Lecture Notes in Computer Science 4782

Commenced Publication in 1973
Founding and Former Series Editors:
Gerhard Goos, Juris Hartmanis, and Jan van Leeuwen

Ronald Perrott Barbara M. Chapman
Jaspal Subhlok Rodrigo Fernandes de Mello
Laurence T. Yang (Eds.)

High Performance Computing and Communications

Third International Conference, HPCC 2007
Houston, USA, September 26-28, 2007
Proceedings

Springer

Volume Editors

Ronald Perrott
Queen's University Belfast
Belfast, UK
E-mail: r.perrott@qub.ac.uk

Barbara M. Chapman
University of Houston
Houston TX 77004, USA
E-mail: chapman@cs.uh.edu

Jaspal Subhlok
University of Houston, Houston
TX 77204, USA
E-mail: jaspal@cs.uh.edu

Rodrigo Fernandes de Mello
University of São Paulo
CEP 13560-970 São Carlos, SP, Brazil
E-mail: mello@icmc.usp.br

Laurence T. Yang
St. Francis Xavier University
Antigonish, NS, Canada
E-mail: lyang@stfx.ca

Library of Congress Control Number: 2007936289

CR Subject Classification (1998): D.2, F.1-2, C.2, G.1-2, H.4-5

LNCS Sublibrary: SL 1 – Theoretical Computer Science and General Issues

ISSN 0302-9743
ISBN-10 3-540-75443-1 Springer Berlin Heidelberg New York
ISBN-13 978-3-540-75443-5 Springer Berlin Heidelberg New York

Springer is a part of Springer Science+Business Media

springer.com

Printed in Germany

Typesetting: Camera-ready by author, data conversion by Scientific Publishing Services, Chennai, India
Printed on acid-free paper SPIN: 12169436 06/3180 5 4 3 2 1 0

Preface

Welcome to the proceedings of the 2007 International Conference on High Performance Computing and Communications (HPCC 2007), which was held in Houston, Texas, USA, September 26–28, 2007.

There have been many exciting developments in all aspects of HPC over the last decade and more exciting developments are on the horizon with, for example, the petaflops performance barrier being targeted in the near future. The rapid expansion in computing and communications technology has stimulated the growth of powerful parallel and distributed systems with an ever increasing demand for HPC in many disciplines. This, in turn, has increased the requirements for more reliable software, better algorithms, more comprehensive models and simulations and represents a challenge to the HPC community to produce better tools, research new areas, etc. Hence conferences, like HPCC 2007, play an important role in enabling engineers and scientists to come together in order to address all HPC-related challenges and to present and discuss their ideas, research results and applications experience.

This year there were 272 paper submissions from all across the world, not only from Europe, North America and South America but also from Asia and the Pacific. All the papers were reviewed by at least three referees from the conference's technical program committee or their colleagues. In order to allocate as many papers as possible and keep the high quality of the conference, we finally decided to accept 69 papers for the conference, which represented the acceptance rate of 25%. We believe that all of these papers and topics not only provide novel ideas, new results, work in progress and state-of-the-art techniques in this field, but will also stimulate future research activities in the area of high performance computing and communications.

This conference is a result of the hard work of very many people such as the program vice chairs, the external reviewers and the program and technical committee members. We would like to express our sincere thanks to everyone involved. Ultimately, however, the success of the conference will be judged by how well the delegates have participated, learnt, interacted and established contacts with other researchers. The committees have provided the venue and created the environment to allow these objectives to be achieved. It is now up to all of us to ensure that the conference is an outstanding success.

We wish you a successful, stimulating and rewarding conference and look forward to seeing you again at future HPCC conferences.

August 2007

Ronald Perrott
Barbara Chapman
Jaspal Subhlok
Rodrigo Fernandes de Mello
Laurence Tianruo Yang

Organization

Executive Committee

General Chairs	Barbara Chapman, University of Houston, USA Jaspal Subhlok, University of Houston, USA
Program Chair	Ronald Perrott, Queens University of Belfast, UK
Program Vice Chairs	Vassil Alexandrov, University of Reading, UK Francois Bodin, University Rennes 1, France Peter Brezany, University of Vienna, Austria Marian Bubak, CYFRONET AGH, Poland Michel Diaz, LAAS, France Edgar Gabriel, University of Houston, USA Laurent Lefevre, École Normale Supérieure de Lyon, France Yunhao Liu, Hong Kong University of Science Technology, China Erik Maehle, University of Luebeck, Germany Allen Malony, University of Oregon, USA Thomas Rauber, University of Bayreuth, Germany Martin Schulz, Lawrence Livermore National Laboratory, USA Alan Sussman, University of Maryland, USA Roland Wismüller, University of Siegen, Germany Hans Zima, California Institute of Technology, USA
Steering Chairs	Beniamino Di Martino, Seconda Università di Napoli, Italy Laurence T. Yang, St. Francis Xavier University, Canada
Publication Chair	Rodrigo Fernandes de Mello, University of São Paulo, Brazil
Web Chairs	Rodrigo Fernandes de Mello, University of São Paulo, Brazil Tony Li Xu, St. Francis Xavier University, Canada Liu Yang, St. Francis Xavier University, Canada
Local Organizing Chair	Rosalinda Mendez, University of Houston, USA

Sponsoring Institutions

Sun Microsystems

Program Committee

David Abramson	Monash University, Australia
Raad S. Al-Qassas	University of Glasgow, UK
Vassil Alexandrov	University of Reading, UK
Henrique Andrade	IBM Thomas J. Watson Research Laboratory, USA
Cosimo Anglano	University of Alessandria, Italy
Irfan Awan	University of Bradford, UK
Frank Ball	Bournemouth University, UK
Purushotham Bangalore	University of Alabama, USA
Ioana Banicescu	Mississippi State University, USA
Alessandro Bassi	Hitachi Europe, France
Alessio Bechini	University of Pisa, Italy
Micah Beck	University of Tennessee, USA
Siegfried Benkner	University of Vienna, Austria
Arndt Bode	Technische Universität München, Germany
Francois Bodin	University of Rennes 1, France
Luciano Bononi	University of Bologna, Italy
George Bosilca	University of Tennessee, USA
Peter Brezany	University of Vienna, Austria
Marian Bubak	CYFRONET AGH, Poland
Wojciech Burakowski	Warsaw University, Poland
Fabian Bustamante	Northwestern University, USA
Armando Caro	BBN Technologies, USA
Calin Cascaval	IBM Thomas J. Watson Research Center, USA
Umit Catalyurek	Ohio State University, USA
Bradford Chamberlain	Cray Inc., USA
Guihai Chen	Nanjing University, China
Lei Chen	Hong Kong University of Science and Technology, China
I-hsin Chung	IBM Research, USA
Toni Cortes	Universitat Politècnica de Catalunya, Spain
Marilia Curado	University of Coimbra, Portugal
Khuzaima Daudjee	University of Waterloo, Canada
Geert Deconinck	University of Leuven, Belgium
Frederic Desprez	INRIA, France
Michel Diaz	LAAS, France
Ivan Dimov	University of Reading, UK
Karim Djemame	University of Leeds, UK
Andreas Doering	IBM Research at Zurich, Switzerland
Werner Dubitzky	University of Ulster, UK

Olivier Dugeon	France Telecom, France
Marc Duranton	NXP Semiconductors, The Netherlands
Ernesto Exposito	University of Toulouse, France
Wu Feng	Virginia Polytechnic Institute and State University, USA
Robert J. Fowler	University of North Carolina at Chapel Hill, USA
Karl Fuerlinger	University of Tennessee, USA
Edgar Gabriel	University of Houston, USA
Alex Galis	University College London, UK
Daniel Rodriguez Garcia	University of Reading, UK
Jean-Patrick Gelas	University Lyon 1, France
Michael Gerndt	Technical University of Munich, Germany
Luc Giraud	ENSEEIHT, France
Olivier Gluck	École Normale Supérieure de Lyon, France
Andrzej M. Goscinski	Deakin University, Australia
Georgios Goumas	National Technical University of Athens, Greece
Anastasios Gounaris	University of Manchester, UK
William Gropp	Argonne National Laboratory, USA
Karl-Erwin Grosspietsch	Fraunhofer Institute for Autonomous Intelligent Systems, Germany
Tao Gu	Institute for Infocomm Research, Singapore
Abdelkader Hameurlain	University of Paul Sabatier, France
Jinsong Han	Hong Kong University of Science and Technology, China
Hermann Hellwagner	University Klagenfurt, Austria
Jano van Hemert	National e-Science Centre, UK
Jeff Hollingsworth	University of Maryland, USA
Chunming Hu	Beihang University, China
Tai-Yi Huang	National Tsing Hua University, Taiwan
Zhiyi Huang	University of Otago, New Zealand
Marty Humphrey	University of Virginia, USA
Toshiyuki Imamura	The University of Electro Communications, Japan
Zhen Jiang	West Chester University of Pennsylvania, USA
Hai Jin	Huazhong University of Science and Technology, China
Helen Karatza	Aristotle University of Thessaloniki, Greece
Karen Karavanic	Portland State University, USA
Constantine Katsinis	Drexel University, USA
Joerg Keller	University of Hagen, Germany
Rainer Keller	University of Stuttgart, Germany
Christoph Kessler	Linköping University, Sweden
Nectarios Koziris	National Technical University of Athens, Greece
Jean-Christophe Lapayre	University of Franche-Comté, France
Jenq-Kuen Lee	National Tsing-Hua University, Taiwan
Wang-Chien Lee	The Pennsylvania State University, USA

Laurent Lefevre	École Normale Supérieure de Lyon, France
Rainer Leupers	Aachen University of Technology, Germany
Xiuqi Li	Florida Atlantic University, USA
Yiming Li	National Chiao Tung University, Taiwan
Jie Lian	University of Waterloo, Canada
Xiaofei Liao	Huazhong University of Science and Technology, China
Wei Lin	Australian Taxation Office, Australia
Yunhao Liu	Hong Kong University of Science and Technology, China
David Lowenthal	University of Georgia, USA
Janardhan Lyengar	Connecticut College, USA
Erik Maehle	University of Luebeck, Germany
Allen Malony	University of Oregon, USA
Muneer Masadah	University of Glasgow, UK
John May	Lawrence Livermore National Laboratory, USA
Eduard Mehofer	University of Vienna, Austria
Piyush Mehrotra	NASA AMES Research Center, USA
Alba C. M. A. de Melo	University of Brasilia, Brazil
Xiaoqiao Meng	University of California at Los Angeles, USA
Edmundo Monteiro	University of Coimbra, Portugal
Shirley Moore	University of Tennessee, USA
Matthias Mueller	Technical University of Dresden, Germany
Henk Muller	University of Bristol, UK
Lenka Novakova	Czech Technical University, Czech Republic
John O'Donnell	University of Glasgow, UK
Gabriel Oksa	Slovak Academy of Sciences, Slovak Republic
Vincent Oria	New Jersey Institute of Technology, USA
Mohamed Ould-Khaoua	University of Glasgow, UK
Béatrice Paillassa	Centre National de la Recherche Scientifique, France
Lei Pan	California Institute of Technology, USA
Stylianos Papanastasiou	University of Glasgow, UK
Steve Parker	University of Utah, USA
Rubem Pereira	Liverpool John Moores University, UK
Cong-Duc Pham	University of Pau, France
Jean-Marc Pierson	University Paul Sabatier, France
Gilles Pokam	University of California at San Diego, USA
Balakrishna Prabhu	VTT Technical Research Centre of Finland, Finland
Isabelle Puaut	University of Rennes, France
Khaled Ragab	Ain Shams University, Egypt
Massimiliano Rak	Seconda Università di Napoli, Italy
Raul Ramirez-Velarde	Monterrey Tech, Mexico
Andrew Rau-Chaplin	Dalhousie University, Canada
Thomas Rauber	University of Bayreuth, Germany
Paul Roe	Queensland University of Technology, Australia
Philip C. Roth	Oak Ridge National Laboratory, USA

Gudula Ruenger	Technische Universität Chemnitz, Germany
Silvius Rus	Google, USA
Miguel Santana	STMicroelectronics, France
Erich Schikuta	University of Vienna, Austria
Martin Schulz	Lawrence Livermore National Laboratory, USA
Assaf Schuster	Israel Institute of Technology, Israel
Stephen L. Scott	Oak Ridge National Laboratory, USA
Hadi S. Shahhoseini	Iran University of Science Technology, Iran
Jackie Silcock	Deakin University, Australia
Peter Sobe	University of Luebeck, Germany
Matt Sottile	Los Alamos National Laboratory, USA
Jeff Squyres	Cisco Systems, USA
Thomas Sterling	California Institute of Technology, USA
Vaidy Sunderam	Emory University, USA
Bronis R. de Supinski	Lawrence Livermore National Laboratory, USA
Alan Sussman	University of Maryland, USA
Martin Swany	University of Delaware, USA
Domenico Talia	Università della Calabria, Italy
Kun Tan	Microsoft Research, China
David Taniar	Monash University, Australia
Jie Tao	Universität Karlsruhe, Germany
Patricia J. Teller	The University of Texas at El Paso, USA
Nigel A. Thomas	University of Newcastle, UK
Bernard Tourancheau	University of Lyon 1, France
Shmuel Ur	IBM Haifa Labs, Israel
Sudharshan Vazhkudai	Oak Ridge National Laboratory, USA
Pascale Vicat-Blanc	École Normale Supérieure de Lyon, France
Luis J. G. Villalba	Complutense University of Madrid, Spain
Xiaofang Wang	Villanova University, USA
Greg Watson	Los Alamos National Laboratory, USA
Josef Weidendorfer	Technische Universität München, Germany
Andrew L. Wendelborn	University of Adelaide, Australia
Roland Wismüller	University of Siegen, Germany
Alexander Woehrer	University of Vienna, Austria
Wolfram Woess	Johannes Kepler University of Linz, Austria
Felix Wolf	RWTH Aachen University, Germany
Joachim Worringen	Dolphin Interconnect Solutions, Germany
Dan Wu	University of Windsor, Canada
Tao Xie	San Diego State University, USA
Baijian Yang	Ball State University, USA
Kun Yang	University of Essex, UK
Wai Gen Yee	Illinois Institute of Technology, USA
Hao Yin	Tsinghua University, China
Hao Yu	IBM Thomas J. Watson Research Center, USA

Hongbo Zhou	Slippery Rock University, USA
Hans Zima	California Institute of Technology, USA
Anna Zygmunt	University of Science and Technology in Krakow, Poland

Additional Reviewers

Sarala Arunagiri
Urtzi Ayesta
Amitabha Banerjee
Puri Bangalore
Alessandro Bardine
Véronique Baudin
Andrzej Beben
Daniel Becker
Ehtesham-ul-haq Dar
Ibrahim Elsayed
Colin Enticott
Nick Falkner
Mauro Gaio
Antoine Gallais
Lei Gao
Karl-Erwin Grosspietsch
Karim Guennoun
Yuzhang Han
Ivan Janciak
Fakhri Alam Khan
Yann Labit
Rui Li
Rodrigo Fernandes de Mello
Olivier Mornard
Philippe Owezarski
Wei Peng
Kathrin Peter
Vincent Roca
Charles Ross
Barry Rountree
A.B.M. Russel
Jean-Luc Scharbarg
Hanno Scharwächter
Weihua Sheng
Ke Shi
Abdellatif Slim
Shaoxu Song
Matthew Sottile
Jefferson Tan
Cédric Tedeschi
Carsten Trinitis
Nicolas Van Wambeke
Richard Vuduc
Qiang Wang
Song Wu
Pingpeng Yuan
Jinglan Zhang
Cammy Yongzhen Zhuang
Deqing Zou

Table of Contents

Keynote Speech

Cluster Computing

Data Mining, Management and Optimization

Distributed, Mobile and Pervasive Systems

Embedded Systems

Grid Computing

High-Performance Scientific and Engineering Computing

Languages and Compilers for HPC

Networking: Protocols, Routing, Algorithms

Parallel/Distributed Architectures

Peer-to-Peer Computing

Performance, Evaluation and Measurements Tools

Reliability and Fault-Tolerance

Web Services and Internet Computing

Programming Challenges for Petascale and Multicore Parallel Systems

Vivek Sarkar

Rice University
vsarkar@cs.rice.edu

Abstract. This decade marks a resurgence for parallel computing with high-end systems moving to petascale and mainstream systems moving to multi-core processors. Unlike previous generations of hardware evolution, this shift will have a major impact on existing software. For petascale, it is widely recognized by application experts that past approaches based on domain decomposition will not scale to exploit the parallelism available in future high-end systems. For multicore, it is acknowledged by hardware vendors that enablement of mainstream software for execution on multiple cores is the major open problem that needs to be solved in support of this hardware trend. These software challenges are further compounded by an increased adoption of high performance computing in new application domains that may not fit the patterns of parallelism that have been studied by the community thus far. In this talk, we compare and contrast the software stacks that are being developed for petascale and multicore parallel systems, and the challenges that they pose to the programmer. We discuss ongoing work on high productivity languages and tools that can help address these challenges for petascale applications on high-end systems. We also discuss ongoing work on concurrency in virtual machines (managed runtimes) to support lightweight concurrency for mainstream applications on multicore systems. Examples will be give from research projects under way in these areas including PGAS languages (UPC, CAF), Eclipse Parallel Tools Platform, Java Concurrency Utilities, and the X10 language. Finally, we outline a new long-term research project being initiated at Rice University that aims to unify elements of the petascale and multicore software stacks so as to produce portable software that can run unchanged on petascale systems as well as a range of homogeneous and heterogeneous multicore systems.

R. Perrott et al. (Eds.): HPCC 2007, LNCS 4782, p. 1, 2007.

Towards Enhancing OpenMP Expressiveness and Performance

Haoqiang Jin

NASA Ames Research Center
hjin@nas.nasa.gov

Abstract. Since its introduction in 1997, OpenMP has become the de facto standard for shared memory parallel programming. The notable advantages of the model are its global view of memory space that simplifies programming development and its incremental approach toward parallelization. However, it is very challenge to scale OpenMP codes to tens or hundreds of processors. This problem becomes even more profound with the recent introduction of multi-core, multi-chip architectures. Several extensions have been introduced to enhance OpenMP expressiveness and performance, including thread subteams and workqueuing. In this talk, we describe applications that expose the limitation of the current OpenMP and examine the impact of these extensions on application performance. We focus on exploiting multi-level parallelism and dealing with unbalanced workload in applications with these extensions and compare with other programming approaches, such as hybrid. Our experience has demonstrated the importance of the new language features for OpenMP applications to scale well on large shared memory parallel systems.

R. Perrott et al. (Eds.): HPCC 2007, LNCS 4782, p. 2, 2007.

Bandwidth-Aware Design of Large-Scale Clusters for Scientific Computations

Mitsuhisa Sato

Center for Computational Sciences, University of Tsukuba
msato@cs.tsukuba.ac.jp

Abstract. The bandwidth of memory access and I/O, network is the most important issue in designing a large-scale cluster for scientific computations. We have been developing a large scale PC cluster named PACS-CS (Parallel Array Computer System for Computational Sciences) at Center for Computational Sciences, University of Tsukuba, for wide variety of computational science applications such as computational physics, computational material science, etc. For larger memory access bandwidth, a node is equipped with a single CPU which is different from ordinary high-end PC clusters. The interconnection network for parallel processing is configured as a multi-dimensional Hyper-Crossbar Network based on trunking of GigabitEthernet to support large scale scientific computation with physical space modeling. Based on the above concept, we are developing an original mother board to configure a single CPU node with 8 ports of Gigabit Ethernet, which can be implemented in the half size of 19 inch rack-mountable 1U size platform. PACS-CS started its operation on July 2006 with 2560 CPUs and 14.3 TFops of peak performance. Recently, we have newly established an alliance to draw up the specification of a supercomputer, called Open Supercomputer Specification. The alliance consists of three Japanese universities: University of Tsukuba, University of Tokyo, and Kyoto University (T2K alliance). The Open Supercomputer Specification defines fundamental hardware and software architectures on which each university will specify its own requirement to procure the next generation of their supercomputer systems in 2008. This specification requests the node to be composed of commodity multicore processors with high aggregated memory bandwidth, and the bandwidth of internode communication to be 5 GB/s or more in physical link level and 4 GB/s or more in MPI level with Link aggregation technology using commodity fabric. We expect several TFlops in each system. In order to support scalable scientific computations in a large-scale cluster, the bandwidth-aware design will be important.

R. Perrott et al. (Eds.): HPCC 2007, LNCS 4782, p. 3, 2007.

OpenMP 3.0 – A Preview of the Upcoming Standard

Larry Meadows

Intel Corporation
lawrence.f.meadows@intel.com

Abstract. The OpenMP 3.0 standard should be released for public comment by the time of this conference. OpenMP 3.0 is the first major upgrade of the OpenMP standard since the merger of the C and Fortran standards in OpenMP 2.5. This talk will give an overview of the new features in the OpenMP standard and show how they help to extend the range of problems for which OpenMP is suitable.

Even with multi-core, the number of hardware cores in an SMP node is likely to be relatively small for the next few years. Further, a number of users want to use OpenMP, but need to scale to more cores than fit in a node, and do not want to mix OpenMP and MPI. Intel supports a production quality OpenMP implementation for clusters that provides a way out. This talk will briefly introduce Intel Cluster OpenMP, provide an overview of the tools available for writing programs, and show some performance data.

R. Perrott et al. (Eds.): HPCC 2007, LNCS 4782, p. 4, 2007.

Manycores in the Future

Robert Schreiber

HP Labs
rob.schreiber@hp.com

Abstract. The change from single core to multicore processors is expected to continue, taking us to manycore chips (64 processors) and beyond. Cores are more numerous, but not faster. They also may be less reliable. Chip-level parallelism raises important questions about architecture, software, algorithms, and applications. I'll consider the directions in which the architecture may be headed, and look at the impact on parallel programming and scientific computing.

R. Perrott et al. (Eds.): HPCC 2007, LNCS 4782, p. 5, 2007.

The Changing Impact of Semiconductor Technology on Processor Architecture

Ray Simar

Texas Instruments
r-simar@ti.com

Abstract. We stand on the brink of a fundamental discontinuity in silicon process-technology unlike anything most of us have seen. For almost two decades, a period of time spanning the entire education and careers of many engineers, we have been beneficiaries of a silicon process-technology which would let us build almost anything we could imagine. Now, all of that is about to change. For the past five years, capacitive loading of interconnect has grown to be a significant factor in logic speed, and has limited the scaling of integrated-circuit performance. To compound the problem, recently interconnect resistance has also started to limit circuit speed. These factors can render obsolete current designs and current thinking as interconnect-dominated designs and architectures will become increasingly irrelevant. Given these fundamental interconnect challenges, we must turn to architecture, logic design and programming solutions. The background on these dramatic changes in semiconductor technology will be discussed in the hopes that the solutions for the future may very well come from the attendees of HPCC 2007!

R. Perrott et al. (Eds.): HPCC 2007, LNCS 4782, p. 6, 2007.

A Windows-Based Parallel File System

Lungpin Yeh, Juei-Ting Sun, Sheng-Kai Hung, and Yarsun Hsu

Department of Electrical Engineering,
National Tsing Hua University HsinChu, 30013, Taiwan
{lungpin,posh,phinex}@hpcc.ee.nthu.edu.tw, yshsu@ee.nthu.edu.tw

Abstract. Parallel file systems are widely used in clusters to provide high performance I/O. However, most of the existing parallel file systems are based on UNIX-like operating systems. We use the Microsoft .NET framework to implement a parallel file system for Windows. We also implement a file system driver to support existing applications written with Win32 APIs. In addition, a preliminary MPI-IO library is also developed. Applications using MPI-IO could achieve the best performance using our parallel file system, while the existing binaries could benefit from the system driver without any modifications. In this paper, the design and implementation of our system are described. File system performance using our preliminary MPI-IO library and system driver is also evaluated. The results show that the performance is scalable and limited by the network bandwidth.

1 Introduction

As the speed of CPU becomes faster, we might expect that the performance of a computer system should benefit from the advancement. However, the improvements of other components in a computer system (i.e. memory system, data storage system) cannot catch up with that of CPU. Although the capacity of a disk has grown with time, its mechanical nature limits its read/write performance. In this data-intensive world, it is significant to provide a large storage subsystem with high performance I/O[1]. Using a single disk with a local file system to sustain this requirement is impossible nowadays. Disks combined either tightly or loosely to form a parallel system provide a possible solution to this problem. The success of a parallel file system comes from the fact that accessing files through network can have higher throughput than fetching files through local disks. This could be attributed to the emergence of high-speed networks such as Myrinet [2], InfiniBand [3], Gigabit Ethernet, and more recently 10 Gigabit Ethernet.

A parallel file system can not only provide a large storage space by combining several storage resources on different nodes but also increase the performance. It could provide high-speed data access by using several disks at the same time. With suitable striping size, the workload in the system can be distributed among these disks instead of being centralized in a single disk. For example, whenever a write happens, a parallel file system would split these data into a lot of small

R. Perrott et al. (Eds.): HPCC 2007, LNCS 4782, pp. 7–18, 2007.

chunks, which are then stored on different disks across the network in a round-robin fashion.

Most of parallel file systems are based on Unix or Linux. As far as we know, WinPFS[4] is the only parallel file system based on Microsoft Windows. However, it does not allow users to specify the striping size of a file across nodes. Furthermore, it does not provide a user level library for high performance parallel file access. In this paper, we implement a parallel file system for Microsoft Windows Server 2003 allowing users the flexibility to specify different striping size. Users can specify the striping size to satisfy the required distribution or using the default striping size provided by the system. We have implemented a file system driver to trap Win32 APIs such that existing binaries can access files stored on our parallel file system without recompilation. Besides, some MPI-IO functions (such as noncontiguous accesses) are also provided for MPI jobs to achieve the best performance. We have successfully used our parallel file system as a storage system for VOD (Video On Demand) services, which can deliver the maximum bandwidth and demonstrate the successful implementation of our parallel file system.

This paper is organized as follows: Section 2 presents some related works. Design and implementation will be discussed in section 3, with the detailed description of our system driver. Section 4 depicts the results of performance evaluation of our windows-based parallel file system, along with the prototype VOD system. Finally, we would make some conclusions and provide some directions in section 5.

2 Related Works

PVFS[5,6] is a parallel file system publicly available in the Linux environment. It provides both user level library for performance and a kernel module package that makes existing binaries working without recompiling.

WinPFS [4] is a parallel file system for Windows and integrated within the Windows kernel components. It uses the existing client and server pairs in the Windows platform (i.e. NFS [7], CIFS [8], ...) and thus no special servers are needed. It also provides a transparent interface to users, just like what does when accessing normal files. The disadvantage is that the user can not specify the striping size of a file across nodes. Besides, its performance is bounded by the slowest client/server pairs if the load balancing among servers is not optimal. For example, if it uses NFS as one of the servers, the overall performance may be gated by NFS. This heterogeneous client/server environment helps but it might also hurt when encountering unbalanced load.

Microsoft adds the support of dynamic disks starting from Windows 2000. Dynamic disks are the disk formats in Windows necessary for creating multi-partition volumes, such as spanned volumes, mirrored volumes, striped volumes, and RAID-5 volume. The striped volumes contain a series of partitions with one partition per disk. However, only up to 32 disks can be supported, which is not very scalable[9].

3 Design and Implementation

The main task of the parallel file system is to stripe data or split files into several small pieces. Files are equally distributed among different I/O nodes and can be accessed directly from applications. Applications can access the same file or different files in parallel rather than sequentially. The more I/O nodes in a system, the more bandwidth it could provide (only limited by the network capacity).

3.1 System Architecture

Generally speaking, our parallel file system consists of four main components: Metadata server, I/O daemons (Iod), a library and a file system driver. Metadata server and I/O daemons set up the basic parallel file system architecture. The library provides high performance APIs for users to develop their own applications on top of the parallel file system. It communicates with the metadata server and Iods, and does the tedious work for users. The complexity behind the parallel file system is hidden by the library and users do not need to concern about how the metadata server and Iods co-operate. With the help of file system driver, we can trap I/O related Win32 API calls and provide transparent file accesses. Most of the user mode APIs have the kernel mode equivalent implementation. The overall architecture is shown in Fig. 1.

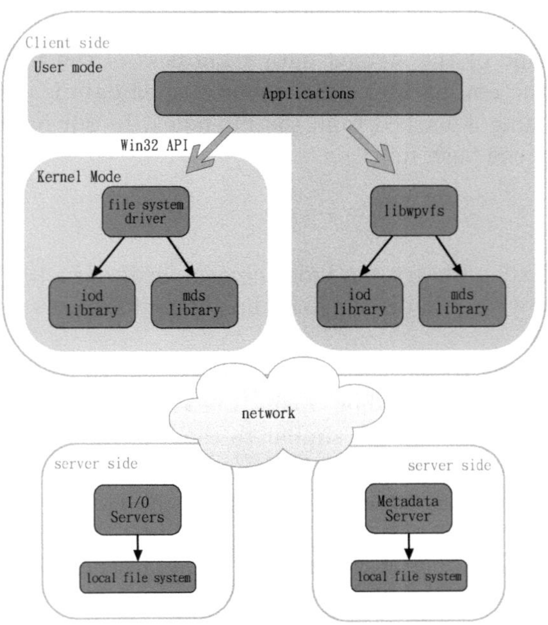

Fig. 1. The overall system architecture

Metadata Server. Metadata means the information about a file except for the contents that it stores. In our parallel file system, metadata contains five parts:

- File size: It describes the size of a file.
- File index: It is a 64-bit number, which uniquely identifies the file stored on the metadata server. Its uniqueness is maintained by the underlying file system, such as the inode number of the UNIX operating systems. It is used as the filename of the striped data stored on I/O nodes.
- Striping size: The size that a file is partitioned.
- Node count: The number of I/O nodes that the file is spread across.
- Starting I/O node: The I/O node that the file is first stored on.

The metadata server runs on a single node, managing the metadata of a file and maintaining the directory hierarchy of our parallel file system. It does not communicate with I/O daemons or users directly, but only converses with the library, `libwpvfs`. Whenever a file is requested, users may call the library to connect with the metadata server and get the metadata of that file. Before a file can be accessed, its metadata must be fetched in advance.

I/O Daemons. The I/O daemon is a process running on each of the I/O nodes responsible for accessing the real data of a file. It can run on a single node or several nodes, and you can run several I/O daemons on an I/O node if you want. After users get the metadata of a file, the library could connect to the required I/O nodes, and the Iods would access the requested file and send stripes back to the client.

Each of the I/O nodes maintains a flat directory hierarchy. The file index is used as the filename of the striped data regardless of the file's real filename. No matter what the real path of a file is, the striped data is always stored in a directory whose name is hashed from the file index. In our implementation, we use modulation as the hash function.

3.2 Library

As mentioned before, a library can hide the complexity of a parallel from users. In this subsection, we would discuss how the different libraries are implemented.

User Level Library. We provide a class library that contains six most important file system methods, including `open`, `create`, `read`, `write`, `seek`, and `close`. These methods are mostly similar to those of the `File` class in C# but with more capabilities support. Users can specify the striping size, starting Iod, and Iod counts when accessing a file. The library separates the users from the Iods and the metadata server. All the tedious jobs will be handled by the library. With the help of the library, users only need to concern how to efficiently partition and distribute the file.

Kernel Level File System Driver. In the Windows operating system, NT I/O Manager, which is a kernel component, is responsible for the I/O subsystem. To allow I/O Manager and drivers to communicate with other components in the operating system, a data structure called I/O Request Packet (IRP) is frequently used. An IRP contains lots of information to describe requests and parameters. Most import of all is the major function code and the minor function code. These two integers contained in an IRP precisely indicate the operation that should be performed. I/O related Win32 APIs will eventually be sent to the I/O Manager, which then allocates an IRP sent to the responsible driver.

With the help of a virtual disk driver, a file system driver, and a mount program, our parallel file system can be mounted as a local file system for Windows. Fig. 2 illustrates the mounting process of our parallel file system. The virtual disk driver presents itself as a normal hard disk to Windows when it is loaded into the system. The mount program invokes the `DefineDosDevice` function call to create a new volume on the virtual disk. After the new volume is created, the mount program tries to create a file on the volume. This request will be routed to the NT I/O Manager. Upon receiving this request, the I/O Manager finds that this volume is not handled by any file system driver yet. Thus, it sends an IRP containing a mount request to each of registered file system drivers in the system. File system drivers check the on disk information when they receive such a request to determinate if it recognizes this volume.

We implement a crafted read function in the virtual disk driver. The driver returns a magic string "-pfs-" without quotes when a file system driver tries

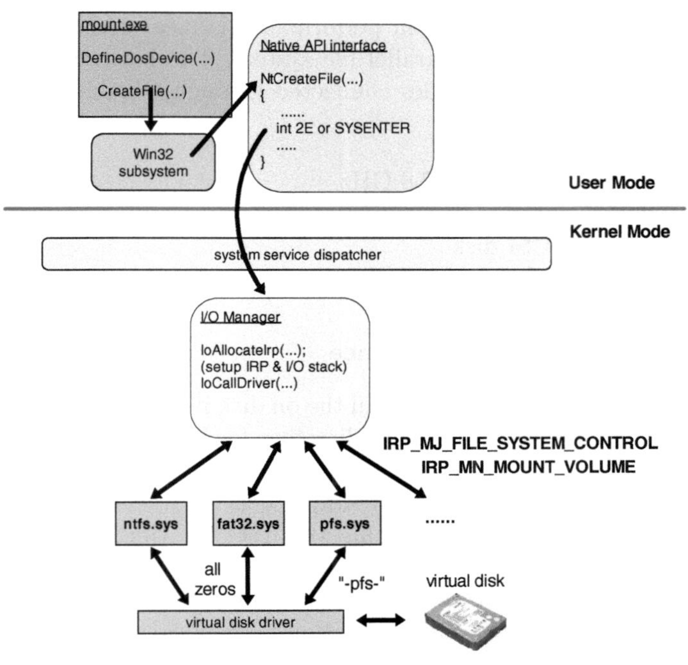

Fig. 2. The process of mounting our parallel file system

to read 6 bytes from the disk. Otherwise, it returns zeros. In this way, while any other file system drivers try to check the on disk information, they do not recognize the volume. "-pfs-" is the magic string that only our parallel file system driver recognizes. When the mount request is sent to our file system driver, it reads 6 bytes from the disk, recognizes the magic string, and tells the I/O Manager that this volume is under our control. The mount operation completes and all I/O operations targetting at this volume will be routed to our file system driver hereafter.

On loading the file system driver into the system, persistent connections are established to all I/O daemons. The connection procedures are performed once at the loading time, and all operations are made through these sockets. This eliminates the connection overhead of all I/O operations from user mode applications. The file system driver effectively does the same thing as the user mode library when it receives a read or write operation.

MPI-IO Library. MPI-IO[10] is the parallel I/O part of MPI and its objective is to provide high performance parallel I/O interface for parallel MPI programs. A great advantage of MPI-IO is the ability to access noncontiguous data with a single function call, which is known as collective I/O. Our parallel file system is built on .NET framework using C#.

4 Performance Evaluation

In this section, the local file system performance is measured along with read and write performance of our parallel file system. The hardware used is IBM eServer xSeries 335 with five nodes connected through Gigabit Ethernet, each housing:

- One Intel Xeon processor at 2.8 GHz
- 512 MB DDR memory
- 36.4 G Ultra 320 SCSI disk
- Microsoft Windows Server 2003 SP1

4.1 Local File System Performance

Our parallel file system doesn't maintain the on disk information itself, but relies on the underlying file system. The root directory for Iods or the metadata server is set in an NTFS partition. To test I/O performance of the local file system and the .NET framework, we write a simple benchmark using C#. The tests are performed on a single node, running the tests ten times and averaged the results.

A 64 KB buffer is filled with random data and written to the local file system continuously until the number of bytes written to the local file system reaches the file size. Note that the write operations are carried out by the Microsoft .NET Framework and the NTFS file system driver which has some caching mechanism

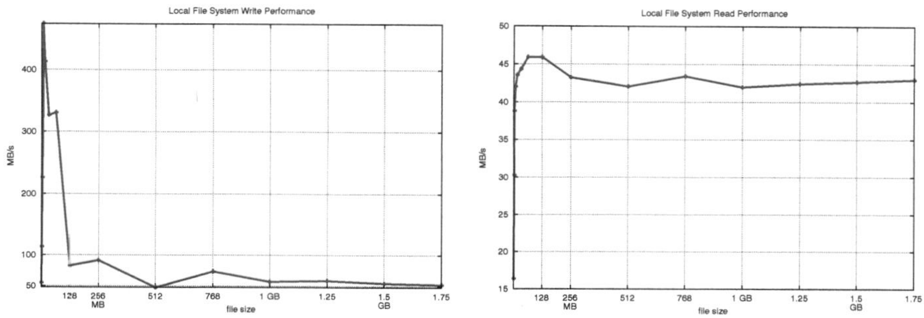

(a) Write Performance (b) Read Performance

Fig. 3. Performance evaluation of local file system

internally. In Fig. 3(a), we observe that write performance of local file system converges to about 55 MB/s when the file size is larger than 768 MB, but the performance varies when the file size is smaller than 512 MB. We think this is the effect of the caching mechanism.

To make sure that the files written are not cached in memory, the system is rebooted before measuring the read performance. The same file is read from the disk into a fixed-size buffer and the buffer is used over and over again. The data read is ignored and overwritten by later reads. As you can see from Fig. 3(b), read performance converges to about 43 MB/s.

4.2 Performance Evaluation Using User Level Library

The performance of our parallel file system are evaluated on five nodes. One of them is served both as a metadata server and a client which runs our benchmark program written with our library. The other four nodes are running I/O daemons, one for each.

Again, a fixed-size memory buffer is filled with random data. After that, a create operation is invoked, and the buffer content is written to the parallel file system continuously until the number of bytes written reaches the file size. The test program then waits for the acknowledgements sent by the I/O daemons to make sure all the data sent by the client are received by all I/O daemons. Note that though the Iods have written received data to their local file systems, this does not guarantee that the data is really written to their local disks. They may be cached in the memory by the operating system and written back to the physical disks later. We ran the tests ten times and averaged the results.

In Fig. 4(a), we measure write performance with various file sizes and various number of I/O nodes. The striping size is 64 KB. The size of the memory buffer used equals to the number of I/O nodes multiplied by the striping size. Write performance converges to about 53 MB/s when only one I/O node is used. We consider that the write performance is bounded by the local file system in this case, since this is almost equal to the local file system write performance as shown in the previous test. The performance of writing to two I/O nodes is about twice of writing to only one node when the file size is large enough.

However, write performance reaches a peak of 110 MB/s when writing to three or four I/O nodes. For these two cases, they almost have the same performance since the bottleneck is the network bandwidth rather than the physical disks. Since all the cluster nodes are connected by a Gigabit Ethernet which has the theoretical peak bandwidth of 125 MB/s, it is conceivable that a client can not write out faster than 125 MB/s due to protocol overhead. The same behavior has been observed in PVFS[11] and IBM vesta parallel file system[12]. However, we expect the write performance to be scalable if a higher bandwidth network is available in the future.

The size of the memory buffer for read is also the number of I/O nodes multiplied by the striping size. The data read into the memory buffer is ignored and overwritten by later data. As you can see in Fig. 4(b) , read performance is not as good as write performance. But when we increase the number of I/O nodes, the performance increases too. For four I/O nodes, read performance reaches a peak of 75 MB/s. With the use of more than two I/O nodes, read performance of our parallel file system is better than that of a local disk.

We have made some tests to figure out why the read performance can not fully utilize the theoretical network bandwidth. The Iod program is modified such that when it receives a read request, it does not read the data from the local file system, but just sends the contents of a memory buffer to the client directly. The contents in the memory buffer are non-deterministic. In this process of measuring read performance, the behavior is exactly the same as previous tests except that no local file system operations are involved. We run the test several times. The results show that when only one I/O node is used, the curves are almost identical and the performance reaches a peak of 90 MB/s. In the case of two I/O nodes, it has the same behavior but the performance reaches a peak of 93 MB/s. For three and four I/O nodes, the curves are desultory and the average performance reaches a peak of around 78 MB/s which is lower than the performance of using only one or two I/O nodes. We think this is due to network congestion and packet collision. Whenever multiple I/O nodes try to send large

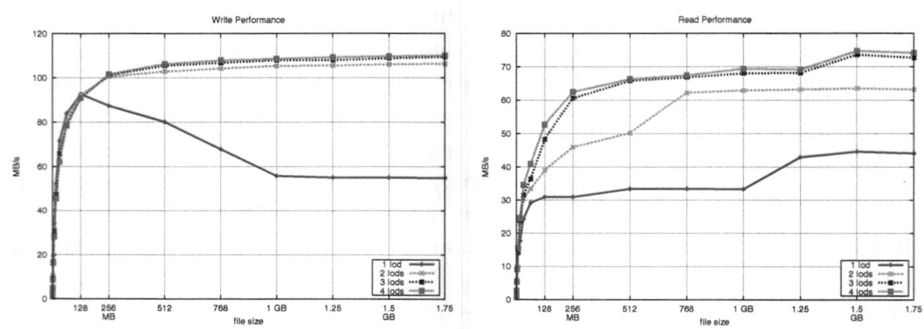

(a) Write Performance (b) Read Performance

Fig. 4. Performance evaluation using variable I/O nodes

amount of data to a client simultaneously, the receiving speed of the client can not catch up with the overall sending speed of I/O nodes. Therefore some packets may collide with others and get dropped. The I/O nodes have to back off and resend packets as required by the protocol design of Ethernet architecture. This explains why the read performance is not as good as the write performance and saturated around 75 MB/s when three or four I/O nodes are used. In Fig. 4(b), the read performance of one I/O node and two I/O nodes are bounded by the local file system. In the case of three and four I/O nodes, it is bounded by the network due to network congestion. Again, as in write performance, we expect the read performance to be improved significantly when a higher performance network is available in the future.

4.3 Performance Evaluation Using Kernel Driver

To measure the performance of our parallel file system when using the file system driver, we write a simple benchmark program which has the same functionality as the one written in C#. But this benchmark uses Win32 APIs directly to create, read and write files. We also repeat the tests ten times and average the results.

Fig. 5(a) shows the write performance with various number of I/O nodes. The striping size is 64 KB and the user supplied memory buffer is 1 MB. When we increase the number of I/O nodes, the performance increases too, but it is worse than that of the local file system even when four I/O nodes are used. As observed from Fig. 5(b), the read performance is much better compared with the write performance. The performance increases with the number of I/O nodes when more than two I/O nodes are used, but the performance with only one I/O node is between that of four I/O nodes and three I/O nodes.

Windows is a commercial product and the source codes are not available. Therefore, the detail operations are opaque. Furthermore the file system driver resides in kernel mode and needs a socket library to communicates with daemons in the parallel file system. Microsoft doesn't provide a socket library for kernel

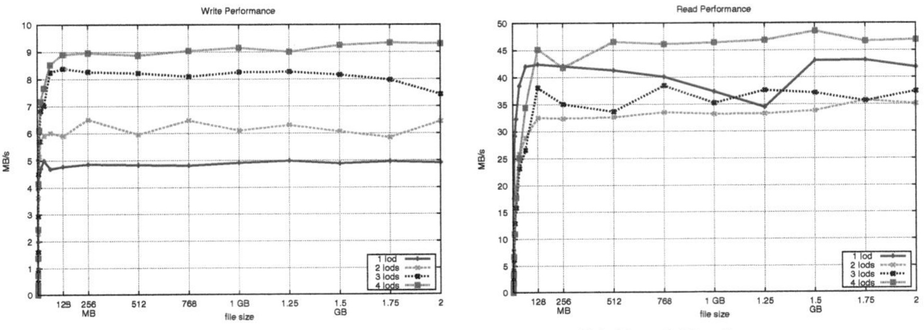

(a) Write Performance (b) Read Performance

Fig. 5. Performance evaluation using the system driver with 64KB striping size and 1MB user buffer mode programmers. The lack of a sophisticated kernel mode socket library also makes it difficult to write high performance programs.

The main purpose of implementing a kernel driver is to enable existing binaries to run over this parallel file system. However the parallel file system is developed mainly for high performance applications. We expect users to write high performance applications using our new APIs created especially to take advantage of this parallel file system.

4.4 Performance Evaluation Using MPI-IO

We use the MPI-IO functions that we have implemented to write a benchmark program. In this case, we set the size of etype[10] to 64 KB which is equal to the striping size of the previous three cases. This is an obvious selection, since an etype (elementary datatype) is the basic unit of data access and all file accesses are performed in units of etype. The visible portion of the filetype is set to an etype and the stride[10] (i.e. the total length of the filetype) is set to the number of I/O nodes multiplied by the etype. Finally, the displacement is set to the rank of the I/O node multiplied by the etype. All the others are set based on the previous settings.

We measure the performance by varying the number of I/O nodes from one to four. The buffer size is set to be the number of I/O nodes multiplied by the etype. Each test is performed ten times and averaged to get the final result. The write and read performance are shown in Fig. 6(a) and Fig. 6(b) respectively.

The trends of the write and read performance resemble those which we have discussed above. Compared with the library of our parallel file system, `libwpvfs`, the MPI-IO functions have some added function calls and operations, but they do not influence the performance deeply. Consequently, the MPI-IO functions are provided without suffering serious overhead.

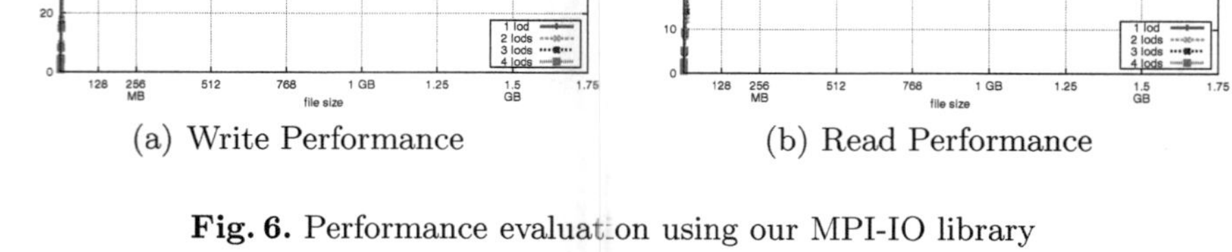

(a) Write Performance (b) Read Performance

Fig. 6. Performance evaluation using our MPI-IO library

4.5 VOD Prototype System

Besides, we have set up a distributed multimedia server on top of our parallel file system. Microsoft DirectShow is used to build a simple media player. Using DirectShow with `libwpvfs`, we build a media player, which could play multimedia files distributed across different I/O nodes. Since DirectShow can only play media files stored on disks or from a URL, we establish a web server as an agent to gather striped files from I/O nodes. This web server is inserted between the media player and our library, `libwpvfs`, and it is the web server that uses the library to communicate with the metadata server and I/O nodes.

The data received by the web server from I/O nodes is passed to the media player. The media player plays a media file coming from the http server through a URL rather than from the local disk. Both the media player and the web server run on the local host. The web server is bound with our media player and transparent to the end user. A user is not aware of the existence of the web server and could use our media player as a normal one.

Any existing media player programs which support playing media files from an URL, such as Microsoft Media Player, can take advantage of our parallel file system by accessing the video file on our web server. In this way, we may provide a high performance VOD service above our parallel file system.

5 Conclusions and Future Work

PC-based clusters are getting more and more popular these days. Unfortunately almost all of the parallel file systems are developed in UNIX-based clusters. It is hard to implement a Windows-based parallel file system because Windows is a commercial product and the source codes are not available. In this paper, we have implemented a parallel file system which provides parallel I/O operations for PC clusters running Windows operating system. A user mode library using .NET framework is also developed to enable users writing efficient parallel I/O programs. We have also successfully implemented a simple VOD system to demonstrate the feasibility and usefulness of our parallel file system. In addition we have implemented key MPI-IO functions on top of our parallel file system and found that the overhead of implementing MPI-IO is very minimal. The performance of MPI-IO is very close to the performance provided by the parallel file system. Furthermore, we have also implemented a file system driver which provides a transparent interface for accessing files stored on our parallel file system so that existing programs written with Win32 APIs can still run on our system.

We have found that both write and read performance are scalable and only limited by the performance of the Ethernet network we use. We plan to further evaluate the performance of this parallel file system when we can obtain a higher performance network such as Infiniband under Windows and believe that our parallel file system can automatically achieve much better performance.

The prototyping VOD system proves the usability of our parallel file system in the Windows environment. Varying the striping size in the VOD system under different load conditions may have distinct behavior. The impact of the striping size may hurt or help the VOD system under different load conditions. We would perform some detailed experiments and analysis in the near future. This would help us develop a more realistic and high performance VOD system that can benefit from our parallel file system.

References

1. Adiga, N.R., Blumrich, M., Liebsch, T.: An overview of the BlueGene/L supercomputer. In: Proceedings of the 2002 ACM/IEEE Conference on Supercomputing, Baltimore, Maryland, pp. 1–22 (2002)
2. Myricom: Myrinet, `http://www.myri.com/`
3. InfiniBand Trade Association: Infiniband, `http://www.infinibandta.org/`
4. Pérez, J.M., Carretero, J., García, J.D.: A Parallel File System for Networks of Windows Worstations. In: ACM International Conference on Supercomputing (2004)
5. Carns, P.H., Ligon III, W.B., Ross, R.B., Thakur, R.: PVFS: A parallel file system for linux clusters. In: 4th Annual Linux Showcase and Conference, Atlanta, GA, pp. 317–327 (2000)
6. Ligon III, W. B., Ross, R.B.: An Overview of the Parallel Virtual File System. In: 1999 Extreme Linux Workshop (1999)
7. Kleiman, S., Walsh, D., Sandberg, R., Goldberg, D., Lyon, B.: Design and implementation of the sun network filesystem. In: Proc. Summer USENIX Technical Conf., pp. 119–130 (1985)
8. Hertel, C.R.: Implementing CIFS: The Common Internet File System. Prentice-Hall, Englewood Cliffs (2003)
9. Russinovich, M.E., Solomon, D.A.: Microsoft Windows Internals, Microsoft Windows Server 2003, Windows XP, and Windows 2000, 4th edn. Microsoft Press, Redmond (2004)
10. Corbett, P., Feitelson, D., Fineberg, S., Hsu, Y., Netzberg, W., Prost, J., Snir, M., Traverset, W., Wong, P.: 32. In: Overview of the MPI-IO Parallel IO Interface. IEEE and Wiely Interscience, Los Alamitos (2002)
11. Ligon III, W.B., Ross, R.B.: Implementation and performance of a parallel file system for high performance distributed applications. In: Proceedings of the Fifth IEEE International Symposium on High Performance Distributed Computing, pp. 471–480. IEEE Computer Society Press, Los Alamitos (1996)
12. Feitelson, D.G., Corbett, P.F., Prost, J.-P.: Performance of the vesta parallel file system. In: 9th Internationl Parallel Processing Symposium, pp. 150–158 (1995)

PARMI: A Publish/Subscribe Based Asynchronous RMI Framework for Cluster Computing

Heejin Son and Xiaolin Li

Scalable Software Systems Laboratory
Computer Science Department,
Oklahoma State University
Stillwater, OK 74078, U.S.A.
{hees,xiaolin}@cs.okstate.edu

Abstract. This paper presents a publish/subscribe based asynchronous remote method invocation framework (PARMI) aiming to improve performance and programming flexibility. PARMI enables high-performance communication among heterogeneous distributed processors. Based on publish/subscribe paradigm, PARMI realizes asynchronous communication and computation by decoupling objects in space and time. The design of PARMI is motivated by the needs of scientific applications that adopt asynchronous iterative algorithms. An example scientific application based on the Jacobi iteration numerical method is developed to verify our model and evaluate the system performance. Extensive experimental results on up to 60 processors demonstrate the significant communication speedup using asynchronous computation and communication technique based on the PARMI framework compared to a baseline scheme using synchronous iteration and communication.

1 Introduction

Large scale distributed computing is complicated and poses a significant challenge due to its scale, dynamics, and heterogeneity. Traditional solutions like RPC (remote procedure call) gained popularity due to widely available built-in libraries. But RPC offers limited capabilities because it supports only C/C++ and Fortran programming languages and primitive point-to-point communication. The goal of this study is to provide a flexible programming tool based on publish/subscribe paradigm for parallel and distributed applications with decoupled communication and asynchronous computation.

Our contributions include: (1) design and develop the PARMI framework to enable scalable asynchronous computation and communication; (2) adopt the publish/subscribe paradigm to enhance the existing remote method invocation (RMI) in Java by enabling asynchronous RMI; (3) make use of the recently introduced generics mechanism in Java to enable flexible interfaces to support dynamic applications and parameterized classes and objects.

The rest of this paper is organized as follows. Section 2 presents the motivation and related work. Section 3 presents the system architecture of the overall design of

R. Perrott et al. (Eds.): HPCC 2007, LNCS 4782, pp. 19–29, 2007.

the PARMI including main features. Section 4 presents and analyses the experimental results and evaluates its performance. Finally, Section 5 concludes this paper.

2 Related Work

Java has grown into a main-stream programming language since its inception. With the increasing adoption of Java for parallel and distributed computing, the built-in remote method invocation (RMI) mechanism is one of the most popular enabling solutions for parallel and distributed computing using Java [1]. Building on the popularity of RMI, we further enhance it with asynchronous communication, which is desirable to decouple sending/receiving objects and facilitate overlapping computation and communication. In PARMI, the publish/subscribe paradigm is used to decouple communicating parties in both time and space. Following the object-oriented programming paradigm, PARMI aims to provide a simple platform to support complex large-scale scientific applications.

The challenging issues are as follows: (1) how to overcome a synchronous and point-to-point communication nature of RMI, (2) how to provide a scalable framework for dynamic applications, and (3) how to maintain a strong decoupling of participants in both time and space.

2.1 Shortcomings of Java RMI

The RMI system allows an object running in one Java virtual machine (JVM) to invoke methods on an object running in another JVM. A user can utilize a remote reference in the same manner as a local reference. However, the synchronous nature of RMI leads network latency and low-performance. Thus, several implementations have been developed that support extended protocols for RMI. These include Manta, NinjaRMI, and JavaParty by changing the underlying protocols such as the serialization protocols [2]. In 1997, Object System presented a communication system called Voyager, providing several communication modes allowing for synchronous invocation, asynchronous invocations with no reply (one way), and asynchronous invocation with a reply (future/promise) [3-6].

Asynchronous RMI with a Future Object. The RMI with a future object is the latest and most efficient means for providing asynchronous communication between a client and a server. A future object allows a client to continue computation after the incomplete communication without being blocked. Thanks to built-in classes and interfaces for the future object which holds a result of an asynchronous call, we don't need to spend time to implement the future object after Java version 1.5 or later. The previous asynchronous RMI studies have manipulated the stub class, which was generated automatically by an rmic compiler. Many difficulties in maintenance have occurred from this approach. For example, if a method that is invoked remotely by an object is modified, the corresponding classes and interfaces should be changed accordingly. After a stub class was generated by a RMIC compiler, we must change the stub class manually. Therefore, we focus on adding codes to access the FutureTask on the client side, not changing the stub class.

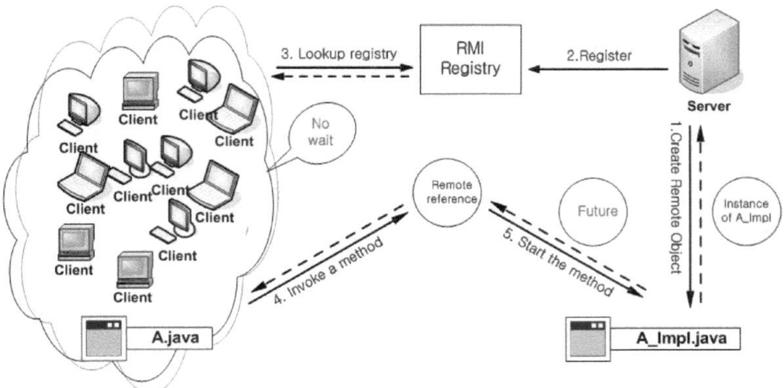

Fig. 1. Operation of the asynchronous RMI with a future object

Fig. 1 shows the operations of the asynchronous RMI with a future object. Several client-side processes send signals to invoke server-side methods. *A_Impl* class has all business logics for methods of *A* interface. *A* interface has all method names available for remote invocation. Server program creates an instance of the *A_Impl* class and binds the service with the RMI registry. Now, a client is able to get a remote reference from the RMI registry using lookup service. Once the client invokes a method, it proceeds with the remaining works without waiting because a future object is returned instantly when it is called.

This model overcomes synchronization but it is still tightly coupled in space and time [7].

2.2 Publish/Subscribe Communication Model

In the publish/subscribe model, the information bus requires operating constantly and tolerating for dynamic system evolution and legacy system. Providers publish data to an information bus and consumers subscribe data they want to receive [8]. Providers and consumers are independent to each other and need not even know of their existence in advance. In general, the provider is also called the publisher, the consumer is the subscriber, and the information bus is the middleware or broker. In systems based on the publish/subscribe interaction paradigm, subscribers register their interests in an event or pattern of events, and are subsequently asynchronously notified of events generated by publishers [7].

As distributed systems on wide area networks grow, the demands of flexible, efficient, and dynamic communication mechanisms have increased. The publish/subscribe communication paradigm provides a many-to-many data dissemination. It is an asynchronous messaging paradigm that allows for better scalable and more dynamic network topology. The publish/subscribe interaction is an asynchronous messaging paradigm, characterized by the strong decoupling of participants in both time and space [9].

The classical categories of publish/subscribe communications are *topic-based* and *content-based* systems. In the topic-based system, messages are published to topics or named logical channels, which are hosted by a broker. Subscribers obtain all messages published to the topics to which they subscribe and all subscribers to the topic will receive the same messages. Each topic is grouped by keywords. In the content-based system, messages are only delivered to a subscriber if the attributes or content of those messages match constraints defined by one or more of the subscriber's subscriptions. This method is based on tuple-based system. Subscribers can get a selective event using filter in form of name-value pairs of properties and basic comparison operators (=, >, ≥, ≤, <) However, these classical approaches have limited in designing the object-oriented system. They have considered the different models for the middleware and the programming language. Consequently, the object-oriented and message-oriented worlds often claimed to be incompatible [10]. Patrick T. Eugster presented *a type-based* publish/subscribe which can provide type safety and encapsulation [11].

3 PARMI System Architecture

3.1 PARMI Framework

As shown in Fig. 2, PARMI consists of four main components: *publishers, subscribers, adapter,* and *items*. Publishers and subscribers preserve the stub/skeleton and the remote reference layer, because they use the same objects and methods from the RMI system. However, the adapter and items are designed particularly for the PARMI framework that enable the publish/subscribe communication.

The adapter is a central entity that keeps a set of all available items for publishers and subscribers in a hierarchical manner. The set is represented as a hashtable €= $\{x_1 \ldots x_m\}$, where x_i is an item. The set has a unique topic as a key mapping with an item x_i for a value as hashtable. The adapter collects subscriptions and forwards events to subscribers. The adapter executes the same role as a server in the existing RMI system.

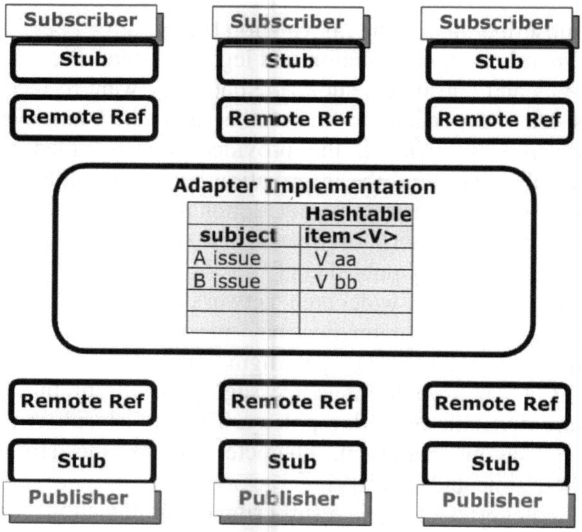

Fig. 2. PARMI organization

An item x_i is an element in the adapter holding the mapping information in three attributes: a topic as a String type, a value as a generic form <V>, and a collection of subscribers' remote references £= $\{S_1 \ldots S_m\}$ interested in the topic. Generics enables type-safety and simple interfaces to create automated data objects[12]. <V> can be any type of object such as Integer, Boolean, Double, String, Arrays, or even Class. Whenever a corresponding event occurs, an item keeps adding or removing three attributes. When *notifySubscribers* method is locally invoked, it notifies all *subscribe* methods of each subscriber registered. And this mechanism was inspired by the observer-observable design pattern, which is also available in java.util.* libraries in the form of interface *Observer* and class *Observable*.

Fig. 3 shows operations in the PARMI system. If we have a set of processes, $\Pi= \{p_1 \ldots p_n\}$ that communicate by exchanging information items, then each process p_i executes three operations: $publish_i(v,t)$, $register_i(S_i,t)$, and $unregister_i(S_i,t)$, where v is a value of an information item, t is a topic for publish and subscribe, and S is a subscriber itself.

Each process p_i can be either a publisher or a subscriber. If p_i is *a publisher*, then it sends the data to the adapter when it is ready. When $publish_i(v, t)$ method is invoked by process p_i, the adapter checks its Hashtable € whether it contains the item with the key t. If the item x already exists, the adapter updates the item's value v. If not, the adapter creates x with v. After updating with the value, the adapter notifies to all subscribers £= $\{S_1 \ldots S_m\}$ who registered themselves to x.

On the contrary, p_i is *a subscriber* S_i, it registers itself with an item in the adapter to receive the data interested in, or releases itself from an item in the adapter if the subscriber doesn't want to subscribe anymore. p_i interacts with other processes by

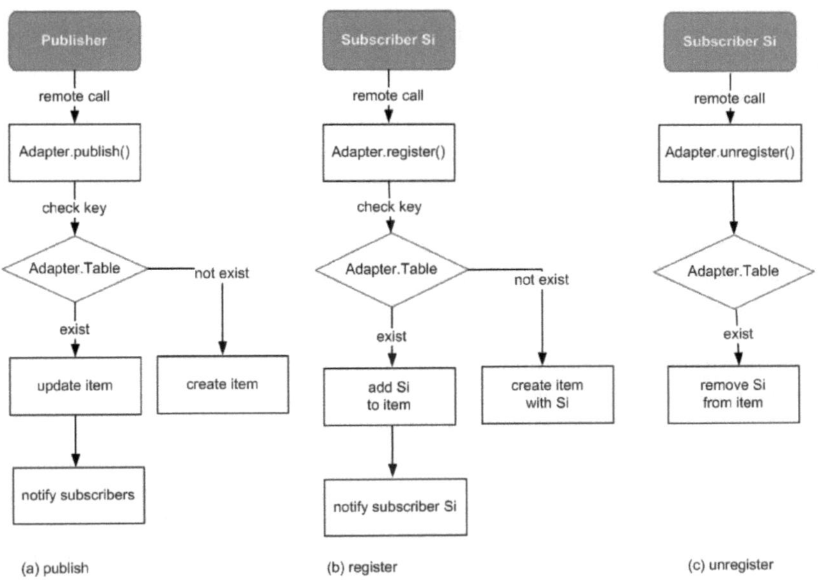

Fig. 3. The operations in PARMI

subscribing the class of events interested in and by publishing event notification. p_i receives a value v submitted by other processes by executing an upcall to *subscribe*$_i$*(v, t)*. Subscriptions are respectively installed and removed at each process by calling *register*$_i$*(S, t)* and *unregister*$_i$*(S, t)* operations. That is p_i receives a value v only after registering its interest by calling *register*$_i$*(S, t)*. After that, if p_i carries out *unregister*$_i$ *(S, sub)*, then it is excluded from the subscription list. The adapter checks Hashtable € with a key t for *register*$_i$*(S, t)* method which is invoked by a subscriber S_i. If an item x with t already exists, the adapter adds S_i to x and notifies only to the current added S_i who wants to receive x at this instant. If not, the adapter creates x with adding S_i to subscription collection £. On the other hand, the adapter removes S_i after *unregister*$_i$*(v, t)* method is called. In the PARMI, publishers and subscribers carry out the same role as clients in the existing RMI system.

Because all three methods are in the adapter and their return types are void, p_i invokes each method using RMI with a future object and passes the control to the adapter. All reference data are handed over to the adapter, but p_i does not need to wait until the method finishes its work.

3.2 Application of PARMI with Jacobi Iteration

To provide a parallel grid computing with the Jacobi method, we use *centralized summation algorithm* that is composed of a central *master* and *N workers*[13]. *The master* partitions a matrix into N number, assigns a unique id to each process, keeps the information item which is provided by workers, and collects all convergence rates from each worker. *The workers* are labelled with a unique id which indicates the position of the matrix. *Each worker* calculates interior points of the matrix, exchanges boundaries of the matrix, and computes the maximum difference between the old matrix and the new matrix. Fig. 4 and 5 show the detailed Jacobi iteration processes on the PARMI between the master and workers.

Fig. 4 shows communications between a master and workers before workers start their own calculations. (1) A master registers itself with the RMI registry on its host name using a unique name, "Jacobi". A worker requests the remote reference of the master to the RMI registry using lookup service. (2) The master returns the remote reference of itself. Now, each worker is ready to invoke methods of the master. (3) The master assigns ids to all the workers at the same time. If there is a worker that did not request its id, then the rest of the workers wait until the entire workers request their ids. As soon as the last worker requests its id, all the workers are notified and their ids are simultaneously assigned. (4) After that, the workers can start their jobs.

Fig. 5 illustrates the remaining communications between a master and workers after workers start their own jobs. (1) A worker registers itself with a topic in which is interested, i.e. the topic is the boundaries of the assigned matrix. And it calculates two different matrix values and computes the maximum difference between them. (2) A worker publishes the boundary values of the matrix. (3) If a worker registered its topic and the topic's value was published, then it automatically subscribes the topic's value from the master. Because each worker has a different speed, information items in the Hashtable is created by either *register* or *publish* method. Each worker repeats its process until the maximum difference reaches the convergence rate. Then the worker sends its convergence rate to the master. The master notifies the worker to stop its job if all the workers have reached the local convergence.

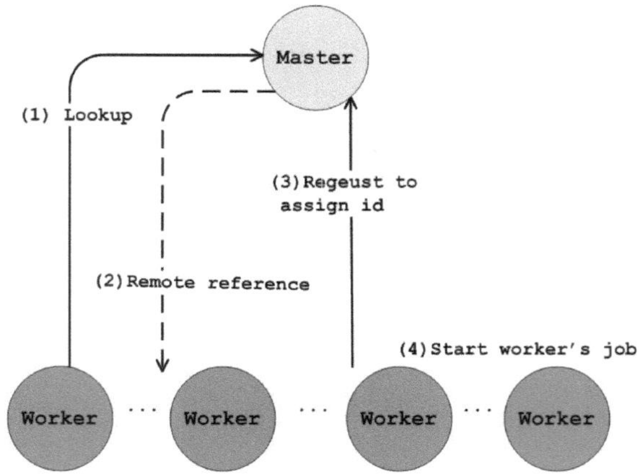

Fig. 4. The Jacobi iteration processes between a master and workers: part1

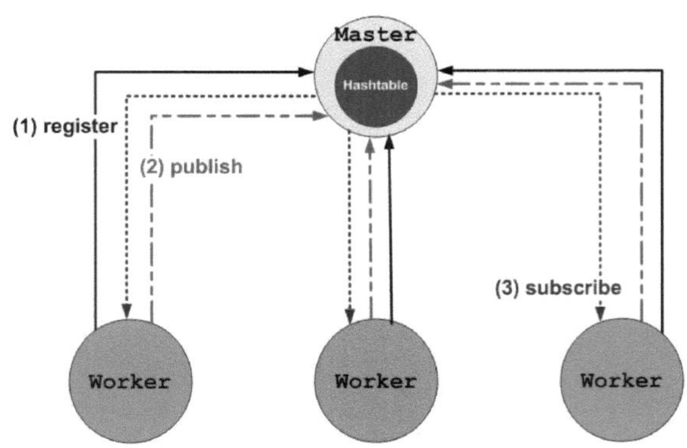

Fig. 5. The Jacobi iteration processes between a master and workers: part2

Fig. 6 is the schematic diagram of the asynchronous Jacobi implementations. The PARMI framework provides the package, *s3lab.parmi.**. The things to use PARMI are implementing *Adapter* and *Subscriber* interface. For the first time user, we provided *AdapterTask* class which already implemented three methods in *Adapter* interface using *Item* class. For Jacobi iteration, we made three different packages, *jacobi.master.**, *jacobi.commnon.**, and *jacobi.worker.** according to the RMI syntax. The master implemented *Adapter* interface and the worker did *Subscriber* respectively. For remote site calls, the *jacobi. commnon.** package is located in both master and worker sides. A worker in jacobi.worker.* package has the roles: registers its interested parts; calculates its own part; publishes the edges of its own matrix; and sends its id, value, and iteration time to the master node when it reaches the convergence rate. A master in jacobi.master* package plays the following roles: divides the whole matrix into the number of workers; allocates the matrix to each worker; collects the convergence rate; and send its decision to stop worker if all worker reach the convergence rate.

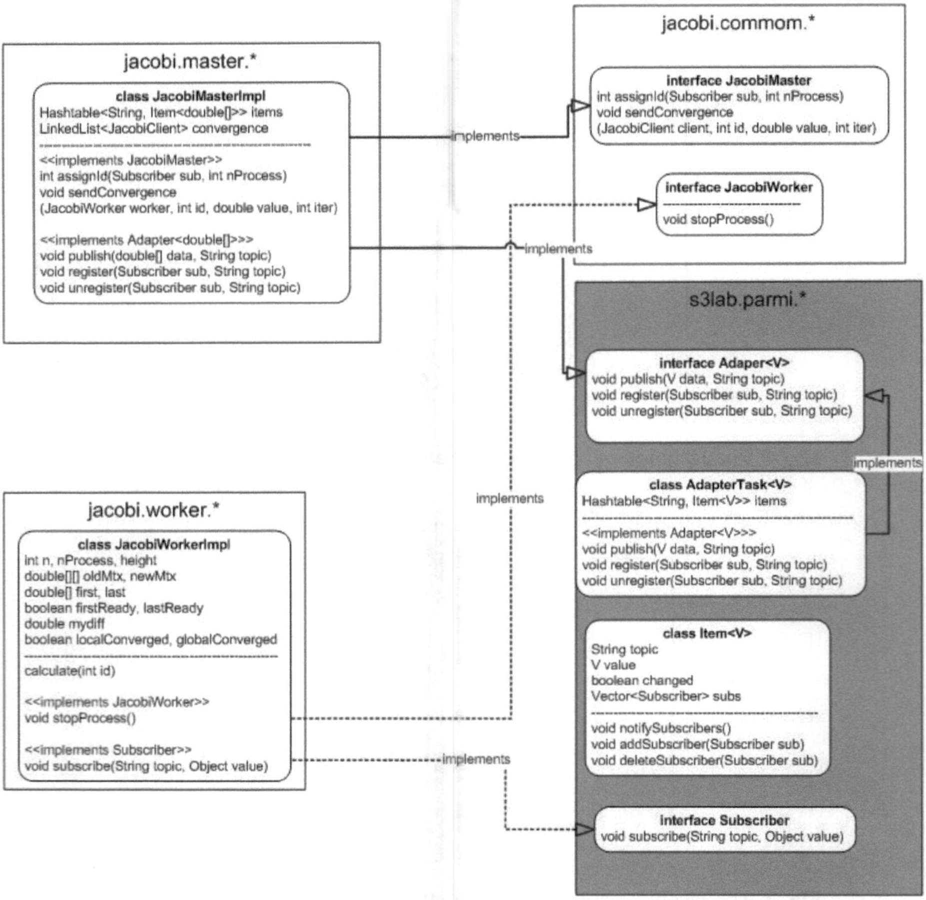

Fig. 6. Implementation of Jacobi application using PARMI framework

4 Performance Results

All experiments were conducted on a cluster with 60 processors combined with a server and 4 racks: The server machine has dual 2.4 GHz processor with 2 GB memory; and the racks have one or dual processors and their speeds are from 1.5 to 3.0 GHz with 0.25 to 1 GB memory. We used the server as a master and the 4 racks as workers. All of the machines run Linux 2.6.9 kernel. All experiments were conducted using SUN's JDK version 1.5.0. All machines were connected to each other by GB interconnection. Measurements were performed using the *System.currentTimeMillis()* method in Java with convergence threshold of 0.03, and grid size of 18,000 by 18,000.

We did not count the initial response time of each worker. All workers wait until the last worker requests its id, and complete their connections to the master. We could not predict the time between the workers and the master because each machine has a different time interval between these processes. Due to this unpredictability, we did not include the time before all processes connected to the master to be assigned ids. The synchronous version was implemented using the current existing RMI communication with synchronous iterative algorithm, and the asynchronous one was done using the PARMI framework with the asynchronous iterative algorithm [14].

Fig. 7 (a) shows the execution costs for the synchronous and asynchronous versions in the master in order to analyse the overall performance. After the total processors for workers reached at a certain point, the asynchronous versions were faster than the synchronous one. In our experiments, the number of processors was 15. Fig. 7 (b) shows, the iteration numbers for asynchronous version has irregular because of asynchronous algorithm.

Fig. 8 shows that asynchronous versions are faster than the synchronous counterparts for communication time. In synchronous version, workers become idle until the process of a master is completed and they get the data needed for their next iteration. Furthermore, the convergence detection is done with a gather-scatter operation at each iteration, which takes longer time than a totally asynchronous detection. On the other hand, synchronous version has less computation time than asynchronous one because its iteration number is always less than the asynchronous one.

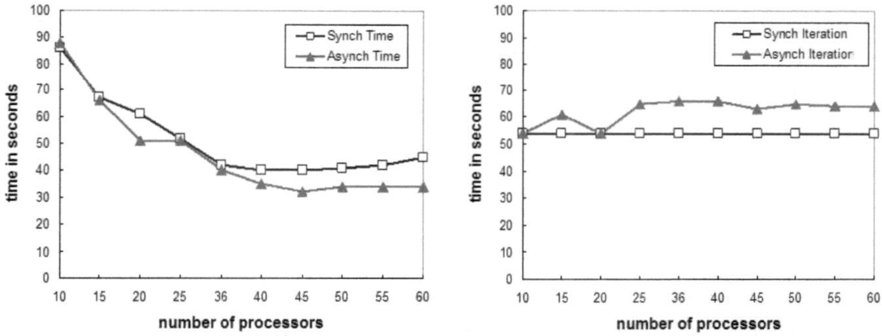

(a) (b)

Fig. 7. Experimental results on a master, (a) Total execution cost for sync/async versions, (b) Iteration number for sync/async versions

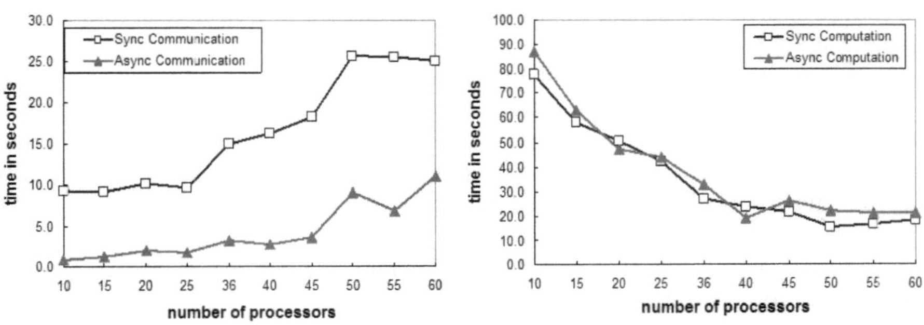

(a) (b)

Fig. 8. Experimental results on workers, (a) Average communication cost for sync/async versions, (b) Average computation cost for sync/async version

Table 1. Comparisons of speedup, (t): total execution cost, (c): average communication cost, Speedup is the time on synchronous versus asynchronous versions

N	10	15	20	25	36	40	45	50	55	60
Sync(t)	86	67	61	52	42	40	40	41	42	45
Async(t)	88	66	51	51	40	35	35	34	34	34
Speedup(t)	1.0	1.0	1.2	1.0	1.1	1.1	1.1	1.2	1.2	1.3
Sync(c)	9.3	9.1	10.2	9.7	15.1	16.3	18.3	25.6	25.5	25.1
Async(c)	0.9	1.2	2.0	1.8	3.2	2.7	3.6	9.0	6.8	11.0
Speedup(c)	10.3	7.4	5.1	5.5	4.7	6.0	5.0	2.8	3.8	2.3

PARMI produces a dramatic performance increase and high throughput by the improvement of communication time. We desynchronized the communication using asynchronous algorithms and asynchronous RMI method invocations. These suppressed all the idle time, and so reduced the whole execution times.

Finally, Table 1 summarizes all the results above. These results demonstrate that the speedup of synchronous versus asynchronous communication improves over 2 times. In these experiments, we can see results that when the ratio of computation time to communication time increases, the ratio of synchronous time to asynchronous time decreases because computation time takes a majority compared to communication times. Hence, for very large problems, it requires more processors to both reduce computation times and preserve an efficient ratio of synchronous time to asynchronous time.

5 Conclusion

We have investigated how RMI can be made asynchronous and suitable for dynamic parallel and distributed systems. Our goal was to design a framework that offers efficient communication for scientific computing, preferably using communication models that integrate cleanly into Java and are easy to use. For this reason, we have taken the existing RMI model as a starting point in our work.

We have given a description of the RMI model and investigated asynchronous RMI implementations on their suitability for high performance parallel programming. This analysis showed that these existing RMI implementations are not efficient to fully utilize a high-performance network because of point-to-point and synchronous communication nature.

To address the communication bottleneck, we designed and implemented PARMI, a high-performance RMI framework that is specifically optimized for parallel programming in a heterogeneous cluster computing environment. To overcome point-to-point and asynchronous communication, we adopted a publish/ subscribe communication model. We also used Generics to provide a flexible and scalable object-oriented platform.

A scientific application using the Jacobi iteration method has been developed to demonstrate the performance gain using PARMI communication framework compared to the conventional RMI mechanism. To enhance the performance improvement, we chose synchronous and asynchronous iterative algorithms. The synchronous version was implemented using the current existing RMI communication and synchronous iterative algorithm. The asynchronous version was implemented using PARMI framework following the asynchronous iterative algorithm. We showed that the asynchronous application using PARMI significantly improved the performance compared to the synchronous one. We have also showed that the performance improvement was mainly owing to reduced communication overhead. Our ongoing research is to realize asynchronous communication for parallel applications in grid environments.

References

1. Sun Microsystems: Java Remote Method Invocation: Distributed Computing for Java (1994)
2. Sysala, T., Janecek, J.: Optimizing Remote Method Invocation in Java, pp. 29–33 (2002)
3. Izatt, M., Chan, P., Brecht, T.: Ajents: Towards an Environment for Parallel, Distributed and Mobile Java Applications. Concurrency: Practice and Experience 12, 667–685 (2000)
4. Kurzyniec, D., Sunderam, V.: Semantic Aspects of Asynchronous RMI: the RMIX Approach, 157 (2004)
5. Raje, R., Williams, J., Boyles, M.: An Asynchronous Remote Method Invocation (ARMI) Mechanism for Java. Concurrency: Practice and Experience 9, 1207–1211 (1997)
6. Walker, E.F., Floyd, R., Neves, P.: Asynchronous Remote Operation Execution in Distributed Systems, 253–259 (1990)
7. Eugster, P.T., Felber, P.A., Guerraioui, R., Kermarrec, A.-M.: The Many Faces of Publish/Subscribe. ACM Computing Surveys 35, 114–131 (2003)
8. Oki, B., Pfluegl, M., Siegel, A., Skeen, D.: The Information Bus: an Architecture for Extensible Distributed Systems. In: Proceedings of the fourteenth ACM symposium on Operating systems principles Asheville, North Carolina, United States, pp. 58–68 (1993)
9. Wikipedia: Publish/Subscribe - Wikipedia, the Free Encyclopedia (2006)
10. Eugster, P.T., Guerraoui, P., Sventek, J.: Type-Based Publish/Subscribe. Technical Report (2000)
11. Eugster, P.T., Guerraoui, R.: Distributed Programming with Typed Events. IEEE Software (2004)
12. Bracha, G.: Generics in the Java Programming Language (2004)
13. Foster, I.: Designing and Building Parallel Programs 2006 (1996)
14. Bahi, J., Contassot-Vivier, S., Couturier, R.: Coupling Dynamic Load Balancing with Asynchronism in Iterative Algorithms on the Computational Grid. In: 17th IEEE and ACM int. conf. on International Parallel and Distributed Processing Symposium, p. 9. IEEE computer society press, Nice, France (2003)

Coarse-Grain Time Slicing with Resource-Share Control in Parallel-Job Scheduling

Bryan Esbaugh and Angela C. Sodan

University of Windsor, Computer Science
401 Sunset Ave., Windsor, Ontario, N9B3P4, Canada
{esbaugh,acsodan}@uwindsor.ca

Abstract. We present a parallel job scheduling approach for coarse-grain timesharing which preempts jobs to disk and avoids any additional memory pressure. The approach provides control regarding the resource shares allocated to different job classes. We demonstrate that this approach significantly improves response times for short and medium jobs and that it permits controlling different desirable resource-shares for different job classes at different times of the day according to site policies.

1 Introduction

User satisfaction in regards to getting jobs run on parallel machines is often not sufficiently satisfactory because short and medium jobs may under certain conditions have long wait times. Our goal is to provide an approach which can provide better service to short and medium jobs, while permitting explicit control over how much resource share is allocated to medium and short jobs vs. long jobs at different times of the day.

We address this problem by providing a practically well feasible approach of time-shared execution. If time sharing is used in parallel-job scheduling, the typical approach is gang scheduling which creates multiple virtual machines and switches synchronously between different time slices via central control, applied to all machine nodes or hierarchically organized to subsets of them. Gang scheduling provides higher probabilities to get short and medium jobs scheduled quickly even in the presence of wide long-running jobs because of having multiple virtual machines available in which to place the jobs. Time slices under gang scheduling are in the range of millisecond or a few seconds. However, gang scheduling increases the memory pressure because gang scheduling can be efficiently implemented only by keeping all active jobs in memory, meaning that the jobs from multiple slices have to share the memory [11][14]. In addition, no explicit control regarding backfilling is provided.

Another option is to preempt jobs to disk by suspension or checkpointing and to support coarse-grain time sharing (switching between jobs in long time intervals like minutes or hours) [11] or take off jobs in special cases only [5]. However, all existing preemptive approaches lack explicit control of scheduling options for individual jobs.

R. Perrott et al. (Eds.): HPCC 2007, LNCS 4782, pp. 30–43, 2007.

Our approach (called SCOJO-PECT as an extension of our earlier SCOJO scheduler by preemption and control) employs coarse-grain time sharing with explicit control (specifying desirable resource share parameters by the system administrator), based on suspension of jobs to disk. In detail, our approach

- uses explicitly controlled preemption in certain long-range time slices to improve response times for short and medium jobs,
- applies safe backfilling,
- supports varying resource allocation policies over the day,
- uses a share-based control without priorities to drive the scheduling of jobs.

In [2], we have presented an initial solution and evaluated it in a grid context. In this paper, we refine our approach for local clusters, experiment with the possible further optimization of merging medium and long jobs under certain conditions in the job context, and evaluate our approach for different configurations and with both workloads generated by the Lublin-Feitelson [6] model and real job traces taken from the Feitelson Workload Archive [3]. We demonstrate that our approach improves overall response times and bounded slowdown significantly, therefore being competitive to gang scheduling without the memory pressure involved in gang scheduling. We also show that short and medium jobs are served very well, independent of the other jobs in the system.

2 Related Work

Preemption has been found useful in providing good utilization and response times, though only if accompanied with migration, i.e. being able to select new resources when rescheduling the job [9][15]. Migration is only possible with checkpointing, but checkpointing should be done at suitable application-specific points in the execution (which may occur in the order of hours based on personal communication with Sharcnet system administrators). Suspension to swap disk requires continuing the job on the same resources.

Gang scheduling, which switches between jobs in a coordinated manner via global time slices, is also a kind of preemption but keeps jobs in memory to reduce switching costs. Gang scheduling is known to provide better average response times and bounded slowdowns [8]. Similar benefits as from gang scheduling were found to be obtainable in [11] [16] via coarse-grain time sharing. The approach in [1] finds benefits if preempting jobs after a maximum of 1 hour runtime provided that they are over a certain size limit (this approach ignores preemption cost) but is combined with migration. These approaches do not provide any fair-share considerations. The approach in [5], however, demonstrates benefits in both average and worst-case slowdowns for all job classes with suspension only. The approach considers the relative slowdown of preemptor and preemptees and imposes limits on possible slowdown and relative sizes between preemptor and preemptees to avoid that long-running jobs suffer disadvantages.

Job schedulers normally apply backfilling which permits jobs to move ahead vs. their normal scheduling position to better utilize space. Conservative backfilling only permits this to happen if no other job in the queue is delayed. EASY backfilling only guarantees the first job in the waiting queue not to be delayed. Preemption may also be applied in conjunction with backfilling: Jobs may be backfilled even if their estimated runtimes are longer than the available backfill window and be terminated/restarted or preempted if their actual runtimes create a conflict [10] [13].

Fair share scheduling was first proposed by Maui [4] though now other schedulers like LSF have included fair-share ideas, too. Maui maintains shares per user or group over time and adjusts them by considering the recent past time intervals and weighing them exponentially. Maui combines a number of factors such as political priorities (like group), system priorities, and user priorities and translates them into priorities. Moab [7] is a commercial extension of Maui with more differentiated optimization and priority approaches and also support of preemption, which can optionally be applied upon failure to meet fair share targets, upon backfilled jobs running into reservations, upon expected response times becoming too long, or upon jobs with higher priority arriving. However, the actual algorithms are not revealed.

3 Preemption and Share Control

3.1 Problems in Standard Job Schedulers

There exist a couple of problems in standard job schedulers which lead to problems which cannot be resolved without preemption or different partitions for different service requirements:

- a wide long-running job may block a short or medium job from being scheduled
- wide short or medium jobs, though occurring infrequently, are hard to serve well and–as all wide jobs–require the machine to be drained
- the machine may be well utilized by medium and long-running jobs, leaving no chance for a short job to be filled or backfilled quickly

In addition, prevention of starvation requires some aging of long jobs but then they may end up being scheduled at an unsuitable time and EASY backfilling can push jobs unfairly backward.

The above reasons can especially affect maximum response times but also average response times. Of course, there exists also a dependence on the current system load—which is unavoidable, i.e. jobs wait longer if there is much work on the machine and in the waiting queue ahead of them.

3.2 Our Basic SCOJO-PECT Approach

The basic idea of our approach is coarse-grain time sharing with time slices in the minute (or potentially hour) range. This permits suspension of jobs to disk, keeping all memory available for the next running job, and making the overhead tolerable, while accomplishing similar benefits as gang scheduling. The approach is feasible in most cluster environments: for example, most clusters in the Sharcnet project [12], have a local disk per node and 8 Gbyte of memory. If all memory is used by the application this makes 2 * 8 Gbyte (in and out) of data to be transferred. Assuming a bandwidth of 150 Mbyte/sec, this results in 109 sec. With 3 time slices (short, medium, and long-running jobs) per time interval but short jobs likely to use little memory, this makes a maximum of 4 min swap time per time interval.

The time slices are primarily designated for certain job classes, i.e. each slice has a specific dominant job type. This permits for example a differentiation into short-job, medium-job, and long-job time slices. This approach provides a basis to perform controlled resource allocation for the different job classes.

In addition, we permit definition of certain resources shares (ratios of the overall resource utilization) to be defined for different job classes in different time intervals of the day. These shares are mapped to corresponding lengths of the time slices. Shares express which jobs should run at each time of the day but also indirectly determine the priority given to a job class. Thus, it makes sense to allocate higher possible shares to a job class than its average usage is (as we have done in our experiments for medium jobs). The expected benefits of our overall approach are

- to make sure that short jobs can be scheduled, independent of the currently running medium and long jobs and independent of their own size,
- consequently also to be able to serve medium jobs well in the presence of wide long-running jobs and being able to control how certain job classes are served over the day, e.g. to provide less share for long jobs at daytime and more at night time,
- to avoid problems like stranded jobs which may result from preemption of individual jobs by preempting the whole group of running jobs.

In addition, we have a more explicit handle to control resource allocation to jobs than using the typical priority approach which keeps the effects hard to understand and can lead to undesirable cases like a wide long-running job finally ending up being scheduled during the day.

A potential problem is that the differentiation of job classes into different time slices leads to fragmentation. However, we permit limited backfilling of other job classes and merging of slices as described below.

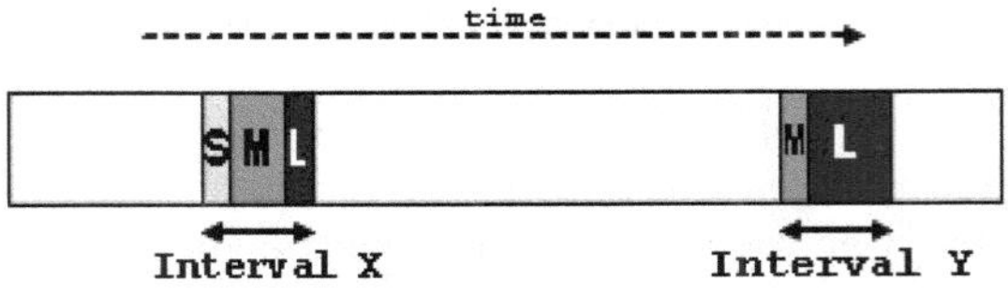

Fig. 1. Share allocation in different intervals during different times of the day. S are short, M medium, and L long-running jobs.

3.3 Shares and Time Slices

We split the 24 hours of the day into a number of equidistant time intervals and permit target ratios for different job classes to be defined for every time interval. This task may be done by system administrators according to site policies, while incorporating feedback from job traces regarding the overall typical job mixes.

To keep slice handling manageable, we define slices per each interval, i.e. we keep fixed boundaries for the slices, see Fig. 1. Thus, we may have one short-job slice, one medium-job time slice, and one long-job time slice per 60 minute interval. We define the slices at the beginning of the time interval, while deciding the slice lengths for the different job classes on the basis of the target ratios. However, short jobs are handled separately and get a slice length according to the longest waiting short job. Thus, short jobs are not preempted except as a result of being backfilled into a non-short time slice. The remaining time of the time interval is split according to the ratios defined among the other job classes. If no jobs of a certain class are available, the share is attributed to another class for the corresponding interval. If using short, medium, and long job classes, the ratio defines how the time is split among medium and long jobs.

Accounting considers all used shares from the different job classes per time interval, with the sum representing the machine utilization for the time interval. The accounting considers jobs backfilled into a slice that is basically designated for a different class. In addition, it can happen that in certain time intervals no jobs of a specific class are available. Thus, we apply an adjustment of share allocation, based on target shares and past usage, with the latter being considered for a certain past time window and weights declining exponentially with time distance. Note that this corresponds to the fair-share idea as introduced by the Maui job scheduler [4]. We use m different weights, calculated as $A * B^m$ with $0 < B < 1$. B determines how quickly the weights decline and $A * \sum_{i=1,m} B^i$ determines the impact given to the past. Then, the share allocation is adapted to variations in job submission within the range of the typical job mix.

3.4 Scheduling with Controlled Backfilling into Time Slices

At the beginning of each time interval, time slices of the different types for the corresponding job classes are obtained, according to the description in Section 3.3. Time slices of different types (designated to a specific job class) are currently always scheduled in the same order for each time interval, i.e. from short to long. The initial allocation for each time slice is the set of jobs that were preempted when the corresponding time slice (of same type) in the previous time interval ended. This permits jobs to be re-started on the same resources. Then, the scheduler attempts to allocate jobs of the job class which corresponds to the slice type from the head of the preemption and waiting queues. The resource-allocation approach is either first-fit or a smart node-selection heuristic as described below. Afterwards, the scheduler attempts to backfill as described below. The scheduling order per job class is FIFO. This is feasible because we no longer need priorities to give shorter-running jobs better service.

Our basic backfilling approach is EASY, adjusted to work with different job classes and our time-slice restrictions. We first try to EASY backfill from the same job class. Then we backfill from other job classes, while searching the other job classes in increasing order of their runtime ranges. This step is important because often jobs of same type cannot be packed well enough or because there may not be sufficient jobs of each type at a certain time. Note that this backfilling is safe, i.e. does not create any resource conflicts but rather permits jobs to start/continue running in the other slice if there are no conflicts. Then, this backfilling stage applies the following rules:

1. Any preempted jobs from other slices that can fit onto the free resources are candidates for backfilling (they are handled before any waiting jobs).
2. Any waiting jobs of other classes can be backfilled if not delaying the first waiting job of their own class.

In both cases, jobs either need to finish before the end of the slice or run on resources which are not yet allocated in the slice of their own corresponding type and therefore can continue running in their own slice. The abstract code is shown in Fig.2.

In addition, we experiment with two further possible optimizations: 1) smart node selection (hoping to increase the options for backfilling of other-type jobs, and 2) merging of medium/long type slices to increase utilization. The smart node-selection heuristic tries to allocate jobs on resources which are not yet allocated to any job in any of the slices. This makes it more likely that jobs can backfill. The heuristic counts the jobs allocated per node and then selects the nodes with the lowest count. This heuristic is less important for highly loaded systems but can play a role in cases of only sporadic arrival of certain job types.

The idea behind merging of medium/long slices is that keeping different job classes can potentially provide poor utilization, e.g. if there is only one job of each class, each using only some of the resources. Thus, the scheduler merges slices if all of the below conditions apply:

- slices (currently medium and long) are merged if the utilization is poor,
- there are no very wide jobs in any of the slices (which would lead to serious resource conflicts),
- merging slices does not create too many resource conflicts (two jobs using fully or partially the same resources).

While running in merged mode, the scheduler attempts to maintain the current target shares and predicted target shares for the next interval (within some tolerance range) when selecting medium and long jobs to schedule. If there are any resource conflicts under merging, the heuristic to solve this problem is to schedule the job first which bests meets the target shares. The backfilling is modified to exclude nodes which are needed to schedule waiting preempted jobs.

Conversely, we need to decide when to split a merged slice again into different slices per class. Fortunately, this does not involve any resource problems: the

```
// POTENTIAL OPTIMIZATION for merging medium/long slices
//     schedule preempted jobs of slice type, avoid collisions if merging
for (all job in preemptionQueue[sliceType])
    if (!collision) scheduleJob(job);
    else scheduleJob(bestFitTargetShareJob());

// try to schedule waiting jobs (avoid collisions with preempted jobs)
if (scheduledPreemptedJobs)
    {  // POTENTIAL OPTIMIZATION for smart node selection
       excludedResources = collectResourcesUsedByScheduledPreemptedJobs();
       for (job in waitingQueue[sliceType] )
            if (jobSchedulableWithExcludedNodes(job,excludedResources))
                scheduleJob(job);  else break;  }
else for (job in waitingQueue[sliceType])
            if (jobScheduable(job) scheduleJob(job); else break;

tryEasyBackfill (sliceType);

// try restrictive backfilling with other job types
for (queue in preemptionQueue) // sorted by increasing runtime class
    for (job in queue)
         if (jobFits() && noCollision(job,excludedResources)
             scheduleJob(job);
for (queue in waitingQueue) // sorted by increasing runtime class
   {limit = findShortestRemainingRuntime(runningQueue);
    for (job in queue)
         if (runtime(job) <= limit && jobFits() &&
           noCollisionInOwnSlice(job.jobType)   scheduleJob(job); }
```

Fig. 2. Core scheduling algorithm

jobs are simply sorted into their corresponding time slices according to their type. We use the following criteria for splitting:

- There are wide jobs which otherwise cannot be scheduled.
- The required share cannot be maintained if keeping the slice merged.

3.5 Properties of SCOJO-PECT Scheduler

As long as conservative scheduling is applied, the following property applies for response times R for a certain job class C, $maxR$ maximum response times, a time span T under consideration with $T(C)$ being the shares allocated to C; $Prio$ being the standard priority scheduler, $Conly$ a FCFS schedule for only the jobs of C (dropping all other jobs from the schedule), and $PECT$ is our SCOJO-PECT time-sharing scheduler:

$R(C, PECT) \leq R(C, Conly) * T/T(C) \leq R(C, Prio) * T/T(C)$ and

$maxR(C, PECT) \leq maxR(C, Conly) * T/T(C) \leq maxR(C, Prio) * T/T(C)$

The second part of the inequality is obvious: scheduling only jobs of the same class must perform better than a mix with other jobs which may delay jobs of

C. The first part of the inequality considers that our time-sharing scheduler only allocates certain time intervals to the job class C. The time $R(C, PECT)$ may be lower than adjustment by the corresponding factor because we may be able to safely backfill into the other slices (gaining additional runtime) and better filling options may arise if stretching the scheduling time by time slices. However, if we apply EASY scheduling, occasionally new opportunities may arise to push back jobs which may affect the response times of C negatively.

Larger time intervals for C do not necessarily provide better response times because few or no jobs of the class may be available at certain times. Better packing may be accomplished if letting jobs queue up a little (giving more choices).

4 Experimental Evaluation

4.1 Experimental Setup

We have used the Lublin-Feitelson model [6] for the workload generation and traces from the Feitelson Workload Archive [3]. The Lublin-Feitelson model is a complex statistical workload description, derived from real traces and considering job sizes, job runtimes, and job interarrival times. The model includes correlations between sizes and runtimes, fractions of sequential jobs, fractions of power-of-two sizes, and differing inter-arrival times according to day/night-cycles. The model can be adjusted to different machine sizes. For details of the workload parameters, see Table 1. We have slightly modified the interarrival times and changed the α parameter from 10.23 to 10.33. This change reduces the average work creation related to available resources from 91% to 84%. The reasons are explained below. The selected real traces are SDSC SP2 1998 and SDSC Blue 2000 (in both cases, the first 10,000 jobs of the cleaned version). Statistics of the workload as obtained with these settings are included in Table 1. What is interesting to observe from the workload characteristics is that though the number of short jobs is very high, the work of the short jobs is $\leq 1\%$. The different workloads have different characteristics: Lublin-Feitelson has the highest percentage of medium-job work. We currently schedule with the knowledge of actual runtimes.

Table 1. Workload parameters and workload statistics

	Lublin-Feitelson	SDSC SP2	SDSC Blue
Machine size	128	128	1152
Number of jobs	10,000	10,000	10,000
N_{short}	63.7%	40.9%	73.75%
N_{medium}	19.3%	35.1%	17.7%
N_{long}	17.0%	24.0%	8.5%
$Work_{short}$	0.5%	0.2%	1.0%
$Work_{medium}$	26.5%	3.7%	15.0%
$Work_{long}$	73.5%	96.1%	84.0%

Table 2. Scheduler parameters

Parameter	Value
Interval	30 min and 60 min
Job classes supported	short, medium, long
Classification short jobs	*runtime* $<$ 10 min
Classification medium jobs	10 min sec $\leq$ *runtime* $<$ 3 hours
Classification long jobs	*runtime* $\geq$ 3 hours
Classification narrow jobs	*size* $\leq$ 10% machine size
Classification medium-size jobs	10% machine size $<$ *size* $\leq$ 50% machine size
Classification wide jobs	*size* $>$ 50% machine size
Bound	15 min
Switch overhead	6 sec and 60 sec
Medium jobs' daily shares	28% over night, 44% over day
Weights, past window	$A = 2, B = 0.81, m = 6$, window is 24 hours $\rightarrow A * \sum_{i=0,m} B^m = 1$

We compared our job scheduler (PECT) to the standard priority scheduling (Prio). In both cases, EASY backfilling was applied. We evaluated response times, bounded slowdowns (response times relative to runtimes with a bound for very short jobs), and utilization. For scheduler parameters, see Table 2. We also tested sensitivity to the scheduling overhead and to the interval time. Furthermore, we explored the effectiveness of different optimizations in the scheduler. Thus, we tested 30 minute intervals with 6 sec switching overhead and all optimizations except merging (PECT 30), 60 min intervals with 60 sec switching overhead and all optimizations except merging (PECT 60), and 60 minute intervals with 60 sec switching overhead without backfilling on non-type jobs (PECT 60N). Additionally, we checked 60 min intervals with 6 sec overhead and merging but do not show the corresponding data in detail.

Summing up the work created by the overall workload and considering the average statistical parameters from the Lublin-Feitelson model regarding the distribution over different time intervals, we obtained the medium shares as shown in Fig. 1. These shares would provide optimum resources to the medium jobs created in this time interval. However, these shares did not provide the best results as actual creation rates show peaks and significant ups and downs.

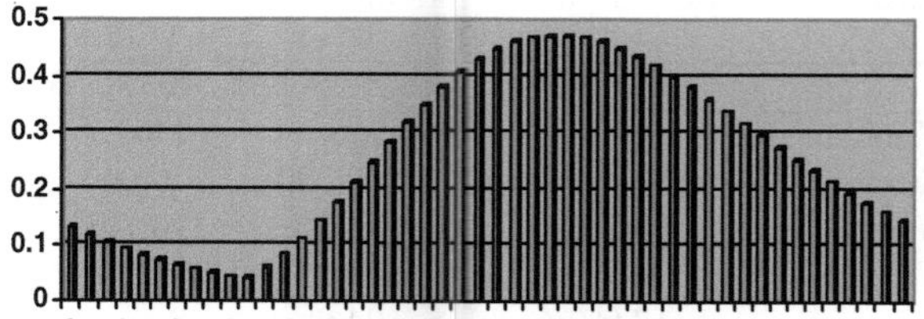

Fig. 3. Shares required to optimally support medium jobs (work created per time interval and corresponding resource share required to schedule this work with 90% utilization of the share)

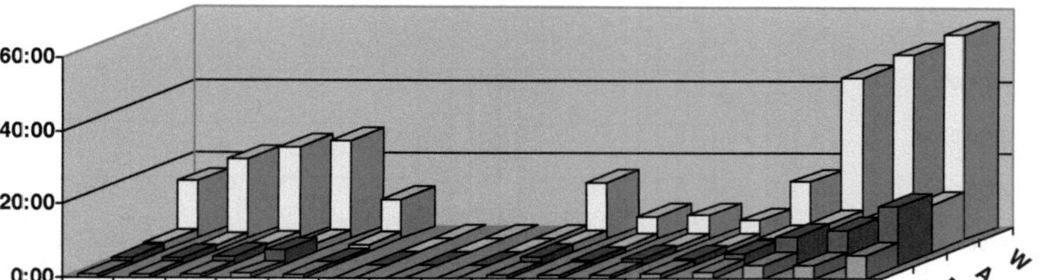

	All Prio	All PECT 30	All PECT 60	All PECT 60N	Short Prio	Short PECT 30	Short PECT 60	Short PECT 60N	Med Prio	Med PECT 30	Med PECT 60	Med PECT 60N	Long Prio	Long PECT 30	Long PECT 60	Long PECT 60N
N	6:09	4:33	4:32	7:11	4:57	0:12	0:22	0:29	6:33	5:45	6:08	7:26	11:17	22:16	22:08	38:26
MS	10:29	12:29	15:24	23:57	6:47	0:13	0:28	0:30	13:46	8:27	10:35	11:51	20:30	48:34	58:46	96:25
A	14:38	9:54	11:08	15:29	11:26	0:12	0:24	0:29	15:21	8:17	9:11	9:56	22:14	48:01	53:19	77:39
W	93:47	130:0	149:1	161:0	65:33	0:34	1:11	0:55	91:57	37:39	38:29	30:20	93:47	265:0	303:2	337:4

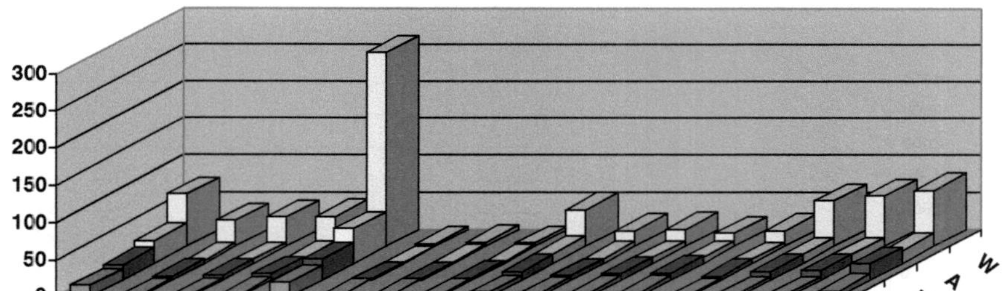

	All Prio	All PECT 30	All PECT 60	All PECT 60N	Short Prio	Short PECT 30	Short PECT 60	Short PECT 60N	Med Prio	Med PECT 30	Med PECT 60	Med PECT 60N	Long Prio	Long PECT 30	Long PECT 60	Long PECT 60N
N	15.60	2.69	3.20	4.01	20.50	1.39	2.12	2.45	7.45	5.45	5.98	6.45	2.59	5.61	4.83	8.30
MS	20.70	4.69	5.97	8.25	27.80	1.46	2.41	2.52	11.40	6.74	8.96	9.68	4.73	11.51	12.25	21.68
A	30.20	4.04	4.87	6.05	46.30	1.42	2.20	2.48	13.30	6.66	7.74	7.92	5.36	11.47	11.52	17.21
W	70.10	34.43	39.55	40.71	262.0	2.84	5.03	4.21	49.40	20.72	22.72	16.96	20.80	61.40	67.21	75.43

Fig. 4. Average response times (top) and average slowdowns (bottom) for Lublin-Feitelson workload. Results are shown for the different job classes short, medium, and long, for the different sizes narrow (N), medium-size (MZ), wide (W), and all (A), and for different scheduler configurations.

We have obtained better results with shares set higher than the actual usage percentages (cf. Table 2).

4.2 Experimental Results

Results for response times and bounded slowdowns are shown in Fig. 4 and Fig. 5. If applying the non-type backfilling optimization, the PECT scheduler served medium and short jobs very well regarding overall response times in most cases and regarding relative response times in all cases. Looking into details for PECT 60 and the Lublin-Feitelson workload, average response times improved by 23% vs. priority scheduling and bounded slowdowns by 84%. This is similar to the results we obtained in [14] for gang scheduling (27% improvement in response

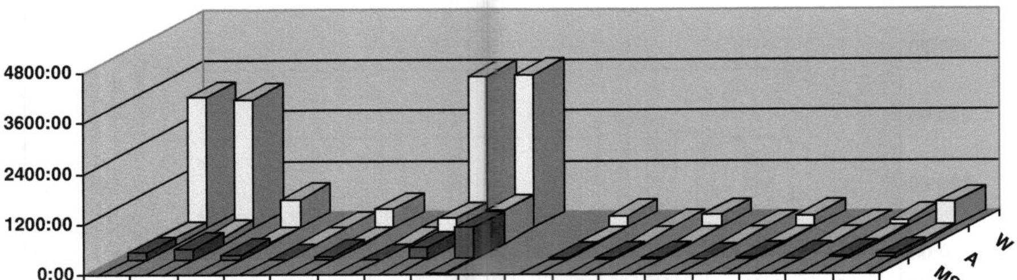

	All Prio	All PECT 60	Short Prio	Short PECT 60	Med Prio	Med PECT 60	Long Prio	Long PECT 60		All Prio	All PECT 60	Short Prio	Short PECT 60	Med Prio	Med PECT 60	Long Prio	Long PECT 60
N	6:14	9:23	1:36	0:15	3:27	2:32	24:33	48:20		3:12	2:18	1:13	0:18	3:42	1:53	27:53	29:06
MS	167:0	268:0	120:2	0:31	80:49	3:12	292:0	757:2		35:12	29:00	32:08	0:30	26:02	15:09	61:52	111:2
A	162:3	187:1	25:26	0:18	26:46	2:43	592:1	772:0		9:33	6:22	3:55	0:20	18:44	6:59	40:14	59:36
W	3099:	3048:	675:4	1:39	427:4	5:02	3584:	3617:		275:5	27:49	292:4	1:01	275:5	8:51	164:5	581:3

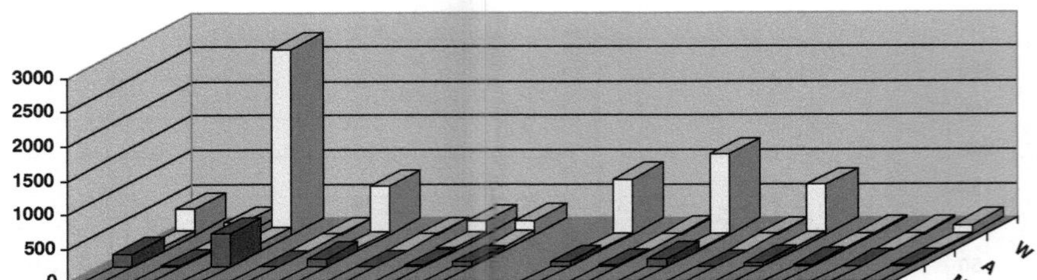

	All Prio	All PECT 60	Short Prio	Short PECT 60	Med Prio	Med PECT 60	Long Prio	Long PECT 60		All Prio	All PECT 60	Short Prio	Short PECT 60	Med Prio	Med PECT 60	Long Prio	Long PECT 60
N	5.52	2.77	7.05	1.67	4.90	2.60	2.79	6.04		5.57	2.09	5.57	1.86	6.77	2.89	3.10	3.53
MS	195.6	24.35	482.0	2.65	127.6	3.92	25.10	62.85		73.34	19.35	129.0	2.54	58.83	33.13	8.72	17.08
A	65.23	14.39	102.3	1.87	42.20	3.06	36.93	52.05		20.39	4.72	16.33	1.91	44.05	14.50	5.14	8.66
W	378.9	172.3	2703.	7.12	706.1	15.24	200.2	202.1		809.6	19.00	1171.	4.54	731.3	17.90	24.00	135

Fig. 5. Average response times (top) and average slowdowns (bottom) for SDSC SP2 (left) and SDSC Blue (right) workload. Results are shown for the different job classes short, medium, and long, for the different sizes narrow (N), medium-size (MZ), wide (W), and all (A)

times and 84% in bounded slowdowns). For the Blue workload, response times decreased by 33% and relative response times by 77%. The SP2 workload looks a little worse: response times increased by 15% but the relative response times again decreased by 78%.

Consistently all results show that short and medium were served very well. Short jobs were served on average within 1/3 of the time interval, and their maximum response times were 3h 53min (Lublin-Feitelson), 5h 40min (SP2), and 2h 13min (Blue). Wallclock runtimes of medium jobs were increased by about a factor of 2.7 in all size classes (because of the time slicing) but wait times significantly reduced (all medium wait times are reduced by 80%), with most reduction for wide medium jobs. However, long jobs currently suffer (due to overall 1.5 times longer wait times), especially long wide jobs. To some extent, longer response times for long jobs are expected because long jobs get less

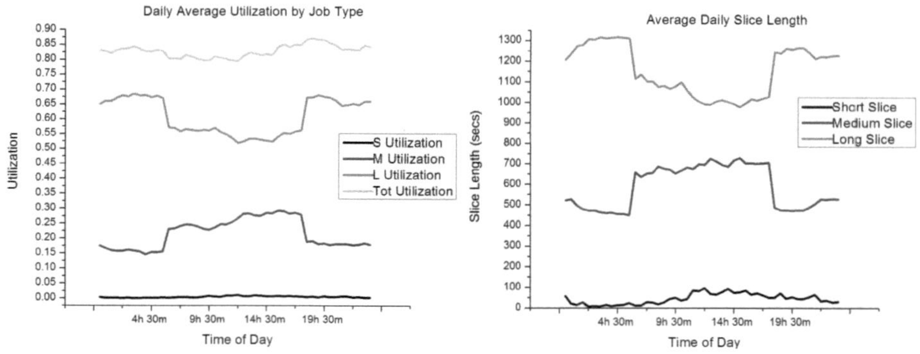

Fig. 6. Daily utilization (left) and daily slice times (right) as average over the whole workload execution

machine share over the day. Since in the SP2 workload long jobs dominate, this explains why the average response times slightly decreased.

For the Lublin-Feitelson workload, we compared using/not using the non-type backfilling optimization. We find that the non-type backfilling optimization cut overall average response times 27% and improved especially results for short and medium jobs. Thus, this is an important optimization. If comparing different time intervals/overheads, 30 min intervals improved overall average response times by 11%, mostly due to improvement for short jobs. This is obvious as they have now a chance every 30 minutes to be scheduled. Otherwise, the difference was not too significant, i.e. higher overheads worked reasonably well. Checking the number of switches, we found that about 2/3 of the possible switches took place. If looking at the response times over the day, PECT gave medium jobs consistently good service (average response times varied only by 26%), whereas the priority scheduler showed a variation of 62%.

Tests showed that using smart node selection and merging medium/long slices (under certain conditions) did not provide any additional benefit (the results are almost the same). If the utilization is high, there is a low chance that smart node selection can help; if the utilization is low, apparently there are not many potential conflicts arising anyway. Regarding merging, backfilling with non-type jobs apparently mimics merging very effectively. We can keep jobs running over several slices of different types (our overhead estimation is therefore pessimistic as on the corresponding nodes, no swapping is necessary), i.e. can say we perform partial slicing for jobs with node conflicts.

Our scheduling approach was capable of maintaining a high utilization: 85.92% with our PECT 30 scheduler and 84.1% for our PECT 60 vs. 88.96% with priority scheduling, i.e. the reduction in utilization was only 3% to 5%. Fig. 6 shows that utilization is consistently high over the whole day and that the share control is effective, reducing the utilization of different job classes at different times of the day, while keeping the machine busy.

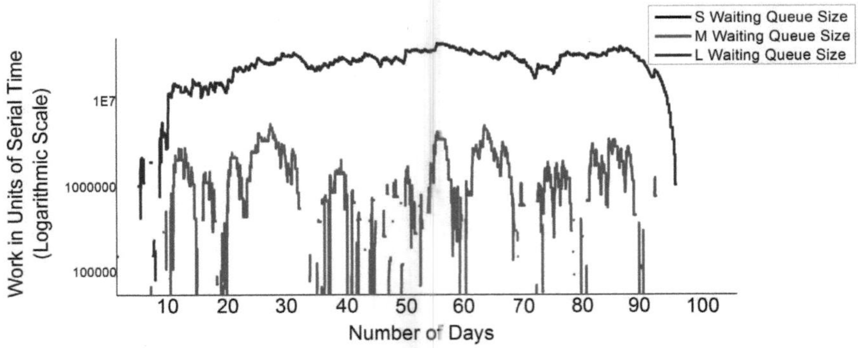

Fig. 7. Average work in waiting queue over makespan, logarithmic scale

Fig. 7 demonstrates that there are ups and downs in the waiting-queues but that no back logging occurs with the current workload settings, i.e. the scheduler is able to keep up with the work generated.

Finally, we have run the scheduler (PECT 30) with a different fine-tuning of the share control. We gave medium jobs the maximum share if the medium workload is high and compare the past usage to the overall averages rather than to the shares set by the scheduler. The medium jobs were served significantly better though at the expense of increasing response times for long jobs. The utilization was 82.2%, i.e. slightly decreased by 3.7%. This means we have a tradeoff between serving medium jobs well and keeping decent response times for long jobs and a good utilization.

5 Summary

We have presented a job-scheduling approach which employs coarse-grain time slicing by suspension to disk and explicit control over how much time share is allocated to different job classes over the day. Such time slicing is more feasible than checkpointing and easier to handle than individual job preemption if the job allocation situation is complex (many jobs to be preempted with very different runtimes). The approach improves overall average response times and average bounded slowdowns, to a similar extent as gang scheduling, and serves especially short and medium jobs well. In future work, we intend to improve our approach by special handling of wide long-running jobs and a combination of time slicing with individual job preemption.

References

1. Chiang, S.-H., Fu, C.: Benefit of Limited Time Sharing in the Presence of Very Large Parallel Jobs. In: Proc. IPDPS (2005)
2. Esbaugh, B., Sodan, A.: Preemption and Share Control in Parallel Grid Job Scheduling. In: Proc. CoreGrid Workshop, Springer, Dresden, Germany (2007)

3. Feitelson Workloads, http://www.cs.huji.ac.il/labs/parallel/workload/logs.html
4. Jackson, D., Snell, Q., Clement, M.: Core Algorithms of the Maui Scheduler. In: Feitelson, D.G., Rudolph, L. (eds.) JSSPP 2001. LNCS, vol. 2221, Springer, Heidelberg (2001)
5. Kettimuthu, R., Subramani, V., Srinivasan, S., Gopalasamy, T.: Selective Preemption Strategies for Parallel Job Scheduling. In: Proc. ICPP (2002)
6. Lublin, U., Feitelson, D.G.: The Workload on Parallel Supercomputers: Modeling the Characteristics of Rigid Jobs. J. of Parallel and Distributed Computing 63(11), 1105–1122 (2003)
7. Moab Workload Manager Administrator's Guide, Version 5.0.0. Cluster Resources, available at http://www.clusterresources.com/products/mwm/docs/index.shtml
8. Moreira, J.E., Chan, W., Fong, L.L., Franke, H., Jette, M.A.: An Infrastructure for Efficient Parallel Job Execution in Terascale Computing Environments. In: Proc. ACM/IEEE Supercomputing (November 1998)
9. Parsons, E.W., Sevcik, K.C.: Implementing Multiprocessor Scheduling Disciplines. In: Feitelson, D.G., Rudolph, L. (eds.) Job Scheduling Strategies for Parallel Processing. LNCS, vol. 1291, Springer, Heidelberg (1997)
10. Perkovic, D., Keleher, P.J.: Randomization, Speculation, and Adaptation in Batch Schedulers. In: Proc. ACM/IEEE Supercomputing, Dallas/TX (November 2000)
11. Setia, S., Squillante, M., Naik, V.K.: The Impact of Job Memory Requirements on Gang-scheduling Performance. Perf. Evaluation Review 26(4), 30–39 (1999)
12. Sharcnet network of clusters (2007), http://www.sharcnet.ca
13. Snell, Q., Clement, M.J., Jackson, D.B.: Preemption Based Backfill. In: Feitelson, D.G., Rudolph, L., Schwiegelshohn, U. (eds.) JSSPP 2002. LNCS, vol. 2537, Springer, Heidelberg (2002)
14. Sodan, A.C., Doshi, C., Barsanti, L., Taylor, D.: Gang Scheduling and Adaptation to Mitigate Reservation Impact. In: Proc. IEEE CCGrid, Singapore (May 2006)
15. Sodan, A.C.: Loosely Coordinated Coscheduling in the Context of Other Dynamic Approaches for Job Scheduling-A Survey. Concurrency & Computation: Practice & Experience 17(15), 1725–1781 (2005)
16. Wong, A.T., Oliker, L., Kramer, W.T.C., Kaltz, T.L., Bailey, D.H.: ESP: A System Utilization Benchmark. ACM/IEEE Supercomputing, Dallas/TX (November 2000)

Quality Assurance for Clusters: Acceptance-, Stress-, and Burn-In Tests for General Purpose Clusters

Matthias S. Müller, Guido Juckeland, Matthias Jurenz, and Michael Kluge

Technische Universität Dresden
Center for Information Services and High Performance Computing (ZIH)
01062 Dresden, Germany
{Matthias.Mueller,Guido.Juckeland}@tu-dresden.de,
{Matthias.Jurenz,Michael.Kluge}@tu-dresden.de

Abstract. Although common sense says that all nodes of a cluster should behave identically since they consist of exactly the same hardware parts and are running the same software, experience tells otherwise.

We present a collection of programs and tools that were gathered over several years during various cluster installations at different sites with clusters from various vendors. The collection contains programs to check for the setup, functionality, and performance of clusters. Components like CPU, memory, disk, network, MPI and file system are checked. Together with the short description of the tools we describe our experiences using them.

1 Introduction

Cluster like systems are in wide use today. This is mainly due to the good price/performance ratio they offer, but also due to the flexibility achieved by combining and selecting from a wide range of components of the shelf (COTS). Since the introduction of the concept in the 90s [1,2] at lot of work has been published how to create and operate such systems [3,4]. Whole collection of tools [5] or complete distributions [6] for clusters have been created. Standard HPC benchmarks can be applied to a cluster if they are suitable for distributed memory machines. Examples are Linpack [7], the NAS Parallel Benchmarks [8,9] and the HPC Challenge benchmark [10]. Acceptance tests used in procurements often comprise micro-benchmarks, standard HPC benchmarks, and different applications from the end users of the system. Despite the wide use of clusters there seems to be a lack of a collection of standard methods, software and tools to help testing a cluster installation from the bottom up. Such a collection would not only help to demonstrate the functionality and performance of a cluster, but also help to identify the cause if the cluster does not work as expected. In this paper we describe the collection of benchmark, programs and tools that we used during the installation, acceptance and early production phase of our latest cluster installation.

R. Perrott et al. (Eds.): HPCC 2007, LNCS 4782, pp. 44–52, 2007.

The next sections briefly describes the cluster that recently was installed at the Center of Information Services and High Performance Computing in Dresden. Although the experiences hold for any larger cluster installation, it should set the expectations and provide some background information about the used technology. Section 3 describes the tests that were used to check the correct configuration of the installed cluster and functionality of various software components. The following sections describe test and benchmarks in a bottom-up approach: First, we describe different component tests, afterwards tests for the complete node, followed by cluster wide tests.

2 Description of the Cluster Environment

The tests and experiences described within this paper have been collected during several installations (and most probably will keep growing with future installations as well). We limit the system description to the most recent and major installation. It is a cluster with 728 nodes with a total of 2592 cores. The nodes are a mixtures of single, dual and quad CPU systems with dual core AMD Opteron 2.6 GHz CPUs. Two Infiniband interconnects (running OFED1.0) are installed, one for message passing with MPI, the other for I/O. The filesystem is Lustre (currently in Version 1.4.10). Two MDS and eight OSS act as file servers. The operating system is SLES 10. For the basic cluster installation, maintenance, and operation the tools delivered by Linux Networx are used. With a revision controlled network installation (Clusterworx), remote power management and serial console access (Icebox, powerman, and conman), OS independent temperature and system monitoring (Icecards), and the BIOS settings and version control offered by LinuxBios [11] the installation can be considered state of the art.

3 Functionality Tests

Before more extensive tests can be performed the basic operation and configuration of the system has to be checked. With hundreds of nodes the tests have to rely on a correctly working batch- and execution environment.

3.1 Execution Environment

With the large number of nodes you need an automated test to check that the system configuration conforms to the contract. Important items include memory and disc capacity on all nodes, and CPU speed.

A collection of simple scripts are sufficient to check that the execution environment is identical on all nodes. This should hold for the interactive sessions on the different login nodes, but also for the batch environment. The execution environment consists of the shell environment and shell limits. A synchronized and correct time is another aspect. Last but not least, all nodes should have the same BIOS version and settings. Sources of informations are the output command of standard unix commands like `limit` and `env` as well as the `proc` file system. The BIOS settings are gathered with the `cmos_util` command from LinuxBios.

3.2 MPI Conformance

Since the dominant programming model on a cluster is message passing using MPI, a working MPI is one of the most important contributions to a cluster. Most open source implementations like Open MPI [12] or MPICH [13] include a test suite. However, to extract them for use outside of their suite is cumbersome. They also tend to check for specific properties of the implementation rather than MPI standard conformance. We decided to use a test program [14] that is easy to use, yet provides a good coverage of the MPI standard and also checks for MPI-2 features.

3.3 OpenMP Conformance

With the introduction of multi-core CPUs the typical number of cores found in a cluster node is increasing. Therefore OpenMP gains importance, either as a standalone solution or in combination with MPI. The availability of various compilers for Fortran and C/C++ raises the need to verify the support for OpenMP in its latest flavor (version 2.5). Since OpenMP has much fewer constructs there is a freely available test suite that provides good coverage of all language features [15].

3.4 Batch System Functionality

Unless a cluster is used only by a small group of users, a batch system environment is an absolute necessity. The batch system should be capable of dispatching sequential, OpenMP, and MPI jobs – especially, running MPI jobs so that they use exactly the nodes and CPU cores they were assigned to. It should be clear that a fully functional batch system should also take care of job cleanup and accounting. For testing these functions of the batch system the MPI and OpenMP tests previously mentioned were used. Additionally, a large number of dummy jobs were used to test maximum submission rates and maximum job throughput.

4 Node Component Tests

With the functionality tests described above it is now guaranteed that it is possible to compile and execute serial and parallel programs using MPI and/or OpenMP via the batch system. This is the pre-condition to execute the following tests on the system.

4.1 Memory

In addition to the correct amount of memory each node should deliver the expected memory bandwidth. Benchmarks like `stream`[16] provide well accepted numbers for the streaming performance of the memory system. In addition, the memory system must not produce any non-correctable bit errors or an exceeding number of single bit errors. We wrote a test program that exercises the memory by writing, reading and verifying random bit patterns to and from memory.

Together with the error detection and correction (EDAC/bluesmoke) tools from the Linux kernel it will detect dysfunctional memory. The program takes the amount of memory and total runtime as an argument and reports any errors to `stderr`. On servers with ECC it should not report any errors, because they should be caught and reported by the kernel. However, in the past we observed one server where two bit errors were silently produced and thus observed by this program.

4.2 Disk

The performance of local discs can be measured with programs like `bonnie`[17]. It is again important to measure the result on each node, Fig. 1 shows the result on a system that had a problem with slow disks. In this case, the major cause was vibrations caused by a bad charge of fans and insufficient insulation of the discs.

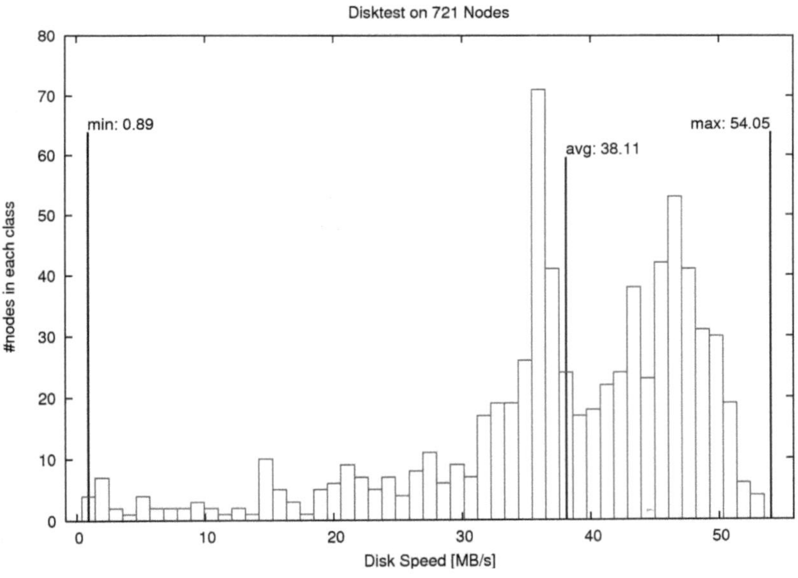

Fig. 1. Disk performance across all nodes of a large cluster. The distribution is unusually wide, ranging from 1 MB/s to 55 MB/s.

5 Node Tests

The best way to test the complete node and not just single components is to run various applications. The results should be checked for correctness and the performance of each node should be checked. We decided to use the SPEC.

OMPM2001 benchmarks [18,19]. SPEC OMPM2001 focuses on systems with less than 16 cores. Since OMPL2001 is based on OpenMP, this benchmark suite

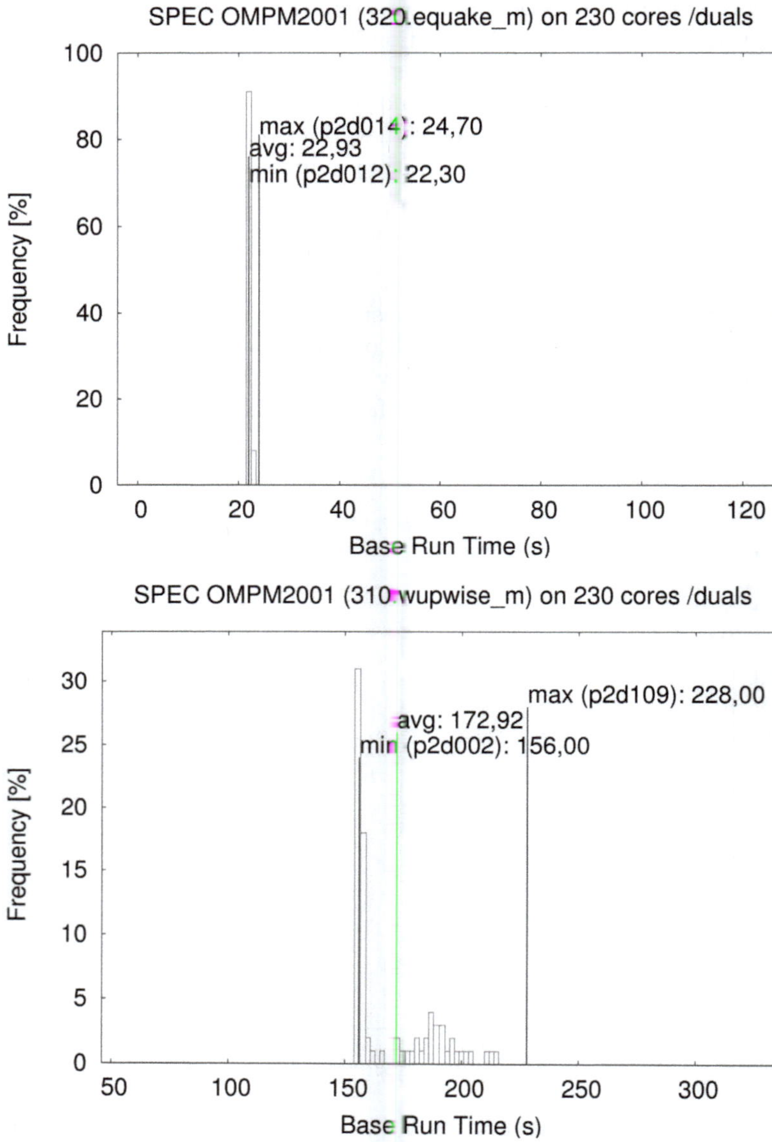

Fig. 2. Runtime distribution of two application across all nodes of a cluster. The application on the top has an almost identical runtime on all nodes, a different application on the bottom shows huge variation of the execution time.

is very suitable for todays cluster nodes. The suite consists of 11 applications written in C and Fortran. The result of the program is validated for correctness. This makes it a perfect candidate to check the build and runtime execution environment. Again, we submitted the benchmark to all nodes in the system. Since we had three different node types and 11 application this resulted in 33 histograms, where each outlier had to be identified and the cause of the problem isolated.

Fig. 2 shows how important it is to run a range of applications on a cluster. With one application there is no indication of problems, however, with another there are nodes that perform significantly worse than others. In this case, it was an undocumented difference in two BIOS versions that caused some applications to run slower.

6 Interconnect and Cluster Tests

After the nodes demonstrated to work properly the final tests can focus on the overall system. Important issues are the interconnect as well as the global file system. Parallel application tests are not covered here.

6.1 Point to Point Measurements

The two major performance criteria for an interconnect are latency and bandwidth between two nodes. Often minimum requirements for these values are fixed in the contract and measured during acceptance. We used an application where both values were not simply measured between a pair of nodes, but the result of every possible pair was measured. Originally designed to visualize hierarchies of interconnects it demonstrated its value to clean up and debug networks. Potential problems include broken or misplaced cables, host channel adapters that did not run at full PCI speed or Infiniband connections that did run at 1X speed instead of 4X. Additionally a test like this produces a very detailed layout of the interconnection network. Fig. 3 shows a visualization of the output. Having good point-to-point connection is the key for the parallel tests done later.

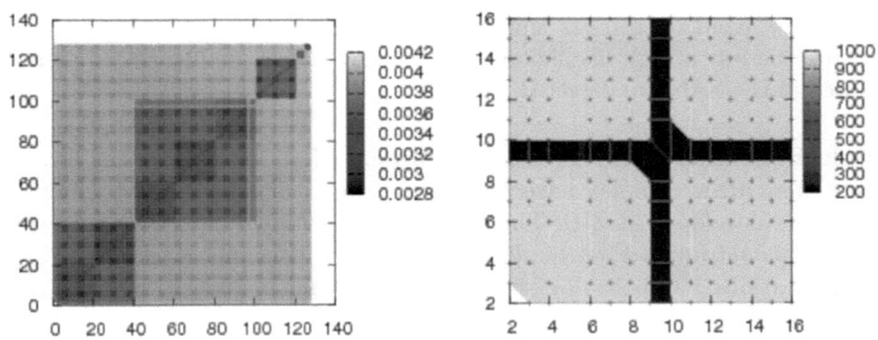

Fig. 3. Latency (in ms) and bandwidth (in MB/s) between pairs. The left pictures shows the expected latency pattern of a system with a hierarchical switch infrastructure. On the right the bandwidth is measured. One of the hosts has a very low bandwidth to all partners, caused by an IB HCA running on 1X instead of 4X.

6.2 Filesystem

Parallel, distributed file systems are powerful and scalable, but also complex. Also, today's systems are often not simple stand alone installations, but embedded in a larger infrastructure. This is also true for the system described in section 2: It has an NFS gateway consisting of 16 NFS servers. NFS is done using IP over IB, each NFS server is a CXFS client. The CXFS Filesystem is 68 TB in size and serves an SGI Altix 4700 system with 2048 cores. The nodes on the cluster can also access a HSM system (SUN STK8500) via the same gateway using standard DMF commands like `dmget`. Altogether this resulted in an infrastructure were many different hard- and software components had to be checked for functionality, reliability and performance.

We used various test programs in different settings to test the SAN and file system infrastructure.

Stress Tests and Content Validation. A basic building block was a program that wrote pseudo-random bit patterns to disk and verified the content in a subsequent reading step. To create the bit patterns a pseudo-random number generator (RNG) was used. Each file contains a small header were the seed of the RNG is stored. The size and number of files written by the program can be selected. Instead of fixing the number of files a total runtime can be provided. This is especially useful during stress tests, when the load on the file system and, thus, the run time can differ significantly from run to run. An MPI parallel version of the code can be used for stress testing. Each process writes and reads a file and acts as an independent client. By choosing small files the required metadata rate can be increased and scenarios like many file in one file system or folder can be tested.

A special version of the program was designed to run in an endless loop. All even processes constantly write files, all odd processes verify the content after the writer has finished. By placing odd and even processes on different nodes caching problems are avoided even for small file sizes. Since data is constantly read or written there is a high likelihood that transient errors or incorrect handled interrupts in the SAN are detected.

Flexible Benchmark and Test Environment. Lots of tests on the filesystems for the HRSK complex have been done using a flexible I/O benchmark, designed and developed at ZIH. There have been two main objectives that this framework should fulfill. First of all, we wanted this benchmark to imitate any given applications I/O behavior and second, it should be able to perform any artificial I/O load.

The project is divided into three sections. The main component is an execution layer that is traversing a list of I/O commands and executing the appropriate I/O calls. The list of commands is generated from an XML file. Main part of this file are XML tags that cause the main program to call libc I/O functions. Around those XML tags other tags are placed that allow to define repetitions and arbitrary patterns. As applications often read a lot of data from files that were already present at the start of the program, we included a concept called file pools. The file pools allow us to create/read/write/unlink files on different directory structures.

The last part within this software is the analysis backend. Each I/O call is reported to the backend and can either be processed or ignored. Processing can include any action within the backend, as the backend itself is a C++ class. Ignoring an I/O call within the backend will cause no overhead within the main code. Evaluations done in the backend can be either on time (during the run) or after the benchmark.

7 Summary and Conclusion

We described a set of tests that we used to check components like CPU, memory, disk, network, MPI and filesystem of a cluster. Each test found at least one issue in at least one cluster installation and contributed to the quality of this specific installation. Despite the fact that all clusters had the technology to provide completely identical nodes, it turned out to be extremely important to run the node component and node tests on all nodes. Histograms of the results highlighted problems and pointed in the direction of the root cause. For a general purpose cluster it was necessary to run a large number of applications, since sometimes only a small subset of applications had a problem. Techniques like a revision controlled network installation, remote power management and complete control over BIOS settings were essential to analyze, isolate and remove the detected problems. Finally we also conclude that a smoothly running cluster is less a hard- and software issue, but a question of system integration. Integration is not only needed for the system components, but also into the environment were the cluster is operated.

References

1. Sterling, T., et al.: How to build a beowulf. In: Cluster Computing Conference, Emory University, Atlanta, GA (1997)
2. Sterling, T., Savarese, D., Becker, D., Dorband, J., Ranawake, I., Packer, C.: Beowulf: A parallel workstation for scientific computation. In: International Conference on Parallel Processing, Oconomowoc, WI, pp. 11–14 (1995)
3. Buyya, R. (ed.): High Performance Cluster Computing, vol. 1. Prentice-Hall, Englewood Cliffs (1999)
4. Pfister, G.: Seach For Clusters. Prentice-Hall, Englewood Cliffs (1998)
5. (2007) `http://oscar.openclustergroup.org/`
6. Papadopoulos, P.M., Katz, M.J., Bruno, G.: Npaci rocks: Tools and techniques for easily deploying manageable linux clusters. Concurrency and Computation: Practice and Experience Special Issue: Cluster 2001 (2001)
7. Dongarra, J.: The linpack benchmark: An explanation. In: Houstis, E.N., Papatheodorou, T.S., Polychronopoulos, C.D. (eds.) Proceedings of the 1st International Conference on Supercomputing, Athens, Greece, pp. 456–474. Springer, Heidelberg (1988)

8. Bailey, D., Barton, J., Lasinski, T., Simon, H.: The NAS Parallel Benchmarks. Technical Report RNR-91-002, NASA Ames Research Center, Moffett Field, CA (1991)
9. Bailey, D., Harris, T., Saphir, W.: van der Wijngaar t, R., Woo, A., Yarrow, M.: The NAS Parallel Benchmarks 2.0. Technical Report NAS-95-020, NASA Ames Research Center, Moffett Field, CA (1995), http://www.nas.nasa.gov/Software/NPB
10. Dongarra, J., Luszczek, P.: Introduction to the hpcchallenge benchmark suite. ICL Technical Report ICL-UT-05-01, ICL (2005)
11. Minnich, R., Hendricks, J., Webster, D.: The linux bios. In: The Fourth Annual Linux Showcase and Conference, Atlanta, GA (2000)
12. Gabriel, E., Fagg, G.E., Bosilca, G., Angskun, T., Dongarra, J.J., Squyres, J.M., Sahay, V., Kambadur, P., Barrett, B., Lumsdaine, A., Castain, R.H., Daniel, D.J., Graham, R.L., Woodall, T.S.: Open MPI: Goals, concept, and design of a next generation MPI implementation. In: Proceedings, 11th European PVM/MPI Users Group Meeting, Budapest, Hungary, pp. 97–104 (2004)
13. (2005) MPICH homepage, http://www-unix.mcs.anl.gov/mpi/mpich/
14. Keller, R., Resch, M.: Testing the correctness of MPI implementations. In: ISPDC 2006, Timisoara, Romania (2006)
15. Müller, M.S., Niethammer, C., Chapman, B., Wen, Y., Liu, Z.: Validating OpenMP 2.5 for fortran and c/c++. In: Sixth European Workshop on OpenMP, KTH Royal Institute of Technology, Stockholm, Sweden (2004)
16. McCalpin, J.D.: Memory bandwidth and machine balance in current high performance computers. IEEE TCCA (1995), http://www.cs.virginia.edu/stream
17. Bray, T.: Bonnie (1990), http://www.textuality.com/bonnie/
18. Saito, H., Gaertner, G., Jones, W., Eigenmann, R., Iwashita, H., Lieberman, R., van Waveren, M., Whitney, B.: Large system performance of SPEC OMP2001 benchmarks. In: Zima, H.P., Joe, K., Sato, M., Seo, Y., Shimasaki, M. (eds.) ISHPC 2002. LNCS, vol. 2327, Springer, Heidelberg (2002)
19. Müller, M.S., Kalyanasundaram, K., Gaertner, G., Jones, W., Eigenmann, R., Lieberman, R., van Waveren, M., Whitney, B.: SPEC HPG benchmarks for high performance systems. International Journal of High Performance Computing and Networking 1, 162–170 (2004)

Performance Evaluation of Distributed Computing over Heterogeneous Networks

Ouissem Ben Fredj and Éric Renault

GET / INT — CNRS UMR 5157 SAMOVAR
9, rue Charles Fourier, 91011 Évry, France
Tel.: +33 1 60 76 45 73, Fax: +33 1 60 76 47 80
{ouissem.benfredj,eric.renault}@int-edu.eu

Abstract. RWAPI is a low-level communication interface designed for clusters of PCs. It has been developed to provide performance to higher applications on a wide variety of architectures. We implemented RWAPI on top of the modular software architecture called GRWA. RWAPI supports Ethernet, InfiniBand and Myrinet network interconnects. This paper introduces RWAPI and the design of its network component on top of both InfiniBand and Myrinet interconnects. We obtained a very low latency and high throughput compared to MPI results.

1 Introduction

High-speed network interconnects that offer low latency and high bandwidth have been one of the main reasons attributed to the success of commodity cluster systems. Some of the leading high-speed networking interconnects include Gigabit-Ethernet, InfiniBand [1], Myrinet [2] and Quadrics [3]. Two common features shared by these interconnects are User-level networking and Direct Memory Access (DMA). The best suited communication protocols that use efficiently these new features are one-sided protocols. It means that the completion of a send (resp. receive) operation does not require the intervention of the receiver (resp. sender) process to take a complementary action. RDMA should be used to copy data to (from) the remote user space directly. Suppose that the receiver process has allocated a buffer to hold incoming data and the sender has allocated a send buffer. Prior to the data transfer, the receiver must have sent its buffer address to the sender. Once the sender owns the destination address, it initiates a direct-deposit data sending. This task does not interfere with the receiver process. On the receiver side, it keeps on doing computation tasks, testing if new messages have arrived, or blocking until an incoming message event arises.

At the network layer, many manufacturers have built RDMA features that ease the implementation of one-sided paradigms. For example, the HSL [4] network uses the PCI-Direct Deposit Component (PCI-DDC) [5] to offer a message-passing multiprocessor architecture based on a one-sided protocol. InfiniBand [1] and Quadrics [3] proposes native one-sided communications. Myrinet [2,6] and QNIX [7] do not provide native one-sided communications. But these features may be added (as for example in GM [8] with Myrinet since Myrinet NICs are programmable).

In the past, remote-write has been implemented in generic message-passing libraries like MPI-2 [9] or dedicated message-passing libraries like the PUT interface [10,11]

R. Perrott et al. (Eds.): HPCC 2007, LNCS 4782, pp. 53–61, 2007.

for MPC-OS, PAPI [12]. However, these implementations were dedicated to a given interconnect and no effort has been made to provide a generic or minimalist API to the remote-write. Gas-Net [13] and MP_Lite [14] provide one-sided communication primitives but without any associated programming model. In this context, we defined RWAPI [15] (the Remote-Write API) as a generic and minimalist API for the remote-write and GRWA (the Global Remote-Write Architecture) which aims at providing a set of independent modules to implement RWAPI on top of any interconnects, any CPU models... This articles describes our implementations of RWAPI for both Myrinet and InfiniBand interconnects and compares performance with other available message-passing libraries (especially MPI).

The document is organized as follows: first we introduce RWAPI [16]; section 3 develops the implementation of RWAPI for Myrinet and InfiniBand; section 4 describes the performance benchmark; the last section before the conclusion provides some performance measurements.

2 RWAPI

A previous study [15] of both hardware and software requirements for high-speed network protocols has led to design the Remote Write protocol. Remote Write is a one-sided communication protocol based on the remote write primitive. It requires the sender of a message to provide all the information needed to copy a contiguous memory area from one node to another node.

RWAPI (which stands for Remote-Write Application Programming Interface) is a lightweight interface designed to provide a single remote-write primitive. The goal we are trying to achieve is to provide the smallest set of functions that enables to write any parallel programs. This way, we expect to achieve the best performance for communications while requiring as less development as possible to port our interface to new architectures.

There are two kinds of messages in RWAPI. The first message type requires the destination node identification, both local and remote addresses and the size of the message. Messages in this case can be of any length. The second message type just requires the destination node identification and the message content; the size is limited to 16 bytes. They may be helpfully used to transfer small amounts of information of any kind from one node to another. However, even if they are not limited to this specific use, they are especially useful to exchange addresses before the other message type transfers can occur.

The API is as follows:

- int rwapi_init (int, char **) must be called before any other RWAPI functions in order to set up the communication interface.
- int rwapi_finalize ( ) should be called after all RWAPI functions and before exiting the program. This function ensures that all FIFOs are flushed before leaving.
- int rwapi_rank ( ) returns the rank of the local node in the virtual parallel machine.
- int rwapi_size ( ) returns the number of nodes in the virtual machine.
- void * rwapi_alloc (size, net *) allocates a contiguous memory block of the given size. If the underlying network interface requires the use of contiguous physical

memory, it is attached to the application transparently. The value returned by this function is the virtual address in the virtual address space of the process where the contiguous memory block has been attached. The second parameter is the address where the "network" address will be stored when returning from the function. This address is the one that must be used for sending data.
- int rwapi_free (void *) deallocates the memory area provided as a parameter.
- int rwapi_send (node, small) sends a small message to another node. The value returned by this function is an error code.
- int rwapi_receive (node *, small *) returns information about the oldest incomming message that has not been taken into account yet. The value returned by the function is 0 if there is no message pending and 1 if a message has been taken into account. In this case, the node and the small message are stored at the addresses provided as parameters.
- sid rwapi_write (node, net, net, size) sends an arbitrary-long message to another. Both local and remote "network" addresses must be provided together with the size of the message. The Send ID (SID) returned by the function can be latter used in order to determine if the message as been sent or not (this is useful to reuse a memory area).
- int rwapi_issent (sid) checks wether the message identified by the SID has been sent or not.

Note that, rwapi_write, rwapi_send, rwapi_issent and rwapi_receive are non blocking functions.

3 Implementation Details

In order to launch application's processes according to the SPMD model, we use an SSH-based spawner which creates a *master process* on the current host and one process on each host provided as an argument. Then, each process can communicate with the *master* to get information such as the process rank, the number of processes, the application's arguments and other control informations. Then, processes perform collective operations on top of socket-based connections to initialize and finalize the application transparently to the user.

To maintain the non-blocking semantic of RWAPI operations (rwapi_send, rwapi_write, rwapi_receive, and rwapi_issent), we used a host memory receive list (HMRL). Thus, when a receive message event arises while the process is not waiting for messages, the interface copies the content of the message event in the HMRL. Note that message events do not contain user data except those corresponding to rwapi_send operations. In this case, the length of the user data is limited to 128 bits and thus a copy of this message is not expensive.

RWAPI over Myrinet Interconnect: There are many ways to implement the Remote-Write protocol on top of Myrinet. One solution would be a native implementation which would consist in developing a new MCP, a new kernel driver and a new user library. This solution is under development and good performance are expected. However, this would not be portable since an MCP should be provided for every network card version. To overcome this limitation, we developed the Remote-Write protocol on top of GM. This way, RWAPI is automatically available for all architectures and NIC versions that GM supports.

In order to associate a process rank to its GM ID, processes perform an exchange of GM IDs using the *exchange* operation set up by the spawner's *master process*.

The rwapi_send function insures that GM is ready to send messages before calling the gm_send_to_peer_with_callback function. The later function is the fastest send function provided by GM since it uses the same port number as the sender at the receiver side. We set up GM to copy the data in the send descriptor to avoid an expensive DMA operation performed by the MCP to copy data from the host memory to the NIC memory. We also use Write-Combining feature instead of PIO or DMA to copy the send descriptor from the host memory to the NIC memory. This feature combines many PIO operations and is adapted to medium message size.

Similar to the rwapi_send function, the rwapi_write function insures that GM is ready to send messages before calling the gm_put function. The gm_put function is suited to the remote-write paradigm since it garantees the minimum number of copies while transferring the data. Indeed, it copies the data from the host memory directly to the NIC memory without any system calls.

RWAPI over InfiniBand Interconnect: RWAPI uses the SSH-based spawner to launch processes in both InfiniBand-based and a Myrinet-based clusters. The first step of the initialization is the creation of a bidirectional channel between each pair of processes. The InfiniBand mechanism that allows the creation of such a channel is the QP (Queue Pair). Each QP is configured for a particular type of service independent from the other. These service types provide different levels of service and different error recovery characteristics. The available transport service types include: Reliable Connection (RC), Unreliable Connection (UD), Reliable Datagram (RD), Unreliable Datagram (UD) and Raw. The transport type used is RC which provides the highest level of reliability and predictability. RC requires for each process an explicit connection with all other processes. Thus, $(N - 1)$ QPs (N being the number of processes in the parallel application) are created by each process.

The second step of the initialization sets up the size of the different comunication queues for each QP, including the send queue (SQ), the receive queue (RQ), the send completion queue (SCQ), and the receive completion queue (RCQ). This step finishes by creating a local process ID (LID) and $(N - 1)$ QP IDs. These IDs are broadcasted to the other processes in order to build the channels between local and remote QP.

4 Performance Benchmark and Testbeds

We compared our implementation with the version of MPI, the other existing library available for both Myrinet-2000 Technology (developed on top of GM) and InfiniBand interconnect (developed on top of VAPI [17]). For both MPI and RWAPI we developed our own benchmarks. We use three separate benchmarks (see figure 1). The first benchmark (figure 1(a)) is the classic ping-pong in which a message can only be sent once the previous one has been received. The second one (figure 1(b)) is a bidirectional

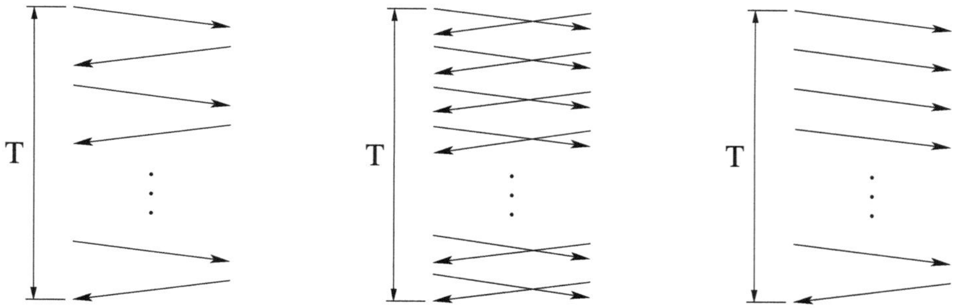

(a) Round-trip (b) Bidirectional RT (c) One-way

Fig. 1. Performance benchmarks

ping-pong which is used to highlight the capability of a library to take benefits of bidirectional links. The last benchmark (figure 1(c)) is the burst which aims at sending as many messages as possible regardless they have been received or not by the counter part.

The measurement protocol is as follows: for each message size, each benchmark is run ten times. The duration of a run is one minute (this ensures a high consistency in results and we have determined that all confidence intervals are greater than or equal to 90%). The system time is registered before the first message is sent (t_1) and after the last message is received on the same node (t_2). Let the elapsed time t be the time difference between both. For a given run, let s be the size of messages and n be the number of effectively transmitted messages.

Let the end-to-end latency L (in the following we use the term latency) be the ratio between the elapsed time t and the number of effectively transmitted messages n. And let the user throughput T (in the following we use the term throughput) be the ratio between the amount of data (number of effectively transmitted messages times the size of a message) and the elapsed time.

$$t = t_2 - t_1 \qquad L = \frac{t}{n} \qquad T = \frac{n \times s}{t}$$

Since the performance for both RWAPI and MPI are almost the same for messages size greater than or equal to 1 MB, we do not include data for larger messages.

Our Myrinet-based cluster is POETS, one of the clusters of the Institut National des Télécommunications. POETS is composed of eight nodes connected using both a Myrinet interconnection network for data and a Gigabit Ethernet interconnect as a control network. Myrinet adaptors are 133-MHz LANai-9 (M3S-PCI64B) with a PCI connector. The host adaptor is equipped with 2 MB of memory. The firmware used for testing was GM 2.0.9. Apart from GM, the port of MPICH on top of GM called MPICH-GM version 1.2.6..14b for Linux x86 was also installed. Each node includes a 800-MHz Intel Pentium III processor with 1.2 GB of memory. The front-end of the cluster is an extra node with almost the same characteristics except that it includes no Myrinet NIC.

The InfiniBand-based cluster is MUSES. It is composed of 16 nodes connected using both an InfiniBand 4× interconnection network for data and a Gigabit Ethernet interconnection network as a control network. Each node includes a 1.8-GHz AMD Opteron processor with a 1-MB cache and 2 GB of memory. For mass storage, each node is associated a 40-GB IDE hard drive, except for the first node (the front-end node) which is associated a 80-GB IDE hard drive; note that users accounts are stored on the front-end. We compared our implementation with two other existing libraries on top of the InfiniBand Technology: VAPI [17], the native interface available on top of InfiniBand, and MPICH developed on top of VAPI for InfiniBand.

The MPI benchmark uses the send/receive primitives instead of put with both Mpich-GM and Mpich-VAPI. Send/receive is based on the rendez-vous model which requires that the receive request should be posted before the send request. Otherwise, a blocking wait or a message buffering is performed on the receiver side. Moreover, the eager-message size is kept to the default value. Finally, the RWAPI benchmark and the MPI benchmark use the same buffer for all the iterations of the communications which allows that the memory pages are pinned only one time.

5 Performance Analysis

Figure 2 presents the latency of RWAPI-GM and MPI-GM for various message sizes. Note that for readability reasons, a logarithmic scale have been used for latency curves. All these graphs highlight that, in the general case, performance with RWAPI are usually better than those with MPI. Exceptions are for the One-Way benchmark for messages which size ranges from 1 kB to 16 kB and for the Round-Trip benchmark for messages which size ranges from 4 bytes to 64 bytes. This may be due to the fact that MPI is using Write-Combining to copy data from the host memory to the NIC memory for small messages avoiding the overhead of the DMA start-up. Regarding the latency, an interesting result is that for the Round-Trip benchmark (see figure 2(a)), the minimum latency for RWAPI is as low as 5.1 μs. As a comparison, the minimum latency for MPI is 7.9 μs and is achieved using the Bidirectional Round-Trip (see figure 2(b)).

Figure 3 shows the throughput of RWAPI and MPI for various message sizes. All these graphs highlight that, in the general case, performance with RWAPI are usually better than those with MPI.

These graphs are highlighting three very important results. First, RWAPI is able to achieve a maximum throughput of up to 1.78 Gb/s for large messages (see figure 3(a)

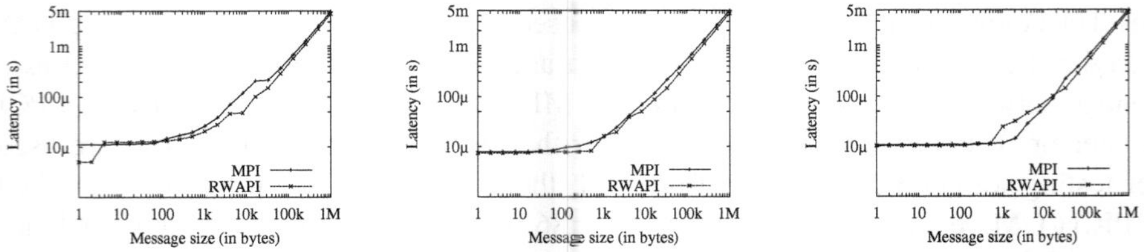

(a) Round-trip (b) Bidirectional RT (c) One-way

Fig. 2. Myrinet Latency

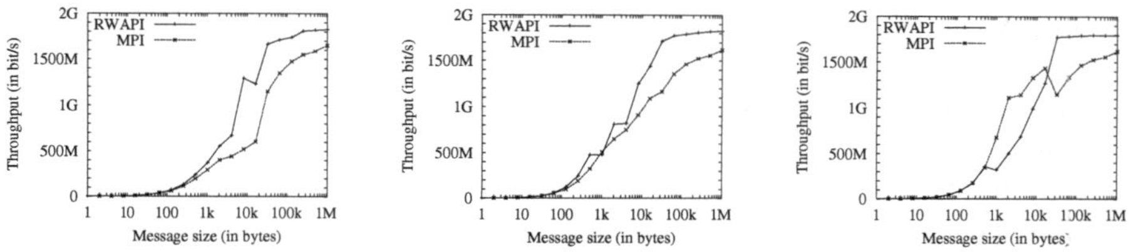

(a) Round-trip (b) Bidirectional RT (c) One-way

Fig. 3. Myrinet Throughput

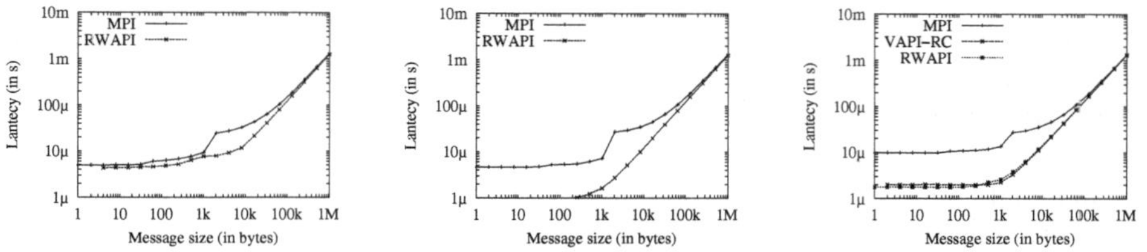

(a) Round-trip (b) Bidirectional RT (c) One-way

Fig. 4. InfiniBand Latency

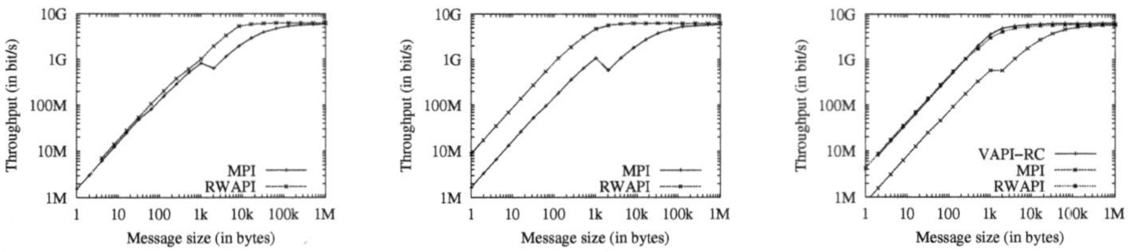

(a) Round-trip (b) Bidirectional RT (c) One-way

Fig. 5. InfiniBand Throughput

and figure 3(b) for Round-Trip and Bidirectional Round-Trip benchmarks respectively). As a comparison, the maximum throughput provided by MPI is 1.62 Gb/s. This means that MPI cannot offer more than 91% of the maximum throughput offerred by RWAPI. In other words, RWAPI is able to deliver large messages 9.8% faster than MPI (in fact, this is not highlighted with latency graphs on figure 2 as a logarithmic scale is used for messages larger than 1 kB).

RWAPI provides a larger part of the bandwidth for smaller messages than MPI. Typically, RWAPI is able to achieve 90% of the maximum throughput for 32-kB messages as MPI requires messages of 256 kB to do the same.

With InfiniBand libraries, in a general way, figure 5 and figure 4 show that RWAPI-VAPI performance are always better than MPI-VAPI performance. More specifically, the maximum ratio between the minimum latency achieved by RWAPI and the minimum latency achieved by MPI is up to 5.5 x for small messages (ie. 1.76 μs for RWAPI and 9.71 μs for MPI using the one-way benchmark).

Both RWAPI and MPI are able to achieve the maximum user throughput for long messages. However, RWAPI is able to provide this maximum user throughput for messages as short as 4 kB while MPI cannot do the same for messages smaller than few hundreds kilo-bytes.

Finally, the curves on figure 5 and figure 4 show that there is an important difference in the management of short and long messages for MPI represented with a knee between 1 and 2 kB.

6 Conclusion

A previous study led us to design our own implementation of the remote write. In this paper, we have presented the design and implementation of RWAPI over Myrinet-2000 and InfiniBand Interconnects. This design takes full advantages of the network hardware such as OS-Bypass and RDMA, thus eliminating the involvement of the operating system and the receive process. In addition, it allows the overlap between communications and computations.

To decrease the latency of small messages, we used the Programmed-IO facility instead of RDMA to copy data to the network card, removing one long-startup-time DMA transaction.

RWAPI over Myrinet achieves a low latency of 5.1 μs and a high user throughput of 2.43 Gb/s even for relatively short messages (2.26 Gb/s is available for 32-kB messages). On the same platform, the lowest latency provided by MPI is 7.9 μs and the maximum user throughput provided by MPI represents 90% of the maximum user throughput provided by RWAPI (this means that message transfer with RWAPI is up to 9.8% faster than with MPI).

RWAPI over InfiniBand design can achieve a low latency (about 1.76 μs) and a high user throughput (more than 6.3 Gb/s, ie the maximum user bandwidth) even for short messages. As a comparison, the lowest latency provided by MPI over the same platform is 4.96 μs and the maximum user throughput cannot be achieved for messages smaller than several hundreds of kilo-bytes.

Note that the lowest InfiniBand communication layer (VAPI) let the user choose between producing events for transfer operation completion or not. This does not suit RWAPI as disabling events does not allow the user to be informed about the completion of the send and enabling events adds an extra overhead due to the unnecessary receive completions.

Currently, RWAPI uses RC as the type of service with InfiniBand packet management. RC requires a connection between each remote HCA and thus consumes much HCA memory resources. Consumed memory is mainly used to store data reassembly informations for each connection. To achieve better scalability, we are working on applying the RD type of service which bypasses any connection management and maintains a reliable communication.

As a short-term, we have planned to optimize the rwapi_write operation by using either PIO, Write-Combining or DMA to copy user data from the host memory to the NIC memory; PIO operations for very small messages, Write-combining for medium messages and DMA for large messages since it involves an expensive start-up overhead.

As a mid-term, we have planned to implement another version of RWAPI which can run simultaneously over heterogeneous architectures composed of different network types and different machine characteristics.

References

1. InfiniBand Trade Association: The InfiniBand Architecture, Specification Volume 1 & 2, Release 1.0.a (2001)
2. Boden, N., Cohen, D., Flederman, R., Kulawik, A., Seitz, C., Selzovic, J., Su, W.: Myrinet - A Gigabit-per-Second Local-Area Network. 15, 29–36 (1995)
3. Beecroft, J., Addison, D., Petrini, F., McLaren, M.: Quadrics QsNet II: A network for Supercomputing Applications. In: Hot Chips 15, Stanford University, Palo Alto, CA (2003)
4. Whitby Strevens, C., et al.: IEEE Draft Std P1355 — Standard for Heterogeneous Interconnect — Low Cost Low Latency Scalable Serial Interconnect for Parallel System Construction (1993)
5. Greiner, A., Desbarbieux, J., Lecler, J., Potter, F., Wajsbürt, F., Penain, S., Spasevski, C.: PCI-DDC Specifications. Laboratoire MASI, Université Paris VI (1996)
6. ANSI/VITA 26-1998: Myrinet-on-VME Protocol Specification (1998)
7. Vivo, A.D.: A Light-Weight Communication System for a High Performance System Area Network. PhD thesis, Università di Salerno - Italy (2001)
8. Myricom: GM: A message-passing system for myrinet networks 2.0.12 (1995)
9. Message Passing Interface Forum MPIF: MPI-2: Extensions to the Message-Passing Interface. Technical Report, University of Tennessee, Knoxville (1996)
10. Fenyo, A.: Conception et rèalisation d'un noyau de communication bâti sur la primitive d'écriture distante, pour machines paralléles de type grappe de PCs. Thése de doctorat, UPMC LIP6, Paris, France (2001)
11. Desbarbieux, J.L., Gluck, O., Zerrouki, A., Fenyo, A., Greiner, A., Wajsbürt, F., Spasevski, C., Silva, F., Dreyfus, E.: Protocol and performance analysis of the mpc parallel computer. In: IPDPS '01: Proceedings of the 15th International Parallel & Distributed Processing Symposium, p. 52. IEEE Computer Society, Washington, DC (2001)
12. Renault, E., David, P.: PAPI message passing library: Comparison of performance in user and kernel level messaging. In: Sakellariou, R., Keane, J.A., Gurd, J.R., Freeman, L. (eds.) Euro-Par 2001. LNCS, vol. 2150, pp. 717–721. Springer, Heidelberg (2001)
13. Bonachea, D.: Gasnet specification, v1.1. Technical report, Berkeley, CA, USA (2002)
14. Turner, D., Selvarajan, S., Chen, X., Chen, W.: The MP_Lite Message-passing Library. In: Akl, S.G., Gonzalez, T. (eds.) the proceedings of the Parallel and Distributed Computing and Systems. PDCS 2002, Cambridge, USA, pp. 373–235 (2002)
15. Ben Fredj, O., Renault, E.: A qualitative analysis of the critical's path of communication models for next perform implementations of high-speed interfaces. In: Sobh, I.T., Khaled Elleithy, E. (eds.) Advances in Systems, Computing Sciences and Software Engineering: Proceedings of SCSS 2005, pp. 70–77. Springer, Secaucus, NJ (2006)
16. Ben Fredj, O., Renault, E.: Rwapi over infiniband: Design and performance. In: 5th International Symposium on Parallel and Distributed Computing (ISPDC 2006), July 6-9, 2006, pp. 50–57. IEEE Computer Society, Timisoara, Romania (2006)
17. Mellanox Technologies Inc: Mellanox ib-verbs api (vapi) (2001)

Hybrid Line Search for Multiobjective Optimization

Crina Grosan[1,2] and Ajith Abraham[1]

[1]Faculty of Information Technology, Mathematics and Electrical Engineering
Centre for Quantifiable Quality of Service in Communication Systems
Centre of Excellence
Norwegian University of Science and Technology
Trondheim, Norway
[2] Department of Computer Science
Babes-Bolyai University, Cluj-Napoca, Romania
cgrosan@cs.ubbcluj.ro, ajith.abraham@ieee.org

Abstract. The aggregation of objectives in multiple criteria programming is one of the simplest and widely used approach. But it is well known that these techniques sometimes fail in different aspects for determining the Pareto frontier. This paper proposes a new line search based approach for multicriteria optimization. The objectives are aggregated and the problem is transformed into a single objective optimization problem. Then the line search method is applied and an approximate efficient point is lacated. Once the first Pareto solution is obtained, a simplified version of the former one is used in the context of Pareto dominance to obtain a set of efficient points, which will assure a thorough distribution of solutions on the Pareto frontier. In the current form, the proposed technique is well suitable for problems having multiple objectives (it is not limited to bi-objective problems). the functions to be optimized must be continuous twice differentiable. In order to assess the effectiveness of this approach, some experiments were performed and compared with two recent well known population-based metaheuristics ParEGO [8] and NSGA II [2]. When compared to ParEGO and NSGA II, the proposed approach not only assures a better convergence to the Pareto frontier but also illustrates a good distribution of solutions. From a computational point of view, of the line search converge within a short time (average about 150 milliseconds) and the generation of well distributed solutions on the Pareto frontier is also very fast (about 20 milliseconds). Apart from this, the proposed technique is very simple, easy to implement and use to solve multiobjective problems.

1 Introduction

The field of multicriteria programming abounds in methods for dealing with different kind of problems. Nevertheless, there is still space for new approaches, which can better deal with some of the difficulties encountered by the previous approaches. There are two main classes of approaches suitable for multiobjective optimization: scalarization methods and nonscalarizing methods. These

R. Perrott et al. (Eds.): HPCC 2007, LNCS 4782, pp. 62–73, 2007.

approaches convert the Multiobjective Optimization Problem (MOP) into a Single Objective Optimization Problem (SOP), a sequence of SOPs, or into another MOP. There are several scalarization methods reported in the literature: weighted sum approach, weighted t-th power approach, weighted quadratic approach, ε-constraint approach, elastic constraint approach, Benson approach, etc. are some of them [5]. Since the standard weighted sum encounters some difficulties, several other methods have been proposed to overcome the major drawbacks of this method. These include: Compromise Programming [4], Physical Programming [9], Normal Boundary Intersection (NBI) [1], and the Normal Constraint (NC) [9] methods. There is also a huge amount of work reported on population-based mataheuristics for MOP [5].

In this paper, we propose a new approach which uses a scalarization of the objectives in a way similar to the weighted t-th power approach (where t is 2 and the coefficients values are 1).

A line search based technique is used to obtain an efficient solution. Starting with this solution, a set of efficient points are further generated, which are widely distributed along the Pareto frontier using again a line search based method but involving Pareto dominance relationship.

Empirical and graphical results and illustrations obtained by the proposed approach are compared with two well known population based metaheuristics namely ParEGO [8] and NSGA II [2].

2 Line Search Generator of Pareto Frontier

The line search [6] is a standard and well established optimization technique. The standard line search technique is modified so that it is able to generate the set of non-dominated solutions for a MOP. The approach proposed is called **L**ine search **G**enerator of **P**areto frontier (LGP) and it comprises of two phases: first, the problem is transformed into a SOP and a solution is found using a line search based approach. This is called as *convergence phase*. Second, a set of Pareto solutions are generated starting with the solution obtained at the end of convergence phase. This is called as *spreading phase*. The *convergence* and *spreading* phases are described below.

Consider the MOP formulated as follows:

Let $\Re^m$ and $\Re^n$ be Euclidean vector spaces referred to as the decision space and the objective space. Let $X \subset \Re^m$ be a feasible set and let f be a vector-valued objective function f: $\Re^m \rightarrow \Re^n$ composed of n real-valued objective functions $f=(f_1, f_2,\ldots, f_n)$, where f_k: $\Re^m \rightarrow \Re$, for $k=1,2,\ldots, n$. A MOP is given by:

$$\min\ (f_1(x),\ f_2(x),\ldots,\ f_n(x)),$$

$$\text{subject to } x \in X.$$

2.1 Convergence Phase

The MOP is transformed into a SOP by aggregating the objectives using an approach similar to the weighted t-th power approach. We consider t= 2 and the values of weights equal to 1. The obtained SOP is:

$$\min F = \sum_{i=1}^{n} f_i^2(x)$$
$$\text{subject to } x \in X.$$

As an important note, we would like to mention that the value 2 for t works fine if the objective functions are positive (which is the case of our experiments). But if at least one objective function is negative, then an odd value (for instance 3) must me used for t. Any of these values (2 or 3) works fine for our examples and will not influence the results.

A modified line search method is used to find the optimum of this problem. The modification proposed in this paper for the standard line search technique refers to direction and step setting and also the incorporation of a re-start procedure. To fine tune the performance, the first partial derivatives of the function to optimize are also made use of. The proposed modifications refer to:

- the setting of the direction and step
- the re-starting of the line search method

After a given number of iterations, the process is restarted by reconsidering other arbitrary starting point which is generated by taking into account the result obtained at the end of previous set of iterations.

Direction and step setting. Initially, several experiments were performed in order to set an adequate value for the direction. The standard value +1 or -1 was used and for some functions the value -1 was favorable to obtain good performance. Some experiments were also performed by setting the direction value as being a random number between 0 and 1. It was found that the usage of random number helped to obtain overall very good performance for the entire considered test functions. But usage of the value -1 for direction, obtains almost the same performance similar to that obtained with a random value. So, either of these values (the random one and the value -1) may be used for better performance.

The step is set as follows:

$\alpha_k = 2 + \frac{3}{2^{2k}+1}$ (1)

where k refers to the iteration number.

The modified *line search* technique is summarized as follows:

Line_search()

```
Set k=1 (Number of iterations)
Repeat
    for i=1 to No of variables
        p_k=random; //or p=-1;
```

$\alpha_k = 2 + \frac{3}{2^{2k}+1}$
$x_i^{k+1} = x_i^k + p_k \cdot \alpha_k$
endfor
if $F(x_{k+1}) > F(x_k)$ then $x^{k+1} = x_k$.
k=k+1
Until k=Number of iterations (a priori known).

Remarks

(i) The condition:
if $F(x_{k+1}) > F(x_k)$ then $x_{k+1} = x_k$
allows to move to the new generated point only if there is an improvement in the quality of the function.
(ii) Number of iterations for which line search is applied is apriori known and is usually a small number. For the experiments reported in this paper, the number of these iterations was set to 10.
(iii) When restarting the line search method (after the insertion of the re-start technique) the value of the iterations number starts again from 1 (this should not be related to the value of α after the first set of iterations (and after each of the following iterations)).

Several experiments were attempted to set a value for the step, starting with random values (until a point is reached for which the objective function achieves a better value); using a starting value for the step and generating random numbers with Gaussian distribution around this number, etc. As a result of the initial experiments performed, it was decided to use equation (1) to compute the step size. But, of course, there are also several other ways to set this.

Incorporation of re-start procedure. In order to restart the algorithm the result obtained in the previous set of iterations (denote it by x) is taken into account and the steps given below are followed:

For each dimension $i$of the point x, the first partial derivative with respect to this dimension is calculated. This means the gradient of the objective function is calculated which is denoted by g. Taking this into account, the bounds of the definition domain for each dimension are re-calculated as follows:

if $g_i = \frac{\partial F}{\partial x_i} > 0$ then *upper bound* $= x_i$;
if $g_i = \frac{\partial F}{\partial x_i} < 0$ then *lower bound* $= x_i$

The search process is re-started by re-initializing a new arbitrary point between the newly obtained boundaries.

2.2 Spreading Phase

At the end of the convergence phase, a solution is obtained. This solution is considered as an efficient (or Pareto) solution. During this phase and taking into account of the existing solution, more efficient solutions are to be generated so as to have a thorough distribution of all several good solutions along the Pareto frontier. In this respect, the line search technique is made use of to generate one solution at the end of each set of iterations. This procedure is applied several times in order to obtain a larger set of non-dominated solutions. The following steps are repeated in order to obtain one non-dominated solution:

Step 1. A set of nondominated solutions found so far is archived. Let us denote it by *NonS*. Initially, this set will have the size one and will only contain the solution obtained at the end of *convergence phase.*

Step 2. We apply line search for one solution and one dimension of this solution at one time. For this:

Step 2.1. A random number i between one and |NonS| ($|\cdot|$ denotes the cardinal) is generated. Denote the corresponding solution by $nonS_i$.

Step 2.2. A random number j between one and the number of dimensions (the number of decision variables) is generated. Denote this by $nonS_{ij}$.

Step 3. Line search is applied for $nonS_{ij}$.

Step 3.1. Set p=1 (the random value also works fine).

Step 3.2. Set α (which depends on the problem, on the number of total nondominated solutions which are to be generated, etc.).

Step 3.3. The new obtained solution *new_sol* is identical to $nonS_i$ in all dimensions except dimension j which is:
$new\_sol_j = nonS_{ij} + \alpha \cdot p$

Step 3.4. if ($new\_sol_j >$ *upper bound*) or ($new\_sol_j <$ *lower bound*)
then new_sol_j = *lower bound* + random $\cdot$ (*upper bound* − *lower bound*).

Step 4. if $F(new\_sol) > F(nonS_1)$
then discard *new_sol*
else if *new_sol* is nondominated with respect to the set *NonS*
then add *new_sol* to *NonS* and increase the size on *NonS* by 1.
Go to step 2.

Step 5. Stop

These steps are repeated until a set on nondominated solutions of a required size is obtained. In our experiments the size of this set is 100.

Note that this procedure it very fast and it takes less than 20 milliseconds to obtain 100 non-dominated solutions.

3 Experiments and Comparisons

In order to assess the performance of LGP, some experiments were performed using some well known bi-objective and three-objective test functions, which are adapted from [3], [7]. These test functions were also used by the authors of ParEGO [8] and NSGA II [2], which are well known in the computational intelligence community as very efficient techniques for multiobjective optimization.

Table 1. Parameters used in experiments by ParEGO and NSGA II. d denotes the number of decision parameter dimensions.

ParEGO		**NSGA II**	
Parameter	*Value*	*Parameter*	*Value*
Initial population in latin hypercube	$11d-1$	Population size	20
Total maximum evaluations	250	Maximum generations	13
Number of scalarizing vectors	11 for 2 objectives 15 for 3 objectives	Crossover probability	0.9
Scalarizing function	Augmented Tchebycheff	Real value mutation probability	$1/d$
Internal genetic algorithm evaluations per iteration	200,000	Real value SBX parameter	10
Crossover probability	0.2	Real value mutation parameter	50
Real value mutation probability	$1/d$		
Real value SBX parameter	10		
Real value mutation parameter	50		

Details about implementation of these two techniques may obtained from [2] and [8]. Parameters used by ParEGO and NSGA II (given in Table 1) and the results obtained by these two techniques are adapted from [8].

A set of 100 non-dominated solutions obtained by LGP, ParEGO, NSGA II is compared in terms of dominance and convergence to the Pareto set. For the first comparison, two indices were computed for each set of two comparisons: number of solution obtained by the first technique which dominate solutions obtained by the second technique and number of solutions obtained by the first technique which are dominated by the solutions obtained by the second technique.

For two sets of A and B of solutions, which are compared, indices are denoted by *Dominate(A, B)* and *Dominated(A, B)* respectively. Visualization plots are used to illustrate the distribution of solutions on the Pareto frontier.

LGP uses only three parameters:

number of re-starts: 20;

number of iteration per each re-start: 10;

$\alpha$for the spreading phase (which is set independent for each test function).

3.1 Test Function DTLZ1a

The test function DTLZ1a is a two objective test function and has 6 variables [8]. It is given by:

$$\textit{minimize } f_1 = \tfrac{1}{2}x_1(1+g)$$
$$\textit{minimize } f_2 = \tfrac{1}{2}(1-x_1)(1+g)$$

$$g = 100\left[5 + \sum_{i=2}^{6}\left((x_i - 0.5)^2 - \cos\left(2\pi\left(x_i - 0.5\right)\right)\right)\right]$$
$$x_i \in [0,\ 1],\ i=1,\ \ldots,\ n,\ n=6.$$

The Pareto set for this function consists of all solutions where all by the first decision variables are equal to 0.5 and the first decision variable may take any value between 0 and 1.

For this test function, the value of α for the spreading phase is set to 0.01. The convergence to the Pareto frontier and the distribution of solutions obtained by LGP, ParEGO and NSGA II for the test function DTLZ1a is depicted in Figure 1. Different sizes of the objective space are illustrated in order to incorporate all solutions obtained by all techniques. It is obvious that LGP assure a very good convergence and distribution for this function. From the results presented in Table 2 it can be observed that none of the solutions obtained by LGP are dominated neither by ParEGO or by NSGA II, while solutions obtained by LGP dominate all 100 solutions obtained by ParEGO and NSGA II. 88 of the solutions obtained by NSGA II are dominated by solutions obtained by ParEGO while 69 of the solutions obtained by ParEGO are dominated by solutions obtained by NSGA II.

Table 2. The dominance between solutions obtained by LGP, ParEGO and NSGA II for test function DTLZ1a

Dominate	ParEGO	NSGA II	***Dominate***	LGP	NSGA II	***Dominate***	LGP	ParEGO
LGP	100	100	ParEGO	0	69	NSGA II	0	88
Dominated	ParEGO	NSGA II	***Dominated***	LGP	NSGA II	***Dominated***	LGP	ParEGO
LGP	0	0	ParEGO	100	88	NSGA II	100	69

3.2 Test Function DTLZ4a

Test function DTLZ4a has three objective functions and 8 decision variables and is given by:

$$minimize\ f_1 = (1+g)\cos\left(\frac{x_1^{100}\pi}{2}\right)\cos\left(\frac{x_2^{100}\pi}{2}\right)$$
$$minimize\ f_2 = (1+g)\cos\left(\frac{x_1^{100}\pi}{2}\right)\sin\left(\frac{x_2^{100}\pi}{2}\right)$$
$$minimize\ f_3 = (1+g)\sin\left(\frac{x_1^{100}\pi}{2}\right)$$
$$g = \sum_{i=3}^{8}(x_i - 0.5)^2$$
$$x_i \in [0,\ 1],\ i=1,\ \ldots,\ n,\ n=8.$$

The Pareto front is 1/8 of the unit sphere centered in origin. The Pareto optimal set consist of all solutions but the first two decision variables are equal to 0.5 and the first two decision variables may take any value between 0 and 1.

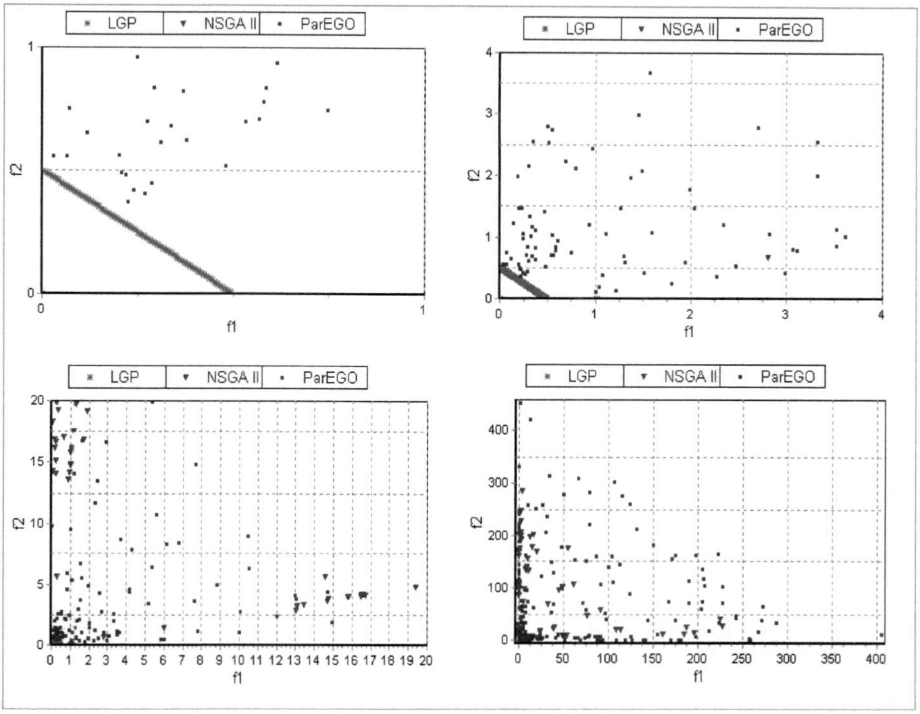

Fig. 1. Distribution of solutions on the Pareto frontier obtained by LGP, ParEGO and NSGA II for test function DTLZ1a

For test function DTLZ4a the value of α is set to 0.2. The distribution of solutions on the Pareto frontier and the convergence to the Pareto frontier for all the three algorithms is depicted in Figure 2. For test function DTLZ4a the value of α is set to 0.2.

From Figure 2 it can be observed that, compared to ParEGO and NSGA II, LGP is assuring a very good convergence. The latter two approaches are not converging very well with the parameters used.

As evident from Table 3 none of the solutions obtained by LGP are dominated neighter by ParEGO or by NSGA II while solutions obtained by LGP dominate all 100 solutions obtained by ParEGO and NSGA II. 87 of the solutions obtained by NSGA II are dominated by solutions obtained by ParEGO while 54 of the solutions obtained by ParEGO are dominated by solutions obtained by NSGA II.

3.3 Test Function DTLZ7a

This test function has 3 objectives and 8 decision variables and it is given by:

$$\begin{aligned}
&minimize\ f_1 = x_1\\
&minimize\ f_2 = x_2\\
&minimize\ f_3 = (1+g)h
\end{aligned}$$

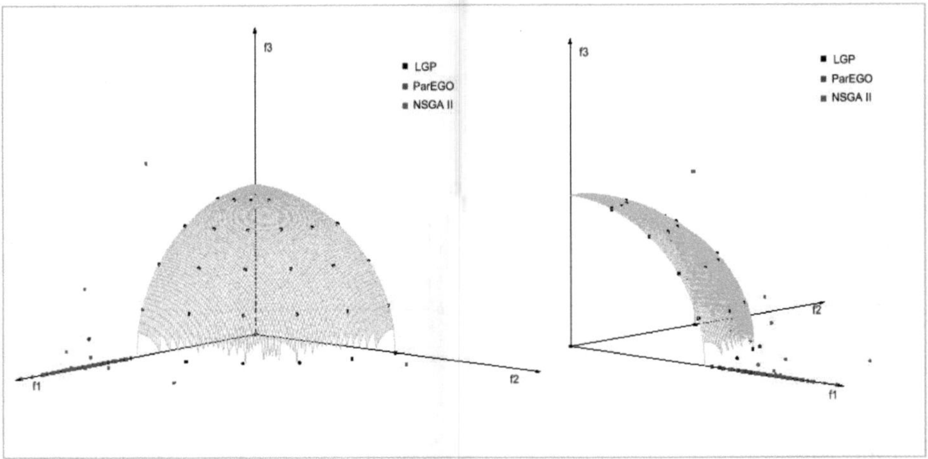

Fig. 2. Convergence to the Pareto frontier and distribution of solutions obtained by LGP, ParEGO and NSGA II on the Pareto frontier for test function DTLZ4a (view from different angles)

Table 3. The dominance between solutions obtained by LGP, ParEGO and NSGA II for test function DTLZ4a

Dominate	ParEGO	NSGA II	***Dominate***	LGP	NSGA II	***Dominate***	LGP	ParEGO
LGP	100	100	ParEGO	0	87	NSGA II	0	54
Dominated	ParEGO	NSGA II	***Dominated***	LGP	NSGA II	***Dominated***	LGP	ParEGO
LGP	0	0	ParEGO	100	54	NSGA II	100	87

$$g = 1 + \frac{9}{6}\sum_{i=3}^{8} x_i$$
$$h = 3 - \sum_{i=1}^{2}\left[\frac{f_i}{1+g}\left(1+\sin\left(3\pi f_i\right)\right)\right]$$
$x_i \in [0, 1]$, i=1, ..., n, n=8.

The Pareto front has four discontinuous regions and the Pareto set consists of all solutions where all by the first two decision variables are equal to 0.

The test function DTLZ7a has 4 discontinuous Pareto regions. LGP is able to converge very well and it is able to spread into the all four disconnected Pareto regions from a single starting point. The value of α used is 1, but there is not much difference between different values of α. As evident from Figure 3, both ParEGO and NSGA II ar far from the Pareto front in terms of convergence. Also, none of the solutions obtained by LGP is dominated by neither ParEGO or NSGA II. 17 solutions obtained by ParEGO are dominated by solutions obtained

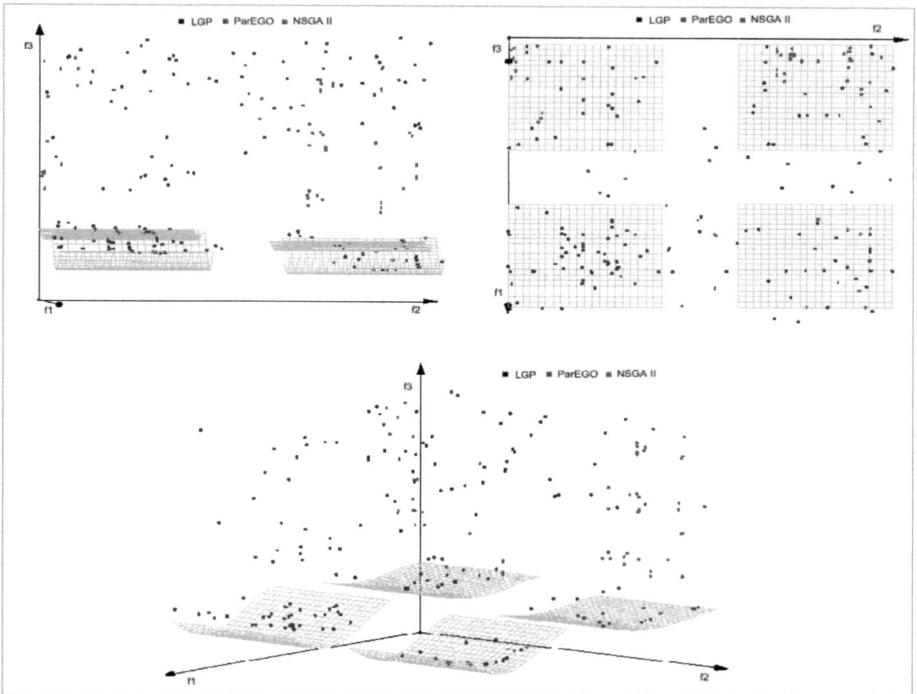

Fig. 3. Convergence to the Pareto frontier and distribution of solutions obtained by LGP, ParEGO and NSGA II on the Pareto frontier for test function DTLZ7a (view from different angles)

Table 4. The dominance between solutions obtained by LGP, ParEGO and NSGA II for test function DTLZ7a

Dominate	ParEGO	NSGA II	***Dominate***	LGP	NSGA II	***Dominate***	LGP	ParEGO
LGP	100	100	ParEGO	0	80	NSGA II	0	17
Dominated	ParEGO	NSGA II	***Dominated***	LGP	NSGA II	***Dominated***	LGP	ParEGO
LGP	0	0	ParEGO	100	17	NSGA II	100	80

by NSGA II while 80 of the solutions obtained by NSGA II are dominated by solutions obtained by ParEGO.

4 Conclusions

The paper proposes a new approach for multiobjective optimization which uses an aggregation of objectives and transforms the MOP into a SOP. A line search based technique is applied in order to obtain one solution. Starting from this solution a simplified version of the initial line search is used in order to generate solutions with a well distribution on the Pareto frontier. Numerical experiments performed show that the proposed approach is able to converge very fast and provide a very good distribution (even for discontinuous Pareto frontier) while compared with state of the art population based metaheuristics such as ParEGO and NSGA II.

Compared to NSGA II and ParEGO, LGP has only few parameters to adjust. It is computationally inexpensive, taking less than 200 milliseconds to generate a set of nondominated solutions well distributed on the Pareto frontier.

The only inconvenience is that LGP involves first partial derivatives which makes it be restricted to a class of problems which are continuous twice differentiable. But almost all practical engineering design problems are continuous differentiable.

One of the further work ideas is to find a better way to set the value of α. In this paper, we considered different α values until we achieved a satisfactory distribution. Also, we would like to extend LGP to deal with constraint multi-objective optimization problems.

References

1. Das, I., Dennis, J.: A Closer Look at Drawbacks of Minimizing Weighted Sums of Objective for Pareto Set Generation in Multicriteria Optimization Problems. Structural Optimization 14, 63–69 (1997)
2. Deb, K., Pratap, A., Agarwal, S., Meyarivan, T.: A fast and elitist multi-objective genetic algorithm: NSGA-II. IEEE Transaction on Evolutionary Computation 6(2), 181–197 (2002)
3. Deb, K., Thiele, L., Laumanns, M., Zitzler, E.: Scalable multi-objective optimization test problems. In: Proceedings of the Congress on Evolutionary Computation (CEC-2002), USA, pp. 825–830 (2002)
4. Chen, W., Wiecek, M.M., Zhang, J.: Quality Utility – A Compromise Programming Approach to Robust Design. Journal of Mechanical Design 121, 179–187 (1999)
5. Ehrgott, M., Wiecek, M.M.: Mutiobjective Programming, In Multiple Criteria Decision Analisys: State of the art surveys. In: Figueira, J., et al. (eds.) International Series in Operations Research & Management Science, vol. 78, Springer, New York (2005)
6. Gould, N.: An introduction to algorithms for continuous optimization, Oxford University Computing Laboratory Notes (2006)
7. Huband, S., Hingston, P., Barone, L., While, L.: A Review of Multiobjective Test Problems and a Scalable Test Problem Toolkit. IEEE Transactions on Evolutionary Computation 10(5), 477–506 (2006)
8. Knowles, J.: ParEGO: A hybrid algorithm with on-line landscape approximation for expensive multiobjective optimization problems. IEEE Transactions on Evolutionary Computation 10(1), 50–66 (2006)

9. Messac, A., Mattson, C.A.: Generating Well-Distributed Sets of Pareto Points for Engineering Design using Physical Programming. Optimization and Engineering 3, 431–450 (2002)
10. Messac, A., Ismail-Yahaya, A., Mattson, C.A.: The Normalized Normal Constraint Method for Generating the Pareto Frontier. Structural and Multidisciplinary Optimization 25(2), 86–98 (2003)

Continuous Adaptive Outlier Detection on Distributed Data Streams

Liang Su, Weihong Han, Shuqiang Yang, Peng Zou, and Yan Jia

School of Computer Science National University of Defense Technology Changsha 410073, China
liangsumail@gmail.com, 13808419839@hnmcc.com, sqyang9999@126.com, zpeng@nudt.edu.cn, jiayanjy@vip.sina.com

Abstract. In many applications, stream data are too voluminous to be collected in a central fashion and often transmitted on a distributed network. In this paper, we focus on the outlier detection over distributed data streams in real time, firstly, we formalize the problem of outlier detection using the kernel density estimation technique. Then, we adopt the *fading strategy* to keep pace with the transient and evolving natures of stream data, and *mico-cluster technique* to conquer the data partition and "one-pass" scan. Furthermore, our extensive experiments with synthetic and real data show that the proposed algorithm is efficient and effective compared with existing outlier detection algorithms, and more suitable for data streams.

1 Introduction

Following with the fast improvement of hardware and communication technologies, many modern data acquisition systems are essentially automatic, distributed and continuous. For example, networking applications (multiple web/ blog crawlers, intrusion detection, network monitoring), financial services (distributed fraud detection, financial monitoring, click stream analysis) and military application(soldier location streams), etc. In these environments, all stream data generally have these characters: continuous and unbounded, distributed and evolvable over time. So, distributed data stream model was provided, which is suitable for these applications very well.

In this paper we study a quintessential monitoring problem on continuously changing distributed data sources, namely, *outlier detection*, which is an important part of data mining. And the outliers may point out some surprising and suspicious activities or observations which appear to be inconsistent with the remainder data. Formally, we have $m+1$ distributed nodes (one leader node and m child nodes), illustrated in Figure 1. Each child node i has a changing source of data S_t^i at time t, and individual data items in the sources may be high dimensional including numerical values, text, audio or video. We are more concerned about monitoring some desirable properties of the union $\bigcup_{i=1}^{m} S_i^t$ of data on all nodes in real time. In order to understand the challenge in monitoring outliers over distributed nodes, there are two straightforward "solutions":

R. Perrott et al. (Eds.): HPCC 2007, LNCS 4782, pp. 74–85, 2007.

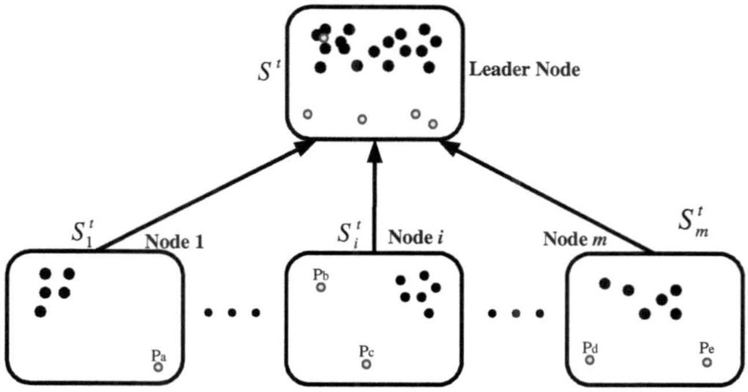

Fig. 1. Continuous Distributed Outlier Detection (red points are local outlers)

1. Each node sends the newly gathered data to the central node. The central node updates the outliers of $\bigcup_{i=1}^{m} S_i^t$ whenever a new point is received.

2. Periodically, each node sends the newly gathered data to the central node, every τ seconds. The central node updates the outliers every τ seconds.

The first method has communication and central processing bottleneck because all data are aggregated to central node; The second solution, which is similar to a batch processing of the first method, will have latency because of the time τ and also does not overcome the communication problem. Furthermore, the outliers would change radically within the time period. So these two solutions are not suitable for monitoring applications. Therefore, we develop efficient online outlier detection techniques. And our contributions are as follows:

1. We propose two novel outlier measures, using *kernel density estimation technique* [1] to approximate the stream data distribution, which are compatible with distance-based outlier and density-based outlier (see Definition 2 and Definition 3).

2. We adopt the *fading strategy* to keep pace with the transient and evolving natures of stream data, and *mico-cluster technique* (commonly used for data compression or summarization) to deal with the data partition.

3. Finally, extensive experiments including synthetic and real data sets, demonstrate that our proposed methods are efficient and effective.

The rest of this paper is organized as follows. Section 2 introduces the related researches of outlier detection. Section 3 presents some preliminaries and the problem definitions. Section 4 provide the effective algorithm (**MOD** algorithm). Section 5 evaluates the effectiveness and efficiency of the algorithm compared with existing algorithms. Section 6 concludes the paper.

2 Related Work

Methods for outlier detection are drawing increasing attention. The salient approaches can be classified as either distribution-based, clustering, distance-based, or density-based.

The distribution-based approach [2,3] assumes the data following a distribution model (e.g. Normal) and flags as outliers those objects which deviate from the model. Thus, such approaches do not work well in moderately high-dimensional (multivariate) spaces, and have difficult to find a right model to fit the evolving data stream. To overcome these limitations, researchers have turned to various non-parametric approaches (clustering, distance-based, and density-based).

Most clustering algorithms (e.g. CLARANS [4], DBSCAN [5], BIRCH [6], WaveCluster [7], CLIQUE [8]), are to some extent capable of handling exceptions. However, since the main objective is to find clusters, they are developed to optimize clustering, and regard outliers as "by-products".

Distance based approaches [9,10,11,12,13,14,15], first proposed by E.M. Knorr and R.T. Ng [9], attempt to overcome limitations of distribution-based approach and they detect outliers by computing distances among points. A point p in a data set T is a distance-based outlier ($DB(\rho, r)$-outlier) if at most a fraction ρ of the points in T lie within distance r from p. It has an intuitive explanation that an outlier is an observation that is sufficiently far from most other observations in the data set. However, it will be no effect when the data points exhibit different densities in different regions of the data space or across time, because this outlier definition is based on a single, global criterion determined by the parameters ρ and r, so more robust density-based techniques were provided.

Density-based approaches proposed originally by M. Breunig, et al. [16] which defines a local outlier factor (LOF) for each point depending on the local density of its neighborhood. In general, points with a high LOF are flagged as outliers. It has attracted considerable attention [17,18,19], and a large number of algorithms have been developed which concern how to define the density or accelerate the efficiency, such as, LOCI [17] method.

In the distributed data stream environments, Babcock and Olston presented an original algorithm for distributed top-k monitoring [20]. Amit Manjhi, et al. [21] thought of finding recently frequent items. S. Subramaniam, et al. [22] cared about the outlier over sensor stream data. Graham Cormode, et al. [23] considered continuous clustering.

3 Preliminaries and Problem Statement

In this section, first, we introduce the two preliminaries: distributed data stream model and kernel density estimation. Then, we propose two novel outlier measures which are compatible with distance-based and density-based outlier.

3.1 Distributed Data Stream Model

A data stream consists of an unbounded sequence $D_1, D_2, \cdots$ of numeric values which $D_i = (D_i^1, D_i^2, \cdots, D_i^d)$ and $D_i^j \in \mathbb{R}$ for $i, j \in \mathbb{N}, j \in [1, d]$. Then, distributed data stream model composes of $m+1$ nodes: a central *leader node* N_c, and m remote stream monitor nodes $N_1, N_2, \cdots, N_m$. Each node just monitors one stream data, illustrated in Figure 1. Except otherwise stated, we assume the stream data as *independent and identical distributed*(***iid***) observations of an unknown continuous random variable which is reasonable for most distributed data stream applications. On the other hand, in many data stream applications, only the most recent elements are important in data mining, while the old ones are very little or not. For example, in a stream of stock market data, people are more concerned about the moving average of the stock price over all observations made in the last hour. So, we only care about the data in the *sliding window* (W) which contains the most recent N elements of stream.

3.2 Kernel Density Estimation

Kernel density estimation is to reveal the unknown density of a distribution by selecting a suitable density estimator and has been successfully applied in diverse application scenarios [1]. Theoretically, kernel density function $\widetilde{f_h}(x)$ is sure to converge to the real density $f(x)$ function for arbitrary distribution. More formally, assume that we have a data set $\mathcal{D}$, containing n points whose values are $\overline{X_1}, \overline{X_2}, \cdots, \overline{X_n}$. We can approximate the underlying distribution $f(x)$ using the following kernel function ($\mathcal{K}(x)$, *kernel function*, is a function of random variable X. h, *kernel width*, determines the smoothing level of kernel function):

$$\widetilde{f_h}(x) = \frac{1}{nh}\sum_{i=1}^{n}\mathcal{K}\left(\frac{x-\overline{X_i}}{h}\right), \quad x, \overline{X_i} \in \mathbb{R}, \quad \int_{x\in\mathbb{R}} \mathcal{K}(x)dx = 1 \ . \tag{1}$$

$$\mathcal{K}\left(\frac{x-\overline{X_i}}{h}\right) = \begin{cases} \frac{3}{4}\cdot\frac{1}{h}\left(1-\left(\frac{x-\overline{X_i}}{h}\right)^2\right), & \left|\frac{x-\overline{X_i}}{h}\right| < 1 \\ 0 & otherwise \end{cases} \tag{2}$$

Common used kernel functions are Epanechnikov, Gaussian, Quartic and Triweight kernel, etc. Since the Gaussian kernel is unbounded ($x \in (-\infty, +\infty)$), it exacerbates the cost of computing integral. Quartic and Triweight kernel are fourth and sixth function each other. So, we choose the Epanechnikov kernel (see Equation 2) which is a square function, more easy to integrate and has bound.

3.3 Novel Outlier Measures

Intuitively, outliers can be defined as given by Hawkins [24] in Definition 1. There are several formal definitions of an outlier in Section 2. In our work, combining the meaning of Definition 1 with kernel density estimation, naturally we can come to a conclusion — the probability of one point is more close to zero, the point is more likely to be an outlier. So, we use the following Equation 3 to measure the outlier degree in this setting. Then, two variations based on the commonly-used outlier definitions are presented.

Definition 1 (Hawkins-Outlier). *An outlier is an observation that deviates so much from other observations as to arouse suspicion that it was generated by a different mechanism.*

$$\Phi(p,r) = P[p-r, p+r] = \int_{p-r}^{p+r} \widehat{f_h}(\mathbf{x})d\mathbf{x} \ . \quad (3)$$

Distance-based Kernel Estimation Outlier: It is a variation of *Distance-based Outlier* [9]. Naturally, we present the definition of *Distance-based Kernel Estimation Outlier* (DisKE-Outlier) (see Definition 2). However, it also has the same limitations as distance-based methods. So we provide a more robust outlier measure: *DenKE-Outlier* in Definition 3.

Definition 2 (DisKE-Outlier). *A point is a "DisKE-Outlier" if* $\Phi(p,r) \leq \rho$.

Density-based Kernel Estimation Outlier: It is a variation of *Density-based Outlier* [16]. Spiros Papadimitriou, et al. [17] provided an outlier metric — Multi-Granularity Deviation Factor(MDEF). For any given point, MDEF is a measure of how the neighborhood count of p (in its counting neighborhood) compares with that of the points in its sampling neighborhood. A point is flagged as an outlier, if its MDEF is significantly different from that of the local averages. r is the sampling neighborhood distance and αr ($\alpha \in (0,1)$) is the range over which the neighborhood counts are considered. Correspondingly, we can define a *Density-based Kernel Estimation Outlier*(DenKE-Outlier) as follows:

Definition 3 (DenKE-Outlier). *A point is defined as a "DenKE-Outlier" that* $\frac{\Phi(p,\alpha r)}{\sum_{q\in\Omega_r}\Phi(q,\alpha r)/|\Omega_r|} < \xi$, *$\xi$ is a real value parameter to define how significant the point p is to the average of its neighbors. Ω_r is a point set that contains all points which distances are below r to point p. $|\Omega_r|$ is the count of Ω_r.*

4 Micro-cluster Based Outlier Detection Algorithm

In this section, firstly, we discuss the selection of kernel width which significantly affects the accuracy of algorithm. Secondly, the fading strategy is proposed to conquer the evolving nature of stream data. Thirdly, we provide the approximate JS-divergence technique to save the traffic between the leader node and other child nodes. Finally, we use the micro-cluster CF-tree [6] data structure to condense the stream data and meet the "one-pass" scan.

4.1 Selecting Kernel Width

The kernel density estimation in the Figure 2 indicate that the *kernel width* significantly affects the shape of a kernel function. If the width is chosen too high, the estimation is over-smoothed and hides important details. Otherwise, the estimation is under-smoothed and brings illusive details, resulting in heavy computation. A widely used rule for approximating the kernel width is the Scott's rule [1] (h in Equation 4). However, these strategies depend on the complete sample, which is impracticable in data stream scenario. To overcome this problem, we adopt an approximate solution, only considering the data in sliding window. The number of sliding window (N) is more large, the approximate kernel width

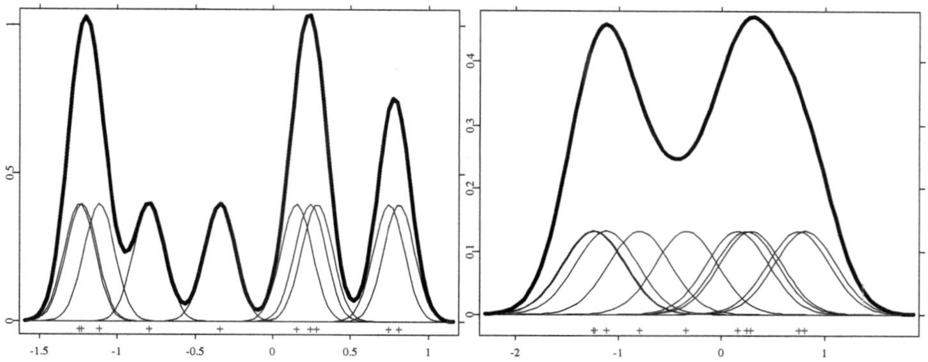

Fig. 2. Kernel density estimation, underlying standard normal kernel and kernel width with $h = 0.1$ (left) and $h = 0.3$ (right)

$\widetilde{h}$ is more close to h. σ is the standard deviation of whole stream data and $\widetilde{\sigma}$ is the sample standard deviation of sliding window.

$$h \approx 2.345\sigma n^{-1/5} \approx \widetilde{h} \triangleq 2.345\widetilde{\sigma}(N)^{-1/5}, \quad \widehat{h} = \min\left\{2.345\widetilde{\sigma}n^{-1/5}, D_x^k\right\} \ . \quad (4)$$

Definition 4 (k-nearest neighbor set). *D_p^k is the distance between point p and its k-nearest neighbor. Δ_p, named " k-nearest neighbor set of point p", is defined as a set that contains all points which distance is not larger than D_p^k. Δ_S, named " k-nearest neighbor set of set S", is defined as $\bigcup_{p\in S}\Delta_p$.*

The $\widetilde{h}$ width will increase very fast following by the $\widetilde{\sigma}$ and n in Equation 4. In order to balance the quality and efficiency, we consider the distance between the point and its k-nearest neighbors. So an improved kernel width $(\widehat{h})$ is listed in Equation 4. Extensive experiments demonstrate its efficiency.

4.2 Capturing Evolution over Data Stream

Because of the dynamic nature of stream data, it is the inherent obstacle that must been conquered by an effective and robust stream analysis technique. To capture the evolution, we are conscious of the problem in Equation 1: each kernel entry is equally weighted with constant $\frac{1}{n}$. So we present a *fading strategy* that couple the kernel density estimation with exponential smoothing [25]. The basic idea of fading strategy is to give older data less weight and to gradually discount the history data, which is widely used in the area of time series analysis and forecasting. Then, Equation 1 is adjusted to Equation 5. Notice that the sum of weights in Equation 5 is $\sum_{i=1}^{N}\frac{\omega^{N-i}}{\sum_{j=1}^{N}\omega^{j-1}} = 1$, which is equal to $\sum_{i=1}^{n}\frac{1}{n} = 1$ in Equation 1. When $\omega = 1$, Equation 5 comes back to Equation 1. The *fading factor*(ω) determines the rate of fading. The higher the value of ω the lower importance of the historical data compared to more recent data. The sliding window becomes an *evolving sliding window*. And also, we can tune the impact degree of past and current data through ω to match different applications.

$$\widehat{f}_h(x) = \frac{1}{h} \sum_{i=1}^{N} \left(\frac{\omega^{N-i}}{\sum_{j=1}^{N} \omega^{j-1}} \cdot \mathcal{K}\left(\frac{x - \overline{X_i}}{h}\right)\right), \quad x \in \mathbb{R}, \quad \omega \in (0,1] \ . \tag{5}$$

4.3 Comparison for Different Distributions

The difference of two distributions can seriously influence the traffic between the leader node and other child nodes. Intuitively, the less change of distribution in one node, the less traffic between this node and the leader node. So it is very important to compare the difference between two distributions. Several methods have been proposed to quantify the difference between probability density distributions [1]. One widely used measure is Ali-Silvey divergences, which is more natural than ℓ_p norms. $p(x)$ and $q(x)$ are probability distribution functions over random variable x, Ω is the value set of x. Two most common divergences are the asymmetric KL-divergence and the symmetric JS-divergence in Equation 6. However, because kernel density estimation method may assign probability of zero to some regions, KL-divergence is undefined and meaningless for the ln function. So, we choose the JS-divergence which is meaningful to any $x \in \Omega$.

$$D_{JS}(p,q) = \sum_{i=1}^{n} \left(p_i \ln \frac{2p_i}{p_i + q_i} + q_i \ln \frac{2q_i}{p_i + q_i} \right) \ . \tag{6}$$

$$\widehat{D}_{JS}(p,q) = \sum_{i \in \Delta_{S_{in} \cup S_{out}}} \left(p_i \ln \frac{2p_i}{p_i + q_i} + q_i \ln \frac{2q_i}{p_i + q_i} \right) \ . \tag{7}$$

Theoretically, a new data incoming the sliding window or an old data being dropped out will change the kernel density for all data points in the whole sliding window. Obviously, it is too inefficient and unscalable to suit the evolving data stream over time. The incoming data and dropped data are a very small proportion to the capacity of sliding window. The points are more close to the changeable data, the influence is more serious. To balance the accuracy and complexity, we provide an approximate JS-divergence($\widehat{D}_{JS}(p,q)$ in Equation 7) which is only concerned about the changeable data (S_{in} and S_{out}) and their k-nearest neighbors set. As the number $\left|\Delta_{S_{in} \cup S_{out}}\right|$ is much less than n, time complexity is reduced to $O\left(d\left(\left|\Delta_{S_{in} \cup S_{out}}\right| + \log n\right)\right)$ which $\log n$ is the time spending to search k-nearest neighbor set. $\widehat{D}_{JS}(p,q)$ is non-negative real number and unbounded. In practical, we can select a parameter λ (*divergence threshold*) as the upper-bound of $\widehat{D}_{JS}(p,q)$. If $\widehat{D}_{JS}(p,q) > \lambda$, there is a significant change and corresponding node transfer its kernel estimation function to the leader node. Otherwise, no transformation.

4.4 Micro-cluster Definition and MOD Algorithm

In this section, we describe the ***M**icro CF-Tree Based Distributed **O**utlier **D**etection Algorithm* (**MOD** Algorithm) that can deal with the two outlier categories (DisKE-Outlier and DesKE-Outlier), in a distributed manner. We

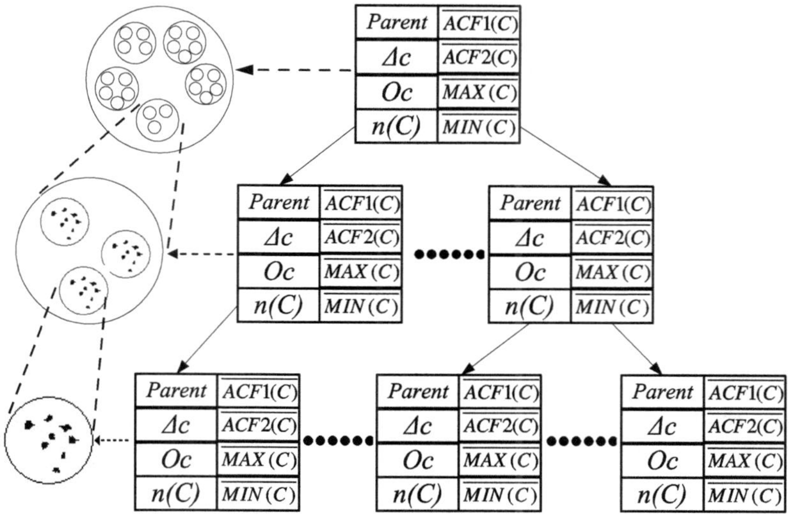

Fig. 3. Extended Micro-Cluster Feather Tree

want to identify outliers in the leader node. By observation of the definition of DisKE-Outlier, it is easy to find that the leader node need only to examine the values that have been marked as outliers by its child nodes. All the other data values can be safely ignored, since they cannot possibly be outliers. However, DesKE-Outliers are non-decomposable because the neighbors of one point may be distributed in different child nodes. A feasible solution is that all child nodes have a copy of the kernel density function of leader node. The child node transfers its kernel estimation function to the leader node when its $\widehat{D}_{JS}(p,q) > \lambda$, and the leader node broadcasts its kernel estimation function to its child nodes when $\widehat{D}_{JS}(p,q) > \lambda$ in N_c. We define the concept of micro-cluster-based outlier more precisely in Definition 5 which is a *CF-Tree* [6]. It is very easy to create and maintain the clusters in a single pass. The whole algorithm is described in algorithm (1).

Definition 5. *An extended micro CF-tree for a set of d-dimensional points* $D_{i_1}, D_{i_2}, \cdots, D_{i_n}$ *with time-stamp* T_{i_1}, T_{i_2}, $\cdots$, T_{i_n} *and each point* $D_j = (d_j^1$, d_j^2, $\cdots$, $d_j^d)$, *is defined as the* $(4d+1+|\Delta_C|+|O_C|)$*-tuple* $(\overline{ACF1(C)}$, $\overline{ACF2(C)}$, $\overline{MIN(C)}$, $\overline{MAX(C)}$, Δ_C, O_C, $n(C))$. $(\overline{ACF1(C)}$, $\overline{ACF2(C)}$, $\overline{MIN(C)}$, $\overline{MAX(C)}$ *are four vectors of d entries. The definition of each of these entries is as follows, Figure 3 is the whole data structure:*

- *The p-th entry of* $\overline{ACF1(C)}$ *is equal to* $\sum_{j=1}^{n(C)} d_{i_j}^p$, *The p-th entry of* $\overline{ACF2(C)}$ *is equal to* $\sum_{j=1}^{n(C)} (d_{i_j}^p)^2$;
- *The p-th entry of* $\overline{MIN(C)}$ *is equal to* $\min_{j=1}^{n(C)} d_{i_j}^p$, *The p-th entry of* $\overline{MAX(C)}$ *is equal to* $\max_{j=1}^{n(C)} d_{i_j}^p$;

Algorithm 1. CF-Tree Based Distributed Outlier Detection Algorithm

Input: m data streams $S_1, \cdots, S_m$ in $N_1, \cdots, N_m$, the divergence threshold λ, ρ is the outlier proportion in DisKE-Outlier, the distance r, α is the distance proportion in DesKE-Outlier.
Output: the outlier set O_c

Function OutlierDetetion()

$BDisKE \Leftarrow$;// variable $BDisKE = true$ indicates using DisKE-Outlier detection method, otherwise using DesKE-Outlier detection method.
LeaderProcess(); // initiate the process for the leader node;
for$(i = 1$ **to** $m)$ **do**
  ChildProcess(); // initiate the process for each child node;

Function LeaderProcess()

find the closest micro-cluster from the CF-Tree to the point p which a new point from a child node, assumed the cluster as GC which is a leaf node in the CF-Tree;
if$(p \notin O_{GC})$ **then isOutlier**(p, GC);
update $\widetilde{f_{\widehat{h}}}(x)$; // using Equation 1 and kernel width in Equation 4;
if$(\widehat{D}_{JS}(\widehat{f}^{new}_{\widehat{h}}, \widehat{f}^{old}_{\widehat{h}}) > \lambda)$ **then** broadcast $\widehat{f}^{new}_{\widehat{h}}$ to the all child node;
if$(p \in O_{GC}$, or is marked as an outlier$)$ **then** report point p as an outlier;
else insert p to micro-cluster GC;

Function ChildProcess()

find the closest micro-cluster from the CF-Tree to the new point p, assumed as LC which is a leaf node in the CF-Tree;
if$(p \notin O_{LC})$ **then isOutlier**(p, LC);
update $\widetilde{f_{\widehat{h}}}(x)$; // using Equation 1 and kernel width in Equation 4;
if$(\widehat{D}_{JS}(\widehat{f}^{new}_{\widehat{h}}, \widehat{f}^{old}_{\widehat{h}}) > \lambda)$ **then** send the point p and $\widehat{f}^{new}_{\widehat{h}}$ to the leader node;
if$(p \notin O_{GC}$, and is not marked as an outlier$)$ **then** insert p to LC;

Function isOutlier(point p, micro cluster MC)

if$(BDisKE =)$
  if$(\Phi(p, r) < \rho)$ **then** mark p as an outlier and add to O_{MC};
else sum up to $\Omega_r = \bigcup_{q \in MC} \Delta_q$;
if$(\frac{\Phi(p,\alpha r)}{\sum_{q \in \Omega_r} \Phi(q,\alpha r)/|\Omega_r|} < \xi)$ **then** mark p as an outlier and add to O_{MC};
update $\widetilde{f_{\widehat{h}}}(x)$; // using Equation 1 and kernel width in Equation 4;

- *The Δ_C is defined in Definition 4. O_C holds the outliers in cluster $\mathcal{C}$, $n(\mathcal{C})$ is the count number in Micro-Cluster $\mathcal{C}$.*

5 Experiments

In this section, we will present the experimental results for our algorithms comparing with two typical algorithms. One is the distanced-based algorithm — NL algorithm [9] which is proposed by E.M. Knorr and R.T. Ng, and the other is an density-based algorithm — LOCI algorithm [17] which is presented by Spiros Papadimitriou, et al. There are three primary purposes: (1) Comparing the precision and recall ratio in four real data sets. (2) Verifying the algorithm

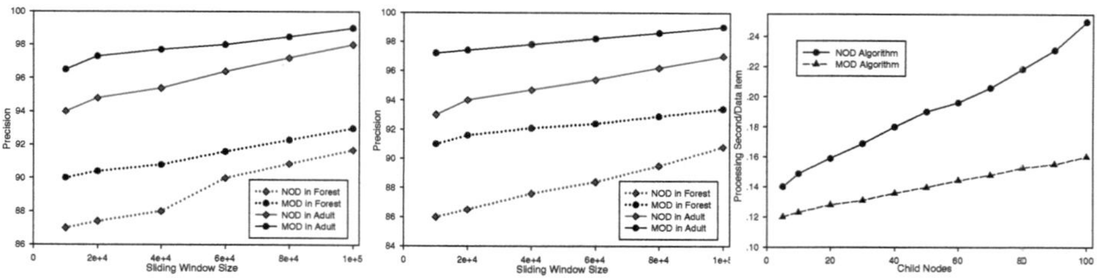

(a) $\sigma = 0.2$ (b) $\sigma = 0.4$ (c) Child Nodes Scalability

Fig. 4. Precision when varying the sliding window size, and nodes scalability

scalability in high dimensions and large numbers of child nodes. (3) Checking the efficiency of two kernel widths ($\widetilde{h}$ and $\widehat{h}$ in Equation 4).

All experiments were conducted on a 3.2 GHz PentiumIV PC with 1GB memory, running Microsoft Windows XP. To evaluate the efficiency and scalability of our two algorithms, both real and synthetic data sets are used. Real data sets include adult, KDD cup 99, forest and stock price series. The last data set consists of the price for a single stock taken at frequent intervals over a six year period. Total count is $330K$ values. The other three data sets come from the UCI machine learning repository. All algorithms are implemented by Microsoft Visual C++ 6.0. The **N**aïve **O**utlier **D**etection Algorithm (NOD Algorithm) use the $\widetilde{h}$ kernel width and C++ STL (NOT the CF-tree), which is to proof the MOD algorithm scalability.

Selecting the Kernel Width: Kernel width is a virtual parameter in kernel density estimation. An proper width can save computing time and improve the accuracy very large. In this experiment, we select two very large data sets (adult and forest cover). We consider the precision change with the sliding window size. The sliding window varied from 10000 to 100000. Unless particularly mentioned, the nearest neighbor $k = 8$, divergence threshold $\lambda = 0.1$ and fading factor $\omega = 0.2$. In Figure 4 (a) the $\sigma = 0.2$ in the real data selected and Figure 4 (b) is 0.4. The results show the MOD algorithm has more better precision than the NOD algorithm, especially in Figure 4 (b). The precision in MOD algorithm is less sensitive than NOD algorithm following with the change of sliding window. This owes to the kernel width. In MOD algorithm, we considered the influence of k-nearest neighbors which can prune effectively off the kernel width.

Efficiency of MOD Algorithm: Our first set of experiments focused on the efficiency of three algorithms. We use two measures, namely *precision* and *recall*, defined as follows. *Precision* represents the fraction of the values reported by our algorithms as outliers that are true outliers. *Recall* represents the fraction of the true outliers that our algorithms identified correctly. We set the sliding window size $N = 10000$. We selected five real value columns in the real data sets (The stock price series data set only has one column and we constructed five columns by simply duplicating the column). The experiment results in Fig. 5 (a) and (b) indicate that our NOD algorithm is better than the corresponding distance-based

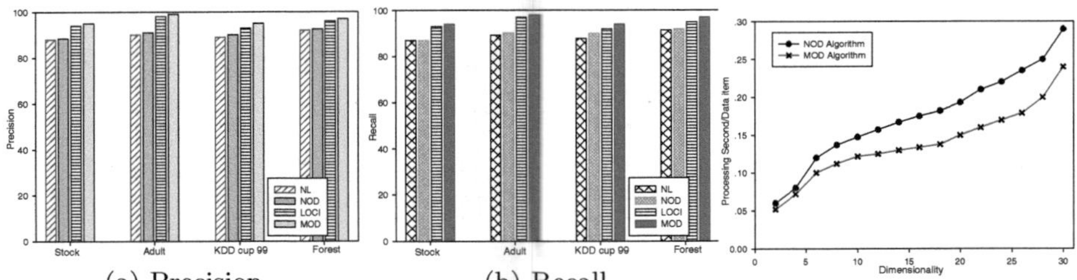

(a) Precision (b) Recall (c) Dimensional Scalability

Fig. 5. Precision and Recall in four real datasets, and dimensional scalability

NL algorithm, the MOD algorithm also exceeds the LOCI, although only $0.2 \sim 3.1\%$ improvement. It is not surprising that our NOD and MOD algorithms are essentially improved from the distance-based and density-based algorithms.

Algorithm Scalability: The scalability is an important criteria to judge whether an algorithm is suitable for distributed environment or not. Our synthetic data were random sample columns from adult data sets.we used ten PCs to simulate the 100 child nodes (each node simulate ten child nodes at most) and one PC as the leader node in Figure 1. each In Figure 4(c) and Figure 5(c), we change the dimension(d) and child nodes(N) to check the precision of NOD and MOD algorithms. The results show that: NOD algorithm is exponential proportional to the dimension, but the MOD is linear to it. This phenomenon is more obvious in Figure 5(c). The processing time per data item of MOD algorithm in Figure 5(c) only varied range of $0.12 \sim 0.16$ second, but $0.14 \sim 0.25$ second in NOD algorithm. The ratio of fluctuation difference ($\frac{0.25-0.14}{0.16-0.12} \approx 3$) approximate to three. So, the MOD algorithm has more scalability than NOD algorithm, mainly because of the micro-cluster compressed data structure.

6 Conclusions and Future Work

In this paper, we study the problem of continuous adaptive outlier detection on distributed data streams. We propose two novel outlier measures and an algorithm which can deal with the distributed and evolving natures of stream data. The mathematical analysis and extensive experiments show that the proposed methods are both efficient and effective comparing with existing outlier detection algorithms. In the future, we will study cluster over distributed data streams.

Acknowledgements

This work is supported by the National High Technology Development 863 Program of China under Grant No.2004AA112020, and the National Grand Fundamental Research 973 Program of China under Grant No.2005CB321804.

References

1. Scott, D.W.: Multivariate Density Estimation: Theory, Practice, and Visualization. Wiley-Interscience, Chichester (2001)
2. Barnett, V., Lewis, T.: Outliers in statistical data, 3rd edn. Wiley, Chichester (2001)
3. Eskin, E.: Anomaly detection over noisy data using learned probability distributions. In: ICML (2000)
4. Ng, R.T., Han, J.: Efficient and effective clustering methods for spatial data mining. In: VLDB (1994)
5. Ester, M., Kriegel, H.P., Xu, X.: A database interface for clustering in large spatial databases. In: KDD (1995)
6. Zhang, T., Ramakrishnan, R., Livny, M.: Birch: an efficient data clustering method for very large databases. In: SIGMOD (1996)
7. Sheikholeslami, G., Chatterjee, S., Zhang, A.: Wavecluster: A multi-resolution clustering approach for very large spatial databases. In: VLDB (1998)
8. Agrawal, R., Gehrke, J., Gunopulos, D., Raghavan, P.: Automatic subspace clustering of high dimensional data for data mining applications. In: SIGMOD (1998)
9. Knorr, E.M., Ng, R.T.: Algorithms for mining distance-based outliers in large datasets. In: VLDB (1998)
10. Knorr, E.M., Ng, R.T.: Finding intensional knowledge of distance-based outliers. In: VLDB (1999)
11. Ramaswamy, S., Rastogi, R., Shim, K.: Efficient algorithms for mining outliers from large data sets. In: SIGMOD (2000)
12. Bay, S.D., Schwabacher, M.: Mining distance-based outliers in near linear time with randomization and a simple pruning rule. In: KDD (2003)
13. Ghoting, A., Parthasarathy, S., Otey, M.E.: Fast mining of distance-based outliers in high-dimensional datasets. In: Ng, W.-K., Kitsuregawa, M., Li, J., Chang, K. (eds.) PAKDD 2006. LNCS (LNAI), vol. 3918, Springer, Heidelberg (2006)
14. Wu, M., Jermaine, C.: Outlier detection by sampling with accuracy guarantees. In: KDD (2006)
15. Tao, Y., Xiao, X., Zhou, S.: Mining distance-based outliers from large databases in any metric space. In: KDD (2006)
16. Breunig, M.M., Kriegel, H.P., Ng, R.T., Sander, J.: Lof: identifying density-based local outliers. In: SIGMOD (2000)
17. Papadimitriou, S., Kitagawa, H., Gibbons, P.B., Faloutsos, C.: Loci: Fast outlier detection using the local correlation integral. In: ICDE (2003)
18. Lazarevic, A., Kumar, V.: Feature bagging for outlier detection. In: KDD (2005)
19. Jin, W., Tung, A.K.H., Han, J., Wang, W.: Ranking outliers using symmetric neighborhood relationship. In: Ng, W.-K., Kitsuregawa, M., Li, J., Chang, K. (eds.) PAKDD 2006. LNCS (LNAI), vol. 3918, Springer, Heidelberg (2006)
20. Babcock, B., Olston, C.: Distributed top-k monitoring. In: SIGMOD (2003)
21. Manjhi, A., Shkapenyuk, V., Dhamdhere, K., Olston, C.: Finding (recently) frequent items in distributed data streams. In: ICDE (2005)
22. Subramaniam, S., Palpanas, T., Papadopoulos, D., Kalogeraki, V., Gunopulos, D.: Online outlier detection in sensor data using non-parametric models. In: VLDB (2006)
23. Cormode, G., Muthukrishnan, S., Zhuang, W.: Conquering the divide: Continuous clustering of distributed data streams. In: ICDE (2007)
24. Hawkins, D.: Identification of Outliers, 1st edn. Springer, Heidelberg (1980)
25. Gijbels, I., Pope, A., Wand, M.: Automatic forecasting via exponential smoothing: Asymptotic properties (1997)

A Data Imputation Model in Sensor Databases

Nan Jiang

The University of Oklahoma
School of Computer Science
Norman, OK 73019, USA
nan_jiang@ou.edu

Abstract. Data missing is a common problem in database query processing, which can cause bias or lead to inefficient analyses, and this problem happens more often in sensor databases. The reasons include power outage at the sensor node, sensors time synchronization, occurrences of local interferences, unstable wireless network communication, etc. Therefore, in sensor database applications, there is a need for data imputation, especially for those applications in which the query response time is tight, and the accuracy of the query results is important. In this paper, we present a data imputation application based on association rule mining of closed frequent itemsets. They are subsets of all frequent patterns but provide complete and condensed information since they do not include redundant patterns. Experimental results compared with the existing techniques using real-life sensor data show that our proposed technique effectively imputes missing sensor data as well as achieves time and space efficiency.

1 Introduction

A wireless sensor network (WSN) is a wireless network consisting of spatially distributed autonomous devices using sensors to cooperatively monitor physical or environmental conditions, such as temperature, sound, vibration, pressure, motion or pollutants, at different locations [15]. Recent advances in sensor technology have made possible the development of relatively low cost and low-energy-consumption micro sensors which can be integrated in a wireless sensor network. These devices - Wireless Integrated Network Sensors (WINS) - will enable fundamental changes in applications spanning the home, office, clinic, factory, vehicle, metropolitan area, and the global environment [3].

Many researches have been conducted by different organizations regarding wireless sensor networks to address many different issues such as power awareness, security, routing protocol, heterogeneous sensor networking and so on, but few of them discuss how to impute the missing data when data is lost or corrupted. In this paper we present a data imputation model to impute data tuples in a sensor database using a data mining technique based on closed pattern mining for association rules. Its goal is to derive imputed values that are not only accurate but also timely. This is significant to many applications where exact data may not be necessary and certain approximate data is acceptable, such as traffic management, intrusion detection, and

R. Perrott et al. (Eds.): HPCC 2007, LNCS 4782, pp. 86–96, 2007.

network routing. This research also reduces the chance that real-time applications would miss their deadlines due to lack of data.

The data imputation model using association rule mining on stream data based on closed frequent itemsets (CARM) [11] discovers relationships between sensors and use them to compensate for missing and corrupted data. The derived association rules provide complete and non-redundant information, since closed pattern mining provides complete and condensed information and thus they do not include redundant patterns [21]. Experimental results compared with the existing techniques using real-life sensor data show that our proposed model effectively imputes missing sensor data as well as achieves time and space efficiency.

The remainder of this paper is organized as follows. We review the existing data imputation solutions in Section 2. We discuss the definitions of terms used in the paper in Section 3. In Section 4, we present our proposed online data imputation application based on the closed pattern mining. Section 5 depicts the performance evaluation of the proposed data imputation model comparing with the existing techniques using real-life traffic data. Finally, Section 6 concludes the paper.

2 Related Works

Many articles have been published to deal with the missing data problem, and a lot of software has been developed based on these methods. Some of the methods totally delete the missing data before analyzing them, like listwise and pairwise deletion [21] while some other methods focus on imputing the missing data based on the available information. The most popular statistical imputation methods include mean substitution, imputation by regression [4], hot deck imputation [9], cold deck imputation, expectation maximization (EM) [13], maximum likelihood [2, 12], multiple imputations [16, 18], and bayesian analysis [6].

Also, there are some mathematical approaches to perform the imputation, like SVDimpute, which is a singular value decomposition based method, called SVDimpute. This method is applied for imputing missing values in DNA microarrays. A DNA microarray is a matrix m rows of which are expression levels of genes and n columns are different experimental conditions. The missing values occur for diverse reasons, including insufficient resolution, image corruption, or simply due to dust or scratches on the slide [20].

But none of these approaches is specially suited for wireless sensor network environments, where streams of data constantly sent from the sensors to the servers, due to several reasons. First, how much old information should be based on to get the associated information for the missing data imputation? Using all of the old readings to perform the imputation is unreasonable, especially when using an iteration procedure until convergence to get the imputation like in the EM algorithm. On the other hand, using only the previous round of sensor readings to perform the imputation is also not a good choice since data streams often have a changing data distribution. Some of the statistical methods use all of the available data points in a database to construct the best possible results, like the Maximum Likelihood. In the wireless sensor networks, the missing sensor data may or may not be related to all of the available information, thus using all of the available information to process the result is not an optimal choice and would use more time and memory space than necessary when it comes to the implementation stage.

Second, which information should be used to perform the missing data imputation? In a wireless sensor network, data is collected within certain scopes and reported to the servers during a certain period of time. Different sensors have different readings at different time periods, and the current readings of one sensor may relate not only to its previous readings, but also to other sensors' previous or current readings. Therefore, replacement of missing values with randomly selected values present in a pool of similar complete cases or by a value which is independent of the data set like in the hot/cold deck imputation is difficult to implement. This is because even though we may get the complete set of information of a certain wireless sensor network, it is not easy to decide which information is similar to the current round of missing sensor's information. In other words, it is hard to draw the pool for a certain sensor's certain round of readings when the application needs to perform the data imputation.

Third, the missing data may or may not miss at random (MAR), while most of the statistical techniques, such as maximum likelihood [2, 12] and multiple imputations [16, 18], are based on the MAR assumption. According to the definition in [12], Data on Y are Missing At Random (MAR) if the probability that Y is missing does not depend on the value of Y after controlling other observed variables X. For example, we are modeling weight (Y) as a function of gender (X). One gender may be less likely to disclose its weight, that is, the probability that Y is missing depends only on the value of X. Such data are MAR.

As there are more and more stream data applications emerge, proper data estimation algorithms for stream data are needed. In the prediction model of both TinyDB and BBQ, multivariate data modeling techniques are used [6]. Such models can be learned from historical data using standard algorithms. For a specific model, training data needs to be collected for some period of time before predicting values. These models need to be updated over time to reflect the most recent changes. Choosing the best model for the given query workload and environment is an important issue in this case. While our proposed technique catches the relationship among sensors using association rule mining, this can be applied to a broad applications without model updating.

In [14], the authors propose using pattern discovery in multiple time-series to estimate missing data, but it's not well suited for sensor networks, where the relationships between sensors decided not only by the time trends, but also some other factors, like locations and so on.

In [8], the authors proposed the WARM (Window Association Rule Mining) algorithm for imputing missing sensor data. WARM uses association rule mining to identify sensors that report the same data for a number of times in a sliding window, called related sensors, and then imputes the missing data from a sensor by using the data reported by its related sensors. WARM has been reported to perform better than the average approach where the average value reported by all sensors in the window is used for imputation. However, there exist some limitations in WARM. First, it is based on 2-frequent itemsets association rule mining, which means it can discover relationships only between two sensors and ignore the cases where missing values are related with multiple sensors. Second, it finds those relationships only when both sensors report the same value and ignores the cases where missing values can be imputed by the relationships between sensors that report different values.

In view of the above challenges, in this paper we present a data imputation model using CARM (Closed Itemsets based Association Rule Mining) [11], which can derive the most recent association rules between sensors based on the closed itemsets in the current sliding window. The definition of closed itemsets is given in Section 3.

3 Definitions

In this section, we describe the notations and definitions that are used throughout this paper.

Let D = $\{d_1, d_2,\ldots, d_n\}$ be a set of n item ids, and V = $\{v_1, v_2,\ldots, v_m\}$ be a set of m item values. An item I is a combination of D and V, denoted as I = D.V. For example, $d_n.v_m$ means that an item with id d_n has the value v_m. A subset $X \subseteq I$ is called an itemset. A k-subset is called a k-itemset. Each transaction t is a set of items in I. Given a set of transactions T, the support of an itemset X is the percentage of transactions that contain X. A frequent itemset is an itemset the support of which is above or equal to a user-defined support threshold.

Let T and X be subsets of all the transactions and items appearing in a data stream D, respectively. The concept of closed itemset is based on the two following functions, f and g: $f(T) = \{i \in I \mid \forall t \in T, i \in t\}$ and $g(X) = \{t \in D \mid \forall i \in X, i \in t\}$. Function f returns the set of itemsets included in all the transactions belonging to T, while function g returns the set of transactions containing a given itemset X. An itemset X is said to be closed if and only if $C(X) = f(g(X)) = f\bullet g(X) = X$ where the composite function $C = f\bullet g$ is called Galois operator or closure operator [19].

For example, let I = {A, B, C, D} be a set of 4 items, and T = {CD, AB, ABC, ABC} be a set of transactions in data streams, then the closed itemsets and their support counts are {(C, 3), (AB, 3), (CD, 1), (ABC, 2)}. Each of the closed itemsets X satisfies $C(X) = f(g(X)) = f\bullet g(X) = X$. Take AB as an example, g(AB) = {AB, ABC, ABC}, $f\bullet g(AB)$ = AB, so C(AB) = f(g(AB)) = $f\bullet g(AB)$ = AB. Closed frequent itemsets are those closed itemsets that have support equal to or greater than the user-defined minimum support. If the user defined the absolute support to be 2, then the closed frequent itemsets are {(C, 3), (AB, 3), (ABC, 2)}. The frequent itemsets are {(A, 3), (B, 3), (C, 3), (AB, 3), (AC, 2), (BC, 2), (ABC, 2)}, from which we can see that closed frequent itemsets are smaller subsets of frequent itemsets and contain all itemsets and support information in the frequent itemsets.

From the above discussion, we can see that a closed itemset X is an itemset the closure C(X) of which is equal to itself (C(X) = X). The closure checking is to check the closure of an itemset X to see whether or not it is equal to itself, ie., whether or not it is a closed itemset.

An association rule $X \rightarrow Y$ (s, c) is said to hold if both s and c are above or equal to a user-specified minimum support and confidence, respectively, where X and Y are sensor readings from different sensors, s is the percentage of records that contain both X and Y in the data stream, called support of the rule, and c is the percentage of records containing X that also contain Y, called the confidence of the rule. The task of mining association rules then is to find all the association rules among the sensors which satisfy both the user-specified minimum support and minimum confidence.

4 Data Imputation Model Based on Closed Pattern Mining

The data imputation model uses association rule mining [1] on stream data to compensate for missing and corrupted data. An association rule is an implication of the form X => Y (s, c), where X and Y are frequent itemsets in a database and $X \cap Y = \varnothing$, s is the percentage of records that contain both X and Y in the database, called support of the rule, and c is the percentage of records containing X that also contain Y, called the confidence of the rule. An association rule is said to hold if both s and c are above or equal to a user-specified minimum support and confidence. An itemset is frequent if its support is above or equal to a user-defined support threshold. A k-frequent itemset is a frequent itemset with k items. Our goal is to find the relationships between sensors, and later use these relationships (or association rules) to impute the values of missing sensor readings.

When a transaction arrives or leaves the current data stream sliding window, the CFI-Stream algorithm [10] checks each itemset in the transaction on the fly and updates the associated closed itemsets' supports. The current closed itemsets are maintained and updated in real time in the DIU tree. The closed frequent itemsets can be output at any time at users' specified thresholds by browsing the DIU tree.

A lexicographical ordered direct update tree is used to maintain the current closed itemsets. Each node in the DIU tree represents a closed itemset. There are k levels in the DIU tree, where each level i stores the closed i-itemsets. The parameter k is the maximum length of the current closed itemsets. Each node in the DIU tree stores a closed itemset, its current support information, and the links to its immediate parent and children nodes. Figure 1 illustrates the DIU tree after the first four transactions arrive. The support of each node is labeled in the upper right corner of the node itself. The figure shows that currently there are 4 closed itemsets, C, AB, CD, and ABC, in the DIU tree, and their associated supports are 3, 3, 1, and 2.

We assume in this paper that all current closed itemsets are already derived, and based on these closed itemsets, we generate association rules for data imputation. Please refer to [10] for the detailed discussion of the update of the DIU tree and the closure checking procedure for addition and deletion operations.

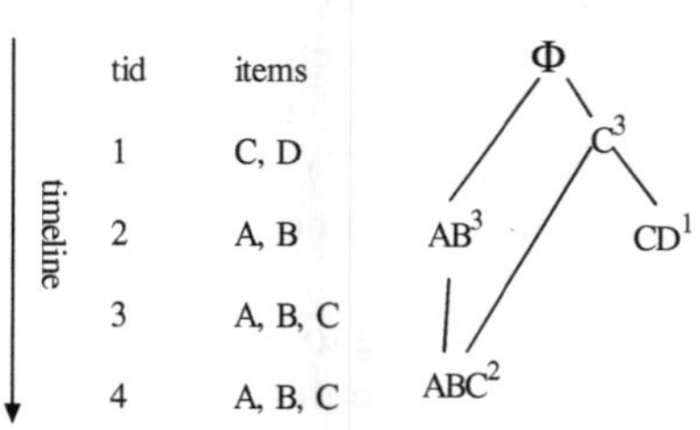

Fig. 1. The lexicographical ordered direct update tree

Based on the current closed itemsets in the DIU tree, instead of generating all possible association rules, we generate the rules that have strong relationships with the current round of sensor readings where one or more readings are missing. We achieve this through browsing the DIU tree, which stores all of the closed itemsets. Based on the users' specified support and confidence thresholds, we find out rules through paths (links) of closed itemsets that suit the users' needs, i.e., satisfy the users' specified support and confidence thresholds. For example, if the user's specified support and confidence threshold is 0.3 and 0.6 respectively, it means that we find out all the closed itemsets whose appearances are equal or greater than 30% of all transaction sets, and we find out all the association rules that related with the specified closed itemset whose possibilities to happen are equal or greater than 60%. The mining process is online and incremental, which is especially beneficial when users have different specified thresholds in their online queries. Please refer to [11] for a detailed discussion of the procedures of the CARM algorithm.

The data imputation model provides the following functionalities as shown in figure 2. It first preprocesses the data received from the input module, then judges if it contains missing or corrupted value. If yes, it performs data imputation, outputs the imputed value and stores it in the database; otherwise, it stores the value in the database directly. The users can thus query the database and get the query results in real time.

Below is an example of data imputation using closed itemsets based association rule mining. Assume a wireless sensor network consists of four sensors S_1, S_2, S_3, and S_4 that send their readings to a server in a certain time interval. Each sensor detects the temperature of a room during a certain period of time. The different values of the observed temperature from each sensor are represented as follows: $S_1.value_1$, $S_2.value_2$, $S_3.value_3$, $S_4.value_4$ and so on, where the value of each sensor may or may not be the same. The sensor node sends the generated tuple to the server using its radio unit.

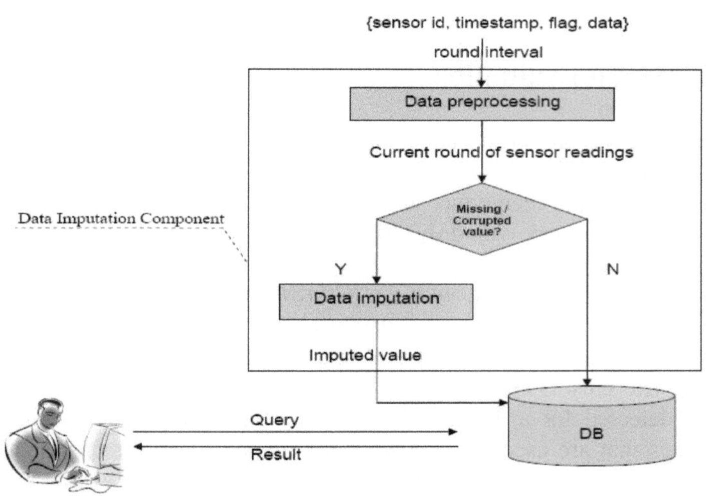

Fig. 2. Data imputation model

Assume the sensors have reported the following values as shown in figure 3 for the last 4 rounds of sensor readings, where round number 1 is the oldest round and round number 4 is the newest one. Assume minimum support = 50%, and minimum confidence = 50%. From the figure we can see all closed frequent itemsets from the current readings are {S_1.70, S_3.72, S_4.60} = 2, {S_1.70, S_4.60} = 3, {S_2.60, S_4.60} = 2, and {S_4.60} = 4. Based on the non-redundant association rules derived from the closed frequent itemsets, we can derive {S_1.70, S_4.60} → S_3.72, support = 1/2, confidence = 2/3 and S_4.60 → S_2.60, support = 1/2, confidence = 1/2, we can impute the missing value S_3.72 and S_4.60. Compared with WARM, CARM can find out the relationship between multiple sensors instead of pairs of sensors; it can also derive the relationship among sensors with different values instead of only same value S_4.60 → S_2.60, therefore it increases the number of missing values that can be imputed.

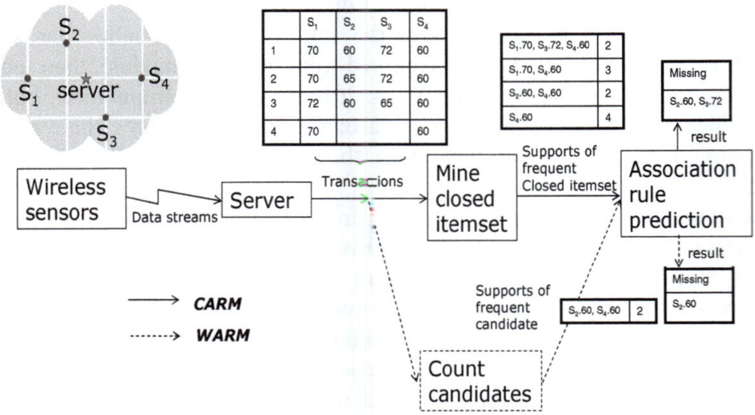

	S1	S2	S3	S4
1	70	60	72	60
2	70	65	72	60
3	72	60	65	60
4	70			60

Fig. 3. An example of data imputation using closed itemsets mining

5 Experimental Evaluations

The performance of our proposed data imputation model is studied by means of simulation. Several different simulation experiments are conducted in order to evaluate the proposed technique and compare it with four existing statistical techniques: the Average Window Size (AWS) approach, the Simple Linear Regression (SLR) approach, the Curve Regression (CE) approach, and the Multiple Regression (MR) approach, and with the WARM approach, the current state-of-the-art data imputation algorithm in sensor database [8].

The simulation model consists of 108 sensor nodes. All sensor nodes report to a single server. The sensors are deployed on city streets, collect and store the number of the vehicles detected for a given time interval. The actual vehicle counts taken as sensor readings that are used as input for our simulation experiments are traffic data provided by [1]. The data was collected in year 2000 at various locations throughout the city of Austin, Texas. The data represents the current location, the time interval, and the number of vehicles detected during this interval. From this set we generated four different input data sets corresponding to the different numbers of the possible sensor states used for the simulation experiments.

The evaluation of the achieved accuracy of an imputation of the missing values is done by using the average Root Mean Square Error (RMSE): $RMSE = \frac{1}{numStates}\sqrt{\frac{\sum_{i=1}^{\#imputations}(Xa_i - Xe_i)^2}{\#imputations}}$

where Xa_i and Xe_i are the actual value and the imputed value, respectively; #imputations is the number of imputations performed in a simulation run; and numStates is the number of subsets, in which the actual readings are distributed. The expression $\sqrt{\frac{\sum_{i=1}^{\#imputations}(Xa_i - Xe_i)^2}{\#imputations}}$ represents the standard error and is an imputation of the standard deviation under the assumption that the errors in the imputed values (i.e. Xa_i - Xe_i) are normally distributed. Thus, the RMSE allows the construction of confidence intervals describing the performance of different candidate missing value estimators. The smaller the RMSE (the standard deviation), the better the estimator. The calculated RMSE for each different set of input data (e.g. Set10 means that the sensor readings are split into 10 subsets) is divided by the number of subsets and the result is the average standard deviation in each case. This is done to keep the measure comparable across experiments.

From figure 4 we can see that CARM gives the best result of the above approaches regarding to the accuracy, follows by the WARM and AWS approaches. The regression approaches performs no better than WARM, CARM and AWS approaches, the main reason might be that it only considers the relationship between the neighbor nodes, while CARM and WARM find out all of the relationship between the existing sensors. Also from figure 4, we can see that the proposed CARM approach provides better imputation accuracy than the WARM approach does. This is because CARM performs the imputation based on the association rules derived by a compact and complete set of information, while WARM performs the imputation based on the association rules derived only by 2-frequent itemsets in the current sliding window.

Figure 5 illustrates the Total Main Memory Access Time per round (TMMAT) in milliseconds of AWS, SLR, CE, MR, WARM and CARM approaches. The TMMAT is defined as the time for performing all main memory accesses required for updating

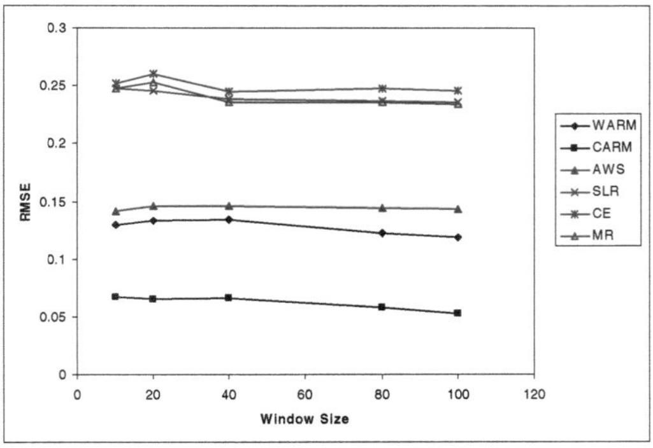

Fig. 4. RMSE for AWS, SLR, CE, MR, WARM and CARM approaches

the associated data structures and impute missing values per round of sensor readings. The experimental results show that in terms of TMMAT, the proposed CARM approach is outperformed by all other four statistical approaches, but it's still very fast comparing with the resend cases. The CARM approach is faster than the WARM technique. As shown in this figure, the TMMAT of WARM increases slightly when the window size increases since the information in WARM stores in the cube data structures, and the time needs to process this information increases when the size of the cube increases. For the CARM approach, the TMMAT first increases as the number of transactions increases since the number of closed itemsets that newly discovered increases; however, the average processing time decreases after the number of newly discovered closed itemsets reaches a threshold. This is because the number of closed itemsets which exist in the DIU tree increases, and they do not need to be processed; only their supports need to be updated incrementally.

Figure 6 illustrates Memory Space (in Kbytes) of AWS, SLR, CE, MR, WARM and CARM approaches. The experimental results show that in terms of Memory Space, the proposed CARM approach is outperformed by all other four statistical approaches, but it's still requires far less than the main memory provided in a contemporary computer. The results of the simulation experiments show that for 108 sensors, using WARM, the needed memory space is much higher than that using

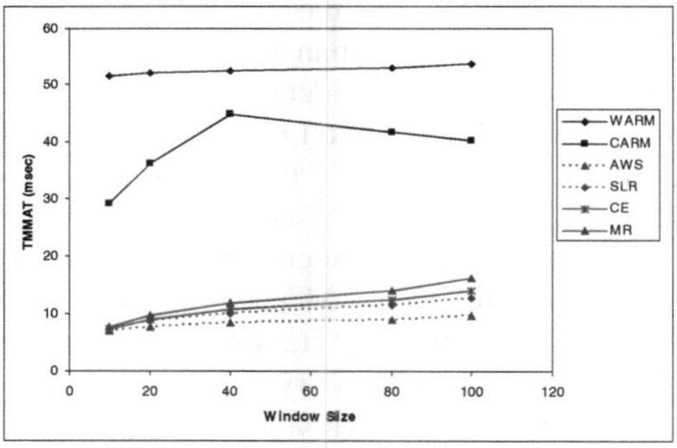

Fig. 5. TMMAT for AWS, SLR, CE, MR, WARM and CARM approaches

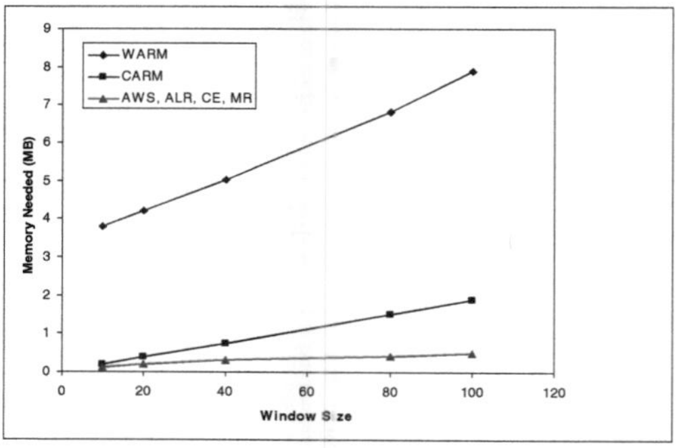

Fig. 6. Memory Needed for AWS, SLR, CE, MR, WARM and CARM approaches

CARM. This is because the DIU tree data structure uses much less memory space than the cube data structures, and it only stores the condensed closed itemsets information.

There is a possibility that the imputation algorithm alone will not be able to impute a missing value. The reasons for that are the following: no association rules between the sensor with the missing value (MS) and other sensors can be derived based on the current information. The percentage of cases in which a missing value cannot be imputed by the imputation algorithm alone (*PCE*) is computed for each simulation run using the following formula: $PCE = \frac{\#casesValueCannotBeImputed}{totalNumberOfAttemptsToImpute} * 100\%$ where *#casesValueCannotBeImputed* is the number of cases in which a missing value cannot be imputed using the imputation algorithm alone, and *totalNumberOfAttemptsToImpute* is the total number of attempts to impute a missing value in a given simulation run. From figure 7 we can see the CARM algorithm greatly reduces the number of cases that can not be imputed directly by the association rules derived. This is because, compared with WARM, it considers the relationships between multiple sensors and enables us to find more associations between multiple sensors even when they have different values.

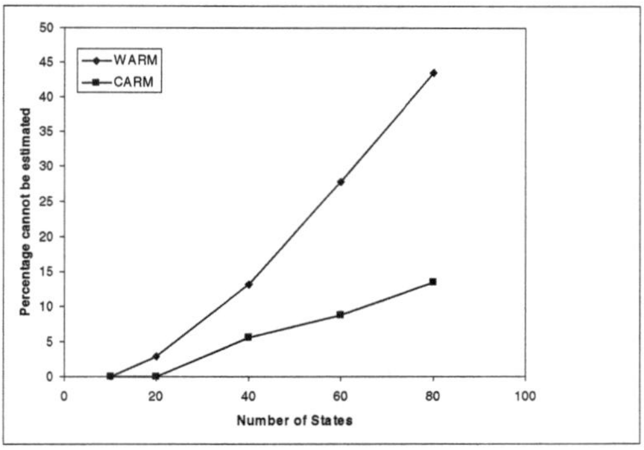

Fig. 7. PCE for WARM and CARM approaches

6 Conclusions

In this paper we present a novel data imputation model in sensor network databases based on closed pattern mining. This model integrates an incremental method to perform data imputation from the derived association rules based on closed pattern mining. Our performance study shows that it is able to impute missing sensor data online with both time and space efficiency, greatly improves the imputation accuracy and reduces the number of cases that cannot be imputed.

References

1. Austin Freeway ITS Data Archive: (accessed 2003), http://austindata.tamu.edu/default.asp
2. Allison, P.D.: Missing data. Thousand Oaks, CA Sage (2002)

3. Asada, G., Dong, M., Lin, T.S., Newberg, F., Pottie, G., Kaiser, W.J., Marcy, H.O.: Wireless Integrated Network Sensors: Low Power Systems on a Chip. In: European Solid State Circuits Conference (1998)
4. Cool, A.L.: A review of methods for dealing with missing data. Paper presented at the Annual Meeting of the Southwest Educational Research Association, Dallas, TX (2000)
5. Laird Dempster, N., Rubin, D.: Maximum Likelihood from Incomplete Data via the EM Algorithm. Journal of the Royal Statistical Society (1977)
6. Deshpande, A., Guestrin, C., Madden, S., Hellerstein, J., Hong, W.: Model-driven data acquisition in sensor networks. In: VLDB (2004)
7. Carlin Gelman, J., Stern, H., Rubin, D.: Bayesian Data Analysis. Chapman & Hall, Sydney (1995)
8. Halatchev, M., Gruenwald, L.: Estimating Missing Values in Related Sensor Data Streams. Int'l. Conf. on Management of Data (2005)
9. Iannacchione, V.G.: Weighted sequential hot deck imputation macros. In: Proceedings of the SAS Users Group International Conference (1982)
10. Jiang, N., Gruenwald, L.: CFI-Stream: Mining Closed Frequent Itemsets in Data Streams. In: ACM SIGKDD international conference on knowledge discovery and data mining (2006)
11. Jiang, N., Gruenwald, L.: Estimating Missing Data in Data Streams. In: 12th International Conference on Database Systems for Advanced Applications (2007)
12. Little, R.J.A., Rubin, D.B.: Statistical analysis with missing data. John Wiley and Sons, New York (1987)
13. McLachlan, G., Thriyambakam, K.: The EM Algorithm and Extensions. John Wiley & Sons, New York (1997)
14. Papadimitriou, S., Sun, J., Faloutsos, C.: Streaming Pattern Discovery in Multiple Time-Series. In: VLDB (2005)
15. Romer, K., Mattern, F.: The Design Space of Wireless Sensor Networks, IEEE Wireless Communications (2004)
16. Rubin, D.: Multiple Imputations for Nonresponce in Surveys. John Wiley & Sons, New York (1987)
17. Rubin, D.: Multiple Imputations after 18 Years. Journal of the American Statistical Association (1996)
18. Shafer, J.: Model-Based Imputations of Census Short-Form Items. In: Proceedings of the Annual Research Conference, Bureau of the Census, Washington, DC (1995)
19. Taouil, R., Pasquier, N., Bastide, Y., Lakhal, L.: Mining Bases for Association Rules Using Closed Sets. In: International Conference on Data Engineering (2000)
20. Troyanskaya, O., Cantor, M., Sherlock, G., Brown, P., Hastie, T., Tibshirani, R., Botstein, D., Altman, R.B.: Missing Value Estimation Methods for DNA Microarrays. In: Bioinformatics (2001)
21. Wilkinson & The APA Task Force on Statistical Inference (1999)
22. Zaki, M.J.: Generating non-redundant association rules. In: ACM SIGKDD international conference on knowledge discovery and data mining (2000)

An Adaptive Parallel Hierarchical Clustering Algorithm

Zhaopeng Li, Kenli Li*, Degui Xiao, and Lei Yang

School of Computer and Communication, Hunan University, Changsha, 410082, P.R. China
lzpd@163.com, LKL510@263.net, jt_dgxiao@hnu.cn,
leideyang@tom.com

Abstract. Clustering of data has numerous applications and has been studied extensively. It is very important in Bioinformatics and data mining. Though many parallel algorithms have been designed, most of algorithms use the CRCW-PRAM or CREW-PRAM models of computing. This paper proposed a parallel EREW deterministic algorithm for hierarchical clustering. Based on algorithms of complete graph and Euclidean minimum spanning tree, the proposed algorithms can cluster n objects with $O(p)$ processors in $O(n^2/p)$ time where $1 \le p \le \frac{n}{\log n}$. Performance comparisons show that our algorithm is the first algorithm that is both without memory conflicts and adaptive.

1 Introduction

Hierarchical clustering[2,3] techniques are widely applied in diversified areas such as gene categorization in biology, image processing, and network intrusion detection. Cluster analysis [1] is the process of classifying objects into subsets that have meaning in the context of a particular problem.

The basic principle behind hierarchical clustering is as follows: if there are n input points or data items, we start with n clusters where each cluster has a single point. From there on, the "closest" two clusters are identified. The two closest clusters are merged, resulting in a reduction in the number of remaining clusters is q where q is the target number of clusters and could be a part of the input.

The distance between two clusters can be defined in many ways. The commonly employed metric is the single link metric, as others did [2-4], and we also employ the single link metric as well.

Efficient sequential and parallel clustering algorithms have been studied extensively from researchers. By far the runtime of $O(n^2)$ is the lowest to reach the goal of clustering n points under the single link metric [2,3]. Rasmussen and Willett [5] discuss the parallel hierarchical clustering using the single link metric and the minimum variance metric on a SIMD array processor. Although a significant constant factor speedup is achieved, their parallel algorithm can not decrease the $O(n^2)$ time required by the serial implementation. Li and Fang [6] describe the parallel algorithms for hierachical clustering using the single link metric on an n-node SIMD supercube

* Corresponding author.

R. Perrott et al. (Eds.): HPCC 2007, LNCS 4782, pp. 97–107, 2007.

and an n-node butterfly respectively. They affirmed that their algorithms run in $O(n\log n)$ time on the hypercube and $O(n\log^2 n)$ on the butterfly. But in fact it was pointed out that the times required by their algorithms must be $O(n^2)$ [2]. Olson has presented $O(n\log n)$-time $\frac{n}{\log n}$-processor algorithms on the Concurrent-Read-Concurrent-Write (CRCW) Parallel Random Access Machine (PRAM), butterfly, and tree models [2]. An $O(n^2)$-time n-processor SIMD shuffle-exchange network algorithm has also been given by Li [7]. E. Dahlhaus [3] develop an efficient parallel algorithm for single linkage clustering in $O(\log n)$ time with $O(n)$ processors on a Concurrent-Read-Exclusive-Write (CREW) PRAM, given that a Minimum Spanning Tree (MST) is known. More recently, Rajasekaran [4] has presented a SIMD algorithm that runs in $O(\log n)$-time using $\frac{n^2}{\log n}$ CRCW PRAM processors, and a Random algorithm that run in an expected $O(\log^2 n)$ cycles on a $1\times n$ AROB.

However, as pointed out in literature [9,10,11], most of the parallel algorithms for hierarchical clustering presented so far have an obvious drawback: they are all impractical, and thus are of only theoretic importance. For example, if a large database has 10000 objects (this is possible in such cases as in bioinformics and in intrusion detection), to perform the algorithms in [2] and [4], then at least 1000 and 10^7 processors which shared a common memory must be needed respectively. However, it is very difficult to build such a shared memory machine by the present techniques [9,10]. Another drawback of these algorithms resides in the parallel models they used. Since concurrent read or write are required in these algorithms, it is not applicable for the EREW (Exclusive-Read-Exclusive-Write) PRAM.

To overcome these two drawbacks, we propose an adaptive parallel hierarchical algorithm based on EREW. The contribution of our algorithm is as follows:

- Using p processors where p can be adjustable in the range from 1 to $1\le p\le\frac{n}{\log n}$, our proposed algorithm run in $O(n^2/p)$ time, according to the definition in [9,10,12], our algorithm is adaptive.
- An improvement from CRCW to EREW is made in this algorithm. Although any CRCW or CREW algorithm can be converted into EREW algorithm in a straightforward way [10], the time needed must increase by $O(\log n)$ times.

The rest of this paper is organized as follows. Section 2 introduces our algorithm , in Section 3, the performance comparisons follow. Finally, some concluding remarks are given in Section 4 as well as some future research directions in this field.

2 The Proposed Algorithm

As we know[2,3,4], Hierarchical clustering may be represented by a dendrogram that can be easily constructed from a Euclidean minimum spanning tree of the n input points [2]. So the main wok for hierarchical clustering is to finding EMST and identifying the connected components of a graph[2,3,4]. However, before the EMST can be constructed for n given points, one auxiliary thing have to be done is to obtain the distance of every pair of points and save them (i.e. to construct a completed graph for n points). Now we at first describe the parallel algorithms for this step.

Suppose that in SIMD-PRAM model we have p processors where $1 \le p \le \frac{n}{\log n}$ and $O(n^2)$ shared memory units, while each processors have $O(1)$ local memory.

2.1 A Complete Graph Constructed in Parallel

Let the initial input points have m dimensions. At first we have to obtain all distances d_{ij} for all pairs of points V_i and V_j where $1 \le i, j \le n$. There are many methods to denote a completed graph $G(V, E)$ [10]. Here, for the convenience of the following process, adjoin matrix M is used. To construct an adjoin matrix for a given graph G, there are several parallel algorithms [11], but they are all based on SIMD-CRCW or SIMD-CREW models. Thus we use the elegant ideas[9,13] to design parallel EREW algorithm for constructing an adjoin matrix M for completed graph G.

Algorithm 1. Parallel algorithm for computing all distances d_{ij} of edges e_{ij}, $1 \le i, j \le n$

The n input points $V_1(x_{11}, x_{12},\ldots,x_{1m})$, $V_2(x_{21}, x_{22},\ldots,x_{2m}),\ldots$, $V_n(x_{n1}, x_{n2},\ldots,x_{nm})$ are given

begin
 for $l = p$ to 1 step by – 1 **do**
  for all processors P_i where $1 \le i \le l$ **do**
  for $k = \left\lceil \frac{n}{p} \right\rceil (i - 1 + p - l) + 1$ to $\left\lceil \frac{n}{p} \right\rceil \times (i + p - l)$ **do**
   for $j = \left\lceil \frac{n}{p} \right\rceil (i - 1) + 1$ to $\left\lceil \frac{n}{p} \right\rceil \times i$ **do**
    begin
    compute the Euclidean distance d_{kj} where $d_{kj} = \sqrt{\sum_{s=1}^{m} (x_{ks} - x_{js})^2}$
    write d_{kj} to the shared memory
    end
end

Here, as figure 2 shows, when $p|n$, algorithm 1 must perform well. However, it is trivial to process similarly in algorithm 1 when $p|n$ is not the case.

Lemma 1. Algorithm 1 can be efficiently executed on SIMD-PRAM without memory conflicts in $O(\frac{n^2}{p})$ time and $O(n^2)$ work.

Proof. Assume that the time for computing the Euclidean distance for each PRAM processor is $O(1)$ (in fact it must be $O(m)$, in this paper we let m be a constant), the conclusion can be shown from the figure 2. (i) if $p|n$. As the algorithm 1 shows, there are 3 *for* cycle to perform in order to execute algorithm. The time needed in the proposed parallel algorithm is $p\times\frac{n}{p}\times\frac{n}{p}=\frac{n^2}{p}$, and the total work (cost) is thus equal to $p\times\frac{n^2}{p}=n^2$. When any processor has obtained the Euclidean distance d_{ij}, it write it to the different location of the shared memory. Because every d_{ij} which different processors write are different from each other, there are no write conflicts among different processor. As figure 1 shows, in any time for different processors P_i and P_j, if $i < j$, then the subscripts of d computed by processor i will be totally different with that read by processor j, which means that the memory units processor i and j read are also different in any time. Thus there are no read conflicts among different processors, either. And algorithm 1 can be executed on EREW computation model. (ii) if $p|n$ does not hold. In this case, except the final processor and in final cycle for l, the situations for other processors and other cycle are as same as that in case 1, and the time and work bound will not change.

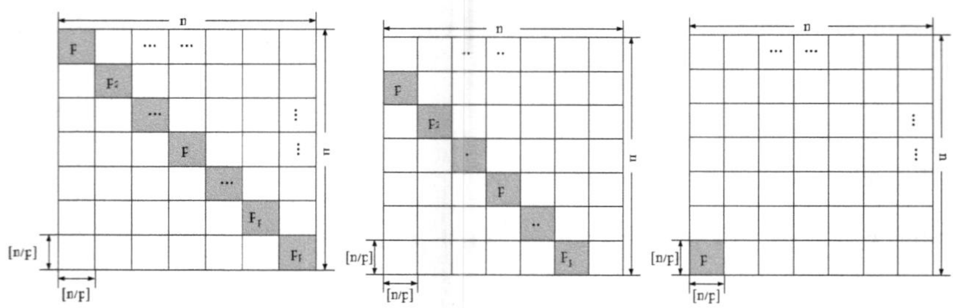

a. when $l = p$ *b*. when $l = p - 1$ *c*. when $l = 1$

Fig. 1. The data allocation in different processor as algorithm 1 executes

2.2 Generation of MST

Now we have known all distances d_{ij} from points V_i and V_j, $1 \le i, j \le n$, which make up a symmetry adjoin matrix. Based on this matrix, the next task for clustering is to generate a MST of completed graph G. parallel algorithms for generating MST have been extensively researched in past twenty years, and there are many different parallel algorithms to get a MST from a known graph [12,14]. However, because of our objective is to cluster n input points (items), which require pruning some edges (section 2.3) after getting a MST, we use the parallel algorithms proposed by Akl, etc[12]. This parallel algorithm is an adaptive and cost-optimal parallel algorithm for MST, and is without memory conflicts.

Algorithm 2. The parallel algorithm for producing MST[12]
Limited by the space, for more details on the parallel algorithms, one can refer to [10,12].

Lemma 2. Algorithm 2 can be execute in PRAM-EREW, and the total time and work for execute the algorithm 2 is (n^2/p), and $O(n^2)$, respectively, if $1 \le p \le \frac{n}{\log n}$.

Proof. Because the number of processors in the proposed algorithm is p where $1 \le p \le \frac{n}{\log n}$, the conclusions are validated [12].

2.3 Pruning Edges and Finding Connected Components

Now we have obtained the MST of the n input points. The q clusters from the n input points are in need, so $q - 1$ longest edges must be pruned from the MST T. if this is done, then the left q subtrees (components) are the q clusters of interest. Thus there is still two parts of work to do, one is pruning edges from T, and the other is finding the needed connected components or subtrees.

At first, we introduce the parallel pruning edges algorithm on EREW-PRAM.

Algorithm 3. Pruning edges algorithm

begin
for $k = 1$ to $q - 1$ **do**
  for all processor P_i where $1 \le i \le p$ **do**
    find the longest edges from the edges set of TREE[$(i-1)[\frac{n-1}{p}]$ + 1] to TREE[$(i[\frac{n-1}{p}])$] **do**
    call Procedure MAXIMUM
    delete the longest edge
    call Procedure BROADCAST in [10,12]
end
Procedure MAXIMUM($A[1,\cdots,p]$)

begin
  for all p_i where $1 \le i \le \left\lceil \frac{p}{2} \right\rceil$ **do**
  for $j = 1$ to $\lceil \log n \rceil$ **do**
   if $(i \bmod 2^{j-1}) = 0$ and $2i + 2^{j-1} \le p + 1$ **then**
    if $A[2i-1] < A[2i-1+2^{j-1}]$ **then** $A[2i-1] = A[2i-1+2^{j-1}]$ **end if**
   end if
 end

Lemma 3. Algorithm 3 can be executed on EREW-PRAM in $O((\frac{n}{p}+2\log p)\times(q-1))$ time.

Proof. It is known that finding the maximum can be finished in linear time with the number of items. In our algorithm, each processor takes $O([(n\text{-}1)/p])$ time to find the longest edge from its $[(n\text{-}1)/p]$ edges. The time needed for running subprocedure BROADCAST [12] and MAXIMUM are both $O(\log p)$. To obtain q clusters from the n input points, the circle will have to be run for $q-1$ times. Thus the total time for running the algorithm 3 is $O((\frac{n}{p}+2\log p)\times(q-1))$. Obviously, as we almost evenly divide the array TREE[$n-1$] into p blocks, and the subprocedure MAXIMUM is designed to perform on EREW, there is no memory conflicts in algorithm 3.

After pruning edges from the MST, the final step to cluster n objects is to decide which points are in the same cluster. In fact, there left q connectivity components or subtrees after performing pruning procedure. So the task of this step is finding the q connectivity components. Parallel connectivity algorithms had once been extensively researched in past years, and there are many such parallel algorithms [15,16]. For more details on the parallel algorithms for this problem, one can refer to [15,16].

However, none of these algorithms are of our need in the case that the graph has become a forest which is saved as an array TREE[$n-1$], either for their not having adaptivity, or for their time complexity. Therefore, we designed a new parallel connectivity algorithm to our need, which is both adaptive and without memory conflicts.

In our algorithm for finding the q connectivity components or clusters, an array $D(n)$ is defined, which is used to store the number of the connectivity components, and at first it is initialized as $D(i) = i$. After the algorithm is performed, if node i and j are in the same cluster k, then $D(i) = D(j) = k$. for the convenience of describing the algorithm, assume that all edges of the forest are saved in array TREE[$n-1$] from TREE[0] to TREE[$n-q-1$]. If this condition is not satisfied, one can easily do this with $O(n)$ time.

Algorithm 4. Algorithm for finding the q connected components of graph G

begin
  for all processor P_i where $1 \le i \le p$ **do**
    for $u = \left[\frac{n}{p}\right]\times(i-1)+1$ to $\left[\frac{n}{p}\right]\times i$ **do**
     $D(u) = u$
   for $i = 0$ to $n-q-1$ **do**
    v_l = TREE[i][1] and v_m = TREE[i][2]
    if $l > m$ **then** $D(l) = D(m)$
    else $D(m) = D(l)$
    end if
    call Procedure **FIND**
end

Procedure FIND(j, k)
for all processor P_i where $1 \leq i \leq p$ **do**
find all $D(t)$ from $D[n/\mathrm{p} \times i]$ to $D[n/\mathrm{p} \times i + n/\mathrm{p} \times (i+1) - 1]$ where $D(t) = j$
$D(t) = k$

In algorithm 4, after the first cyclic is performed, the array $D(n)$ is initialized, and the corresponding relation of array $D(n)$ with the vertexes of the graph $G(V, E)$ is depicted in Figure 2.

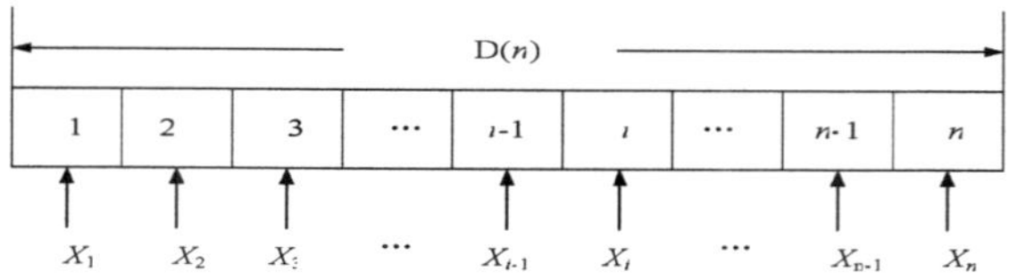

Fig. 2. The corresponding relation of the initialized array $D(n)$ with all vertexes

While in the second cyclic, if two vertexes connected by the current input edge are $x_{i\text{-}1}$ and x_i (ie., TREE[k][1] = $x_{i\text{-}1}$, and TREE[k][2] = x_i), then the value of ith element of array D, $D(i)$ will change into $i - 1$, as Figure 3 shows.

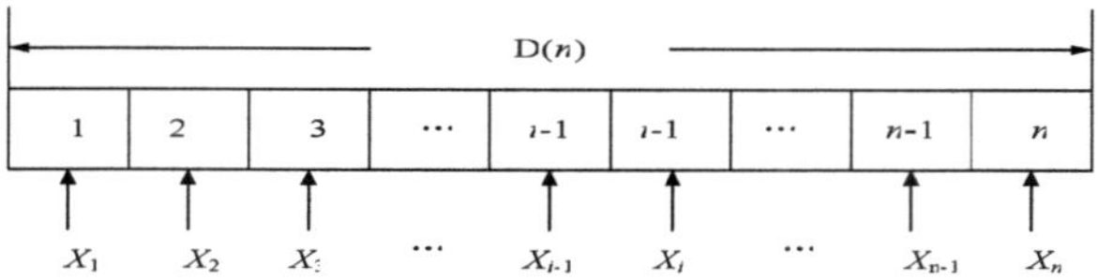

Fig. 3. If an edge ($x_{i\text{-}1}$, x_i) appears in the forest, the value of array D will be altered

If this is the case happened, the Procedure FIND is called to alter the values of all array elements whose value is being i to $i - 1$, for it is possible that one edge in the forest connecting v_i and v_j ($j > i$) is processed before edge ($x_{i\text{-}1}$, x_i). There are only $n - q$ edges in forest, so after these codes are performed for $n - q$ times, the label number stored in array $D(n)$ is the number of cluster. While for certain cluster, all sequence number of array $D(n)$ having same array element are the corresponding vertexes of this cluster.

Lemma 4. Algorithm 4 can be performed in $O(n^2/p)$ time with p processors on PRAM-EREW where $p \leq n$.

Proof. The first cyclic can be finished in $O(n/p)$ time. Since the subprocedure FIND takes at most $O(n/p)$ time, the second cyclic will take $O((n - q) \times n/p)$ time. Thus the total time to perform algorithm 4 is $O(n/p) + O((n - q) \times n/p) = O(n^2/p)$. As in the whole performing stage, all processor only read data from its memory units, and at the same time there are no different processors write into the same memory units. It is obvious that the algorithm can be performed on PRAM-EREW model.

2.4 The Proposed Parallel Hierarchical Clustering Algorithm

So far, we have finished the three steps of hierarchical clustering. Now combined the above four algorithms, one can obtain the parallel hierarchical cluster algorithm, which is based on the EREW model, and totally without memory conflicts.

Algorithm 5. Parallel algorithm for hierarchical cluster based on EREW

Step 1. perform the parallel algorithm for computing the adjoin matrix.
Step 2. perform the algorithm for producing the MST.
Step 3. perform the parallel pruning edge algorithm.
Step 4. perform the algorithm for recognizing connected components.

Theorem 1. Clustering n input points can be finished in $O(n^2/p)$ time with p processors on EREW-PRAM model.

Proof. Following the lemmas 1 to 4, the total time needed to perform the proposed parallel cluster algorithm is $T = n^2/p + n^2/p + ((\frac{n}{p} + 2\log p)\times(q-1)) + n^2/p$. Notice that in algorithm 2 and 4, the number of processors are limited in the range of $1 \le p \le \frac{n}{\log n}$ and $1 \le p \le n$ respectively. At the same time, the number of the needed clusters q satisfies that $q \le n$. Thus the total time to perform the parallel clustering algorithm is bounded by $O(n^2/p)$. Since algorithms 1 to 4 are all based on SIMD-EREW model, it is obvious that the proposed parallel algorithm can be performed on EREW-PRAM. And it is also shown that in this algorithm, all memory conflicts which may happen among different processors are avoided.

3 Performance Comparisons

For the past years, hierarchical clustering has been extensively researched, and there are many parallel algorithms for it on different models, for example the algorithms in literature [2,3], [7-8,17]. These algorithms are mainly based on SIMD parallel computation models. Following the previous researches, the performance comparison will be described in terms of time-processor tradeoff, i.e., the cost of the parallel algorithm[9,10]. Rasmussen and Willett [5] were the first to discuss parallel implementations of clustering using the single link metric on a SIMD array processor. But they can not decrease the $O(n^2)$ time required by the best serial implementation. Li and Fang [6] also presented parallel algorithms for hierarchical clustering using the single link metric on an n-node hypercube and an n-node butterfly. However, The time needed in their parallel algorithm is $O(n^2)$, too [6]. Later, an $O(n^2)$-time n-processor SIMD shuffle-exchange network algorithm has been given by Li [7]. Olson [2] showed that parallelism could accelerate to solve larger instances of this problem. Their algorithm runs in $O(n\log n)$ time by allowing $\frac{n}{\log n}$ processors to currently access $O(n^2)$ units in shared memory. E. Dahlhaus's efficient parallel algorithm [3] for single linkage clustering runs in $O(\log n)$ time with $O(n)$ processors on CREW-PRAM under condition that the a Minimum Spanning Tree of the n input points is given. Another PRAM algorithm recently presented by Rajasekaran [4] runs in $O(\log n)$-time using $\frac{n^2}{\log n}$ CRCW processors, resulting to an $O(n^2)$ computation cost.

The common characteristic of most of the above algorithms is that they are all based on the CRCW or CREW models, and thus there exist memory conflicts among different processors when they access the same memory units concurrently. Although It has been shown that every CREW or CRCW algorithms that require $T(L)$ time using $P(L)$ processors (where L denotes the size of input or output data) can be transformed into an EREW algorithm which requires time $O(T(L)\log L)$ still using $P(L)$ processors [10]. Since our proposed algorithm is designed for EREW model, hence. In practical, in according to the definition of cost for parallel computation [10], the cost of our algorithm is only $1/(\log n)$ of the algorithms in [2,3,4]. Secondly, like the algorithms in [9,12,13], the number of processors in our algorithm can be adjusted from 1 to $1/(\log n)$ in accordance to the scale of the problem and actual computation condition, while keeping the computation cost unchanged. It can provide the probability for clustering very large datasets. Therefore, according to the definition in [9,10,12], our algorithm is adaptive or scalable.

For the purpose of clarity, the comparisons of the mentioned parallel algorithms which are based on SIMD model for hierarchical clustering are depicted in Table 1. It is obvious that our parallel algorithm outtakes undoubtedly other parallel algorithms in the overall performance.

Recently, M. Dash etc [17] present a parallel algorithm for hierarchical clustering, and this algorithm has good theoretical and experimental performance as compared with the relevant algorithms, but that algorithm is not based on SIMD models and

Table 1. Comparisons of the parallel algorithms for Hierarchical Clustering

algorithms	Model	Time	Processor	Adaptability	Cost
Sequential [2]	sequential	$O(n^2)$	1	*no*	$O(n^2)$
Rasmussen and Willett[5]	CRCW	$O(n^2)$	$n/\log n$	*no*	$O(n^2)$
S.Rajasekaran[4]	CRCW	$O(\log n)$	$O(n^2/\log n)$	*no*	$O(n^2)$
Li and Fang[6]	SIMD hypercube	$O(n\log n)$	n	*no*	$O(n^2\log n)$
Li[7]	SIMD shuffle-exchang	$O(n^2)$	n	*no*	$O(n^3)$
Olson[2]	CRCW	$O(n\log n)$	$n/\log n$	*no*	$O(n^2)$
Dahlhaus[3]	CREW	$O(n)$	n	*no*	$O(n^2)$
Ours	EREW	$O(n^2/p)$	p	*yes*	$O(n^2)$

their metrics for distance are centroid method. Although their method is valuable on improving the practical performance of clustering algorithms, theoretically, it is not comparable with our proposed algorithm.

4 Conclusions

Based on the EREW-SIMD machine with shared memory model, a new parallel algorithm for hierarchical clustering is proposed. The proposed algorithm use p, $1 \le p \le \frac{n}{\log n}$, processors to clustering n objects in $O(n^2/p)$ time, resulting to $O(n^2)$ computation work which is by far the lowest in serial way. Since the number of processors in our algorithm can be adjusted in the range from 1 to $n/\log n$ according to the available computation resources and the scale of the problem to be processed, our algorithm is adaptive[9,10]. To our knowledge, it is the first parallel algorithm in the sense of both without memory conflicts and adaptive for the hierarchical clustering on PRAM model.

On the other hand, although the number of processors in the proposed algorithm can be adjusted, we still can't rely on this kind of algorithms to solve the large scale clustering instances, for it is not possible to construct such shared memory computers with a large number of computers. Although presently M. Dash etc s' algorithm are also performed in shared memory multiprocessor system (MIMD) [17], the number of processors is not large enough to clustering large scale datasets, especially to cluster gene data clustering. Since the supercomputer systems are mainly based on cluster technology now, it is a worthwhile work to investigate on feasible parallel algorithms based on present main distributed memory system.

Acknowledgments. This research was supported by the National Natural Science Foundation of China under Grant No. 60603053 and the Key Project of Education Ministry of China under Grant No.105128.

References

1. Han, J., Kamber, M.: Data Mining: Concepts and Techniques. Morgan Kaufmann Publishers, San Francisco (2000)
2. Olson, C.F.: Parallel Algorithms for Hierarchical Clustering. Parallel Computing 21, 1313–1325 (1995)
3. Dahlhaus, E.: Parallel Algorithms for Hierarchical Clustering and Applications to Split Decomposition and Parity Graph Recognition. Journal of Algorithms 36, 205–240 (2000)
4. Rajasekaran, S.: Efficient Parallel Hierarchical Clustering Algorithms. IEEE transactions on parallel and distributed systems 16(6), 497–502 (2005)
5. Rasmussen, E.M., Willett, P.: Efficiency of hierarchic agglomerative clustering using the ICL Distributed Array Processor. Journal of Documentation 45, 1–24 (1989)
6. Li, X., Fang, Z.: Parallel Clustering Algorithms. Parallel Computing 11, 275–290 (1989)
7. Li, X.: Parallel Algorithms for Hierarchical Clustering and Clustering Validity. IEEE Trans. Pattern Analysis and Machine Intelligence 12, 1088–1092 (1990)

8. Tsai, H.R., Horng, S.J., Lee, S.S., Tsai, S.S., Kao, T.W.: Parallel Hierarchical Clustering Algorithms on Processor Arrays with a Reconfigurable Bus System. Pattern Recognition 30, 801–815 (1997)
9. Akl, S G.: Optimal parallel merging and sorting without memory conflicts. IEEE Trans, Comput. 36(11), 1367–1369 (1987)
10. chen, G.: Design and analysis of parallel algorithm. higher education press, Beijing (2002)
11. Datta, A., Soundaralakshmi, S.: Fast Parallel Algorithm for Distance Transform. IEEE Transactions on Systems, Man, and Cybernetics 33(5), 429–434 (2003)
12. Akl, S.G.: An adaptive and cost-optimal parallel algorithm for minimum spanning trees. Computing 3, 271–277 (1986)
13. Li, K.L., Li, Q.H., Li, R.F.: Optimal parallel algorithm for the knapsack problem without memory conflicts. Journal of Computer Science and Technology 19(6), 760–768 (2004)
14. Jun, M., Shaohan, M.: Effcient Parallel Algorithm s for Some Graph Theory Problems. J?of Comput?Sci?Technol 8(4), 362–366 (1993)
15. Nath, D., Maheshwari, S.N.: Parallel algorithms for the connected components and minimal spanning tree problems. Inf: Proc. Lett. 14(1), 7–11 (1982)
16. Chong, K.W., Han, Y.J.: Concurrent Threads and Optimal Parallel MinimumSpanning Trees Algorithm. Journal of the ACM 48(2), 297–323 (2001)
17. Dash, M., Petrutiu, S., Scheuermann, P.: pPOP: Fast yet accurate parallel hierarchical clustering using partitioning. Data & Knowledge Engineering 61(3), 563–578 (2007)

Resource Aggregation and Workflow with Webcom

Oisín Curran[1], Paddy Downes[1], John Cunniffe[1], Andy Shearer[2], and John P. Morrison[3]

[1] Dept of Information Technology, National University of Ireland, Galway, Galway, Ireland
oisin.curran@geminga.it.nuigalway.ie, {paddy,jc}@.it.nuigalway.ie
[2] Dept of Physics, National University of Ireland, Galway, Galway, Ireland
andy.shearer@nuigalway.ie
[3] Centre for Unified Computing, University College Cork, Cork, Ireland
j.morrison@cs.ucc.ie

Abstract. Efficient exploitation of the aggregate resources available to a researcher is a challenging and real problem. The challenge becomes all the greater when researchers who collaborate across functional or administrative domains need to pool their disjoint heterogeneous resources to achieve their objectives. Support tools are becoming available for these ad hoc resource integration and sharing scenarios. The focus of this paper is the identification of suitable deployment and usage strategies when using the workflow approach with these tools. In particular, we present a novel two-level peer-to-peer model for dynamic resource aggregation that operates entirely at the user level. This sidesteps the need for system level middleware and administrative support. This paper presents our strategy, the underlying framework and the workflow expression and evaluation semantics.

1 Introduction

Much research is conducted in environments where there exist multiple independent resources and multiple dynamic groups of occasionally collaborating users with varying levels of access to subsets of those resources. Resources may be widely distributed and have very different architectures and administrative policies. Such environments evolve naturally within and among academic and industrial research facilities. These sites are typically funded departmentally yet collaborations may span departmental or other administrative or functional borders.

Without a large-scale, ongoing and predictable pattern of collaboration there is little motivation to implement and maintain a computational grid [1] solution to the problem of resource aggregation. Even if such a solution were provided it may not be optimal for many classes of problem; the grid model of distributed computing is largely focused on high throughput batch processing and not easily tuned to support other priorities such as deadline or interactive (steered) processing.

R. Perrott et al. (Eds.): HPCC 2007, LNCS 4782, pp. 108–119, 2007.

In such scenarios there is a need for lightweight user-level tools that facilitate dynamic resource aggregation while ensuring the efficient use of resources. Such tools are emerging, yet the availability of tools that facilitate resource aggregation is not in itself enough to guarantee efficient resource usage and sharing.

This report presents our use of the Webcom system [2], [3], [4] to aggregate and abstract distributed heterogeneous resource sets. These resources are typically under the control of different administrators and in active use by different research groups for a diverse range of application types. We demonstrate our use of Webcom as a tool for resource aggregation using a novel two-level peer-to-peer *Overlay Metacomputer* model.

This paper makes the following contributions,

1. We propose a methodology for the dynamic aggregation of disjoint independent resources. The aggregate resource set behaves as a distributed virtual machine that interprets a common workflow expression format on all platforms, thereby abstracting site differences. Our model provides a locality-aware peer based system for load balanced workflow execution. This approach relies solely on user-level access to resources and does not depend on any particular middleware or uncommon site policies.
2. We present our workflow expression and execution tools built on top of the Webcom graph evaluation engine. With these tools we can make use of Webcom's unique semantics and task distribution capabilities to achieve efficient workflow evaluation.
3. We present results of tests of our model, using both synthetic and real application examples in a real and very heterogeneous environment consisting of distinct and independently administered resources of different architectures with only partially overlapping user communities.

The rest of this paper is organised as follows. In section 2 some background details are presented to put our current report in context. The approach taken is detailed in section 3. A motivating example is presented and discussed in section 4. Initial results of experimental evaluations are presented in section 5. Related work is discussed in 6. We close this report by presenting conclusions and outlining future work targets in section 7.

2 Background

Webcom is based on the Condensed Graph (CG) model of computing [5]. A CG is a directed acyclic graph (DAG) with enhanced semantics. Each node in a CG is an executable entity. A node may represent a single instruction or an entire graph that has been *condensed* (abstracted) to node form. Certain nodes affect the flow of control (e.g., the conditional node) and, used in conjunction with CG edge semantics, direct the order of graph evaluation. The CG model is intended to give the programmer the benefits of both the control flow and data flow models of computing in a single system. Webcom is implemented as a peer-based CG evaluation engine; a distributed virtual machine that interprets the CG language. A detailed description of Webcom is presented in [4]. Webcom is implemented entirely in Java to increase its portability. Our work focuses on implementing workflow via Webcom, and analysing suitable deployment strategies for different application and environment types, to support the development of real applications with Webcom.

3 Approach

This section presents the runtime organisation and workflow expression strategies of our model. These have been designed to leverage the capabilities of the Webcom framework and the features of the CG model to support application deployment in heterogeneous computing environments.

3.1 Runtime Topologies

Our approach to resource aggregation consists of constructing a two level overlay network across the available independent resources. When combining mixed resources including stand-alone machines, Condor pools [6], [7], and clusters controlled in space sharing mode, a top level ring links the sites. This is in keeping with the work presented in [8] where it was shown that a peer-to-peer ring of Condor pools proved an efficient, scalable and fault tolerant method to connect sites. This top level ring is set up by starting the Webcom service at each site using the site's task submission interface. These Webcom instances connect to each other forming the top level *Webcom world*.

Systems have been reported in the literature that use Ring (e.g., Chord [9] and Tapestry [10]), and Balanced Tree [11] based topologies. The Webcom system is in this sense 'model free'; machines can be connected in any of these structures or combinations thereof. We exploit this fact to group peers in the topology best suited to their characteristics. Overlay strategies for connection management are well established, with many efficient and robust solutions reported in the literature [12], [11], [13], so new topologies are not a strong focus of our current research. Hence, our illustrative example uses a ring at the top level, rather than any more sophisticated connection management structure.

Similarly, a site-internal Webcom world is set up at each location using the processors allocated by the local resource manager as appropriate. The topology chosen at each site depends on the nature of the site and the number of processors to be used. In a mixed and widely disbursed or very heterogeneous Condor pool, a tree topology is suitable as this means less traffic over the network and less impact to the Webcom world when an individual Webcom is preempted by Condor [14], [11]. At a dedicated space sharing site where preemption is not a factor and all processors are connected to the same switch, a star topology is more efficient. There remains the risk of losing connection with the entire cluster, but the loss of an individual node is far less likely than at a cycle scavenging site. Webcom can also be used to group any number of independent workstations as it does not rely on the existence of a batch job manager for resource access.

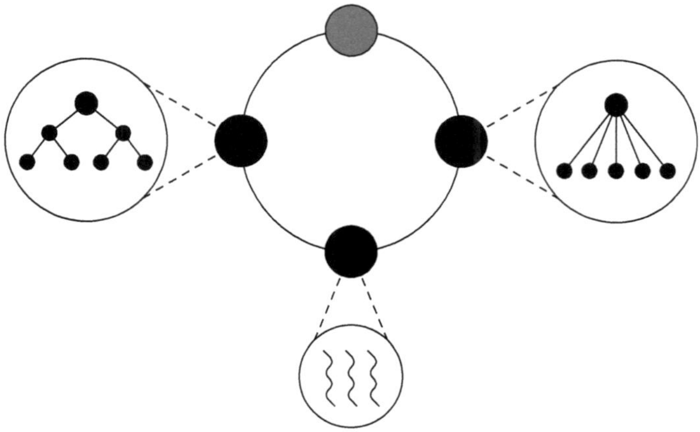

Fig. 1. An example of a Webcom overlay metacomputer. This example scenario shows the aggregation of resource from three sites: a Condor pool shown on the left, a PBS queue on the right, and an SMP machine at the bottom. A top level ring is formed with these and a stand-alone workstation (at the top of the circle). The latter may act as the point of submission for the user's workflow and also contribute to task execution. With this overlay in place, workflow graphs can be evaluated across the aggregate resource set, with Webcom providing load balancing across the individual sites and abstracting underlying differences in platform and topology.

A schematic of the approach is presented in figure 1. This shows a scenario wherein three sites form a top level Webcom world. Each site hosts an independent Webcom world.

3.2 Workflow and Job Description Languages

Expression languages for scientific workflow have been presented in a range of projects for various systems [15], [16], [17]. Our workflow and job description languages make use of the Condor DAGMan syntax and offer the potential to go beyond its semantics. DAGMan workflows and CGs are both DAG based, therefore every DAGMan workflow has an equivalent CG representation. The workflow structure, expressed in the DAGMan language as a collection of jobs in a parent/child hierarchy maps to an equivalent shape CG. As Webcom facilitates flow control with conditional execution, iteration, and explicit recursion, it can be seen as a superset of the DAGMan language. Mapping the semantic concepts of Condor/DAGMan to the Webcom equivalent, as the two systems have radically different deployment architectures, requires the use of specific Webcom *nodes* designed to handle particular Condor job types. Figure 2 shows a synthetic example application.

Our Webcom based workflow strategy uses three layers of condensation in each workflow. The top layer is the workflow shape; the middle layer provides exception handling. This is built into the graph expression making exception handling a fully distributed part of workflow execution. The lowest layer shows

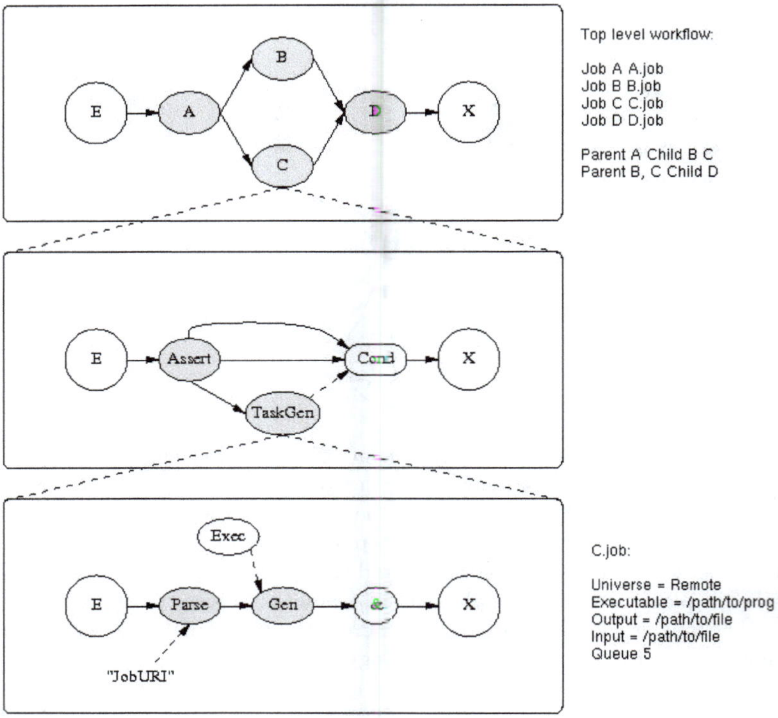

Fig. 2. Workflows are implemented as a set of graphs with this pattern of condensation. The top level graph expresses the logical sequence of tasks in the workflow. The middle layer provides exception handling using Webcom's flow control operators and lazy evaluation semantics. The lowest level graph uses Webcom's ability to add nodes to a graph at runtime, using an enumeration node, dynamically creating a cluster of one or more jobs.

the mapping from workflow task expression (attribute/value pairs) to graph. When one of these graphs is executed on a Webcom worker, the "Parse" node downloads the job definition from the submission site and generates a set of parametrised job entities. These are passed to the "Gen" node who generates a set of work execution nodes (instances of "Exec"), feeding each its own version of the job description. This allows for site specific and task specific information to be calculated when and where appropriate. This deferred approach to task interpretation means that task definitions can change after submission of the initial workflow. In the event of a task's failure, its children must not attempt to execute and the workflow should be aborted. In the absence of a central point of control (fully decentralised workflow execution), where the tasks have been submitted to the workflow system and are pooled for execution, conditional execution of child nodes is used. The condition for execution of a child node is the successful completion of all parents. Failure of a parent results in the propagation of trace data through the graph that facilitates fault detection and re-running of the outstanding tasks. This behaviour is guaranteed by the lazy evaluation semantics of the CG model. The nodes containing the task execution logic are only evaluated when needed. If all parents are not fully successful, the child task is not needed and so needless resource consumption is minimised.

4 Example Application

As part of the Webcom project we are collaborating with the Medical Physics department of the University College Hospital Galway (UCHG) in an ongoing effort to develop a distributed Radiation Treatment Planning (RTP) workflow solution based around the use of a Monte Carlo (MC) radiation transport simulation package.

The use of the MC technique in medical physics has become increasingly common. In radiotherapy, it is widely accepted that MC is the most accurate and general calculation method available [18]. A number of review articles exist on this topic [19], [20], [21]. The main drawback of this method is the time involved. The more traditional, but less accurate [22], solutions such as the pencil beam [23] or collapsed cone [24] calculation will take minutes to perform on a standard desktop. A MC calculation can take days to complete. Our objective is to provide MC based results in a time-frame that makes MC a viable option for clinical use.

The full workflow for this application involves data transport, the acquisition and generation of simulation model input data, preparation of executables for distribution based on currently available resources, the simulation itself, result accumulation and verification, format translation, and visualisation. This application and our approach to it are described fully in a forthcoming paper. The purpose of this overview is to demonstrate the applicability of our resource aggregation approach and workflow model to very real and computationally intensive problems.

4.1 Application Requirements

The main computational step in the RTP workflow is the MC phase. However, there are significant data input and output issues. An accurate run of the simulation results in the transport, analysis and visualisation of approximately 20 GB of data. Runtimes are dependent on a number of user specified parameters; a usable result can be generated in 70 to 100 CPU hours. A clinical time-frame is in the order of one to three hours. Providing medical rather than computational science experts with the necessary computational power in a flexible, reliable and accessible manner is the central challenge of this application. Our approach answers that challenge by facilitating the harnessing of disjoint resources *on-demand*, and ensuring the efficient use of those resources.

4.2 Application Environment

Due to the interdepartmental nature of our research group, a mixed collection of resources are accessible for use. These include a 64 processor cluster (32 dual Intel Xeon 2.4 GHz processor machines with two GB of RAM per machine) managed by Torque [25] [26] and Maui [27] in space sharing mode, a very heterogeneous Condor pool featuring high end graphics workstations (two dual-core AMD Opteron processors with 16GB of RAM), more modest Linux workstations, and general purpose laboratory machines (Intel Pentium III and Pentium IV based machines with between 128MB and 384MB of RAM running Windows 2000 or Windows XP). Added to this there is a 16 CPU AMD Opteron SMP machine with 32GB of RAM, partially managed by Torque/Maui and donating spare cycles to the Condor pool. Other independent (typically midrange) workstations are managed directly as Webcom clients.

4.3 Discussion

None of these resources alone can provide the computational needs of the application due to queue lengths, policy restrictions, or a combination of these and other factors at individual sites. The benefit of our approach is the ease and consistency with which the available resource subsets at each site can be harnessed into an efficient and powerful metacomputer suitable for the processing of the application workflow. Users can drive the application from a portal that supports parametrisation of the workflow and the upload of patient- and treatment plan-specific inputs. Progress can then be steered based on the expert user's interpretation of intermediate results.

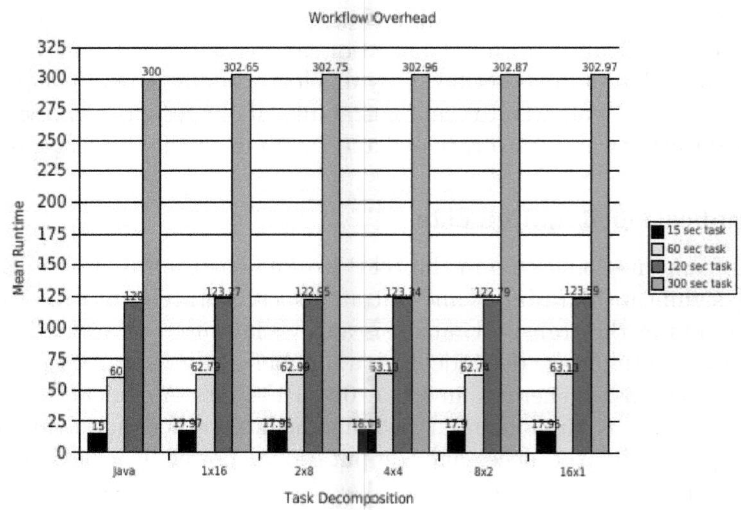

Fig. 3. A sequence of decompositions for a parallel application. The 'Java' timings show the basic performance of a 16 way parallel application. The '1x16' arrangement refers to a one node workflow version of the application. The '2x8' arrangement uses two nodes, each of which executes eight threads at the application level. The '16x1' set of timings show the performance when the parallelism of the application is entirely exposed at the workflow level. In all cases, for all task durations, the cost of workflow expression is approximately two or three seconds.

5 Evaluation

The basic performance of Webcom as an efficient resource manager and graph execution engine is shown in figure 3. This experiment shows Webcom's performance when executing different workflow expressions of a parallel application. The test application simulates a scalable job capable of being decomposed into different arrangements. As the decomposition becomes increasingly fine grained with more of the parallelism exposed at the workflow level (rather than the application logic level), the overhead of graph management remains reasonably constant, with execution times as short as 15 seconds.

The RTP application is scalable and can be tuned dynamically to the available resources such that each node in the MC phase has approximately the same runtime. Figure 4 shows the results of executing the MC phase of the RTP workflow over a dynamically aggregated collection of 165 processors from different sites.

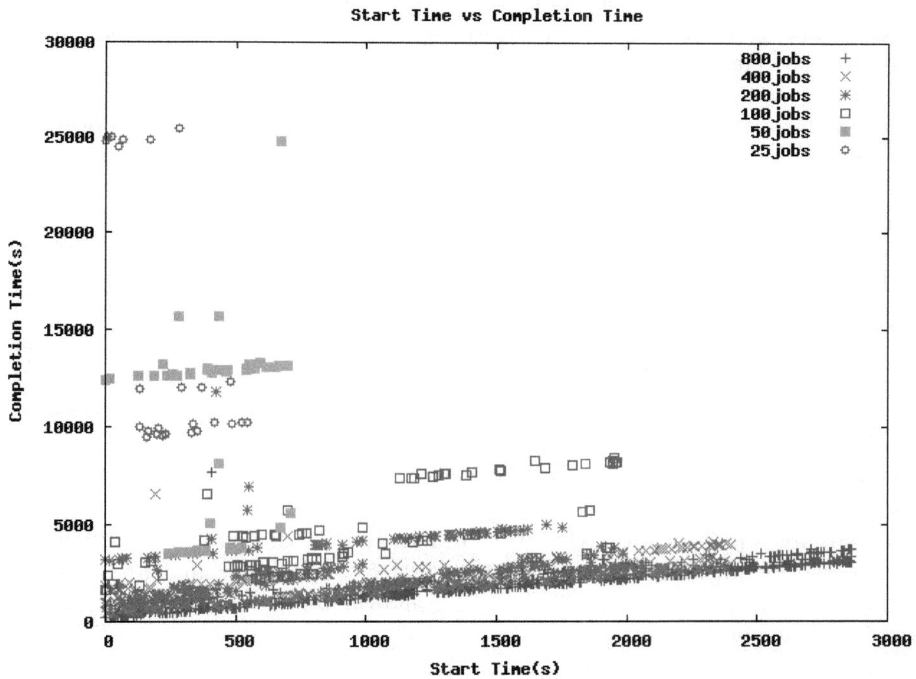

Fig. 4. A plot of start times to finish times for different decompositions of the radiotherapy workflow executed over a collection of 165 machines using our resource aggregation strategy to combine resources controlled by different managers including Condor and PBS, in conjunction with stand-alone machines running the Webcom service. The arrangement of 800 jobs shows the best performance as it is capable of making use of all available resources. The smaller decompositions expose less parallelism at the workflow level and consequently fewer machines are used, leading to longer execution times for each job and consequently longer workflow execution times.

This shows that the more fine grained decompositions of the task, which equate to a larger number of shorter jobs, results in better completion times. One reason for this is that the overall impact of jobs that are delayed on non-dedicated machines is greatly reduced as each individual job represents a smaller fraction of the overall workflow. The efficient handling of large parallel workflows over mixed machines, demonstrates the utility of our approach.

6 Related Work

This work builds on investigations conducted in many areas related to task and resource management. Pinchak [28] describes an approach to building a load balancing meta-queue over a set of independent batch processing sites using the concept of a *placeholder* task, inserted into each site's queue, whose purpose is to pull down instructions from a work server and execute tasks using the queue's resources. Our work is conceptually similar in that we use multiple independent sites to form a metacomputing platform by overlay. However, the work in [28] and [29] (the latter presents a DAG oriented extension to server) achieves load balancing via self-scheduling workers that pull tasks from a single master site following a fixed eager strategy. The Master/Worker (MS) model of work distribution, as used in [28], is popular and well suited to certain classes of problems. However it is not necessarily the most scalable model as the master's ability to dispatch work and collect results can prove a bottleneck [30]. The use of hierarchical MS topologies has been studied as a step towards dealing with the scalability issue [31]. A fundamental difference between [28] and this project is our use of the peer-to-peer model [32] to achieve greater scalability.

Other systems offer a single package solution for resource aggregation and workflow execution, e.g., the Java CoG kit [33] offers the ability to use a suite of resource access protocols including Globus [34], SSH [35], and web services, for workflow execution. Java CoG kit is built on the client/server model with pre-packaged components for interacting with various resource managers. Workflow is defined and managed using graphical or script languages that are translated to an XML-based intermediate representation. Our work differs as we use a multi-level P2P overlay abstraction for resource aggregation to increase scalability and robustness. In addition, as our approach is based on a graph-oriented model of computing that provides a strong theoretical basis from which to reason about the execution of workflow graphs. This model also provides conditional and iterative execution, recursive graph definitions, and combinations of lazy and eager evaluation strategies, which we leveraged to achieve decentralised exception handling.

The use of heterogeneous resources introduces the problem of platform dependence; we have tackled this problem using concepts developed within the Condor project [6], [7]. Condor supports cycle harvesting across heterogeneous non-dedicated resources to implement a distributed batch system focused on high throughput computing [36], [37]. Our technique uses the job and workflow description languages developed in Condor and incorporates ClassAd [38] matching with novel worker filtering techniques to achieve task to resource matching.

Support for the expression and correctly ordered submission of dependency constrained job collections, or workflows, is implemented in Condor via the DAGMan tool [39]. DAGMan is a user level tool whose purpose is to ensure that jobs are submitted in dependency-preserving order; it exerts no influence on Condor's placement and scheduling behaviour. Condor is batch-oriented and assumes that jobs are independent of each other; DAGMan proceeds as a Condor client, working on behalf of a user, submitting jobs when all their dependencies have been satisfied. Blythe et al. [40] discuss the impact of this lack of workflow awareness at the scheduler or task allocation level; excessive or redundant input and output transfer is identified as an issue with the Condor/DAGMan approach. Our approach is built on the Webcom graph evaluation engine and so the scheduler has access to all the inter-task dependency information in the workflow expression.

7 Conclusions and Future Work

In this paper we have described our approach and methodology for resource aggregation and collaboration support. The problem of integrating the diverse and distributed heterogeneous resources available to researchers has been tackled with a user-level overlay metacomputer. In addition, we have described the usage of our tools designed to support workflow execution in a dynamic peer-to-peer context. Initial results from both simulation and real applications have been presented to illustrate the efficiency of our model.

The purpose of this work is to support the development of solutions to real problems; these problems exhibit significant data and computational requirements. The solutions proposed herein are capable of exploiting the aggregate computing power of disjoint resource sets. Our future work will involve further optimisation of the tools discussed and support for dynamic task priority management, as well as enhancements to the automation of topological structuring and restructuring of Webcom overlays.

References

1. Foster, I., Kesselman, C. (eds.): The Grid: Blueprint for a New Computing Infrastructure. Morgan Kaufmann, Los Altal, CA (1999)
2. Morrison, J.P., Power, D.A., Kennedy, J.J.: Webcom: A volunteer-based metacomputer. The Journal of Supercomputing 18(1), 47–61 (2001)
3. Morrison, J.P., Power, D.A., Kennedy, J.J.: An evolution of the Webcom metacomputer. The Journal of Mathematical Modelling and Algorithms 23(2), 263–276 (2003) (Special Issue on Computational Science and Applications)
4. Morrison, J., Coghlan, B., Shearer, A., Foley, S., Power, D., Perrot, R.: WebCom-G: A candidate middleware for Grid-Ireland. International Journal of High Performance Computing Applications 20(3), 409–422 (2006)
5. Morrison, J.P.: Condensed Graphs: Unifying Availability-Driven, Coercion-Driven and Control-Driven Computing. PhD thesis, Technische Universiteit Eindhoven (1996)

6. Litzkow, M., Livny, M., Mutka, M.: Condor - a hunter of idle workstations. In: Proceedings of the 8th International Conference of Distributed Computing Systems (June 1988)
7. Thain, D., Tannenbaum, T., Livny, M.: Distributed computing in practice: the Condor experience. Concurrency - Practice and Experience 17(2-4), 323–356 (2005)
8. Butt, A.R., Zhang, R., Hu, C.Y.: A self-organising flock of Condors. Journal of Parallel and Distributed Computing 66, 145–161 (2006)
9. Stoica, I., Morris, R., Karger, D., Kaashoek, M.F., Balakrishnan, H.: Chord: A scalable peer-to-peer lookup service for internet applications. In: Proceedings of the ACM SIGCOMM 2001 Conference, San Diego, California (August 2001)
10. Zhao, B.Y., Stribling, L.H., Rhea, J., Joesph, S.C., Kubiatowicz, A.D.J.D.: Tapestry: A resilient global-scale overlay for service deployment. IEEE Journal on Selected Areas in Communications 22(1) (2004)
11. Celaya, J., Arronategui, U.: Scalable architecture for allocation of idle CPUs in a P2P network. In: Gerndt, M., Kranzlmüller, D. (eds.) HPCC 2006. LNCS, vol. 4208, pp. 240–249. Springer, Heidelberg (2006)
12. Stioca, I., et al.: Chord: a scalable peer-to-peer lookup protocol for internet applications. IEEE/ACM Trans. Netw. 11(1), 17–32 (2003)
13. Rowstron, A., Druschel, P.: Pastry: scalable, decnetralized object location and routing for large-scale peer-to-peer systems. In: Proceedings of the Middleware 2001, IFIP/ACM International Conference on Distributed Systems Platforms, pp. 329–350 (2001)
14. Jagadish, H., Ooi, B., Vu, Q.: Baton: A balanced tree structure for peer-to-peer networks. In: Proceedings of the 31st VLDB Conference, pp. 661–672 (2005)
15. Fahringer, T., Qin, J., Hainzer, S.: Specification of grid workflow applications with AGWL: An abstract grid workflow language. In: EEE International Symposium on Cluster Computing and the Grid 2005 (CCGrid 2005) (May 2005)
16. Alt, M., Hoheisel, A., Pohl, H.W., Gorlatch, S.: Using high level petri-nets for describing and analysing hierarchical grid workflows. In: Proceedings of the CoreGRID Integration Workshop 2005 (2005)
17. Hoheisel, A., Der, U.: An XML-based framework for loosely coupled applications on grid environments. In: Sloot, P., et al. (eds.) International Conference on Computational Science, pp. 245–254. Springer, Heidelberg (2003)
18. Mohan, R.: Why Monte Carlo? In: Leavitt, D.D., Starkschall, D. (eds.) Proc. XII Int. Conf. on the Use of Computers in Radiation Therapy, vol. XII, pp. 16–18. Medical Physics Publishing, Salt Lake City, Madison (1997)
19. Andreo, P.: Monte Carlo techniques in medical radiation physics. Physics in Medicine and Biology 36(7), 861–920 (1991)
20. Ma, C.M., Jiang, S.B.: Monte Carlo modelling of electron beams from medical accelerators. Physics in Medicine and Biology 44(12), R157–R189 (1999)
21. Rogers, D.W.O.: Fifty years of Monte Carlo simulations for medical physics. Physics in Medicine and Biology 51(13), R287–R301 (2006)
22. Krieger, T., Sauer, O.A.: Monte Carlo- versus pencil-beam-/collapsed-cone-dose calculation in a heterogeneous multi-layer phantom. Physics in Medicine and Biology 50(5), 859–868 (2005)
23. Ahnesjo, A., Saxner, M., Trepp, A.: A pencil beam model for photon dose calculation. Medical Physics 19(2), 263–273 (1992)
24. Ahnesjo, A.: Collapsed cone convolution of radiant energy for photon dose calculation in heterogeneous media. Medical Physics 16(4), 577–592 (1989)

25. Bayucan, A., Henderson, R.L., Lesiak, C., Mann, B., Proett, T., Tweten, D.: Portable Batch System: External reference specification. Technical report, MRJ Technology Solutions, 2672 Bayshore Parkway, Suite 810, Mountain View, CA 94043 (July 2005)
26. Henderson, R.L.: Job scheduling under the portable batch system. In: IPPS '95: Proceedings of the Workshop on Job Scheduling Strategies for Parallel Processing, pp. 279–294. Springer, London (1995)
27. Jackson, D., Snell, Q., Clement, M.: Core algorithms of the Maui scheduler. In: Feitelson, D.G., Rudolph, L. (eds.) JSSPP 2001. LNCS, vol. 2221, pp. 87–102. Springer, Heidelberg (2001)
28. Pinchak, C., Lu, P., Goldenberg, M.: Practical heterogeneous placeholder scheduling in overlay metacomputers: Early experiences. In: Feitelson, D.G., Rudolph, L., Schwiegelshohn, U. (eds.) JSSPP 2002. LNCS, vol. 2537, pp. 205–228. Springer, Heidelberg (2002)
29. Goldenberg, M., Lu, P., Schaeffer, J.: TrellisDAG: A system for structured DAG scheduling. In: Feitelson, D.G., Rudolph, L., Schwiegelshohn, U. (eds.) JSSPP 2003. LNCS, vol. 2862, pp. 21–43. Springer, Heidelberg (2003)
30. Aida, K., Natsume, W., Futakata, Y.: Distributed computing with hierarchical master-worker paradigm for parallel branch and bound algorithm. In: Proceedings of the Third IEEE/ACM International Symposium on Cluster Computing and the Grid (CCGRID'03), pp. 156–163 (May 2003)
31. Iverson, M.A., Özgüner, F.: Hierarchical, competitive scheduling of multiple DAGs in a dynamic heterogeneous environment. Distributed Systems Engineering 6, 112–120 (1999)
32. Milojicic, D.S., et al.: Peer-to-peer computing. Technical Report HPL-2002-57, HP Laboratories Palo Alto (March 2002)
33. von Laszewski, G., Hategan, M.: Workflow concepts of the Java CoG Kit. Journal of Grid Computing 3(3-4), 239–258 (2006)
34. Foster, I.: Globus toolkit version 4: Software for service-oriented systems. In: IFIP International Conference on Network and Parallel Computing, pp. 2–13. Springer, Heidelberg (2005)
35. Ylonen, T.: The Secure Shell (SSH) protocol architecture (January 06) Status: Proposed Standard
36. Livny, M., Basney, J., Raman, R., Tannenbaum, T.: Mechanisms for high throughput computing. SPEEDUP Journal 11(1) (June 1997)
37. Basney, J., Livny, M.: Deploying a high throughput computing cluster. In: Buyya, R. (ed.) High Performance Cluster Computing: Architectures and Systems, vol. 1, Prentice Hall PTR, Englewood Cliffs (1999)
38. Raman, R., Livny, M., Solomon, M.: Resource management through multilateral matchmaking. In: Proceedings of the Ninth IEEE Symposium on High Performance Distributed Computing (HPDC9), Pittsburgh, PA, pp. 290–291. IEEE Computer Society Press, Los Alamitos (2000)
39. Tannenbaum, T., Wright, D., Miller, K., Livny, M.: Condor – a distributed job scheduler. In: Sterling, T. (ed.) Beowulf Cluster Computing with Linux, MIT Press, Cambridge (2001)
40. Blythe, J., Jain, S., Deelman, E., Gil, Y., Vahi, K., Mandal, A., Kennedy, K.: Task scheduling strategies for workflow-based applications in grids. In: Proceedings of CCGrid 2005. Cardiff, Wales, vol. 2, pp. 759–767 (May 2005)

Performance Evaluation of View-Oriented Parallel Programming on Cluster of Computers

Haifeng Shang[1], Jiaqi Zhang[1], Wenguang Chen[1], Weimin Zheng[1], and Zhiyi Huang[2]

[1] Institute of High Performance Computing, Department of Computer Science and Technology, Tsinghua University, Beijing, China
[2] Department of Information Science, University of Otago, Dunedin, New Zealand
smickers@gmail.com, zation99@gmail.com, cwg@mail.tsinghua.edu.cn, zwm-dcs@tsinghua.edu.cn, hzy@cs.otago.ac.nz

Abstract. View-Oriented Parallel Programming(VOPP) is a novel programming style based on Distributed Shared Memory, which is friendly and easy for programmers to use. In this paper we compare VOPP with two other systems for parallel programming on clusters: LAM/MPI, a message passing system, and TreadMarks, a software distributed shared memory system. We present results for ten applications implemented and optimized using all the three systems. Experimental results demonstrate that VOPP is almost as efficient as Message Passing Interface when running on up to 32 processors, which means there is significant performance improvement compared with TreadMarks. The factors contributing to the performance of VOPP are discussed and analyzed. VOPP is still slower than MPI when the number of processes is large because of extra messages for separate synchronization and lack of bulk transfer mechanisms.

1 Introduction

Software distributed shared memory systems provide a shared memory abstraction on top of the native message passing facilities. It leaves the chore of message passing to the underlying DSM systems so that it is easier to program. However, DSM systems tend to generate more communication and therefore be less efficient than message passing systems [14].

View-Oriented Parallel Programming(VOPP) [3] is a programming style based on distributed shared memory. Traditional DSM systems like TreadMarks [1] are far from efficient compared with Message Passing Interface [7]. The reason is that programs written in MPI can be finely tuned by reducing unnecessary message passing. As we know, message passing is a significant cost for parallel applications, which is also true for DSM programs. Programmers cannot help reduce messages with traditional DSM systems as consistency maintenance of the underlying systems deals with the whole shared memory space.

We propose a novel VOPP programming style for DSM applications which optimizes DSM performance by introducing a new consistency maintenance protocol

R. Perrott et al. (Eds.): HPCC 2007, LNCS 4782, pp. 120–131, 2007.

VOUPID [4] and allows programmers to participate in performance tuning by wise optimization of data allocation. VOPP programs perform much more efficiently than TreadMarks. The performance gain is two-fold. First, the consistency maintenance for views reduces unnecessary messages and unnecessary data. The consistency maintenance protocol VOUPID solves the diff accumulation problem which limits the performance of TreadMarks. Second, VOPP enables programmers to tune the performance by wisely partitioning views.

Our ultimate goal is to make VOPP performs as efficiently as MPI. So far, the performance of VOPP is comparable with MPI and they are almost as efficient as each other while running on less than 32 processors. In this paper, we are going to compare VOPP with TreadMarks and MPI in terms of performance and investigate the factors contributing to the performance of VOPP.

Contributions of this paper include:

1) We use 10 applications for performance evaluation. The result presented in this paper is more convincible than the previous research which only used 4 applications.[6]

2) We perform a detailed performance analysis between VOPP, MPI and TreadMarks and reveals that VOPP could achieve comparable performance with MPI when the number of processes is up to 32. The performance of VOPP is not as good as MPI when the number of processes is larger than 32. Then main reasons are extra messages for separate synchronization and lack of bulk transfer mechanisms.

The rest of the paper is organized as following. In section 2 we briefly describe the VOPP programming style. In section 3 we evaluate the performance of VOPP by comparing it with TreadMarks and MPI. The reasons of the results are also discussed and analyzed. Finally, we conclude our work and summarize the usability and programmability of VOPP.

2 View-Oriented Parallel Programming

In the View-Oriented Parallel Programming(VOPP) style, programmers should divide shared data into views according to the memory access pattern of the parallel algorithm. A view [3] consists of data objects that require consistency maintenance as a whole body. Views are indicated through primitives such as acquire_view and release_view. Acquire_view means acquiring exclusive access to a view, while release_view means finishing the access. In addition, acquire_Rview and release_Rview are provided for read-only accesses. The read-only access primitives can be called in a nested style while other access primitives cannot. By using these primitives, programmers are now focusing on the access of shared data rather than synchronization and mutual exclusion.

Views are defined implicitly by programmers in their mind. Therefore, it is convenient for programmers to use the shared data and optimize the programs by wisely partitioning views. Views cannot overlap each other and they are unchangeable once they're defined. A view can only be accessed when the primitives are used.

Programming in VOPP style, programmers can enjoy the convenience of accessing shared data with almost ordinary read and write operations. Programmers don't have to determine what to communicate as the message passing is left to the underlying system. Furthermore, by dividing shared data into views, VOPP allows programmers to participate in performance optimization.

Much work has been done in the previous papers [3,5,4] to introduce optimizations in VOPP compared with traditional distributed shared memory systems. In the following section, we will describe our recent work on VOPP.

3 Performance Comparison

In this section, we present our experimental results of 10 applications. All applications in our experiment are coded in VOPP, TreadMarks and MPI respectively. Our DSM system, optimized VODCA[15], is used to run the VOPP programs, Traditional DSM system, TreadMarks, is used to run the TreadMarks programs and LAM/MPI V7.1 is used to run the MPI programs.

The applications used in our experiment include Gauss, Integer Sort(IS), Successive Over-Relaxation(SOR), Neural network(NN), Water, Traveling Salesman Problem(TSP), Barnes-Hut, BT, CG and MG, which are mostly chosen from SPLASH-2 benchmark suite [12] and NAS Parallel Benchmarks(NPB) [13].

Our experiments were carried out on a cluster with Infiniband interconnections, running Linux 2.6. Each node has two 1.6GHz processors and 4 Gbytes memory. The page size of the virtual memory is 32 KB.

3.1 Improvement on VODCA

As we know, barriers incur many messages, especially when the number of processes increases. Furthermore, every process often waits for the slowest process as there is synchronization when barrier is called. It significantly slows down the parallel program if barriers are frequently called.

The source files of VODCA V1.0.1 can be downloaded from the web site http://vodca.otago.ac.nz/. Barriers in this VODCA totally rely on the work of the Vdc_barrier_manager, which is running on proc 0. When a barrier function is called, every process sends a barrier requirement to proc 0 and then waits for an end-of-barrier reply from Proc 0. We therefore could conclude that the barrier manager has too much burden that it is the bottleneck of the whole barrier process.

We implement the barrier with the binomial-tree model[11] whose complexity is just O(logN) instead of the O(N) of the original linear model. As a result the barrier time on 64 processes in the cluster we mentioned above is 623.8 microseconds, which is 17% faster than the original implementation. When the number processes increases, it is expected that our new barrier implementation could outperform the original VODCA implementation more significantly due to the complexity difference.

3.2 Performance Overview

Gauss implements the gauss elimination algorithm in parallel. The matrix size of Gauss is 8000*8000 and the number of iterations is 1024 in our tests. The TreadMarks Gauss program has the false sharing effect. The VOPP version has significantly improved the performance by removing the false sharing effect with local buffers. The speedup of Gauss is shown in Figure 1.

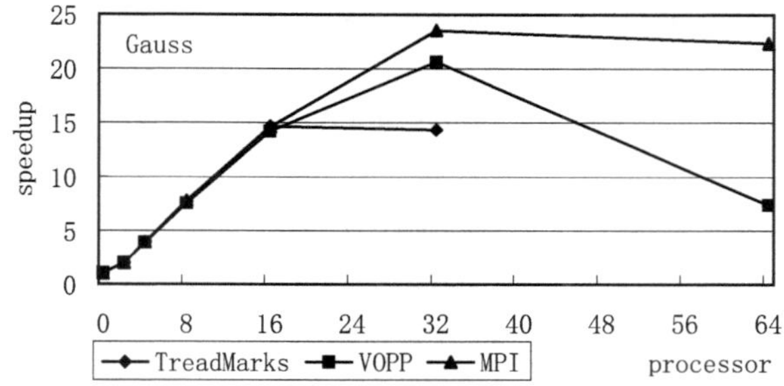

Fig. 1. Speedup of Gauss

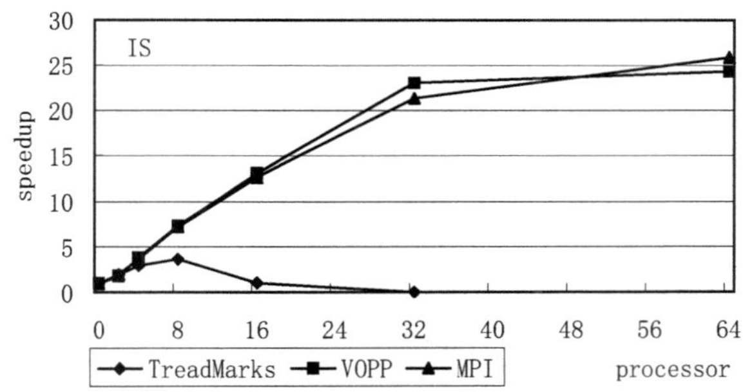

Fig. 2. Speedup of IS

IS ranks an unsorted sequence of N keys using bucket sort and the problem size in our tests is (2^{27}*2^{17},40). The speedup of IS is shown in Figure 2.

SOR uses a simple iterative relaxation algorithm with a two-dimensional grid as input. Every element is updated to a function of its neighbors' values in each iteration. We use local buffers for those infrequently-shared data in our VOPP programs. In contrast, we use shared memory (a set of views) for those frequently-shared data such as the border elements. In our tests SOR processes a matrix with size of 8192*1024 in 100 iterations. The speedup of SOR is shown in Figure 3.

NN trains a back-propagation neural network in parallel using a training data set. The VOPP version of NN uses local buffers for infrequently-shared data and acquire_Rview for read-only data. The acquire_Rview for read-only data is

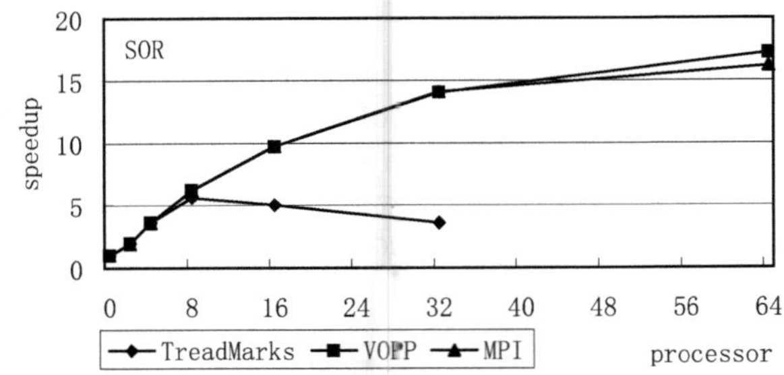

Fig. 3. Speedup of SOR

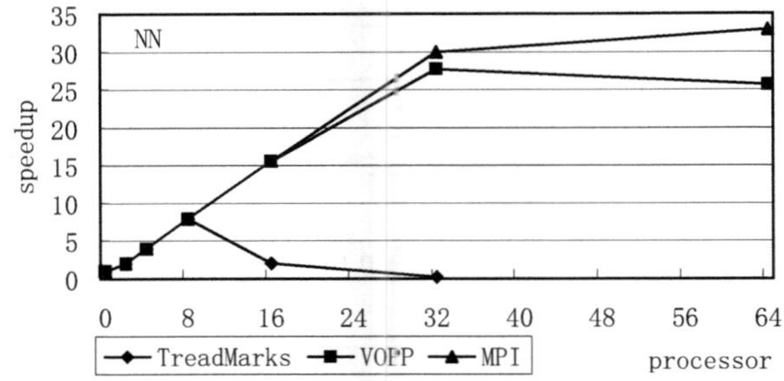

Fig. 4. Speedup of NN

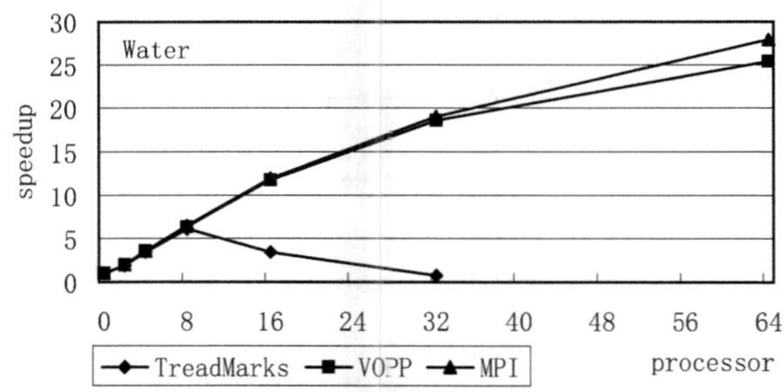

Fig. 5. Speedup of Water

very important for the VOPP program. The size of the neural network in NN is 9*40*1 and the number of epochs taken for the training is 1024. The speedup of NN is shown in Figure 4.

Water from the SPLASH-2 benchmark suite is a molecular dynamics simulation program. The bulk of the inter-processor communication happens during the force computation phase. Each processor computes and updates the intermolecular force between each of its molecules and each of n/2 molecules following

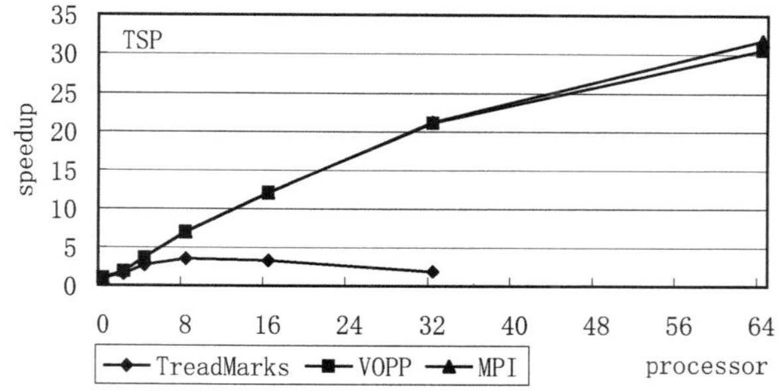

Fig. 6. Speedup of TSP

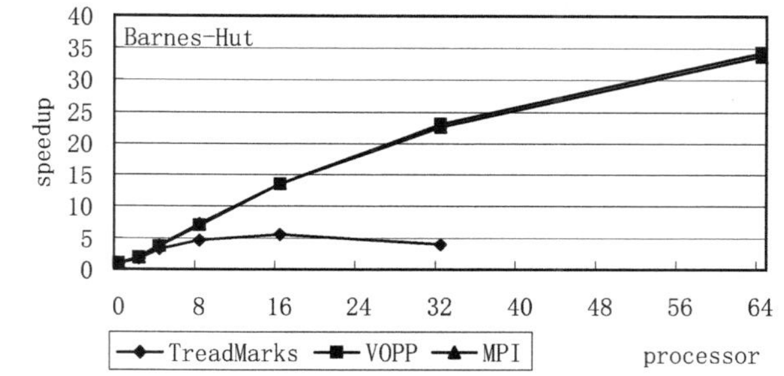

Fig. 7. Speedup of Barnes-Hut

it in the array in wrap-around fashion. The data set we used is 1728 molecules and 5 steps. The speedup of Water is shown in Figure 5.

TSP solves the traveling salesman problem using a branch and bound algorithm. We solve a 19-city problem with a recursive-solve threshold of 12. The speedup of TSP is shown in Figure 6.

Barnes-Hut from the SPLASH-2 benchmark suite is an N-body simulation using the hierarchical Barnes-Hut Method. We run Barnes-Hut with 32768 bodies in our experiment. The speedup of Barnes-Hut is shown in Figure 7.

BT creates a binary tree with a depth of 9. It keeps all those unexpanded nodes by using a task queue. The speedup of BT is shown in Figure 8.

CG from NAS Parallel Benchmarks implements conjugate gradient algorithm. The data set we used in the experiment is 15 niter and 11 nonzer and 14000 nn, which is defined as LARGE problem size. The speedup of CG is shown in Figure 9.

MG, which comes from NAS Parallel Benchmarks, solves a poisson problem on a 128 by 128 by 128 grid, using 20 multigrid iterations. The speedup of MG is shown in Figure 10. Speedup results for these applications are shown in Figure1 - Figure10. These figures show the relative speedup of the 10 applications written in different styles running in the same hardware environment while the number of processes increases. Programs of TreadMarks slow down significantly when

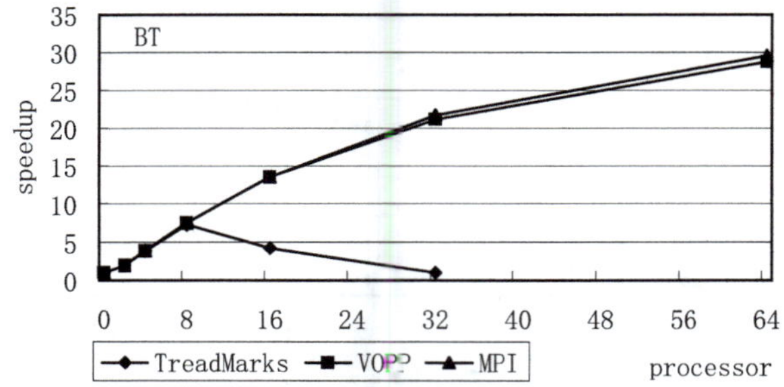

Fig. 8. Speedup of BT

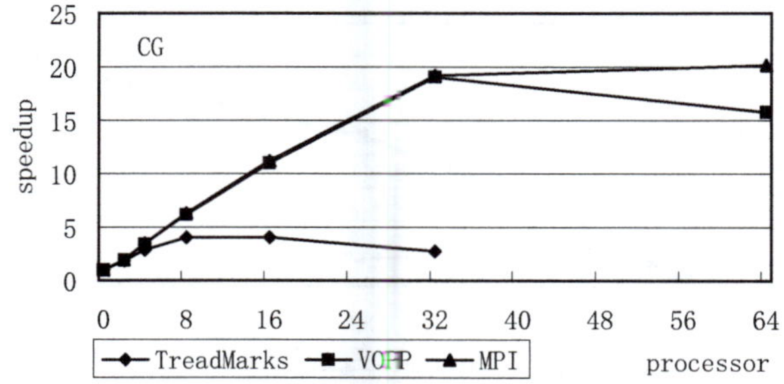

Fig. 9. Speedup of CG

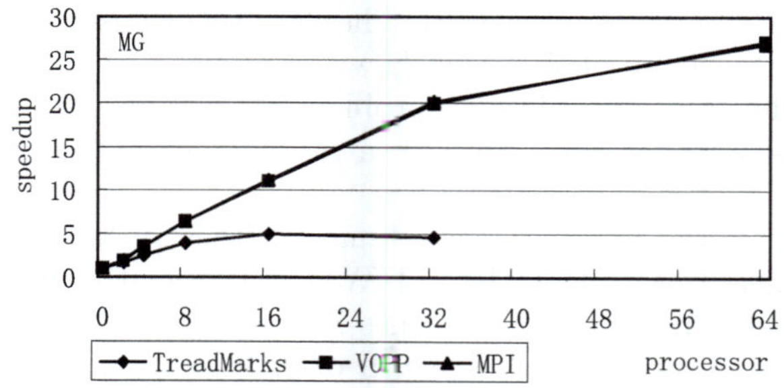

Fig. 10. Speedup of MG

running on more than 8 processors. Comparing the speedup of TreadMarks, VOPP and MPI on 32 processors, we can find that the performance of VOPP is much better than TreadMarks and it is nearly the same efficient as MPI. However, when running on more than 32 processors, some of VOPP programs including Gauss, NN and CG slow down and there are obvious performance gaps between VOPP and MPI. In contrast, SOR and Barnes-Hut of VOPP perform better. For the rest five programs, there are very small performance gaps between VOPP and MPI.

3.3 Factors Contributing to the Performance of VOPP

Most of the differences in speedup and communication requirement between TreadMarks and MPI are a result of TreadMarks' lack of bulk transfer, extra messages for separate synchronization(barriers), false sharing and diff accumulation [9,10,14]. Contrastively, there is no false sharing in VOPP programs, as the views access are not nested and VOPP has succeeded in solving the problem of diff accumulation by introducing the new consistency maintenance protocol VOUPID [4,8].

Table 1. 32-Processor Total Messages for TreadMarks, VOPP and MPI

program	TreadMarks	VOPP	MPI
Gauss	2138734	1849612	247969
IS	239082	247682	39680
SOR	57850	49380	12400
NN	717741	696551	37355
Water	72800	65200	3760
TSP	74269	52276	5736
Barnes-Hut	604004	579006	1040
BT	52700	48200	7800
CG	84468	74849	12480
MG	125782	117601	21920

Table 1 and Table 2 provide figures for the number of messages and the amount of data exchanged when the programs are running on 32 processors. For TreadMarks and VOPP programs, we count the total number of messages and the total amount of data transferred. While for the MPI programs, we count the number of user-level messages and the amount of the user data sent in each run.

Table 2. 32-Processor Data Transferred(KB) for TreadMarks, VOPP and MPI

program	TreadMarks	VOPP	MPI
Gauss	23853491	7994304	7936991
IS	4012643	1318835	1300234
SOR	179728	52890	50840
NN	1603279	500457	282970
Water	91476	36942	36758
TSP	13925	171	168
Barnes-Hut	296320	208843	202560
BT	2383	816	761
CG	2884	1153	982
MG	12460	3752	3740

In Table 1, the two DSM systems, TreadMarks and VOPP require comparable amount of messages in ten different applications. They do not support bulk transfer either. In TreadMarks and VOPP systems, messages are incurred when updating shared data which is based on page fault. The messages occur in MPI programs are not comparable. In table 2, we can easily find the reason VOPP programs perform more efficiently than those of TreadMarks. By reducing false sharing and solving diff accumulation problem [3], VOPP programs send significantly less data when running on parallel systems. There are few unnecessary data transferred in VOPP programs, compared with those of MPI.

Although performance of VOPP has significantly improved and it is as efficient as MPI when running on less than 32 processors, we can find some programs of VOPP slow down when running on more processors. The main factors contributing to the performance slowdown are overheads of barriers and lack of bulk transfer.

As we point out in 3.1, barriers incur large amount of messages, especially when the number of processors increases. However, if we compare the VOPP programs with the MPI programs, we find there are much more barriers in the VOPP programs [7]. Gauss calls 15998 barriers; IS calls 120 barriers; SOR calls 200 barriers; NN calls 2410 barriers; Water calls 72 barriers; TSP calls 3 barriers; Barnes-Hut calls 19 barriers; BT calls 2 barriers; CG calls 1186 barriers and MG calls 484 barriers. In contrast, there are almost no barriers in MPI programs. Accordingly the performance gaps for Gauss, NN and CG are larger, while for other programs they are much smaller.

The reason why there have to be more barriers in VOPP programs has been explained in [5]. Barriers have to be used to make sure the sequential consistency. In contrast, in the MPI code, there is no need to use a barrier for synchronization, since the receive primitive is synchronized with the send primitive and is always finished after the send primitive. To verify the overhead of barriers is a significant contributor, we run our VOPP programs on two versions of VODCA and compare them with the MPI version programs. As the largest performance gap occurs in Gauss program, we choose Gauss for our verification. Firstly, we run Gauss on our optimized VODCA system with improved barrier in our previous tests, relatively, we now run Gauss again on original VODCA without improved barrier. Secondly, we intentionally make the number of barriers in VOPP Gauss the same as that in MPI version. We run the new Gauss on LAM/MPI, optimized VODCA with improved barrier and original VODCA without improved barrier.

In Figure 11, VOPP_ori represents VOPP program running on original VODCA without improved barrier. We find the performance gap is obviously larger between VOPP_ori and MPI when running on 64 processors. Compared with the performance gap between optimized VODCA and MPI, we can find that there is obvious performance gain by optimization of barrier.

In Figure 12, the performance gaps of the modified Gauss become very small. The MPI curve represents the MPI Gauss with unnecessary barriers. The VOPP curve represents the Gauss program with less barriers running on optimized VODCA with improved barrier, of which the execution result is wrong. The VOPP_ori represents the modified Gauss running on original VODCA without

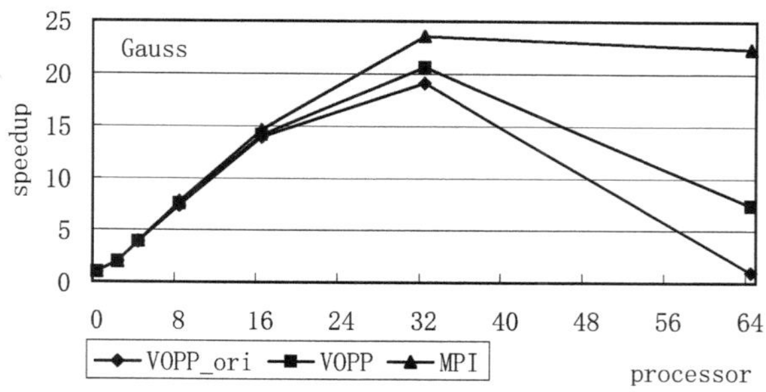

Fig. 11. Speedup of Gauss with/without improved barrier

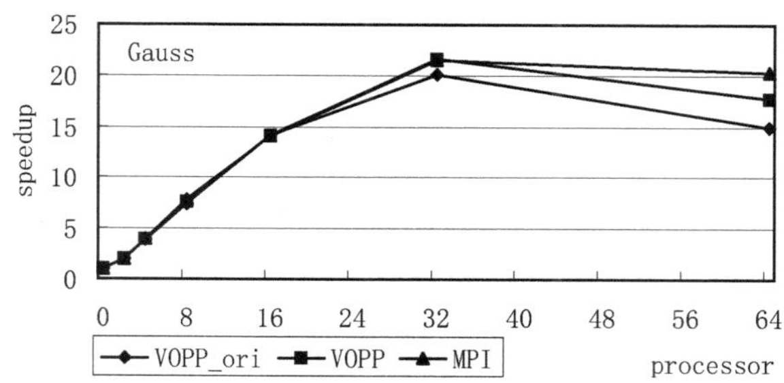

Fig. 12. Speedup of Gauss with same number of barriers

improved barrier. The numbers of barriers in each of them are nearly the same. For 64 processors, the speedup gap between VOPP and MPI is reduced from 15 to 3 and the gap between VOPP and VOPP_ori is also reduced, from 7 to about 3.

From the two figures above, we can conclude that the performance slowdown of VOPP programs is a result of the overheads from barriers. Although our optimized barrier performs more efficiently, the overheads of barriers are still the source contributing to the performance slowdown when the number of processors increases. We are considering replace barriers with some light weight synchronization primitives.

Like TreadMarks, VOPP does not support bulk transfer. The update of views is based on pages, which means for large amount of data to be updated, there will be extra messages to handle access misses. In contrast, MPI is able to aggregate large amounts of data in a single message. To simulate the effect of bulk transfer, we define the VOPP page size as a multiple of the hardware page size. By increasing the VOPP page size, a view is likely to be updated with less page faults, which means there will be less messages for handling the access miss.

As shown in Table 3, by increasing the VOPP page size, the total number of messages has been reduced greatly. And the speedup gap of Gauss between

Table 3. 64-processor messages, data transferred and speedup of Gauss

Gauss	VOPP_32KB	VOPP_64KB	MPI
messages	3921256	2248807	531362
data transferred	17149791	17144611	17026841
speedup	7.376	10.218	22.343

VOPP and MPI becomes smaller. The extra messages incurred by lack of bulk transfer is another source contributing to the performance of VOPP.

In summary, by reducing false sharing and solving the problem of diff accumulation, VOPP performs much more efficiently than TreadMarks. There are significantly less data to be transferred. However, the extra messages incurred by separate synchronization and lack of bulk transfer limit the performance of VOPP when the number of processors increases.

4 Conclusions

This paper evaluates a novel DSM programming style VOPP for cluster computers. Ten applications of all kinds are converted and optimized under three different parallel systems, VOPP, TreadMarks and LAM/MPI. Our experimental results demonstrate that, on a large variety of programs, there is significant performance advantage of VOPP against TreadMarks and VOPP is now comparable with MPI. The performance of VOPP is nearly as efficient as MPI on less than 32 processors. We also analyze the factors contributing to the performance gap between VOPP and MPI. VOPP is still slower than MPI when the number of processes is large because of extra messages for separate synchronization and lack of bulk transfer mechanisms.

As a software distributed shared memory system, VOPP is convenient for programmers to use and it is easy to achieve correctness and efficiency. It only requires programmers to insert primitives when a view is accessed. Programmers do not have to determine what to communicate but focus on the implementation of the parallel algorithm. Besides, programmers are allowed to participate the performance optimization by wisely partitioning views. For programs with complicated communication patterns, especially for those with complicated or irregular array accesses or with data structures accessed through pointers, VOPP shows better programmability than MPI. Our experience indicates that it is convenient to port traditional DSM programs to VOPP and it is easier to implement parallel algorithm using VOPP than MPI. Furthermore, the performance of VOPP is as efficient as MPI for 32 processors and we therefore have another choice for parallel programming on cluster of computers.

References

1. Keleher, P., Dwarkadas, S., Cox, A.L., Zwaenepoel, W.: TreadMarks: Distributed shared memory on standrd workstations and operating systems. In: Proceedings of the 1994 Winter Usenix Conference, pp. 115–131 (1994)

2. Amza, C., Cox, A.L., Dwarkadas, S., Keleher, P., Lu, H., Rajamony, R., Yu, W., Zwaenepoel, W.: TreadMarks: Shared memory computing on networks of workstations. IEEE Computer 29, 18–28 (1996)
3. Huang, Z., Purvis, M., Werstein, P.: View-Oriented Parallel Programming and View-based Consistenc. In: Proc. of the Fifth International Conference on Parallel and Distributed Computing, Applications and Technologies, pp. 505–518 (2004)
4. Huang, Z., Purvis, M., Werstein, P.: View-Oriented Update Protocol with Integrated Diff for View-based Consistency. In: Proc. of the IEEE/ACM Symposium on Cluster Computing and Grid (2005)
5. Huang, Z., Purvis, M., Werstein, P.: Performance Evaluation of View-Oriented Parallel Programming. In: Proc. of the 2005 International Conference on Parallel Processing, IEEE Computer Society, Los Alamitos (2005)
6. Gropp, W., Lusk, E., Skjellum, A.: A high performance, portable implementation of the MPI message passing interface standar in Parallel Computing, 789–828 (1996)
7. Werstein, P., Pethick, M., Huang, Z.: A Performance Comparison of DSM, PVM and MP. In: Proc. of the Fourth International Conference on Parallel and Distributed Computing, Applications and Technologies, pp. 476–482 (2003)
8. Keleher, P.: Lazy Release Consistency for distributed shared memor. In: Ph.D. Thesis (Rice Univ) (1995)
9. Klaiber, A.C., Levy, H.M.: A comparison of message passing and shared memory architectures for data parallel language. In: Proc. of the 2th Annual International Symposium on Computer Architecture, pp. 94–106 (1994)
10. Lu, H., Dwarkadas, S., Cox, A.: Zwaenepoel: Quantifying the performance differences between PVM and TreadMark. Journal of Parallel and Distributed Computing, 65–78 (1997)
11. Hensgen, F.R., Manber, U.: Two algorithms for barrier synchronizatio. International Journal of Parallel Programming, 1–17 (1988)
12. Singh, J.P., Webber, W.-D., Gupta, A.: SPLASH: Stanford parallel applications for shared memor. In Technical Report, Department of Computer Science, Stanford University (1991)
13. Bailey, D., Barton, J., Lasinski, T., Simon, H.: The NAS parallel benchmark. In Technical Report 103863, NASA (1993)
14. Lu, H., Dwarkadas, S., Cox, A.L., Zwaenepoel, W.: Message Passing versus distributed shared memory on networks of workstation. In: Proc. of SuperComputing 95 (1995)
15. Huang, Z., Chen, W., Purvis, M., Zheng, W.: VODCA: View-Oriented, Distributed, Cluster-based Approach to parallel computin. In: Proc. of the IEEE/ACM Symposium on Cluster Computing and Grid (2006)

Maximum-Objective-Trust Clustering Solution and Analysis in Mobile Ad Hoc Networks*

Qiang Zhang, Guangming Hu, and Zhenghu Gong

School of Computer, National University of Defense Technology,
Changsha 410073, Hunan, China
powerfbi007@sina.com, {gmhu,zhgong}@nudt.edu.cn

Abstract. In mobile-AdHoc networks (MANETs), many applications need the support of layer-structure. Clustering solution is the most widely used layer-structure and the choosing of clusterheads is the key problem of all. Traditional ways of clustering lack of instructions of trust mechanisms. This paper presents a maximum-objective-trust-based clustering solution (MOTBCS), which aims at the opinion of maximum stable links and energy viewpoint and gives nodes their objective trust estimation. This solution can be better suitable for the realistic working environments for MANETs. We make necessary simulations for our design and results show that MOTBCS can generate more stable clustering groups. It also has less communication costs and better efficiency than other clustering algorithms.

1 Introduction

Mobile ad hoc networks (MANETs) are self-organized wireless networks that are formed by mobile nodes through distributed protocols. MANETs can work without the support of communication infrastructure and such networks are being widely used in more and more fields. The features of dynamic topology and non-existence of central facilities ensure widespread application prospects of them, but at the same time these features bring about many new problems and challenges. Among them clustering of nodes is one of the biggest challenges that MANETs are facing with and it is also a hot spot in the research areas nowadays. Suited clustering solutions can greatly enhance the practicability and performance of MANETs[1].

Till now, researchers have raised a lot of clustering algorithms and some of them have been practically used, such as the Lowest-ID Algorithm[2,3], the Highest Connectivity Degree Algorithm[4], Distributed Clustering Algorithm(DCA)[5], Weighted Clustering Algorithm(WCA)[6,7] and k-hop Clustering Algorithm[8,9]. These algorithms each have different peculiarities and working environments.

* This research was supported by the National Grand Fundamental Research 973 Program of China under Grant No. 2003CB314802 and the National 863 Development Plan of China under Grant No. 2006AA01Z401.

R. Perrott et al. (Eds.): HPCC 2007, LNCS 4782, pp. 132–143, 2007.

2 Clustering

2.1 Definition of Clustering

MANETs nodes and links can be shown as an undirected graph G(V,E). V denotes the set of nodes and E denotes the set of links. If there is a link (*u*, *v*) □E, that means node *u* and *v* are 1-hop neighbors and they can communicate with each other directly. What we want to do is to choose a group of cluster-head nodes. They and their 1-hop nodes make up of the whole network.

Definition 1. The number of node Vi´s 1-hop neighbors is called Vi´s degree, and it is denoted by D (Vi).

Definition 2. If there is a node $u \in V$ and *u* is in the range of Vi´s communication ability, and $(u,Vi) \in E$, then we called them neighbors. We use *N*(Vi) to denote node Vi´s neighbor set, and $u \in N(Vi)$.

Definition 3. There are three node states in MANETs. They are pending nodes, member nodes and cluster-head nodes.

2.2 Stability of Clustering

In wireless networks, the signal intensity that nodes receive is tightly correlative with the distance between nodes. In the pure Friis free space model, T_xP (signal sending power) and R_xP (signal receiving power) have the following relation: (*d* is the distance between nodes)

$$\frac{R_xP}{T_xP} \propto \frac{1}{d^2}$$

While in actual environments, the expression $R_xP = \frac{c}{d^{\alpha}} T_xP$ is more accurate. But the value of α is not easy to set exactly.

Although it is not applicable to judge nodes´ distance directly, we can estimate nodes´ relative mobility with the consecutive message packets such as periodic "HELLO" packets. Assume that node *u* has received three periodic "HELLO" packets from node *v*, and the power is respective shown as $RxP^{(1)}_{v \to u}$, $RxP^{(2)}_{v \to u}$ and $RxP^{(3)}_{v \to u}$, then we can define node *v*´s relative mobility degree as:

$$M_u^{rel}(v) = 10\lg \frac{R_xP^{(2)}_{v \to u}}{R_xP^{(1)}_{v \to u}} + 10\lg \frac{R_xP^{(3)}_{v \to u}}{R_xP^{(2)}_{v \to u}} \quad (1)$$

This way is somewhat like the definition of MOBIC[10]. If $R_x P^{(2)}_{v\to u} < R_x P^{(1)}_{v\to u}$, then $\lg \frac{R_x P^{(2)}_{v\to u}}{R_x P^{(1)}_{v\to u}} < 0$. That means node v is leaving away from node u. If $R_x P^{(2)}_{v\to u} > R_x P^{(1)}_{v\to u}$, then $\lg \frac{R_x P^{(2)}_{v\to u}}{R_x P^{(1)}_{v\to u}} > 0$. That means node v is moving towards node u.

Link Stability Estimation Rules

If node u and v meet the following demands, then we regard the link between them stable[13]:

1. The times that node u and v continuously send HELLO packets to each other equals or is more than 3;
2. If $M_u^{rel}(v) < 0$, then $|M_u^{rel}(v)| < M_{\min}^{Threshold}$;
3. If $M_u^{rel}(v) > 0$, then $M_u^{rel}(v) < M_{\max}^{Threshold}$

Here $M_{\min}^{Threshold}$ and $M_{\max}^{Threshold}$ are thresholds of relative mobility. Of course, $M_{\min}^{Threshold} < M_{\max}^{Threshold}$, and their actual values can be set according to the actual network environments.

3 Objective Trust

We mentioned the notion of node stability ahead, now we introduce the definition of objective trust.

Definition 4. Objective trust is more extensive than the notion of node stability. The decay of objective trust is a time-associated function[11]. We will combine it with the evaluation of node stability.

Definition 5. We mark the original objective trust with T_0, and T_t after a period of time t.

Definition 6. $\theta(t)$ denotes the decay factor of objective trust. Then

$$T_t = \theta(t) T_0 \quad (t \geq 0) \tag{2}$$

Definition 7. We use λ to denote the "suspicion parameter" of objective trust.

Definition 8. Function of $f(t)$ can be denoted as:

$$f(t) = \begin{cases} \lambda e^{-\lambda t}, & \text{when } t>0 \\ 0, & \text{when } t \leq 0 \end{cases} \tag{3}$$

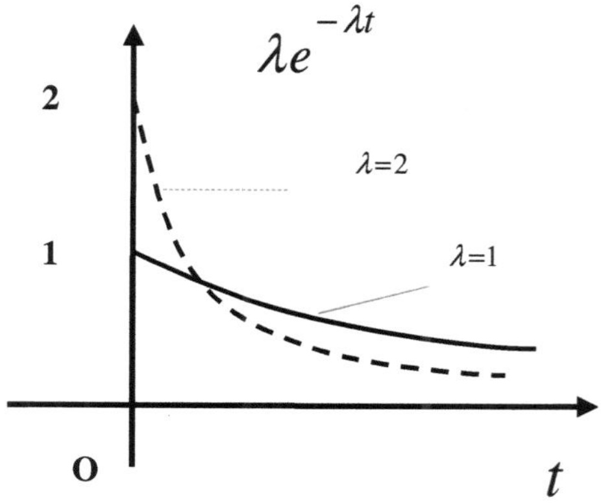

Fig. 1. Function of $f(t)$

Definition 9. We use $TSF(t)$ to denote the cumulation of decayed objective trust.

$$TSF(t) = \int_{-\infty}^{t} f(t')dt' = \begin{cases} 1-e^{-\lambda t} & , t \geq 0 \\ 0 & , t < 0 \end{cases} \tag{4}$$

Then $\theta(t) = 1 - TSF(t)$

$$\theta(t) = e^{-\lambda t} \ (t \geq 0) \tag{5}$$

The proofs of expression (3) and (5) are shown in Appendix I.

The objective trust and stability of nodes are relational with the ability of nodes in some degree, such as energy remains, computing speed and memory size. We use Cap_u to denote the integrative ability of node *u*. It is a discrete value after modifying.

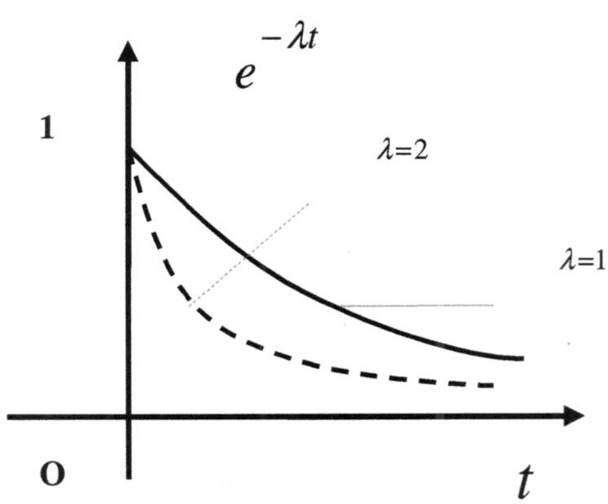

Fig. 2. Function of $\theta(t) = 1 - TSF(t)$

According to the upper objective trust model, the parameter λ_u can be set as follows. We first calculate the value of stable links N_u^{slink} and we set a quite big λ_{upper} beforehand.

$$\lambda_u = \begin{cases} 1\Big/\sqrt{\left(N_u^{slink} \cdot Cap_u\right)} & N_u^{slink} \neq 0 \\ \lambda_{upper} & N_u^{slink} = 0 \end{cases} \tag{6}$$

Assume that node u has k neighbors, and their relative mobility values to u are $M_u^{rel}(n_1)$, $M_u^{rel}(n_2)$, …, $M_u^{rel}(n_k)$, then we define $\overline{M}_u$ as node u's integrative relative mobility.

$$\overline{M}_u = \sum_{i=1}^{k} \frac{\left|M_u^{rel}(n_i)\right|}{M} \tag{7}$$

In the upper formula, when $M_u^{rel}(n_i) < 0$, then M equals $M_{\min}^{Threshold}$; when $M_u^{rel}(n_i) \geq 0$, $M = M_{\max}^{Threshold}$. Of course, the less $\overline{M}_u$ is, the more believable node u is.

4 Reconstruction of HELLO Message

The realization of MOTBCS is executed by reconstructing the message of HELLO. Since MANETs nodes send out HELLO packets periodically and they need brief computing, we can draw the following assumptions.

1. We can only consider λ (objective trust suspicion parameter) instead of complicated value of objective trust.
2. If no signal is received after a period of waiting time T_w (generally is twice larger than the cycle of HELLO packets), we can regard the link disabled referring to the tendency in Fig 2.

The new HELLO message with trust information can be shown as:

$$HELLO_i = (ID_i \mid Status \mid ntable \mid \overrightarrow{M}_i^{rel} \mid \overline{M}_i \mid Cap_i \mid timestamp \mid sk_i(hash))$$

Where $\boldsymbol{ID_i}$: The ID number of node i ;

Status: This field has two choices of *ClusterID* and NULL. If it is NULL, that means the status of node is "Pending". If it is the ID of some node else, that means the node is node i's clusterhead. If it is the ID number of node i itself, that means node i is a clusterhead.

ntable: the table of 1-hop neighbors to node *i*. Nodes can set up 2-hop topology by acquiring neighbors´ 1-hop neighbors.

$\overrightarrow{M}_i^{rel}$ *:* The relative mobility vector of node i . The components of $\overrightarrow{M}_i^{rel}$ can be stored in the table of *ntable* .

$\overline{M}_i$ *:* Nodes´ integrative relative mobility and it is computed from $\overrightarrow{M}_i^{rel}$

Cap_i *:* The integrative ability of node *i*, including its energy remains, computing speed and memory size. In this paper, we mainly concern nodes´ energy.

timestamp: Current time.

sk_i(hash): The abstract information[12] to sign the message using the private key sk_i of node *i*.

We can divide the HELLO packets into three types for the difference of node *ID* and *Status* [13]:

Neighbor-search-message: its *Status*= NULL;
Clusterhead message: its $Status = ID_i \neq$ NULL;
Cluster-join-message: its $Status = ID_j \neq$ NULL and $i \neq j$

Each node sets up a neighbor table and a 2-hop topology table. By using the table information and the periodic HELLO packages, we can work out the variables $\overrightarrow{M}_i^{rel}$, $\overline{M}_i$, N_i^{slink} and Cap_i . After that, the objective trust "suspicion parameter" λ_i can be computed, then the solution of MOTBCS showed below will select out the clusterhead of the group.

5 MOTBCS Algorithm

A pending node *u*(*Status* = NULL) chooses its clusterhead with the following algorithm MOTBCS.

ClusterHead(*u*)

IF (!initialized(G(V,E))) initialize(G(V,E))

IF (!assign(*u*.ID)) assign(ID) to *u*

IF (*u.Status* !=NULL) return ERROR;

$N'(u) = N(u) \cup \{u\}$ // $N(u)$: neighbor of node *u*

FOREACH node *j* ($j \in N'(u)$) **DO**

computer $\overrightarrow{M}_i^{rel}$, $\overline{M}_i$, N_i^{slink} and λ_j ; // N_i^{slink} : number of stable links of *i*

ENDFOR

FOREACH neighbor node *j*($j \in N'(u)$) **DO**

select node with minimum λ_j ;

```
        ENDFOR
    IF nodes with minimum λ_j is only THEN
        node j becomes Clusterhead;
        u.Status= j;
        return;
    ELSE
    FOREACH node j with minimum λ_j DO
        select node with minimum M̄_j;
    ENDFOR
    IF node j with minimum M̄_j is only THEN
        node j becomes Clusterhead;
        u.Status= j;
        return;
    ELSE
    FOREACH node k with minimum M̄_k DO
        select node with minimum ID;
        node k with minimum ID becomes Clusterhead;
        u.Status= k;
    ENDFOR
    return;
  ENDIF
  ENDIF
```

6 Simulation and Experiments

We use network simulator ns-2 to do our experiments. Two widely used models are simulated in our tests. They are Random Waypoint Mobility Model(RWMM) and Reference Point Group Mobility model(RPGM). The behaviors and setting of the experiments are similar with that of [13].

6.1 Estimation Criterions

1. Clusterhead-Changing Frequency(CCF)

The status changing times of clusterheads in one second. A small CCF means a good clustering solution and a stable mobile adhoc network.

2. Node-Changing Frequency(NCF)

The status changing times of nodes in one second. A small NCF always means a stable mobile adhoc network.

3. Number of Clusterheads(NC)

The number of clusterheads changes when nodes move around in the network. Fewer clusterheads mean that a node can communicate with another node through fewer hops.

4. Communication Costs(CC)

The number of communication packages in one second. For simplification, we don't take the package encrypting into account.

6.2 Simulation Setting and Figures

We simulate the Lowest-ID Algorithm, Max-Degree Algorithm, Weighted Clustering Algorithm(WCA) and MOTBCS. The simulation parameters are shown in the following table.

Table1. Simulation Parameter Setting

Parameter	Description	Default Value
T_h	Cluster Message Cycle	2 s
T_w	Link Max-valid-time	4.2 s
T_c	Clusterhead Competing Interval	5s
Node number	Number of Mobile Nodes	60
Sim-time	Simulation time	600 s
Pause time	Pause Time When Moving in the Waypoint Model	4 s
Max-speed	Max Moving Speed	20 m/s
Tx-range	Range of Radio Transmission	30 to 180 m
Head-Energy-Use	Clusterhead Energy Use	0.03%/s
Member-Energy-Use	Normal Node Energy Use	0.01%/s
Length	Area Length	Sqrt (Num*3.14*150*150/8)
Width	Area Width	Sqrt (Num*3.14*150*150/8)

Simulation 1: Node-Changing-Frequency(NCF) vs. Communication Range(CR)

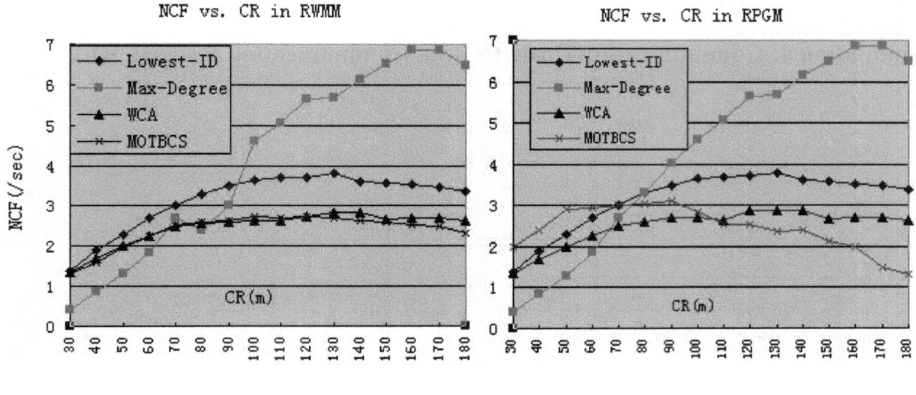

Fig. 3. NCF vs. CR

Simulation 2: Clusterhead-Changing-Frequency(CCF) vs. CR

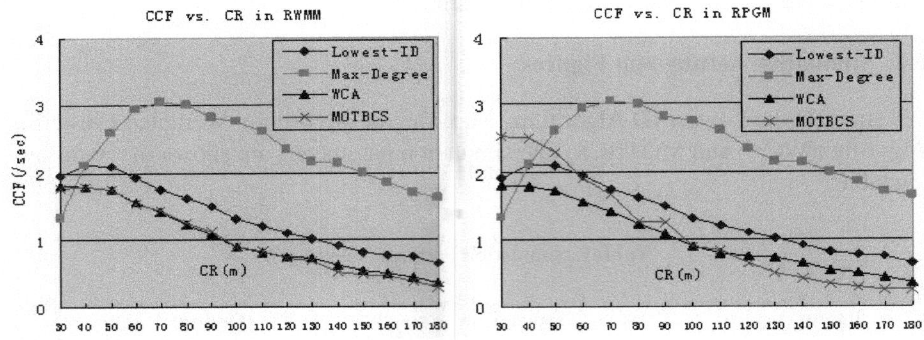

Fig. 4. CCF vs. CR

Simulation 3: Node-Changing- Frequency(NCF) vs. Node Speed(NS)

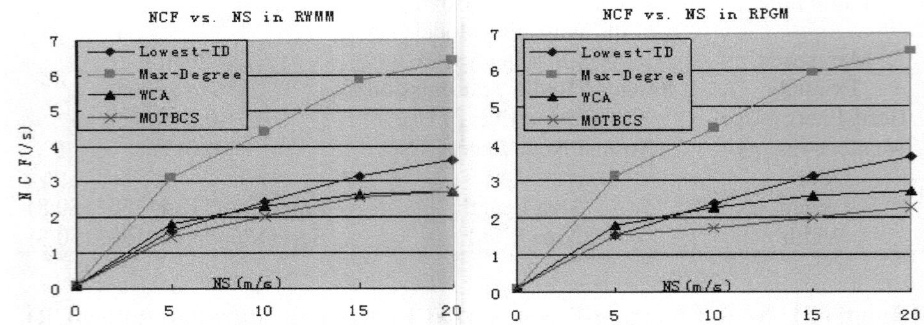

Fig. 5. NCF vs. NS

Simulation 4: Communication Costs(CC) vs. Communication Range(CR)

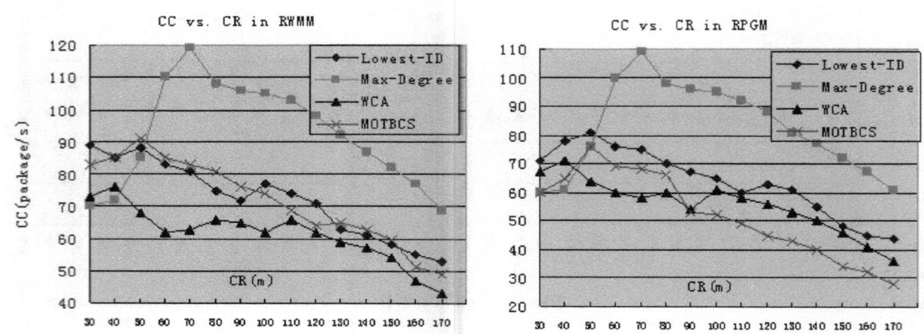

Fig. 6. CC vs. CR

Simulation 5: Communication Costs(CC) vs. Node Speed(NS)

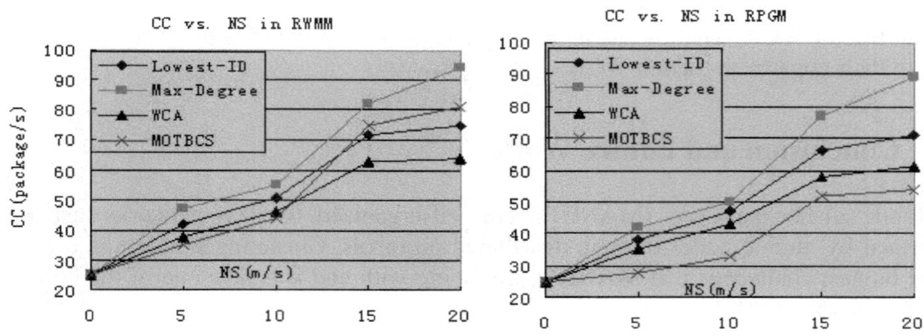

Fig. 7. CC vs. NS

Simulation 6: Number of Clusterheads in Experiments

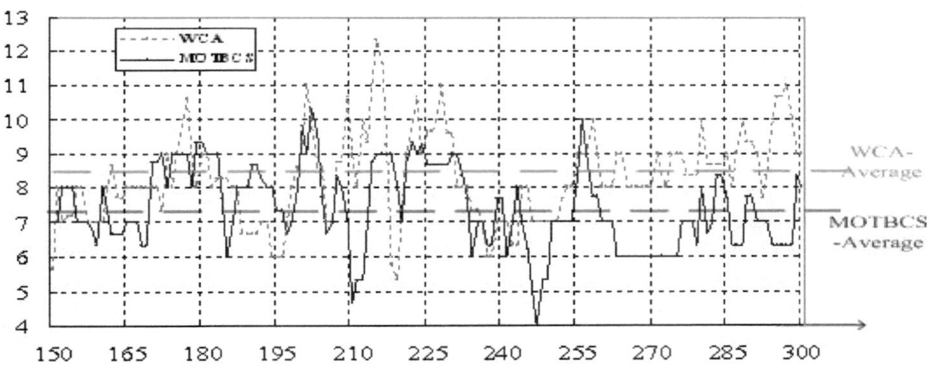

Fig. 8. Number of Clusterheads

6.3 Experiment Analysis

Fig.3 shows that NCF increases gradually when CR gets large. The reason is that when CR increases, the number of nodes in a cluster also increases. The performance of MOTBCS is better than the other three especially when $r > 100$m. That is because MOTBCS chooses the best objective trust nodes as clusterheads.

Fig.4 shows that when CR is about 50-70m, CCF is comparatively large. For at that time, the number of clusterheads is rather big and nodes are easy to compete to act as clusterheads.

We know from Fig.5 that NCF increases greatly when Node Speed gets large. But apparently, MOTBCS is more steady than the other solutions.

Fig.6 and Fig.7 compare MOTBCS's communication and control costs with the other three algorithms. Generally speaking, MOTBCS has better performance because

MOTBCS extends "HELLO" message to choose clusterheads instead of constructing new package, which reduces extra costs greatly.

Fig.8 shows that the average clusterhead number of MOTBCS is obviously less than that of WCA. That means nodes in MOTBCS need fewer hops to communicate with their copartners than in WCA.

7 Conclusion and Future Work

Mobile ad hoc networks (MANETs) are self-organized wireless networks that are formed by mobile nodes through distributed protocols. Clustering problem is one of the biggest challenges that MANETs are facing with and it is also one of the hottest spots in the research areas. This paper presents a maximum-objective-trust-based clustering solution(MOTBCS). It well takes into account objective trust and the relative mobility of nodes on the criteria for clusterhead selection. In this way, it overcomes the drawbacks of over-idealization assumptions about node behavior models in similar research, and thus is more applicable to realistic environments. Simulation results show that the MOTBCS solution can generate more stable clustering structure and has lower communication overheads, compared with other homologous algorithms. So MOTBCS can be a valuable solution for clustering with further realization in the future.

References

1. Varadharajan, V., Shankaran, R., Hitchens, M.: Security for Cluster Based Ad hoc Networks. Computer Communications 27(5), 488–501 (2004)
2. Ephrem ides, A., Wieselth ier, J.E., Baker, D.J.: A design concept for reliable mobile radio networks with frequency hopping signaling[A]. Proceedings of IEEE[C] 75(1), 56–73 (1987)
3. Gerla, M., sai, T., Multicluster, J.T.C.: mobile, multimedia radio network [J]. Wireless Networks 1, 255–265 (1995)
4. Parekh, A.K.: Selecting Routers in Ad-Hoc Wireless Networks. In: Proceeding of the SBT/IEEE International Tele. Symposium (1994)
5. Basagni, S.: Distributed clustering for ad hoc networks[C]. In: Proceedings of International Symposium on Parallel Architectures[C], Algorithms and Networks, pp. 310–315 (1999)
6. Chatterjee, M., Das, S.K., Turgut, D.: WCA: a weighted clustering algorithm for mobile Ad Hoc networks [J]. Journal of Cluster computing, Special issue on Mobile Ad hoc Networking, 193–204 (2002)
7. Chatterjee, M., Das, S.K., Turgut, D.: An on-demand weighted clustering algorithm (WCA) for ad hoc networks[A]. In: Proceedings of IEEE GLOBECOM 2000 [C], San Francisco, pp. 1697–1701. IEEE Computer Society Press, Los Alamitos (2000)
8. Krishna, P., Vaidya, N.H., Chatterjee, M., et al.: A cluster based approach for routing in Ad Hoc networks[J]. ACM SIGCOMM Computer Communication Review (CCR) (1997)
9. Amis, A.D., Prakash, R., Vuong, T.H., Huynh, D.T.: Max-M in D-Cluster formation in wireless Ad Hoc networks [C]. In: Proceedings IEEE INFOCOM 2000, Israel (2000)
10. Basu, P., Khan, N., Little, T D C.: Amobility based metric for clustering in mobile Ad Hoc networks [C]. In: Proceedings of IEEE ICDCS 2001 Workshop on Wireless Networks and Mobile computing, Phoenix, A Z, pp. 413–418. IEEE Computer Society Press, Los Alamitos (2001)

11. Zhang, Q., Li, D., Gong, Z., et al.: EDDTSTM: A Time-Nonlinear-Sensitive Trust Model Based on Human Mental Peculiarity in Virtual Computing Environments. Guangzhou, China. In: Proceeding of CIS. vol. 1, pp. 62–67 (2006)
12. Kuang, X.-H.: Research of Group Key Management in Mobile Ad hoc Networks. PhD thesis, National University of Defense Technology, Changsha, Hunan, P.R. China (2003)
13. Hu, G.-M.: Research on Key Security Issues in Clustered Mobile Ad hoc Networks. PhD thesis, National University of Defense Technology, Changsha, Hunan, P.R. China (2007)

Appendix I Proof of Expression (3) and (5)

Since λ is an objective trust suspicion constant, according to the mathematic fluxional way, we can conclude that:

$$\forall t > 0, \quad \lim_{\Delta t \to 0} \frac{[TSF(t+\Delta t) - TSF(t)]\big|_{conditional}}{\Delta t} = \lambda$$

According to the conditional probability, the formula just equals to

$$\forall t > 0, \quad \lim_{\Delta t \to 0} \frac{[TSF(t+\Delta t) - TSF(t)]/(1-TSF(t))}{\Delta t} = \lambda$$

Then we get the formula $TSF'(t) \big/ (1-TSF(t)) = \lambda$

So we can get $TSF(t) = 1 - Ce^{-\lambda t} \quad (t \geq 0)$

We consider that $TSF(t)$ =0 when $t \leq 0$, according to the formula $TSF(0)$ =0, then we can know that $C = 1$, so we can conclude $TSF(t) = 1 - e^{-\lambda t} \quad (t \geq 0)$. And as $f(t)$ is the probability density function of $TSF(t)$, so formula(3) can be proved subsequently.

As $TSF(t) = 1 - e^{-\lambda t} \quad (t \geq 0)$ is proved, and $\theta(t) = 1 - TSF(t)$, so expression (5) of $\theta(t) = e^{-\lambda t} \ (t \geq 0)$ is also proved.

A New Method for Multi-objective TDMA Scheduling in Wireless Sensor Networks Using Pareto-Based PSO and Fuzzy Comprehensive Judgement

Tao Wang[1,2], Zhiming Wu[2], and Jianlin Mao[2]

[1] School of Electrical and Computer Engineering
Georgia Institute of Technology, USA
twang31@mail.gatech.edu
[2] Department of Automation
Shanghai Jiao Tong University, 200240, Shanghai, China
{reno_wang0823,ziminwu,jlmao}@sjtu.edu.cn

Abstract. In wireless sensor networks with many-to-one transmission mode, a multi-objective TDMA (Time Division Multiple Access) scheduling model is presented, which concerns about the packet delay and the energy consumed on node state transition. To realize the scheme, a mapping between the problem and evolutionary algorithm is reasonably set up. A multi-objective particle swarm optimization based on Pareto optimality (PAPSO) is then proposed to solve such multi-objective optimization problem and find a better tradeoff between time delay and energy consumption. The simulation results validate the effectivity of PAPSO algorithm and also show that PAPSO outperforms other techniques in the literature.

1 Introduction

Large-scale networks of wireless sensors are becoming a hot topic of research due to their potential usage in defense, pervasive commercial and scientific applications. In such networks, sensors are units with sensing, processing, and wireless networking capability. They can automatically collect information and report the results to an access point. However, as the sensors are usually battery-powered, saving energy becomes an essential problem in sensor networks.

The medium access control (MAC) method is a major consumer of sensor energy [2]. Different medium access methods result in different time and energy efficiencies. Among those proposed MAC protocols, including contention based access and contention free access, TDMA is a suitable access method for wireless sensor networks. First, TDMA can save energy by eliminating collisions, avoiding idle listening, or entering inactive states until being allocated time slots. Secondly, as a collision-free access method, TDMA can bound the delays of packets and guarantee reliable communication.

R. Perrott et al. (Eds.): HPCC 2007, LNCS 4782, pp. 144–155, 2007.

However, allocating time slots to each sensor to finish a couple of data collection tasks is an NP-complete problem [3]. Additionally the energy constraint makes the problem more difficult to solve. Therefore, in sensor networks, the main challenge of TDMA is how to allocate time slots to each node to minimize the energy consumption and time delay.

As to the aspect of TDMA scheduling against time performance, some references have studied how to minimize packet delay [4], how to improve fairness [5], how to maximize parallel operation [6], and how to shorten the total slot cost to finish a set of transmission tasks [7]. By setting up a convex optimization model, Cui et al. [4] adopted relaxation methods to solve the problem, and got a pareto optimal energy-delay curves, where the power consumption on switching was neglected. Sridharan et al. [4] developed a linear programming formulation and presented a distributed solution, which outperformed than random MAC in terms of fairness and delay etc. To minimize the overall transaction time, which was modeled as a graph partitioning problem, Gandham et al. [5] proposed a distributed edge-coloring algorithm. In order to save time for data collection, Ergen et al. [6] proposed three algorithms based on coloring method in graph theory. However, these two references did not consider the energy saving problem in sensor networks.

In this paper a practical hierarchical solution approach is proposed to solve the multi-objective TDMA scheduling problem. Given the strong search ability in combinatorial optimization, particle swarm optimization (PSO) is introduced. And to reach multi-objective optimality, Pareto optimization serves as evaluation criterion of candidate solutions during the evolutionary process of PSO, by means of which we can get Pareto optimal solutions to TDMA scheduling problem. In this sense, the whole framework of the method can deal with several constraints flexibly and reach a multi-objective optimal slot-allocation scheme.

The remainder of the paper is organized as follows: the problem statement is introduced in section 2; the part of optimization algorithm, including coding method and evaluation system of PSO solutions, is expatiated in section 3; and the computational results are given in section 4; and section 5 gives some final conclusions.

2 Problem Statement

2.1 Network and Scheduling Model

From a viewpoint of network, a sensor network can be represented by an undirected graph $G = (V, E)$, where V represents the set of all sensors in the network and $E \subset V \times V$ represents the set of communication links between a pair of nodes. There is one access point (AP) in V. All traffic generated at sensors are destined for AP, composing a routing tree. Such a network is called many-to-one sensor network.

The distance $d(i, j)$ between nodes i and j is defined as the minimum number of edges to go from one to the other. From this definition, the topology of sensor network can be described by an $N \times N$ symmetric connectivity matrix C, which is defined as $C_{ij} = 1$, if $d(i,j) = 1$; else $C_{ij} = 0$. In addition, the conflict relationship in the network can be described by an interference matrix $I_{N\times N}$, where if $d(i,j) \leq 2$, $I_{ij} = 1$; else $I_{ij} = 0$. As a result, node i and j can transmit data at the same time if the communication distance $d(i,j)$ is larger than 2.

In TDMA scheduling problem, time is splitted into equal intervals called time slots. Each time slot is designed to accommodate a single packet to be transmitted and received between pairs of nodes in the network. And once the routes are established, the allocation of time slots directly influences the performance of transmission in network. Moreover, TDMA also ensures collision-free communication when several transmission tasks run simultaneously. So what we need to do is to find a schedule of time slots to reach our requirements of network transmission performances, such as energy consumption and time delay.

In many-to-one sensor networks, as the sensor data flows toward AP from respective source nodes, the TDMA scheduling problem can be formulated as follows. There are a set of sensor data packets, forwarding to AP, over the routing trees, which are established using GPSR [7]. Each data-collection process that a packet flows to AP from its source node is called a task. On the established routes, each task consists of a sequence of transmission actions called subtasks, where one subtask needs one slot occupation. The aim of the problem is to determine a slot-allocation sequence of subtasks so that collision would not happen, and some optimization criteria could be satisfied.

2.2 Description of Optimization Objectives

In this paper, two optimization objectives are considered : the average energy consumption and the average time delay of data packets of sensor nodes.

To save energy, a common idea is just to switch off the radio when it is neither transmitting nor receiving. However, frequently turning on/off the radio also consumes large amounts of power, especially when the packet is small. Hence, we take into account this part of energy, which is ignored in many TDMA researches. According to Ref. [3], the formula of average consumed energy of sensor nodes is as follows:

$$AvgEngy = \frac{1}{N}\sum_{i=1}^{N}[P_i^{tx} \cdot (t_i^{tx} + t_i^{s-tx}) + P_i^{rx} \cdot (t_i^{rx} + t_i^{s-rx})$$

where N denotes the number of nodes in the network, P_i^{tx} (P_i^{rx}) is the power consumption of transmitter (receiver) at node i. $t_i^{tx}(t_i^{rx})$ is the total work time of the transmitter (receiver) at node i . $t_i^{s-tx}(t_i^{s-rx})$ is the total transition time consumed between the sleep and active states.

As a performance index, the average time delay of data packets should be as small as possible, which can help those sensor nodes to increase their sampling rate to the maximum possible level. In other words, the smaller the average time delay of data packets is, the more data those nodes can collect in every unit time and more efficiently they can communicate with each other.

3 PSO-Based Energy-Delay Pareto Optimization

3.1 Standard PSO

Particle swarm optimization (PSO) is a new swarm intelligence technique proposed by Eberhart and Kennedy [8], inspired by social behavior of bird flocking or fish schooling.

The main idea of PSO is as follows. There is a population of random solutions. Each potential solution, called particle, flies through the problem space by following the current optimal particle. Flying in the search space, each particle has a velocity which is dynamically adjusted according to the experiences of its own and its colleagues. This makes the swarm have an intelligent ability of flying towards the optimal position.

The global model equations of PSO are:

$$V_{id}(t+1) = W \cdot V_{id}(t) + C_1 \cdot rand_1() \cdot (p_{id}(t) - X_{id}(t)) + \\ + C_2 \cdot rand_2() \cdot (p_{gd}(t) - X_{id}(t)) \quad (1)$$

$$X_{id}(t+1) = X_{id}(t) + V_{id}(t) \quad (2)$$

$$|V_{id}| \leq V_{\max} \quad (3)$$

where V_{id} and X_{id} are respectively the velocity and position of particle. p_{id} and p_{gd} respectively represent the best position of i_{th} particle and the swarm. W called inertia weight, is a user-specified parameter. A large inertia weight pressures towards global exploration while a smaller inertia weight pressures towards fine-tuning the current search area. Proper selection of the inertia weight and acceleration coefficients can provide a balance between the global and the local search. C_1 and C_2 are acceleration factors, which are usually set to 2 . $rand_1$ and $rand_2$ are random numbers between (0,1). And during the recursive process, the selections of p_{id} and p_{gd} depend on the evaluation system, which will be expatiated later.

According to our TDMA problem, some parameters in PSO are defined as follows. There are N tasks and each task i includes M_i hops. The dimension of particles is set to M, the number of the total subtasks, i.e., $\sum_{i=0}^{N-1} M_i$. The border of the searching space is defined by X_{max}, which is set to $N-1$. Parameter $V_{\max}$ determines the maximum change one particle can take during one iteration, which is set as $N-1$. Under this setting, V_{id} is a value in the range $[-(N-1), N-1]$, thus the positions of the flying particles are under control.

3.2 Pareto-Based Evaluation System of PSO Solutions

In PSO, there are concepts of individual and population. During the process of iterations of the algorithm, the flying direction of particle derives from its own optimal position p_{id} and global optimal position p_{gd} of population. Consequently, the selection of p_{id} and p_{gd} directly influences the searching performance of PSO. Thus, the evaluation system of candidate solutions is critical to the whole optimization algorithm.

However, there exists confliction between time delay and energy consumption in TDMA scheduling, i.e., pursuing the optimization of power consumption on switching inextricably damages the performance of time delay index. It is mainly because to lessen state transitions and thus save energy, the good working continuity of nodes is required, which means that after collecting data from its child nodes, the sensor node better wait to transmit data packets to its own parent node instead of switching off. As a result, the performance of time delay would be damaged, and vice versa. Consequently, as to such multi-objective optimization problem in which the objectives cannot be optimized simultaneously, the concept of Pareto optimality was introduced into the evaluation system.

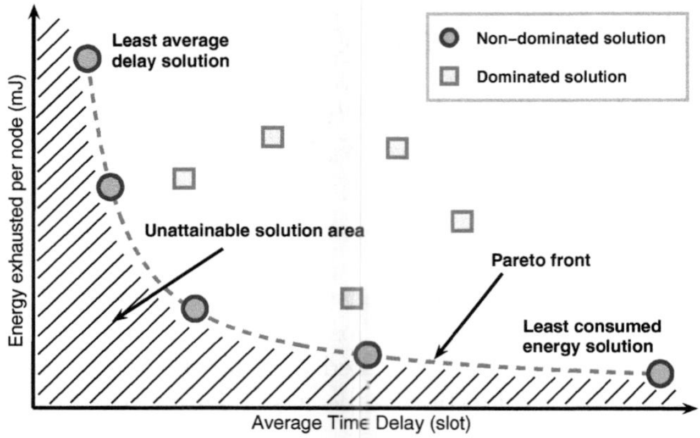

Fig. 1. Pareto frontier of candidate scheduling solutions

3.3 Pareto Optimality

In general, the multi-objective optimization problem is described as follows:

$$\min f(x) = \big(f_1(x), f_2(x), \cdots, f_k(x)\big)$$
$$s.t. \quad g_i(x) \le 0,\ i = 1, 2, \ldots, m$$

Where $x \in R^n$ is the decision vector belonging to the feasible region S, which is described as follows: $S = \{x \in R^n | g_i(x) \le 0,\ i = 1, 2, \ldots, m\}$

A decision vector $x_1 \in S$ is said to *dominate* a decision vector $x_2 \in S$ (denoted $x_1 \prec x_2$) iff:

- The decision vector x_1 is not worse than x_2 in all objectives, or $f_i(x_1) \le f_i(x_2) \quad \forall i = 1, 2, \ldots, q$.
- The decision vector x_1 is strictly better than x_2 in at least one objective, or $f_i(x_1) < f_i(x_2)$ for at least one $i = 1, 2, \ldots, q$.

If any of above conditions is not met, then x_1 does not dominate x_2. All the solutions that do not dominate each other compose nondominated solution set. If the comparative space of picked solutions is the global space, then the nondominated solution set is called *Pareto-optimal* set.

In this paper, the objective functions of our TDMA scheduling algorithm are:

$$f_1 = AvgEngy,\ f_2 = AvgDelay \tag{4}$$

which are used in the Pareto dominance comparison as Fig.1.

Crowding-Measure-Based Maintenance of Pareto Archive. Since Pareto optimal set is often infinite, Pareto archive is used to offer available optimal solutions representing the Pareto frontier. In order to make the nondominated solutions in the archive uniformly distribute on the Pareto frontier, here crowding-measure is introduced into the maintenance of archive.

$$d_i = (d_i^1 + d_i^2)/2 \tag{5}$$

where d_i^1 and d_i^2 are the minimum two Euclidean distances between individual i and other members of archive.

The crowding measure d_i reflects the distribution of other individuals around i. The smaller d_i is, the more the number of individuals surrounding i is. Consequently during the evolutionary process, when the Pareto archive is full, we filter the member with the minimum crowding measure. By means of the maintenance, the retained members of archive would evenly distribute on the Pareto frontier.

Evaluation of Global Optimal Solution. The selection of global best position p_{gd} is crucial to the whole PSO algorithm. In multi-objective problem, the population generates several nondominated solutions, all of which can be p_{gd}. Then how to assign suitable p_{gd} for every single particle is one important task. Here, combined with the maintenance of Pareto archive, we select the global best position p_{gd} as follows:
To every newly-generated nondominated solution x_i:

- If x_i dominates some members of archive, then x_i takes the place of all the dominated ones, and let x_i be the new p_{gd} of those particles whose previous p_{gd} are the replaced;
- Else if the current archive is full, let x_i replace the least-crowding-measure solution x_l as well as its position as a p_{gd}.
- Else if the archive is not full, directly insert x_i first, and use a newly-defined variable $np(x_i)$, the number of particles whose p_{gd} is x_i, to make every member of Pareto archive at least be p_{gd} of some particles, thus ensuring the diversity of solutions:
  a For all the solutions x_k in the archive, define $s = \min\{np(x_k)\}, k = 1, 2, \ldots, M (M$ is the swarm size),
  if $s \geq g$, then $s = g$ and $np(x_i) = 0$; where $g = (0.025 \sim 0.05)M$, an index to prevent a solution from being p_{gd} of too many particles;
  b Define $F = \{x_k | np(x_k) > s\}$, $u = |F|$, $v = 1$; "||" means the size of set.

c Find x_k in F, closest to x_i, and let x_i be p_{gd} of one of the particles whose previous p_{gd} is x_k. $F = F|\{x_k\}$, $np(x_i) = np(x_i) + 1$, $v = v + 1$;
d If $np(x_i) < s$ and $v < u$, go to step **c**; If $np(x_i) < s$ and $v = u$, go to step **b**; If $np(x_i) = s$, then add x_i to the archive.

After the process of PSO flying iterations, the Pareto solution archive represents the Pareto-optimal solutions of the multi-objective TDMA scheduling.

Evaluation of Local Optimal Solution. As to the selection of local optimal solution p_{id} in every iteration, we take the method proposed by Coello Coello [10]:
every time the particle j gets the new position x_j and the current local optimal position p_{jd} after the j iterations, compare the dominance between p_{jd} and x_j. If $x_j \succ p_{jd}$, then the updated $p_{jd} = x_j$; else if $x_j \prec p_{jd}$, then p_{jd} does not change; else if x_j and p_{jd} do not dominate each other, then randomly pick one between them two.

3.4 PAPSO Optimization

The procedure of PAPSO optimization is as follows:

1. Initialize parameters of PSO, including max_generation_PSO, W, C_1, C_2, M and M', and make Pareto archive empty;
2. Initialize Pareto solution archive:
   As to all the initial solutions, calculate their fitness $f_1, f_2, \ldots, f_i$ one by one. Then according to Pareto optimality, judge their dominance with each other, add all non-inferior solutions to P', and form an initial Pareto solution archive.
3. While (max_generation_PSO is not reached)
   { generation=generation+1;
   generate next generation of swarm by eq.(1)~(3);
   evaluate the new swarm according to the evaluation system, and update the local optimal position P_{id} of individual particle and the global optimal position P_{gd} of swarm (i.e. Pareto solution archive) }
4. When the evolutionary process is over, output the result of Pareto optimization.

3.5 Encoding and Decoding Scheme

To apply our evolutionary algorithm on TDMA time slot scheduling, encoding is a key step. The aim is to express a solution as a sequence code, which is an individual in population-based algorithms.

According to the description of the scheduling problem in section 2.1, a data collection task can be divided into several hops called subtasks in sequence. Hence, a set of data collection tasks can be viewed as a combination of all the subtasks. Therefore, we first denote a subtask as (TaskID,HopNo.), where the TaskID points out which task the subtask belongs to, and the HopNo. gives the sequence number of this subtask, showing its position in the whole task TaskID. For example, if task 1 includes 2 hops, then the two corresponding subtasks can be denoted as (1,0) and (1,1). Observing this representation method, we take the TaskIDs out of a subtask sequence to form an individual.

To expatiate the above encoding approach, an example shown in Fig. 2 is given. There are three tasks, i.e., transmitting a packet from node 0 to AP, another from node 1 to AP and the third one from 4 to AP. According to their routes, each task contains 4 subtasks, i.e., (0,0)(0,1)(0,2)(0,3), (1,0)(1,1)(1,2)(1,3), and (4,0)(4,1)(4,2) respectively. There is a random combination of above subtasks, such as,

$$(0,0)\,(1,0)\,(0,1)\,(4,0)\,(0,2)\,(4,1)\,(4,2)\,(0,3)\,(1,1)\,(1,2)\,(1,3)$$

then taking the TaskIDs out, the above sequence can be encoded as follow: 01040440111

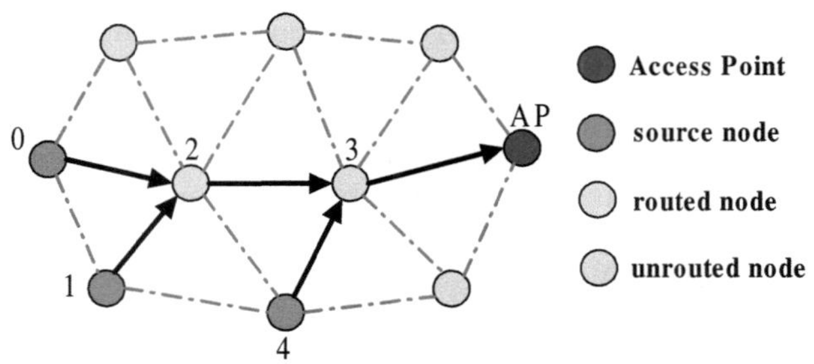

Fig. 2. Example of Network

To summarize the above process, the encoding rule is concluded as follows. There are N tasks and each task i includes M_i hops. An individual is a sequence composed of all the task ID numbers from 0 to $N-1$, where each task number i appears M_i times. Obviously, the length of the individual is $\sum_{i=0}^{N-1} M_i$. According to this rule, an individual can be generated at random.

The decoding method of an individual is denoted as follows. First, an individual is transformed to a sequence of subtasks. Then, from the first subtask to the last one, we assign slots in sequence at the same time, trying to assign those subtasks to one slot without rousing collisions. Take an individual from the above example network, i.e., 00001111444, the corresponding sequence of subtasks is (0,0)(0,1)(0,2)(0,3)(1,0)(1,1)(1,2)(1,3)(4,0)(4,1)(4,2). Then the slot allocation scheme is shown in Fig.3.

From the decoding scheme, it can be seen that an individual can finally be decoded to a slot scheduling, which allows parallel operations. The mapping between an individual and a TDMA scheduling solution is now completely established.

Slot No.	1	2	3	4	5	6	7	8	9
Subtask Sequence	(0,0)	(0,1)	(0,2)	(0 3) (1 0)	(1,1)	(1,2)	(1,3) (4,0)	(4,1)	(4,2)
Execution Node	0	2	3	AP [illegible]	2	3	AP 4	3	AP

Fig. 3. Slot allocation of the example

4 Simulation Results

In the simulation, to validate the viability of our proposed method, four networks with different numbers of nodes were deployed, randomly generating the network topology. The access point is located in the center of the area. The capacity of the channel is set to 500kbps and the packet lengths are 1kbits. Some parameters in energy aspect are as follows: similar to Ref. [2], the transition time between the sleep and active states is assumed to 470s. The power consumed in transmission and reception of a packet is set to 81 and 180mW, respectively. The power consumed in idle state is set as same as the reception state. In addition, we assume that the clock drift can be ignored by lengthening the slot time slightly.

To illustrate the effectiveness and performance of our PAPSO optimization algorithm, two other algorithms are used as comparisons: Max Degree First coloring algorithm (MDFCA, which is a common 2-distance coloring algorithm), and Node Based Scheduling Algorithm (NBSA) proposed by Ergen and Varaiya [6]. In the initialization of PAPSO, the swarm size M was set as 50 and the maximal generation was 200. And the task amount was that every sensor node generated a data packet and sent it to the access point. For the sake of fair comparison, all these algorithms run in energy-efficient mode, i.e. a node with no packet to send keeps its radio off during its allocated time slots.

Table 1 to 3 respectively illustrates the results of average time delay, average energy consumption and total time cost of three scheduling algorithms in the four networks. Among the three algorithms, MDFCA performs the worst on all the three indice. It is mainly because in MDFCA, many slots are allocated to those nodes which do not have the transmission task, thus wasting a lot of slots and decreasing the time performance of transmission. Moreover, in MDFCA, the node could only work on one slot in each coloring period, which makes the node have to switch its state in every single transmission. Consequently, graph coloring is not suitable to the data collecting sensor networks.

Although NBSA is still an algorithm based on coloring idea, its optimization performance for sensor networks is much better. This is because NBSA excludes such empty slots assigned to those nodes without packets, which can help the algorithm to save time and energy. However, both of the two algorithms lack the ability to adjust energy consumption of network.

The seven PAPSO solutions were continuous 7 members of the Pareto archive, which size is set as 7 to make data table succinct. According to Pareto optimality, the results of NBSA and PAPSO dominated the ones of MDFCA obviously

Table 1. Performance of average delay (slot)

No. of Nodes	25	49	121	169
MDFCA	16.67	44.06	136.13	198
NBSA	12.21	28.58	76.68	108.25
PAPSO(1)	9.08	24.38	70.32	100.75
PAPSO(2)	9.11	24.57	71.18	101.32
PAPSO(3)	9.53	24.77	71.92	101.92
PAPSO(4)	10.28	25.52	73.15	103.25
PAPSO(5)	11.82	26.65	74.82	105.38
PAPSO(6)	13.50	27.95	76.80	107.98
PAPSO(7)	14.58	29.35	78.82	110.91

Table 2. Performance of average energy consumption (mJ)

No. of Nodes	25	49	121	169
MDFCA	0.419	1.211	4.656	7.353
NBSA	0.378	1.057	4.186	6.624
PAPSO(1)	0.394	1.069	4.194	6.565
PAPSO(2)	0.387	1.049	4.138	6.520
PAPSO(3)	0.374	1.035	4.075	6.483
PAPSO(4)	0.365	1.023	4.030	6.450
PAPSO(5)	0.361	1.015	3.990	6.432
PAPSO(6)	0.357	1.009	3.982	6.426
PAPSO(7)	0.355	1.005	3.975	6.425

as table 1 and 2 showed. To illustrate the advantage of PAPSO over NBSA, we picked data in the 121-node network and compared the two algorithms as Fig. 4 showed. The solution of NBSA was dominated by solution (2)(3)(4)(5) of PAPSO. Furthermore, since during the maintenance of archive, we fixed two extreme cases as the border of archive and solutions are evenly distributed, so we can easily find a tradeoff by picking the middle one, i.e. PAPSO(4). , the performance of total time cost was not sacrificed. From table 3, we can see that the performance of total time cost of PAPSO was comparative to NBSA. Even when a tradeoff is met, the performance of total time cost was not sacrificed.

In a whole, the optimization effect of PAPSO outperforms the other two algorithms. It is mainly because: 1. Effective idea of solving the scheduling problem. As to the scheduling of TDMA time slots, we assigned time slots to subtasks sequentially, which avoids generating idle slots and thus ensures the good time performance. And time slots of subtasks can be flexibly adjusted, which leads to good working continuity of sensor nodes and decreases the energy consumption. 2. Good search algorithm. As an efficient evolutionary algorithm, PSO not only owns good search ability in the solution space but also is good at solving multi-objective optimization problem. Combined with Pareto optimality, PSO can search the solution space more thoroughly and find out a better tradeoff between energy consumption and time delay.

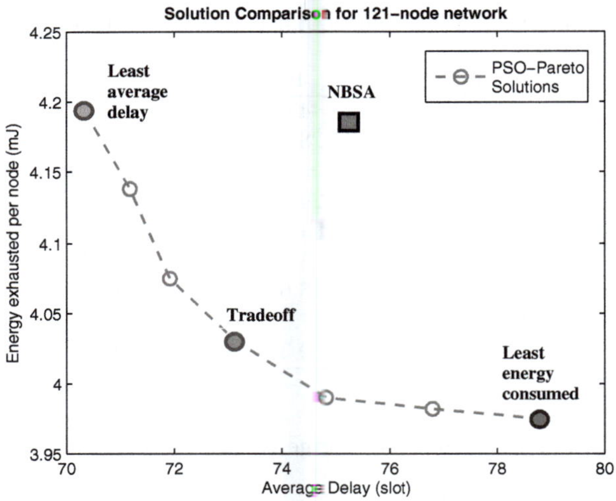

Fig. 4. Comparison of results of NBSA and PAPSO in 121-node network

Table 3. Performance of total time cost(slot)

No. of Nodes	25	49	121	169
MDFCA	46	123	397	602
NBSA	30	62	154	212
PAPSO(1)	28	59	149	205
PAPSO(2)	31	61	155	207
PAPSO(3)	29	59	151	210
PAPSO(4)	31	62	153	209
PAPSO(5)	31	61	153	213
PAPSO(6)	30	60	155	209
PAPSO(7)	32	61	156	215

5 Conclusions

In this paper, according to the characteristics of many-to-one data transmission in wireless sensor networks, we combined Pareto optimality with PSO and proposed an effective PAPSO optimization to solve the TDMA scheduling problem. And the optimization has the following advantages: 1. Under the framework of evolutionary search algorithm, it is easy to deal with multi-objective optimization problem and set up the model of multi-objective optimization; 2. We can make the best of the problem-solving ability of evolutionary search algorithm over NP problem; 3. The introduction of Pareto optimality makes the evaluation system of multi-objective PSO solutions more reasonable and effective, thus offering more choices of network performance to decision-makers.

In wireless sensor networks, multi-objective optimization problem widely exists and there is usually confliction among such objectives. Consequently, our proposed multi-objective optimization algorithm can serve as a good idea to such kind of problem.

References

1. Jolly, G., Younis, M.: An Energy-Efficient, Scalable and Collision-Free MAC layer Protocol for Wireless Sensor Networks. Wireless Communications and Mobile Computing (2005)
2. Ergen, S.C., Varaiya, P.: TDMA: Scheduling Algorithms for Sensor Networks, Technical Report, Department of Electrical Engineering and Computer Sciences U.C. Berkeley (July 2005)
3. Cui, S., et al.: Energy-Delay Tradeoffs for Data Collection in TDMA-based Sensor Networks. In: The 40th annual IEEE International Conference on Communications. Seoul, Korea (May 16-20, 2005)
4. Sridharan, A., Krishnamachari, B.: Max-Min Fair Collision-free Scheduling for Wireless Sensor Networks. In: Workshop on Multihop Wireless Networks (MWN 2004) (April 2004)
5. Gandham, S., Dawande, M., Prakash, R.: Link scheduling in sensor networks: Distributed edge coloring revisited. In: INFOCOM, pp. 2492–2501 (2005)
6. Ergen, S.C., Varaiya, P.: TDMA: scheduling algorithms for sensor networks, Technical Report, Department of Electrical Engineering and Computer Sciences University of California, Berkeley (July 2005)
7. Karp, B., Kung, H.T.: GPSR: Greedy perimeter stateless routing for wireless networks. In: Proc. ACM/IEEE MobiCom (August 2000)
8. Eberhart, R., Kennedy, J.: A new optimizer using particle swarm theory. In: Proceedings of the sixth international symposium on micro machine and human science, pp. 39–43 (1995)
9. Deb, K.: Evolutionary Algorithms for Multi-Criterion Optimization in Engineering Design [A]. In: Proceedings of Evolutionary Algorithms in Engineering and Computer Science (EUROGEN-99), pp. 135–161 (1999)
10. Coello Coello, C.A., Salazer Lechuga, M.: MOPSO: A Proposal for Multi Objective Particle Swarm Optimization. In: Congr. on Evolutionary Computation, Piscataway, New Jersey. vol. 2, pp. 1051–1056 (2002)

Energy-Aware Online Algorithm to Satisfy Sampling Rates with Guaranteed Probability for Sensor Applications*

Meikang Qiu[1] and Edwin H.-M. Sha[2]

[1] University of New Orleans,
New Orleans, LA 70148, USA
mqiu@uno.edu

[2] University of Texas at Dallas,
Richardson, TX 75083, USA
edsha@utdallas.edu

Abstract. Energy consumption is a major factor that limits the performance of sensor applications. Sensor nodes have varying sampling rates since they face continuously changing environments. In this paper, the sampling rate is modeled as a random variable, which is estimated over a finite time window. We presents an online algorithm to minimize the total energy consumption while satisfying sampling rate with guaranteed probability. An efficient algorithm, EOSP (*Energy-aware Online algorithm to satisfy Sampling rates with guaranteed Probability*), is proposed. Our approach can adapt the architecture accordingly to save energy. Experimental results demonstrate the effectiveness of our approach.

1 Introduction

Sensor networks are emerging as a main technology for many applications, such as national security and health care. In this context, a key problem that ought to be tackled is that of devising embedded software and hardware architectures that can effectively operate in continuously changing, hard-to-predict conditions [1].

In sensor network applications, such as moving objects tracking, tasks may not have fixed sampling rate. The sampling rate may vary, depending on the rate the object moves. At same time, the time window of sensor systems is fix. Hence, how to select the proper number of *processing engine* PEs and adopt different sampling rates becomes an important problem. The existing methods are not able to deal with the uncertainty of sampling rate. Therefore, either worst-case or average-case sampling rates for these tasks are usually assumed. Such assumptions, however, may not be applicable for the hard-to-predict conditions and may result in an inefficient task assignment. In this paper, we models the sampling rate in each time window as a random variable [2].

* This work is partially supported by TI University Program, NSF EIA-0103709, Texas ARP 009741-0028-2001, NSF CCR-0309461, NSF IIS-0513669, and Microsoft, USA.

R. Perrott et al. (Eds.): HPCC 2007, LNCS 4782, pp. 156–167, 2007.

In sensor applications, such as camera based sensors, same tasks can be processed under different voltage levels with different energy consumptions. DVS (Dynamic voltage scaling) is one of the most effective low power system design techniques. We want to select proper number of PEs and assign a proper voltage to each task of a sensor such that minimize the total energy consumption while satisfying the total sampling rate constraint with a guaranteed confidence probability.

This paper use adaptive approach for online customization of embedded architectures that function in non-stationary environments [1]. we design an algorithm, EOSP (*Energy-aware Online algorithm to satisfy Sampling rates with guaranteed Probability*), to select the proper number of PEs and assign different voltage levels to the PEs to minimize total energy consumption. The obtained voltages will affect the adaptation thresholds of control policies.

The experimental results show that EOSP achieves a significant reduction in total energy consumption on average. For example, with 4 voltages, compared with Method 1 (worse-case scenario without DVS), EOSP shows an average 33.8% reduction in total energy consumption while satisfying sampling rate constraint 360 with probability 0.80.

In the next section, we introduce necessary background, including basic definition and models. The algorithm is discussed in Section 3. We show our experimental results in Section 4. Related work and concluding remarks are provided in Section 5 and 6, respectively.

2 Basic Concepts and Models

In this section, we introduce some basic concepts and models which will be used in the later sections. First the sensor model is given. Next, two kinds of PE processing mode are introduced. Then we give the basic formula of DVS. Finally, we give the formal definition of VAP problem. An example is given throughout the whole section.

2.1 The Sensor Model

Sensor usually works in highly continue changing environments. For example, consider a camera based sensor, which is tracking a moving object, such as a person or vehicle [3,1]. The tracking granularity requirement demands one image sample per meter of distance traveled by the object, in order to have a trace of the object's trajectory accurate to within one meter distance of the object's actual location at all times. If the object is traveling at a speed of 20 m/s this sampling speed would translate to having 20 samples/s for the camera. We modeled the system dynamics with discrete events formulated over a fixed time window used to sample the future performance requirements of the system.

For the adaptive architecture, the working procedures are as follows. During look ahead for the next window, the incoming data is buffered. The controller uses the inputs from the sampling rate look ahead to update its control policy.

The controller makes changes to the pool of hardware with the updated policy after each window.

The time *window length* (WL) is a design parameter, and will have to be decided by the designer based on empirical data obtained from simulations of the particular application. In the tracking example the sampling rate may vary, depending on the rate the object moves. If sampling rate is high then the number of hardware resources turned on is larger in order to meet the tighter timing constraint. There is an upper limit to sampling rate based on the number and working mode of hardware processing engines (PEs) being made available for it by an architecture.

Problem Definition: For a given adaptive architecture, how to select the proper number of processing engine (PE) and assign voltage levels to the PEs for each time window of a sensor such that the total energy consumption can be minimized with a guaranteed confidence probability satisfying sampling rates.

2.2 Sampling Rate Estimator and PE Processing Method

The sampling rate is a random variable, which is decided by the moving rate of objects. We use a fixed time window to collect the data. The method we estimate the sampling rate for each window is based on the previous four average sampling rates. For example, if the average sampling rates are S_{n-1}, S_{n-2}, S_{n-3}, and S_{n-4}, for S in previous 4 time windows, then for current window, $S_n = S_{n-1} + 0.5 * (S_{n-1} - S_{n-2}) + 0.3(S_{n-2} - S_{n-3}) + 0.2(S_{n-3} - S_{n-4})$. We use several similar functions to estimate S_n. The estimated S_n is also a random variable. For example, the S shown in Figure 1 (a) is a random variable. "P1" represent the corresponding probability of every sampling rate value.

S	P1
80	0.10
60	0.10
50	0.20
40	0.20
30	0.40

(a)

S	P2
80	1.00
60	0.90
50	0.80
40	0.60
30	0.40

(b)

Fig. 1. (a) Distribution function of sampling rate. (b) Cumulative distribution function of sampling rate.

When the sampling rate is high, we need use more PEs to process the collected data. The energy consumption of each PE includes two parts.

$$E_{total} = E_{base} + E_{process} \tag{1}$$

There are two cases about PE processing modes.

1. Case 1. Each PE works under same voltage and frequency. There is no processing time and energy difference.
2. Case 2. $E_{process}$ can be tuned with *dynamic voltage scaling* (DVS). There are several different voltage levels for each PE to work on with.

In case 1, since S is a random variable, we want to find a configure of PE such that the energy is minimized with a guaranteed probability. For example, the distribution of random variable S is shown in Figure 1 (a). We first compute the cumulative distribution function (CDF) of S, which is shown in Figure 1 (b). "P2" represent the cumulative probability of each sampling rate value. Assume the processing rate R of each PE is 20, and the energy consumption E of each PE is 50. Then we need 4 PEs with 100% confidence while satisfying timing latency. The minimum total energy consumption is 200. But, this is the worst-case scenario. We can minimize energy while satisfying timing constraint with 90% confidence by using only 3 PEs. In this scenario, the total energy consumption is 150. We will discuss case 2 after introducing DVS.

2.3 Dynamic Voltage Scaling

DVS (Dynamic voltage scaling) is a technique that varies system's operating voltages and clock frequencies based on the computation load to provide desired performance with the minimum energy consumption. It has been demonstrated as one of the most effective low power system design techniques and has been supported by many modern microprocessors. Examples include Transmeta's Crusoe, AMD's K-6, Intel's XScale and Pentium III and IV, and some DSPs developed in Bell Labs [4].

Dynamic power, which is the dominant source of power dissipation in CMOS circuit, is proportional to $N \times C \times V_{dd}^2$, where N represent the number of computation cycles, C is the effective switched capacitance, and V_{dd} is the supply voltage [5,6,7]. Reducing the supply voltage can result in substantial power and energy saving. Roughly speaking, system's power dissipation is halved if we reduce V_{dd} by 30% without changing any other system parameters. However, this saving comes at the cost of reduced throughput, slower system clock frequency, or higher cycle period time (gate delay). The cycle period time T_c is proportional to $\frac{V_{dd}}{(V_{dd}-V_{th})^\alpha}$, where V_{th} is the threshold voltage and $\alpha \in (1.0, 2.0]$ is a technology dependent constant.

Let t represent the computation time and E represent energy, they are calculated as follows:

$$T_c = \frac{k \times V_{dd}}{(V_{dd} - V_{th})^\alpha} \tag{2}$$

$$t = N \times T_c = N \times \frac{k \times V_{dd}}{(V_{dd} - V_{th})^\alpha} \tag{3}$$

$$E = N \times C \times V_{dd}^2 \tag{4}$$

In Equation (2), k is a device related parameter. From Equations (3) and (4), we can see that the lower voltage will prolong the execution time of a node but reduces its energy consumption.

Low power design is of particular interest for the soft real time multimedia systems and we assume that each PE has multiple voltages available on the chip such that the system can switch from one level to another.

2.4 VAP Problem

For multi-voltage systems, assume there are maximum M different voltages in a voltage set $V = \langle V_1, V_2, \cdots, V_M \rangle$. An assignment is to assign a voltage to each PE that has been selected. Define an *assignment* A to be a function from domain U to range V, where U is PE set and V is the voltage set.

We define F to be the *cumulative distribution function* of the random variable S (abbreviated as *CDF*), where $F(t) = P(S < t)$. When S is a discrete random variable, the CDF $F(t)$ is the sum of all the probabilities associating with the computation times that are less than or equal to t. If S is a continuous random variable, then it has a *probability density function* (PDF). If assume the pdf is f, then $F(t) = \int_0^t f(s)ds$. Function F is nondecreasing, and $F(-\infty) = 0$, $F(\infty) = 1$.

We define the *voltage assignment with guaranteed probability (VAP)* problem as follows: Given M different voltage levels: V_1,V_2,$\cdots$,V_M, a sampling rate constraint S and a confidence probability P, find the proper number of PEs and an assignment of voltage level for each of the selected PEs to gives the minimum total energy consumption E with confidence probability P under sampling rate constraint S.

	V1	V2	V3
R	10	20	30
E	20	45	100

Fig. 2. The processing speed and energy consumption of each PE under different voltage levels

We will use dynamic programming to solve the VAP problem. Our algorithm *Volt_Ass* is shown in the next section. Here we give an example first. Figure 2 shows the processing rate R and energy consumption E of each PE under different voltage levels. The base energy E_{base} is 12. Suppose the confidence probability we need is 0.90. Then we check the table in Figure 1 (b), and find that the sampling rate S will not higher than a value, i.e., 60 samples/s. That is, $S \leq 60$. We denote the corresponding value to be L. Let N to be the number of PEs needed, we list the constraint equations to make it clear.

$$R_1 + R_2 + \cdots + R_N \geq L \quad (5)$$

We want to get

$$Min(E_1 + E_2 + \cdots + E_N) \tag{6}$$

After running our dynamic programming, we the number of PEs and the voltage assignment for the PEs. We need 3 PEs, the voltage level assignment of PEs is shown in Ass_1 of Table 1. The total energy consumption is $135 + 12 * 3 = 171$. For the case with no DVS, i.e., we only use the maximum processing speed, we will choose using only 2 PEs. The total energy is $200 + 12 * 2 = 224$. The assignment is shown in Ass_1 of Table 1. There is 23.7% difference. Hence, the solution is: use 3 PEs and let them all work under V_2, then the total energy consumption is 171, while satisfying the sampling rate with 0.90 confidence probability.

Table 1. The assignments with energy consumption 171 and 224 with sampling rate constraint 60

	Num.	**Speed**	**Volt.**	$E_{proc.}$	E_{base}
	1	20	V_2	45	12
Ass_1	2	20	V_2	45	12
	3	20	V_2	45	12
	Total	60		135	36
Total Energy = 171					
	N	**Speed**	**Volt.**	$E_{proc.}$	E_{base}
Ass_2	1	30	V_3	100	12
	2	30	V_3	100	12
	Total	60		200	24
Total Energy = 224					

3 The Algorithms

In this section, we will propose our algorithms to solve the energy-saving problem of sensor applications. The basic idea is to obtain the minimum total energy consumption by selecting proper number of PEs and doing voltage assignment. We consider two cases: Case 1: there is no voltage change for each PE. This case is simple, and we use *EOSP_Simple* to solve it. We will focus on case 2. Case 2: PE may work under different voltages. We proposed an algorithm, EOSP (*Energy-aware Online algorithm to satisfy Sampling rates with guaranteed Probability*). In this algorithm, we use *Volt_Ass* sub-algorithm to give the best PE number selection and voltage assignment for each selected PE.

3.1 The *EOSP_Simple* Algorithm for Case 1

The EOSP_Simple algorithm for case 1 is shown in Algorithm 3.2. In EOSP_Simple algorithm, we use the adaptive model [1] to solve energy saving problem for camera based sensors. The adaptive approach includes three steps: First, look ahead on performance parameters, such as image sampling rate and system

Algorithm 3.1. EOSP_Simple Algorithm for Case 1

Require: Several PEs, the sampling rate S, the guaranteed probability P.

Ensure: The number of PEs N with minimum total energy consumption E_{min} while satisfying S with P.

1: Look ahead on performance parameters: $S \leftarrow$ sampling rate.
2: Use fixed time window to collect data that will be processed.
3: Estimate the sample rate with different estimator, get the random variable S.
4: Build the cumulative distribution function (CDF) based on the distribution function of S.
5: $L \leftarrow$ the corresponding sampling rate value that has $Prob.(S \leq L) \geq P$ by looking CDF of S.
6: $N \leftarrow$ the number of PEs with $\sum_{i=1,\cdots,N} R_i \geq L$.
7: $E_{min} \leftarrow \sum_{i=1,\cdots,N} E_i$.
8: Output results: N and E_{min}.
9: Use online architectural adaptation to reduce energy consumption while satisfying timing constraints with guaranteed probability.

latency, by buffering input data coming in a given time window. Second, dynamic processing requirements prediction using a high efficient estimator activated at the end of every window period. Third, use an on-line architecture adaptation control policy. Since our design is for a non-stationary environments, the control policy varies with the environment but is stationary within a time window.

3.2 The EOSP Algorithm for Case 2

3.3 Volt_Ass Algorithm

To solve the voltage assignment problem, we use dynamic programming method. In our algorithm, table $D_{i,j}$ will be built. Each entry of table $D_{i,j}$ will store $E_{i,j}$, which is the minimum energy for i number of PEs with sampling rate j.

Lemma 1. *Given $E^1_{i,j}$ and $E^2_{i,j}$ with same sampling rate j, then $Min(E^1_{i,j}, E^2_{i,j})$ is selected to be kept.*

In every step of our algorithm, one more PE will be included for consideration. The (R, E) pair of each PE is stored in a local table B, which is shown in Figure 2. We first sort (R, E) pairs in B according to R in an ascending order, and represent them in the form: $(R_1, E_{R_1}), (R_2, E_{R_2}), \cdots, (R_M, E_{R_M})$. Hence, E_j is the energy consumption only for one PE with sampling rate j, The algorithm to compute $D_{i,j}$ are shown in Algorithm 3.3.

Theorem 1. *The energy consumption in $D_{I,L}$ obtained by algorithm* Volt_Ass *is the minimum total energy consumption with sampling rate L.*

Proof: By induction. **Basic Step:** When $i = 1$, there is only one PE and $D_{1,j} = B_j$. Thus, when $i = 1$, Theorem 1 is true. **Induction Step:** We need to show that for $i \geq 1$, if $D_{i,j}$ is the minimum total energy consumption of i PEs, then

Algorithm 3.2. The EOSP Algorithm

Require: M different voltages, Several PEs, the sampling rate S, the guaranteed probability P.
Ensure: The number of PEs N with minimum total energy consumption E_{min} while satisfying S with P.

1: Look ahead on performance parameters: $S \leftarrow$ sampling rate.
2: Use fixed time window to collect data that will be processed.
3: Estimate the sample rate with different estimator, get the random variable S.
4: Build the cumulative distribution function (CDF) based on the distribution function of S.
5: $L \leftarrow$ the corresponding sampling rate value that has $Prob.(S \leq L) \geq P$ by looking CDF of S.
6: Use algorithm *Volt_Ass* to obtain PEs assignment A with minimized E for the schedule graph.
7: $N \leftarrow$ the number of PEs with $\sum_{i=1,\cdots,N} R_i \geq L$.
8: $E_{min} \leftarrow \sum_{i=1,\cdots,N} E_i$.
9: Output results: N, A, and E_{min}.
10: Use online architectural adaptation to reduce energy consumption while satisfying timing constraints with guaranteed probability.

Algorithm 3.3. Volt_Ass algorithm to compute $D_{i,j}$

Require: Sampling rate L, M different voltage levels.
Ensure: $D_{i,L}$

1: Build a local table B according to M different voltage levels;
2: $R \leftarrow$ processing speed, $E \leftarrow$ energy;
3: Sort (R, E) pairs in B according to R in an ascending order, and represent them in the form: $(R_1, E_{R_1}), (R_2, E_{R_2}), \cdots, (R_M, E_{R_M})$;
4: $B_k \leftarrow E_k$ in each pair of B;
5: Start from the first PE, $D_{1,j} \leftarrow B_k$;
6: $I \leftarrow L/R_1$;
7: **while** $i \leq I$, **do**
8:     **for all** sampling rate j, **do**
9:         Compute the entry $D_{i,j}$ as follows:
10:         **for all** k in B_k, $j > k$ **do**
11:             $D_{i,j} = D_{i-1,j-k} + B_k$;
12:             Insert $D_{i,j-1}$ to $D_{i,j}$ and remove redundant energy value using Lemma 1;
13:         **end for**
14:     **end for**
15: **end while**
16: The energy in $D_{I,L}$ is the minimum total energy consumption with sampling rate L and the assignment can be obtained by tracing how to reach $D_{I,L}$;
17: Output $D_{I,L}$;

$D_{i+1,j}$ is the minimum total energy consumption of $i+1$ PEs. In the algorithm, since $j = k + (j-k)$ for each k in B_k, we try all the possibilities to obtain j. Then we use add the energy consumptions of two tables together. Finally, we using Lemma 1 to cancel the conflict energy values. The new energy consumption obtained in table $D_{i,j}$ is the energy consumption of i PEs at sampling rate k plus the energy consumption in $D_{i-1,j-k}$. Since we have used Lemma 1 to cancel redundant energy values, the energy consumption in $D_{I,L}$ is the minimum total energy consumption for sampling rate L. Thus, Theorem 1 is true.

From Theorem 1, we know $D_{I,L}$ records the minimum total energy consumption of the whole path within the timing constraint L. We can record the corresponding voltage assignment of each PE when computing the minimum system energy consumption in the algorithm *Volt_Ass*. Using these information, we can get an optimal assignment by tracing how to reach $D_{I,L}$.

It takes $O(M^2)$ to compute one value of $D_{i,j}$, where M is the maximum number of voltage levels. Thus, the complexity of the algorithm *Volt_Ass* is $O(S * M^2)$, where S is the given sampling rate constraint. Usually, S is upper bounded by a constant. In this case, *Volt_Ass* is polynomial.

4 Experiments

In this section, we conduct experiments with the EOSP algorithm on a set of benchmarks including 8-Stage Lattice filter,4-Stage Lattice filter, FFT1, Laplace, Karp10, Almu, Differential Equation Solver, RLS-Laguerre Lattice filter, Elliptic filter, and Voltera filter. The distribution of sampling rates is Gaussian. For each benchmark, we conduct a set of the experiments based on different configurations. K different voltages, $V_1, \cdots, V_K$, are used in the system, in which a voltage with level V_K is the quickest with the highest energy consumption and a voltage with level V_1 is the slowest with the lowest energy consumption.

We conduct the experiments using three methods. Method 1: Use the hard real-time and we need to guarantee to satisfy the varying sampling rates. Method

Table 2. The comparison of total energy consumption for Method 1, Method 2 and EOSP while satisfying sampling rate $S = 360$ for various benchmarks

4 voltages, S = 360										
Benchmarks	Num.	M1	M2	EOSP 0.8			M2	EOSP 0.9		
		E	E(0.8)	E(0.8)	% M1	% M2	E(0.9)	E(0.9)	% M1	% M2
8-stage Lattice	42	2655	2154	176[illegible]	33.6%	18.2%	1921	1467	44.8%	23.6%
4-stage Lattice	26	1576	1264	102[illegible]	35.1%	19.1%	1163	853	45.9%	26.7%
FFT1	28	1644	1312	107[illegible]	34.8%	18.3%	1228	937	43.0%	23.7%
Laplace	16	1032	852	689	32.6%	19.1%	736	531	48.1%	27.9%
Karp10	21	1305	1054	86[illegible]	33.7%	17.9%	942	712	45.5%	24.4%
Almu	17	1053	827	68[illegible]	35.2%	17.5%	762	583	44.6%	23.5%
Diff. Equ. Solver	11	767	624	50[illegible]	33.7%	18.6%	567	425	44.6%	25.0%
RLS-laguerre	19	1255	1062	84[illegible]	32.7%	20.4%	897	682	45.6%	24.0%
Elliptic	34	2384	1935	158[illegible]	33.4%	17.9%	1802	1345	43.6%	25.4%
Voltera	27	1668	1358	110[illegible]	33.6%	18.4%	1147	892	46.5%	22.2%
Average Reduction (%)					33.8%	18.5%	–	–	45.2%	24.6%

2: Use soft real-time and model the sampling rates as random variables. Method 3: Combine soft real-time and DVS together and use EOSP algorithm. We compare the results from EOSP algorithm with those from Method 1 and Method 2. The experiments are performed on a Dell PC with a P4 2.1 G processor and 512 MB memory running Red Hat Linux 9.0. In the experiments, the running time of EOSP on each benchmark is less than one minute.

The experimental results for Method 1 ("$M1$"), Method 2 ("$M2$"), and our EOSP algorithm with 4 voltages, are shown in Table 2. Column "Num." represents the number of nodes of each filter benchmark. Column "E" represents the minimum total energy consumption (μJ) obtained from the three different methods. Labels "E(0.8)" and "E(0.9)" represent the minimum total energy consumption when the guaranteed probability is 0.8 and 0.9, respectively. Column "% M1" and "% M2" under "EOSP" represents the percentage of reduction in total energy consumption, compared to Method 1 and Method 2, respectively. The average reduction is shown in the last row of the table.

The results show that our algorithm EOSP can significantly improve the performance of multi-voltage sensor nodes. For example, with 4 voltages, compared with Method 1, EOSP shows an average 33.8% reduction in total energy consumption while satisfying sampling rate constraint 360 with probability 0.80. The experimental results show that when the number voltages increases, the percentage of reduction on total energy increases correspondingly.

Through the experimental results from Table 2, we found that Method 1 doesn't explore the larger solution space for total energy consumption with soft real-time. Our EOSP algorithm combined both soft real-time and DVS, and can significantly reduce total energy consumption while satisfying sampling rates with guaranteed probability. It is efficient and provides overview of all possible variations of minimum energy consumptions comparing with the worst-case scenario generated by Method 1.

5 Related Work

Dynamic voltage and frequency scaling. Many researchers have studied on DVS [8,5,7,9,10,11]. Semeraro et al. designed a multiple clock domain (MCD) system to support fine-grained dynamic voltage scaling within a processor [12]. Yao et al. [13,14] and Ishihara et al. [6] studied the optimal schedule for DVS processors in the context of energy-efficient task scheduling. Both showed that it is most energy efficient to use the lowest frequency that allows an execution to finish before a given deadline. Ishihara et al. also showed that when only discrete frequencies are allowed, the best schedule is to alternate between at most two frequencies.

The International Technology Roadmap for Semiconductors [15] predicts that the future system will feature multiple supply voltages (V_{dd}) and multiple threshold voltages (V_{th}) on the same chip This enables the DVS, which varies the supply voltage according to workload at run time [13, 6]. The highest energy efficiency is achieved when voltage can be varied arbitrarily [16]. However, physical constraints of CMOS circuit limit the applicability of having voltage varying continuously. Instead, it is more practical to make multiple discrete voltages simultaneously available for the system [17].

Sensor networks. Sensor networks have wide applications in areas such as health care, military, environmental monitoring, infrastructure security, manufacturing automation, collecting information in disaster prone areas and surveillance applications [18]. In fact, the vision is that sensor networks will offer ubiquitous interfacing between the physical environment and centralized databases and computing facilities [19]. Efficient interfacing has to be provided over long periods of time and for a variety of environment conditions, like moving objects, temperature, weather, available energy resources and so on.

In many sensor networks, the DSP processor consumes a significant amount of power, memory, buffer size, and time in highly computation-intensive applications. However, sensor node can only be equipped with a limited power source (≤ 0.5 Ah, 1.2 V) [20]. In some application scenarios, replenishment of power resources might be impossible. Therefore, power consumption has become a major hurdle in design of next generation portable, scalable, and sophisticated sensor networks. In computation-intensive applications, an efficient scheduling scheme can help reduce the power consumption while still satisfying the performance constraints. This paper focuses on reducing the total energy of sensor applications on architectures with multiple PEs and multiple voltage levels.

6 Conclusion

In this paper, we studied the proper PE number selection and voltage assignment problem that minimizes the total energy while satisfying sampling rates with guaranteed probability for sensor applications. We proposed a high efficient algorithm, EOSP (*Energy-aware Online algorithm to satisfy Sampling rates with guaranteed Probability*), in this paper. Our algorithm can give a much larger solution space of total energy minimization to be used for online adaptation control. A wide range of benchmarks has been tested on the experiments and the experimental results showed that our algorithm significantly improved the energy-saving of applications on computation-intensive sensor nodes.

References

1. Kallakuri, S., Doboli, A.: Energy conscious online architecture adaptation for varying latency constraints in sensor network applications. CODES+ISSS 2005, Jesey City, New Jersey, pp. 148–154 (September 2005)
2. Tongsima, S., Sha, E., Chantrapornchai, C., Surma, D., Passos, N.: Probabilistic Loop Scheduling for Applications with Uncertain Execution Time. IEEE Trans. on Computers 49, 65–80 (2000)
3. Weng, Y., Doboli, A.: Smart sensor architecture customized for image processing applications. IEEE Real-Time and Embedded Technology and Embedded Applications, 336–403 (2004)

4. Hua, S., Qu, G., Bhattacharyya, S.S.: Exploring the Probabilistic Design Space of Multimedia Systems. In: IEEE International Workshop on Rapid System Prototyping, pp. 233–240. IEEE Computer Society Press, Los Alamitos (2003)
5. Zhang, Y., Hu, X., Chen, D.Z.: Task Scheduling and Voltage Selection for Energy Minimization. DAC 40, 183–188 (2002)
6. Ishihara, T., Yasuura, H.: Voltage scheduling problem for dynamically variable voltage processor. In: ISLPED, pp. 197–202 (1998)
7. Shin, D., Kim, J., Lee, S.: Low-Energy Intra-Task Voltage Scheduling Using Static Timing Analysis. In: DAC, pp. 438–443 (2001)
8. Stan, M.R., Burleson, W.P.: Bus-Invert Coding for Low-Power I/O. IEEE Trans. on VLSI Syst. 3(1), 49–58 (1995)
9. Saputra, H., Kandemir, M., Vijaykrishnan, N., Irwin, M.J., Hu, J.S., Hsu, C.-H., Kremer, U.: Energy-conscious compilation based on voltage scaling. In: LCTES 2002 (June 2002)
10. Sakurai, T., Newton, A.R.: Alpha-power law MOSFET model and its application to CMOS inverter delay and other formulas. IEEE J. Solid-State Circuits SC-25(2), 584–589 (1990)
11. Chandrakasan, A., Sheng, S., Brodersen, R.: Low-Power CMOS Digital Design. IEEE Journal of Solid-State Circuits 27(4), 473–484 (1992)
12. Semeraro, G., Albonesi, D., Dropsho, S., Magklis, G., Dwarkadas, S., Scott, M.: Dynamic Frequency and Voltage Control for a Multiple Clock Domain Microarchitecture. In: 35th Intl. Symp. on Microarchitecture (November 2002)
13. Yao, F., Demers, A., Shenker, S.: A scheduling model for reduced cpu energy. In: 36th symposium on Foundations of Computer Science (FOCS), Milwankee, Wisconsin, pp. 374–382 (October 1995)
14. Li, M., Yao, F.: An efficient Algorithm for computing optimal discrete voltage schedules. SIAM J. Comput. 35(3), 658–671 (2005)
15. ITRS: International Technology Roadmap for Semiconductors. International SEMATECH, Austin, TX (2000), `http://public.itrs.net/`
16. Burd, T.B., Pering, T., Stratakos, A., Brodersen, R.: A dynamic voltage scaled microprocessor system. IEEE J. Solid-State Circuits 35(11), 1571–1580 (2000)
17. Intel: The Intel Xscale Microarchitecture. Technical Summary (2000)
18. Im, C., Kim, H., Ha, S.: Dynamic Voltage Scheduling Technique for Low-Power Multimedia Applications Using Buffers. In: Proc. of ISLPED (2001)
19. Rahimi, M., Pon, R., Kaiser, W., Sukhatme, G., Estrin, D., Srivastava, M.: Adaptive sampling for environmental robots. In: International Conference on Robotics and Automation (2004)
20. Berman, P., Calinescu, G., Shah, C., Zelikovsly, A.: Efficient energy management in sensor networks. Ad Hoc and Sensor Networks (2005)

A Low-Power Globally Synchronous Locally Asynchronous FFT Processor

Yong Li, Zhiying Wang, Jian Ruan, and Kui Dai

National University of Defense Technology, School of Computer,
Changsha, Hunan 410073, P.R. China
liyong.nudt@163.com, {zywang,ruanjian,daikui}@nudt.edu.cn

Abstract. Low-power design became crucial with the widespread use of the embedded systems, where a small battery has to last for a long period. The embedded processors need to efficient in order to achieve real-time requirements with low power consumption for specific algorithms. Transport Triggered Architecture (TTA) offers a cost-effective trade-off between the size and performance of ASICs and the programmability of general-purpose processors. The main advantages of TTA are its simplicity and flexibility. In TTA processors, the special function units (SFUs) can be utilized to increase performance or reduce power dissipation. This paper presents a low-power globally synchronous locally asynchronous TTA processor using both asynchronous function units and synchronous function units. We solve the problem that use asynchronous circuits in TTA framework, which is a synchronous design environment. This processor is customized for a 1024-point FFT application. Compared to other reported implementations with reasonable performance. our design shows a significant improvement in energy-efficiency.

1 Introduction

In recent years, special-purpose embedded systems have become one very important area of the processor market. Digital signal processor (DSP) offer flexibility and low development costs, but it has limited performance and typically high power dissipation. Field programmable gate arrays (FPGA) combine the flexibility and speed of application specific integrated circuit (ASIC), but it cannot compete with the energy efficiency of ASIC implementations. For a specific application, TTA can provide both flexibility and configurability during the Application Specific Instruction Processor (ASIP) design process. In TTA processor, special function units can be utilized to increase performance or reduce power dissipation.

Fast Fourier Transform (FFT) is one of the most important tools in the field of digital signal processing. When operating in embedded environment, the devices are usually sensitive to power dissipation. So the energy-efficient FFT implementation is needed. There are many FFT implementations and low power architectures were described in paper [1], [2], [3], [4] and [5] et al. They all emphasized the importance of the low power consumption in embedded FFT applications. In CMOS circuits, power dissipation is proportional to the square of the supply voltage [6]. Reducing the supply voltage can achieve a good energy-efficiency, but this will result in circuit performance [7].

R. Perrott et al. (Eds.): HPCC 2007, LNCS 4782, pp. 168–179, 2007.

There is a problem for designers to solve that how to make systems have low power dissipation without performance loss.

Since the early days, asynchronous circuits have been used in many interesting applications. There are many successful examples of asynchronous processors, which were described in [8], [9], [10], [11] and [12]. In these papers, asynchronous circuits have advantages of low power dissipation and high performance. The asynchronous circuits are very suitable for systems that are sensitive to power consumption and performance. In order to take advantage of flexibility and configurability of TTA and low power consumption of asynchronous circuits, we attempt to use asynchronous circuits in TTA. In this paper, we solve the problem that use asynchronous circuits in TTA that is synchronous design environment. This novel architecture may be viewed as a globally synchronous locally asynchronous (GSLA) implementation. Based on this architecture, a low power TTA processor is customized for a 1024-point FFT application. Asynchronous function units are utilized to obtain the low power dissipation. The performance and power dissipation are compared against the synchronous version. The results showed that GSLA implementation based on TTA can can offer a low power solution for some application specific embedded systems.

This paper is organized as follows. Section 2 briefly describes the radix-4 FFT algorithm. Section 3 describes the Transport Trigger Architecture. Section 4 describes the implementation of the GSLA TTA processor. Next, the power and performance analysis results are presented. The last section gives the conclusion.

2 Radix-4 FFT Algorithm

There are several FFT algorithms and, the most popular FFT algorithms are the Cooley-Turkey algorithms [13]. It has been shown that the decimation-in-time (DIT) algorithms provide better signal-to-noise-ratio than decimation-in-frequency algorithms when finite word length is used. In this paper, a radix-4 DIT FFT approach has been used since it offers lower arithmetic complexity than radix-2 algorithms.

The N-point DFT of a finite duration sequence $x(n)$ is defined as

$$X(k) = \sum_{n=0}^{N-1} x(n) W_N^{nk} \tag{1}$$

Where $W_N = e^{-j(2\pi/N)}$, $k = 1, 2, ..., N-1$, known as the twiddle factor. The direct implementation of the DFT have a complexity of $O(N^2)$. Using the FFT, the complexity can be reduced to $O(N \log_2(N))$.

For example, one may formulate the radix-4 representation of the FFT in the following manner:

Let, $N = 4T$, $k = s + Tt$ and $n = l + 4m$, where $s, l \in \{0, 1, 2, 3\}$ and $m, t \in \{0, 1, ..., T-1\}$. Applying these values in equation (1) and simplifying results in,

$$X(s+Tt) = \sum_{l=0}^{3} W_4^{lt} \left[W_{4T}^{sl} \sum_{m=0}^{T-1} x(l+4m) W_T^{sm} \right] \tag{2}$$

For $N = 1024$ and $T = 256$, the 1024-point FFT can be expressed from equation (2) as

$$X(s + 256t) = \sum_{l=0}^{3} W_4^{lt} \left[W_{1024}^{sl} \sum_{m=0}^{255} x(l + 4m) W_{256}^{sm} \right] \quad (3)$$

Equation (3) is obvious and suggests that the 1024-point DFT can be computed by first computing an 4-point DFT on the appropriate data slot, then multiplying them by 765 non-trivial complex twiddle factors and computing the 4-point DFT on the resultant data with appropriate data reordering.

Because the architecture is determined by the characteristics of the application set, the first step of the architecture design is to analyze the application [14]. By the analysis of FFT application, we find that the major operations of FFT are add, multiply, et al. According to the type of the major operations, designer can quickly decide what function unit to implement; similarly, the amount of the function units is decided to the proportion of the equivalent operations. Furthermore, we expect to customize special function unit to complete these operations in order to achieve a good power-efficiency.

3 Transport Triggered Architecture

As compared to conventional processor architectures, one of the main features of TTA [15] processor is that all the actions are actually side-effects of transporting data from one place to another. Thus, the processor has only one instruction *move*, which moves data. The programming is oriented on communications between FUs and, therefor, the interconnection network is visible to the software level, when developing TTA applications. A TTA processor consists of a set of function units and register files are connected to an interconnection network, which connects the input and output ports. The architecture is flexible and new function units (FUs), buses, and registers can be added without any restrictions. Naturally, the approach is well suited for application-specific processors. In addition, application specific support is provided by implementation of user-defined function units customized for a given application. The advantages of TTA processors are, also, short cycle time, and fast and application specific processor design. The MOVE software toolset [16] enables an exhaustive design-space exploration.

The entire hardware-software co-design flow is presented on Fig. 1. We modify the MOVE tools in order to make it possible that the special asynchronous function units library can be also included in MOVE design flow. The applications can be described by high level languages (HLLs), such as the C/C++ programming language. In our case, hardware design starts with a C language description of the FFT algorithm. By using modified MOVE tools and the library of designed special asynchronous function units we are able to generate description of power efficient processors. After that, the processor description file can be converted into a hardware description language (HDL) representation of the processor core by using MOVEGen tool. Automatically generated HDL code for the processor core together with HDL predesigned components (instruction dan data memories and other peripherals) can be used by the ASIC synthesis tools. As mentioned, HDL code of our special asynchronous function units cannot be directly

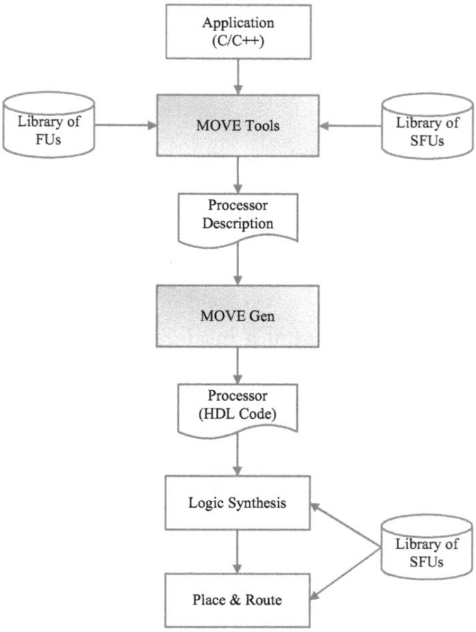

Fig. 1. Design flow with specifical function units form high level language code to hardware implementation

used by the synchronous synthesis tools. They should be synthesized and implemented according to the asynchronous circuits design flow. As predesigned components, the special asynchronous units can be used by IC design platform in order to obtain layout of the target processor.

4 Implementation of FFT Processor

4.1 Special Function Units

We propose a 32-bit wide ASIP architecture (buses and ports of FUs are 32 bit wide). The design of ASIP architecture is optimized by implementing several special function units. The design of the SFUs is started with the most frequent operation, i.e., ADD, MUL. Asynchronous adder unit (AADD) is one of the designed SFUs. Another SFU, the asynchronous multiplier unit for arithmetic operations with sub-word parallelism is also implemented (sub-word multiply operation between two 32-bit numbers represents two parallel multiply operations on packed 16-bit operands, for example). The implementation of these SFUs help to achieve high energy-efficiency. By implementing all of these SFUs we are able to significantly reduce the power dissipation, and to optimize the overall architecture design.

Asynchronous Adder Unit. In this work, we implemented a 32-bit asynchronous adder. Pipelining is a standard way of decomposing an operation into concurrently

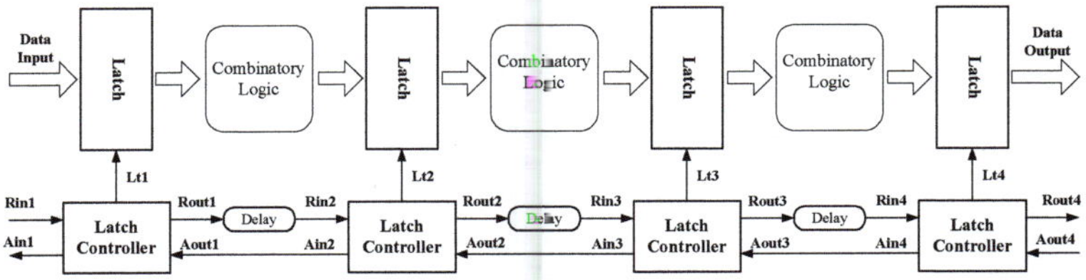

Fig. 2. The pipeline architecture of asynchronous adder unit that has four pipeline stages

operating stages to increase throughput at a moderate increase in area. The adder employs four pipeline stages, and the architecture can be illustrated in Fig. 2.

As shown in Fig. 2, the adder unit is composed of tradition combinatorial circuit and control circuit. The tradition combinatorial circuit is the same as synchronous circuit. In synchronous circuit, the communication of data between pipeline stages is regulated by the global clock. It is assumed that each stage takes no longer than the period of the clock and data is transferred between consecutive stages simultaneously. In asynchronous pipeline, the communication of data between the stages is regulated by local communication between stages, therefor, the global clock is replaced by a local communication protocol. The communication protocol employed in pipeline can be either a 2-phase or 4-phase signaling. The control circuit, called handshake circuit, realizes the communication protocol locally inside a pipeline between adjacent stage. When one stage has data which it would like to send to a neighboring stage, it sends a request ($Rout$) to that stage. If that stage can accept new data, it accepts the new data and returns an acknowledgment (Ain). The pipeline designed with request and acknowledgement signal that governs the transfer of data between stages.

This pipeline circuit is similar to Sutherland's Micropipeline [17], except that it uses latches and relies on a simple set of timing constraints for correct operation. Between signal $Rout$ and Rin, the matching delay elements provides a constant delay timing constraint that matches the worst case latency of combinatory logic in each stage. Latch controllers can generate the clock, $Lt1$, $Lt2$, $Lt3$, $Lt4$, to control the operation of pipeline circuit.

Based on our asynchronous design flow in [18], it is very simple to design an asynchronous function unit quickly. We employe ripple-carry adder, which performance is limited but this implementation can satisfy the requirement and save layout area. The combinatorial circuit can be described by VHDL/Verilog and synthesized by Synopsys Design Compiler. The control circuit can be described as signal transition graphs (STG) [19] and synthesized by Petrify [20]. Then the Verilog description that comes from Petrify also can be further synthesized by Design Compiler. The logic gates of control circuit that is synthesized by Petrify and implemented by C-element [21] can be illustrated in Fig. 3. The control circuit will increase extra cost in terms of area.

In addition to the combinatorial circuit itself, the delay element represents a design challenge: to a first order the delay element will track delay variations that are due to the fabrication process spread as well as variations in temperature and supply voltage. On the other hand, wire delays can be significant and they are often beyond the designer's control. So some design policy for matched delays is obviously needed.

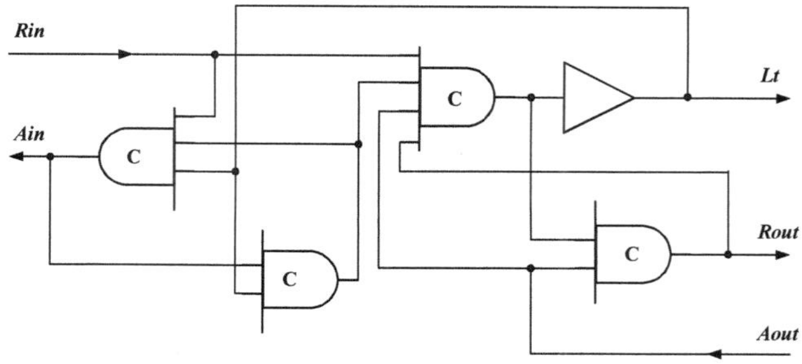

Fig. 3. The control circuit of one pipeline stage

In our procedure, the post-layout timing analysis has been done and custom delay cells is implemented to provide the constant delay.

Asynchronous Multiplier Unit. Any multiplier can be divided into three stages: partial products generation stage, partial products addition stage, and the final addition stage. The second stage is the most important, as it is the most complicated and determines the speed of the overall multiplier. In high-speed designs, the Wallace tree construction method is usually used to add the partial products in a tree-like fashion in order to produce two rows of partial products that can be added in the last stage.

According to the multiply stages of multiplier, our multiplier is divided into three stages execution pipeline. The multiplier supports arithmetic operations with sub-word parallelism. While being similar to the asynchronous adder unit, the multiplier employs three stages pipeline architecture and matching delay elements.

4.2 Address Generation

In the radix-4 FFT operation, the address of any operand can be defined as follows [22]:

$$A = 4 \times \sum_{i=1}^{r-1} 4^{i-1} \times A(i) + A(0) \tag{4}$$

Let,

$$B = 4 \times \sum_{i=1}^{r-1} 4^{i-1} \times A(i) + \left(\sum_{i=1}^{r-1} A(i) \right) \pmod 4 \tag{5}$$

For any address $A(0 \leq A < N)$, the A and B are in one-to-one correspondence.

If $m = \sum_{i=1}^{r-1} 4^{i-1} \times A(i)$, $b = \left(\sum_{i=0}^{r-1} A(i) \right) \pmod 4$, so

$$B = 4m + b \tag{6}$$

Based on the equation (6), the address A can be mapping to the memory. The memory can be divided into four memory banks. The b is the bank number and the m is the inner address.

At the p stage of FFT, the m of the four operands that required by the calculation can be described as follows:

$$\begin{cases} m_0 = \sum_{i=p+1}^{r-1} 4^{i-1} \times A(i) + 4^{p-1} \times 0 + \sum_{i=1}^{p-1} 4^{i-1} \times A(i) \\ m_1 = m_0 + 4^{p-1} \\ m_2 = m_0 + 2 \times 4^{p-1} \\ m_3 = m_0 + 3 \times 4^{p-1} \end{cases} \quad (7)$$

Such an operation can be easily implemented with the aid of counter, which represents the $\sum_{i=p}^{r-2} 4^{i-1} \times A(i+1) + \sum_{i=1}^{p-1} 4^{i-1} \times A(i)$. Then the m can be produced.

4.3 Globally Synchronous Locally Asynchronous Wrapper

The GSLA methodology aims to combine the advantages of asynchronous design with the convenience of standard synchronous design methodologies.

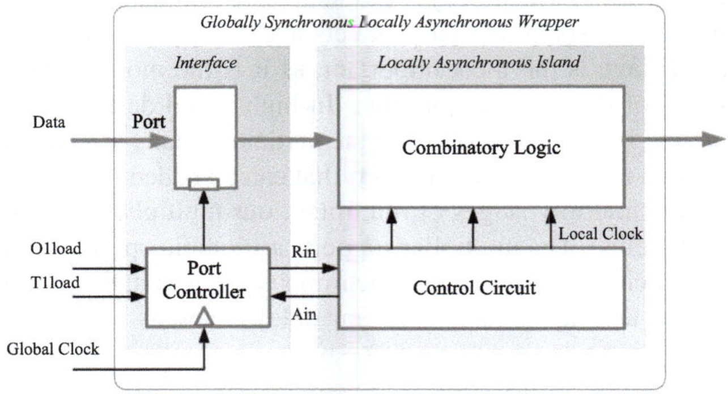

Fig. 4. Function unit with locally asynchronous island surrounded by the GSLA wrapper

Fig. 4 depicts a block level schematic of a GSLA module with its wrapper surrounding the locally asynchronous island. The wrapper contains an arbitrary number of ports, a local asynchronous unit, and port controller.

Each function unit in TTA has one or more operator register (O), only one trigger register (T) and one or more result registers (R). In GSLA wrapper, the operator and triggered registers are both controlled by global clock. When the signal $O1load$ is high, the operand will be send to operator register. When signal $T1load$ is high, the another operand will be send to trigger register and the function unit will be triggered to work. Locally asynchronous island is driven by local clock. The control circuit uses two or four phase handshaking protocol to control the data flow between adjacent pipeline stages.

In our modified MOVE framework, the latency of specifical asynchronous function unit should be converted to cycles according to global clock period. For example, if the latency of our asynchronous multiplier unit is 14ns and the global clock period is 5ns, the cycles of this unit should be defined as 3. It means that asynchronous function unit can complete one operation and other unit can read the result register after three clocks periods. This method can solve the problem that how to use asynchronous function units in synchronous TTA environment. In fact, the latency of asynchronous is only 14ns, which is less than three clock periods. The latency conversion may cause performance loss, but it a worthy trade-off between performance and power to some systems that are sensitive to power dissipation.

4.4 General Organization

The processor is composed of nine separate function units and a total of eight register files (RF) containing 32 general-purpose registers. The function units include one asynchronous multiplier unit (AMUL), two asynchronous adder units (AADD), one comparator unit (COMP), one data address generator (AG), one I/O unit (I/O), one shift unit (SH) and two Load/Store units (LSU). These function units and register files are full connected by interconnection network consisting of 6 buses. The 32-bit buses are used to transport data. In addition, the processor contains instruction and data memories. The general organization of proposed GSLA TTA processor tailored for FFT (GSLAFFT) processor is presented in Fig. 5.

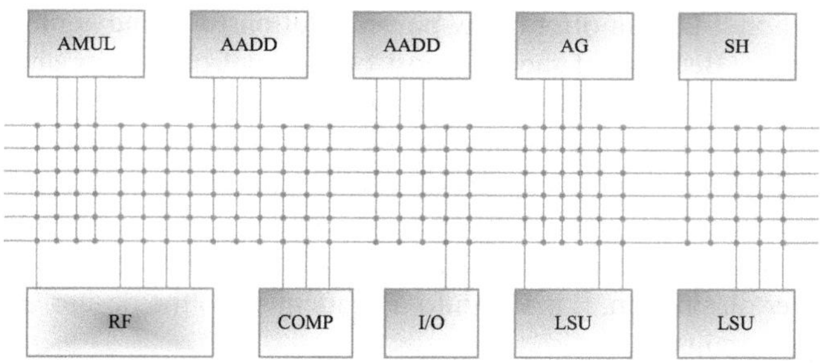

Fig. 5. Architecture of the proposed GSLAFFT processor core

The structural description of the FFT processor core was obtained with the aid of the hardware subsystem of the modified MOVE framework, which generated the Verilog description. The structures of the address generator unit, comparator unit, shift unit, I/O unit, and Load/Store unit were described manually in Verilog. The predesigned asynchronous function units library including Verilog description and layout is implemented according our asynchronous circuits design flow.

5 Simulation Results

In this work, we also implemented synchronous multiplier unit and adder unit using the same data path as their asynchronous versions. The synchronous FFT processor core (SFFT) replaced the asynchronous function units of GSLAFFT with their synchronous versions. The SFFT and GSLAFFT were implemented in $0.18\mu m$ 1P6M CMOS standard cell ASIC technology. The layouts were both implemented in standard cell automatic place and route environment. The Mentor Graphics Calibre was used for the LPE (Layout Parasitic Extraction) and the Synopsys Nanosim was used for performance and power analysis. In performance and power analysis, the simulation supply voltage was 1.8V, the temperature was 25°C, and the device parameters used the typical values that comes from the foundry. The clock frequency of these processor was 200MHz. The obtained results are listed in Table 1. It should be noted that the power dissipation of instruction and data memories are not taken into account.

Table 1. Characteristics of 1024-point FFT on SFFT and GSLAFFT

Design	Clock Frequency [MHz]	Execution Time [μs]	Area [mm^2]	Power [mW]	Energy [μJ]
SFFT	200	26.09	0.8543	51.24	1.34
GSLAFFT	200	26.09	0.9001	43.41	1.13

Due to characteristics of fine-grain clock gating and zero standby power consumption, the total power of GSLAFFT is less than SFFT. It shows that the GSLA implementation based on TTA can offer a low power solution for some application specific embedded systems. Because of the area cost of the control circuits and independent power rings layout, the area of asynchronous function unit is lager than its synchronous version. The area cost is a disadvantage of the asynchronous circuits implementation without any area optimization. Designers should seek the trade-off between power consumption, performance and area. In different applications, optimization techniques for different targets should be used [23]. Taking into account of the improvement in performance and power dissipation, it is worth to pay attention to the design and application of asynchronous circuits.

Table 2 presents how many 1024-point FFT transforms can be performed with energy of $1mJ$. The results are presented for some different implementations of the 1024-point FFT. The 1024-point FFT with radix-4 algorithm can be computed in 6002 cycles in TI C6416 when using 32-bit complex words [24]. The Stratix is an FPGA solution with dedicated embedded FFT logic using Altera Megacore function [25]. The MIT FFT uses subthreshold circuit techniques [26]. The FFTTA is a low power application-specific processor for FFT [1].

Compared to other FFT implementation, the proposed GSLAFFT processor shows significant energy-efficiency. The MIT FFT outperforms the GSLAFFT. However, due to its long execution time, the MIT FFT is not suitable for high performance design. The FFTTA also shows a significant improvement in energy-efficiency. It should be noted that the instruction and data memories take 40% of the total power consumption of FFTTA. If the power consumption of instruction and data memories is not included, the FFTs per mJ of FFTTA should be 1076 and outperforms our GSLAFFT. However, the performance of our processor can be scaled, i.e., the execution time can be halved by doubling the resources. On the other hand, if we use advanced process technology, such as 130nm, the clock frequency will be increase and the supply voltage will be decrease, moreover, the power consumption will be reduced. So the FFTs per energy unit will be

Table 2. Statistics of some 1024-point FFTs

Design	Technology [nm]	Clock Frequency [MHz]	Supply Voltage [V]	Execution Time [μs]	FFT/mJ
GSLAFFT	180	200	1.8	26.09	884
	130	720	1.2	8.34	100
TI C6416	130	600	1.2	10.0	167
	130	300	1.2	21.7	250
	130	275	1.3	4.7	241
Stratix	130	133	1.3	9.7	173
	130	100	1.3	12.9	149
MITFFT	180	0.01	0.35	250000	6452
	180	6	0.9	430.6	1428
FFTTA	130	250	1.5	20.9	645

increased. Simply, if the clock frequency of GSLAFFT on 130nm technology can be increased to 250MHz, the FFTs per mJ will be increased to 1104 even that the power consumption is not reduced.

6 Conclusion

Because TTA is very suitable for embedded systems for its flexibility and configurability. Supported by special low-power function units, the architecture is easy to be modified for different embedded applications that are sensitive to power. In this paper, a low-power application-specific globally synchronous locally asynchronous processor for FFT computation has been described. The resources of the processor have been tailored according to the need of the application. Asynchronous circuits have been used for reducing the power consumption of the processor. We implemented GSLA wrapper, which make it be possible to use asynchronous function units in TTA. The processor was implemented on 180nm 1P6M CMOS standard cell ASIC technology. The results showed that GSLA implementation based on TTA can can offer a low power solution for some application specific embedded systems. Although there is some area cost, it is worth to pay attention to the design and application of asynchronous circuits in embedded systems that are sensitive to power dissipation.

The described processor has limited performance but the purpose of our experiment was to prove the feasibility and potential of the proposed approach. However, the performance can be improved by introducing additional function units and optimizing the code of the application.

Acknowledgment

This work has been supported by the National Natural Science Foundation of China (90407022). Thank Andrea Cilio and his colleagues in Delft Technology University of Netherlands for their great help and support to our research work.

References

1. Pitkanen, T., Makinen, R., Heikkinen, J., Partanen, T., Takala, J.: Low–power, high–performance tta processor for 1024–point fast fourier transform. In: Vassiliadis, S., Wong, S., Hämäläinen, T.D. (eds.) SAMOS 2006. LNCS, vol. 4017, pp. 227–236. Springer, Heidelberg (2006)
2. Wang, A., Chandrakasan, A.P.: Energy-aware architectures for a real-valued FFT implementation. In: Proceedings of the 2003 international symposium on low power electronics and design, pp. 360–365 (2003)
3. Lee, J.S., Sunwoo, M.H.: Design of new DSP instructions and their hardware architecture for high-speed FFT. VLSI Signal Process 33(3), 247–254 (2003)
4. Takala, J., Punkka, K.: Scalable FFT processors and pipelined butterfly units. VLSI Signal Process 43(2-3), 113–123 (2006)
5. Zhou, Y., Noras, J.M., Shepherd, S.J.: Novel design of multiplier-less FFT processors. Signal Process 87(6), 1402–1407 (2007)
6. Weste, N.H.E., Eshraghian, K.: Principles of CMOS VLSI design: a systems perspective. Addison-Wesley Longman Publishing Co., Inc, Boston, MA, USA (1985)
7. Chandrakasan, A., Sheng, S., Brodersen, R.: Low-power CMOS digital design. IEEE Journal of Solid-State Circuits 27(4), 473–484 (1992)
8. Werner, T., Akella, V.: Asynchronous processor survey. Computer 30(11), 67–76 (1997)
9. Furber, S.B., Garside, J.D., Temple, S., Liu, J., Day, P., Paver, N.C.: AMULET2e: An asynchronous embedded controller. In: Proceedings of the International Symposium on Advanced Research in Asynchronous Circuits and Systems, pp. 290–299 (1997)
10. Garside, J.D., Bainbridge, W.J., Bardsley, A., Clark, D.M., Edwards, D.A., Furber, S.B., Lloyd, D.W., Mohammadi, S., Pepper, J.S., Temple, S., Woods, J.V., Liu, J., Petlin, O.: AMULET3i - an asynchronous System-on-Chip. In: Proceedings of the 6th International Symposium on Advanced Research in Asynchronous Circuits and Systems, pp. 162–175 (2000)
11. Furber, S.B., Edwards, D.A., Garside, J.D.: AMULET3: a 100 MIPS asynchronous embedded processor. In: Proceedings of the 2000 IEEE International Conference on Computer Design, pp. 329–334. IEEE Computer Society Press, Los Alamitos (2000)
12. Kawokgy, M., Andre, C., Salama, T.: Low-power asynchronous viterbi decoder for wireless applications. In: Proceedings of the 2004 international symposium on Low power electronics and design, pp. 286–289 (2004)
13. Cooley, J., Turkey, J.: An algotithm for the machine calculation of complex fourier series. Math Computer 19, 297–301 (1965)
14. Jain, M.K., Balakrishnan, M., Kumar, A.: ASIP design methodologies: Survey and issues. In: Proceedings of the The 14th International Conference on VLSI Design (VLSID '01), pp. 76–81 (2001)
15. Corporaal, H.: Microprocessor Architecture: from VLIW to TTA. John Wiley & Sons Ltd. Chichester (1998)
16. Corporaal, H., Arnold, M.: Using Transport Triggered Architectures for embedded processor design. Integrated Computer-Aided Engineering 5(1), 19–37 (1998)
17. Sutherland, I.E.: Micropipelines. Communications of the ACM 32(6), 720–738 (1998)
18. Gong, R., Wang, L., Li, Y., Dai, K., Wang, Z.Y.: A de-synchronous circuit design flow using hybrid cell library. In: 8th International Conference on Solid-State and Integrated Circuit Technology, pp. 1860–1863 (2006)
19. Piguet, C., Zahnd, J.: STG-based synthesis of speed-independent CMOS cells. In: Workshop on Exploitation of STG-Based Design Technology (1998)

20. Cortadella, J., Kishinevsky, M., Kondratyev, A., Lavagno, L., Yakovlev, A.: Petrify: a tool for manipulating concurrent specifications and synthesis of asynchronous controllers. IEICE Transactions on Information and Systems E80-D(3), 315–325 (1997)
21. Shams, M., Ebergen, J., Elmasry, M.: A comparison of CMOS implementations of an asynchronous circuitsprimitive : the C-element. In: International Symposium on Low Power Electronics and Design, vol. 12(14), pp. 93–96 (1996)
22. Xie, Y., Fu, B.: Design and implementation of high throughput FFT processor. Computer Research and Development 41(6), 1022–1029 (2004)
23. Zhou, Y., Sokolov, D., Yakovlev, A.: Cost-aware synthesis of asynchronous circuits based on partial acknowledgement. In: Proceedings of the 2006 IEEE/ACM international conference on Computer-aided design, pp. 158–163. ACM Press, New York (2006)
24. Texas Instruments: TMS320C64x DSP Library Programmer Reference (2002)
25. Lim, S., Crosland, A.: Implementing FFT in an FPGA co-processor. In: The International Embedded Solutions Event (GSPx), pp. 27–30 (2004)
26. Wang, A., Chandrakasan, A.: A 180-mV subthreshold FFT processor using a minimum energy design methodology. IEEE Journal of Solid-State Circuits 40, 310–319 (2005)

Parallel Genetic Algorithms for DVS Scheduling of Distributed Embedded Systems

Man Lin* and Chen Ding

Department of Mathematics, Statistics and Computer Science,
St. Francis Xavier University
{mlin,x2002rah}@stfx.ca

Abstract. Many of today's embedded systems, such as wireless and portable devices rely heavily on the limited power supply. Therefore, energy efficiency becomes one of the major design concerns for embedded systems. The technique of dynamic voltage scaling (DVS) can be exploited to reduce the power consumption of modern processors by slowing down the processor speed. The problem of static DVS scheduling in distributed systems such that the energy consumption of the processors is minimize while guaranteeing the timing constraints of the tasks is an NP hard problem. Previously, we have developed a heuristic search algorithm: Genetic Algorithm (GA) for the DVS scheduling problem. This paper describes a Parallel Genetic Algorithm (PGA) that improves over Genetic Algorithm (GA) for finding better schedules with less time by parallelizing the GA algorithms to run on a cluster. A hybrid parallel algorithm is also developed to further improve the search ability of PGA by combining PGA with the technique of Simulated Annealing (SA). Experiment results show that the energy consumption of the schedules found by the PGA can be significantly reduced comparing to those found by GA.

1 Introduction

With the growing mobile technology, nowadays our life is heavily relied on mobile and portable devices with built in CMOS processors. The biggest limitation of current electronic devices is the battery life. The technique of dynamic voltage scaling(DVS) has been developed for modern CMOS processors, which can effectively lower the power consumption and enable a quieter-running system while delivering performance-on-demand(DVS) [1]. Many of today's advanced processors, such as AMD and Intel, have this technology.

The CPU power consumed per cycle in a CMOS processor can be expressed as $P = C_L f V_{DD}^2$, where C_L is the total capacitance of wires and gates, V_{DD} is the supply voltage and f is the clock frequency. It is obvious that a lower voltage level leads to a lower power consumption. The price to pay for lowering the voltage level is that it also leads to a lower clock frequency and thus slows

* The author would like to thank NSERC (National Science Engineering Research Council, Canada) and CFI for supporting this research.

R. Perrott et al. (Eds.): HPCC 2007, LNCS 4782, pp. 180–191, 2007.

down the execution of a task. The relation between the clock frequency and the supply voltage is: $f = K * (V_{DD} - V_{TH})^2 / V_{DD}$. As a result, exploiting DVS may hurt the performance of a system (If we reduce the voltage by half, the energy consumed will be one-quarter and the execution time will be double.) When using DVS in a hard real-time system where tasks have deadlines, we can not lower the voltage levels of the processors too much as we also need to guarantee the deadlines of the tasks be met.

In the past few years, there have been a number of algorithms proposed for applying DVS to hard real-time systems. Many previous works of DVS-based scheduling either focus on single processor power conscious scheduling or consider independent tasks only [2,3]. In this paper, we focus on static DVS-based scheduling algorithm for distributed real-time systems consisting of dependent tasks. We consider dicrete-voltage DVS processors instead of variable-voltage DVS as real DVS processors show only a limited number of supply voltage levels at which tasks can be executed.

The problem of optimally mapping and scheduling tasks to distributed systems has been shown, in general, to be NP-complete [4]. Because of the computational complexity issue, heuristic methods have been proposed to obtain optimal and suboptimal solutions to various scheduling problems. Genetic algorithms, inspired by Darwin's theory of evolution, have received much attention as searching algorithms for scheduling problems in distributed real-time systems [5,6,7]. The appeal of GAs comes from their simplicity and elegance as robust search algorithms as well as from their power to discover good solutions rapidly for difficult high-dimensional problems [8]. Our previous work has adopted Genetic Algorithms for DVS-based scheduling algorithm [9]. The method is different from other approaches of DVS scheduling of distributed systems [10,11,12,13,14] in that it does not just construct one schedule once, it constructs populations of schedules and iterates many generations to find one near optimal schedule. Different from another GA algorithm [15], our GA algorithm integrates task assignment (to which processor the tasks will be assigned), task scheduling (when the tasks will be executed) and voltage selection (at which voltage level the tasks will run) at the same phase.

When the task size becomes bigger, the search space becomes larger. In order to find better DVS schedules more efficiently, we design parallel genetic algorithms (PGA) to improve our previous GA. The PGA is implemented on a SUN cluster with 33 computation nodes. The PGA cuts down the computation time of genetic algorithm by reducing the population to smaller size in order to achieve reasonable speed up with good result. And it also expands the search space, increases the probability to obtain better result. A hybrid parallel algorithm which combines PGA and Simulating Annealing (SA) is also developed to further improve the search process.

The paper is organized as follows. First, the energy model and the task model are described in Section 2. The Parallel Genetic Algorithm (PGA) for DVS schedling will be described in Section 3 and the experimental results will be shown in Section 4. Finally the conclusions are presented in section 5.

2 Task and Schedule Models

We consider the energy aware scheduling problem for a set of task $T_1, T_2, \ldots, T_N$ on a set of heterogeneous processors $P_1, P_2, \ldots, P_M$ where N is the number of tasks and M is the number of processors. Each processor has a number of voltage levels. The power dissipation for a processor p running at level l is denoted as $POW_{p,l}$.

There are precedence constraints among tasks which can be represented by a directed acyclic graph (DAG). If there is an edge from task T_i to T_j in the DAG, then T_j can only start after T_i finishes execution.

- Each task has a Worst Case Execution Time (WCET) on each processor for a given voltage level. The worst case execution time of task t on processor p at voltage level l is represented as $w_{t,p,l}$.
- Each task has a deadline. d_i is used to denote the deadline of task i.
- Suppose task t is assigned to processor p and run at voltage level l. Then the energy used to execute task t is given by $POW_{p,l} * w_{t,p,l}$.
- Assume T_i is mapped to processor $P(i)$ and runs at level $L(i)$. Then the total energy consumptions can be easily calculated as follows: $E_{total} = \sum_{i=1}^{N}(POW_{P(i),L(i)} * w_{T_i,P(i),L(i)})$.

To simplify the energy model, we assume that it takes no time to switch from one voltage level to another and therefore no energy consumed accordingly.

2.1 DVS Scheduling Example

Let's consider a very simple scheduling example with 3 tasks to be mapped into 2 processors as shown in Fig. 1(a).

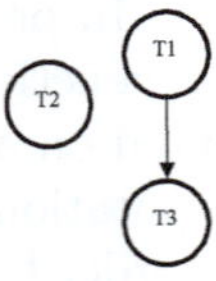

	Length	Deadline
T1	10	20
T2	10	40
T3	10	40

(Voltage = 5.0V)

(a) Task Set

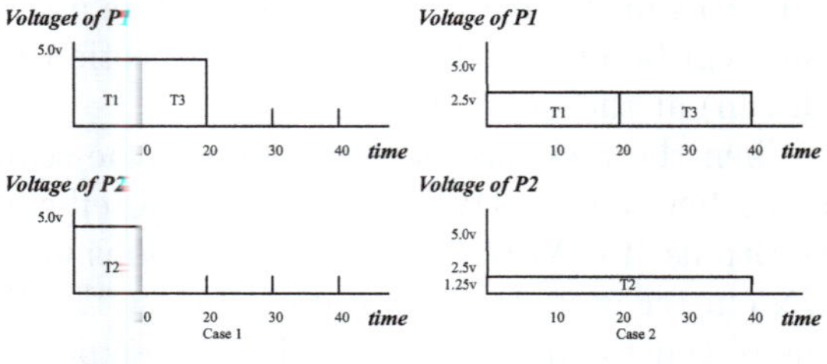

(b) Without and With DVS

Fig. 1. An Example

Case 1 and case 2 in Fig. 1(b) shows the schedule without DVS and with DVS, respectively. With DVS, task T_1 and T_3 can be slowed down to run at 1/2 speed and consume only 1/4 energy. Task T_2 can save more energy consumption as it can slow down more.

3 Parallel Genetic Algorithms

GA approach can find better solutions for most scheduling problems in terms of the feasibility of the solutions and the energy saving [9]. However, when the problem size increases, it takes very long time for the solution to converge due to the bigger population size needed. Parallel computations reduce the computation time significantly when using multiple processors. To find better solution faster, we developed an Parallel Genetic algorithm (PGA) for the energy aware scheduling problem.

The chromosome representation and the genetic operator are adopted from the GA developed previously [9].

3.1 Chromosome Representation

Each chromosome is a schedule represented by an ordered list of genes and each gene contains three data items: Task (the task number), Proc (the processor number) and Level (the voltage level). The chromosome can be viewed as an $N * 3$ array where N is the total number of tasks. The first row of a chromosome indicates the tasks ordered from left to right. The second row indicates the corresponding processor that each task will be assigned to. And the third row is the voltage level selected for the corresponding processor for each task.

In the example shown in table 1, tasks t_1, t_2, t_3 and t_4 are to scheduled onto processor p_1 and p_2 where both of the processor have two voltage levels. Task t_2 and t_3 are assigned to process p_2 and runs at voltage level 1 and 2 respectively. Task t_4 and t_1 are assigned to process p_1 and runs at level 2 and 1 respectively.

Table 1. A Schedule Example

Task	2	3	4	1
Proc	2	2	1	1
Level	1	2	2	1

A random order of tasks may result in infeasibility because the precedence constraints might be violated. To avoid such problem, we only allow chromosomes that satisfy topological order [7]. A topological ordered list is a list in which the elements satisfy the precedence constraints. To maintain all the individuals in the populations of any generation to be topological ordered, we adopted the techniques in [7].

- Generate only topological ordered individuals in the initial population;
- Use carefully chosen genetic operators (see section 3.2).

Note that the deadline constraint may still be violated even when a schedule satisfies a topological order.

3.2 Genetic Operators

Genetic Algorithms generate better solutions by exploring the search space by genetic operators. New individuals are produced by applying crossover operator or mutation operator to the individuals in the previous generation. The probability of the operators indicates how often the operator will be performed. We choose 0.6 as the probability for the crossover and 0.4 as the probability for the mutation.

The mutation operator creates a new individual with a small change to a single individual. In our approach, we use Processor and/or Level Assignment Mutation. To perform the mutation, we randomly select a range of genes from a chromosome and then randomly change the processor number and/or the level number within this range. Obviously, the mutation operator does not change the order of the tasks. Therefore, the new individual also satisfy the topological order.

The crossover operator creates new individuals by combining parts from two individuals. To perform crossover operation, we first randomly pick two chromosomes (as parents) from the current population. Then we randomly select a position where the crossover is going to occur. The first part of child 1 uses the schedule of parent 1 up to the chosen position. The second part of child 1 is constructed by selecting the rest tasks (the tasks not in the first part) from parent 2 in order. The same mechanism also applies to child 2. Below is an example of our crossover operator. Assume we have two individuals as shown in Fig. 2(a). Suppose the selected position is 2, then the children will be shown as in Fig. 2(b).

Task #	1	5	2	4	3
Processor #	1	2	1	2	1
Level #	1	2	1	2	1
Parent 1					

Task #	5	3	2	1	4
Processor #	2	2	1	1	1
Level #	1	1	2	2	2
Parent 2					

(a) Crossover Parents

Task #	1	5	3	2	4
Processor #	1	2	2	1	1
Level #	1	2	1	2	2
Child 1					

Task #	5	3	1	2	4
Processor #	2	2	1	1	2
Level #	1	1	1	1	2
Child 2					

(b) Crossover Children

Fig. 2. Crossover operator

The crossover operator will produce two offsprings that are topological ordered if the parents are topological ordered. The detailed proof can be found in [7].

3.3 Parallel Environment

Our parallel hardware environment is a Sun Netra X1 Cluster Grid of 33 Sun servers (nodes). Each node has a Ultra SPARC II 64bit CPU, capable of 1 Gflops with 1 Gb main memory.

Based on the characteristic of coarse-grain PGA, a C language with MPI (Message Passing Interface) library has been chosen for our PGA programming.

We choose to implement a coarse-grain PGA to solve our problem because it is the most popular method and others have reported good results with this method in literature. First it cuts down the computation time of genetic algorithm by reducing the population to smaller size in order to achieve reasonable speed up with good result. Second improvement is that it expands the search space, increases the probability to obtain better result. And most it is obvious the large that amount of nodes used the better chance to get good result.

3.4 The Basic PGA Algorithm

In our PGA, subpopulation is employed in each node instead of the whole population as shown in Fig. 3. Assume the population of PGA is denoted as *POPSIZE* and there are n computation nodes. Each node performs GA with *POPSIZE*/n number of individuals in the subpopulation. Note that when a larger number of nodes being used, the subpopulation size may become too small. To avoid subpopulation exiguity problem, we set a minimal value for subpopulation. Typically in our case, 40 is the minimal population size. Each node randomly creates initial schedules that satisfy the partial order constraints.

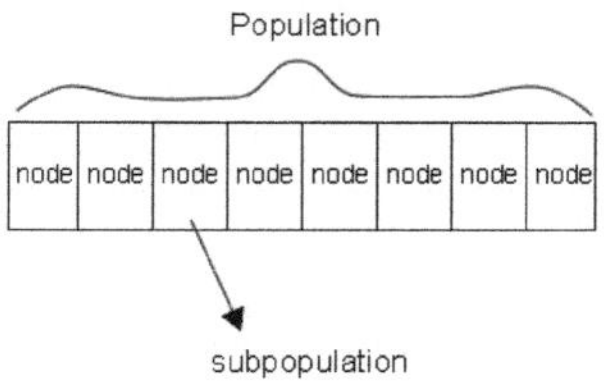

Fig. 3. Subpopulation Division

After initialization, PGA goes into a loop with many iterations, each iteration contains two steps: 1) basic GA operations in each node including selection, crossover and mutation and 2) communications among nodes. Step 1) in each iterations allows each node generate a new subpopulation. Each node then finds its own best chromosome out of the new subpopulation. In Step 2), one node sends its best solution to the rest of nodes. Note that the nodes take turns to send the best solution, one node in each iteration. Each node compares the best solution that it receives with its own best solution and determines whether to replace or discard it following the replacing policy.

The iterations will keep repeating until the stopping criteria are satisfied. Finally,the master node gathers all best chromosome from all the nodes and finds the best one among all.

The cost of communication between two MPI processors can be approximately divided into two parts: (i) Start-up time $T_{startup}$ and (ii) Transmission Time $T_t =$ (Data size)/Bandwidth. Start-up time is the duration needed for MPI to set up the communication links. Transmission time is the time needed to send the message to its destination. The most commonly used model of the cost of communication is linear: $T_{comm} = T_{startup} + (Datasize)/Bandwidth$.

When PGA broadcasts the best chromosome, a large amount of communication is required. Since the chromosome contains many pointers and MPI is not able to send non-continuous memory block at once, we defined a MPI communication data type called ChromosomeType that contains all the information in a chromosome. Therefore each MPI communication only need one start up time plus one transmission time to broadcast the best chromosome.

3.5 PGA Strategies

Initial Schedules Using EDF with TopologicalOrder. To increase the efficiency of our genetic algorithm, during the initialization of PGA, the generalization of schedules include two schedule that uses EDF scheduling policy while conforming to the topological order of tasks, where one has minimal energy level and the other one has maximal energy level.

Evaluation and Punishment of Infeasible Solutions. The aim of our optimization problem is to minimize the energy consumption of the tasks. The evaluation function is defined as the energy consumption of all the tasks.

However, the individual schedule with smaller E_{total} will not always be considered better. This is because we need to consider one more factor: the deadline constraints (precedence constraints are already encoded into the schedule enforced by the topological order.). Our algorithm give penalty to the individuals violating the deadline constraints so that they have less chance in getting into the next generation.

The Policy of Replacing the Best and Worst Solution. The best feature of parallel programming is that one computation node can communicate its result to other nodes to help them get close to optimal solution in the search process. So exchanging results becomes the most important issues in our PGA. The nodes send their best result sequentially to other nodes. When a node receive a solution from others nodes, it will determine whether to use it or discard it.

The replacing best policy works as follow: before communication, each node find both the best and worst chromosome indexes. After communication, the local best index will be compared to the received best solution. If the local best is better than the received one, the node abandons the received one, and replaces the local worst index with the received best. Otherwise, the node replaces both the local best and local worst solution with the receive best solution.

Cost Slack Stealing. In a schedule, it is possible that some tasks have slacks. Our cost slack stealing strategy is to determine whether the scheduling has slacks and try to reuse these slacks by extending the task execution time as much as possible by using the feasible lowermost energy level, so the total tasks execution time can be reduced. Considering the speed efficiency of PGA, this cost clack stealing method only performed on the final best result in each node to gain last enhancement of the PGA.

3.6 Hybrid Parallel Algorithm

This section, we present a hybrid parallel algorithm: parallel simulation annealing and genetic algorithm(PSAGA), to further improve the performance.

Simulated annealing (SA) is a generic probabilistic meta-algorithm for the global optimization problem, namely locating a good approximation to the global optimum of a given function in a large search space [16].

In analogy with the physical process of annealing in metallurgy, each step of the SA algorithm replaces the current solution by a random "nearby" solution, chosen with a probability that depends on the difference between the corresponding function values and on a global parameter T (called the temperature), that is gradually decreased during the process. The allowance for "uphill" moves saves the method from becoming stuck at local minimum as would the greedier methods.

GA lacks of focus in the search due to the crossover operation. The combination of SA with GA allows finding a better solution in a particular search point and search area. PSAGA preforms an extra SA process besides those of PGA in each iteration. After performing replacing the best and worst solution, an SA will be started using the best schedule currently found. The SA tries to find the best neighbor in each iteration.

4 Experimental Results

In this section, we describe the simulation experiments preformed and the results.

First we describe the features of the generated scheduling problems in the experiments. Fig. 4 shows the number of task and the number of constraints we have considered in our experiments. For each category, we compared GA, PGA, and PSAGA. The results you see later in the section are average result of the 5 problems in each category. Beside the task graph, each scheduling problem has one more variable: the number of processors. We have chosen 3, 5, and 10 processors for each task graph but we mainly compared 10 processors cases in this paper.

Number of Tasks	Number of Constraints
50	0, 10, 25
100	0, 10, 25, 50
200	0, 10, 25, 50, 100

Fig. 4. Test Parameters

We first demonstrate the efficiency of energy reduction and the corresponding runtime of the PGA algorithm. Fig. 5(a) shows the average energy consumption of the solutions found by GA and PGA using various computation nodes under different test categories. When the number of node equals one, it means that it

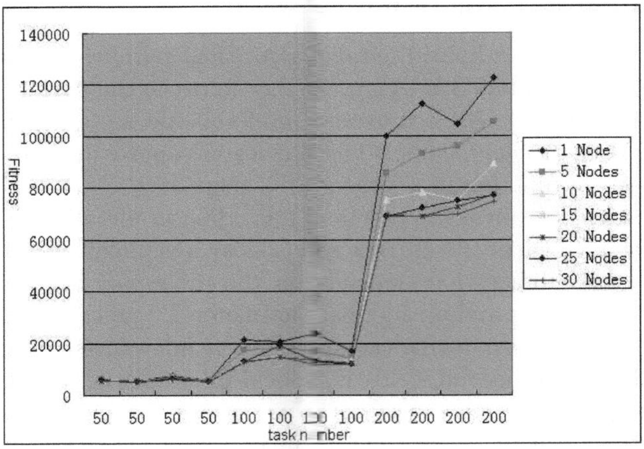

(a) Energy Consumption

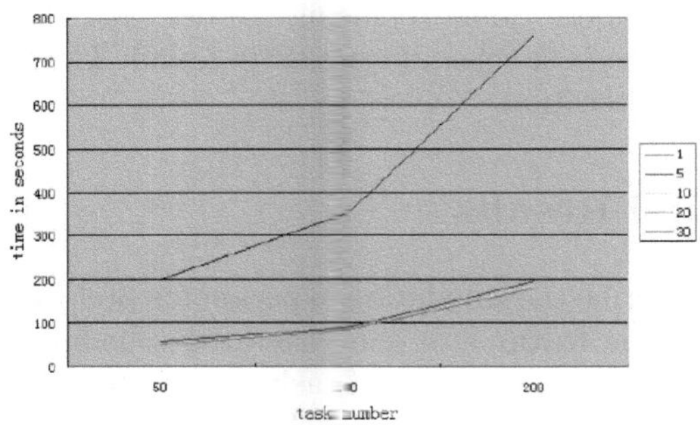

(b) Computation Time

Fig. 5. The Parallelization Effect

is a serial genetic algorithm. That is, the GA program here is represented by a 1 node PGA program. For small task size 50, GA and PGAs almost have the same result, and it is hard to see any advantage of PGA in this case. PGA's results get better when the computation nodes used for PGA increase. The improvement is particularly obvious when the task size is above 100 where the energy consumption can be reduced more than 50% with 8 times less run time. With task number being 500, the difference seems to be more obvious. PGA has better solution as the size of nodes is increased and clearly the results show that the 30 node PGA has the best result overall. With more nodes PGA can find a schedule closer to the optimal solution. Fig. 5(b) shows the corresponding average computation time used for the GA and PGA. under different computation nodes. The run time speed up of PGA with 5 nodes is 4. As the node increases to more than 5, we use a fixed time for PGA as we aim to find better solutions and the computation time is tolerable.

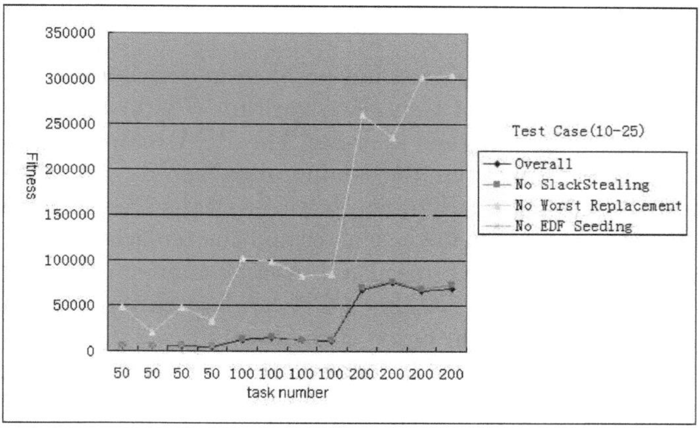

Fig. 6. PGA Strategies

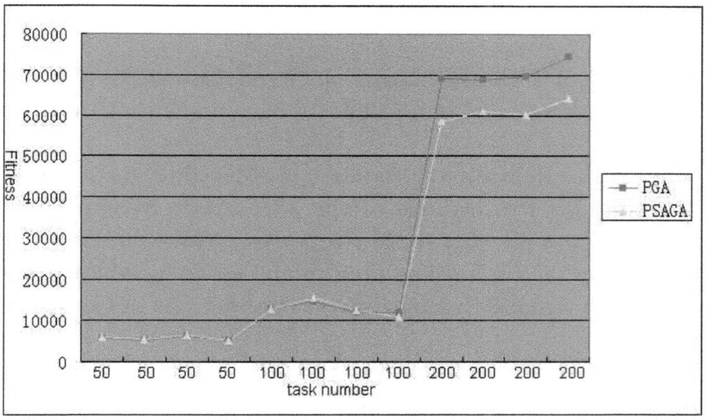

Fig. 7. PSAGA vs. PGA

Next, we show the effect of the typical strategies in PGA in Fig. 6 We found that policy of the best and worst replacement results in the best fitness than any other strategies. Including initial schedules using EDF becomes more efficient when the task number becomes lager. And slack stealing contributes to some improvement.

For the PSAGA, we only compare it with the PGA with 30 nodes as an example. From Fig. 7, for task size being 50 to 100, PSAGA and PGA have similar results. But when the task size increases to 200, the difference become obvious. PSAGA can save around 15% more energy than PGA. As the results are similar when using different number of nodes, we can expect that when the schedule problem size gets larger, the PSAGA performs better than PGA. When task size increases to 200, PSAGA can reduce 15% more energy than PGA.

5 Conclusion

We have addressed energy minimization problem in distributed embedded systems in this paper. A Parallel Genetic Algorithm (PGA) has been described for static DVS scheduling in such systems. The PGA improves over GA both on the optimization results and the computation speed. The PGA has been implemented in a SUN cluster. Various strategies of the PGA, such as the communication between the nodes in the cluster and the replacing strategies of best and worst individuals, have been discussed. A large number of experiments were conducted. The experiments show that PGA can find better solutions than GA for most scheduling problems in terms of the feasibility of the solutions and the energy saving. The improvement is particularly obvious when the task size is above 100 where the energy consumption can be reduced more than 50% with 8 times less run time. PSAGA (Parallel hybrid Simulating Annealing and Genetic Algorithm) is also employed to achieve the further improvement of PGA.

References

1. Burd, T.D., Pering, T.A., Stratakos, A.J., Brodersen, R.W.: A dynamic voltage scaled microprocessor system. IEEE Journal of Solid-State Circuits 35, 1571–1580 (2000)
2. Pillai, P., Shin, K.G.: Real-time dynamic voltage scaling for low-power embedded operating systems. In: Proceedings of the 18th ACM Symposium on Operating Systems Principles, pp. 89–102. ACM Press, New York (2001)
3. Demers, A., Shenker, F.Y.: A scheduling model for reduced CPU energy. In: Proceedings of the 36th Annual Symposium on Foundations of Computer Science, pp. 374–382 (1995)
4. Garey, M.R., Johnson, D.S.: Computers and Intractability: A Guide to the Theory of NP-Completeness. Freeman Publisher, San Francisco (1979)
5. Hou, E.S.H., Ansari, N., Ren, H.: A genetic algorithm for multiprocessor scheduling. IEEE Transactions Parallel and Distributed Systems 5, 113–120 (1994)
6. Lin, M., Yang, L.T.: Hybrid genetic algorithms for partially ordered tasks in a multi-processor environment. In: Proceedings of the 6th International Conference on Real-Time Computing Systems and Applications (RTCSA-99), Hongkong, pp. 382–387 (1999)
7. Oh, J., Wu, C.: Genetic-algorithm-based real-time task scheudling with multiple goals. Journal of Systems and Software, 245–258 (2004)
8. Glover, F., Laguna, M.: Modern Heuristic Techniques for Combinatorial Problems. Blackwell Scientific Publications (1993)
9. Lin, M., Ng, S.: Energy Aware Scheduling for Heterogeneous Real-Time Embedded Systems Using Genetics Algorithms. In: New Horizons of Parallel And Distributed Computing, vol. ch. 8, pp. 113–127. Kluwer Academic Publisher, Dordrecht (2005)
10. Luo, J., Jha, N.K.: Static and dynamic variable voltage scheduling algorithms for real-time heterogeneous distributed embedded systems. In: ASP-DAC/VLSI Design 2002, Bangalore, India, pp. 719–726 (2002)
11. Zhang, Y., Hu, X., Chen, D.: Task scheduling and voltage selection for energy minimization. In: DAC 2002, New Orleans, Louisiana, USA, pp. 183–188 (2002)

12. Zhu, D., Melhem, R., Childers, B.: Scheduling with dynamic voltage/speed adjustment using slack reclamation in multi-processor real-time systems. In: Proceedings of the 22nd IEEE Real-Time Systems Symposium (RTSS-01), pp. 84–94. IEEE Computer Society Press, Los Alamitos (2001)
13. Mishra, R., Rastogi, N., Zhu, D., Mosse, D., Melhem, R.: Energy aware scheduling for distributed real-time systems. In: International Parallel and Distributed Processing Symposium (IPDPS'03), Nice, France (2003)
14. Ruan, X.J., Qin, X., Zong, Z.L., Bellam, K., Nijim, M.: An energy-efficient scheduling algorithm using dynamic voltage scaling for parallel applications on clusters. In: Pcoc. 16th IEEE International Conference on Computer Communication and Networks (ICCCN) (2007)
15. Schmitz, M., Al-Hashimi, B., Eles, P.: Energy-efficient mapping and scheduling for DVS enabled distributed embedded systems. In: Proceedings of 2002 Design, Automation and Test in Europe (DATE2002), pp. 514–521 (2002)
16. Kirkpatrick, S., Gelatt, C.D., Vecchi, M.P.: Optimization by simulated annealing. Science 4598, 671–680 (1983)

Journal Remap-Based FTL for Journaling File System with Flash Memory

Seung-Ho Lim, Hyun Jin Choi, and Kyu Ho Park

Computer Engineering Research Laboratory,
Department of Electrical Engineering and Computer Science,
Korea Advanced Institute of Science and Technology,
Daejon, South Korea
{shlim,hjchoi}@core.kaist.ac.kr, kpark@ee.kaist.ac.kr

Abstract. Constructing flash memory based storage, FTL (Flash Translation Layer) manages mapping between logical address and physical address. Since FTL writes every data to new region by its mapping method, the previous data is not overwritten by new write operation. When a journaling file system is set up upon FTL, it duplicates data between the journal region and its home location for the file system consistency. However, the duplication degrades the performance. In this paper, we present an efficient journal remap-based FTL. The proposed FTL, called $JFTL$, eliminates the redundant data duplication by remapping the journal region data path to home location of file system. Thus, our $JFTL$ can prevent from degrading write performance of file system while preserving file system consistency.

1 Introduction

Flash Memory has become one of the major components of data storage since its capacity has been increased dramatically during last few years. It is characterized by non-volatile, small size, shock resistance, and low power consumption [1]. The deployment of flash memory is spreading out ranging from consumer electronic devices to general purpose computer architecture. In nonvolatile memories, NOR flash memory provides a fast random access speed, but it has high cost and low density compared with NAND flash memory. In contrast to NOR flash memory, NAND flash memory has advantages of a large storage capacity and relatively high performance for large read/write requests. Recently, the capacity of a NAND flash memory chip became gigabyte stage, and the size will be increased quickly. NAND flash memory chips are arranged into blocks, and each block has a fixed number of pages. Currently, typical page size of NAND is 2 KB, and the block size is 128 KB.

For flash memory, many operations are hampered by its physical characteristics. First, the most important feature, is that bits can only be cleared by erasing a large block of flash memory, which means that in-place updates are not allowed. It is called out-of-place update characteristics. When data are modified, the new

R. Perrott et al. (Eds.): HPCC 2007, LNCS 4782, pp. 192–203, 2007.

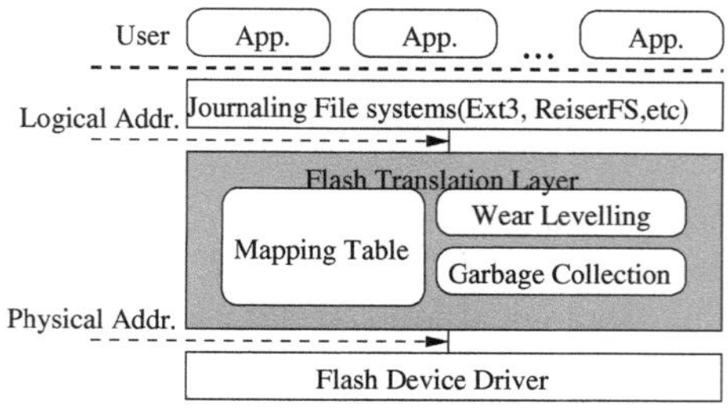

Fig. 1. Layered Approach with FTL

data must be written to an available page in another position, and old page becomes dead page. As time passes, a large portion of flash memory is composed of dead pages, and the system should reclaim the dead pages for writing operations. The erase operation makes dead pages become available again. Second, erasing count of each block has a limited number and erase operation itself gives much overhead to system performance. Thus, the write operation is more important than read operation when designing flash memory based systems.

Due to its limitations, data robustness and consistency is a crucial problem. Since every data should be written to new place in flash memory, a sudden power failure of system crash may do harm the file system consistency. There are two approaches to address these problems. One approach would be possible by designing a file system itself to support specific flash memory characteristics. There are several native flash file system, such as JFFS [8] and YAFFS [9]. These are log-structured based file system that sequentially stores the nodes containing data and metadata in every free region, although their specific structures and strategies are different from each other. These approach are available only on board flash memory chips, and if it is desired to be used in PC systems, a flash memory interfaces should be integrated into computer architecture.

The other approach is to introduce a special layer, called a Flash Translation Layer (FTL), between existing file system and flash memory [4][5][6]. The FTL remaps the location of the updated data from one page to another and manages the current physical location of each data in the mapping table. FTL gives a block device driver interface to file system by emulating a hard disk drive with flash memory through its unique mapping algorithm. Figure 1 shows the layered approach with FTL in flash memory based systems. This layered approach can offload flash memory from processing unit by utilizing commonly used interfaces such as SCSI or ATA between processor and flash memory chips. The removable flash memory cards such as USB mass storage and compact flash are using this approach. The advantage of it is that existing file systems can be used directly on the FTL. When existing file systems use upon a FTL, journaling technique is essential element to prevent from corrupting user data and enhance system reliability, data consistency and robust. However, journaling method existing file system itself degrades the system performance. Generally, journaling file systems write a journal record of important file system data before processing write operation. Then this journal is re-written to real file system home location to make clear consistent state. Due to out-of-place update characteristics of flash memory, the journaling affects flash memory based system more than disk based system.

In this paper, we present an efficient journaling interface between file system and flash memory. The proposed FTL, called *JFTL*, eliminates the redundant data copy by remapping the logical journal data location to real home location. Our proposed method can prevent from degrading system performance while preserving file system consistency. Our approach is layered approach, which means that any existing file system can be setup upon our method with little modification to provide file system consistency. In the implementation, we consider Ext3 file system among many kinds of journaling file systems because it is the main root file system of Linux. The remainder of this paper is organized as follows. Section 2 describes motivation and background. Section 3 describes the design and implementation of our proposed technique, and Section 4 show experimental results. We conclude in Section 5.

2 Background

Ext3 [13] is a modern root file system of Linux that aims to keep complex file system structures in a consistent state. All important file system updates are first written to a log region called a *journal*. After the write of journal are safely done, the normal write operations occur in its *home locations* of file system layout. The journal region is recycled after the normal updates is done. A central concept when considering a Ext3 file system is the transaction. A transaction is the update unit of Ext3 file system, which contains the sets of changes grouped together during an interval. A transaction exists in several phases over its lifetime, running, commit, checkpoint and recovery. The detailed operation of transaction is omitted due to the page limit. Please refer to other papers.

The Ext3 file system offers three journaling operation modes: *journaled mode*, *ordered mode*, and *writeback mode*. In *journaled mode*, all data and metadata are written to the journal. This provides high consistency semantics including user level data consistency. However, all are written twice to file system; first to the journal region, second to its home location. Thus, its write performance is extremely degraded. In *ordered mode*, only file's metadata are written to the journal, and file's data are written directly to its home location. To provide data consistency without double writes for file's data, *ordered mode* guarantees a strict ordering between two writes. The file's data are should be written to its home location before the corresponding metadata are written to the journal when a transaction commits. This can guarantee that file metadata never indicate a wrong data before it has been written. After the metadata logging to the journal is done, they are re-written to its home location. In *writeback mode*, as does *ordered mode*, only metadata are written to the journal, and file's data are

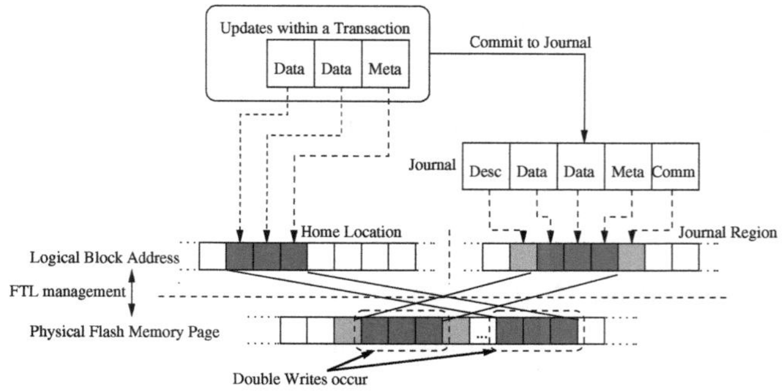

(a) Conventional FTL cannot eliminate double write

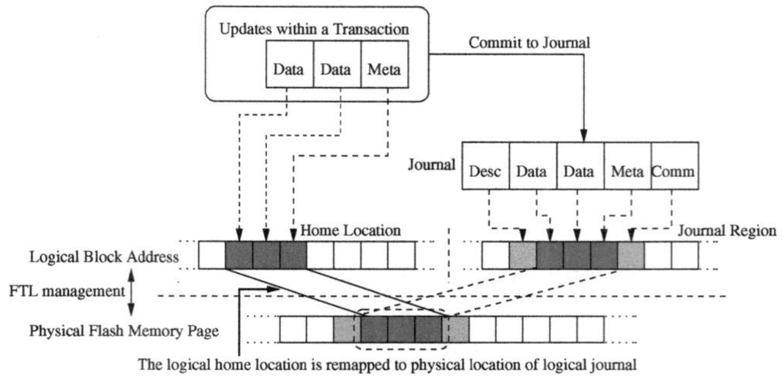

(b) JFTL can eliminate double write by redirecting logical path of journal region to home location

Fig. 2. Double writes of Journaling

written directly to its home location. However, unlike *ordered mode*, there is no strict write ordering between these two. A metadata indexes can point to the wrong data block which wasn't written to storage. It cannot guarantee data consistency, but the journal can restore metadata.

The journaling operation of Ext3 can preserve file system consistency, at some level for each mode. Regardless of the consistency level of each mode, the key concept is that important information is first written to the journal, and then written to home location. During these journaling operations, double writes occur. If Ext3 is mounted in flash memory based storage system, it is set up upon FTL since FTL presents a block device interface to file system. The issue we should take notice is system performance. Since write operation occurs twice, write performance is degraded, since double writing occurs by a log manner in the flash memory. The Figure 2(a) represents the problem of double writing in FTL. However, if we make use of the FTL mapping management for journaling technique, we can eliminate double writing overhead. Figure 2(b) shows the $JFTL$ that remaps the logical home location to the physical location of journal that was written before. It can eliminate the redundant writes of home location.

3 Design and Implementation

In this section, we describe our design and implementation of *JFTL*. The design goal is to prevent from degrading system performance, while providing file system consistency semantics. Our *JFTL* can be applied to all modes in Ext3 journaling, we explain based on Ext3 *journaled mode*.

3.1 JFTL: Journaling Remap-Based FTL

In the Ext3 *journaled mode*, all updates including file data and metadata are first written to *journal region*, then copied to *home location*. Transaction provides system consistency by writing all together or not included in a transaction. This atomicity is possible to manage duplication between journal region and home location, because original data remained in home location are not corrupted by operations during transaction running. The duplication is crucial feature in journaling. In view of this, if we look into FTL, FTL already uses duplication by the mapping management. Every written blocks from file system is written to another new flash pages by FTL not over-written. The problem is that FTL has no idea about journaling. Specifically, it does not know what data are associated with transactions, and the relationship between home location and journal region. If we can manage these information within FTL, we can eliminate double writing. In Ext3, these important information is stored in block called *descriptor*. For a transaction, *descriptor* block contains list of addresses of home location for each subsequent blocks to be written to journal region. The list represents the blocks to be written to home location after the transaction commit. During transaction commit phase, these are locked by journal transaction. The *pdflush* daemon of Linux may write journaled data to flash after commit ends. Thus, all updated data are duplicated.

In the *JFTL*, blocks in the list of *descriptor* is not written again to flash memory actually, but the *JFTL* remaps the physical location of journal region to home location. For this, we make a dedicated interfaces between journaling module and *JFTL* to pass these information. The lists recorded in *descriptor* are passed through the interface from Ext3 journaling module to *JFTL*, and *JFTL* uses these for updating the mapping table associated with blocks recorded in the list. The overall write process of journal including our remapping of duplicated blocks is as follows.

At first, when a new transaction is started, all updates are added and clustered to some buffers, and these are linked and locked. In the Figure 3, the updates, whose block numbers are 1 and 3, are clustered into running transaction.

Next, when the transaction expires, it enters commit phase where the clustered buffers, made in running phase, are inserted into list, log list. These data are written to journal region with two additional special blocks, *descriptor* and *commit*. In the Figure 4, Journal region represents the list to be written to journal region. In the journal, both *descriptor* and *commit* block have a magic header and a sequence number to identify associated transaction. The *descriptor* block records the home addresses of the list by checking their address of journal

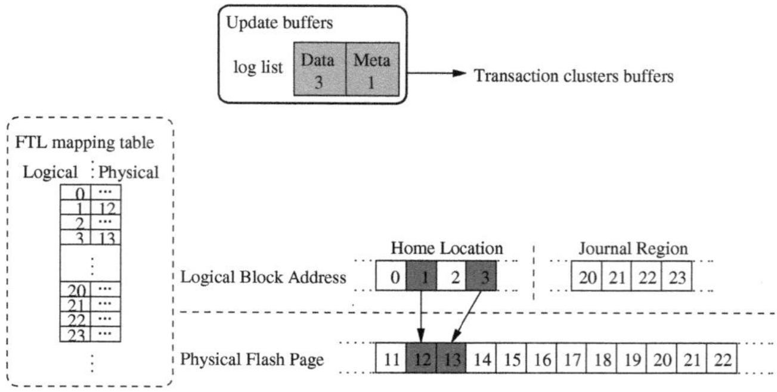

Fig. 3. In transaction running phase, buffers are clustered

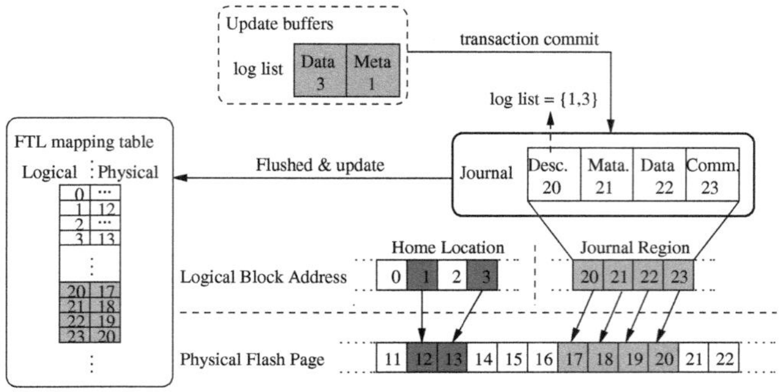

Fig. 4. Transaction commit phase makes journal and writes

region, according to the order of subsequent blocks that are written to journal region. For example, the block number of logging data within journal sequence is 1 and 3, so *descriptor* records the block number 1 and 3, which is associated to the journal logging sequence. When *descriptor* block is made with the list, blocks included in this transaction are flushed to block device layer. The submitted blocks are written to flash memory through the $JFTL$, and the $JFTL$ updates the mapping table. The region are logically journal region, but physically new free region. Up to now, journal data are flushed to storage.

After journal flushing is done, journaling module of Ext3 sends the list to $JFTL$ through the dedicated interfaces for remapping. $JFTL$ makes the remap list, *remap list*, from using information sent from journaling module. the *remap list* has not only home address but also journal address associated with it. For one element of the list is composed of home address and journal address to be updated, as shown in Figure 5. Specifically, the block number 1 of home location is related to the block number 21 of journal region, and the block number 3 of home location is related to the block number 22 of journal region. For each element, $JFTL$ finds out the physical address of the journal region, and updates mapping table of home location of the element with the physical address. Then,

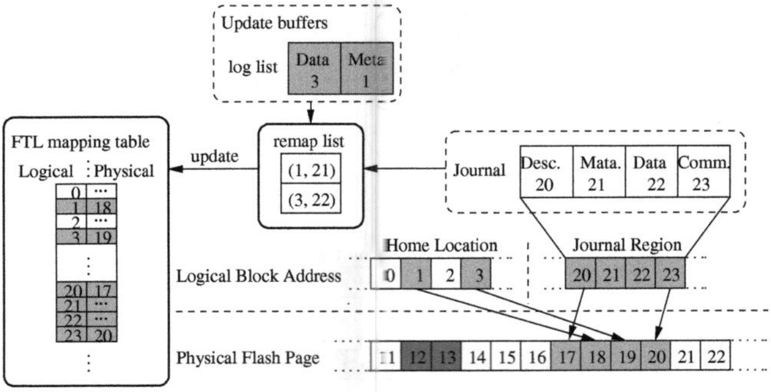

Fig. 5. JFTL remaps the home addresses of blocks to currently updated physical addresses

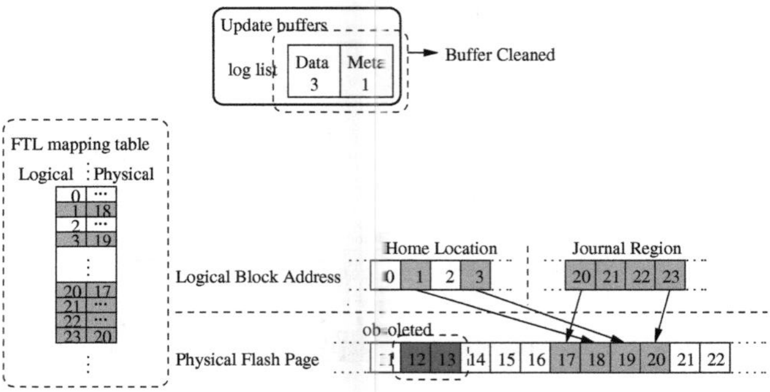

Fig. 6. Buffers are cleaned not to be flushed

JFTL clears the physical address of journal region. This clearing indicates that this region is obsoleted so it can be included in erase operation. In the Figure, in case of element (1,21), the physical address of journal region (21) is remapped to (18), so we link home location (1) to the physical address (18). Then we clear the physical address of journal region (21). When crash occurs before the remapping, this transaction journal data should be revoked and the file system status should be rolled back to the older version. This can be properly operated by recovering older version of mapping table. When crash occurs after the remapping, the new version is pointed by home location although this transaction is not properly completed. We don't worry about the consistency since the remapping can guarantee that all the journaling data in this transaction are written to storage.

When updating of mapping table associated to this transaction is done by remapping, the journaled data don't need to be written to flash memory any more, because all journal region are written out properly and the home location of that now points to the newly written data by remapping of these. The remaining job is to clear the buffers related to blocks of home location. This is possible by simply setting the BUFFER_CLEAN flag of *buffer head* structure of the blocks. By doing this, these are not flushed to flash by the *pdflush* daemon because it only flushed the dirty blocks by its mechanism. After all is done, the transaction is removed. Lastly, in *JFTL*, previous address of updated blocks now can be considered as obsoleted, so these are marked as dead in the table and can be included in erase operation.

3.2 Recovery

When system crashes, Ext3 performs checking of its superblock and journal superblock, and scans the journal region to recover system corruption. When crash occurs during journal writes, we should provide rollback mechanism for improperly logged transaction. The improperly logged transaction represents that crash occurs during this outstanding transaction write. The recovery of Ext3 with *JFTL* proceeds as follows. When Ext3 is mounted, it checks the mount flag in the superblock, and it finds that the system was crashed last. In this case, Ext3 checks journal superblock and identifies its consistency was corrupted by crash. After finding crashes, it identifies the committed transactions with their sequence number by scanning journal region. For the properly logged transactions, it reads *descriptor* block that is marked as same sequence number of superblock, checks the logged block addresses, and makes *remap list* for remapping these. The *remap list* is sent to *JFTL*, and *JFTL* updates mapping table, as commit does during runtime. Then, Ext3 scans more to find logged transactions and repeatedly performs the updating processes as described above. During the checking of journal region, if Ext3 finds improperly committed transaction, it stops the recovery process and returns. Lastly, Ext3 resets the journal region to be cleared and used later for journaling.

Different from previous recovery process, in case of properly logged transaction, our recovery policy doesn't perform copies of logged data to permanent home location, only remaps their addresses to the physical position of that by updating *JFTL* mapping table, just like that of run-time commit policy. Even though the amount of blocks to be recovered is small, our recovery mechanism is efficient because we only remap the committed transactions without copies to home locations.

4 Evaluation

The proposed Ext3 journaling mode with *JFTL*, will be compared to original Ext3 *ordered mode* and *journaled mode*. we exclude the *writeback mode* since it gives similar results of *ordered mode*. The experimental environment we use is NAND flash memory simulator based PC system. We use *nandsim* simulator [6] for the experiments. The nandsim can be configured with various NAND flash devices with associated physical characteristics. The physical characteristics of simulated the flash memory is as follows. Page size and block size of configured flash memory is 512Bytes and 16KBytes, respectively. One page read time and

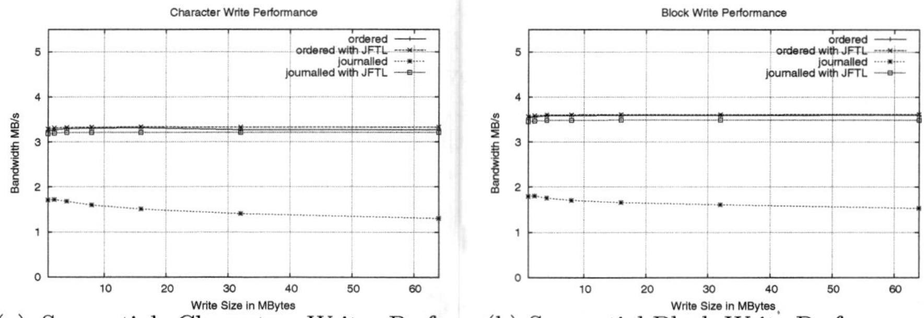

(a) Sequential Character Write Performance (b) Sequential Block Write Performance

Fig. 7. Bonnie Benchmark Performance. **The left graph plots character write performance, and the right graph plots block write performance as the amount of data written increases along the x-axis.**

programming time is configured to 25us and 100us, respectively. Block erase time is set to 2ms. Those configured settings are accepted among many flash memories. The reason we chose 512Bytes page size is to match with the sector size of disk. The capacity of nandsim is set to 128MB, which is allocated in main memory. For the performance evaluation, we used two benchmark programs, Bonnie [15] and IOzone [16]. For the benchmark tests, we have tested as follows, Ext3 ordered, Ext3 ordered with $JFTL$, Ext3 journaled, and Ext3 journaled with $JFTL$. Among many file system operations, we are interested in write performance of file system since journaling gives overheads for write performance. During all the experiments, we guarantee all the data are written to flash memory, not be cached to main memory.

We begin our performance evaluation with Bonnie benchmark [15]. Bonnie performs a series of tests one file of known size. The series of tests are composed of character write/read, block write/read/rewrite, and random write. Among these, we plots character write and block write performance as file size increases. The Figure 7 shows their results. First of all, we can identify throughout the graphs that the bandwidth of each mode does not changed largely as the written file size increases. It is from distinct feature of flash memory. Flash memory is electrical device and does not affected by any geometry structure. Thus, flash memory is free from complex scheduling overhead like disk scheduling. Instead of, flash memory is affected by garbage collection issue.

The left graph of Figure 7 plots character write performance, and the right graph of Figure 7 represents block write performance of Bonnie. The character write operation indicates that file is written using the *putc*() stdio. In block writes, file is written using *write*() calls with the amount of block size, 4KB. They show similar results of write trends without average bandwidth rate values. In the graphs, among two ordered modes, the ordered with $JFTL$ slightly outperforms the ordered mode. It is the effect of remapping of metadata not real writing to home location. However, compared to large amount of data, the

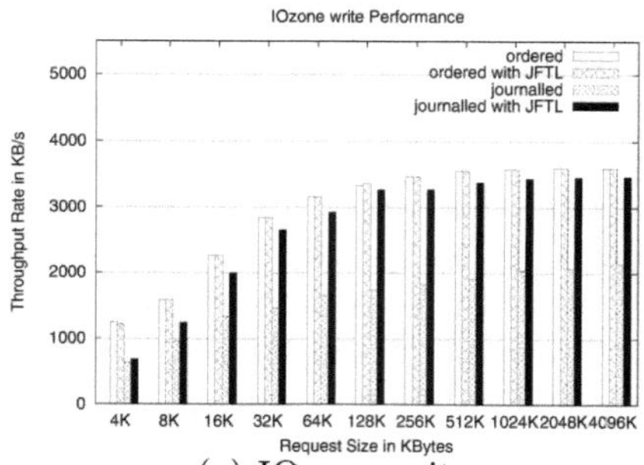

(a) IOzone write

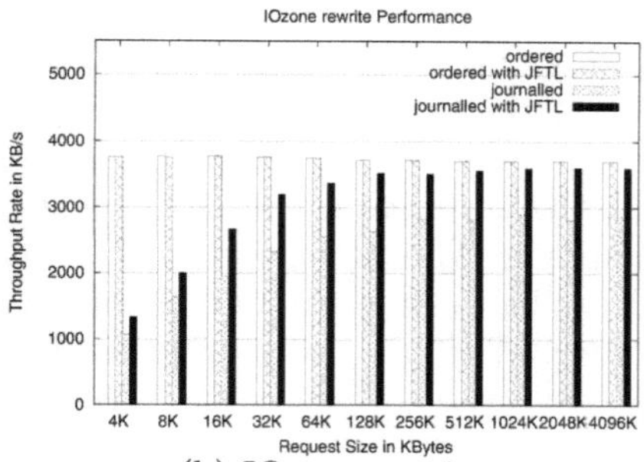

(b) IOzone rewrite

Fig. 8. IOzone Benchmark Performance. **The Left graph plots write performance, and the right graph plots rewrite performance as write request size per write operation increases along the x-axis.**

amount of metadata is negligible, so the performance gap is small. If we compare two journaled modes, journaled mode with $JFTL$ greatly outperforms the pure journaled mode, as we expect. In case of journaled mode, bandwidth is slowly down as file size increases, which is due to the number of descriptors in the journal transactions. As the written data is increased within a transaction interval, number of descriptors is also increased. If journaled with $JFTL$ is compared with ordered modes, bandwidth degradation of the journaled with $JFTL$ is not a serious problem. The journaled with $JFTL$ mode can give similar performance to ordered mode while preserving strict data consistency.

Next benchmark program we tested is IOzone benchmark [16]. It generates and measures a variety of file operations including write/rewrite, read/reread, aio read/write, and so on. Among these, we plot write performance verifying that data is actually written to flash memory. we have experimented write operation while varying the write request size per write operation from 4KBytes to 4096KBytes when file, whose size is 4096KBytes, is created and written. It means that 1024 write operation occurs for 4KBytes write size, whereas only one write operation occurs for 4KBytes write size, when 4096KBytes file is written. In the Figure 8, left graph plots write performance, and right graph plots rewrite performance as write size per write operation increases along the x-axis. The difference between write and rewrite is that write means new data is written and rewrite means the update of previous data. From the left graph, we identify that throughput of new data increases as the write size increases, which is due to the metadata update between each write intervals, such as free block allocation. For small write request size, performance is much degraded in both ordered mode and journaled mode. It results from the fact that free block allocation and metadata operation is performed for each write operation. As write request size increases, throughput is increased since metadata operation number is decreased. However, it is not required to allocate free blocks when the data is rewritten, which means metadata operation decreases. Thus, throughput of ordered mode is constant through the request sizes, as we see from the right graph. In all experiments, we can identify that the performance of ordered with $JFTL$ mode and journaled with $JFTL$ outperforms its comparison.

From the experiments, we note several results. First, journaling itself gives overhead to system while providing consistency. Second, when *ordered mode* is used, the performance of $JFTL$ is slightly better than original with the help of remapping. Third, the *journaled mode* downs system performance extremely, at least half of the throughput and latency, even if it provides strict data consistency. Whereas, the journaled with $JFTL$ does not harm largely, and gives little overhead to the system, while providing strict data consistency.

5 Conclusion

Flash Memory has become one of the major components of data storage since its capacity has been increased dramatically during last few years. Although it has many advantages for data storage, many operations are hampered by its physical characteristics. The major drawbacks of flash memory are that bits can only be cleared by erasing a large block of memory and the erasing count of each block has a limited number. Considering a journaling file system on the FTL, the duplication of data between the journal region and its home location is crucial for the file system consistency. However, the duplication degrades the file system performance.

In this paper, we present an efficient journaling interface between file system and flash memory. The proposed FTL, called $JFTL$, eliminates the redundant data copy by remapping the logical journal data location to real home location. Our proposed method can prevent from degrading system performance while preserving file system consistency. Our approach is layered approach, which means that any existing file system can be setup upon our method with little modification to provide file system consistency. In the implementation, we consider Ext3 file system among many kinds of journaling file systems because it is the main root file system of Linux.

References

1. Douglis, F., Caceres, R., Kaashoek, F., Li, K., Marsh, B., Tauber, J.A.: Storage alternatives for mobile computers. In: Proc. of the 1st Symposium on Operating Systems Design and Implementation(OSDI), pp. 25–37 (1994)
2. Best, S.: JFS Overview (2004), `http://www.ibm.com/developers-orks/library/l-jfs.html`
3. Sanmsung Electronics Co., NAND Flash Memory & SmartMedia Data Book (2002), `http://www.samsung.com/`
4. Ban, A.: Flash file system. U.S. Patent 5404485 (Arpil 4, 1995)
5. Intel Corporation: Understanding the flash translation layer(FTL) specification, `http://developer.intel.com/`
6. Memory Technology Device (MTD) subsystem for Linux, `http://www.linux-mtd.infradead.org`

7. Gal, E., Toledo, S.: Mapping Structures for Flash Memories: Techniques and Open Problems. In: Proceedings of the IEEE International Conference on Software-Science, Technology and Engineering (2005)
8. Woodhouse, D.: JFFS: The Journalling Flash File System. Ottawa Linux Symposium (2001)
9. Aleph One Ltd: Yaffs: A NAND-Flash File system, `http://www.aleph1.co.uk/yaffs/`
10. Lim, S.-H., Park, K.H.: An Efficient NAND Flash File System for Flash Memory Storage. IEEE Transactions on Computers 55(7), 906–912 (2006)
11. Kawaguchi, A., Nishioka, S., Motoda, H.: A Flash-Memory Based File System. Usenix Technical Conference (1995)
12. Rosenblum, M., Ousterhout, J.K.: The Design and Implementation of a Log-Structured File System. ACM Transactions on Computer Systems 10(1) (1992)
13. Ts'o, T., Tweedie, S.: Future Directions for the Ext2/3 File system. In: Proceedings of the USENIX Annual Technical Conference(FREENIX Track) (June 2002)
14. Timothy, E., et al.: Journal-guided Resynchronization for Software RAID. In: Proceedings of the 4th USENIX Conference on File And Storage Technologies(FAST) (December 2005)
15. Tim Bray: Bonnie Benchmark, `http://textuality.com/bonnie/`
16. William, D.: Norcott: Ioznoe File system Benchmark, `http://www.iozone.org/`

A Complex Network-Based Approach for Job Scheduling in Grid Environments

Renato P. Ishii[1], Rodrigo F. de Mello[2], and Laurence T. Yang[3]

[1] Federal University of Mato Grosso do Sul – Department of Computer and Statistics
Campo Grande, MS, Brazil
renato@dct.ufms.br
[2] University of São Paulo – Institute of Mathematics and Computer Sciences
Department of Computer Science – São Carlos, SP, Brazil
mello@icmc.usp.br
[3] St. Francis Xavier University – Antigonish, NS, Canada
lyang@stfx.ca

Abstract. Many optimization techniques have been adopted for efficient job scheduling in grid computing, such as: genetic algorithms, simulated annealing and stochastic methods. Such techniques present common problems related to the use of inaccurate and out-of-date information, which degrade the global system performance. Besides that, they also do not properly model a grid environment. In order to adequately model a real grid environments and approach the scheduling using updated information, this paper uses complex network models and the simulated annealing optimization technique. The complex network concepts are used to better model the grid and extract environment characteristics, such as the degree distribution, the geodesic path, latency. The complex network vertices represent grid process elements, which are generalized as computers. The random and scale free models were implemented in a simulator. These models, associated with Dijkstra algorithm, helps the simulated annealing technique to find out efficient allocation solutions, which minimize the application response time.

1 Introduction

The evolution of the computer electronics have made feasible the production of low cost microprocessors and the computer networks dissemination. This motivated the development of distributed systems. Applications executing on such systems have been presenting a complexity growth. Moreover the high cost of specialized resources, such as parallel machines and clusters, demonstrate the need of a technology able to provide high scale distributed resources. Those needs have been motivating the development of grid computing. The resource allocation, in such systems, motivated the study of new job schedulers addressed to large scale environments.

The job scheduling problem is an important subject in distributed systems, which aims at taking decisions based on policies such as [1,2,3,4,5]. Policies distribute processes, which compose distributed parallel applications, on the available processing elements (PEs), according to different objectives such as: load balancing, reduce application response time and improve the resource usability. Such scheduling policies,

R. Perrott et al. (Eds.): HPCC 2007, LNCS 4782, pp. 204–215, 2007.

proposed for conventional systems (parallel machines, clusters and NOWs[1]) do not provide good solutions for grids, which bring some new functional requirements and objectives.

Grid resources are heterogeneous and of unpredictable availability. Common decisions to be taken by the scheduler, in such environments, are: which applications will have access to the computational resources, the amount and localization of resources to be used by applications. These decisions are affected by different factors, such as the imposed system workload, the resource heterogeneity and user requirements. Besides that, the scheduling has to deal with specialized applications which have different resource usage behaviors [6].

The computational grids can better be represented, and also the scheduling problem, by using complex networks with optimization techniques. Complex networks are a knowledge representation theory which use topological structures (graphs) to store information. They allow diverse connection types to be established among entities, better representing the real system. In this work, the network vertices represent computers and the edges, the connectivity among them. Relevant grid characteristics and properties can be mapped in complex network models, such as the degree distribution, distance and latency between vertices. Such information can be used by the scheduling optimization technique.

By observing the advantages of complex networks, Zhao *et al.* [7] and Costa *et al.* [8] proposed specific solutions to computing problems. In the first work, the congestion control problem, in Internet gateways, was modeled by using complex networks. Important metrics were defined to prevent traffic chaotic situations in communication channels. In the second work, complex network models characterize grid environments. Experiments provide model comparisons and conclude about how they can improve parallel application performance.

Motivated by such works, this paper adopts complex network models and the simulated annealing technique to optimize the job scheduling problem in computational grids. The complex network usage is convenient to model grid environment characteristics, such as: statistical metrics to remove and insert computers, the distance and connectivity among computers. The random and scale free models and the simulated annealing technique were implemented in a simulator. Results demonstrate the scheduling behavior in environments following the two complex network models.

This paper is organized as follows: section 2 reviews the related work; concepts on complex networks are described in section 3; section 4 presents the simulated annealing technique; the job scheduling model is proposed in section 5; section 6 presents the results; conclusions are described in section 7.

2 Related Work

The grid computing research is mainly concentrated in the coordination of geographically distributed resources aiming high performance. Some of the main approaches are [9,10,1,2,3].

In [9] the UGSA (Unified Grid Scheduling Algorithm) algorithm is proposed, which considers the information on the system workload and the communication latency for

[1] NOWs – Network of Workstations.

job scheduling. Applications are mapped in DAGs (Directed Acyclic Graph) which define how processes will be allocated in the grid processing elements. In this work, a genetic algorithm module is also defined, which is executed when processes arrive at the system.

In [10] a stochastic algorithm, named stochastic greedy, is proposed which prevents degradation in the system performance. Dynamic hash tables and data replication techniques are used to keep the system workload updated and, in this way, to consider the last information in the job scheduling. The proposed algorithm is compared to the original greedy algorithm indicating higher performance and lower overload.

In [2] the TDS policy [1] is extended by solving some of its limitations. Optimizations are defined to determine whether the scheduling result is good or not, according to the policy objective. The main idea is to duplicate processes before placing in the processing elements. The TDS is conceptually compared to the TDS extended, although no real results are presented to quantify such assumptions.

In [3] the opportune cost policy is proposed, which considers applications in which processes may cause communication overhead. A marginal cost function [11] is adapted to allow process reallocation on processing elements. The routing concepts of virtual circuits are used to determine the CPU overhead in communication terms. Each system processing element has multiple resources: CPU, memory, input/output devices, etc. The objective is to minimize the overload caused by using such resources. The proposed overhead evaluation model does not consider the impact that each process cause in the network communication. Besides that, the model does not evaluate the overhead weight in a measure which allows to determine the CPU delay, such as the number of MIPS (millions of instructions for second) consumed when sending and receiving network messages.

3 Complex Networks

A network is composed of a set of interconnected elements [12] called vertices or nodes, which are linked by edges. Many kinds of systems can be modeled by using this concept, such as social, information, technological and biological networks.

Newman [12] has defined three topics to be prioritized when studying complex networks. The first aims at finding topological properties such as the path length among vertices and the degree distribution, what characterizes the network structure and behavior. For this, we may define mechanisms to measure and collect network properties. The second aims at creating a complex network model which helps to understand such properties, for instance, the interaction degree among elements. The third objective aims at predicting the network behavior based on measured properties and local rules, such as how the network structure affects the Internet traffic, or how the web server structure influences on its performance.

3.1 Complex Networks Models

In the last years, lots of attention has been paid to complex networks what is evident by studies in diverse areas: social, information, technological, and biological networks.

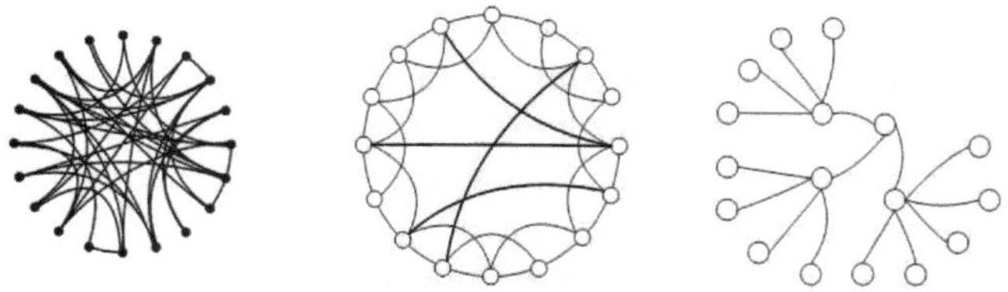

(a) Random model (b) Small world model (c) Scale free model

Fig. 1. Examples of complex network models

This motivated the design of evaluation models to explain the area characteristics and properties. The main models are:

Random Model. The random model, defined by Erdös and Rény [13], is considered the most basic complex network. This model generates random graphs with N vertices and K edges, called random graph ER. Having N initially detached vertices, the model ER is obtained by randomly connecting selected vertices (there is a probability for an edge binding each two vertices) until the number of graph edges is equal to K (techniques may be used to avoid multiple connections).

In this network model, the vertex expected degree is defined by the equation 1.

$$G^{ER}_{N,K} \quad (1)$$

The figure 1(a) presents an example of the random model. By considering that the generation of network connections is random, Erdös and Rényi concluded that all vertices, in a specific network, have approximately the same number of connections, or the same probability to have new connections [14].

Small World Model. Duncan Watts and Steven Strogatz [15] discovered that networks present high interconnection patterns, tending to form clusters of vertices. They had considered a model similar to the one by Erdös and Rényi, where the edges are established between adjacent vertices while part of them are randomly defined. This model demonstrates that the average distance between any two vertices, in large networks, does not exceed a small number of hops [16].

In the Watts and Strogatz model each vertex knows the localization of several others. In large scale, such connections guarantee few separation hops among the network vertices. In a large network, few edges among clusters are enough to form a small world, transforming the network into a large cluster [16]. An example of this network model is presented in figure 1(b).

Scale Free Model. Barabási and Albert demonstrate in [14] that some networks are not randomly built. They define the existence of an order in the network structure dynamics with some specific characteristics. One of such characteristics is called preferential attachment: a new vertex tends to connect to any other, although there is a high probability to attach it to the highly connected vertices. This implies in another basic premise: networks are not formed by degree-equitable vertices.

On another hand, such networks present few highly connected vertices (hubs), being most of them with few connections. Hubs always tend to receive more connections. Networks with such characteristics (example in the figure 1(c)) are named scale free.

The name scale free comes from the mathematical representation characteristics of the connection degree presented by the network vertices, which follows a very particular distribution, named power law. This is a logarithmic distribution which decreases abruptly and has a long tail. This distribution implies that most of the vertices have just few connections and a significant minority of them has a large number of connections. More specifically, the degree distribution follows a power law for a large k according to the equation 2.

$$P(k) \sim k^{-\gamma} \tag{2}$$

4 Simulated Annealing

The simulated annealing (SA) technique aims at finding a global minimum for an energy function [17]. Its name comes from an analogy to the metallurgy annealing, which consists of the controlled heating and cooling of materials aiming at reducing the physical defects.

Algorithm 1. Simulated Annealing

```
Let the initial solution S0, the initial temperature T0, the cooling rate α and the maximum number of iterations to
reach stability Iter_max
N(S) // possible solutions space
S ← S0; // current solution
S' ← S; // better solution until the current moment
T ← T0; // current temperature
IterT ← 0; // iteration number T
while T > 0 do
    while IterT < Iter_max do
        IterT ← IterT + 1;
        Generate a neighbor S' ∈ N(S)
        Δ ← f(S') − f(S)
        if Δ < 0 then
            S ← S';
            if f(S') < f(S*)) then
                S* ← S'; // best solution
            end if
        else
            Generate a random x ∈ [0, 1];
            if x < e^(−Δ/T) then
                S ← S';
            end if
        end if
    end while
    T ← α * T;
    IterT ← 0;
end while
Return S*;
```

This technique introduces the system temperature concept, which makes possible to find global minima. In this way, for each iteration, the algorithm considers a possible neighbor state to the current S, which is named S'. During this evaluation, the transition probability from the current state to the new one is randomly determined, trying to pass the system to less energy states. The neighborhood is explicitly defined, being able to present some variations for specific problems.

The state transition probability is calculated by the function $P(E, E', T)$, where E and E' represent the energies $E(S)$ and $E(S')$ for the states S and S', respectively, and the parameter T is the system temperature, modified as the execution flows. The essential property of this function is that the transition probability P must be different from zero, with $E_0 > E$ which indicates a transition to a worse state (i.e., it presents higher energy). This system characteristic is used as a way to prevent the system stagnation in a local minimum.

On the other hand, when the system temperature T is next to zero, the probability function must tend to zero if $E_0 > E$, otherwise return a positive value. This guarantees, for low values of T, that the system tends to find lower energy solutions. In this way, the simulated annealing technique can be used to determine behavior changes, trying to offer a global adequate solution for the optimization problem. However, to apply this technique, we must know the system in order to determine the temperature variation when searching the global minimum. The simulated annealing solution adopted in this work is presented in the algorithm 1.

5 Model for Job Scheduling in Grid Computing

Motivated by results presented in [7][8] and by the need of wide scale systems, this paper adopts a model based on complex networks to represent a grid computing topology and study the job scheduling problem. From this model it is possible to determine important characteristics to take scheduling decisions such as: the connection degree among vertices, the communication channel latency and the overhead. Such characteristics are included in the proposed job scheduling algorithm which is based on the simulated annealing technique. This algorithm aims at reducing the application response time and increasing the system throughput.

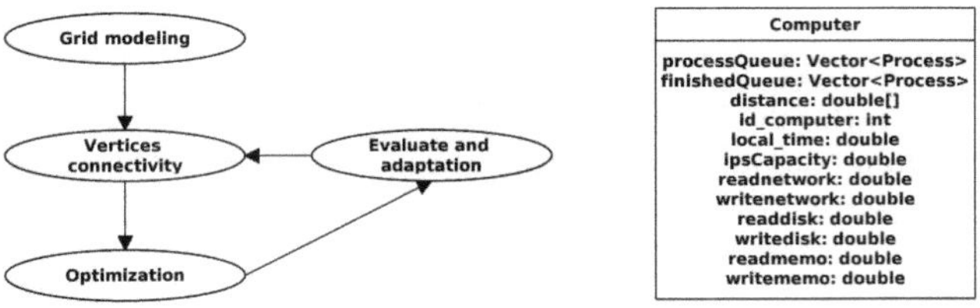

(a) The model architecture **(b) A grid processing element**

Fig. 2. Modeling the job scheduling problem

The proposed model architecture is presented in the figure 2(a), which is composed of the following modules:

1. **Grid computing modeling** – In this module, complex network models are used to characterize the grid processing elements, i.e., the computers. Initially, the random and scale free models were studied and implemented in a simulator;

2. **Vertices connectivity** – This module is responsible for the complex network configuration which characterizes the relationships among vertices. In the simulator, each network vertex represents a processing element and its attributes are (figure 2(b)):
   (a) processQueue and finishedQueue – they represent the ready and finished process queues, respectively;
   (b) distance – this vector stores the costs to reach each network vertex. The costs are calculated using the Dijkstra algorithm;
   (c) id_computer – it identifies each grid computer;
   (d) local_time – it represents the computer local clock;
   (e) other fields – there are other fields to represent the processing capacity, memory read and write operations, network read and write operations, which are defined by probability distribution functions (PDF). For example, the ipsCapacity is defined as an exponential PDF with average 1500, defining the capacity of processors. Other attributes are defined in the same way.

   The failure of a communication channel is represented, in the complex network, as the edge elimination. In the same way, the failure of a processing element is represented by the vertex elimination. The elimination as well as the insertion of new vertices requires the model reconfiguration, in order to adjust the costs of better paths between vertices. This is obtained by using the Dijkstra algorithm, i.e., for each vertex elimination/insertion in the environment, the algorithm is re-executed. In the proposed simulator, the insertion as well as the removal of grid processing elements is modeled by PDFs, as demonstrated in the section 6.
3. **Scheduling optimization** – This paper considers complex network models and the simulated annealing optimization technique to improve the application performance running on grids. After defining the system model and calculating all the costs to reach any vertex, from any origin, we have to allocate processes in grid computers. In order to solve this problem, optimization techniques can be applied to generate solutions. In this work we adopt the simulated annealing technique (SA) to improve process allocations. The problem solution is represented by a matrix $M = (m_{pc})$, where p is the process number (matrix rows) and c is the computer number (matrix columns). The allocation of the process p in a computer c is represented by $Proc = m_{pc}$.

   An example is presented in the figure 1, where each row represents a process and each column a computer. Matrix elements with value 1 represent the process

Table 1. A solution matrix

	Computers				
Processes	1	2	3	4	5
1	0	0	0	1	0
2	0	1	0	0	0
3	1	0	0	0	0
4	0	0	1	0	0
5	0	0	0	0	1
6	0	1	0	0	0
7	0	0	0	1	0
8	1	0	0	0	0
9	0	0	0	1	0

and computer association, i.e., which computer (column) will execute which process (row) according to the suggested solution. A process can be only placed in a computer.

The SA simulates a method analogous to the cooling of atoms in thermodynamics, known as annealing. The algorithm procedure (algorithm 1) is invoked whenever there are processes to be allocated. The adopted energy function (equation 3, where $getProcessET_{pc}$ returns the estimated execution time of the process p when placing it in the computer c) aims at finding solutions which minimize the total process response time.

$$\frac{1}{\sum getProcessET_{pc}} \tag{3}$$

The proposed optimization algorithm is decomposed in two overlapping searches:

- The internal search contains the optimization process. For each iteration, the solution neighborhood is explored;
- The external search controls the process finish. It is based on cooling state concepts and considers that a state is cooled when there is no possibility to reach a better solution.

The algorithm starts searching based on an initial solution. For instance, consider a parallel application launched in the computer c with $\mathsf{id\_computer} = 1$. In this way, all the application processes are located in the column corresponding to that computer in the initial solution matrix (example in table 2).

Table 2. Initial matrix solution

	Computers				
Processes	1	2	3	4	5
1	1	0	0	0	0
2	1	0	0	0	0
3	1	0	0	0	0
4	1	0	0	0	0
5	1	0	0	0	0
6	1	0	0	0	0

In each loop iteration, which characterizes the main procedure, a solution S', which is a neighbor of the current one S, is randomly generated. The energy function value variation is evaluated to each neighbor solution generated. In this evaluation the equation 4 is calculated. When $\Delta < 0$, S' starts to be the new current solution. When the generated neighbor is worse, in a quantity $\Delta > 0$, it can be accepted with a probability equals to $e^{-\Delta/T}$, where T is the temperature parameter.

$$\Delta = f(S') - f(S) \tag{4}$$

This process is repeated until T is too small that no movement can be accepted, i.e., the system reaches stability. The obtained solution, in this stability instant, evidences a local minimum. The probability to accept movements, which degrade the energy function value, decreases proportionally to the temperature.

4. **Evaluation and adaptation** – After the optimizer execution, i.e. the simulated annealing procedure, a new solution (matrix) for a specific application is obtained. Then, the processes are migrated to the respective computers, i.e., the system is adapted to the new solution, until receiving new applications (what launches the optimizer again).

A simulator was developed to analyze the job scheduling performance under different complex network models for grid computing. This simulator, developed in the Java language, implements the random and scale free complex network models. Besides that, it also implements the simulated annealing technique (section 4) to optimize the job scheduling problem in grids.

The main module of this simulator is the class *Simulator*, responsible for all simulation process control. Computers are represented by the class *Computer*, and its characteristics presented in the figure 2(b). After creating all computers, they are associated to one of the two possible complex network models: *Random* or *ScaleFree*. After that, the communication costs are calculated by using the Dijkstra algorithm (class *Dijkstra*).

The application submission is simulated by attributing processes (object of the class *Process*) to computers and calling the optimizer (class *SA*), which applies the simulated annealing algorithm to look for job scheduling solutions. Process characteristics are obtained by a trace file (class *Log-Feitelson*) [6].

This trace follows the Feitelson's workload model[2] [6] which is based on the analysis of six execution traces of the following production environments. This model defines that the process arrival follows an exponential PDF with average 1500 seconds. The other simulator parameters are also defined by PDFs[3]: computer capacity in MIPS; read and write network throughput in KBytes/s; read and write memory throughput in MBytes/s; read and write disk throughput in MBytes/s; process workload in MIPS; arrival interval of parallel applications; process memory occupation in MBytes: used to define the cost to transfer the process in case of migration; probability distributions to define when vertices (i.e. computers) are inserted and removed from the complex network.

6 Results

Experiments were conducted to determine which environment configuration presents better results for the job scheduling, i.e., lower parallel application response time. Other simulator parameters had to be defined for such experiments. Some of those parameters were defined in accordance with previous experiments on real environments, others in accordance with our experience and expectations on grid computing environments. The parameters are:

– application arrival time – it is represented by an exponential PDF with average 1500 seconds, which follows the behavior detected by Feitelson [6];

[2] Feitelson's workload model – available at http://www.cs.huji.ac.il/labs/parallel/workload/models.html.

[3] This simulator uses the probability distribution functions from the PSOL library, available at http://www.math.uah.edu/psol/objects/PSOL.jar.

- addition of computers in the environment – the interval between consecutive additions is represented by an exponential PDF with average 2500 seconds and another exponential PDF with average 20 for the number of computers to be added;
- removal of computers from the environment – the interval between consecutive removals is represented by an exponential PDF with average 3500 seconds and another exponential PDF with average 10 for the number of computers to be removed;
- number of changes made by the SA algorithm – 5 and 10 changes when searching for a global minimum;
- complex network average degree – 3, 5 and 7 which represent the average number of connections for each network vertex;
- process size – exponential PDF with average 2MB which represents the process memory occupation. This is used by the SA algorithm to calculate the migration cost when transferring the process over the communication channel;
- network read and write operations – they are represented by an exponential PDF with average 100KB, i.e., each process, on average, reads or writes 100KB in the network;
- disk read and write operations – they are represented by an exponential PDF with average 100KB;

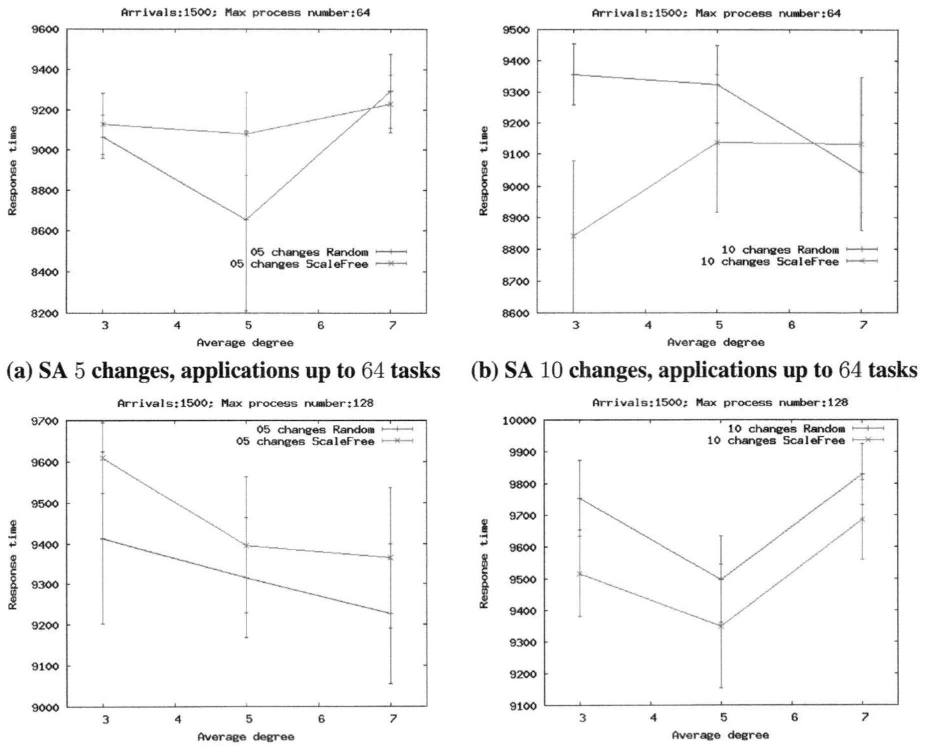

(a) **SA** 5 **changes, applications up to** 64 **tasks** (b) **SA** 10 **changes, applications up to** 64 **tasks**

(c) **SA** 5 **changes, applications up to** 128 **tasks** (d) **SA** 10 **changes, applications up to** 128 **tasks**

Fig. 3. Scheduling results

- memory read and write operations – they are represented by an exponential PDF with average 500KB;

The figures 3(a), 3(b), 3(c) and 3(d) present the application response times. The lower the response time is, the higher is the performance.

The first case (figure 3(a)) presents the obtained results for the SA algorithm with 5 changes executing parallel applications composed of up to 64 tasks. We observe that the response time for the random complex network model had better results. Only for the last case, with larger average degree of the vertices, the scale free model outperforms the random.

The second case (figure 3(b)) considers 10 changes for the SA, executing applications of up to 64 tasks. In this situation the largest number of changes increases the scale free performance, which, only in the last case, is surpassed by the random model.

The figure 3(c) presents results considering 5 changes for the SA and parallel applications composed of up to 128 tasks. The figure 3(d) considers 10 changes for the SA and applications with up to 128 tasks. In these two cases, we observe that by increasing the number of changes, the scale free model outperforms the random one.

By knowing the Internet follows the scale free model [18,19] and computational grids are implemented on such infrastructure, we can conclude that algorithms which explore a large solution space provide better results on such model. This becomes evident when, by using a large number of changes in the SA algorithm, better solutions are obtained. Situations where the algorithm does not provide a large solution space, the inherently characteristics of the complex network random model assists in the process distribution, however this is not the case when considering grid environments.

7 Conclusion

In order to better model a real grid computing environment and approach the scheduling problem using updated information, this paper considers a scheduling model applying two complex network models and the simulated annealing optimization technique. Complex networks assist in the grid modeling and also in the extraction of the environment characteristics such as: interconnection degree, distance and latency between pairs of vertices (which represent computers).

Experiments were conducted using a simulator which allowed to observe that the inherently Internet characteristics, which follows the scale free model, needs an optimization algorithm with a large search space to provide good scheduling solutions. The complex network random model assists in the process distribution, as the search space is amplified by the inherently interconnection of vertices. Therefore, the interconnection among vertices defines the optimization approach to be adopted.

Acknowledgments

This paper is based upon work supported by CAPES, Brazil under grant no. 032506-6. Any opinions, findings, and conclusions or recommendations expressed in this material are those of the authors and do not necessarily reflect the views of CAPES.

References

1. Ranaweera, S., Agrawal, D.P.: A task duplication based scheduling algorithm for heterogeneous systems. In: Parallel and Distributed Processing Symposium - IPDPS 2000, pp. 445–450 (2000)
2. Choe, T., Park, C.: A task duplication based scheduling algorithm with optimality condition in heterogeneous systems. In: International Conference on Parallel Processing Workshops (ICPPW'02), pp. 531–536. IEEE Computer Society Press, Los Alamitos (2002)
3. Keren, A., Barak, A.: Opportunity cost algorithms for reduction of i/o and interprocess communication overhead in a computing cluster. IEEE Transactions on Parallel and Distributed Systems 14, 39–50 (2003)
4. Wang, Q., Zhang, L.: Improving grid scheduling of pipelined data processing by combining heuristic algorithms and simulated annealing. In: Computer and Computational Sciences, 2006. IMSCCS '06. First International Multi-Symposiums on, vol. 1, pp. 583–588 (2006)
5. Pop, F., Dobre, C., Godza, G., Cristea, V.: A simulation model for grid scheduling analysis and optimization. In: Parallel Computing in Electrical Engineering, 2006. PAR ELEC 2006. International Symposium on, pp. 133–138 (2006)
6. Feitelson, D.G.: Packing schemes for gang scheduling. In: IPPS '96: Proceedings of the Workshop on Job Scheduling Strategies for Parallel Processing, London, UK, pp. 89–110. Springer, Heidelberg (1996)
7. Zhao, L., Lai, Y.-C., Park, K., Ye, N.: Onset of traffic congestion in complex networks. Phys. Rev. E Stat. Nonlin. Soft Matter Phys. 71, 26–125 (2005)
8. Costa, L.F., Travieso, G., Ruggiero, C.A.: Complex grid computing. The European Physical Journal B - Condensed Matter 44(1), 119–128 (2005)
9. Aggarwal, A.K., Aggarwal, M.: A unified scheduling algorithm for grid applications. In: High-Performance Computing in an Advanced Collaborative Environment, 2006. HPCS 2006. 20th International Symposium on, pp. 1–1 (2006)
10. Zheng, Q., Yang, H., Sun, Y.: How to avoid herd: a novel stochastic algorithm in grid scheduling. In: High Performance Distributed Computing, 2006 15th IEEE International Symposium on, pp. 267–278. IEEE Computer Society Press, Los Alamitos (2006)
11. Aspnes, J., Azar, Y., Fiat, A., Plotkin, S., Waarts, O.: On-line routing of virtual circuits with applications to load balancing and machine scheduling. J. ACM 44, 486–504 (1997)
12. Newman, M.E.J.: The structure and function of complex networks. SIAM Review 45, 167–256 (2003)
13. Erdos, P., Renyi, A.: On random graphs. Publicationes Mathematicae 6, 290–297 (1959)
14. Barabasi, Albert.: Emergence of scaling in random networks. Science 286, 509–512 (1999)
15. Watts, D.J., Strogatz, S.H.: Collective dynamics of 'small-world' networks. Nature 393, 440–442 (1998)
16. Buchanan, M.: Nexus - Small Worlds and the Groundbreaking Science of Networks. W. W. Norton & Company (2002)
17. Kirkpatrick, S., Gelatt, C.D., Vecchi, M.P.: Optimization by simulated annealing. Science, Number 4598, 220 (4598), 671–680 (1983)
18. Albert, R., Barabasi, A.L.: Statistical mechanics of complex networks. Reviews of Modern Physics 74, 47–101 (2002)
19. Faloutsos, M., Faloutsos, P., Faloutsos, C.: On power-law relationships of the internet topology. In: SIGCOMM '99: Proceedings of the conference on Applications, technologies, architectures, and protocols for computer communication, pp. 251–262. ACM Press, New York (1999)

Parallel Database Sort and Join Operations Revisited on Grids

Werner Mach and Erich Schikuta

University of Vienna
Department of Knowledge and Business Engineering
Research Lab Computational Technologies and Applications
Rathausstraße 19/9, A-1010 Vienna, Austria
werner.mach@univie.ac.at, erich.schikuta@univie.ac.at

Abstract. Based on the renowned method of Bitton et al. (see [1]) we develop a concise but comprehensive analytical model for the well-known Binary Merge Sort, Bitonic Sort, Nested-Loop Join and Sort Merge Join algorithm in a Grid Environment.

We concentrate on a limited number of characteristic parameters to keep the analytical model clear and focused. Based on these results the paper proves that by smart enhancement exploiting the specifics of the Grid the performance of the algorithms can be increased and some results of Bitton et al. for a homogenous multi-processor architecture are to be invalidated and reversed.

1 Introduction

Today the Grid gives access to viable and affordable platforms for many high performance applications hereby replacing expensive supercomputer architectures.

Basically the Grid resembles a distributed computing model providing for the selection, sharing, and aggregation of geographically distributed autonomous resources dynamically at runtime depending on their availability, capability, performance, cost, and users quality-of-service requirements (see [2]).

We believe that the Grid delivers a suitable environment for parallel and distributed database systems. There is an urgent need for novel database architectures due to new stimulating application domains, as high energy physics experiments, bioinformatics, drug design, etc., with huge data sets to administer, search and analyze. This situation is reflected by a specific impetus in distributed database research in the Grid, starting with the Datagrid project [3] and now manly carried by the OGSA-DAI project [4]. After a few years where the research was mainly in the area of distributed data management, now a new stimulus on research on parallel database operators focusing Grid architectures (see [5], [6]) can be noticed.

Sorting and Joining are extremely demanding operation in a database system. They are the most frequently used operators in query execution plans generated

R. Perrott et al. (Eds.): HPCC 2007, LNCS 4782, pp. 216–227, 2007.

by database query optimizers. Therefore their performance influences dramatically the performance of the overall database system [7], [8]. Generally these algorithms can be divided into internal (main memory based) and external (disk based) algorithms [9]. An external algorithm is necessary, if the data set is too large to fit into main memory. Obviously this is the common case in database systems.

In this paper we present an analysis and evaluation of the most prominent parallel sort and join algorithms, Binary Merge Sort and Bitonic Sort, and Nested-Loop and Sort Megre Join in a Grid architecture based on the well known analysis and findings of Bitton et al. [1]. Now these algorithms are reviewed under the specific characteristics of the Grid environment. The surprising fact, justified by the findings of this paper and resulting from the characteristic situation of the Grid, is that the some results on the general performance of these parallel algorithms are invalidated and reversed, i.e. in a Grid environment the performance of these algorithms can be improved choosing an adapted workflow layout on the Grid taking smartly into account the specific node characteristics.

The paper is organized as follows. In the next section the architectural framework of the Grid environment is laid out. This is followed by the description of the parallel algorithms, their specific characteristics and the definition of the basic parameters of the analysis. In section 5 a comprehensive analytical formulation of the parallel algorithms is given, followed by the evaluation of the algorithms compared and discussion of the findings. The paper is closed by a conclusion and a presentation of topics for further research.

2 Architectural Framework

2.1 A Generalized Multiprocessor Organization

The work of Bitton et al. [1] is based on a so called generalized multiprocessor organisation, which basically comprises a *homogenous* classical supercomputer architecture, where every working node shows the same characteristics in processing power, every disk node in I/O performance and the network interconnect bandwidth is the same for all node. So basically for the evaluation of the parallel operations on the generalized multiprocessor organization only the following components are considered:

1. a set of general-purpose processors,
2. a number of mass storage devices,
3. an interconnect device connecting the processors to the mass storage devices via a high-speed cache

Such an organization is depicted by Figure 1.

2.2 A Static Simplified Grid Organization

The big difference of a Grid environment, which is the focus of this paper, to the architecture laid out above is the *heterogeneity* of all comprising elements,

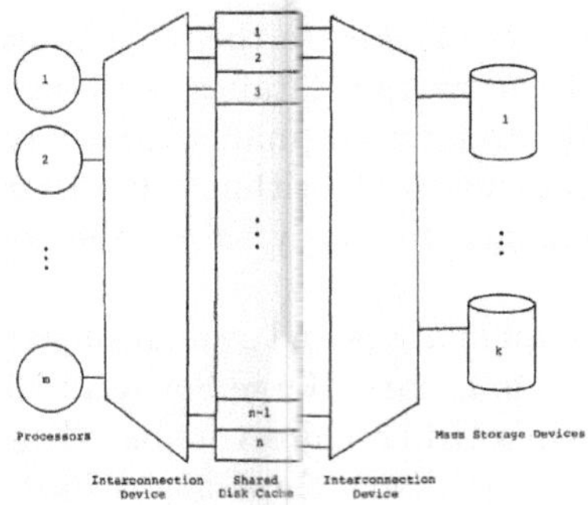

Fig. 1. Generalized Multiprocessor Organization (from [1])

which are processing nodes, interconnect bandwidth, disk performance. For the analytical comparison of the parallel algorithms in focus we restrict our approach to a simplified Grid computing organization focusing on the sensitive parameters of the model in focus.

We use the term *Static Simplified Grid Organization*, which describes an organization to perform a distributed query on a loosely coupled number of heterogeneous nodes. There is no logical order or hierarchy. That means, there is no logical topology of the nodes (e.g. no master/slave nor a equivalent order). Each node has a fixed number of properties with defined values. The term *Static* is used to describe that the values of each node and also the speed of the network are fixed and not changeable during the execution of a query. The sketch of such an organization is shown in Figure 2. These assumptions build the basis for our approach to analyze the operations in a simplified Grid.

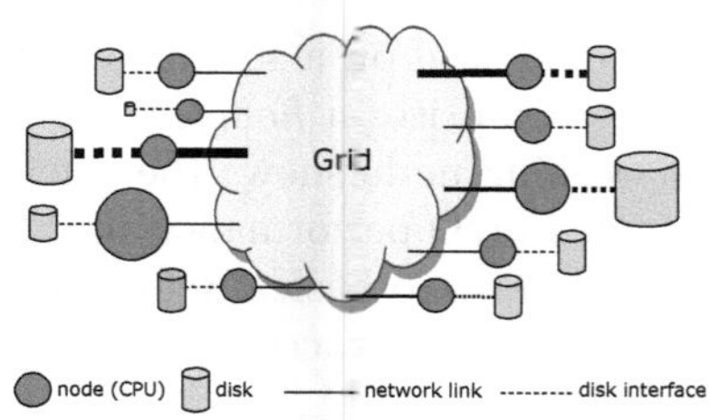

Fig. 2. Static Generalized Grid Organization

2.3 General Architectural Overview

The general layout of our architecture is as follows: A number of nodes are connected via a network. A node consists of one ore more CPUs (Central Processing Unit), a disk or disk-array and a network interface to the interconnect they are connected.

The actual configuration of a node is transparent (that means not "seen" by the outside user). However there exists a database to describe the system at the time of the query execution. All relevant data (parameters) describing the architecture are stored in this database. It is assumed that the following architectural characteristics hold during the operation:

- Nodes can perform a dedicated operation (compare, sort and merge two tuples).
- each node has it own mass storage device and the nodes are connected over a network.
- Nodes can send and receive messages.
- The performance of a node will not drop below a given value during query execution.
- The availability of each node during the query execution also is guaranteed.

3 Parallel Sorting Algorithms

In this paper we concentrate on two parallel sorting algorithms, the parallel binary merge sort and the block bitonic sort algorithm and analyze their performance in a generalized multiprocessor and a static simplified grid organization. These two algorithms are broadly accepted and the algorithms of choice in parallel database systems, because of their clear design and well studied properties. Due to the size restriction of the paper only a short description of the two well known algorithms is given in the following. A more detailed and comprehensive discussion can be found in the cited literature.

3.1 Parallel Binary Merge Sort

Binary Merge-Sort uses several phases, to sort mk tuples, where m is the number of pages containing the data set and k denotes the number of tuples per page. We assume that there are much more pages m than processing nodes p (i.e. $m >> p$) and that the size of the data set is much larger than the available main memory of the nodes.

We assume the very general case that the pages are not distributed equally among the mass storage media of the available nodes and that the tuples are not presorted according to the sorting criteria in the pages. Therefore the algorithm starts with a prepare phase, which distributes the pages equally across all p nodes, sorts the tuples inside every page according to the sort criterion and writes the pages back to disk. After the prepare phase m/p (respectively $m/(p-1)$) pages are assigned to each node and the tuples of each page are sorted. The algorithm continues with the suboptimal phase by merging pairs of longer and longer runs[1]. In every step the length of the runs is twice as large as in the preceding run. At the beginning each processor reads two pages, merges them into a run of 2 pages and writes it back to the disk. This is repeated, until all pages are read and merged into 2-pages-runs. If the number of runs exceeds $2p$, the suboptimal phase continues with merging two 2-page-runs to a sorted

[1] A *run* is an ordered (respective to the tuples contained) sequences of pages.

4-page-run. This continues until all 2-page-runs are merged. The phase ends, when the number of runs is $2p$. At the end of the suboptimal phase on each node 2 sorted files of length $m/2p$ exist. In the suboptimal phase the nodes work independently in parallel. Every node accesses its own data only. During the following optimal phase each processor merges 2 runs of length $m/2p$ and pipelines the result (run of length m/p) to a target node. The number of target nodes is $p/2$. The identification of the target-node is calculated by

$$nodenr_{target} = \frac{p}{2} + nodenr_{source}$$

for even source-node-numbers, and

$$nodenr_{target} = \frac{p}{2} + nodenr_{source} + 1.$$

for odd source-node-numbers.

In the postoptimal (last) phase the remaining $p/2$ runs are merged into the final run of length m. At the beginning of the postoptimal phase, we have $p/2$ runs. During this phase one of the p nodes is no longer used. Each of the other nodes is used only once during this phase. Two forms of parallelism are used. First, all nodes of one step work in parallel. Second, the steps of the postoptimal phase overlap in a pipelined fashion. The execution time between two steps consists of merging the first pages, building the first output-page and sending it to the target-node. During the postoptimal phase every node is used only in one step, that means that every node is idle for a certain time.

Thus the algorithm costs are

$$\underbrace{\frac{n}{2p}log(\frac{n}{2p})}_{suboptimal} \qquad \underbrace{+\frac{n}{2p}}_{optimal} \qquad \underbrace{+logp - 1 + \frac{n}{2}}_{postoptimal} \tag{1}$$

which can be expressed as

$$\frac{nlogn}{2p} + \frac{n}{2} - (\frac{n}{2p} - 1)(logp) - 1. \tag{2}$$

3.2 Block Bitonic Sort

Batchers bitonic sort algorithm sorts n numbers with $n/2$ comparator modules in $\frac{1}{2}logn(logn + 1)$ steps [10]. Each step consist of a comparison-exchange at every comparator module and a transfer to the target-comparator module. The comparator modules are connected by the perfect shuffle [11]. The perfect shuffle uses three types of comparator modules. The comparator module is represented by a node, which merges two pages and distributes the lower page and the higher page to two target-nodes. The target-nodes are defined by using a mask-information (i.e. the information for the perfect shuffle).

It is necessary to build $2p$ equally distributed and sorted runs of length $m/2p$. The prepare-phase and the suboptimal phase produce the runs. The total cost are:

$$\frac{n}{2p}\left(logn + \frac{log^2 2p - log2p}{2}\right) \tag{3}$$

4 Parallel Join Algorithms

In this section we introduce two join algorithms for relational databases, a parallel "nested-loop" and a parallel "sort-merge" algorithm. Like the sort algorithms in section 3, the presented join algorithms are commonly used in database systems.

4.1 Nested-Loop Join

The inner (smaller) relation T, and the outer relation R (larger one) are joined together. The algorithm can be divided in two steps:

1. **Initiate**
   Each of the processors read a different page of the outer relation R
2. **Broadcast and join**
   All pages of the inner relation T are sequentially broadcasted to the processors. After receiving the broadcasted page, each processor joins the page with its page from R.

n and m are the sizes (number of pages) of the relations R and R', and we suppose $n >= m$. To perform the join of R and R' we assign p processors. If $p = n$, the execution time is:

$$\begin{aligned} T_{nested-loop} = T(\text{read a page of R}) \\ +mT(\text{broadcast a page of R'}) \\ +mT(\text{join 2 pages}) \end{aligned} \tag{4}$$

S is the join selectivity factor and indicates the average number of pages produced by the join of a single page of R with a single page of R'. Joining two pages is performed by merging the pages, sorting the output page on the join attribute and write the sorted page to disk.

$$S = \frac{size(\text{R join T})}{mn} \tag{5}$$

If the number of processors p smaller than the number of pages n, step 1) and 2) must be repeated $\frac{n}{p}$ times. Therefore the costs for the Parallel Nested-Loop Join is

$$T_{nested-loop} = \frac{n}{p}\left[C_r + m[C_r + C_m + S(C_{so} + C_w)]\right] \tag{6}$$

4.2 Sort Merge Join

The algorithm processes in two steps. The first step of the algorithm is to sort the two relations on the join attribute (we assume, that the two relations are not already sorted). After sorting, the second step is performed, where the both sorted relations are joined together and the result relation is being produced. If we use the block bitonic sort in the first step described in section 3, the costs of the Sort Merge Join are

$$T = \left[\tfrac{n}{2p}logn + \tfrac{m}{2p}logm + (log^2 2p - log2p)\tfrac{n+m}{4p}\right]C_p^2 \\ +(n+m)C_r + max(nm)C_m + mnS(C_{so} + C_w). \tag{7}$$

Using the Parallel Binary Merge Sort the costs in term of C_2^p costs are

$$T = \left[\tfrac{nlogn}{2p} + \tfrac{n}{2} - (\tfrac{n}{2p} - 1)(logp) - 1\right]C_p^2 \\ +(n+m)C_r + max(nm)C_m + mnS(C_{so} + C_w). \tag{8}$$

4.3 Analysis Parameters

For comparing the algorithm we use the same definitions of the analysis parameter as described in [1].

We define with n the number of pages with k tuples each, and with p the number of processors.

Communication Cost. A processor must request a page, for this purpose a message is necessary, the cost for such an "I/O-related" message is C_{msg}

I/O Cost Parameters

- H ... certain hit ratio for the cache
- H' ... fraction amount of time a free page frame will be available in the cache during a write operation
- R_c ... cost of a cache to processor transfer
- R_m ... cost of a mass-storage transfer
- k is the number of tuples in a page.
- C are costs of a simple operation (scan, compare, add)

The average cost of a read by a processor is

$$C_r = HR_c + (1 - H)(R_c + R_m) + 2C_{msg}$$

The average cost of writing a page is

$$C_w = H'R_c + (1 - H')(R_c + R_m) + 2C_{msg}$$

Merge Cost

- V are costs of moving a tuple inside a page

Thus the costs of merging a page are

$$C_m = 2k(C + V)$$

To group some of the above parameters we define analogously to [1] a "2-page operation" C_p^2 as

$$C_p^2 = 2C_r + C_m + 2C_w$$

5 Analysis

The costs of reading a page C_r in a simplified Grid architecture can be split into two parts:

- C_{ard} average costs of reading a page from the disk to the main memory of the remote node, and
- C_{arn} average costs of transferring a page from the remote node to the local node.

Analogously to reading a page, writing a page C_w can also be split in two parts

- C_{awn} average costs of transferring a page from the local node to the remote node, and
- C_{awd} average costs of writing a page from the main memory to the disk of the remote node.

5.1 Lessons Learned for Grid Workflow Orchestration

Based on a thorough analysis, which is beyond the scope of this paper, we proved the correlation "the higher the network speed, the lower the impact on the costs of a C_p^2 operation", which is also intuitively clear. The costs of disk accesses are therefore much less influencing the overall performance than the network costs. Focusing on the sort algorithms the impact on the performance of the sort merge algorithms depends predominantly on the network bandwidth. Therefore the nodes with the best network-bandwidth numbers should be grouped to perform the last ($postoptimal_{II}$) phase in the binary merge sort. One stage in this last phase consists of a number of sender and a (number of) receiver nodes. The algorithm of defining the layout of the workflow for the postoptimal phase can be described by choosing the nodes with the best network bandwidth starting from the final stage. If the bandwidth is the same for some nodes the ones with the best computational power have to be chosen then. This can be described by the following algorithm:

1. Determine the network bandwidth and processing power for each processing node
2. Sort nodes according to network bandwidth

3. Nodes with equal network bandwidth in the sequence are sorted according to their processing power
4. Identify postoptimal phase as binary tree structure with node creating final run of length m as root
5. Starting from the beginning of sequence (i.e. best node first) map nodes level-wise from right to left beginning from root (root is level 0, successors of root are level 1, etc.).

5.2 Effects on the Performance of the Sort Algorithms

Based on the work of Bitton et al. in a generalized multiprocessor organization the block bitonic sort always outperforms the binary merge sort (see Figure 3 which is based on the analysis in [1]). To analyze the mapping of the algorithms onto a simplified Grid organization we have specifically pay attention to the three phases of the algorithms as laid out in section 3. The last phase (postoptimal) of the Binary Merge Sort algorithm is split into three parts (see equation 9) because only one processor is necessary for $\frac{n}{2}$ costs.

$$\underbrace{[logp - 1 + \frac{n}{2}]}_{postoptimal} \rightarrow \underbrace{logp - 1}_{postoptimal_I} \quad \underbrace{+\frac{n}{2}}_{postoptimal_{II}} \tag{9}$$

If we choose for this "bottleneck" the nodes of the Grid with the best network bandwidth available, the effect on the overall performance has to be at least remarkable. We specifically emphasize that even only one processing node with high network performance is worth while to exploit this effect. It is intuitively clear that this situation can be seen as normal in a heterogenous Grid organization, where nodes with different performance characteristics are the rule.

This leads to the clear policy for orchestration of a Grid workflow for a parallel binary merge sort to use nodes with the highest network performance in the $postoptimal_{II}$ phase as laid out in algorithm of section 5.1.

On the other hand using one high performance node in the bitonic sort gives no performance at all, because this node is slowed done by all other nodes working in parallel in a staged manner.

This effect that the binary merge sort now outperforms the block bitonic sort in a simplified Grid organization is shown in the Figure 4 by the line labeled "binary merge modified" (please notice the logarithmic scale of the values).

This effect can easily be explained by Amdahl's law too, where simply said the performance of a parallel algorithm is dependent on its sequential part. More specific Amdahls's law is stating that the speedup is limited by the sequential part. This is intuitively clear that even the most powerful node will limit the performance increase if the number of used nodes for the whole parallel algorithm is increasing. A speedup and scale-up analysis will clear up this issue.

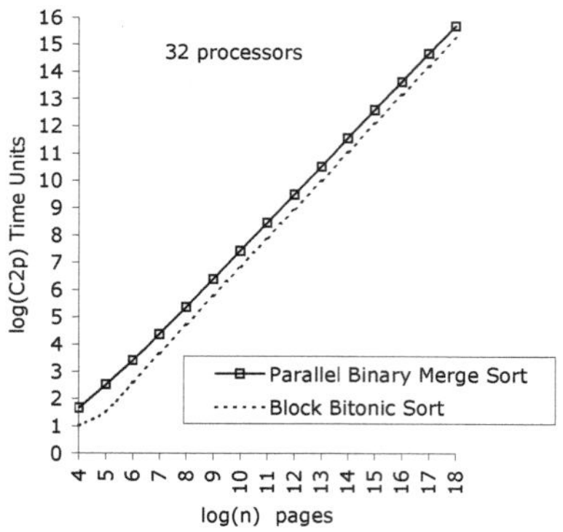

Fig. 3. Sort in a Generalized Multiprocessor Organization

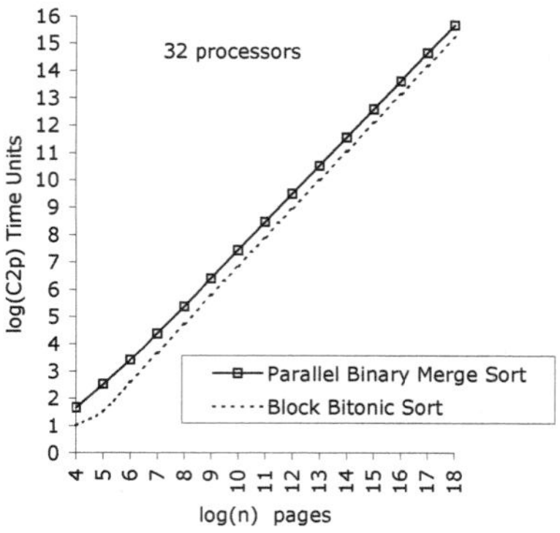

Fig. 4. Sort in a Simplified Grid Organization

5.3 Effects on the Performance of the Join Algorithms in a Generalized Multiprocessor Organization

Similar to the effects on the performance of the Sort Algorithms in a Generalized Multiprocessor Organization, the Merge-Sort Join algorithm with the block bitonic sort outperforms the Nested-Loop Join and also outperforms the Merge-Sort Join based on the binary-merge sort algorithm, unless the number of processors available is close to the larger relation size. Figure 5 shows the join algorithms with a selectivity factor of 0,001. If the ratio between the relation sizes is significantly different from 1, the nested-loop algorithm outperforms the merge-sort (except for a small numbers of processors). For lower selectivity values, the merge-sort algorithm performs better than the nested-loop algorithm because the merge step (handled by a single processor) has to output fewer pages.

In a generalized multiprocessor organization we have analyzed the algorithms with a selectivity factor between 0,001 and 0,9. The effect is, that the difference between the two merge-sort algorithms stays constant (if $S >= 0.05$). The reason is, that the merge costs are only marginal to the overall costs of the algorithm.

5.4 Effects on the Performance of the Join Algorithms in a Simplified Grid Organization

In the simplified grid organization the Merge-Sort join algorithm based on the bitonic-sort outperforms the Merge-Sort join based on the binary merge sort up to 2^6 processors, see "Parallel Merge-Sort Join (Binary Merge Sort) modified" in Figure 6.

As in the multiprocessor organization we have analyzed the join algorithms in the simplified grid organization with a selectivity factor between 0,05 and 0,9. The effect is the same for the generalized multiprocessor organization, the difference between the merge-sort algorithm (Parallel Merge-Sort Join (Bitonic) and Parallel Merge-Sort (Binary Merge) remains constant (if $S >= 0.05$). The reason is, that the merge costs have insignificant influence on the overall costs of the algorithm. The Parallel Merge-Sort (Binary Merge) modified is always faster (in terms of C_2^p costs) than the unmodified version.

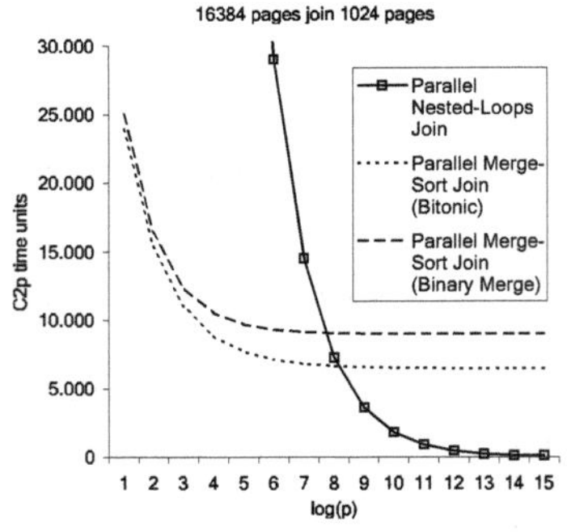

Fig. 5. Join in a Generalized Multiproc. Organization with S=0,001

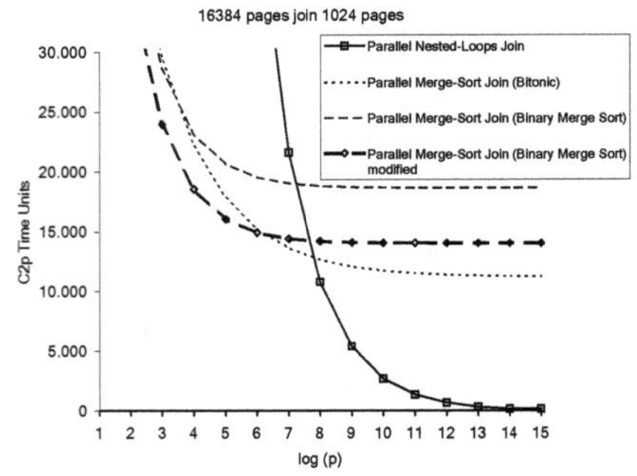

Fig. 6. Join in a Simplified Grid Organization with S =0,001

6 Conclusion and Future Work

In this paper the performance of the well-known parallel Sort-merge, Bitonic sort, Nested-Loop join and Merge-sort join algorithms for parallel database systems was analyzed. We developed an analytical evaluation model based on a simplified Grid architecture.

These are important results for the creation of parallel execution plans for database queries in the Grid. This work is performed as part of a project, which aims towards the development of a query optimizer for database query execution workflows and their orchestration in the Grid. We aim for a general execution time model for parallel query evaluations which can be integrated into a smart broker. This should give us the basis to create (near) optimal bushy query execution plans, which exploit operator- and data-parallelism in a Grid environment. In a further state of this research project we also will take into a account the availability of nodes and will extend our static model towards a more dynamic version.

References

1. Bitton, D., Boral, H., DeWitt, D.J., Wilkinson, W.K.: Parallel algorithms for the execution of relational database operations. ACM Trans. Database Syst. 8, 324–353 (1983)
2. Online: The gridbus project: grid computing info centre (grid infoware). Website (last access: 2007-01-30)

3. Hoschek, W., Jaen-Martinez, J., Samar, A., Stockinger, H., Stockinger, K.: Data management in an international data grid project. In: IEEE/ACM International Workshop on Grid Computing Grid, ACM Press, New York (2000)
4. Antonioletti, M., Atkinson, M., Baxter, R., Borley, A., Hong, N.C., Collins, B., Hardman, N., Hume, A., Knox, A., Jackson, M., Krause, A., Laws, S., Magowan, J., Paton, N.W., Pearson, D., Sugden, T., Watson, P., Westhead, M.: The design and implementation of grid database services in ogsa-dai. Concurrency and Computation: Practice and Experience 17, 357–376 (2005)
5. Gounaris, A., Sakellariou, R., Paton, N.W., Fernandes, A.A.A.: Resource scheduling for parallel query processing on computational grids. In: 5th International Workshop on Grid Computing (GRID 2004), pp. 396–401 (2004)
6. Soe, K.M., Aung, T.N., Nwe, A.A., Naing, T.T., Thein, N.: A framework for parallel query processing on grid-based architecture. In: ICEIS 2005, Proceedings of the Seventh International Conference on Enterprise Information Systems, pp. 203–208 (2005)
7. Iyer, B., Dias, D.: Issues in parallel sorting for database systems. In: Proc. Int. Conf. on Data Engineering, pp. 246–255 (1990)
8. Jarke, M., Koch, J.: Query optimization in database systems. ACM Computing Surveys 16, 111–152 (1984)
9. Bitton, D., DeWitt, D., Hsiao, D., Menon, J.: A taxonomy of parallel sorting. ACM Computing Surveys 16, 287–318 (1984)
10. Batcher, K.E.: Sorting networks and their applications. In: Proc. of the 1968 Spring Joint Computer Conference, Atlantic City, NJ, April 30-May 2, vol. 32 (1968)
11. Stone, H.: Parallel processing with the perfect shuffle. IEEE Trans. Computing C-20 (1971)

Performance Prediction Based Resource Selection in Grid Environments

Peggy Lindner[1], Edgar Gabriel[2], and Michael M. Resch[1]

[1] High Performance Computing Center Stuttgart, University of Stuttgart,
Nobelstr. 19, 70569 Stuttgart, Germany
{lindner,resch}@hlrs.de
[2] Parallel Software Technologies Laboratory,
Department of Computer Science, University of Houston,
4800 Calhoun Road, Houston, TX 77204, USA
gabriel@cs.uh.edu

Abstract. Deploying Grid technologies by distributing an application over several machines has been widely used for scientific simulations, which have large requirements for computational resources. The Grid Configuration Manager (GCM) is a tool developed to ease the management of scientific applications in distributed environments and to hide some of the complexities of Grids from the end-user. In this paper we present an extension to the Grid Configuration Manager in order to incorporate a performance based resource brokering mechanism. Given a pool of machines and a trace file containing information about the runtime characteristics of the according application, GCM is able to select the combination of machines leading to the lowest execution time of the application, taking machine parameters as well as the network interconnect between the machines into account. The estimate of the execution time is based on the performance prediction tool Dimemas. The correctness of the decisions taken by GCM is evaluated in different scenarios.

1 Introduction

During the last couple of years, Grid computing has emerged as a promising technology for science and industry. Computational Grids give users the power to harness spare cycles on PCs or workstations or distribute data and compute intensive applications on a large number of geographically distributed machines. Among the key challenges of current Grid technologies is the problem of resource selection and brokering. The set of resources chosen for a particular job can vary strongly depending on the goals of the researchers, and might involve minimizing the costs, optimizing the location of the computational resources in order to access a large data set or minimizing the overall execution time of the job.

In this paper we present a novel approach for resource selection based on the estimation of the execution time of an application using a simulator. Given a pool of machines, the Grid Configuration Manager (GCM) tests several possible combination of machines fulfilling the requirements of the end user regarding

R. Perrott et al. (Eds.): HPCC 2007, LNCS 4782, pp. 228–238, 2007.

the required computational resources. The estimate for the execution time of the application is based on the performance prediction tool Dimemas [1]. After GCM determined the combination of machines with the lowest predicted execution time, it can automatically generate the required configuration files for PACX-MPI [6] jobs and launch the distributed simulation.

The distribution of an application onto heterogeneous, dynamically changing resources requires the complex coordination of the resources. Currently, there is no general strategy or solution to the scheduling problem in a Grid environment available which meet all the demands [12]. The currently available brokering and scheduling systems are usually very specific for the target systems [15] or tailored for special application scenarios [10]. Work on interoperability includes Grid projects targeting resources running different Grid middlewares [3,14] and projects using (proposed) standard formats, e.g., to describe computational jobs [3,9]. The UniGrids [11] project specifically targets inter-operation between Globus and UNICORE. The difference between the GCM and these projects is that we also target the use of traditional non-Grid resources.

The remainder of the paper is organized as follows: section 2 presents the main ideas behind GCM, section 3 introduces the performance prediction tool Dimemas. In section 4 we detail the integration of Dimemas and GCM. Section 5 evaluates the correctness of the decision taken by GCM for various configurations and application settings. Finally, section 6 summarizes the paper.

2 The Grid Configuration Manager GCM

The Grid Configuration Manager (GCM)[7] is a tool developed to hide some of the complexity of Grid environments from the end-user. The central objective is to ease the handling of scientific, computational jobs in heterogeneous Grid environments, by abstracting the necessary tasks and implementing them for the most widely used Grid software toolkits (Globus[4], UNICORE[13]) and for the most widespread traditional access mechanism, SSH. To hide the complexity of different Grid protocols, the Grid Configuration Manager provides methods to support users in dealing with three main problems: handling of different authentication/authorization methods, a common interface for the job description and steering, and support of file transfer mechanisms for post- and preprocessing. The GCM idea is based on individual models for the resource description of the participating resources. Two different abstract classes were defined: the *host class* and the *link class*. The *host class* describes the properties of a resource such as name of the machine, user authentication information, number of processes to be started on this machine, or the start command and path of the application on this machine. The *link class* contains information about the network connection between two hosts, e.g. number of connections between the hosts and network parameters for each connection (i. e. network protocol and port numbers to be used). The current implementation of the GCM does not include an automatic resource discovery mechanism. This means, that the user has to specify how many nodes are available on the hosts to run his job. In the future this information will be gathered by requesting this information from the batch system for ssh based machines or using information services offered by Unicore and Globus.

A key component to create advanced applications in Grid environments is the file transfer component. File transfer operations are required for staging data files and executables, but also to distribute configuration files required by various tools and libraries. The file transfer component developed for GCM is based on the graphical Network Interface to File Transfer in the Internet (NIFTI). Given its modular architecture, NIFTI is conceptually independent of the implementation of the services it represents, and currently interfaces alternatively to FTP, SCP or an UNICORE file services.

One of the original goals of GCM from the very beginning was to support parallel multi-site jobs, since distributing jobs onto several machines simultaneously is an important part of the Grid concept. The need for doing so mainly comes from applications, which have a high demand on computational resources or for increasing throughput on the machines and reducing turnaround times. The GCM supports the handling of a multi-site job for a communication library called PACX-MPI [6]. While GCM is not restricted to PACX-MPI, the defined abstract interface has been up to now implemented to support the PACX-MPI configuration files and startup procedures.

3 Dimemas

Dimemas [1] is a performance prediction tool for message passing applications develop at the Barcelona Supercomputing Center (BSC). Within the frame of the European DAMIEN project [5], Dimemas has been extended to support performance prediction of distributed applications in Grid environments. Based on trace-files of the application, the tool is capable of estimating the execution time of the very same application for varying networking parameters or processor speeds. In order to generate a trace-file, the user has to re-link its MPI application with a tracing library provided by Dimemas, and run a small number of iterations of the application on the required number of processes. The tracing libraries intercepts each call to a communication routine using the MPI profiling interface [8], and stores relevant information such as the sender and receiver processes of the message, message length and time stamps for the beginning and the end of the data transfer operation. Thus, Dimemas offers a scalable and fully automatic approach for generating the trace-files, which has been used for highly complex, parallel applications.

The Grid architecture supported by Dimemas consists of individual nodes which can be grouped to a machine. Each node can have one or multiple, identical processors. Multiple processors are connected by a network interconnect. Based on a GUI, the user can modify the performance of the network as well as specify the no. and the relative performance of the nodes. The network connection between different machines can be based either on a shared network resources, i.e. the internet, or on dedicated network connections. The first approach is based on a congestion function, a latency respectively flight time and a maximal bandwidth. Values for latency and bandwidth of a message are adapted at runtime based on the congestion function provided by the user. The second approach allows the precise definition of a latency and the bandwidth between the machines. However, the current version of the tool is restricted to the same latency/bandwidth values between all machines within a given Grid configuration.

4 Integration of GCM with Dimemas

Among the most challenging issues the user has to deal with in current Grid environments is the resource selection step. In order to ease the resource selection procedure, GCM has been integrated with the performance prediction tool Dimemas. The goal of this approach is to enable the automatic selection of the most efficient Grid configuration for a particular application within a given set of resources. To encapsulate the required data for Dimemas, GCM has been extended by a new 'Dimemas-model'. This model contains all the configuration details and necessary information to run the Dimemas simulator program. The *host* and *link classes* of GCM have been extended to collect the additional information such as the specification of the available SMP-nodes on each machine, a detailed description of the network connection between the machines, location of the trace file etc. A special 'Mapping-Dialog' allows the user to specify how the application processes should be distributed across the machines. The class *Dimemas* contains all methods to run a Dimemas simulation and store the input data and results for future usage in a special metadata format. Special GUI interfaces help the user to specify all the information. The results of the Dimemas simulation and (error) messages can be retrieved on a special output window in order to allow the usage of GCM as a GUI interface for Dimemas.

To propose an optimal Grid configuration, GCM takes into account the given resources, the currently available nodes on each machine and the communication pattern of the machine respectively between the machines. Therefore, the user has to provide information about the application in form of a trace file which will be used by Dimemas. Based on this information, GCM creates different 'test configurations' for Dimemas, starts the Dimemas simulator, stores and compares the results achieved. The Grid configuration with the lowest estimated execution time will be offered to the user in form of a PACX-MPI configuration. 1 shows the individual steps of the resource selection within GCM.

Since the number of potential process distributions on the available machines can be fairly large, GCM creates test configurations based on a simple heuristic. The distribution of processes onto the hosts follows a simple round robin algorithm. For each available machine, GCM generates a test configuration using that machine as the starting point and allocates all available processors on that machine for the simulation. In case more processors are required in order to run the simulation, GCM takes the required number of processors from subsequent machines in sequential order, etc. Using this approach, n test configurations are generated for a machine pool containing n machines. An advantage of this heuristic is, that since we assign the maximum number of processes on each machine to the simulation, the number of machines used for each test configuration are being minimized. For parallel, distributed applications this kind of configuration usually produces the lowest execution times, as the communication between the machines is fairly time consuming. As an example, lets assume that 16 processes should be distributed onto 4 machines, each of which offer 6 processors. According to the described algorithm, the following test configurations will be created:

- 6 processes each on machines 1 and 2; 4 processes on machine 3
- 6 processes each on machines 2 and 3; 4 processes on machine 4
- 6 processes each on machines 3 and 4; 4 processes on machine 1
- 6 processes each on machines 4 and 1; 4 processes on machine 2

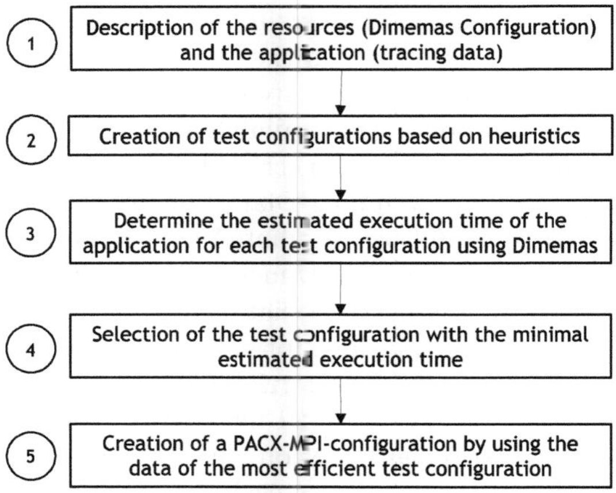

Fig. 1. Resource selection based on the estimated execution time of the application within GCM

The simulations based on Dimemas estimating the execution time of the application for different network and machine parameters are executed on the very same a local PC where GCM is being executed. Depending on the size of the trace file, a single simulation takes at most a couple of seconds on a state-of-the-art PC, and does not require any Grid resources itself.

5 Evaluation

In the following, we would like to verify the correctness and the accuracy of the choices made by the Grid Configuration Manager. The application used throughout this section is a finite-difference code which solves a partial differential equation (PDE), respectively the set of linear equations obtained by discretization of the PDE using center differences. To partition the data among the processors, the parallel implementation subdivides the computational domain into rectangular subdomains of equal size. Thus, each processors holds the data of the corresponding subdomain. The processes are mapped onto a regular three-dimensional mesh. Due to the local structure of the discretization scheme, a processor has to communicate with at most six processors to perform a matrix-vector product.

The Grid configurations used in this subsection consists of a collection of homogeneous resources based on the technical data of local InfiniBand clusters of the participating institutions. This approach allows us to investigate the effect of different parameters on the decision procedure of GCM individually. As an example, a configuration such as shown in table 1 consists of four machines, each of them being from the technical perspective identical. Unless explicitly mentioned otherwise, the communication bandwidth between the machines is set to 12MB/s and the latency is 4 ms for the Dimemas simulations, which are typical values for Fast Ethernet connections.

Table 1. Estimated execution times by Dimemas of the Finite Difference Code for various test configurations. Times are given in seconds.

No. of processors on machine				estimated execution time	No. of processors on machine				estimated execution time
1	2	3	4	(s)	1	2	3	4	(s)
Test 1: 16 processes					Test 3: 16 processes				
16	0	0	0	1.17 *	10	6	0	0	10.08 *
0	6	6	4	7.41	0	6	6	4	12.19
4	0	6	6	7.41	6	0	6	4	12.22
10	0	0	6	5.13	10	2	0	4	12.23
Test 2: 32 processes					Test 4: 32 processes				
6	26	0	0	8.71	10	12	10	0	10.69
0	32	0	0	4.44 *	0	12	20	0	7.99 *
6	14	6	6	10.52	6	0	22	4	11.22
6	20	0	6	10.52	10	12	6	4	10.47

The first set of tests have been set up in order to verify, that the configuration chosen by GCM/Dimemas is using the minimal number of machines to run an application. Since the communication between machines is at least an order of magnitude slower than the communication between processes on the same, minimizing the number of machines used by an application will in the vast majority of scenarios also minimize the overall execution time of the application. Of special importance are scenarios where the application could run on a single machine and thus would completely avoid expensive inter-machine communication. The results of the simulations are shown in table 1. GCM was configured such that its pool of machines consisted of four identical clusters. In the tests 1 and 2 (left) one machine advertises sufficient available resources to run the simulation entirely on that platform, while using any of the other machines would require a combination of multiple machines in order to provide the requested number of processors. Similarly, in the tests 3 and 4 only a single combination of available resources leads to a two-host configuration capable of providing the number of requested processors. Table 1 shows for each test case the combination of resources respectively the number of processes on each platform used by GCM to find the minimal execution time and the estimated execution time by Dimemas. The combination of resources selected by GCM is marked by a star. In all scenarios, GCM/Dimemas was able to spot the combination requiring the minimal number of machines, by either using a single machine (tests 1 and 2), or two machines (tests 3 and 4).

Table 2. Estimated execution times by Dimemas of the Finite Difference Code for various test configurations with varying bandwidth between the machines. Times are given in seconds.

Bandwidth	No. of processors on machine				estimated exec. time	No. of processors on machine				estimated exec. time
	1	2	3	4	(s)	1	2	3	4	(s)
	Test 5: 24 processes					Test 6: 32 processes				
100 MB/s	12	12	0	0	6.601620 *	16	16	0	0	6.937863 *
25 MB/s	0	12	12	0	7.361563	0	16	16	0	6.958964
30 MB/s	0	0	12	12	7.052505	0	0	16	16	6.939202
95 MB/s	12	0	0	12	6.601623	16	0	0	16	6.937885

In the second test we would like to evaluate, whether our approach is correctly determining the set of resources offering the best network connectivity between the machines. Two separate set of tests have been conducted. In the first test, we keep the latency between all machines constant, vary however the available bandwidth between each pair. As shown in table 2, all four machines advertise the same number of processors, namely either 12 in test 5 or 16 in test 6. Since the user requested however 24 respectively 32 processors, GCM has to find the combination of two machines promising the lowest execution time. The bandwidth between the according pair of machines is shown in in the first column of table 2.

As can be seen on the results shown in table 2, this particular application shows only limited sensitivity to the bandwidth between the machines. All execution times estimated by Dimemas end up in the very same range. Nevertheless, in both scenarios GCM did chose the combination of resources with the highest bandwidth, namely machines 1 and 2.

Similarly to the previous scenario, we kept this time the bandwidth between all four machines constant at 12 MB/s. However, we varied the latency between the different machines between 0.6 ms and 5 ms. The first column in table 3 shows the latency value used between the according pair of machines. The results indicate, that this finite difference code is significantly more sensitive to the network

Table 3. Estimated execution times by Dimemas of the Finite Difference Code for various test configurations with varying latency between the machines. Times are given in seconds.

Latency	No. of processors on machine 1	2	3	4	estimated exec. time (s)	No. of processors on machine 1	2	3	4	estimated exec. time (s)
	Test 2: 24 processes					Test 4: 32 processes				
4 ms	12	12	0	0	9.59947	16	16	0	0	7.414597
5 ms	0	12	12	0	10.480678	0	16	16	0	7.948321
0.6 ms	0	0	12	12	6.612363 *	0	0	16	16	5.660996 *
0.8 ms	12	0	0	12	6.787963	16	0	0	16	5.759201

latency between the machines than to the inter-machine bandwidth. GCM could correctly identify the combination of machines with the lowest network latency.

5.1 Using GCM and Dimemas to Evaluate Implementation Options for Grid Environments

An alternative usage scenario for GCM together with Dimemas is to evaluate the performance of different implementation options given a particular execution environment or Grid configuration. The motivation behind this approach is, that many scientific applications have various options given a particular functionality. As an example, there are many different ways to implement the occurring neighborhood communication in scientific code, each of which might lead to the optimal performance for different problem sizes and machine configurations [6]. Similarly, a application often has the choice to use different linear solvers, or multiple algorithms with slightly different characteristics for a particular solver. Using GCM together with Dimemas gives the code developers and end-users a powerful tool to get very realistic estimates about the performance of each option on various machines respectively set of machines, without having to run the actual code on all possible combination of machines.

This approach is being demonstrated in this subsection using different algorithms for the solver used in the finite difference code. Four different solvers namely QMR, QMR_1, $TFQMR$, and $TFQMR_1$ are available within the framework. The main difference between these four solvers are the number of collective operations per iteration and the number of synchronization points per iteration [2]. As a first step, the user has to generate tracefiles for each solver and for the number of processors it would like to utilize. This can be done on a single machine, and does not require any distributed resources per se. In a second step, the user generates in GCM the Grid environment of interest, by specifying the number of machines, the number of nodes on each machine and the network characteristic between the machines. GCM will then call Dimemas for each of the according scenarios/traces and suggest the version of the code delivering the best performance. In case the application chooses the solver at runtime based on a setting in an input file, it is straight forward to add a plug-in to GCM which could alter the input file by choosing the solver which has been determined to deliver the best performance for the given Grid environment.

The main limitation of this approach is due to GCM not having any 'extended' knowledge about the different version being compared. As an example, comparing the execution time of two iterative solvers for a given number of iterations does not include any information about the convergence behavior of the solver. Thus, a decision purely based on the execution time might not necessarily lead to the best solver. However, there are sufficient important scenario, where a useful decision can be made based on the execution time. In the tests shown here, the QMR and the $TQMR$ solvers expose significantly different convergence characteristics from the numerical perspective. It is however safe to compare the execution times between QMR and QMR_1 and between $TFQMR$ and $TQMR_1$ in order to determine, which implementation performs better in a given environment. The results of such a comparison is shown in table 4. Two different scenarios are shown here requiring 16 processes and 32 processes. The configurations provided to GCM have been chosen such that the processes are evenly distributed on two machines.

Table 4. Results scenario 3 - Comparison of different solvers (* indicates the configuration chosen by GCM)

Solver	No. of processors on machine 1	No. of processors on machine 2	measured exec time (s)	simulated exec time (s)	No. of processors on machine 1	No. of processors on machine 2	measured exec time (s)	simulated exec time (s)
	Test 1: 16 processes				Test 3: 32 processes			
QMR	8	8	1.78	1.85 *	16	16	1.42	5.66 *
QMR_1	8	8	2.78	5.62	16	16	2.35	7.90
	Test 2: 16 processes				Test 4: 32 processes			
$TFQMR$	8	8	3.63	4.25	16	16	2.79	8.67
$TFQMR_1$	8	8	3.50	3.50 *	16	16	2.66	5.04 *

Table 4 shows the number of processes used on each machine, the avg. execution time of each solver measured on the given environment and the estimated execution time by Dimemas. In all four scenarios, the estimated execution time and the average measured execution times are pointing to the same solver as being optimal/faster. Please note, that within the context of these tests, the accuracy of the estimated execution time compared to the measured execution time is only of limited interest. More relevant is the factor, that the relative performance between the solvers is determined correctly by Dimemas. Since Dimemas is a powerful tool with many options, a user can optimize the settings of Dimemas in order to close the absolute gap between the estimated and the measured execution time. Clearly, the smaller the gap the wider the range of useful scenarios for GCM. In the tests shown in table 4 most of the parameters of Dimemas were the default values delivered by the tool.

6 Summary

We described in this paper the integration of the performance prediction tool Dimemas with the Grid Configuration Manager GCM. Based on a trace-file of the application, this combination of tools can be used in order to optimize the resource selection procedure for a given pool of resources. Using various test-cases we have shown, that GCM together with Dimemas (i) minimizes the number of machines used for a configuration and (ii) does take network parameters such as network and latency between different machines into account when choosing the set of resources for a particular run. Furthermore, we have demonstrated, that this combination of tools can also be used in order to evaluate various implementation/functional options of an application for a given Grid environment without having to actually run the application on a distributed set of resources.

The limitations of this approach are two-fold: first, in order to use the approach outlined in this paper, the user has to generate a trace file of his application. This can be a restriction for some applications. However, since the trace file can be generated on a single machine, this step should not pose any major challenges. Second, the accuracy of the predictions does inherently have an influence on scheduling decisions. Using Dimemas as basis for the performance prediction step gives GCM the most powerful performance prediction tool currently available - in fact, the only tool known to the authors which can be used out-of-the-box for arbitrary complex MPI applications.

References

1. Badia, R.M., Labarta, J., Giménez, J., Escalé, F.: DIMEMAS: predicting mpi applications behavior in grid environments. In: Workshop on Grid Applications and Programming Tools Held in conjunction with GGF8, Seattle, USA Proceedings, June 25, 2003, pp. 52–62. Seattle (2003)
2. Bücker, H.M., Sauren, M.: A Parallel Version of the Quasi-Minimal Residual Method Based on Coupled Two-Term Recurrences. In: Madsen, K., Olesen, D., Waśniewski, J., Dongarra, J.J. (eds.) PARA 1996. LNCS, vol. 1184, pp. 157–165. Springer, Heidelberg (1996)
3. Open Middleware Infrastructure for Europe. OMII Europe
4. Foster, I., Kesselmann, C.: The globus project: A status report. In: Proc. IPPS/SPDP 1998 Heterogeneous Computing Workshop, pp. 4–18 (1998)
5. Gabriel, E., Keller, R., Lindner, P., Müller, M.S., Resch, M.M.: Software Development in the Grid: The DAMIEN tool-set. In: Sloot, P.M.A., Abramson, D., Bogdanov, A.V., Gorbachev, Y.E., Dongarra, J.J., Zomaya, A.Y. (eds.) ICCS 2003. LNCS, vol. 2659, pp. 235–244. Springer, Heidelberg (2003)
6. Keller, R., Gabriel, E., Krammer, B., Müller, M.S., Resch, M.M.: Efficient execution of MPI applications on the Grid: porting and optimization issues. Journal of Grid Computing 1(2), 133–149 (2003)
7. Lindner, P., Gabriel, E., Resch, M.M.: GCM: a Grid Configuration Manager for heterogeneous Grid environments. International Journal of Grid and Utility Computing (IJGUC) 1(1), 4–12 (2005)
8. Message Passing Interface Forum. MPI: A Message Passing Interface Standard (June 1995), http://www.mpi-forum.org

9. Miura, K.: Overview of japanese science grid project NAREGI. Progress in Informatics 1(3), 67–75 (2006)
10. Natrajan, A., Humphrey, M.A., Grimshaw, A.S.: Grid resource management in Legion. In: Nabrzyski, J., Schopf, J.M., Weglarz, J. (eds.) Grid Resource Management: State of the Art and Future Trends, pp. 145–160. Kluwer Academic Publishers, Dordrecht (2004)
11. Riedel, M., Sander, V., Wieder, P., Shan, J.: Web Services Agreement based Resource Negotiation in UNICORE. In: Arabnia, H.R. (ed.) 2005 International Conference on Parallel and Distributed Processing Techniques and Applications, pp. 31–37 (2005)
12. Tonellotto, N., Yahyapour, R., Wieder Ph.: A proposal for a generic grid scheduling architecture. Technical Report TR-0015, CoreGRID Technical Report, Institute on Resource Management and Scheduling (2006)
13. Joint Project Report for the BMBF Project UNICORE Plus, Grant Number: 01 IR 001 A-D, Duration: January 2000 to December 2002 (2003)
14. Venugopal, S., Buyya, R., Winton, L.: A grid service broker for scheduling e-science applications on global data grids. Concurrency and Computation: Practice and Experience 18, 685–699 (2006)
15. Young, L., McGough, S., Newhouse, S., Darlington, J.: Scheduling Architecture and Algorithms within the ICENI Grid middleware. In: Cox, S. (ed.) UK e-Science Program All Hands Meeting, pp. 5–12 (2003)

Online Algorithms for Single Machine Schedulers to Support Advance Reservations from Grid Jobs

Bo Li and Dongfeng Zhao

School of Information Science and Engineering, Yunnan University,
Kunming, Yunnan, 650091, China
{libo,dfzhao}@ynu.edu.cn

Abstract. Advance Reservations(AR) make it possible to guarantee the QoS of Grid applications by reserving a particular resource capability over a defined time interval on local resources. However, advance reservations will also cause the processing time horizon discontinuous and therefore reduce the utilization of local resources and lengthen the makespan (i.e., maximum completion time) of non-resumable normal jobs. Single machine scheduling is the basis of more complicated parallel machine scheduling. This study proposed a theoretic model, as well as four online scheduling algorithms, for local single machine schedulers to reduce the negative impact on the utilization of local resources and to shorten the makespan of non-AR jobs resulting from advance reservations for Grid jobs. The performances of the algorithms were investigated from both of the worst case and the average case viewpoints. Analytical results show that the worst case performance ratios of the algorithms against that of possible optimal algorithms are not less than 2. Experimental results for average cases suggest that the First Fit and the First Fit Decreasing algorithm are better choices for the local scheduler to allocate precedence-constrained and independent non-AR jobs respectively.

1 Introduction

Grid computing affiliates the sharing of all kinds of resources. Advance Reservations(AR) make it possible to guarantee the QoS of Grid applications by reserving a particular resource capability over a defined time interval on local resources[1]. In general, advance reservations will also cause the processing timeframe discontinuous and therefore reduce the utilization of local resources and lengthen the completion time of non-AR jobs. Up till now, many studies and experiments have been conducted on reservation-based Grid QoS technologies. Most of them have been focused on the benefits for AR jobs coming from advance reservations[2-5]. Few of them have been devoted to evaluate the impact from advance reservations on the performances of local resources and/or normal jobs[6-9]. However, none of them has been published to try to reduce the disadvantages on local resources and/or normal jobs resulting from advance reservations for Grid jobs.

Single machine scheduling is the basis of more complicated parallel machine scheduling. In this study, it was aimed to find a theoretic model, as well as local

R. Perrott et al. (Eds.): HPCC 2007, LNCS 4782, pp. 239–248, 2007.

scheduling algorithms, for single machine schedulers to reduce the disadvantages on utilization of local resources and to shorten the makespan(i.e., maximum completion time) of normal jobs while supporting advance reservations for Grid jobs. The local scheduler can fulfill both targets by allocating non-AR jobs into available intervals with the goal to minimize the makespan of the jobs.

For single machines, the impact of AR on the allocating of local jobs is depicted in Fig.1. Jobs are classified into two types: AR jobs and non-AR jobs. AR jobs are Grid jobs with advance reservations. Once their reservation requirements are accepted, they will be put into the AR-job queue and will be processed within reserved processing intervals. Non-AR jobs are put into the non-AR-job queue and will be processed only within those available intervals left by AR jobs. All reserved intervals for AR jobs are unavailable for non-AR jobs.

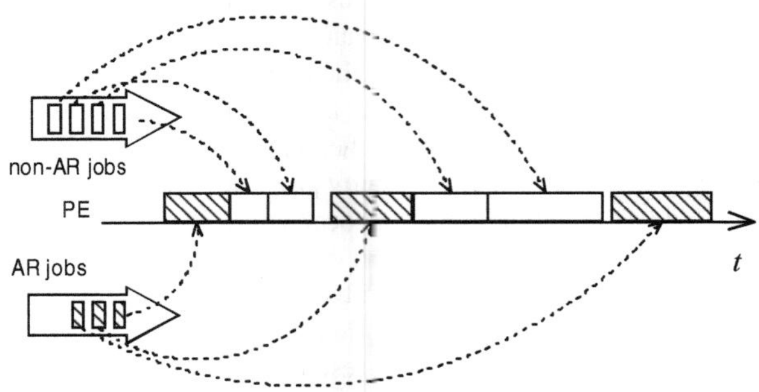

Fig. 1. The allocating of AR and non-AR jobs on single processing element

Because non-AR jobs can be processed only in discontinuous available intervals, the allocating of non-AR jobs into available intervals to minimize the makespan is a kind of availability constraint scheduling problem. With availability constraints, a machine can process jobs only in discontinuous available intervals. This restriction arises in many areas such as manufacturing industry, production scheduling and computer science [10, 11]. Suppose all non-AR jobs are non-resumable, it is NP-hard to allocate them into available intervals to minimize the makespan. When a non-resumable job cannot be finished before the ending of the available interval in which it is processed, it has to restart from the beginning later in another available interval, rather than to continue.

Assume the information (e.g., arrival time, required processing time) on queued AR and non-AR jobs is known before scheduling. If there are more than one AR jobs in queue, in order to meet the reservation requirements of possible arriving AR jobs, it is necessary to allocate non-AR jobs only into existing available intervals made by accepted reservations. Suppose the number of queued AR jobs is one and the scheduler knows the time point when a new available or unavailable interval will end as it arrives, this setting can be referred to as online.

This paper will propose a new theoretic model for the local single machine scheduler to minimize the makespan of non-AR jobs while supporting advance reservations for

Grid jobs, will present four online scheduling algorithms and evaluate their performances from both of the average case and the worst case viewpoints.

2 Problem Definition and Online Algorithms

For the online scheduling problem to allocate non-AR jobs into available intervals, the completion time of the non-AR jobs equals the length of the duration from the starting time point to the end time of the last job. Considering that the makespan comes from both of the available intervals and unavailable intervals, we can divide it into two parts: the one coming from unavailable intervals and the other coming from available intervals. For any deterministic problem instance of which the arrival time and the duration of any unavailable or available interval are fixed, the makespan is dominated by how to allocate the jobs to the available intervals. In other words, the goal to minimize the makespan is equivalent to allocate the jobs into as small number of earlier available intervals as possible. By regarding the jobs and their processing time requirements as items and item sizes respectively, and by regarding the available intervals of the machine and their durations as bins and bin sizes respectively, the deterministic online single machine scheduling problem for nonresumable non-AR jobs can be transformed into a online version of a variant of the standard Variable-Sized Bin Packing(VSBP) problem firstly proposed in [12]: given a list $L=(a_1,a_2,\cdots,a_n)$ of items, each with size $s(a_j),1\le j\le n$, and a list $B=(b_1,b_2,\cdots)$ of bins, each with size $s(b_i),1\le i$, the goal is to pack the n items into the bins with a minimum total size of bins from b_1 to the last one being used (say $b_t,t\ge 1$) in list B. To distinguish the new problem from the standard VSBP[13], in this paper we call it the Variant of the Variable-Sized Bin Packing problem, VVSBP for short.

According the sizes of the bins and the items, we can divide VVSBP into two subcases with feasible solutions:

1. Subcase A: $\min\{s(b_i)\}\ge\max\{s(a_j)\}$;
2. Subcase B: $\min\{s(b_i)\}\ge\min\{s(a_j)\}$ and $\max\{s(b_i)\}\ge\max\{s(a_j)\}$.

where $\min\{s()\}$ and $\max\{s()\}$ denote the minimum and the maximum in $s()$ respectively.

In the online scheduling problem, if the duration of a new available interval is not greater or equal to the maximum processing time requirement of the non-AR jobs in queue, this interval will not be considered for processing jobs. This is equivalent to subcase A. If only those available intervals the durations of which are less than the minimum processing time requirement of the non-AR jobs are ignored, we get subcase B. To guarantee feasible solutions for any problem instance of subcase A, we assume the number of bins is not less than the number of items. Because subcase A can be viewed as a subset of subcase B, to guarantee feasible solutions for any problem instance of subcase B, we can exclude those intervals with sizes less than the maximum item size in the problem instance of subcase B to construct a problem instance of subcase A, and require the number of the bins in the derived problem instance is not less than the number of items.

In the online version of the classic one-dimensional bin packing(BP) problem or the standard VSBP problem, we have all information on the bins but we cannot preview the information of items before they arrive. An online algorithm assigns items to bins in the order of $(a_1, a_2, ..., a_n)$ with item $a_j, 1 \leq j \leq n$ solely on the basis of the sizes of the preceding items and the bins to which they were assigned. Without loss of generality, we can assume that no new item will arrive until its preceding items are packed into bins and no new bin will open to hold items until all the used bins cannot accept the current item. In the online version of the VVSBP problem, we have all information on items but we cannot preview the type of the bins before packing. We must decide which items should be packed into the bin as it arrives. A new bin will arrive after the current bin is closed and a closed bin will never be reopened to accept items even if it is empty.

The differences between the two kinds of online issues make those online algorithms for the BP or the VSBP problem not applicable to the VVSBP problem. It is reasonable to extend the four algorithms in [12] for both of the subcases of the online version of the VVSBP problem. Except NF, FF, NFD and FFD, we can also adapt other bin packing algorithms for the VVSBP problem, such as the Worst Fit(WF) algorithm, the Almost Worst Fit(AWF) algorithm, the Worst Fit Decreasing(WFD) algorithm and the Almost Worst Fit Decreasing(AWFD) algorithm. However, because these four algorithms are not as typical as NF, FF, NFD and FFD in practice, they will not be considered in the following. Without causing confusion, we also use the names of the algorithms in the BP problem to denote their analogues adapted for two subcases of the VVSBP problem.

1. Next Fit(NF): Pack items in the order of $(a_1, a_2, ..., a_n)$, if the current bin b_i has not sufficient space to hold the current item a_j, b_i will be closed and the next bin b_{i+1} will be opened to try to hold a_j
2. First Fit(FF): All items are given in the VVSBP problem and they will be packed in the order of $a_1, a_2, ..., a_n$. When packing $a_j, 1 \leq j \leq n$, put it in the lowest indexed open bin into which it will fit. If such an open bin does not exist, open a new bin until we find the least i such that $b_i, i \geq 1$ is capable of holding a_j.

Presort the items in non-increasing order by size, and then apply FF, BF to the re-ordered items, we will get the First Fit Decreasing(FFD) algorithm and the Best Fit Decreasing(BFD) algorithm respectively.

In the scheduling problem, if the jobs must be processed in index order, we say they are precedence-constrained. On the other hand, if the jobs can be processed in random order, we say they are independent. Of course, only these algorithms without presorting as mentioned above, i.e., FF, BF, WF and AWF, can be used for precedence-constrained jobs.

3 Worst Case Analysis

Competitive ratio is usually used to evaluate the performance of online algorithms, which describes the maximum deviation from optimality that can occur when a specified algorithm is applied within a given problem class. For a list L of items, a list B of bins and an approximation algorithm H for the VVSBP problem, let $H(L,B)$ denote the total size of bins by algorithm H, and let $OPT(L,B)$ denote the total size in optimal packing, the competitive ratio of algorithm H is defined as $R_H^\infty = \lim_{k\to\infty} \sup_{L,B}\{H(L,B)/OPT(L,B) \mid OPT(L,B) \ge k\}$. For instances of subcase B in section 2, if feasible solutions do not exist, we say $R_H^\infty = \varnothing$ for any algorithms; if feasible solutions exist but algorithm H can not get a feasible solution, we say $H(L,B) = \infty$ and thus $R_H^\infty = \infty$.

Let H denote any of the online algorithms defined above, we get their competitive ratios in subcase A and subcase B as follows.

Theorem 1. For worst case instances of subcase A, $R_H^\infty = 2$.

This theorem can be proved by applying the results in [12] in that subcase A is equivalent to the essential assumption in [12] and that the four algorithms here are identical to those in [12].

Theorem 2. For worst case instances of subcase B, $R_H^\infty = \infty$.

Proof. Without lose of generality, let the list L of items with sizes $s(a_1) > s(a_2) > \ldots > s(a_n), s(a_1) < s(a_2) + s(a_3)$, two instances are considered:

Instance 1: If $H \in \{NF, NFD\}$, suppose the sizes of bins are $\begin{cases} s(b_i) = s(a_{n-i+1}), 1 \le i \le n \\ s(b_i) = s(a_n), i > n \end{cases}$, an optimal and feasible solution is to pack a_i into $b_{n-i+1}, 1 \le i \le n$.

Instance 2: If $H \in \{FF, FFD\}$, suppose the sizes of bins are $\begin{cases} s(a_2) + s(a_3) \le s(b_1) < s(a_1) + s(a_3) \\ \qquad < s(a_1) + s(a_2) \\ s(a_1) \le s(b_2) < s(a_2) + s(a_3) \\ s(b_i) = s(a_{i+1}), 3 \le i \le n-1 \\ s(b_i) = s(a_n), i \ge n \end{cases}$, an optimal and feasible solution is to pack a_1 into b_2, a_2 and a_3 into b_1, a_i into b_{i-1} for $4 \le i \le n$.

In the two instances, H cannot get feasible solutions, we have $H(L,B) = \infty$ and $OPT(L,B) = \sum_{i=1}^{n} s(b_i)$, and thus $R_H^\infty = \infty$.

4 Average Case Experiments

Usually, simulations with real workload trace records(e.g. Parallel Workload Archive[14]) are used to investigate the performances of parallel or Grid scheduling algorithms[8, 15-17]. However, up till now, there are not any real workload trace records for single machine schedulers. In this section, we will investigate the average-case performances of the online algorithms only from the bin packing viewpoint.

Theoretical analysis and experimental simulation are two methods to determine the average-case performance of bin packing algorithms. The usual approach for average-case analysis is to specify a density function for the problem data, including the sizes of items and bins, and then to establish probabilistic properties of an algorithm, such as the asymptotic expected ratio, the wasted space expected ratio etc.[18, 19] Among the distributions for average-case analysis, continuous uniform or discrete uniform distributions are widely used. For uniform distributions, the asymptotic expected ratio is defined as $ER_H^\infty = \lim_{n\to\infty} ER_H^n = \lim_{n\to\infty} E[H(L_{n,1})/OPT(L_{n,1})]$ and the wasted space expected ratio is defined as $EW_H^\infty = \lim_{n\to\infty} EW_H^n = \lim_{n\to\infty} E[H(L_{n,1}) - s(L_{n,1})]$, $s(L_{n,u})$ denoting the total size of n items drawn independently from the uniform distribution on the interval $[0,u]$. It is suggested in [18] that it is often easier to analysis the average case results about EW_H^n than about ER_H^n directly and the former typically imply the latter in that for most distributions that have been studied $\lim_{n\to\infty} E[OPT(L_{n,1})/s(L_{n,1})] = 1$ and $ER_{BF}^\infty = ER_{BFD}^\infty = 1$ for the classic one-dimensional bin packing problem. Moreover, when item sizes are drawn independently from the uniform distribution on the interval $[0,u]$, it is proved in [20] that $ER_u^\infty(H) = \lim_{n\to\infty} E[H(L_{n,u})/s(L_{n,u})]$ and $ER_u^\infty(FF) = 1, u \in [0,1]$. When only consider the limitation of $u \to 1$, it is suggested in [18] that $\lim_{b\to 1} R_{FF}^\infty(U[0,b]) = 1$.

For instances of the VVSBP problem with given lists of bins and items, the goal to minimize the total size of bins is not only related to the numbers and sizes of items and bins, but also related to the sequence of bins. This makes it more difficult to analyze the average-case performance of algorithms mathematically. Instead of mathematical analysis, a numerical experiment solution is used in this section to simulate the behaviors of different algorithms for typical problem instances with sizes of items and bins drawn independently from uniform distributions.

In the following, we adapt the definition of $ER_u^\infty(H)$ for algorithms in average-case experiments as $ER_u^\infty(H) = \lim_{n\to\infty} E[H(L_{n,u}, B_{wn,u^*})/s(L_{n,u})]$, in which B_{wn,u^*} denotes the list B of wn bins with sizes drawn from the uniform distribution in the interval $[u^*,1]$. In experiments, for problem instances in subcase A, $u^* = u$ and $w = 1$; For subcase B $u^* = \min\{L_{n,u}\}$ and $w = 10$.

Let us first examine the situation when n is small. Fig. 2(a) and Fig. 2(b) depict the average values of $H(L_{n,u}, B_{wn,u})/s(L_{n,u})$ of algorithms when applying them to problem instances in subcase A and subcase B respectively with $n = 20$ and 100. u ranges from 0.05 to 1 in steps of 0.05 and each data point represents the average of 100 runs. It can be seen from Fig.2(a) and Fig.2(b) that FF and FFD are always better than NF and NFD. As n increases, the improvement of NF and NFD are not as remarkable as that of FF and FFD. Surprisingly, in Fig.2(a), when $u > 0.7$, the performance curves of NF and NFD with $n = 100$ become even worse than that with $n = 20$.

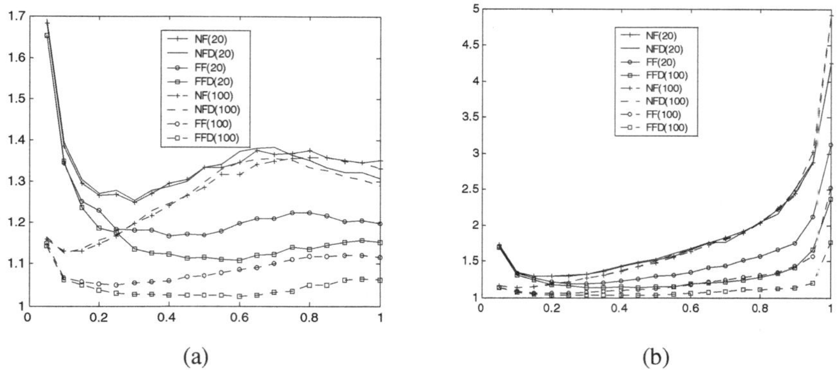

Fig. 2. $ER_u^{\infty}(H)$ for subcase A(a) and B(b) as a function of u with n=20 and 100

To be able to obtain the asymptotical average performances of algorithms, long lists of items are preferred. Experiments with $n = 500$, 1000, 2000, 5000, 10000 and 20000 are carried out to examine the asymptotical performances. Fig.3(a) and Fig.3(b) depict the experimental results with $n = 500$ and 2000 for subcase A and B respectively.

In experiments, with the same value of u, the improvements of the average ratios of NF and NFD coming from the increasing of n are very small: as n increases from 500 to 2000, in subcase B, the improvements of the average ratios of NF for $u = 0.05$ through 1 are within $[-0.8405, 0.0202]$, and the improvements of NFD are within $[-0.9870, 0.0184]$; In subcase A, the improvements of NF are within $[-0.0005, 0.0187]$, and the improvements of NFD are within $[-0.0022, 0.0168]$.

Moreover, by $n = 10000$, both the NF and the NFD algorithm appear to have converged, and both of them seem to be very near their asymptotes: in subcase A, as n increases from 10000 to 20000, the improvements of NF for $u = 0.05$ through 1 are within $[-0.0013, 0.0017]$, and the improvements of NFD are within $[-0.0005, 0.0019]$. Similar convergence happens in subcase B. Due to the convergence properties of NF and NFD in subcase A, it is shown that the results of NF or NFD with different n are overlapped in most cases. Thus in Fig.3, for simplicity, only the results with $n = 2000$ were given out.

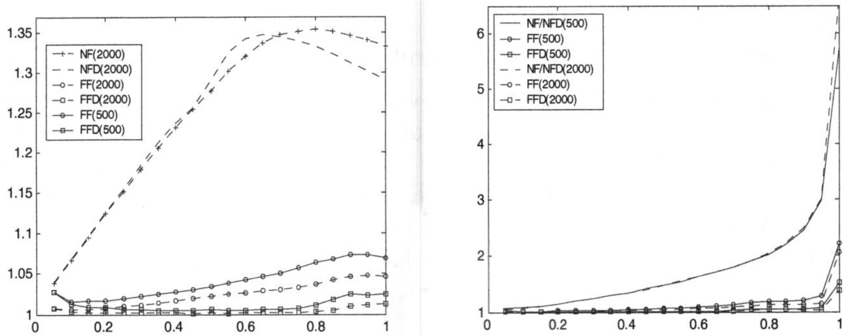

(a) (b)

Fig. 3. $ER_u^\infty(H)$ for subcase A(a) and B(b) as a function of u with n=500 and 2000

5 Conclusions

Although many researches have been reported to investigate the impact on the utilization of local resources, on AR jobs or on non-AR jobs resulting from advance reservations for Grid jobs, none of them has been devoted to try to reduce the disadvantages on local resources and/or normal jobs. This study was aimed to find a theoretic model, as well as local scheduling algorithms, for single machine schedulers to reduce the disadvantages on utilization of local resources and to shorten the makespan(i.e., maximum completion time) of normal jobs while supporting advance reservations for Grid jobs. The performances of the algorithms were investigated from both of the worst case and the average case viewpoints. Analytical results show that the worst case performance ratios of the algorithms against that of possible optimal algorithms are not less than 2. Experimental results for average cases show FF and FFD are better than NF and NFD. Considering that when applying NFD to the non-AR jobs, it's required that the processing order of the jobs can be changed without interfering the execution of the jobs, NFD can be applied only for independent non-AR jobs. Because it is not necessary to change the processing order of the jobs when applying FF, this algorithm can be applied for both of independent of precedence-constrained non-AR jobs. For independent non-AR jobs, the average performance of FFD is better than that of FF.

Acknowledgements

This research was supported by the Natural Science Foundation of China with grant number 60663009, the Application Science and Technology Foundation of Yunnan Province with grant number 2006F0011Q, the Research Foundation of the Education Department of Yunnan Province with grant number 6Y0013D and the Research Foundation of Yunnan University with grant number 2005Q106C.

References

1. MacLaren, J.: Advance Reservations: State of the Art. Global Grid Forum (2003)
2. Noro, M., Baba, K., Shimojo, S.: QoS control method to reduce resource reservation failure in datagrid applications, IEEE Pacific Rim Conference on Communications, Computers and Signal Processing (PACRIM) (IEEE Cat. No. 05CH37690) (2005) pp. 478–481 (2005)
3. Mohamed, H.H., Epema, D.H.J.: The design and implementation of the KOALA co-allocating grid scheduler. In: Sloot, P.M.A., Hoekstra, A.G., Priol, T., Reinefeld, A., Bubak, M. (eds.) EGC 2005. LNCS, vol. 3470, pp. 640–650. Springer, Heidelberg (2005)
4. Sulistio, A., Buyya, R.: A Grid Simulation Infrastructure Supporting Advance Reservation. In: Proceedings of the 16th International Conference on Parallel and Distributed Computing and Systems. Anaheim, pp. 1–7. ACTA Press, Cambridge, Boston (2004)
5. Mateescu, G.: Quality of service on the grid via metascheduling with resource co-scheduling and co-reservation. International Journal of High Performance Computing Applications 17, 209–218 (2003)
6. McGough, A.S., Afzal, A., Darlington, J., Furmento, N., Mayer, A., Young, L.: Making the grid predictable through reservations and performance modelling. Computer Journal 48, 358–368 (2005)
7. Smith, W., Foster, I., Taylor, V.: Scheduling with advanced reservations. In: Proceedings of 14th International Parallel and Distributed Processing Symposium, pp. 127–132. IEEE Computer Society Press, Los Alamitos (2000)
8. Heine, F., Hovestadt, M., Kao, O., Streit, A.: On the impact of reservations from the Grid on planning-based resource management. In: Sunderam, V.S., van Albada, G.D., Sloot, P.M.A., Dongarra, J.J. (eds.) ICCS 2005. LNCS, vol. 3514, pp. 155–162. Springer, Heidelberg (2005)
9. Snell, Q., Clement, M., Jackson, D., Gregory, C.: The Performance Impact of Advance Reservation Meta-Scheduling. In: Feitelson, D.G., Rudolph, L. (eds.) IPDPS-WS 2000 and JSSPP 2000. LNCS, vol. 1911, pp. 137–153. Springer, Heidelberg (2000)
10. Sanlaville, E., Schmidt, G.: Machine scheduling with availability constraints. Acta Informatica 35, 795–811 (1998)
11. Schmidt, G.: Scheduling with limited machine availability. European Journal of Operational Research 121, 1–15 (2000)
12. Zhang, G.: A new version of on-line variable-sized bin packing. Discrete Applied Mathematics 72, 193–197 (1997)
13. Friesen, D.K., Langston, M.A.: Variable sized bin packing. SIAM journal on computing 15, 222–230 (1986)
14. Parallel Workloads Archive, http://www.cs.huji.ac.il/labs/parallel/workload/
15. Edi, S., G, F.D: Backfilling with lookahead to optimize the packing of parallel jobs. Journal of Parallel and Distributed Computing 65, 1090–1107 (2005)
16. Chiang, S.-H., Fu, C.: Re-evaluating Reservation Policies for Backfill Scheduling on Parallel Systems. In: Proceedings of the 16th IASTED International Conference on Parallel and Distributed Computing and Systems, Cambridge, MA, pp. 455–460 (2004)
17. Weng, C., Li, M., Lu, X.: An Online Scheduling Algorithm for Assigning Jobs in the Computational Grid. IEICE Trans Inf Syst E89-D, 597–604 (2006)
18. Coffman Jr., E.G., Garey, M.R., Johnson, D.S.: Approximation Algorithms for Bin Packing: A Survey. In: Hochbaum, D. (ed.) Approximation Algorithms for NP-Hard Problems, pp. 46–93. PWS Publishing, Boston (1996)

19. Coffman Jr., E.G., Galambos, G., Martello, S., Vigo, D.: Bin Packing Approximation Algorithms: Combinatorial Analysis. In: Du, D.-Z., Pardalos, P.M. (eds.) Handbook of Combinatorial Optimization, pp. 1–47. Kluwer Academic Publishers, Dordrecht (1998)
20. Csirik, J., Johnson, D.S.: Bounded Space On-Line Bin Packing: Best is Better than First. Algorithmica 31, 115–138 (2001)

CROWN FlowEngine: A GPEL-Based Grid Workflow Engine

Jin Zeng, Zongxia Du, Chunming Hu, and Jinpeng Huai

School of Computer Science and Engineering, Beihang University, Beijing, China
National Key Laboratory of Software Development Environment, Beijing, China
Trustworthy Internet Computing Lab, Beihang University, Beijing, China
{zengjin,duzx,hucm}@act.buaa.edu.cn,
huaijp@buaa.edu.cn

Abstract. Currently some complex grid applications developing often need orchestrate multiple diverse grid services into a workflow of tasks that can submit for executing on the grid environment. In this paper, we present CROWN FlowEngine—a GPEL-based grid workflow engine for executing grid workflow instances. Besides basic functions of a conventional BPEL4WS-based workflow engine, CROWN FlowEngine has many features including hierarchical processing mechanism, multiple types of task scheduling, transaction processing, etc, which are of paramount importance to supporting workflow instances using GPEL language. CROWN FlowEngine will be adopted and widely deployed in CROWN Grid environment to support a wide range of service grid applications integration. We conduct several experiments to evaluate the performance of CROWN FlowEngine, and the results of comparing our work with GWES are presented as well.

1 Introduction

As the progress of grid[1] technology and the increasing prevalence of grid applications, a large numbers of resources could be accessed by drawing on grid middleware. In particular, with the appearance of OGSA (Open Grid Service Architecture)[2] architecture, some web service technologies have been successfully incorporated into grid computing to deal with resource heterogeneity and other important issues. And the resultant service grid is generally regarded as the future of grid computing. Many specifications, including OGSI, WSRF, WSDL, SOAP, etc, are introduced to standardize service grid technology. In service grid, various resources such as computers, storage, software, and data are encapsulated as services (e.g. WSRF services). As a result, resources can be accessed through standard interfaces and protocols, which effectively mask the heterogeneity.

Our CROWN (China Research and Development Over Wide-area Network)[3, 4] project aims to support large-scale resource sharing and coordinated problem solving by using service grid and other distributed computing technologies. In CROWN development, we find currently many complex grid applications developing often need integrate multiple diverse grid services into a new application. Traditional workflow technology gives this requirement a well solution. Like workflow model,

R. Perrott et al. (Eds.): HPCC 2007, LNCS 4782, pp. 249–259, 2007.

first we need to use a language to orchestrate logic and sequence relation of available grid services and create a grid workflow description document; second the grid workflow description document is deployed a execution engine, and then is scheduled and run according to defined logic and sequence relation.

In this paper, we present CROWN FlowEngine—a GPEL[5]-based grid workflow engine for executing grid workflow instances. Besides basic functions of a BPEL4WS-based workflow engine, CROWN FlowEngine has many features including hierarchical processing mechanism, multiple types of task scheduling, transaction processing, etc, which are of paramount importance to support workflow instances using GPEL language. GPEL is a grid process execution language based on the web service composition language BPEL4WS[6], and a detailed introduction about GPEL were discussed in the reference paper. CROWN FlowEngine will be adopted and widely deployed in CROWN Grid environment to support a wide range of service grid applications integration.

The rest of the paper is organized as follows. In section 2, we analyze related works in the area of grid workflow engine. The system architecture of CROWN FlowEngine is presented in section 3. In section 4, we show the implementation details and design issues. Results of performance evaluation are discussed in section 5. And in section 6, we conclude this paper and pave the way of future works.

2 Related Works

Grid computing offers tremendous benefits in the domains of different industry and science especially Life Sciences, Finance Energy, Biology etc. As the appearance of service grid, we use various resources expediently through grid middleware. But while encountering a complex computing paradigm, we have to integrate and orchestrate multiple single computing units to a new application for the original requirement, so from grid service to grid workflow is a natural evolvement process in grid computing area.

Now there are some existing researches about grid workflow engine in academy domain. Jia Yu and Rajkumar Buyya of University of Melbourne publish a technical report[7] about workflow management systems for grid computing and survey current main grid workflow systems. Condor DAGMan[8] is a service for executing multiple jobs with dependencies in a declarative form. It uses DAG (Directed Acyclic Graph) as the data structure to represent job dependencies. Each job is a node in the graph and the edges identify their dependencies. Each node can have any number of "parent" or "children" nodes. Kepler[9] is a popular scientific workflow system, with advanced features for composing scientific applications, it has been extended to support web service.Taverna[10] is used to assist scientists with the development and execution of bioinformatics workflows on the Grid. FreeFluo is also integrated into Taverna as a workflow enactment engine to transfer intermediate data and invoke services. The Grid Workflow Execution Service(GWES)[11, 12] is the workflow enactment engine, which coordinates the composition and execution process of grid workflows by GWorkflowDL language. GWES supports pure web services and Globus Toolkit 4 (GT4)[13], and it is easily extendible for further execution platforms. In the section 5 we compare the performance of GWES with CROWN FlowEngine.

3 System Architecture

CROWN FlowEngine is a GPEL-based grid workflow engine, and Figure 1 shows its system architecture. Traditional BPEL4WS-based workflow engine provides a basic implementation of a workflow engine; however, these basic functions are not enough to satisfy the requirements for workflow engine in real grid environments.

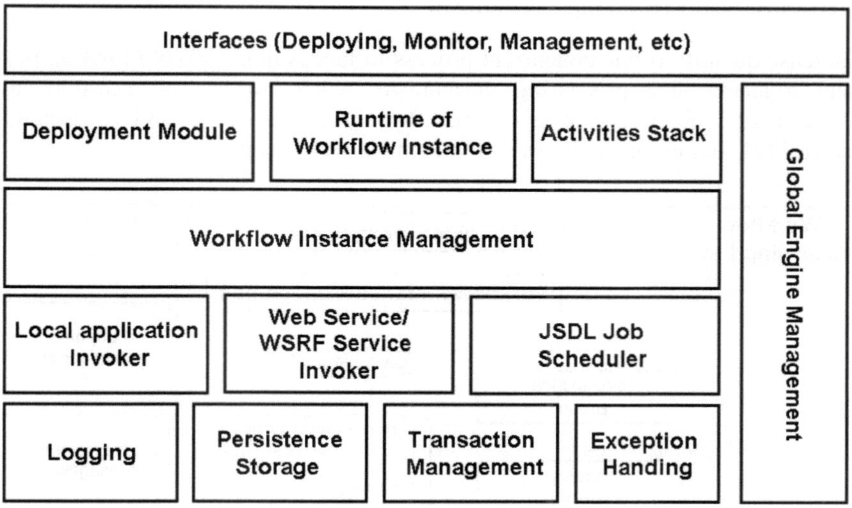

Fig. 1. System Architecture of CROWN FlowEngine

The Grid Workflow Forum[14] thinks grid workflows differ from "normal" workflows, such as business workflows or IT workflows, mainly in the following sense: stateful/stateless, reliability and performance. So when designing and developing CROWN FlowEngine, we fully consider specialties of grid environment, such state of service, reliability, performance, scheduling, etc.

Fist, our CROWN FlowEngine is based on GPEL, the extension of a international factual specification – BPEL4WS version 1.1, so some general workflow model created using BPEL4WS can be explained and executed. We provide stack-based explaining and executing algorithm to explaining GPEL documents and adopt multithread mechanism to executing each process instance for reducing complexity of the CROWN FlowEngine. Especially, in order to solve the problem that process instance waits longer for system resource consumption in asynchronous invoking, the method to persistencing process instance is adopted. Second, an advanced hierarchical processing mechanism enhances concurrent performance of CROWN FlowEngine. And a task processing adaptor supports multiple invocation types, including local application, general stateless web service, stateful WSRF service, and dynamic grid service schedule based on JSDL[15] for various requirements of users. Third, in order to ensure consistency and reliability, besides usual exception handling CROWN FlowEngine implements WS-Reliability[16] protocol for reliability in message layer.

It is significant to adopt a novel distributed transaction processing technology which ensures reliability and consistency of a group of key grid services. In next section we will describe these implementation details of various components and mechanisms.

4 Implementation Experience

4.1 Hierarchical Processing Mechanism

To increase quantity of the concurrent process instances in CROWN FlowEngine, we present a hierarchical processing mechanism which consists of transport layer, message layer, public service layer, workflow processing layer and corresponding interfaces[17], and Figure 2 shows its hierarchical architecture.

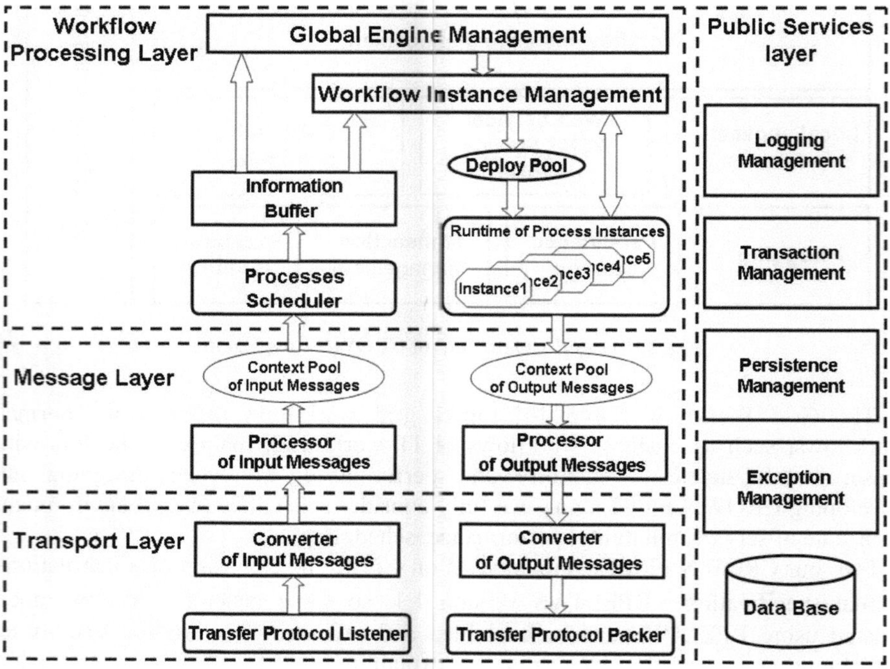

Fig. 2. Hierarchical Processing Mechanism

Transport layer is a foundation layer that can receive and transmit SOAP message bound on various transfer protocols (such as HTTP, SMTP). Therefore, this layer needs to deal with various types of user request and SOAP message transferred on various protocols. With processing content mentioned above, we design converters of in-output messages, protocol listener and packer.

Message layer ensures reliability of transferring message at invoking external application or service. Message layer provides support to WS-Reliability protocol and at least can realize one of following two requirements: first, when fault and exception occur occasionally but not crash, all messages from sender should be delivered to receiver; second, if fault and exception keeps continuously or crash occurs, after fault and exception is eliminated or crash is resumed, all previously sent message should be delivered to receiver. In addition, in order to solve a number of concurrent requests, context pool is set in this layer.

Workflow processing layer is core of CROWN FlowEngine. Its function is to create, explain and execute a process instance, control the lifecycle of this process instance and so on. It consists of following parts:

- Global engine manager: this is the global control component; it is in charge of the management of all external messages and internal information.
- Workflow instance manager: it manages lifecycle of all process instances, including operations of creating, pausing, persistencing and resuming. According to the name and method of request message a new process instance is created.
- Runtime of process instance: it explains and executes activity in GPEL language and processes the activity of a process instance by stack-based explaining and executing algorithm.
- Processes scheduler: it adopts the policy of "First Come First Served" to deal with requests of users or external applications.

Public services layer provides the necessary functions in runtime of CROWN FlowEngine, which includes logging management, transaction management, persistence management, exceptions management, database connectivity and so on.

4.2 Task Scheduling

Conventional workflow engine and workflow engine based on BPEL4WS invokes tasks in a process instance is static, that is it has been decided in orchestrating. But compared with other distributed computing the characteristic of grid computing is

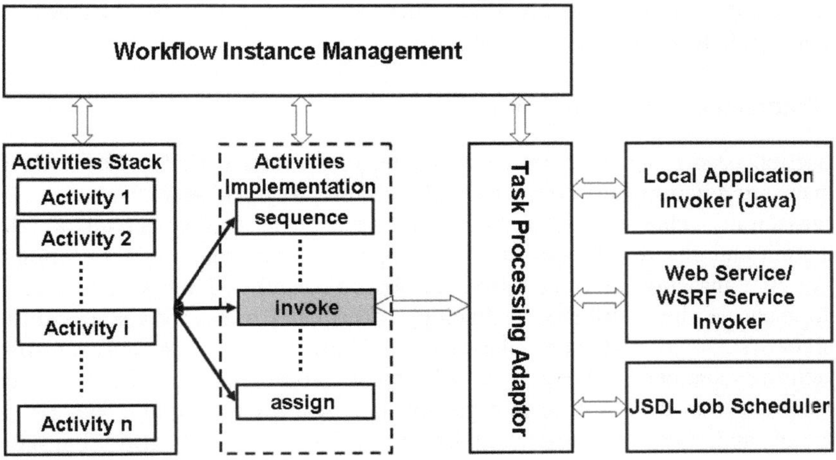

Fig. 3. Multiple Types of Task Scheduling

able to carry out job dynamic scheduling, that is a task can be located to the proper computing node according to user requirement and environment constraint. Shown as figure 3, our CROWN FlowEngine not only support local application and web service/ WSRF service call but also support grid job dynamic scheduling based on JSDL language.

In practical development, we add mostly inner property of cooperative partner description activity (partnerLink) and service invoking activity (invoke). When creating grid workflow model, user may assigns different type for certain external invoking task. In CROWN FlowEngine, workflow instance is explained and executed in activities stack. When external invoking activity (invoke) is executed, task processing adapter selects different call modes by the description at user modeling.

- Local Application Invoker: it may invoke local grid application and support currently Java language programs. This mode is mainly aimed at grid workflow developers, who need not to encapsulate the developed local application into a service in some situations. The local program is more efficient than other.
- Web Service/WSRF Service Invoker: this is a remote method invocation based on SOAP RPC. Invoking web service is basic function provided by CROWN FlowEngine. However with service grid presented and developed stateful service plays more and more important role in grid computing environment. Our GPEL support invoking standardized WSRF service. By being tested, CROWN FlowEngine is completely compatible with Globus Toolkit 4 (GT4).
- JSDL Job Scheduler: it is a remote method invoking based on SOAP Document. JSDL is used to describe the requirements of computational jobs for submission to resources, particularly in Grid environments. Noticeably, CROWN FlowEngine is a light and stand-alone grid workflow engine. CROWN FlowEngine itself has no grid job scheduling function and must together work with a external scheduler (such as CROWN Scheduler) supporting JSDL. At processing a JSDL description task, our JSDL Job Scheduler encapsulates JSDL document into SOAP message and commits it to a real external scheduler that can process JSDL document. Last, JSDL Job Scheduler parses and processes.

4.3 Transaction Processing Model

Transaction processing is an advanced characteristic of the CROWN FlowEngine, which ensure reliability and consistency of a group of key grid services.

Transaction technology is a processing mechanism that constructs a reliable application and ensures identity of all participants' output result in application. The concept of transaction originates from database research. Along with research and development of the distributed technology, distributed transaction across different resources (such as database, message middleware) is presented. Distributed transaction mentioned above can effectively apply in the closely coupled organization, but they are unable to satisfy participant's requirement for local autonomy and long transaction under the relaxed coupled grid environment. Therefore, under CROWN FlowEngine we need a new transaction technology and model – service grid transaction to coordinate a group of key grid services in the workflow instance and ensure reliability and consistency of their result.

CROWN FlowEngine adopts BTP[18] protocol as specification of our transaction processing model. BTP is first XML-based B2B transaction processing protocol that defines the logic sequence and message types of the transaction processing. Purpose of BTP is to permit many participants owned and controlled by heterogeneous organization to coordinate running. BTP uses Two-Phase-Commit in order to ensure that the entire application program obtains the consistent result.

Figure 4 shows our transaction processing model:

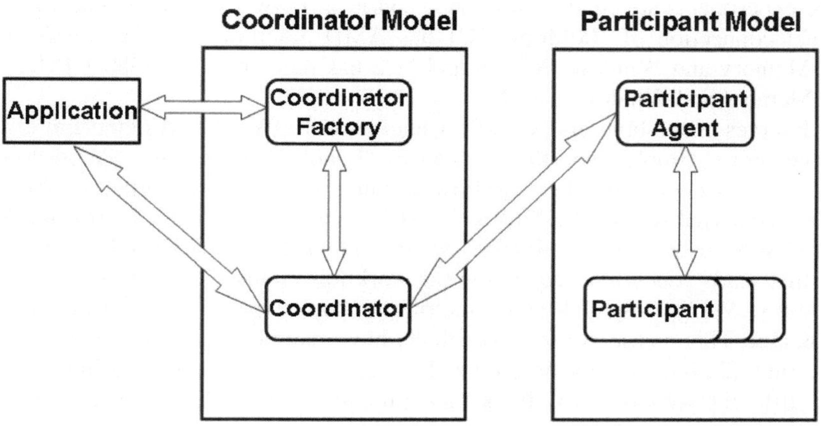

Fig. 4. Transaction Processing Model in CROWN FlowEngine

In this model a grid workflow instance is regarded as the "external" Application. When it need to processes a group of transactional grid services the grid workflow instance sends message to Coordinator Factory. After the coordinator factory initializes an instance of Coordinator, the instance sends a request to the Participant Agent to inform it. The Participant Agent searches this node for the Participant and registers it. The instance of Coordinator commits request message to the Participant Agent and the Participant by Two-Phase-Commit and finally returns transaction result to the Application.

The Coordinator is kernel component created by the Coordinator Factory at generation of the transaction. The Coordinator coordinates the transaction processing throughout all time. During the transaction processing the Coordinator communicates with the Participant Agent time after time. One transaction processing corresponds to one coordinator and is identified with a unique ID.

The Participant Agent is essential component in transaction node. All Participants in distributed node communicate with a Coordinator via the Participant Agent. Actually in course of communication none of Participants communicate with external. The advantage of our implementation is: the Participant Agent simplifies a Participant's work; a lot of messages are centralized by a Participant Agent to control transactions in the node. With Participant Agent, logging and committing Two-Phase-Commit become more convenient. And Participant's task is finishing business operation, but not regarding transaction implementing.

5 Performance Evaluation

In this section, we present the performance evaluation of CROWN FlowEngine. We will show the performance comparison of CROWN FlowEngine and GWES implementation.

The experiments are conducted on PC1 and PC2. Between the two PCs, there is an Ethernet connection of 100Mbps. PC1 has AMD Sempron 1.6 GHz CPUs, 1 GB DDR Memory and Windows XP OS and PC2 has Pentium D 3.4 GHz CPUs, 1 GB DDR Memory and Windows XP OS.

PC1 serves as the both grid workflow engines (CROWN FlowEngine and GWES) and the most simple workflow instance "Echo" (GPEL and GWorkflowDL) respectively is deployed. The workflow instance is a sequence process which can invoke an external service "HelloWorld". There is a stateless "HelloWorld" service in the grid service container (CROWN Node Server), that is deployed on PC2. The client that sends concurrent requests to the workflow engines (PC1) using Java thread technology. We use the following metrics to do the performance evaluation. (1) Throughput. This metric is used to evaluate how many requests can be processed in one second. (2) Average response time. This metric reflects the processing efficiency under different scenarios. (3) Invocation processing time. This metric reflects the processing efficiency of external invocation. (4) Overhead of transaction processing. This metric is used to evaluate the system overhead of transaction processing under CROWN FlowEngine.

In our experiment, we hope to evaluate the metrics mentioned above of CROWN FlowEngine and GWES under different numbers of concurrent requests. All the experiments are repeated 10 times, and we report the average results.

The results of our experiment are presented in Figure 5~8. We vary the number of concurrent requests from 0 to 200 (10, 20, 30 … 130, 140, 150, 175 and 200). Figure 6 plots the throughput of CROWN FlowEngine and GWES, while Figure 7 shows the

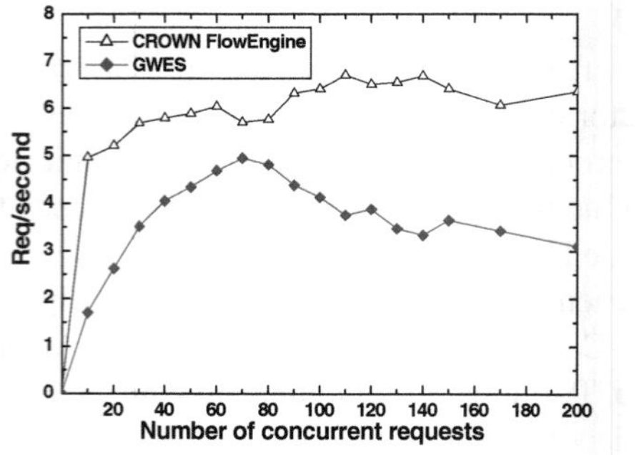

Fig. 5. Throughput vs. Num of concurrent requests (Echo)

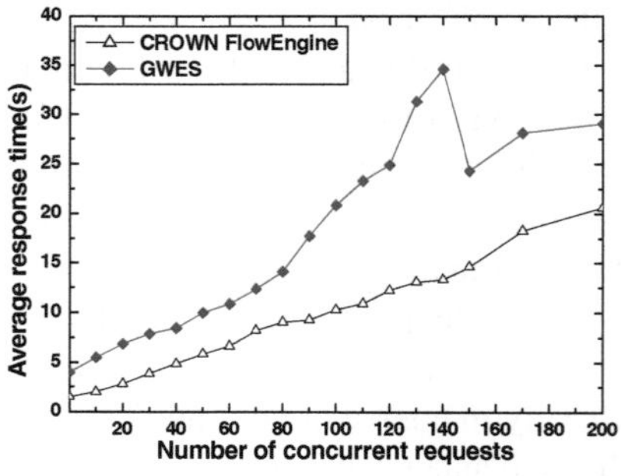

Fig. 6. Average response time vs. Num of concurrent requests (Echo)

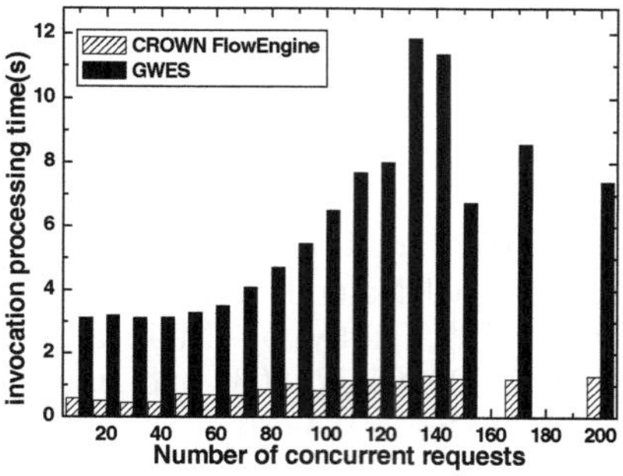

Fig. 7. Comparison of Invocation processing time (Echo)

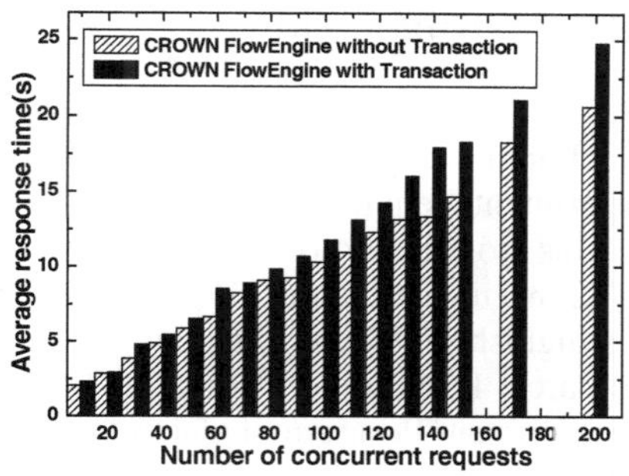

Fig. 8. Performance of transaction processing (Echo)

average response time for processing a request. We can see that CROWN FlowEngine outperforms GWES almost in all cases. We figure out there can be two reasons to explain the results. First, GWES is a service self and depends on third-part Application Server that deals with messages processing, in this experiment GWES is deployed under Tomcat; however, CROWN FlowEngine is a stand-alone server. Second, we design the hierarchical processing mechanism in CROWN FlowEngine to implement some buffers and pools for concurrent requests. In a general way external invocation can consume much time of a process instance processing. Therefore, on grid workflow engine side, we measure the processing time of invocation function, and the result is shown in Figure 8. The difference of invocation processing time further verifies that CROWN FlowEngine performs better than GWES. The result of Figure 9 shows that the transaction processing does not incur too much overhead to CROWN FlowEngine itself.

6 Conclusion and Future Works

In this paper, we present the design and implementation experience of a novel GPEL-based grid workflow engine, CROWN FlowEngine.

In order to solve the difficult problems such as stateful/stateless, reliability and performance in grid workflow, we present many feasible mechanisms and approaches including hierarchical processing mechanism, multiple types of task scheduling, transaction processing, etc. CROWN FlowEngine will include in CROWN 3.0 release that have been tested in CROWN testbed and widely use in real applications as well. We perform extensive experiments to evaluate the performance of our implementation and GWES. The results show that CROWN FlowEngine performs better that GWES under various numbers of concurrent requests.

We will perform more tests on all components of CROWN FlowEngine to make it stable to use. Also we will further design and develop a console for CROWN FlowEngine to monitor, manage and analyze process instances. In addition, we hope to perform further study on performance issues.

Acknowledgement

We thank our colleagues; Bing Bu and Dou Sun implement much programming work, and Jiabin Geng, for his help with client-side programming work and setting up the experimental environment.

This work is part of the results of the project CROWN and is supported by the NSF of China under Grant 90412011 and by China National Natural Science Funds for Distinguished Young Scholar under Grant 60525209, partly by National Basic Research Program of China 973 under grant no. 2005CB321803, and partly by Development Program of China 863 under grant no. 2006AA01A106.

References

1. Foster, I., Kesselman, C.: The Grid: Blueprint for a New Computing Infrastructure, 2nd edn. Morgan Kaufmann, San Francisco (2004)
2. Foster, I., Kesselman, C., Nick, J.M., Tuecke, S.: Grid Services for Distributed System Integration. IEEE Computer 35, 37–46 (2002)
3. CROWN project: http://www.crown.org.cn/en
4. Huai, J., Hu, C., Li, J., Sun, H., Wo, T.: CROWN: A Service Grid Middleware with Trust Management Mechanism. Science in China Series F 49, 731–758 (2006)
5. Wang, Y., Hu, C., Huai, J.: A New Grid Workflow Description Language. IEEE International Conference on Services Computing. IEEE Computer Society Press, Los Alamitos (2005)
6. Andrews, T., Curbera, F., Dholakia, H., Goland, Y., Klein, J., Leymann, F., Liu, K., Roller, D., Smith, D., Thatte, S., Trickovic, I., Weerawarana, S., Thatte, S.: Business Process Execution Language for Web Services version 1.1 (2003), http://www-128.ibm.com/developerworks/library/specification/ws-bpel/
7. Yu, J., Buyya, R.: A Taxonomy of Workflow Management Systems for Grid Computing. Technical Report, GRIDS-TR-2005-1, Grid Computing and Distributed Systems Laboratory, University of Melbourne (2005)
8. Thain, D., Tannenbaum, T., Livny, M.: Condor and the Grid. Grid Computing: Making the Global Infrastructure a Reality. John Wiley & Sons, NJ, USA (2003)
9. Ludäscher, B., Altintas, I., Berkley, C., Higgins, D., Jaeger, E., Jones, M., Lee, E.A., Tao, J., Zhao, Y.: Scientific Workflow Management and the Kepler System. Concurrency and Computation: Practice & Experience, Special Issue on Scientific Workflows (2005)
10. Oinn, T., Addis, M., Ferris, J., Marvin, D., Senger, M., Greenwood, M., Carver, T., Glover, K., Pocock, M.R., Wipat, A., Li, P.: Taverna: a tool for the composition and enactment of bioinformatics workflows. BIOINFORMATICS, vol. 20 (2004)
11. Neubauer, F., Hoheisel, A., Geiler, J.: Workflow-based Grid applications. Future Generation Computer Systems 22, 6–15 (2006)
12. The Grid Workflow Execution Service: http://www.gridworkflow.org/kwfgrid/gwes/docs/
13. Globus Toolkit: http://www.globus.org/toolkit/
14. The Grid Workflow Forum: http://www.gridworkflow.org/
15. Anjomshoaa, A., Brisard, F., Drescher, M., Fellows, D., Ly, A., McGough, S., Pulsipher, D., Savva, A.: Job Submission Description Language (JSDL) (2005), http://forge.gridforum.org/projects/jsdl-wg
16. Iwasa, K., Durand, J., Rutt, T., Peel, M., Kunisetty, S., Bunting, D.: OASIS Web Services Reliability (2004), http://docs.oasis-open.org/wsrm/ws-reliability/v1.1/wsrm- ws_ reliability-1.1-spec-os.pdf

17. Bu, B.: Research and Implementation of a Lightweight Workflow Engine based on Web Service, Master Thesis, in School of Computer Science and Engineering, Beihang University (2006)
18. Ceponkus, A., Dalal, S., Fletcher, T., Furniss, P., Green, A., Pope, B.: OASIS Business Transactions Technical Committee. Business Transaction Protocol?BTP?Version 1.0 (2002), http://www.oasis-open.org/committees/download.php/1184/2002-06-03.BTP_cttee_spec_1.0.pdf

Dynamic System-Wide Reconfiguration of Grid Deployments in Response to Intrusion Detections*

Jonathan Rowanhill[1], Glenn Wasson[1], Zach Hill[1], Jim Basney[2], Yuliyan Kiryakov[1], John Knight[1], Anh Nguyen-Tuong[1], Andrew Grimshaw[1], and Marty Humphrey[1]

[1] Dept of Computer Science, University of Virginia, Charlottesville VA 22094
{jch8f,wasson,zjh5f,yyk5v,jck,an7s,grimshaw,mah2h}@virginia.edu
[2] National Center for Supercomputing Applications (NCSA),
University of Illinois at Urbana-Champaign, IL USA
jbasney@ncsa.uiuc.edu

Abstract. As Grids become increasingly relied upon as critical infrastructure, it is imperative to ensure the highly-available and secure day-to-day operation of the Grid infrastructure. The current approach for Grid management is generally to have geographically-distributed system administrators contact each other by phone or email to debug Grid behavior and subsequently modify or reconfigure the deployed Grid software. For security-related events such as the required patching of vulnerable Grid software, this *ad hoc* process can take too much time, is error-prone and tedious, and thus is unlikely to completely solve the problems. In this paper, we present the application of the ANDREA management system to control Grid service functionality in near-real-time at scales of thousands of services with minimal human involvement. We show how ANDREA can be used to better ensure the security of the Grid: In experiments using 11,394 Globus Toolkit v4 deployments we show the performance of ANDREA for three increasingly-sophisticated reactions to an intruder detection: shutting down the entire Grid; incrementally eliminating Grid service for different classes of users; and issuing and applying a patch to the vulnerability exploited by the attacker. We believe that this work is an important first step toward automating the general day-to-day monitoring and reconfiguration of all aspects of Grid deployments.

1 Introduction

As Grids become increasingly relied upon as critical infrastructure, it is imperative to ensure the highly-available and secure day-to-day operation of the Grid infrastructure. However, as Grid system administrators and many Grid users know, the deployed Grid software remains quite challenging to debug and manage on a day-to-day basis: Grid services can fail for no apparent reason, Grid services can appear to be running

* This material is based upon work supported by the National Science Foundation under Grant No. 0426972 and through TeraGrid resources provided by NCSA. Any opinions, findings, and conclusions or recommendations expressed in this material are those of the author(s) and do not necessarily reflect the views of the National Science Foundation.

R. Perrott et al. (Eds.): HPCC 2007, LNCS 4782, pp. 260–272, 2007.

but users are unable to actually utilize them, information services report seemingly incorrect information, new versions of Grid software can need to be deployed, etc. Whenever a Grid administrator at a site determines that some part of their local Grid services is not running according to plan, and it is not immediately solvable via some local action, Grid administrators must generally contact their peers at other sites via telephone or email to debug overall Grid behavior. Email and telephone are also heavily utilized to disseminate the availability of new versions of deployed Grid software, including those updates that contain important security patches. For such critical security-related events, this *ad hoc* process can take too much time, is error-prone and tedious, and thus is unlikely to remedy the situation.

The broad challenge addressed in this paper is: How can a large collection of Grid services be dynamically managed in near-real-time to provide an acceptable level of functionality without significant human involvement? At the core of our approach is Willow 1, a large-scale feedback control architecture featuring sensors, actuators, and hierarchical control logic that runs side-by-side to the Grid and uniquely facilitates near-real-time monitoring, reconfiguration, and general management of the overall functionality of a Grid. Specifically, in this paper, we narrow our focus to the design and implementation of the integration of ANDREA 2 -- the "actuation" part of Willow -- to ensure the security of large-scale Globus Toolkit v4 deployments (GT4 3) in the presence of an intruder (essentially a person gaining unauthorized access to some service). *We do not pursue ways in which to determine that a Grid intrusion has occurred, rather we assume that one has occurred and subsequently need to holistically react to such a breach.* In simulated experiments involving 11,394 Globus Toolkit v4 deployments, run over 768 real nodes, we show the application and performance of ANDREA for three increasingly-sophisticated reactions: shutting down the entire Grid; incrementally eliminating Grid service for different classes of users; and issuing and applying a patch to the vulnerability exploited by the attacker. We believe that this work is an important first step toward automating the general day-to-day monitoring and reconfiguration of all aspects of Grid deployments.

2 The ANDREA Management Framework

Willow 1 is a comprehensive approach to management in critical distributed applications that uses a combination of (a) fault avoidance by disabling vulnerable network elements intentionally when a threat is detected or predicted, (b) fault elimination by replacing system software elements when faults are discovered, and (c) fault tolerance by reconfiguring the system if non-maskable damage occurs. The key is a reconfiguration mechanism that is combined with a general control structure in which network state is sensed, analyzed, and required changes actuated. ANDREA is the "actuation" component of Willow. Because the focus of this paper is on the integration and use of ANDREA to control Globus installations in the context of intrusion detection, we will not discuss Willow any further. That is, we do not discuss the use of Willow to *detect* the intruder in question here, rather we focus on the use of Willow (specifically ANDREA) to *react* and take corrective actions.

2.1 Distributed, Hierarchical Management Policy in ANDREA

ANDREA is a distributed, hierarchical collection of nodes that integrate at the lowest level with the controlled system, which is in this case the Globus-based Grid. The collection of ANDREA nodes exist in a loosely coupled configuration in that they are not aware of one another's names, locations, or transient characteristics. This is a key point for controlling truly large-scale systems where the cost of any node maintaining a precise and up-to-date view of the remainder of the system is prohibitive (or the task is impossible). An ANDREA node advertises a set of attributes that describe important state of the application node it wraps. This is similar to a management information base (MIB) in traditional management architecture.

ANDREA nodes interact with one another through delegation relationships. Each ANDREA node: (1) states a policy describing constraints on the attributes of other ANDREA nodes from which it will accept delegated tasks; and (2) describes constraints on the kinds of tasks it will accept. This is known as a *delegate policy*. Meanwhile, if an ANDREA node seeks to delegate a task, it states constraints on the attributes of possible delegates. This is called a *delegator policy*. A delegated task is given to all ANDREA nodes in the system and *only* those nodes for which, within a time window specified by the delegator: (1) the delegator's attributes and the tasks's attributes meet the requirements of the delegate's policy; and (2) the delegate's attributes meet the requirements of the delegator's policy. Details of the addressing language and underlying mechanism used by ANDREA, called *Selective Notification*, are provided in the work by Rowanhill et al. 4.

Once an ANDREA node has delegated a task to other ANDREA nodes, there exists a loosely coupled management relationship between the nodes, because the delegator need not know the name/locations of its delegates. It merely uses the "handle" of the management relationship to multicast a communication to its delegates. Likewise, a delegate need not know the name/location of its delegator, and it need not be aware of any other delegates. It can communicate with the delegator using the management relationship as a handle. In addition, temporal coupling between nodes is minimized. Delegates can join a relationship at any time during the specified time window and receive all relevant communications in the correct order upon joining. This management relationship is particularly advantageous for users (or other software components) that are "injecting work" into the ANDREA system.

2.2 Hierarchical Workflow in ANDREA

Management is affected in ANDREA by defining *workflows*. A workflow is a partially ordered graph of task nodes, and a task can define a child workflow within it. A task with an internal workflow represents a composite action. A leaf task is one that contains no sub-workflow. A leaf task may be an action, a constraint, a query, or a periodic query. Each leaf task of a workflow states an *intent*, a token from an enumerated language known to all ANDREA nodes. A task's intent can represent any delegated entity. A common semantic is that each intent represents a requirement on distributed system service state defined as a *constraint* on system variables. Each workflow defines a *reason*, a token from a global enumerated language stating the purpose for the intents of the workflow. Conflicts are detected between the intents of tasks on the same blackboard, and resolved through preemptive scheduling based on the priority of reasons.

An application component can introduce management requirements into a system by defining a local workflow to its ANDREA service interface. This workflow is managed by the ANDREA system so that tasks and constraints are carried out in the right order and delegated to appropriate nodes. Each ANDREA node has a conventional hierarchical workflow and blackboard model.

It is impractical (and non-scaleable) for any single ANDREA node to keep track of all nodes in the system and thus be able to precisely determine *which* nodes have or have not complied with the intended action of a given workflow. Instead, delegates in a workflow keep *aggregate* determinations, in the form of histogram state. When a delegated task is completed (based on this aggregate state), the delegator rescinds the delegation. The management relationship that had linked the delegator to the delegates is terminated and eliminated and thus again fully decoupled.

2.3 Programming ANDREA

ANDREA has a unique programming model in which system configuration tasks (which are part of the workflows) are represented as *constraints*. The set of managed resources to which a constraint applies is described by a set of properties of the resource. Each system administrator (or automated system administration component) can issue "constraint tasks", i.e. system configuration workloads that put the system in a state with certain properties. ANDREA will configure the system and attempt to maintain the given constraints. Conflicting constraints are typically resolved by a priority scheme among administrators. For example, suppose an administrator issued the constraint that no low priority jobs should be run on a set of Grid nodes. ANDREA will reconfigure the selected nodes, and then continue to make sure this constraint is upheld. So, if another administrator issued a command that no medium priority jobs should be run on an overlapping set of nodes, ANDREA can configure the system to maintain both constraints (i.e., no low or medium priority jobs). However, if a command to allow low priority jobs is received, there is a conflict since both constraints cannot be simultaneously met.

3 Using ANDREA to Reconfigure the Grid in Response to Intrusion Detections

While we believe ANDREA can be used for many aspects of day-to-day Grid management, our focus in this effort is to use ANDREA to reconfigure the Grid in response to intrusion detections. We generically describe the scenario as follows: an operator has manually discovered a rootkit on a node in the Grid and relays this information to the Grid control center, which uses ANDREA to put the system into a number of different configurations on a per-node basis[1]:

- *Normal* – nodes are accessible to all user classes
- *Shutdown* – no users are allowed to access nodes (since they are considered infected)

[1] In the future, we will use the sensing capabilities of Willow to discover such rootkits and automatically engage ANDREA (with human confirmations as warranted).

- *No low priority* – no low priority users are allowed access to nodes. Eliminating this class of resource consumption reduces the dynamics of the Grid so that it is easier to assess the extent of the intrusion, without shutting down the entire Grid.
- *Only high priority* – no low or medium priority users are allowed access to nodes. If "*no low priority*" is not sufficient, this configuration further simplifies the Grid environment while still offering some level of service.
- *Patch nodes* – in the case that the rootkit is exploiting a known vulnerability, grid system administrators can use ANDREA to take nodes identified as vulnerable offline, patch them, and bring them back online. This will cause a percentage of the Grid nodes to be unavailable for a time, but will allow other nodes (ones without the vulnerability that allowed the break in) to remain operational.

While this may seem like a simple scenario (and a simple "shutdown" solution), it is an ideal illustrator of the tasks for which ANDREA was designed. Every node in the system must be informed of the vulnerability and (as we shall see) ANDREA can provide a probabilistic description of "the grid's" response (i.e. "the grid is now 90% patched"). While an equivalent effect of avoiding the vulnerable nodes may be achieved via the manipulation of scheduling policies in either a single global queue or via coordination of scheduling policies on multiple queues in the Grid, we believe that this is not a reasonable approach for a number of reasons. First, it is extremely unlikely that there is a single queue, and coordination of multiple queues requires sites to give up site autonomy, which is unlikely (in contrast, as will be discussed in Section 5, ANDREA can be configured to respect site autonomy). Second, we require a mechanism affecting all Grid access, not just access to a job submission queue. Third, ANDREA is part of the larger Willow functionality, which will in the longer term automate some of the sensing and control-rule aspects implied here. Solely manipulating queuing policies offers no automated solution for integrating the discovery of a rootkit and an appropriate reaction.

In the remainder of this section, in the context of intrusion detection and reaction, we describe how to deploy ANDREA for the Grid (Section 3.1), the Grid-specific ANDREA workflows (Section 3.2), and the actual mechanisms executed on the controlled nodes to reconfigure the Globus Toolkit v4 Grid (Section 3.3).

3.1 Deployment of ANDREA for the Grid

An ANDREA deployment consists of a set of ANDREA nodes, which are processes that can receive configuration messages, enforce constraints, and report status. It is up to the ANDREA system deployer to determine how many managed resources are controlled by an ANDREA node, but typical deployments have one ANDREA node per machine. In the experiments reported in this paper, each deployment of GT4 on a machine was managed by a single ANDREA node. Each ANDREA node uses an XML deployment descriptor for the various supported system configurations and the possible conflicts between those configurations. (The mechanism by which ANDREA maps these labels into actual operations on the Grid is discussed in Section 3.3.) From this XML description of the supported configurations, ANDREA can infer that "patch nodes" mode conflicts with the "shutdown", "no low priority" and "only high priority" modes since they are subsets of "normal" mode. Configuration tasks, i.e. "go into normal mode", are issued with an attached priority that is used to resolve conflicts. In this way, configuration tasks can be dynamically prioritized. The current security model in the ANDREA implementation still requires receiver-side authentication and authorization checks to fully ensure that a received command is legitimate.

3.2 Using ANDREA Workflows for Configuring the Grid

System administrators (and potentially automated entities) inject workflows into their ANDREA nodes to manage and configure the Grid. Some of the tasks in a given workflow are delegated from the controller to all ANDREA nodes with particular attributes and acceptance of the controller's attributes, as described previously. The result is the arrival of a task at a set of ANDREA nodes managing Grid components, with the members of the set based on the policies stated at controllers and managers. Once an ANDREA node receives a task, it carries it out by actuation of the managed Grid components.

The status of the action, constraint, query, or periodic query is periodically sent back to the task's delegator on the order of seconds. ANDREA's distributed algorithms perform aggregation of this status information from all other delegates of the task. The command issuer can use these status updates to make determinations about the overall system state based on the percentage of nodes that have responded. ANDREA includes mechanisms to map task status histograms to representation of the task's state at the commander. For example, the Grid may be considered to be in a certain configuration when 90% of the nodes have replied that they are in that configuration, as 100% responses may be unlikely or unnecessary in large systems, depending on the situation.

Once a task is received by an ANDREA node, it must decide what action to take. The actual action an ANDREA node takes depends on what other active tasks are already being enforced. For example, if an ANDREA node is enforcing a "normal" constraint and a task arrives to enforce the "patch nodes" constraint, the net effect will depend on the priority of the "patch" command. If it has a higher priority then it will be enforced in place of the "normal" configuration. However, if it has a lower priority, it will not be discarded. Instead the "patch" command will remain on the ANDREA node's blackboard. If the "normal" configuration is withdrawn by its issuer, then the highest priority commands still on the blackboard that do not conflict with one another will be the ones that are enforced.

3.3 Integration of ANDREA with GT4 (A2GT4)

Although each local ANDREA manager knows which constraints apply to the resources it manages, it requires a resource- (or application-) specific mechanism to enforce those constraints. In our system, enforcement is performed by a system component called ANDREA-to-GT4 (A2GT4). A2GT4 is the mechanism by which ANDREA can alter the behavior of Globus Toolkit v4 and also includes an actuator, plugged into each ANDREA node's blackboard, defining how and when to run the mechanisms based on tasks on the blackboard and their scheduling.

We developed A2GT4 based on particular scenarios. Each new scenario is additive to the existing A2GT4 mechanism (note that ANDREA itself handles conflicts between concurrent re-configurations). With regard to the specific intrusion-detection scenario, this means that we had to map each of the five identified states ("normal", "shutdown", "no low priority", "only high priority", and "patch nodes") into a particular configuration of the GT4 services. In these experiments, the A2GT4 mechanisms manipulate authorization mechanisms to control the overall performance of a GT4 services. We chose this approach because the scripts were easy to debug, easy to confirm correctness, and not prohibitive of more sophisticated approaches necessary for different scenarios (which are under development).

4 Experiments

To evaluate the performance of our Grid management methodology, we deployed 15 instances of the Globus Toolkit, each managed by a local ANDREA node, on 768 nodes of the NCSA IA-64 TeraGrid cluster. Each cluster node had two 1.3-1.5 GHz Intel Itanium 2 processors and 4-12 GB of memory, with a switched Gigabit Ethernet interconnect. Excluding approximately 1.1% of the instances that failed to correctly start up for various reasons, this single NCSA cluster allowed us to simulate a much larger Grid of 11,394 ANDREA nodes and Globus Toolkit instances (Note that it was a “simulation” because it *wasn’t* an actual wide-area Grid deployment, although we truly had 11,394 independently-configurable Globus deployments in our experiments). 101 of the computers each ran a Selective Notification node, forming an overlay network in the form of a tree with a branching factor of 100. Each of the ANDREA nodes was connected as a leaf to this overlay network, such that each lower level Selective Notification dispatcher supported roughly 115 ANDREA nodes. The ANDREA controller ran on a machine at the University of Virginia.

We performed three experiments using the attacker/intruder scenario introduced in Section 3, with normal, no low priority, only high priority, and patch node configurations. For each experiment, we issued configuration commands from the ANDREA controller at UVA and logged the results at UVA and NCSA. At UVA, ANDREA logged the times at which commands were issued to and acknowledged by the ANDREA nodes at NCSA. At NCSA, the A2GT4 scripts logged each state change to the cluster’s file system. Over 200,000 events were recorded.

4.1 Experiment 1: Shutdown

The first experiment evaluated the ability of the ANDREA manager to shutdown all Grid nodes in the event of an attack or break-in. An ANDREA controller issues the command for all nodes to enter the shutdown state, prompting all ANDREA nodes to run the A2GT4 shutdown script, which denies access to all Globus Toolkit services on that node.

In Figure 1, we see that within seconds of issuance of the Shutdown command from UVA, the A2GT4 shutdown scripts begin running at NCSA, with all nodes shutdown 26 seconds after the command was issued. ANDREA confirms the state change of most nodes after 16 seconds, and all 11,394 nodes are confirmed shutdown after 26 seconds.

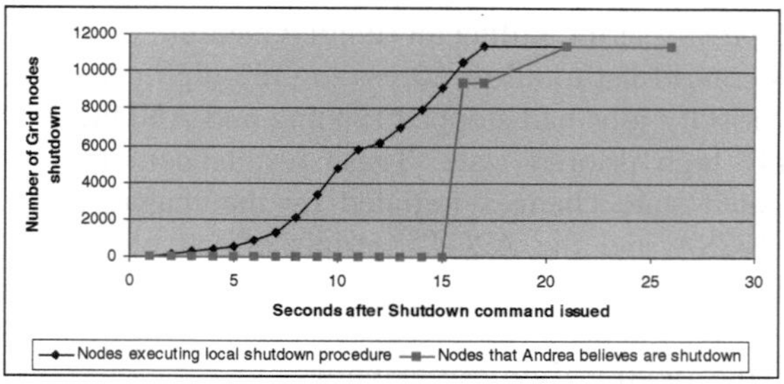

Fig. 1. Testing Basic ANDREA Control of GT4 Grid Nodes

4.2 Experiment 2: Shutting out User Groups/Classes to Enable Better Damage Assessment

The second experiment evaluates the ability of ANDREA state changes to control Grid jobs. We periodically submitted batches of Globus WS-GRAM jobs using three different user credentials, where the users and their jobs are prioritized as low, medium, and high. As the Grid comes under attack, an ANDREA controller decides to limit access to medium and high priority users only, to reduce the number of accounts to monitor while continuing to serve higher priority customers. As the intruder/attack continues, a separate ANDREA controller decides to limit access to only high priority users to further lock down the system.

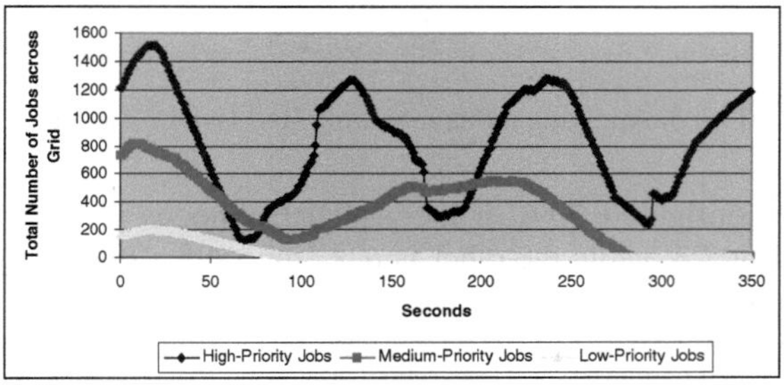

Fig. 2. Incrementally Eliminating Classes of Users

Figure 2 shows the number of high, medium, and low priority jobs running during the experiment. The number of running jobs follows a cyclical pattern, as a new batch of jobs is submitted while the previous batch is completing. The first ANDREA controller issued the command to deny access to low priority jobs 80 seconds into the experiment. Each ANDREA node ran the A2GT4 *NoLow* script, killing all running low priority jobs and updating the Globus authorization mechanism list to deny access to low priority users. 24 seconds later, all low priority jobs had stopped running and ANDREA acknowledged the change to the no low priority state. While the low priority user continued to attempt job submissions, they were all denied due to the access control change. At 275 seconds into the experiment, the ANDREA manager issued the command to deny access to both low and medium priority jobs. ANDREA ran the A2GT4 *HighOnly* script, killing all running medium priority jobs and updating the Globus access control list to allow access to only high priority users. 16 seconds later, all medium priority jobs had stopped running and ANDREA acknowledged the change to the only high priority state. These results demonstrate how ANDREA efficiently propagated state changes initiated by the manager at UVA across the 11,394 nodes at NCSA and the A2GT4 scripts effectively reconfigured the Grid environment on the nodes.

4.3 Experiment 3: Patching Vulnerable Nodes

The first two experiments evaluated the functionality and performance when the ANDREA manager issued constraints defining new states that the nodes should maintain, such as shutdown, no low priority, or only high priority. The third experiment evaluates ANDREA actions, which are commands that are executed once per node, and also demonstrates the use of ANDREA's selective notification.

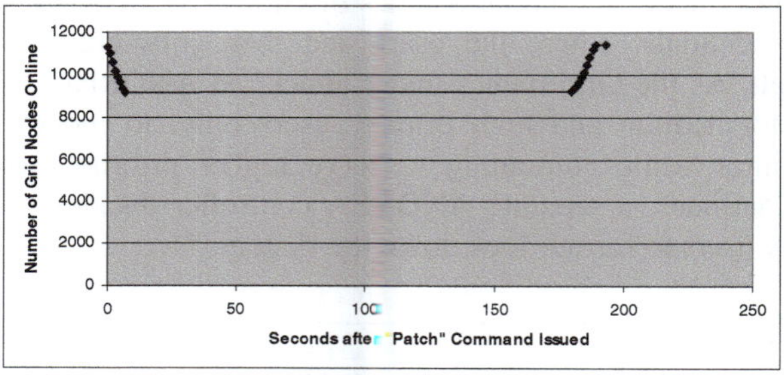

Fig. 3. Applying a Patch to Only Red Hat 7.3 Nodes

We artificially configured 20% of the ANDREA nodes to report their operating system to be RedHat 7.3, while the other nodes report other operating systems. In this scenario, we assume that the operator determines that the specific compromise arose as a result of a vulnerability in Red Hat 7.3, and the operator want to use ANDREA to issue an instruction *to only those vulnerable* nodes to take themselves offline, apply the patch, and then re-register with the Grid (i.e., come back on-line). Figure 3 illustrates that it took about 7 seconds for all 20% of the nodes (approx. 2250 nodes) to initiate the *patch* command within A2GT4. We simulated that the operation of retrieving the patch (the patch itself is not contained in the control signals provided by ANDREA, although a URL could be included), installing the patch, and then rebooting (which subsequently brings the Grid node back on-line) took 3 minutes in its entirety. Hence, these patched Grid nodes began appearing back on-line at 180 seconds into the experiments; all nodes were back on-line and reported to ANDREA within 13 seconds after that. While not directly shown in Figure 6, ANDREA's Selective Notification facilitated the "addressing" of the *Patch* command to only those nodes that published the property of being Red Hat 7.3 -- the actual destination (IP Address, port) of relevant nodes was handled by ANDREA. Only simulated Red Hat 7.3 nodes received such a command.

5 Discussion

We believe that our experiments show a valuable approach to the semi-real-time management of large-scale Grids. However, some discussion of the results is warranted. First, there is a question of the base-line against which these results can be compared. Unfortunately, we are aware of no Grid systems that actuate changes on system components in the automated manner at the scale of which ANDREA is capable. Intuitively, it seems that being able to issue re-configuration commands (such as “shutdown”) to over 11,000 receivers in ~25s seems not unreasonable for many application domains (while our experiments take place in a simulated environment of millisecond-latency between nodes, we believe that in a real wide-area deployment, network propagation delays will not have an overwhelming effect on these durations, as the dominant effect is based on the algorithms and topology of the ANDREA control structure, hierarchy, and fan-out as used in these experiments).

Second, there is the question of the cost of ANDREA’s ability to scale. In other words, ANDREA’s ability to reach large numbers of nodes, addressed by properties, and aggregate responses from those nodes may make it inappropriate for smaller scale communications tasks. We are not claiming that ANDREA should be used as a replacement for current grid communication mechanisms. Instead, we wish to use ANDREA for the task of managing large-scale distributed systems such as Grids and ANDREA’s performance should be viewed in this context. While ANDREA would undoubtedly be slow when compared to a single grid communication (e.g., typically a web service call), its performance is reasonable when compared to 11,394 such calls.

In general, we recognize that there are clearly a number of issues that must be addressed before our approach is ready for an operational Grid like the TeraGrid:

- The scope of the A2GT4 mechanism must be expanded to all functionality of the Grid node beyond those capabilities provided by GT4 (e.g., SSHD).
- A more comprehensive mechanism to put the managed node into each particular state must be developed. For example, the "shutdown" mechanism we implemented is of negligible value in the unlikely event that a vulnerability in the GRAM service allowed an attacker to *bypass* the GT4 authorization mechanism!
- A comprehensive suite of use-cases that cover the majority of day-to-day operations must be developed and reflected in A2GT4 and the ANDREA workflows. For example, we are currently investigating a scenario in which a high-importance operation that relies on the Grid (such as tornado prediction) is not meeting its soft deadlines. The corrective actions to take, particularly *at the same time* as an event such as an ANDREA reaction to intrusion detection, is the subject of on-going research in this project.
- The security of ANDREA must be further analyzed and improved. Certain operations in ANDREA are not mutually authenticated, instead relying on a trust model that in some cases oversimplifies the actual requirements of a TeraGrid-like environment. In addition to analyzing and correcting for vulnerabilities within ANDREA itself, we would like to leverage the authentication infrastructure of the Grid itself, namely GSI 16.

- The entire functionality of Willow (including the sensing and control logic) must be applied to a Grid setting, leveraging existing solutions. For example, regarding the sensing capability/requirement of Willow, we believe that existing tools such as INCA 5 could be leveraged as the foundation for such operations.

In our experiments, for simplicity we had a single control center, which is not unlike the Operations center of the OSG or the TeraGrid. But Willow/ANDREA supports multiple "control loops", such as a "VO controller" and an "Enterprise controller". Ultimately, the decision regarding what to enact (even in the presence of conflicting commands) lies with the controlled nodes. Willow/ANDREA's control mechanism accommodates sites *not* executing commands requested of them.

It is clear that ANDREA can be used to manage the Grid, but one can reasonably wonder about the management of ANDREA itself! Currently, the assumption is that ANDREA is running on a dedicated, reliable network of reliable machines. This is plausible for most situations but clearly not applicable for the broader Internet itself. It is the subject of future work to make ANDREA itself manageable.

6 Related Work

There exist many hierarchic management architectures in the literature, some commercial, others experimental. The size of the managed systems and the scope of management vary with architecture.Applying the model of Martin-Flatin et al 6, we can describe management architectures by the nature of their delegation capabilities. Early management architectures, such as traditional SNMP based systems 7, apply centralized control. SNMP-based systems have evolved to apply management hierarchy (e.g., SAMPA 8, Tivoli 9, and Open Systems Interconnection (OSI) standard 6. In each, the power of late-bound, run-time delegation service is increased, allowing managers to delegate authority through commands over otherwise static infrastructure in static hierarchical configurations. These architectures assume capability divided into static, implicit hierarchy assigned at system design time.

Many of the tools of large-scale Grid deployments can be utilized in Willow/ANDREA. Existing monitoring systems, such as the Globus Toolkit MDS 11, the TeraGrid INCA system 5, and MonALISA 12, could provide sensor data to ANDREA for analysis. For software configuration management of Grid systems, ANDREA could interface with package managers such as Pacman 14 and GPT 15 to (for example) patch vulnerable software versions. ANDREA's delegation policies, matching tasks to nodes, are similar to Condor's ClassAd Matchmaking 16, though we believe ANDREA's mechanisms for resolving conflicting constraints are more powerful. The Hawkeye monitoring system 13 builds on Condor's ClassAd capabilities. Evolving grid/web service standards are important for the management of large-scale grids. WS-Notification 18 and WS-Eventing 19 define message formats for the delivery of asynchronous messages. ANDREA's selective notification system could be modified to send messages that are consistent with these standards. ANDREA could also use WS-Management 20 and WSDM 21, specifications that define a standard language used for management of web services.

7 Conclusion

For security-related events such as the required patching of vulnerable Grid software, the *ad hoc* process of human operators using email and telephones can take too much time, is error-prone and tedious, and thus is unlikely to completely work. The Willow/ANDREA approach provides powerful mechanisms for dynamic reconfiguration of Grid nodes in the presence of intruder detections. We have demonstrated, through the development of A2GT4 and experiments involving 11,394 Globus Toolkit v4 deployments, that ANDREA can effectively enact a variety of different holistic Grid reconfigurations to limit the vulnerabilities. We believe that this work is an important first step toward automating the general day-to-day monitoring and reconfiguration of all aspects of Grid deployments.

References

1. Knight, J.C., Heimbigner, D., Wolf, A., Carzaniga, A., Hill, J., Devanbu, P., Gertz, M.: The Willow Architecture: Comprehensive Survivability for Large-Scale Distributed Applications. In: Intrusion Tolerance Workshop, DSN-2002 The International Conference on Dependable Systems and Networks (2002)
2. Rowanhill, J., Knight, J.C.: ANDREA: Implementing Survivability in Large Distributed Systems. University of Virginia technical report (2005)
3. Foster, I., Kesselman, C.: Globus: A Metacomputing Infrastructure Toolkit. International Journal of Supercomputer Applications 11(2), 115–128 (1997)
4. Rowanhill, J., Varner, P., Knight, J.C.: Efficient Hierarchic Management for Reconfiguration of Networked Information Systems. In: International Conf. on Dependable Systems and Networks (DSN 2004), Florence, Italy (2004)
5. INCA Test Harness and Reporting Framework, http://inca.sdsc.edu/
6. Martin-Flatin, J., Znaty, S., Hubaux, J.: A Survey of Distributed Network and Systems Management Paradigms. Network and Systems Management Journal 7(1), 9–22 (1999)
7. Mountzia, M., Benech, D.: Communication Requirements and Technologies for Multi-Agent Management Systems. In: Eighth IFIP/IEEE International Workshop on Distributed Systems Operations and Management (DSOM'97), Sydney, Australia (1997)
8. Endler, M.: The Design of SAMPA. In: The Second International Workshop on Services in Distributed and Networked Environments, IEEE Press, Los Alamitos (1995)
9. Taylor, S.D.: Distributed System Management Architectures. Masters Thesis, Univeristity of Waterloo, Ontario, Canada (1006)
10. Valetto, G., Kaiser, G.: Using Process Technology to Control and Coordinate Software Adaptation. In: 25th International Conf. on Software Engineering, Portland, OR (2003)
11. Czajkowski, K., Fitzgerald, S., Foster, I., Kesselman, C.: Grid Information Services for Distributed Resource Sharing. In: Proceedings of the Tenth IEEE International Symposium on High-Performance Distributed Computing (HPDC-10), IEEE Press, NJ, New York (2001)
12. Legrand, I., Newman, H., Voicu, R., Cirstoiu, C., Grigoras, C., Toarta, M., Dobre, C.: MonALISA: An Agent based, Dynamic Service System to Monitor, Control and Optimize Grid based Applications. In: CHEP 2004, Interlaken, Switzerland (2004)
13. Hawkeye, A.: Monitoring and Management Tool for Distributed Systems, http://www.cs.wisc.edu/condor/hawkeye/

14. Pacman, http://physics.bu.edu/ youssef/pacman/
15. GPT Grid: Packaging Tools, http://www.gridpackagingtools.org/
16. Raman, R., Livny, M., Solomon, M.: Matchmaking: Distributed Resource Management for High Throughput Computing. In: Proceedings of the Seventh IEEE International Symposium on High Performance Distributed Computing, IEEE Computer Society Press, Los Alamitos (1998)
17. Butler, R., Engert, D., Foster, I., Kesselman, C., Tuecke, S., Volmer, J., Welch, V.: A National-Scale Authentication Infrastructure. IEEE Computer 33(12), 60–66 (2000)
18. OASIS. Web Services Notification (WS-Notification) TC, http://www.oasis-open.org/committees/tc_home.php?wg_abbrev=wsn
19. Box, D., et al.: Web Services Eventing (WS-Eventing) (August 2004),
20. http://ftpna2.bea.com/pub/downloads/WS-Eventing.pdf
21. McCollum, R(ed.): Web Services Management (WS-Management) (2005),
22. http://msdn.microsoft.com/library/en-us/dnglobspec/html/ws-management.pdf
23. OASIS. Web Services Distributed Management TC (2005), http://www.oasis-open.org/committees/tc_home.php?wg_abbrev=wsdm

File and Memory Security Analysis for Grid Systems

Unnati Thakore and Lorie M. Liebrock

Department of Computer Science, New Mexico Institute of Mining and Technology, USA
{unnati,liebrock}@cs.nmt.edu

Abstract. The grid security architecture today does not prevent certain unauthorized access to the files associated with a job executing on a remote machine. Programs and data are transferred to a remote machine for completing staging jobs. Footprints may remain on the machine after the programs and data are deleted from the remote machine. Anyone with super user privileges can access these footprints. In this paper, we explore the problem of unauthorized access to programs and data, supporting it with experiments carried out using the Globus toolkit. After showing that programs and data are at risk, we also discuss ways for secure deletion of data and programs on the remote machine.

1 Introduction

One of the goals of any security architecture is confidentiality, which ensures access to computer resources must be authorized. Grid security research and development currently revolves around developing better solutions to address the following requirements: Authentication, Secure Communication, Effective Security Policies, Authorization, and Access Control [11]. For example, Globus Grid Security Infrastructure (GSI) provides X.509 certificates with the use of TLS (Transport Layer Security) protocol for mutual authentication and delegation. GSI by default provides unencrypted communication but can be configured to use encryption for communication [2, 3, 13].

The policies for access control and authorization as applied to grid computing today affect access to resources by authenticated grid users abiding by authorization policies on the resource [11]. Other policies define the resource usage and security provided by the resource providers in the VO (Virtual Organization) [12]. Access to grid resources, which are spread across geographical and political boundaries, is based on a trust relationship between the resource providers and the users. Thus both have to stick to the security policies agreed upon by both parties. Before a grid user accesses any grid resource, the user needs assurance that the machine is not compromised and his data and programs will not be stolen [10]. However, the users and providers may not be aware of the system being compromised or having internal breaches based on some users having administrator privileges. User data is at risk if there is any significant compromise of any system on the grid.

To our knowledge, no grid security architecture deals with securing a grid user's data and programs on the remote machine from any user that has super user privileges; there are no policies governing this type of confidentiality requirement. In other words, we have found no grid security policies that deal with confidentiality of user data and code on the grid system with respect to super users. Although super user related risks include insider threat, these risks also include risks associated with any

R. Perrott et al. (Eds.): HPCC 2007, LNCS 4782, pp. 273–284, 2007.

hacker that compromises a component of the grid system to an extent that provides super user access. In this paper, we discuss this confidentiality issue and show how the programs and data of a grid user are in jeopardy on a remote machine.

An experimental grid was set up to search for the footprints of a grid user's data and program on the remote machine. Experiments were carried out using the Globus tools for job submission on a grid implemented with the Globus toolkit version 4.0.3. Physical memory and secondary storage on the remote machine were examined for footprints using standard digital forensics tools. The results indicate when a grid user's program and data on the remote machine can be at risk. We then discuss some ways to prevent this unauthorized access in certain scenarios or to at least limit the time frame during which such access could occur.

The paper begins in Section 2 with explanation of what can be extracted by reading secondary storage and physical memory on a grid system. Then the paper describes details of our experimental configuration in Section 3. Section 4 describes the experiments carried out to find footprints of the grid user's data and programs in secondary storage and physical memory. Section 5 explains how a grid user's data and programs can be deleted from secondary storage and outlines a plan for how to remove them from physical memory. The paper finally concludes after a brief discussion of future work.

2 Memory and Storage Analysis

Data Lifecycle Management (DLM) is an important security requirement for Grids [8, 13]. DLM is the process of managing data throughout its lifecycle from generation to its disposal across different storage media for the span of the entire life of the data. The data lifecycle is the time from data creation to its indefinite storage or deletion. There is a need for a security system that protects the grid user's data from all other users, even the super user. In order to determine whether current grid systems provide this protection, we need to analyze the memory and storage on grid systems. Ideally, backups and other storage media should be considered. However, here we focus on the standard memory and storage on grid systems. We recommend encryption of data when it is in storage, so that all backups will also be more secure. Next, we show how such analysis for memory and secondary storage is performed.

2.1 Secondary Storage Analysis

Secondary storage of a machine on a grid system can be analyzed to locate data and programs. A super user can read any grid user's data and programs as the current security systems don't restrict the super user's access. Based on the method used for deleting data and programs (<fileCleanUp> attribute of job description language or the UNIX rm command), files are just deleted from the file system. Deletion of files typically entails just removing the metadata and adding the storage space to the available space list. This deletion is not a secure deletion of files on the remote system. These files could be temporary files, executables, program files, data files, output files, stdout, or stderr.

The normal deletion of files from secondary media is not secure, as the file contents are still stored on unallocated space of the secondary storage. The deleted files can be read by a variety of software or utilities (e.g., dd command of UNIX) that can read raw images of the storage. Due to the nature of grid systems, multiple users can be mapped to a single account on local machine. Therefore, there it may be possible for the users to read footprints left by previous users.

For securely deleting files from secondary storage it is necessary to overwrite that data with some pattern on the secondary storage. The number of times the data of the file is overwritten determines whether the data from the deleted space can be extracted [5]. The required level of confidentiality of the file to be deleted determines the number of times it has to be overwritten. There are many free tools and utilities that securely delete files from secondary storage. An example is the UNIX command shred that deletes files securely.

Another place to look for footprints is swap space, which is used for swapping out inactive jobs when system resources are low. With the size of RAM increasing, there are fewer possibilities to find much on the swap space, unless the system is heavily loaded [9]. Virtual memory provides the same opportunity.

Slack space is another place to look on the disk, which can be used to read information, which was previously written to the disk. Slack space is part of an allocated file that has not yet been written to and thus retains the old data (this is at the end of the last block of a file).

Overwriting is one method to prevent reading deleted data using programs that read raw data. However another analysis method, magnetic force microscopy (MFM), can still read overwritten data. MFM is a technique based on imaging magnetization patterns with high resolution and minimal sample preparation [6]. Although this technique can be used in forensics, it requires access to the physical media and this limitation would protect the data from hackers, at the very least. Further, most super users do not have access to MFM, so we do not consider this a significant risk for grid users unless there is a critical forensic analysis in progress.

Thus, it appears that analysis of hard drives on a grid system can let anyone with super user privileges read data and programs or their footprints. In particular, our experiments assess how much can be found and under what conditions.

2.2 Physical Memory Analysis

Physical memory analysis can be used to extract text and executable files, detailed information about terminated and executing processes, open network connections, and passwords [1]. Some confidential information access may prove to be fatal for a grid user, as this information can be used on behalf of that user on other sites. The physical memory can be analyzed to read raw data and meta data. Raw data is the execution code or data section from a memory mapped file or temporary data structures such as the stack used by the executing processes. Meta data is the data related to memory structures of page tables, processes, etc.

Physical memory (RAM) on Linux machines can be read using /dev/mem or /proc/kcore with administrator privileges. /dev/mem is a device file that mirrors main memory and the byte offsets are interpreted as the memory addresses by the kernel. /proc/kcore is an alias to the RAM. When we read from /proc/kcore the kernel performs the necessary memory reads. The output of kcore is in Executable and Linkable Format (ELF) and can be read using gdb, whereas the output of /dev/mem can be read using an ascii editor or examined by running strings on it. The System.map (the name depends on the kernel version) contents provide a map for addresses of important kernel symbols, which can be used to find the addresses of system calls, the address of memory attached to a process, etc. [1]. The physical memory contents are not erased after the process ends, but are stored as free memory until it is overwritten (as shown in Experiment III) or the power is turned off [9]. Thus an image of physical memory can give details related to the process and its data.

3 Experimental Grid Setup

A grid with two machines connected to each other using Globus Toolkit version 4.0.3 with default settings was configured [3]. Details of the grid environment (machines Grid0 and Grid1) are:

1) Before installing the globus toolkit, the prerequisites were installed on both of the machines. The basic prerequisites installed are j2sdk-1.4.2.13 and Apache Ant 1.6.5.
2) For consistency, Fedora Core 5 was installed on both machines. The Fedora Core 5 kernel (2.6.19) was patched to enable administrator access to /dev/mem.
3) Grid0 has 512 MB of system memory and a 30 GB hard drive. Grid1 has 256 MB of system memory and a 40 GB hard drive.
4) The minimum set of users is created on both machines. Grid0 has globus, postgres, and student in addition to root. Grid1 has globus and student1 in addition to root. The user student1@Grid1 maps to student@Grid0 when executing jobs.
5) The PostgreSQL database was installed on Grid0 to support the RFT (Reliable File Transfer). RFT is a web service that provides scheduling properties for data movement, for which it requires a relational database to storing states [3].
6) Tomcat was installed on Grid1 to support WebMDS (MDS - Monitoring and Discovery System), which is a web based interface for viewing information about grid services.
7) A Simple CA certificate authority was set up for signing user certificates on Grid0 to support grid operation.
8) Firewall settings were changed to allow tcp traffic for ports 8443 (GRAM and MDS), 2811 (GridFTP), and 5432 (PostgreSQL). A range of ports were added on both machines for temporary connections to PostgreSQL for RFT transfers.

3.1 Job Submission and Execution

This project uses WS - GRAM for job submission and management. Job submission to WS-GRAM requires a valid X.509 proxy certificate. WS GRAM has many command line clients to submit jobs, of which we use globusrun-ws, the official command line client, to submit jobs. The jobs submitted are staging jobs specified in a job description file using the standard job description XML schema.

4 Analysis on the Experimental Grid Setup

The following sets of experiments were carried out to find footprints of the grid user's data and programs that reside in the secondary storage and physical memory. Each experiment was run numerous times with results always being consistent with the observations presented below. In the tables that follow for each experiment, the approximate number of times the experiment was run is listed along with the summary of results. Any super user can access any grid user's data and programs. Such files can be accessed (until they are overwritten) even after the files are deleted by a grid user.

4.1 Experiment Set I

The experiments carried out consist of executing a simple Perl script on the remote machine. This Perl script (Appendix I) creates a temporary file and writes easily identifiable strings in the file and then deletes the file using the UNIX rm command. The job was submitted from Grid1 to Grid0 using the job description file (Appendix II). During the execution of that script on Grid0 a dump of the physical memory (/dev/mem) was taken using the standard forensic tool dd on an external hard drive. After the job was executed, an image of the hard disk (/dev/hda) on Grid0 was taken using the dd command. Then the dumps were searched using the UNIX command grep to find the strings that were written in the file.

Observations. The experiments of remote execution were performed many times to observe and verify consistent behavior depending on the size of the file created by the Perl script on the remote machine (Grid0).

Small Files. Small files (with size around 24MB) created by the Perl scripts were tested. A search on the secondary storage dump did not reveal the file created by the Perl script, although a search on the physical memory dump showed the contents of the file.

Large Files. Large files (with size greater than 40MB) created by the Perl script were tested. A search on the secondary storage as well as in physical memory showed the contents of the file.

File Risk Summary. In summary, Experiment Set I shows the following.

File size	# Experiments	Files found in secondary storage	Files found in physical memory dump
~ 24 MB	90	0%	100%
> 40MB	75	100%	100%

As an indication of the degree of security risk detected, let the risk associated with a file being found in storage be 70% risk and the risk associated with a file being found in the physical memory dump be 30%. This is an arbitrary allocation of risk, but it illustrates the point that the risk of information being recovered is higher in storage than it is in memory, as memory is more volatile. Further, when the data is located in both, methods can be used on either to recover the information so the risk of both recovery approaches is present. Note that this discussion does not consider all security risks, e.g., it does not account for the security risk associated with transferring data across the Internet. However, with these caveats, we arrive at the risk analysis graph in Figure 1, where the percentage of risk is shown on the y-axis and risk associated with different security issues are shown are the x-axis.

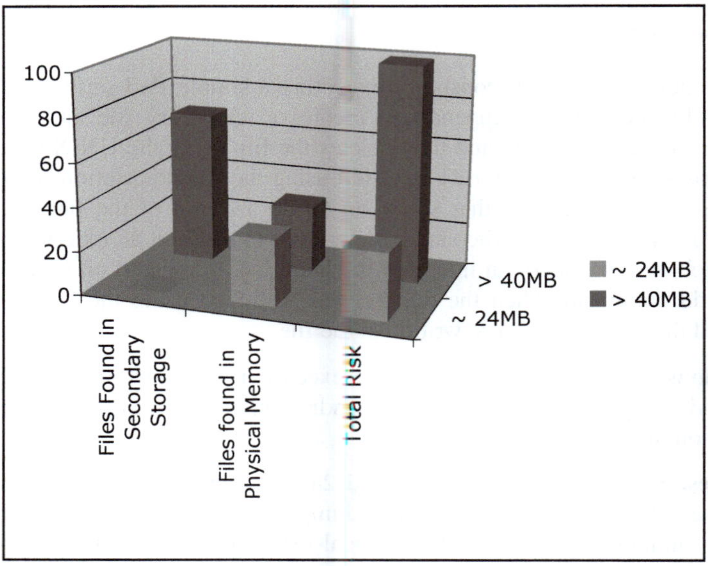

Fig. 1. Security risk using rm for file removal

Other Findings. The Perl script submitted was found in the secondary storage dump of Grid0. The contents of files written by earlier scripts were found on the physical memory and secondary storage dump.

4.2 Experiment Set II

The experiments carried out consist of executing a modified version of the Perl script used in Experiment Set I. The modified Perl script replaces the UNIX rm command with the UNIX shred command to securely delete the file it creates on the remote machine. Note that, as a reviewer pointed out, the secure deletion of files is based on the operating system. Here were we are using a built in UNIX command. If the operating system does not have a command in the standard distribution, then we would recommend an operating system extension or add on.

During the execution of that script on Grid0, a dump of the physical memory (/dev/mem) was taken using the dd command on an external hard drive. After the job was executed, the image of the hard disk (/dev/hda) on Grid0 was taken using the dd command. Then the dumps were searched using UNIX command grep to find the strings that were written in the file.

Observations. A search of the secondary storage dump did not show the file created by the Perl script but a search on the dump of physical memory did find the contents of the file. A search on the secondary storage dump showed the existence of the Perl script that was submitted. The execution of the Perl script takes a long time because of overwriting the storage space many times (25 times as a default number) with the shred command. For many users, the number of overwrite iterations could be significantly lower and still provide adequate security.

File Risk Summary. In summary, Experiment Set II shows the following.

File size	# Experiments	Files found in secondary storage	Files found in physical memory dump
~ 24 MB	60	0%	100%
> 40MB	45	0%	100%

Using the same indications of the degree of security risk detected, we arrive at the risk analysis graph in Figure 2.

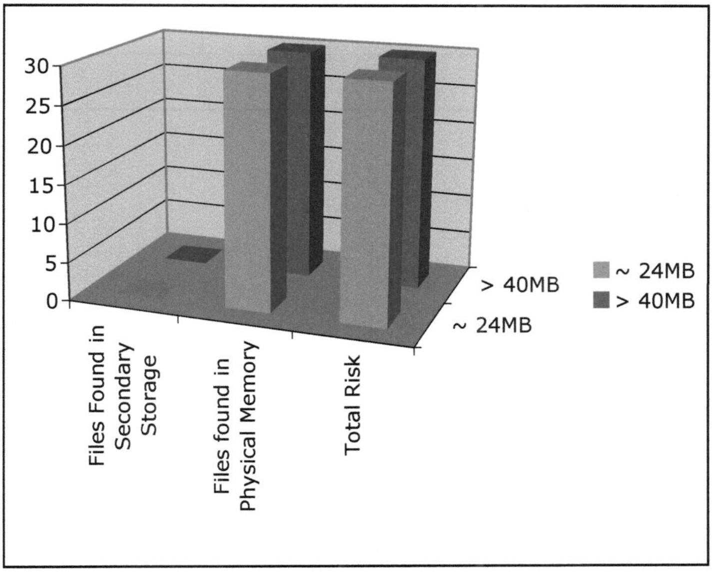

Fig. 2. Security risk using shred for file removal

4.3 Experiment Set III

The experiment carried out consisted of adding the following script [9] to the Perl script used in Experiment Set I and creating a new script with just the following lines of code. This script attempts to overwrite all accessible user memory. These files were submitted for execution on Grid0. After the job was finished, the dump of primary memory was taken on the external hard drive. Then the dump was searched using the UNIX command grep to attempt to find the strings that were written in the original file.

```
for (;;) {
        $buffer[$n++] = '\000' x 4096;
        }
```

Observations. In both cases (the combined scripts and the separate scripts), the process was killed after some time but the physical memory dump showed the memory was overwritten with \000. Thus the strings, which were originally written in the file, were overwritten by \000 preventing the original strings from being shown.

File Risk Summary. In summary, Experiment Set III shows the following.

File size	# Experiments	Files found in secondary storage	Files found in physical memory dump
~ 24 MB	10	0%	0%
> 40MB	10	0%	0%

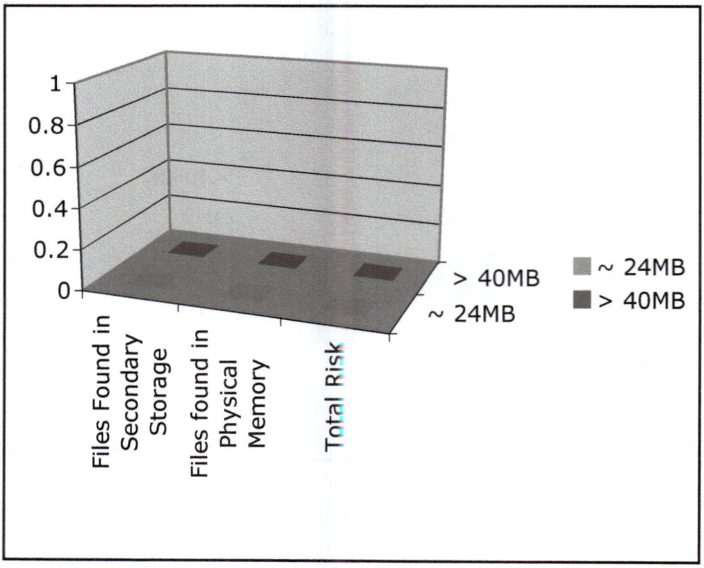

Fig. 3. Security risk using shred and the buffer erasing script for file removal

Using the same indications of the degree of security risk detected, we arrive at the risk analysis graph in Figure 3.

5 Secure Deletion

In order to overcome some of the security risks that have been illustrated in this paper, we now present some simple deletion methods that reduce these dangers.

5.1 Secondary Storage

Secure deletion methods depend on the operating system and file system on the remote machine. Securely deleting the code and data from the hard drive on a Linux machine with a default Linux file system can be is done using the shred command (as shown in the experiments). In the case of Windows machines, there are no such included commands that securely delete files. Thus, in case of remote machines having Windows, the grid user should make sure there is secure file deletion software installed. If no such software is installed, the user should in addition to sending the files for the job, send an executable to be used to securely delete files that were transferred or created. There are free software and utility programs (e.g., sdelete, eraser, cipher, etc.) available for Windows that can be used to securely delete the data and code. Each of these scripts works in the Windows operating system framework, but extends the standard capabilities of the operating system.

5.2 Physical Memory

Removing the data files from physical memory can be dependent on the memory management system of the underlying operating system. One naive way to erase the data/code from memory is shown in Experiment Set III. Even repeated runs of such programs as root only changes about 3/4 of the main memory. Such programs do not overwrite any of anonymous memory, kernel, or file caches [9]. The best way to eradicate everything from physical memory is to reboot the system, but that is not possible for a normal user in a grid environment. The naïve method shown tries to overwrite as much memory as a process can access and when as it exceeds its limit the process is killed by the operating system. This reduces the likelihood of finding the data and programs of the grid user in the physical memory. Since the data was originally written in the user space, we should be overwriting that same memory with this deletion process. However, more experiments are needed to verify this assumption and to ensure that data and code are not sometimes stored in anonymous memory or file caches.

Another approach that may be considered is to overwrite all variables in the program before termination of the simulation or solution code. This, however, is more complicated and we have not experimented with this approach yet.

5.3 Other Storage

Note that we have not addressed any backups or offline storage systems. If the data were encrypted, when not in actual use, then with high probability, the data in other storage systems would be encrypted. The caveat of "with high probability" is needed for the case where the data must be decrypted and temporarily put in storage. If that occurs, then it would be possible, but not probable, that a backup would occur while the data was stored in an unencrypted form.

If the unencrypted data is only in memory, and never put in user storage, then there should not be a backup created as systems typically only create backups of files in storage, not files in memory.

In summary, the presence of other storage systems and backups creates an added level of security risk for access to the data. If the data is not encrypted and securely deleted, it can be recovered from any of the media on which it is stored.

6 Future Work

This work was a side effect of our beginning work on Grid Forensics. In the future, we intend to carry out more detailed memory analysis and extend attribution methods for use in Grid Forensics. In particular, we will extend our experiments on secure file and memory deletion to provide better guidelines for grid user data and code protection. We will also explore the encryption of data while in untrusted file systems.

In addition to memory analysis, other vulnerabilities of the grid systems will be studied to inform grid users of the risks to their data and code, as well as for use in Grid Forensics. For example, network forensics will be applied to grids in order to find other means to collect code and/or data belonging to grid users.

7 Conclusion

Currently, it is the responsibility of the grid user to determine whether they wish to trust their code and data on a remote file / computer system. Here we examined what risks are associated with the decision to extend such trust.

In particular, we analyzed the problem of unauthorized access to deleted data and program files on a grid system. We have shown that both data and code can be found in secondary storage and memory by anyone with super user access. Grid users must be made aware of such security loopholes and provided with methods to avoid them. We have also illustrated how such unauthorized access can be avoided or limited in certain cases.

References

1. Burdach, M.: Digital Forensics of the Physical Memory (2005), http://forensic.seccure.net/pdf/mburdach_digital_forensics_of_physical_memory.pdf
2. Welch, V., Foster, I., Kesselman, C., Mulmo, O., Pearlman, L., Tuecke, S., Gawor, J., Meder, S., Siebenlist, F.: X.509 Proxy Certificates for Dynamic Delegation. In: 3rd Annual PKI R&D Workshop, pp. 42–58 (2004)
3. Globus Toolkit 4.0 Release Manuals (2007), http://www.globus.org/toolkit/docs/4.0/

4. Foster, T., Kesselman, C., Tsudik, G., Tuecke, S.: A Security Architecture for Computational Grids. In: Proc. 5th ACM Conference on Computer and Communications Security Conference, pp. 83–92 (1998)
5. Mallery, J.R.: Secure File Deletion, Fact or Fiction (2006), http://www.sans.org/rr/papers/27/631.pdf
6. Gutmann, P.: Secure Deletion of Data from Magnetic and Solid-State Memory. In: Sixth USENIX Security Symposium Proceedings, San Jose, California, pp. 77–89 (1996)
7. Foster, I.: A Globus primer (2005), www.globus.org/primer
8. Naqvi, S., Arenas, A., Massonet, P.: Scope of Forensics in Grid Computing - Vision and Perspectives. In: Proceedings of the International Workshop on Information Security and Digital Forensics. LNCS, pp. 964–970 (2006)
9. Farmer, D., Venema, W.: Forensic Discovery. Addison Wesley Professional, Boston (2005)
10. Humphrey, M., Thompson, M.R.: Security Implications of Typical Grid Computing Usage Scenarios. In: Proceedings of HPDC, pp. 95–103 (2001)
11. Humphrey, M., Thompson, M.R., Jackson, K.R.: Security for Grids. Proceedings of the IEEE 93(3), 644–652 (2005)
12. Verma, D., Sahu, S., Calo, S., Beigi, M., Chang, I.: A Policy Service for GRID Computing. In: Proceedings of the Third International Workshop on Grid Computing (2002)
13. Sotomayor, B., Childers, L.: Globus Toolkit 4: Programming Java Services. Morgan Kaufmann, San Francisco (2005)

Appendix I: Perl Script for File Creation

```
#!/usr/bin/perl
open ( TMPFILE , ">outputFile" ) || die "cannot open
outputFile for writing : $!" ;
for ( $i =0 ; $i < 10000 ; $i++ ) {
  print TMPFILE "5555aaaa5555aaaa5555aaaa5555aaaa\n" ;
  }
  close ( TMPFILE ) ;
  `rm -rf outputFile` ;
```

Appendix II: Job Description File

```
<job>
  <executable>/usr/bin/perl</executable>
  <directory>${GLOBUS_USER_HOME}</directory>
  <argument>my_echo</argument>
<stdout>${GLOBUS_USER_HOME}/stdout</stdout>
<stderr>${GLOBUS_USER_HOME}/stderr</stderr>
<fileStageIn>
  <transfer>

<sourceUrl>gsiftp://student1@Grid0:2811/tmp/second.pl</so
urceUrl>
      <destinationUrl>
```

```
                  file:///${GLOBUS_USER_HOME}/my_echo
       </destinationUrl>
    </transfer>
  </fileStageIn>
  <fileStageOut>
    <transfer>

<sourceUrl>file:///${GLOBUS_USER_HOME}/stdout</sourceUrl>
       <destinationUrl>
                 gsiftp://student1@Grid0:2811/tmp/stdout
         </destinationUrl>
    </transfer>
  </fileStageOut>
  <fileCleanUp>
      <deletion>

<file>file:///${GLOBUS_USER_HOME}/my_echo</file>
      </deletion>
  </fileCleanUp>
</job>
```

Business Model and the Policy of Mapping Light Communication Grid-Based Workflow Within the SLA Context

Dang Minh Quan[1] and Jörn Altmann[1,2]

[1] International University of Bruchsal, Campus 3, 76646 Bruchsal, Germany
dang.minh@i-u.de
[2] TEMEP, School of Engineering, Seoul National University, South-Korea
jorn.altmann@acm.org

Abstract. In the business Grid environment, the business relationship between a customer and a service provider should be clearly defined. The responsibility of each partner can be stated in the so-called Service Level Agreement (SLA). In the context of SLA-based workflows, the business model is an important factor to determine its job-resource-mapping policy. However, this aspect has not been described fully in the literature. This paper presents the business model of a system handling SLA-based workflow within the business Grid computing environment. From this business model, the mapping policy of the broker is derived. The experiment results show the impact of business models on the efficiency of mapping policies.

1 Introduction

Service Level Agreements (SLAs) [1] are currently one of the major research topics in Grid Computing, as they serve as a foundation for a reliable and predictable job execution at remote Grid sites. The SLA is a business contract between a user and a service provider. Thus, a SLA context represents a business environment. In our system of SLA-based workflows [3,4,5,6,7], the user wants to run a SLA-based workflow, which can be represented in Directed Acyclic Graph (DAG) form. The user has responsibility to specify the estimated runtime of the sub-jobs correlating with the specific resource requirements. He also provides the number of data to be transferred among sub-jobs. In the case of communication workflows with light communication, the data that will be transferred between sub-jobs is very small. The main requirement of the user is finishing the whole workflow within a specific period of time. Figure 1 depicts a sample scenario of running a workflow on the Grid environment. To ensure the deadline constraints, sub-jobs of the workflow must be distributed over many high performance computing centers (HPCCs) in the Grid. The resources in each HPCC are locally managed by the software called Resource Management System (RMS). To ensure that a sub-job can be executed within a dedicated time period, the RMS must support advance resource reservations such as CCS [2]. As HPCCs

R. Perrott et al. (Eds.): HPCC 2007, LNCS 4782, pp. 285–295, 2007.

usually connect with the network through a broadband link, the data transfer of the workflow with light communication can easily be executed in one time slot without the necessity to reserve network capacity. In our system, the time slot that executes the data transfer is right after the slot within which the source sub-job finished.

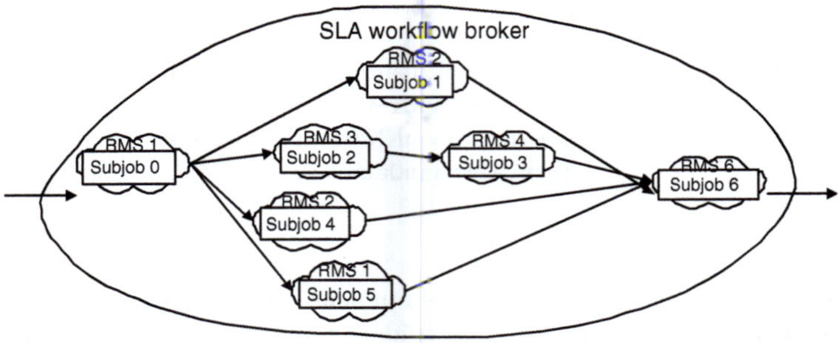

Fig. 1. Sample running workflow scenario

Assigning sub-jobs of the workflow to resources requires considering many constraints such as workflow integrity, on-time conditions, and optimality conditions. To free users from those tedious stuff, it is necessary to have a SLA workflow broker performing the co-operation task of many entities in the Grid. The business relationship of the SLA workflow broker with the users and the Grid service providers will determine the core working mechanism of the broker, the mapping policy. However, this issue has not been fully considered in most of the previous works about SLA-based workflows [9,10,11,12,13,14,15]. This paper, which is belongs to a series of efforts supporting SLA -based workflow [3,4,5,6,7], will analyze this problem to fill the gap. The paper is organized as follows. Section 2 describes the business model and SLOs. Section 3 presents the mapping policy and Section 4 describes the experiment. Related works are described in section 5. Section 6 concludes the paper with a short summary.

2 Business Model and SLOs Analysis

2.1 Business Model

The business relationship between entities in the system running the SLA-based workflow is depicted in Figure 2. There are three main types of entities: end-user, SLA workflow broker and service provider.

The end-user wants to run a workflow within a specific period of time. The user asks the broker to execute the workflow for him and pays the broker for the workflow execution service. The user does not need to know in detail how much he has to pay to each service provider. He only needs to know the total amount. This number depends on the urgency of the workflow and the financial ability

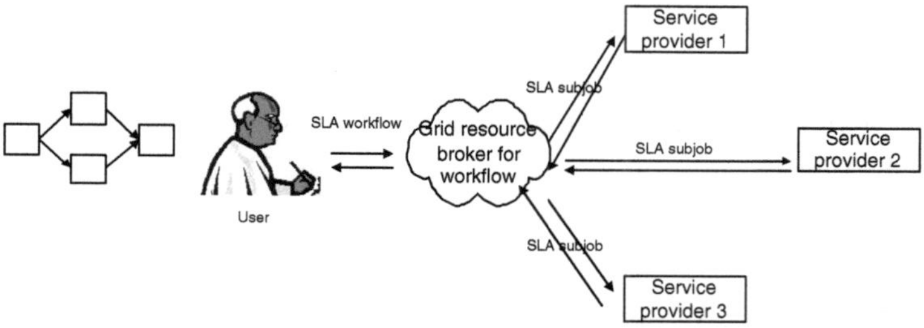

Fig. 2. Stakeholders and their business relationship

of the user. If there are problems with the workflow execution, for example the deadline is violated, the user will ask for compensation from the broker. This is clearly defined in the Service Level Objectives(SLOs) of the SLA.

The SLA workflow broker takes on the task of a user as specified in a SLA with the user. It controls the workflow execution. The SLA workflow broker also has to perform mapping of sub-jobs to resources, signing SLAs with the services providers, monitoring, and error recovery. When the workflow execution has finished, it settles the accounts. It pays the service providers and charges the end-user. The profit of the broker is the difference between boths. The value-add that the broker provides is the handling of all the tasks for the end-user.

The service providers execute the sub-jobs of the workflow. In our business model, we assume that each service provider fix the price for its resources at the time of the SLA negotiation. As the resources of one HPCC usually have the same configuration as well as quality, each service provider has a fixed policy for fining SLA violation, for example $n\%$ of the total cost for each late time slot.

Concluding, all activities between end-users, SLA workflow broker and service providers are specified in legally binding contracts (SLAs). We think that the consideration of business models is important in order to establish incentive structures for users to consume HPCC services at low risk, and for brokers to perform efficiently. That means:

- If the broker does not get any monetary compensation for managing the workflow, there is little incentive for the broker to find a high-quality mapping solution. On contrary, it will encourage the broker to find some unreliable solutions, increasing its income.
- With the proposed business model, the broker takes responsibility for the mapping solution. When a failure happens and the workflow does not finish within the desired period, the user can fine the broker and the broker would not get compensated. This method allows the user to get a guaranteed service.
- The described business model frees the user from the hard work of managing the workflow execution. He only signs an SLA with the SLA workflow broker and then waits for the result.

2.2 SLOs Analysis

Based on the analysis of the kind of entities in our model as described in the previous section, we identified five main SLOs, which ensure a successful execution of the workflow:

SLO1

Between broker - provider: *If the storage usage of the sub-job exceeds the pre-determination space, the sub-job will be canceled and the broker has to pay resource reservation cost.* The provider has to cancel the sub-job, since otherwise, other sub-jobs would be affected through less-available resources.

Between user - broker: *If the storage usage of a sub-job of the workflow exceeds the pre-determination space, the workflow will be canceled and the user has to pay for the cost of the executed sub-jobs.* As the wrong estimated sub-job is canceled, the whole workflow must also be canceled because of the workflow integrity characteristic.

SLO2

Between broker - provider: *If the memory usage of the sub-job exceeds the pre-determined space, the sub-job will be canceled and the broker has to pay resource reservation cost.* If the memory usage of the sub-job exceeds the predetermined space, the data has to be swapped in and out of the harddrive and, thus, slows down the processing speed and exceeds the specified runtime. Therefore, the provider has to cancel it.

Between user - broker: *If the memory usage of a sub-job of the workflow exceeds the pre-determined space, the workflow will be canceled and the user has to pay for the cost of the executed sub-jobs.*

SLO3

Between broker - provider: *If the sub-job cannot be finished because of computing system failure, the sub-job will be canceled and the provider is fined.* This case happens when the whole system is down. For example, the electric power or the network link to the Internet is broken.

Between user - broker: *If the workflow cannot be finished because of computing system failure, the workflow will be canceled and the broker is fined.* This case happens when many RMSs are down and the remaining healthy RMSs cannot execute the sub-jobs of the workflow. In general, the probability of this kind of problem is very small.

SLO4

Between broker - provider: *If the runtime of the sub-job exceeds the limitation because of a wrong estimation, the sub-job can run some more time slots if there* *are sufficient resource free. The broker has to pay the extra cost for computation. If there are insufficient resources available, the sub-job will be canceled and the broker has to pay the cost for the computation.*

Between user - broker: *If the runtime of a sub-job of the workflow exceeds the limit because of customer's wrong estimation and the sub-job is cancelled, the workflow will be cancelled and the user has to pay the computation cost. If the sub-job is not cancelled, the broker will change the mapping solution to ensure the workflow integrity. The user has to pay the extra cost and accept the runtime delay of the entire workflow.*

To check the conditions for this SLO, we have to monitor the entire system. Therefore, if there is no failure in the computing system and the sub-job finished late nonetheless, it can be deduced that the user estimated the resources needed wrongly.

SLO5

Between broker - provider: *If the runtime of the sub-job exceeds the limit because of computing system failure, the sub-job can continue running and the provider has to pay fining for late time slots.*

Between user - broker: *If the runtime of the workflow exceeds the limit because of computing system failure, the workflow can continue running and the broker has to pay fining k\$ for each additional time slot.*

To check the condition of this SLO, we also have to monitor the system status. If there is a failure of the computing system and the sub-job cannot finish on time, we can deduce that any delay detected is caused by the failture. Usually, the latency caused by this error is small. However, if the runtime of the sub-job exceeds a threshold (which includes the extra time for error recovery), we can say that the user had wrong estimation and the sub-job is cancelled.

Each SLO stated above can be implemented by using the structures described in [4].

2.3 The Influence of SLOs on Mapping Policies

In most of the existing systems [9,10,11,12,13,14,15], the broker tries to find a solution which is as inexpensive as possible. As the number of data to be transferred among sub-jobs in the workflow is very small, we can omit the cost of data transfers. Thus, the cost of running a workflow is the sum of three factors: the cost of using: (1) the CPU, (2) the storage and (3) the expert knowledge. The runtime of the sub-job depends on the resource configuration. The runtime estimation mechanism can be based on the one described in [17].

However, within the SLA context, the formal policy is not always effective. First, the deadline of the workflow has different meanings for different people. The importance of a deadline depends on the urgency of the workflow. If the workflow is critical, the user is willing to pay more but also imposes stricter fining rates. Otherwise, he pays less and requires lower fining rates. Thus, having a fixed mapping mechanism is not enough. The broker must have suitable policies to handle different user requirements. In this way, the fining rate, as defined in the SLO, can effect seriously the mapping policy of the broker. Assume that the broker choses a hardware resource with a higher processing power than specified by the user. Assume as well that the broker adjusts the expected runtimes of the sub-jobs according to the processing power. If a sub-job is finishing late (although no failure in any RMS has been detected), the broker cannot impose the responsibility to the user. It is not possible to determine whether the delay is caused by a poor initial estimation of the user or by a faulty adjustment of the broker. Therefore, the broker should pay in this case.

3 Mapping Policies

In our system, we define three types of urgencies: high, medium, and low. It depends on the user to select the level of urgency for his workflow. Depending on the preference of the user, the broker will have different mapping policies. The goal of those mapping policies is securing the deadline of the workflow under different constraints.

The most common cause that affects the deadline of a workflow is the crash of some nodes within a RMS. According to Google system [18], there is a node down every hour. When a node is down, the sub-jobs cannot continue running. In this case, the RMS will restart the sub-job from its checkpoint, an image of the process states from some time in the past [16]. This event will prolong the previously estimated runtime of the sub-job. Consequently, the deadline of the workflow may be violated. A mapping policy could deal with this situation in different ways. For example, a mapping policy could be to introduce spare time slots between sub-jobs of high urgency workflows. In the following, we propose three mapping algorithms correlating to three urgency levels of workflows.

3.1 Mapping Workflow of High Urgency

For high urgency workflows, each sub-job of the workflow should have a spare time period. In case of light communication workflows, we can distinguish two types of spare time. The first type is for sub-jobs having flexible start time and end time. For example, sub-job 1 can start in a wide range of time slots without effecting the deadline of the workflow. Therefore, if this sub-job is scheduled earlier than the latest end time, we could introduce a spare period. The second type of spare time is for dependent, sequential sub-jobs (e.g. sub-jobs in the critical path 0, 2, 3, 6 of figure 1. The spare period can only be introduced if those sub-jobs are assigned to run on a more powerful RMS than initially specified. Thus, the actual runtime will be shorter than originally specified and, therefore, opening up the opportunity for spare time periods.

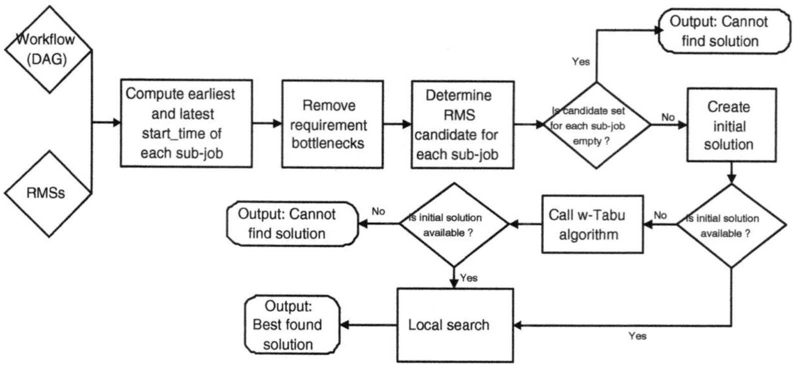

Fig. 3. EL-Map algorithm

From the two above observations, we propose the algorithm called EL-Map (E presents Extreme, L presents Light), which is based on L-Tabu algorithm [3], to do the mapping task. At the step of determining RMS candidates for each sub-job, we refine further the solution space in order to satisfy the spare time period requirements. For critical sub-jobs (i.e. a sub-job on which many subsequent sub-jobs depend), we choose only RMSs having more powerful configuration than required by the user.

3.2 Mapping Workflow of Medium Urgency

For medium-urgent workflows, each sub-job of the workflow does not need to have a spare time period. The broker will reserve the resources according to the user's estimated time period. The mapping algorithm for this case is L-Tabu algorithm [3].

3.3 Mapping Workflow of Low Urgency

For workflow of low urgency, we suggest to use existing algorithms, which reduce the cost by considering the runtime on sub-job-suitable hardware resources. The algorithm that we use is called L-Map and is presented in Figure 4 [5].

Step 0: With each sub-job of the workflow, we sort the RMSs in the candidate set according to the cost of running.

Step 1: We form the first configuration by assigning each sub-job to the RMS that has the lowest cost in the candidate list.

Step 2: We compute the earliest start time and the latest stop time of each sub-job using the conventional graph algorithm.

Step 4: If the reservation profile has conflict periods, we have to move sub-jobs by adjusting the earliest start time slot or the latest end time slot of the sub-jobs.

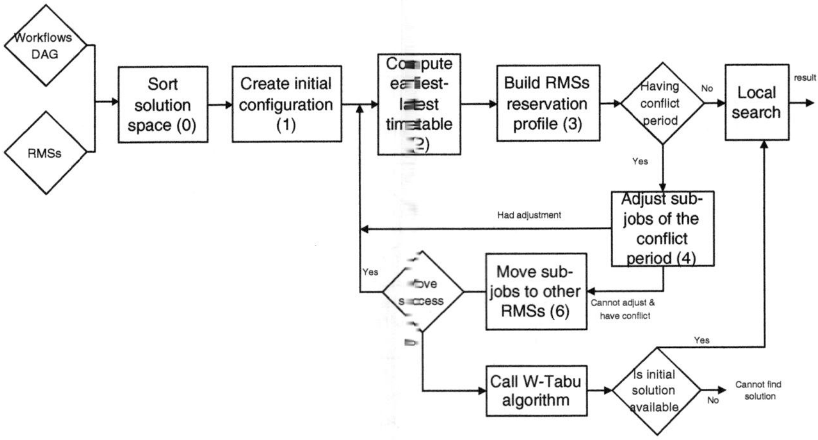

Fig. 4. L-Map algorithm architecture

Step 5: We have to adjust the earliest start time and latest stop time of each sub-jobs that is affected by moving of sub-jobs in step 4 and then repeat step 3 and 4 until we cannot adjust the sub-jobs any more.

Step 6: If after adjusting phase, there are still some conflict periods, we have to move some sub-jobs contributing to the conflict to other RMSs. If a sub-job cannot be moved to other RMS, we can deduce that the Grid resource is busy and the w-Tabu algorithm is invoked. If w-Tabu cannot find an initial solution, the algorithm will stop.

Step 7: The process from step 3 to step 6 is repeated until there is no conflict period or w-Tabu algorithm is invoked. After this phase, we have a feasible candidate solution.

Step 8: A local search procedure is used to improve the quality of the solution as far as possible.

4 Experimental Results

The goal of the experiment is to measure the difference in cost and the influence on business decisions when applying different mapping policies. The experiments are executed with 18 workflows. Each workflow differs in its topology, the number of sub-jobs, the sub-job specifications, and the amount of data to be transferred. These workflows are then mapped to 20 RMSs with different resource configurations and initial resource reservations by 3 algorithms: EL-Map, L-Tabu and L-Map. In the experiment, 30% of all RMSs have a CPU performance equal to the requirement, 60% of RMSs have a CPU performance is twice as powerful than required, 10% of RMSs have a CPU performance that is 200% more powerful than required. Along with the increase in performance, we assume that the price for each CPU class increases linearly, however, slower

Table 1. Performance experiment result

Sjs	EL-Map	L-Tabu	L-Map		
			30-10-60	60-10-30	90-10-0
Simple level experiment					
7	660.18	629.28	624.97	616.80	612.47
8	830.01	753.90	749.29	741.12	733.42
9	828.47	776.87	772.73	763.25	755.55
10	938.15	831.62	829.36	822.30	814.60
11	972.64	886.94	884.14	875.34	870.44
12	1061.01	942.71	940.04	931.23	920.11
13	1136.78	996.76	993.82	987.07	972.58
Intermediate level experiment					
14	1250.05	1052.43	1050.61	1033.62	1025.05
15	1367.41	1176.33	1174.93	1159.48	1146.00
16	1431.28	1301.06	1299.31	1283.80	1270.36
17	1586.94	1355.84	1345.80	1336.27	1322.15
18	1707.27	1480.52	1470.12	1460.59	1439.00
19	1823.63	1503.77	1493.56	1480.60	1462.43
20	1762.85	1558.92	1557.85	1536.56	1515.27
21	1966.77	1582.08	1581.23	1559.94	1535.85
Advance level experiment					
25	2093.66	1775.10	1770.44	1747.02	1729.51
28	2257.40	1875.54	1865.57	1845.57	1824.08
32	2570.20	2202.74	2198.52	2167.76	2160.52

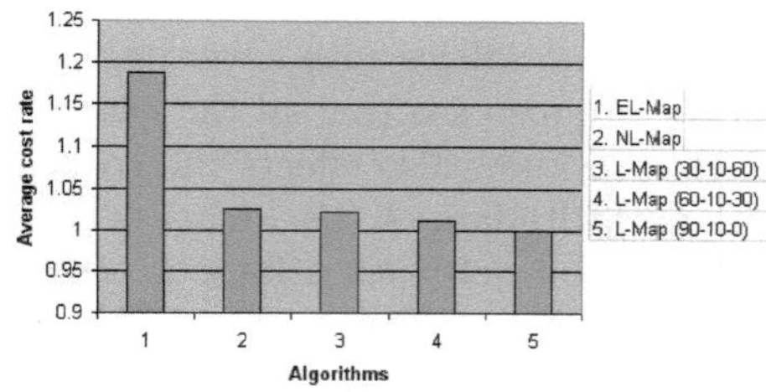

Fig. 5. Cost in average of each algorithm

than f(x)=x. The information about RMS and workflow is available online at http://it.i-u.de/schools/altmann/DangMinh/desc_expe1.txt.

In the experiment, the runtime of each sub-job in each type of RMS, which is used in L-Map algorithm, is assigned by using following formula:

$$rt_j = \frac{rt_i}{\frac{pk_i + (pk_j - pk_i)*k}{pk_i}} \tag{1}$$

pk_i, pk_j is the performance of a CPU in RMS r_i, r_j respectively. rt_i is the estimated runtime of the sub-job with the resource configuration of RMS r_i. k is a speed-up control factor, which depends on the structure of the application. In the experiment, k had three values: 0.5, 0.25, 0.1. We specified the workload configuration as a vector of numbers of sub-jobs in the workflow with different k. For example, the workload configuration 90-10-0 defines that 90% of sub-jobs have a $k = 0.5$, 10% of sub-jobs have $k = 0.25$, 0% of sub-jobs have $k = 0.1$. In this experiment, we use three workload configurations: (1) 30-10-60, (2) 60-10-30, (3) 90-10-0.

The result of the experiment is presented in Table 1. As the runtime of each algorithm is very small in few seconds, we only present the cost of the mapping solution with each algorithm. Figure 5 presents the average cost in relative value of each algorithm. From the experiment results in Table 1 and Figure 5, we can see that the cost of executing a workflow is lower for workflows with a lower level of urgency. This is caused by the fact that the time reserved for each sub-job is lower for workflows with lower urgency.

We can also see that in the case of the L-Map algorithm, the configuration of the workflow also effects the cost of the solution. If the workflow has many sub-jobs with large k (i.e. higher performance), the cost will be lower. The reason is that a sub-job with a large k needs less time periods to finish. In 5, we can see that the cost of using less-powerful resource is higher than the price of more powerful resource multiplied by their runtime.

In 5, we can also notice that the difference in cost between medium-urgent and low-urgent workflows varies depending on the configuration of the workflow. If the workflow has a small number of sub-jobs with big k, the difference in cost is not so much. This fact suggest that NL-Map algorithm can be applied even to low urgency workflows with low number of large k.

The price to pay for preventive actions against failures of nodes within in workflows of high urgency is high. As can be seen in 5, the cost for running the EL-Map algorithm is about 15% higher than the cost of the L-Map algorithm. However, it can be assumed that a user, who has to finish his workflow on time, is willing to pay this extra cost to him. Therefore, this extra cost is acceptable.

5 Related Works

The literature records many efforts supporting QoS for workflow. Imperial College e-Science Network Infrastructure (ICENI) is an end-to-end Grid middleware system developed at the London e-Science Centre [9]. AgFlow is a middleware platform that enables the quality-driven composition of Web services [10]. QoS-aware Grid Workflow is a project, which aims at extending the basic QoS support to Grid workflow applications [11]. The work in [12] focuses on mapping the sweep task workflow to Grid resource with deadline and budget constraints. However, all of them do not define business model for the system. Therefore, they just simply find a solution satisfying the deadline and optimizing the cost. Recently, there are many Grid projects working on SLA issue [13,14,15]. Most of them focus on single job and thus, only the direct relation between user and service provider is considered. The business role of the broker in such systems is not fully evaluated. More over, supporting SLAs for Grid-based workflows is just at the initial phase and is not fully exploited.

6 Conclusion

This paper presents a business model to execute SLA-aware workflows in Grid environments. In particular, we focused on the aspects related to Service Level Objectives. Within the SLA context, end-user can negotiate different fining rates with the service provider. Scoping with different levels of urgency, brokers have to use different mapping algorithm to find solutions with different risk levels. Our measurements have shown that using runtime adaptation gains only benefit in some special cases, namely where workflows have many sub-jobs with a large speed-up factor. For workflows of high urgency, the extra cost of 15% is still acceptable, if we consider that user with high urgency jobs are willing to pay more for getting a higher quality.

References

1. Sahai, A., Machiraju, V., Sayal, M., Jin, L.J., Casati, F.: Automated sla monitoring for web services. In: Feridun, M., Kropf, P.G., Babin, G. (eds.) DSOM 2002. LNCS, vol. 2506, pp. 28–41. Springer, Heidelberg (2002)
2. Hovestadt, M.: Scheduling in hpc resource management systems:queuing vs. In: Feitelson, D.G., Rudolph, L., Schwiegelshohn, U. (eds.) JSSPP 2003. LNCS, vol. 2862, pp. 1–20. Springer, Heidelberg (2003)
3. Quan, D., Kao, O.: Mapping grid job flows to grid resources within sla context. In: Sloot, P.M.A., Hoekstra, A.G., Priol, T., Reinefeld, A., Bubak, M. (eds.) EGC 2005. LNCS, vol. 3470, pp. 1107–1116. Springer, Heidelberg (2005)
4. Quan, D., Kao, O.: On architecture for an sla-aware job flows in grid environments. Journal of Interconnection Networks, World scientific computing, 245–264 (2005)
5. Quan, D., Altmann, J.: Mapping Grid-based workflows with light communication, onto Grid resources within an SLA context. In: GSEM 2007 conference (submitted 2007)
6. Quan, D.: Error recovery mechanism for grid-based workflow within sla context. Accepted by International Journal of High Performance Computing and Networking (IJHPCN) (2006)
7. Quan, D.: Mapping heavy communication workflows onto grid resources within sla context. In: Gerndt, M., Kranzlmüller, D. (eds.) HPCC 2006. LNCS, vol. 4208, pp. 727–736. Springer, Heidelberg (2006)
8. http://en.wikipedia.org/wiki/Service_level_objectives
9. McGough, S., Afzal, A., Darlington, J., Furmento, N., Mayer, A., Young, L.: Making the Grid Predictable through Reservations and Performance Modelling. The Computer Journal 48(3), 358–368 (2005)
10. Zeng, L., Benatallah, B., Ngu, A., Dumas, M., Kalagnanam, J., Chang, H.: QoS-Aware Middleware for Web Services Composition. IEEE Transactions on Software Engineering 30(5), 311–327 (2004)
11. Brandic, I., Benkner, S., Engelbrecht, G., Schmidt, R.: QoS Support for Time-Critical Grid Workow Applications. In: Proceedings of e-Science 2005 (2005)
12. Yu, J., Buyya, R.: Scheduling Scientific Workflow Applications with Deadline and Budget Constraints using Genetic Algorithms, Scientific Programming Journal, ISSN: 1058-9244. IOS Press, Amsterdam, The Netherlands (2006)
13. Surridge, M., Taylor, S., De Roure, D., Zaluska, E.: Experiences with GRIA. In: Proc. 1st IEEE International Conference on e-Science and Grid Computing, IEEE Computer Society Press, Los Alamitos (2005)
14. http://www.eu-egee.org/
15. Hovestadt, M., Kao, O., Vo, K.: The First Step of Introducing Risk Management for Preprocessing SLAs. In: SCC 2006, pp. 36–43 (2006)
16. Garbacki, P., Biskupski, B., Bal, H.: Transparent fault tolerance for grid application. In: Sloot, P.M.A., Hoekstra, A.G., Priol, T., Reinefeld, A., Bubak, M. (eds.) EGC 2005. LNCS, vol. 3470, pp. 671–680. Springer, Heidelberg (2005)
17. Spooner, D.P., Jarvis, S.A., Cao, J., Saini, S., Nudd, G.R.: Local grid scheduling techniques using performance prediction. In: Local grid scheduling techniques using performance prediction, pp. 87–96. IEEE Computer Society Press, Los Alamitos (2003)
18. http://www.renci.org/publications/presentations/LACSI/prelimSlides.pdf

The One-Click Grid-Resource Model

Martin Rehr and Brian Vinter

Department of Computer Science
University of Copenhagen
Copenhagen, Denmark
{rehr,vinter}@diku.dk

Abstract. This paper introduces the One-Click Grid resource, which allows any computer with a Java enabled web browser to safely provide resources to Grid without any software installation. This represents a vast increase of the number of potential Grid resources that may be made available to help public interest research. While the model does make restrictions towards the application writer, the technology provides a real Grid model and supports arbitrary binaries, remote file access and semi-transparent checkpointing. Performance numbers show that the model is usable even with browsers that are connected to the Internet through relatively weak links, i.e. 512 kb/s upload speeds. The resulting system is in use today, and freely available to any research project.

1 Introduction

Grid Computing and Public Resource Computing, PRC, provide increasingly interesting means of obtaining computational resources. Grid Computing is mostly used for connecting university supercomputers, while PRC is predominantly used by research projects for harvesting PC based idle CPU-cycles for a small number of research projects. However, even though the two fields appear closely related, little effort has been made to combine them to a system that offers the flexibility of Grid computing with the resource richness of the PRC model.

1.1 Motivation

Harvesting 'free' cycles through PRC is of great interest since a modern PC is powerful and highly underutilized, and as such cycle harvesting provides a huge calculation potential if one combines millions of them in a computing Grid[1].

Most known Grid systems such as ARC[2] which is based on the Globus toolkit[3] and Condor[4] are unsuitable for PRC computing, as they work under the underlying assumption that the resources are available at anytime, which PRC resources by their very nature are not.

To extend the PRC concept to actual Grid computing, security and installation of software on the donated resource are vital issues. All, to the authors known, PRC projects requires the donor to install software on the resource that should contribute, which alone eliminates users from donating resources from computers that they do have administrative rights on. The software installation also opens

R. Perrott et al. (Eds.): HPCC 2007, LNCS 4782, pp. 296–308, 2007.

for possible exploits and requires the donor to perform updates on that software. This is not desirable and may reduce the amount of donated resources.

Ensuring the safety of a donated resource while it executes a Grid job in a PRC context is an all important topic since all free resources will vanish if the model proves harmful to the hosts. Contrary to standard PRC tasks, a Grid job may take any form and include the execution of any binary. Thus it is necessary to take precautions to ensure that the execution of Grid jobs cannot harm donated resources neither from intention nor by accident.

This paper addresses some of the problems that need to be solved in order to combine PRC computing and Grid Computing. Our goal is to design a Grid PRC secure sandbox model, where Grid jobs are executed in a secure environment and no Grid or application specific software is needed on the donated resource. Furthermore we ensure that resources may be in a typical PRC context, i.e. located behind a Network Address Translation router and a firewall and thus that the model may be used without modifications to firewalls or the routers.

1.2 Related Work

BOINC[5] is a middleware system providing a framework, which has proved the concept of PRC, and is widely used by scientific research projects such as SETI@HOME[6] and FOLDING@HOME[7]. MiG-SSS[8] is a Screen Saver Science model built to combine PRC with Minimum intrusion Grid, MiG[9][10]. Our work differs from BOINC by aiming at a full Grid model, and differs from both BOINC and MiG-SSS by aiming at no Grid specific software installation on the client.

2 Web Browsers and Java

To reach our stated goal of no Grid specific software installation and no modification of the donated machines firewall settings, we are forced to use software which is an integrated part of a common Internet connected resource.

We found that amongst the most common software packages for any PC type platform there is a Java enabled web browser. The web browser provide a common way of securely communicating with the Internet, which is allowed by almost all firewall configurations of the resources we target.[1] The web browser itself provides us with a communication protocol, but it does not by itself, provide a safe execution environment, however all of the most common graphics enabled web browsers have support for Java applets[11], that are capable of executing Java byte-code located on a remote server.

The Java applet security model[12], ASM, prevents the Java byte-code executed in the applet from harming the host machine and thereby provides the

[1] Resources located behind firewalls that do not support outgoing HTTPS is considered out of range for this PRC, however it is not unseen that outbound HTTPS is blocked.

desired sandbox effect for us to trust the execution of unknown binaries on donated resources.

The choice of web browsers and Java applets as the execution framework, results in some restrictions on the type of jobs that may be executed in this environment:

- Applications must be written in Java
- Applications must apply to ASM
- The total memory usage is limited to 64 MB including the Grid framework
- Special methods must be used to catch output
- Special methods must be used for file access

By accepting the limitations described above, a web browser may become a Grid resource simply by entering a specific URL. This triggers the load and execution of an applet which acts as our Grid gateway and enables retrieving and executing a Java byte-code based Grid job. The details of this process is described next.

3 The Applet Grid Resource

Several changes to the Grid middleware are needed to allow Java applets to act as Grid resources. First of all the Grid middleware must support resources which can only be accessed through a pull based model, which means that all communication is initiated by the resource, i.e. the applet. This is required because the ASM rules prevents the applet from initiating listening sockets, and to meet our requirement of functioning behind a firewall with no Grid specific port modifications. Secondly, the Grid middleware needs a scheduling model where resources are able to request specific type of jobs, e.g. a resource can specify that only jobs which are tagged to comply to the ASM can be executed.

In this work the Minimum intrusion Grid[9], MiG, is used as the Grid middleware. The MiG system is presented next, before presenting how the Applet Grid resource and MiG work together.

3.1 Minimum Intrusion Grid

MiG is a stand alone Grid platform, which does not inherit code from any earlier Grid middlewares. The philosophy behind the MiG system is to provide a Grid infrastructure that imposes as few requirements on both users and resources as possible. The overall goal is to ensure that a user is only required to have a X.509 certificate which is signed by a source that is trusted by MiG, and a web browser that supports HTTP, HTTPS and X.509 certificates. A fully functional resource only needs to create a local MiG user on the system and to support inbound SSH. A sandboxed resource, which can be used for PRC, only needs outbound HTTPS[8].

Because MiG keeps the Grid system disjoint from both users and resources, as shown in Figure 1, the Grid system appears as a centralized black box[9] to both users and resources. This allows all middleware upgrades and trouble

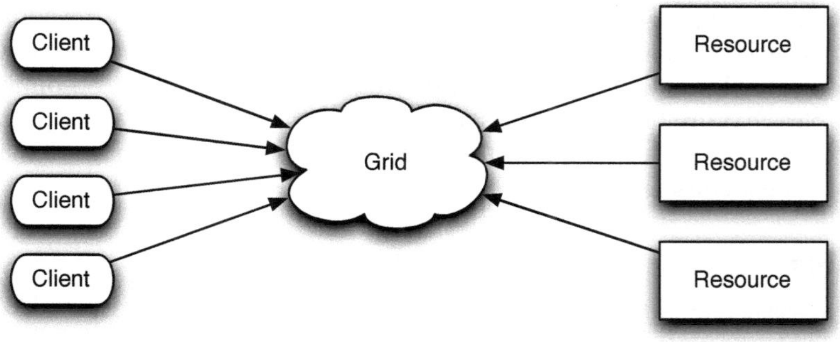

Fig. 1. The abstract MiG model

shooting to be executed locally within the Grid without any intervention from neither users nor resource administrators. Thus, all functionality is placed in a physical Grid system that, though it appears as a centralized system in reality is distributed. The basic functionality in MiG starts by a user submitting a job to MiG and a resource sending a request for a job to execute. The resource then receives an appropriate job from MiG, executes the job, and sends the result to MiG that can then inform the user of the job completion. Since the user and the resource are never in direct constant, MiG provides full anonymity for both users and resources, any complaints will have to be made to the MiG system that will then look at the logs that show the relationship between user and resource.

3.1.1 Scheduling

The centralized black box design of MiG makes it capable of strong scheduling, which implies full control of the jobs being executed and the resource executing them. Each job has an upper execution time limit, and when the execution time exceeds this time limit the job is rescheduled to another resource. This makes the MiG system very well suited to host PRC resources, as they by nature are very dynamic and frequently join and leave the Grid without notifying the Grid middleware.

3.2 The MiG Applet Resource

As explained above, all that is required for a PRC resource to join MiG is a sandbox and support for outgoing HTTPS. However, the previous solution[8] requires installation of non standard software to activate and execute the sandbox.

The Java applet technology makes it is possible to turn a web browser into a MiG sandbox without installing any additional software. This is done automatically when the user accesses "MiG One-Click"[2], which loads an applet into the web browser. This applet functions as a Grid resource script and is responsible for requesting pending jobs, retrieving and executing granted jobs, and delivering the results of the executed jobs to the MiG server.

[2] The URL accessed to activate the web browser as a sandboxed MiG Java resource is called "MiG One-Click", as it requires one click to activate it.

To make the applet work as a resource script, several issues must be addressed. First of all ASM disallows local disk access. Because of this both executables and input/output files must be accessed directly at the Grid storage. Secondly only executables that are located at the same server as the initial applet are permitted to be loaded dynamically. Thirdly text output of the applet is written to the web browser's Java console and not accessible by the Grid middleware.

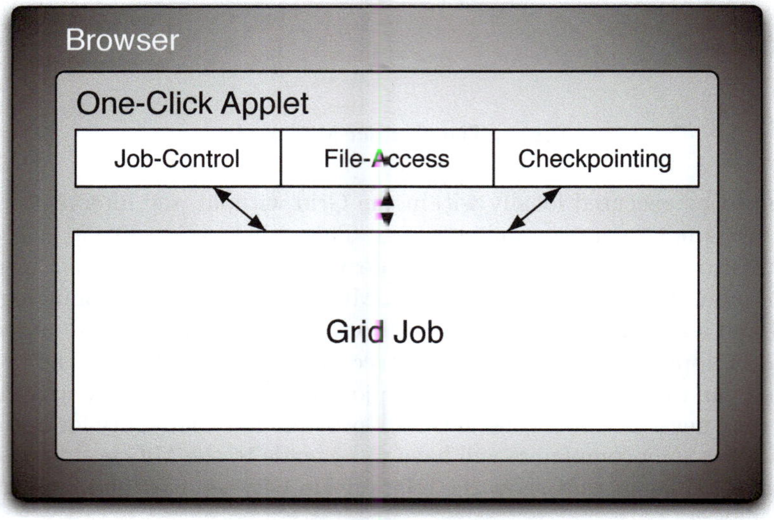

Fig. 2. The structure of an One-Click job

When the applet is granted a job by the MiG server, it retrieves a specification of the job which specifies executables and input/output files. The applet then loads the executable from the Grid this is made possible by the MiG server which sets up an URL from the same site as the resource applet was originally loaded which points to the location of the executables. This allows unknown executables to be loaded and comply with the ASM restrictions on loading executables. Figure 2 shows the structure of an One-Click job. Executables that are targeted for the MiG One-Click model must comply with a special MiG One-Click framework, which defines special methods for writing stdout and stderr of the application to the MiG system[3]. Normally the stdout and stderr of the executing job is piped to a file in the MiG system, but a Java applet, by default, writes the stdout and stderr to the web browsers Java console. We have not been able to intercept this native output path. Input and output files that are specified in the job description must be accessed directly at the Grid storage unit since the ASM rules prohibits local file access. To address this issue the MiG One-Click framework provides file access methods that transparently

[3] The result of a MiG job is the stdout/stderr and the return code of the application that is executed.

provide remote access to the needed files. Note that the MiG system requires input files and executables to be uploaded to the MiG server before job submission which ensures that the files are available at the Grid storage unit.

In addition to the browser applet a Java console version of the MiG resource has been developed, to enable the possibility of retrieving and executing MiG One-Click jobs as a background process. This requires only a Java virtual machine. To obtain the desired security model, a customized Java security policy is used, which provides the same restrictions as the ASM.

3.3 Remote File Access

The One-Click executing framework that was introduced above also provides transparent remote file access to the jobs that are executed. The MiG storage server supports partial reads and writes, through HTTPS, of any file that is associated with a job. When the resource applet accesses files that are associated with a job, a local buffer is used to store the parts of the file that are being accessed. If a file position which points outside the local buffer is accessed, the MiG server is contacted through HTTPS, and the buffer is written to the MiG server if the file is opened in write mode. The next block of data is then fetched from the server and stored into the buffer and finally the operation returns to the user application. The size of the buffer is dynamically adjusted to utilize the previously observed bandwidth optimally.

3.3.1 Block Size Estimation

To achieve the optimal bandwidth for remote file access it is necessary to find the optimal block size for transfers to and from the server. In this case the optimal block size is a trade off between latency and bandwidth. We want to transfer as large a block as possible without excessive latency increment since the chance of transferring data that will not be used increases with the block size.

We define the optimal block size bs_{opt} as the largest block where a doubling of the block size does not double the time to transfer it. This can be expressed the following way:

$$t(x) * 2 > t(x * 2) \qquad \forall x < bs_{opt} \tag{3.1}$$

$$t(x) * 2 < t(x * 2) \qquad \forall x > bs_{opt} \tag{3.2}$$

$$t(x) \; = time\ to\ transfer\ block\ of\ size\ x$$

We do not want block sizes below bs_{opt} as the time t used to transfer a block of size x is less than doubled when the block size is doubled. On the other hand we don't want 'too large' block sizes as we do not know if the retrieved data is going to be used or discarded due to a seek operation beyond the end of the local buffer.

As the One-Click resources can be placed at any sort of connection, and the bandwidth of the connection thus may differ greatly from one resource to another, it is not possible to use a fixed block size and reach a good ratio between bandwidth and latency at an arbitrary type of connection.

The simplest approach would be to use a fixed bs_{opt} based on empirical tests on the most common connections.

A less trivial, but still simple, approach would be to measure the time it takes to connect to the server and then choose a block size which ensures the transfer time of that block to be a factor of x larger than the time to connect, to make sure that the connection overhead does not exceed the time of the actual data transfer.

The chosen approach is to estimate bs_{opt} from the time spent transferring block $x-1$ with the time of transferring block x, starting with an initial small[4] block size bs_0 and then doubling the block size until a predefined cutoff ratio CR is reached. After each data transfer the bandwidth bw_x is calculated and compared to the bandwidth of the previous transfer bw_{x-1}. If the ratio is larger than the predefined CR:

$$\frac{bw_x}{bw_{x-1}} > CR \tag{3.3}$$

then the block size is doubled:

$$bs_{x+1} = bs_x * 2 \tag{3.4}$$

As the block size is doubled in each step the theoretical CR to achieve bs_{opt} should be 2, since there is no incentive to increase block size once the latency grows linearly with the size of the data that is transferred. However in reality, one need to get a CR below 2 to achieve bs_{opt}. This is due to the fact that all used block sizes are powers of 2, and one cannot rely on the optimal block size to match a power of 2.

Therefore to make sure to get a block size above bs_{opt} you need a lower CR. Empirical tests showed that a CR about 1.65 yields good results, see section 5.2

Additional extensions include adapting to the frequency of random seeks in the estimation of the CR. A large amount of random seeks to data placed outside the range of the current buffer will cause new blocks to be retrieved in each seek. Therefore the block size should be lowered in those cases to minimize the latency of each seek.

4 Checkpointing

PRC resources will join and leave the Grid dynamically, which means that jobs with large running time have a high probability of being terminated before they finish their execution. To avoid wasting already spent CPU-cycles a checkpointing mechanism is build into the applet framework. Two types of checkpointing have been considered for inclusion, transparent checkpointing and semi-transparent checkpointing.

[4] An initial small block size gives a good result as many file accesses applies to small text files such as configuration files.

4.1 Transparent Checkpointing

All to the authors known transparent checkpoint mechanisms provided to work with Java, require the JVM to be replacement or access to the /proc file system on Linux/Unix operating system variants, as the default JVM does not support storing program counter and stack frame. Since our goal is to use a web browser with the Java applet as a Grid resource neither of those solutions are satisfactory, since both the replacement of the JVM and access to the /proc file system violates the Java applet security model. Furthermore most PRC resource will be running the Windows operating system which do not support the /proc file system.

4.2 Semi-transparent Checkpointing

Since transparent checkpointing is not applicable to the One-Click model, we went on to investigate what we call semi-transparent checkpointing. Semi-transparent checkpointing covers that the One-Click framework provides a checkpoint method for doing the actual checkpoint, but the application programmer is still responsible for calling the checkpoint method when the application is in a checkpoint safe state.

The checkpoint method stores the running Java object on the MiG server through HTTPS. Since it can only store the object state, and not stack information and program counters, the programmer is responsible for calling the checkpoint method at a point in the application, where the current state of the execution may be restored from the object state only. To restart a previously checkpointed job, the resource applet framework first discovers that a checkpoint exists and then loads the stored object.

To ensure file consistency as part of the checkpoint, the framework also supports checkpointing of modified files, which is done automatically without involving the application writer. Open files are checkpointed if the job object includes a reference to the file.

5 Experiments

To test the One-Click model we established a controlled test scenario. Eight identical Pentium 4, 2.4 GHz machines with 512 MB ram were used for tests.

5.1 One-Click as Concept

The test application used, is an exhaustive algorithm for folding proteins written in Java. This was changed to comply with the applet framework.

A protein sequence of length 26 was folded on one machine, which resulted in a total execution time of 2 hours, 45 minutes and 33 seconds. The search space of the protein was then divided into 50 different subspaces using standard divide and conqueror techniques. The 50 different search spaces were submitted as jobs to the Grid, which provides an average of 6 jobs per execution machine and 2 extra jobs to prevent balanced execution. The search spaces on their own also provide unbalanced execution as the valid protein configurations vary from one search space to another and thus results in unbalanced execution times. The experiment was made without checkpointing the application. The execution of the 50 jobs completed in 29 minutes and 8 seconds, a speedup of 5.7 for 8 machines. While this result would be considered bad in a cluster context it is quite useful in a Grid environment.

To test the total overhead of the model, a set of 1000 empty jobs was submitted to the Grid with only one One-Click execution resource connected. The 1000 jobs completed in 19935 seconds, which translates to an overhead of approximately 20 seconds per job.

5.2 File Access

To achieve the best bandwidth cutoff ratio CR several experiments has been made. In the experiments a 16 MB file was read 100 times by the One-Click resource on a 20 Mb/s broadband Internet connection. All experiments start with an initial block size of 2048 (2^{11}) bytes. The first experiment was run with a CR of 0, which means that the block size is doubled in every transfer. The result is shown in figure 3.

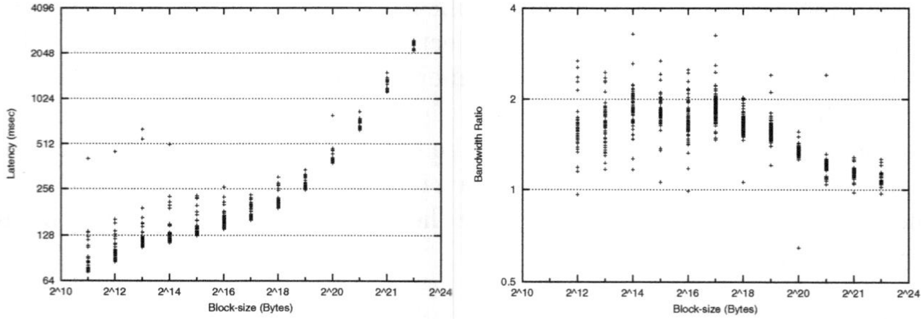

Fig. 3. The upper figure shows the latency as a function of the block size, the lower figure shows the bandwidth ratio $\frac{bw_x}{bw_{x-1}}$ as a function of the block size. Between block size 2^{18} and 2^{20} the latency starts to raise and the bandwidth ratio starts to fall. This is where the cutoff is chosen to avoid excessive raise in latency.

The figure shows how that the latency starts to raise dramatically between block size 2^{18} and 2^{20} and the bandwidth to latency ratio starts to fall at those block sizes. The bandwidth to latency ratio between block size 2^{18} and 2^{20} lies in the interval from 1.25 to 1.75. Based on these observations we performed the same test with a CR of 1.5. The result is shown in figure 4.

This shows that a CR of 1.5 is too low as block sizes of 2^{21} occur and we want the block sizes to be between 2^{18} and 2^{20} to limit the maximum latency.

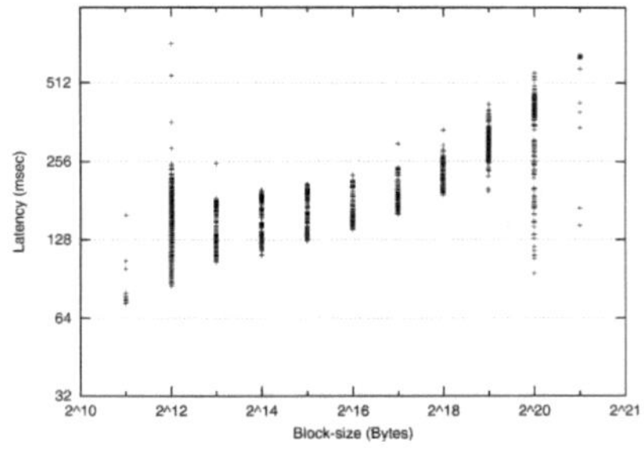

Fig. 4. The latency as a function of the block size with CR 1.5

Therefore the CR must be between 1.5 and 1.75. The test was then run with CR 1.55, 1.60, 1.65, 1.70 and 1.75. The result is shown in figure 5.

We observe that a CR of 1.75 is too high, as only a few block sizes of 2^{19} occur and no block sizes of 2^{20} occurs. A CR of 1.55 results in a few block sizes of 2^{21} which is above the block sizes we want. 1.60 represents the block sizes we want and block size 2^{20} is well represented. A CR of 1.65 represents block size 2^{19} well and a few block sizes of 2^{20} is reached as well, and a CR 1.70 represents block size 2^{20} but no block sizes of 2^{21} are represented. We choose a CR of 1.65 as block size 2^{20} is considered the braking point where the latency starts to grow excessively, therefore we do not want it to be to well represented, but we want it to be represented, which is exactly the case at a CR of 1.65.

To verify the previous finding that the CR value should be 1.65 a test application, which traverses a 16 MB file of random 32 bit integers was developed. First the application was tested against the framework, where fixed block sizes were used, and then the application was tested against the framework, where the dynamic block sizes with a CR of 1.65 were used. The results are shown i figure 6.

The experiment shows, as expected, that the execution time decreases as the block sizes increase in the experiments with static block sizes. The execution time in the experiments with the dynamic block sizes all reside around 256 seconds[5] which are satisfactory, as this shows that compared to largest static buffer size of interest[6], the execution time loss using a dynamic buffer size is at most a factor of four. The reader should note that this type of application is the worst case for dynamic buffer sizing as all the data are read sequentially. If the integers were read in random order, the dynamic buffer size execution would perform much better.

5.3 Checkpointing

The next obvious performance issue is to test the overhead of performing a checkpoint operation within a process. This was tested by submitting jobs that

[5] With the exception of 3 runs, which are classified as outliers.

[6] The largest static buffer of interest is 2^{17} as this is where the time gained by doubling the buffer, levels out.

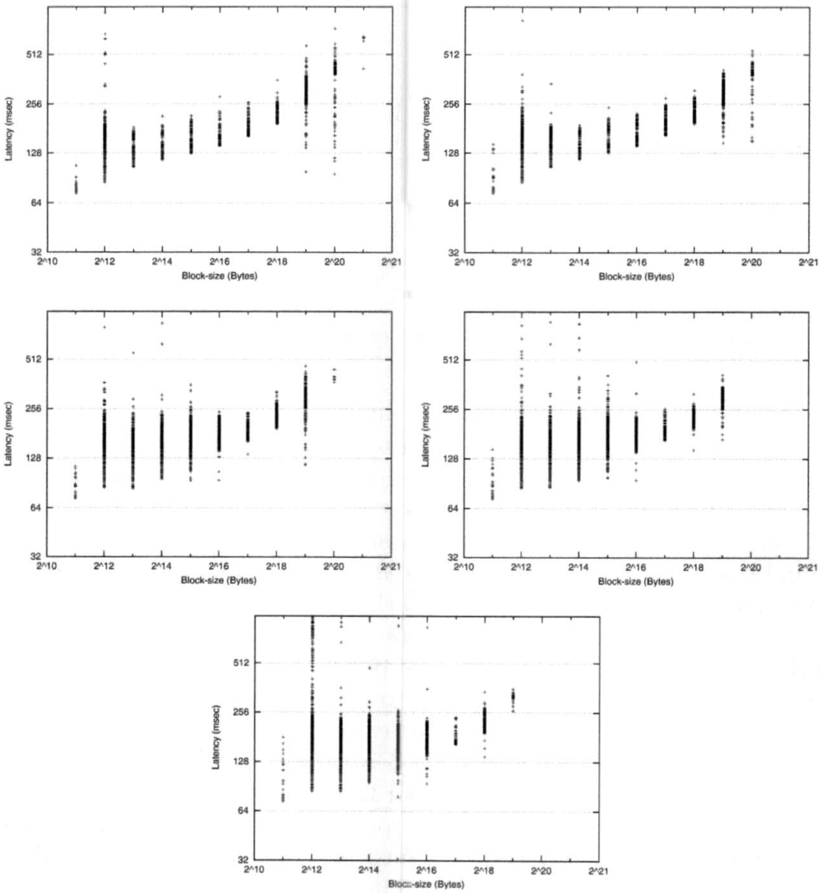

Fig. 5. The latency as a function of the block size with CR 1.55, 1.60, 1.65, 1.70, 1.75

allocate heap memory in the range from 0 kB to 8192 kB. Each job first allocates X kB, where X is in the order power of 2, and does 10 checkpoints, which saves the entire heap space. The performance was first tested on a 20 Mb/s broadband Internet connection. The test was then repeated using a more modest 2048/512 kb/s broadband Internet connection. The result of these tests is shown in figure 7. In the first test, using the 20 Mb/s connection, the checkpoint time is constant as the memory size grows. We can conclude from this, that the overhead of serializing the Java object is dominating compared to the actual network transfer time. The opposite is the case when we examine the results of the 2048/512 kb/s connection. Here we see that the time spent grows linearly with the size of the allocated memory, from which we may conclude that on a 512 kb/s connection the bandwidth is, not surprisingly, the limiting factor.

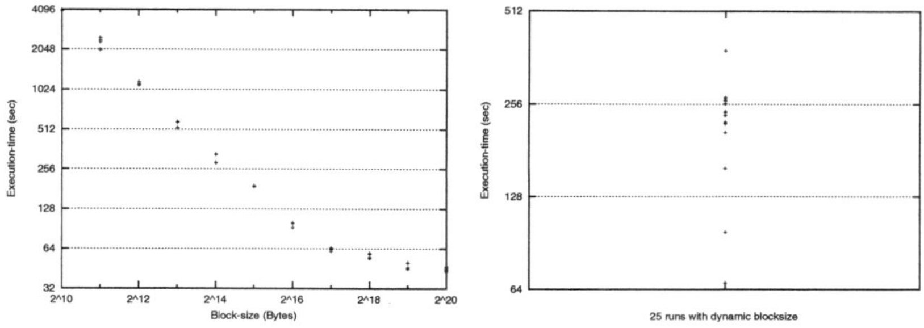

Fig. 6. The execution time as a function of the block size and the execution time with dynamic block sizes and a CR of 1.65

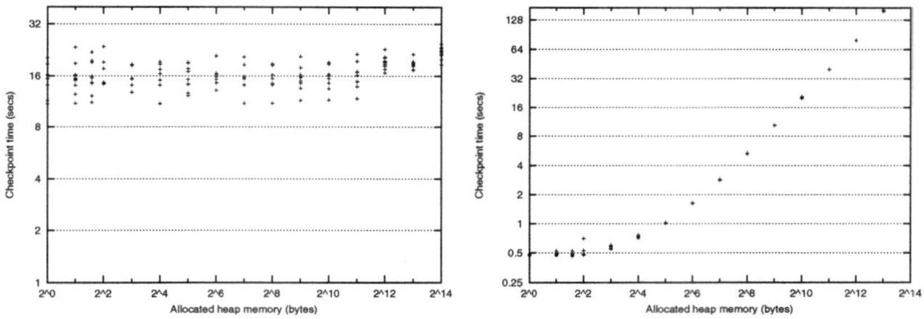

Fig. 7. The time spend checkpointing on a 20 Mb/s and a 2048/412 kb/s Broadband Internet

6 Conclusion

In this work we have demonstrated a way to combine Grid computing with PRC, the One-Click framework, without the need to install any PRC client software on the donating resource.

The use of Java applets provides a secure sandboxed executing environment that prevents the executing Grid jobs from harming the donated machine. The disadvantage of this approach is that all jobs must be written in Java and in addition comply with the presented framework, including the Java Applet Security Model. However the modifications that are needed to port an existing Java application are limited to using special methods for stdout and stderr, applying to the Java applet security model, and using the One-Click framework for remote file access. The One-Click framework also includes the means to provide semi-transparent checkpointing of the applications at runtime.

The Minimum intrusion Grid supports the required pull-job model for retrieving and executing Grid jobs on a resource located behind a firewall without the need to open any ingoing ports. By using the One-Click approach any computer with a web browser that can execute Java applets can become a Grid resource for PRC simply by entering the MiG One-Click URL. Once the user of the donated computer wishes to stop the execution, the browser is simply closed down or pointed to another URL, and the execution stops. The MiG system detects this event, by a timeout, and resubmits the job to another resource, where the job is resumed from the latest checkpoint that was made.

Experiments have been performed to find the optimal block size for the remote file transfer that the framework includes. The experiments show that doubling the block size in each transfer gives the optimal tradeoff between bandwidth and latency as long as the CR is below 1.65.

The experiments also show that the dynamic block sizes approach increases the execution time by of factor of four compared to the execution time reached with the largest static block size in a worst case scenario.

The building checkpointing mechanism has an overhead of 15 seconds per checkpoint on a 2.4GHz P4 and the One-Click framework overall is causing approximately 20 seconds of overhead to each execution, compared to local execution. Despite of this a considerable speedup is reached in the presented protein experiment.

References

1. Foster, I.: The Grid: A New Infrastructure for 21st Century Science. Physics Today 55(2), 42–47 (2002)
2. NorduGrid, http://www.nordugrid.org
3. Globus Toolkit, http://www.globus.org/toolkit
4. Condor Project, http://www.cs.wisc.edu/condor
5. Berkeley Open Infrastructure for Network Computing http://boinc.berkeley.edu
6. SETI@homes, http://setiathome.berkeley.edu
7. Folding@Home, http://folding.stanford.edu
8. Andersen, R., Vinter, B.: Harvesting Idle Windows CPU Cycles for Grid Computing. In: Arabnia, H.R. (ed.) Proceedings of (2006) International Conference on Grid Computing & Applications (GCA'06/ISBN #:1-60132-014-0/CSREA), Las Vegas, USA, pp. 121–126 (2006)
9. Vinter, B.: The architecture of the Minimum intrusion Grid: MiG. In: Roebbers, H., Sunter, J., Welch, P., Wood, D. (eds.) Communicating Process Architectures, pp. 189–201. IOS Press, Amsterdam (2005)
10. Karlsen, H.H., Vinter, B.: Minimum intrusion Grid - The Simple Model, wetice. In: WETICE 2005, pp. 305–310 (2005)
11. Java Applets http://Java.sun.com/applets
12. Java Applet Security http://Java.sun.com/sfaq

Optimizing Performance of Automatic Training Phase for Application Performance Prediction in the Grid

Farrukh Nadeem, Radu Prodan, and Thomas Fahringer

Institute of Computer Science, University of Innsbruck
Technikerstraße 21a, A-6020 Innsbruck, Austria
{farrukh,radu,tf}@dps.uibk.ac.at

Abstract. Automatic execution time prediction of the Grid applications plays a critical role in making the pervasive Grid more reliable and predictable. However, automatic execution time prediction has not been addressed due to the diversity of the Grid applications, usability of an application in multiple contexts, dynamic nature of the Grid, and concerns about result accuracy and time expensive experimental training. We introduce an optimized, low-cost, and efficient yet automatic training phase for automatic execution time prediction of Grid applications. Our approach is supported by intra- and inter-platform performance sharing and translation mechanisms. We are able to reduce the total number of experiments from an polynomial complexity to a linear complexity.

1 Introduction

Automatic execution time prediction of the Grid applications plays a critical role in making the pervasive Grid more reliable and predictable. Application execution time prediction based on historical data is a generic technique used for application performance prediction of scientific and business applications in the Grid infrastructures [14]. Historical data obtained through carefully designed experiments can minimize the need and cost of performance modeling. Controlling number of experiments to get the historical/training data, in order to control the complexity of training phase is challenging. Methods from the field of experimental design have been applied to similar control problems in many scientific and engineering fields for several decades. However, experimental design to support automatic training phase and performance prediction in the Grid has been largely ignored due to diversity of the Grid applications, usability of applications in different contexts and heterogeneous environments, and concerns about the result accuracy and time expensive experimental training. There may be some applications whose scientific phenomenon of performance behavior could be well understood, and useful results including performance models can be developed, but, this is not possible for all applications because of expensive performance modeling, which makes it almost impractical to serve several predictions using these models at run time. Moreover, the size, diversity, and

R. Perrott et al. (Eds.): HPCC 2007, LNCS 4782, pp. 309–321, 2007.

extensibility of heterogeneous resources make it even harder for the Grid. These challenging issues lead to the requirement of a generic and robust experimental design for automatic training phase that maximizes the amount of information gained in minimum number of experiments by optimizing the combinations of independent variables.

To overcome this situation, we present a generic experimental design EXD (Extended Experimental Design) for compute intensive applications. EXD is formulated by modifying the traditional experimental design steps to eliminate time taking modeling and optimization steps that make it inefficient and application specific. We also introduce inter- and intra-platform performance sharing and translation mechanisms to later support EXD. Through EXD, we are able to reduce the polynomial complexity of training phase to a linear complexity, and to show up to a 99% reduction in total number experiments for three scientific applications, while maintaining an accuracy of more than 90%. We have implemented a prototype of the system based on the proposed approach in ASKALON project [12], as a set of Grid services using Globus Toolkit 4.

The rest of the paper is organized as follows: Section 2 describes an introduction to Automatic Training Phase and System Architecture of automatic training phase is presented in Section 3. Section 4 describes the performance sharing and translation mechanisms, which support our design of experiments. Section 5 narrates design and composition of EXD. Section 6 explains the Automatic Training phase based on EXD. Application performance prediction based on the training set is described in Section 7, and experimental results with analysis are summarized in Section 8. Finally we describe related work in Section 9.

2 Training Phase on the Grid

The training phase for an application consists of the application executions for different setups of problem-size and machine-size on different Grid-sites, to obtain execution times (training set or historical data) for those setups. Automatic performance prediction based on historical data needs *enough* amounts of data present in database. The historical data needs to be generated for every new application ported to a Grid environment and/or for every new machine (different from existing machines) added to the Grid. To support the automatic application execution time prediction process experiments should be made against some experimental design and the generated training data be archived automatically. However, to automatize this whole process is a complex problem and we address it in Section 2.1. Here, we describe some of the terms that we use in this paper.

- *Machine-size*: The total number of processors on a Grid site.
- *Grid-size*: The total number of Grid-sites in the Grid.
- *Problem-size*: Set of input parameter(s) effecting performance of application.

2.1 Problem Description

Conducting an automatic training for application execution time predictions on the Grid is a complex problem due to complexity of the factors involved in it.

Generally speaking, automatic training phase is the execution of E experiments, for a set of applications α, on a selected set of Grid sites β each with a scarce capacity (e.g. number of processors p_i), for a set of problem-sizes λ. Execution of each experimental $e \in E$ for a Grid application $\alpha \in \alpha_i$, for a problem-size $r \in \lambda$, on a Grid site $g \in \beta$, at a certain site capacity $p \in \gamma$ yields the execution time interval (duration) $T(e) = endt(e) - startt(e)$. More formally, the automatic training phase comprises of:

- A set of Grid applications: $\alpha = \{\alpha_1, ..., \alpha_n\} | n \in \mathbb{N}$.
- A set of selected non-identical Grid sites:

  $$\beta = \cup_{i=1}^{m} g_i \mid \beta \subseteq G_T, |\beta| = m$$

  where G_T represents set of all the Grid sites. Thus the total Grid capacity in terms of processors is $\gamma = \sum_{i=1}^{m} p_i$, where p_i is number of processor on the Grid-site i.
- A set of different machine-sizes ω, encompassing different machine sizes for z different (heterogeneous) types of CPUs on each of the Grid-sites g_i, where $\omega_i = \bigcup_{k=1}^{z} \{1, ..., S_k\}$. Here S_k represents maximum number of CPUs of type k. We categorize CPUs to be different if their architecture and/or speed is/are different.
- A set of problem-sizes $\lambda = \{r_1, r_2, ..., r_x\} \mid r_i = \cup_{j=1}^{y} param_j, |\lambda| = x$, here $param_j$ represents value of an input parameter j of the Grid application α having y parameters in total, which depends upon the Grid application.

  In this way, for n applications, m Grid-sites, z_i different types of CPUs on a Grid-site i where maximum number of CPUs of category k is S_k, the total number of experiments N is given by:

  $$N = \sum_{h=1}^{n} \sum_{i=1}^{m} \sum_{k=1}^{z_i} \sum_{s=1}^{S_k} p_{k,s}^{i} \times x_h$$

  Where $p_{k,s}^{i}$ denotes s processors of type k on Grid-site i, and x_h denotes the number of different problem sizes of application h.

  The size of each of $\alpha, \beta, \gamma, \lambda$ and ω has a significant effect on the overall complexity of the automatic training phase.The goal is to come up with a set of execution times from E experiments $\psi : \psi = \{t_1, ..., t_E\} || \psi | < N$, such that utility U of execution times $\sum_{e=1}^{E} U(t_i)$ is maximized, N is minimized, and the accuracy ξ is maximized.

3 System Architecture

The Architecture of the system is simple and shown in Figure 1. Application specific information (the different parameters affecting execution time of the application, their ranges and effective step sizes) is provided to Experiment Design Manager, which plans experiments through Extended Experimental Design

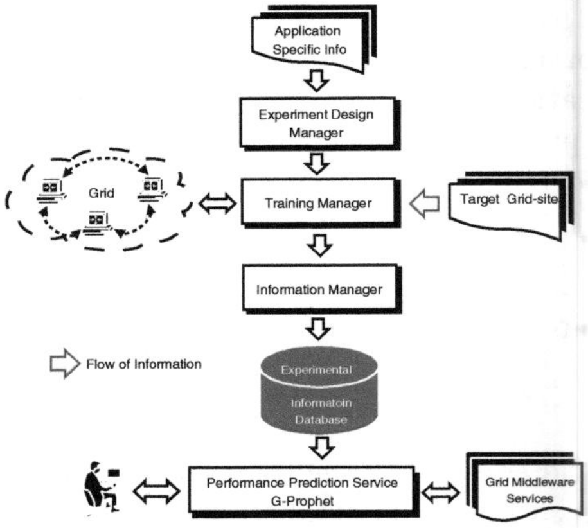

Fig. 1. System Architecture

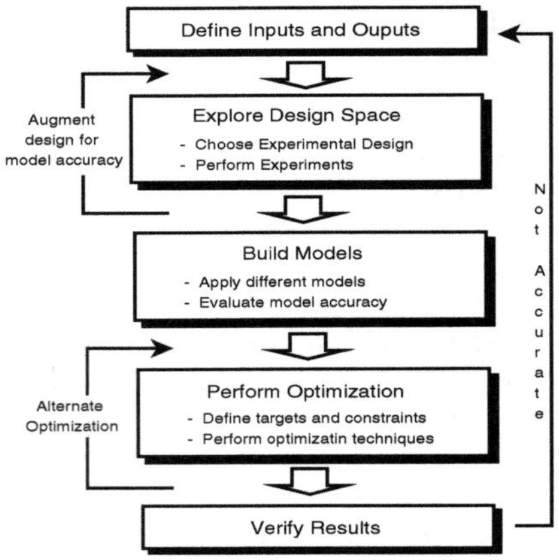

Fig. 2. Sequence of steps in a typical experimental design

(EXD) (see Section 5). The Training Manager executes these experiments on the set of the selected Grid sites. It dynamically decides the detailed execution of minimum number of experiments on different Grids-sites. The information Manager orchestrates the experimental information and, according to configuration policy, either archives this information directly in experimental performance repository or uses the performance sharing and translation mechanisms (see Section 4) to calculate the remaining values (not in the experimental information) before archiving the information. The prediction engine of application execution time prediction service (G-Prophet) gets the archived information from performance repository to serve application execution time predictions to different Grid middleware services.

4 Performance Sharing and Translation (PST)

We introduce PST to share execution times within one Grid-site (for scalability) and among different Gridsites. PST mechanism is based on our experimental observation of inter- and intra-platform performance relativity (rate of change of performance) of compute intensive applications, across different problem-sizes. We explain them as under.

4.1 Inter-platform PST

Inter-platform PST specifies that the normalized execution time ρ of an application for the different problem-sizes in α, is preserved on different Grid sites. More specifically, the normalized execution time ρ for a problem-size r_i relative to another problem-size r_j on a Grid-site g is similar to that on another Grid-site

h. If $\rho_g(\alpha, r_i)$ represents the execution time of application α for problem-size r_i on Grid-site g then:

$$\frac{\rho_g(\alpha, r_i)}{\rho_g(\alpha, r_j)} \approx \frac{\rho_h(\alpha, r_i)}{\rho_h(\alpha, r_j)}, i \neq j \tag{1}$$

This phenomenon is based on the fact that rate of change in execution time of an application across different problem-sizes is preserved on different Gridsites, i.e. the rate of change in execution time of an application $\triangle\rho$ for the problem-size r_i (the target problem-size) with respect to another problem-size r_j (the reference problem-size) on Grid-site g is equal to the rate of change in execution time for the problemsize r_i with respect to the problem-size r_j on Grid-site h.

$$\frac{\triangle\rho_g(\alpha, r_i)}{\triangle\rho_g(\alpha, r_j)} \approx \frac{\triangle\rho_h(\alpha, r_i)}{\triangle\rho_h(\alpha, r_j)}, i \neq j$$

4.2 Intra-platform PST

Intra-platform PST specifies that the normalized execution time ρ of an application α, on a Grid-site g for a machine-size $l \in \omega_g$ (the target machine-size) relative to another machine-size $m \in \omega_g$ (the reference machine-size), for a problem-size r_i is similar that for another problem-size r_j. If $\rho_g(\alpha, r_i, l)$ represents the execution time of an application for problem-size r_i and machine-size l then:

$$\frac{\rho_g(\alpha, r_i, l)}{\rho_g(\alpha, r_j, m)} \approx \frac{\rho_g(\alpha, r_i, l)}{\rho_g(\alpha, r_j, m)}, i \neq j, l \neq m \tag{2}$$

This phenomenon is based on the fact that rate of change in execution time of an application across different problem-sizes is preserved for different machine-sizes, i.e. the rate of change in execution time of an application for the problem-size r_i and machinesize l on Grid-site g with respect to that for machinesize m will be equal to the rate of change in execution time for the problem-size r_j and machine-size l with respect to that for a machine-size m on the same Grid-site:

$$\frac{\triangle\rho_g(\alpha, r_i, l)}{\triangle\rho_g(\alpha, r_j, m)} \approx \frac{\triangle\rho_g(\alpha, r_i, l)}{\triangle\rho_g(\alpha, r_j, m)}, i \neq j, l \neq m$$

Similarly, rate of change in executions time of the application across different machine sizes is also preserved for different problem-sizes. i.e.

$$\frac{\triangle\rho_g(\alpha, r_j, m)}{\triangle\rho_g(\alpha, r_i, m)} \approx \frac{\triangle\rho_g(\alpha, r_j, l)}{\triangle\rho_g(\alpha, r_i, l)}, i \neq j, l \neq m$$

We use this phenomenon to share execution times with in one Grid-site for scalability. The accuracy of inter- and intra-platform similarity of normalized behaviors, for embarrassingly parallel applications, does not depend upon the selection of reference point. However, for the parallel applications exploiting inter-process communications during their executions, this accuracy increases as the reference point gets closer to the target point the closer the reference point, the greater the similarity (of interprocess communication) it encompasses. Thus in case of inter-platform PST the reference problem-size closer to the target problem-size, and in case of intra-platform PST, the reference problem-size closer to the target problem-size as well as the reference machine-size closer to target machine-size. i.e. For inter-platform PST:

$$\lim_{r \to p} \left[\frac{\rho_g(\alpha, r_i)}{\rho_g(\alpha, r_j)} - \frac{\rho_h(\alpha, r_i)}{\rho_h(\alpha, r_j)}\right] = 0$$

Similarly, for intra-platform PST:

$$\lim_{l \to m} \left[\frac{\rho_g(\alpha, r_i, l)}{\rho_g(\alpha, r_j, m)} - \frac{\rho_g(\alpha, r_i, l)}{\rho_g(\alpha, r_j, m)}\right] = 0$$

For normalization from the minimum training set only, we select the maximum problem size (in normal practice of user of the application) and maximum machine size (for which application scales good) as reference point, to incorporate the maximum effects of inter-process communications in the normalization. The distance between the target point and the reference point for inter- and intra-platform PST on one Grid site is calculated respectively as:

$$d = \begin{cases} \sqrt{(\rho(r_i) - \rho(r_j))^2 + (r_i - r_j)^2} \\ \sqrt{(\rho(l) - \rho(m))^2 + (l - m)^2} \end{cases}$$

Note that, in the presented work, $\rho_g(\alpha, p) = \rho_g(\alpha, p, 1)$.

5 Experimental Design

Specifically in our work, the general purpose of the experimental design phase is to set a strategy for experiments to get the execution time of an application to support its performance prediction later on, in minimum number of experiments. A typical experimental design has specific sequence of steps, which is shown in Figure 2. For experimental design, one of our objectives is to eliminate/minimize the modeling and optimization phases in this model to make the training phase and performance prediction system robust. Among other key objectives are, to:

a) reduce/minimize training phase time;
b) minimize/eliminate the heavy modeling requirements after the training phase;
c) develop and maintain the efficient scalability of ED with respect to Grid-size;
d) make it generalizable to a wide range of applications on heterogeneous Grid-sites.

To address these objectives, we design our experimental design EXD in the light of guide lines given by Montgomery et al. in [1]. These are as follows:

a) Recognition of statement of problem: We describe our problem statement as: To obtain maximum execution time information of the application at different problem-sizes on all heterogeneous Grid-sites with different possible machine-sizes in minimum number of the experiments.

b) Selection of response variables: In our work the response variable is the execution time of the application.

c) Choice of factors, levels and ranges: The factors affecting the response variable are the problem-size of the application, the Grid-size and the machine-size. The range of problem-size incorporates ranges of each of input variables and the levels consist of their values at effective step sizes specified by the user of the application. The range of the Grid-size is $\{1, 2, , m\}$, and the levels include numeration of non-identical Grid-sites g_i. The range of machine-size spans over all different subsets of number of (heterogeneous) processors on a Grid-site and its levels include of all these individual subsets. Consider a simple case, for one Grid application with x different problem sizes and m Grid sites with p_i machine sizes. If we include all the respective levels of the above described factors, then the total number of possible experiments N is given by:

$$N = \sum_i^m (p^i \times x)$$

For single processor Grid-sites the above formula reduces to $N = x \times m$. The objective function f of the experimental design is:

$$f : \mathbb{R} \times \mathbb{N}^+ \times \mathbb{N}^+ \rightarrow \mathbb{N}^+$$
$$f(\beta, \gamma, \lambda) = Min_N : \xi \geq 90\%$$

where ξ represents accuracy.

d) Choice/Formulation of Experimental Design: In our Extended Experimental Design (EXD), we minimize the combinations of Grid-size with problem-size and then, combinations of Gridsize with machine-size. By minimizing the combinations of Grid-size with problem-size, we actually minimize number of experiments against different problem-sizes across the heterogeneous Grid-sites. Similarly, by minimizing the Grid-size combinations with machine-size factor, we actually minimize number of experiments against different problem-sizes across different number of processors. To meet these objective, we employ PST [2] mechanisms (see Section 4.2).

In our design, we choose one Grid-site (the fastest one, based on the initial runs) as a base Grid-site and make a full factorial of experiments on it. Later, we use its execution times as reference values to calculate predictions for other platforms (the target Grid-sites) using inter-platform PST and thus minimize problem-size combinations with Grid-size. We also make one experiment on each of other different Grid-sites to make PST work. Similarly, to minimize machine-size combinations with Grid-size, we need a full factorial of experiments with one machine-size, the base machine-size (which we already have while minimizing problem-size combinations with Grid-size), and later use these values as reference values. We make one experiment each for all other machine-sizes, to translate the reference execution times to that for other machine-sizes using intraplatform PST. The approach of making full factorial design of experiments is also necessary because we neither make any assumptions about the performance models of different applications, nor we make their analytical models at run time, because making analytical performance models for individual applications is more complex and less efficient than making full factorial design of experiments for once. In initial phase, we restrict the full factorial design of experiments to the commonly used range of problem-size.

By means of inter-platform PST, the total number of experiments N reduces from a polynomial complexity of $p \times x \times m$ to $p \times x + (m-1)$ for parallel machines, and from a polynomial complexity of $x \times m$ to a linear complexity of $x - 1 + m$ for single processor machines. Introducing intra-platform PST, we are able to reduce total number of experiments for parallel machines (Grid-sites) further to a linear complexity of $p + (x-1) + (m-1)$ or $p + m + x - 2$.

e) Performing of experiments: We address performing of experiments under automatic training phase as described in Section 6.

We design above step d to eliminate the need of next two steps presented by Montgomery et. al. the *statistical analysis & modeling* and *conclusions*, to minimize the serving costs on the fly.

6 Automatic Training Phase Based on EXD

If we consider the whole training phase as a process then, in executing the designed experiments, the process variables are application problem-size, machine-size and Grid-size. Our automatic training phase is a three layered approach; layer 1 for the initial test runs, layer 2 for the execution of experiments designed in experimental design phase, and layer 3 for the PST mechanism. In the initial test runs of the training phase, we need to make some characterizing or screening to establish some conjectures for the next set of experiments. This information is used to decide the base Grid-site. In layer 2, execution of experiments on different Grid-sites is planned according to the experimental design, and the training manager executes these experiments on the selected Grid-sites. The collected information is passed through layer 3, the PST mechanism, to store in a performance repository. First, the training manager makes one initial experiment on all non-identical Grid-sites G from Grid-site sample space G_T for maximum values of problem-sizes. Second, it selects the Gridsite with minimum execution time as base Grid-site and makes a full factorial design of experiments on it. We categorize two Grid-sites to be identical if they have same number of processors, processor architecture, processor speed, and memory. We also exclude those identical Grid-sites with i processors g_i for which $g^i \subset g^j$ for $i \leq j$, i.e. if a cluster or a sub-cluster is identical to another cluster or sub-cluster then only one of them will be included in training phase. Similarly, if a cluster or a sub-cluster is a subset of another cluster or sub-cluster then only the super set cluster is included in the training phase

The information from these experiments is archived to serve prediction process. For applications having a wide range of problem-size, we have support of distributed training phase. In distributed training phase we split the full factorial design of experiments from one machine to different Grid-sites which are included in the training phase. Distributed training phase exploits opportunistic load balancing [15] for executing experiments to harness maximum parallelization of training phase.

7 Application Performance Prediction System: G-Prophet

The ultimate goal of automatic training phase is to support performance prediction service, to serve automatic execution time predictions. To serve application performance prediction prediction system G-Prophet (Grid- Prophet), implemented under ASKALON, utilizes the inter- and intra-platform PST mechanisms to furnish the predictions using Equation 1 as:

$$\rho_g(\alpha, r_i) = \frac{\rho_h(\alpha, r_i)}{\rho_h(\alpha, r_j)} \times \rho_g(\alpha, r_i)$$

and/or Equation 2 as:

$$\rho_g(\alpha, r_i, l) = \frac{\rho_g(\alpha, r_i, l)}{\rho_g(\alpha, r_j, m)} \times \rho_g(\alpha, r_i, m)$$

G-Prophet uses the the nearest possible reference values for serving the predictions, available from the training phase or actual run times. From the minimum training set, we observe a prediction accuracy of more than 90% and a standard deviation of 2% in the worst case.

8 Experiments and Analysis

In this section, we describe the experimental platform and real world applications used for our experiments, the scalability and performance results from our proposed experimental design against the different changes in the factors involved (as defined in Section 5). For our results presented here, each result is taken as an average of five repetitions of the experiment to guard against the anomalous results.

8.1 Experimental Platform

We have conducted all of our experiments on the Austrian Grid environment. The Austrian Grid consortium combines Austria's leading researchers in advanced computing technologies with well-recognized partners in Grid-dependant application areas. The description of Austrian test bed sites for our experiments is given in Table 2. Our experiments are conducted on three real world applications Wien2k [3], Invmod [2] and meteoAG [4]. Wien2k application allows performing electronic structure calculations of solids using density functional theory based on the full-potential augmented planewave ((L)APW) and local orbital (lo) method. Invmod application helps in studying the effects of climatic changes on the water balance through water flow and balance simulations, in order to obtain improved discharge estimates for extreme floods. MeteoAG produces meteorological simulations of precipitation fields of heavy precipitation cases over the western part of Austria with RAMS, at a spatially and temporally fine granularity, in order to resolve most alpine watersheds and thunderstorms.

8.2 Performance and Scalability Analysis

The EXD scales efficiently against the changes in different factors involved in ED. During our experiments we obtain quite promising results. The scalability analysis is shown in Figure 3. We analyzed the scalability of EXD w.r.t. problem-size, by varying the problem-size factor for fixed remaining factors; 10 parallel Grid-sites with machine-size 20 and 50 single processor machines. The problem-size was varied from 10 to 200 and the reduction in the total number of experiments was observed from 96% to 99%. In another setup for analyzing scalability w.r.t machine-size, a reduction of 77% to 97% in the total number of experiments was observed when machine-size was varied from 1 to 80, for fixed factors of 10 parallel machines, 50 single processor Grid-sites and problem-size of 5. From another perspective, we observed that total number of experiments increased from 7% to 9% when Grid-size was increased from 15 to 155, for the fixed factors of 5 parallel machines with machine-size of 10 and problem-size 10. We observed an overall reduction of 78% to 99% when all the factors we varied simultaneously: 5 parallel machines with machine-size from 1 to 80, single processor Grid-sites from 10 to 95, and problem-size from 10 to 95. Performance and normalization results for interplatform PST are shown for Wien2k and Invmod in Figure 4, 5 and 6, 7 respectively. Figure 4 shows the measured performance of Wien2k on 5 different Grid-sites, and Figure 5 shows the normalized performance on these Grid-sites. For Wien2k the normalization is made with the execution time against the problem-size of 9.0. Figure 6 shows the measured execution time of Invmod on 5 different Grid-sites, and Figure 7 shows the normalized performance on these Grid-sites. The normalization for Invmod is made with execution time against the problem-size of 20.0. Automatic training phase made only 49 experiments out of total 492 for Wien2k (approx. 10%), and for Invmod conducted only 24 experiments out of total 192 (approx. 14%) on the test bed described above. Identical curves of normalized execution times exhibit the realization of inter-platform PST for these applications. Performance normalization results for interaplatform PST are shown in Figure 8, 9 and 10, 11. Figure 8 and 10 show the measured performance of MeteoAG for different values of problem-sizes and machine-sizes on two different Grid-sites hcma and zid-cc. Figures 9 and 11 show normalized performance of MeteoAG on these Grid-sites. The normalization is performed with the execution time against machine-size of 1. The training phase conducted only 162 out of total 6669 experiments (approx. 2.2%). The identical normalized performance curves show the realization of intera-platform PST.

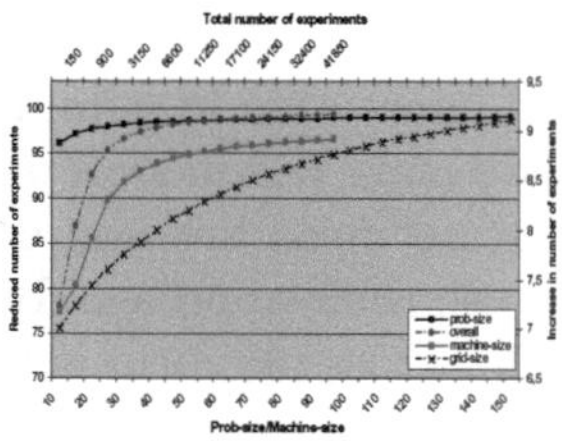

Fig. 3. Scalability analysis of EXD

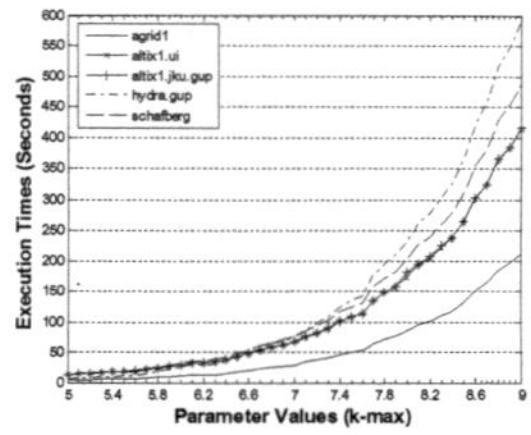

Fig. 4. Wien2k original exe. times

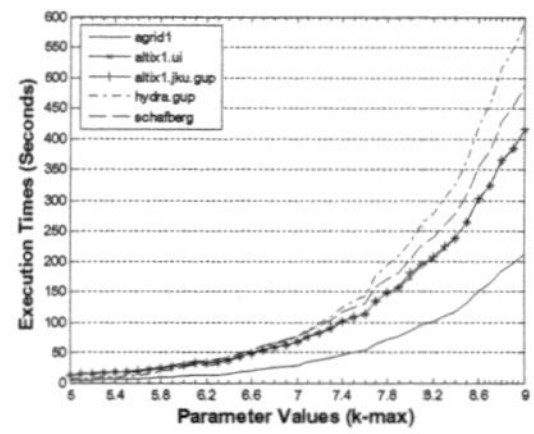

Fig. 5. Wien2k original exe. times

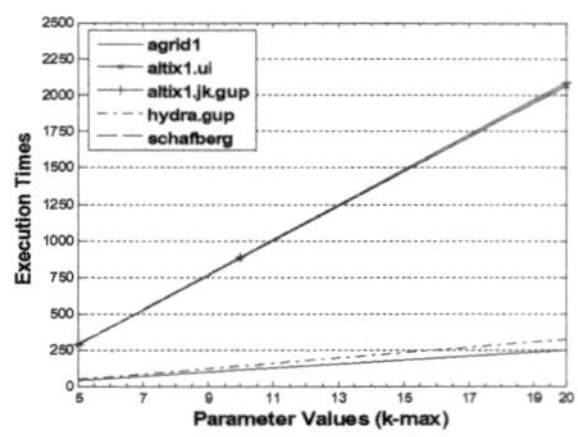

Fig. 6. Invmod original exe. times

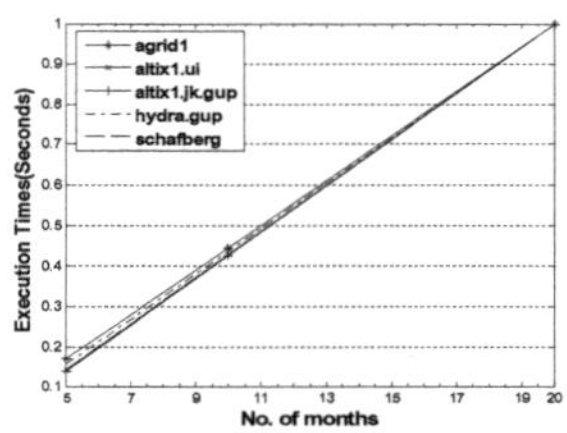

Fig. 7. Invmod normalized exe. times

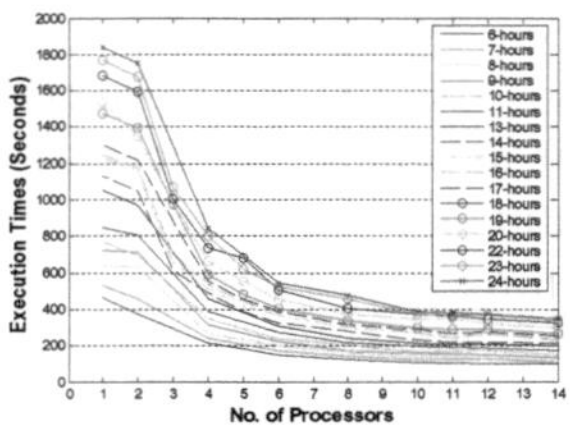

Fig. 8. MeteoAG on *hcma*, original exe. times

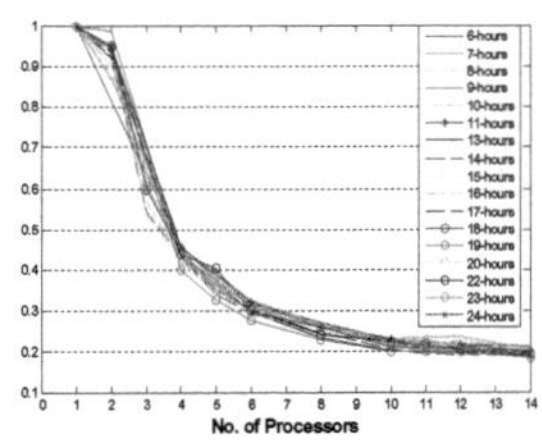

Fig. 9. MeteoAG on *hcma*, normalized exe. times

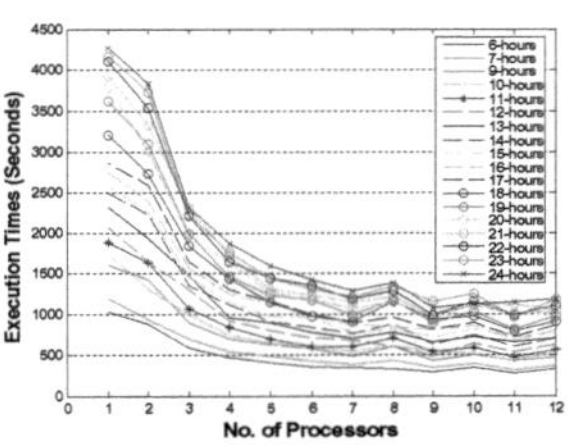

Fig. 10. Invmod normalized exe. times

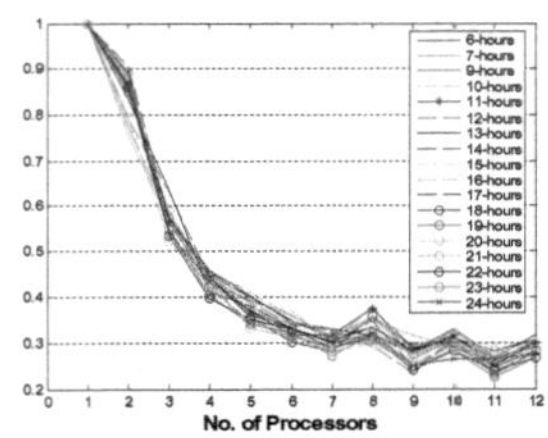

Fig. 11. MeteoAG on *hcma*, original exe. times

9 Related Work

There have been some attempts to get reactive training set for performance prediction of Grid application through Analytical benchmarking and templates [5,6], but to our knowledge we are the first to attempt to make a proactive training phase with proper experiment management and control through a step by step experimental design. Many multi-platform performance studies like [11] evaluated their approaches with data collected at multiple Grid-sites. However, data from each Grid-site are processed individually. Our approach, instead, utilizes benchmark performance results from one platform (base Grid-site) to share this information for other platforms (target Grid-sites). Iverson et al. [5] and

Marin et al. [13] use parameterized models to translate execution times across the heterogeneous platforms but their models need human interaction to parameterize all machines performance attributes. In contrast, we make our models on the basis of normalized/relative performance, which does not need any source code and manual intervention. Moreover it is difficult to use their approach for different languages. Leo et al. [7] use partial executions for cross platform performance translation. Their models need source code instrumentation and thus require human intervention. To contrast, our approach does not require source code and code instrumentation. Systems like Prophesy [9], PMaC [8], and convolution methods to map hand-coded kernels to Grid-site profiles [10] base on modeling computational kernels instead of complex applications with diverse tasks. In contrast, we develop observation-based performance translation, which does not require in-depth knowledge of parallel systems or codes. This makes our approach application- , language- and platform-independent. Reducing the number of experiments in adaptive environments dedicated to applications, by tuning parameters at run time have been applied by couple of works like ATLAS [15] (tuning performance of libraries by tuning parameters at run time). Though this technique improves the performance, yet is very specific to the applications. Unlike this work, we do not do run time modeling/tuning at run time, as modeling individual applications is more complex and less efficient than making the proposed experiments.

References

1. Douglas, C., Montgomery: Design and Analysis of Experiments. ch. 1, vol. 6. John Wiley & Sons, New York (2004)
2. Theiner, D., et al.: Reduction of Calibration Time of Distributed Hydrological Models by Use of Grid Comp. and Nonlinear Optimisation Algos. In: International Conference on Hydroinformatics, France (2006)
3. [3] Blaha, P., Schwarz, K., Madsen, G., Kvasnicka, D., Luitz, J.: WIEN2k: An Augmented Plane Wave plus LocalOrbitals Program for Calculating Crystal Properties, Instituteof Physical and Theoretical Chemistry, TU Vienna (2001)
4. Schller, F., Qin, J., Farrukh N.: Performance, Scalability and Quality of the Meteorological Grid Workflow MeteoAG. In: 2nd Austrian Grid Symposium, Innsbruck, Austria. OCG Verlag (2006)
5. Iverson, M.A., et al.: Statistical Prediction of Task Execution Times Through Analytic Benchmarking for Scheduling in a Heterogeneous Environment. In: Heterogeneous Computing Workshop (1999)
6. Smith, W., Foster, I., Taylor, V.: Predicting Application Run Times Using Historical Information. In: Proceedings of the IPPS/SPDP (1998) Workshop on Job Scheduling Strategies for Parallel Processing (1998)
7. Yang, L.T., Ma, X., Mueller, F.: Cross- Platform Performance Prediction of Parallel Applications Using Partial Execution. In: Supercomputing, USA (2005)
8. Carrington, L., Snavely, A., Wolter, N.: A performance prediction framework for scientific applications,Future Gener. Computing Systems, vol. 22, Amsterdam, The Netherlands (2006)

9. Taylor, V., et al.: Prophesy: An Infrastructure for Performance Analysis and Modeling of Parallel and Grid Applications. ACM Sigmetrics Performance Evaluation Review 30(4) (2003)
10. Bailey, D.H., Snavely, A.: Performance Modeling: Understanding the Present and Predicting the Future. In: Euro-Par Conference (August 2005)
11. Kerbyson, D.J., Alme, H.J., et al.: Predictive Performance and Scalability Modeling of a Large-Scale Application. In: Proceedings of Supercomputing (2001)
12. Fahringer, T., et al.: ASKALON: A Grid Application Development and Computing Environment. In: 6th International Workshop on Grid, Seattle, USA (November 2005)
13. Marin, G., Mellor-Crummey, J.: Cross-Architecture Performance Predictions for Scientific Applications Using Parameterized Models. In: Joint International Conference on Measurement and Modeling of Computer Systems (2004)
14. Nadeem, F., Yousaf, M.M., Prodan, R., Fahringer, T.: Soft Benchmarks-based Application Performance Prediction using a Minimum Training Set. In: Second International Conference on e-Science and Grid computing, Amsterdam, Netherlands (December 2006)
15. Whaley, R.C., Petitet, A.: Petitet: Minimizing development and maintenance costs in supporting persistently optimized BLAS. Software: Practice and Experienc 35(2):101Ž013121 (2005)

Multiobjective Differential Evolution for Mapping in a Grid Environment

Ivanoe De Falco[1], Antonio Della Cioppa[2], Umberto Scafuri[1], and Ernesto Tarantino[1]

[1] ICAR–CNR, Via P. Castellino 111, 80131 Naples, Italy
{ivanoe.defalco,ernesto.tarantino,umberto.scafuri}@na.icar.cnr.it
[2] DIIIE, University of Salerno, Via Ponte don Melillo 1, 84084 Fisciano (SA), Italy
adellacioppa@unisa.it

Abstract. Effective and efficient mapping algorithms for multisite parallel applications are fundamental to exploit the potentials of grid computing. Since the problem of optimally mapping is NP–complete, evolutionary techniques can help to find near–optimal solutions. Here a multiobjective Differential Evolution is investigated to face the mapping problem in a grid environment aiming at reducing the degree of use of the grid resources while, at the same time, maximizing Quality of Service requirements in terms of reliability. The proposed mapper is tested on different scenarios.

1 Introduction

Grid [1] is a decentralized heterogeneous multisite system which aggregates geographically dispersed and multi–owner resources (CPUs, storage system, network bandwidth,...) and can be profitably used to execute MPI computational intensive applications [2]. In fact, a grid represents a collaborative computational–intensive problem–solving environment in which MPI applications, each made up of multiple message–passing subtasks, can be submitted without knowing where the resources are or even who owns these resources.

When the execution of an MPI grid application must satisfy user–dependent requirements [3], such as performance and Quality of Service (QoS) [4], single sites resources could be insufficient for meeting all these needs. Thus, a multisite mapping tool, able to obtain high throughput while matching applications needs with the networked grid computing resources, must be conceived.

In this work we deal with the mapping problem from the grid manager's point of view, so our aim is to find the solution which uses at the minimum the grid resources it has to exploit at the most and respects the user QoS requests. Naturally, in absence of information about the communication timing, the job execution on grid is possible only if the co–scheduling of all its subtasks is guaranteed [5]. In this hypothesis, we distribute the application subtasks among the nodes minimizing execution time and communication delays, and at the same time optimizing resource utilization. QoS issues are investigated only in terms

R. Perrott et al. (Eds.): HPCC 2007, LNCS 4782, pp. 322–333, 2007.

of reliability. This means that our mapper maximizes reliability by preferring solutions which make use of devices, i.e. processors and links connecting the sites to internet, which only seldom are broken.

As the mapping is an NP–complete problem [6], several evolutionary–based techniques have been used to face it in a heterogeneous or grid environment [7,8,9,10,11]. To provide the user with a set of possible mapping solutions, each with different balance for use of resources and reliability, a multiobjective version of Differential Evolution (DE) [12] based on the Pareto method [13] is here proposed. Unlike the other existing evolutionary methods which simply search for one site onto which map the application, we deal with a multisite approach. Moreover, as a further distinctive issue with respect to other approaches proposed in literature [14], we consider the nodes making up the sites as the lowest computational unit taking into account its reliability and its actual load.

Paper structure is as follows: Section 2 illustrates our multiobjective mapper, Section 3 reports on the test problems experienced and shows the results achieved and Section 4 contains conclusions and future works.

2 DE for Mapping Problem

The technique. Given a minimization problem with q real parameters, DE faces it starting with a randomly initialized population consisting of $\mathcal{M}$ individuals each made up by q real values. Then, the population is updated from a generation to the next one by means of different transformation schemes. We have chosen a strategy which is referenced as *DE/rand/1/bin*. In it for the generic i–th individual in the current population three integer numbers r_1, r_2 and r_3 in $[1, \ldots, \mathcal{M}]$ differing one another and different from i are randomly generated. Furthermore, another integer number s in the range $[1, q]$ is randomly chosen. Then, starting from the i–th individual a new trial one i' is generated whose generic j–th component is given by $x_{i'_j} = x_{r_{3_j}} + F \cdot (x_{r_{1_j}} - x_{r_{2_j}})$, provided that either a randomly generated real number ρ in $[0.0, 1.0]$ is lower than the value of a parameter CR, in the same range as ρ, or the position j under account is exactly s. If neither is verified then a simple copy takes place: $x_{i'_j} = x_{i_j}$. F is a real and constant factor which controls the magnitude of the differential variation $(x_{r_{1_j}} - x_{r_{2_j}})$. This new trial individual i' is compared against the i–th individual in the current population and, if fitter, replaces it in the next population, otherwise the old one survives and is copied into the new population. This basic scheme is repeated for a maximum number of generations g.

Definitions. To focus the mapping problem in a grid we need information on the number and on the status of both accessible and demanded resources. We assume to have an application task subdivided into P subtasks (demanded resources) to be mapped on n nodes (accessible resources) with $n \in [1, \ldots, N]$, where P is fixed *a priori* and N is the number of grid nodes. We need to know *a priori* the number of instructions α_i computed per time unit on each node i. Furthermore, we assume to have cognition of the communication bandwidth β_{ij} between any couple of nodes i and j. Note that β_{ij} is the generic element of an $N \times N$

symmetric matrix $\boldsymbol{\beta}$ with very high values on the main diagonal, i.e., β_{ii} is the bandwidth between two subtasks on the same node. This information is supposed to be contained in tables based either on statistical estimations in a particular time span or gathered tracking periodically and forecasting dynamically resource conditions [15,16]. In the Globus Toolkit [4], which is a standard grid middleware, the information is gathered by the Grid Index Information Service (GIIS) [16].

Since grids address non dedicated–resources, their own local workloads must be considered to evaluate the computation time. There exist several prediction methods to face the challenge of non–dedicated resources [17,18]. For example, we suppose to know the average load $\ell_i(\Delta t)$ of the node i at a given time span Δt with $\ell_i(\Delta t) \in [0.0, 1.0]$, where 0.0 means a node completely discharged and 1.0 a node locally loaded at 100%. Hence $(1 - \ell_i(\Delta t)) \cdot \alpha_i$ represents the power fraction of the node i available for the execution of grid subtasks.

As regards the resources requested by the application task, we assume to know for each subtask k the number of instructions γ_k and the number of communications ψ_{km} between the k–th and the m–th subtask $\forall m \neq k$ to be executed. Obviously, ψ_{km} is the generic element of a $P \times P$ symmetric matrix $\boldsymbol{\psi}$ with all null elements on the main diagonal. All this information can be obtained either by a static program analysis, or by using smart compilers or by other tools which automatically generate them. For example the Globus Toolkit includes an XML standard format to define application requirements [16].

Finally, information must be provided about the degree of reliability of any component of the grid. This is expressed in terms of fraction of actual operativity π_z for the processor z and λ_w for the link connecting to internet the site w to which z belongs. Any of these values ranges in $[0.0, 1.0]$.

Encoding. In general, any mapping solution should be represented by a vector $\boldsymbol{\mu}$ of P integers ranging in the interval $[1, N]$. To obtain $\boldsymbol{\mu}$, the real values provided by DE in the interval $[1, N+1[$ are truncated before evaluation. The truncated value $\lfloor \mu_i \rfloor$ denotes the node onto which the subtask i is mapped.

For all mapping problems in which also communications ψ_{km} must be taken into account, the allocation of a subtask on a given node can cause that the optimal mapping needs that also other subtasks must be allocated on the same node or in the same site, so as to decrease their communication times and thus their execution times, taking advantage of the higher communication bandwidths existing within any site compared to those between sites. Such a problem is a typical example of *epistasis*, i.e. a situation in which the value taken on by a variable influences those of other variables. This situation is also *deceptive*, since a solution $\boldsymbol{\mu_1}$ can be transformed into another $\boldsymbol{\mu_2}$ with better fitness only by passing through intermediate solutions, worse than both $\boldsymbol{\mu_1}$ and $\boldsymbol{\mu_2}$, which would be discarded. To overcome this problem we have introduced a new operator, named *site mutation*, applied with a probability p_m any time a new individual must be produced. When this mutation is to be carried out, a position in $\boldsymbol{\mu}$ is randomly chosen, let us suppose its contents refers to a node in site C_i. Then another site, say C_j, is randomly chosen. If this latter has N_j nodes, the position chosen and its next $N_j - 1$ are filled with values representing consecutive nodes of C_j, starting from the first one of C_j. If the right end of the chromosome is reached, this process continues circularly. If $N_j > P$ this operator stops after modifying the P alleles. If *site mutation* does not take place, the classical transformations typical of DE must be applied.

Fitness. Given the two goals described in Section 1, we have two fitness functions, accounting one for the time of use of resources and the other for their reliability.

Use of resources. Denoting with τ_{ij}^{comp} and τ_{ij}^{comm} respectively the computation and the communication times requested to execute the subtask i on the node j it is assigned to, the generic element of the execution time matrix $\boldsymbol{\tau}$ is computed as:

$$\tau_{ij} = \tau_{ij}^{\text{comp}} + \tau_{ij}^{\text{comm}}$$

In other words, τ_{ij} is the total time needed to execute the subtask i on the node j. It is evaluated on the basis of the computation power and of the bandwidth which remain available once deducted the local workload. Let τ_j^s be the summation on all the subtasks assigned to the j–th node for the current mapping. This value is the time spent by node j in executing computations and communications of all the subtasks assigned to it by the proposed solution. Clearly, τ_j^s is equal to zero for all the nodes not included in the vector $\boldsymbol{\mu}$.

Considering that all the subtasks are co–scheduled, the time required to complete the application execution is given by the maximum value among all the τ_j^s. Then, the fitness function is:

$$\Phi_1(\boldsymbol{\mu}) = \max_{j\in[1,N]} \{\tau_j^s\} \tag{1}$$

The goal of the evolutionary algorithm is to search for the smallest fitness value among these maxima, i.e. to find the mapping which uses at the minimum, in terms of time, the grid resource it has to exploit at the most.

Reliability. In this case the fitness function is given by the reliability of the proposed solution. This is evaluated as:

$$\Phi_2(\boldsymbol{\mu}) = \prod_{i=1}^{P} \pi_{\lfloor\mu_i\rfloor} \cdot \lambda_w \tag{2}$$

where $\lfloor\mu_i\rfloor$ is the node onto which the i–th subtask is mapped and w is the site this node belongs to.

It should be noted that the first fitness function should be minimized, while the second should be maximized. We face this two–objective problem by designing and implementing a multiobjective DE algorithm based on the Pareto–front approach. It is very similar to the DE scheme described in Sect. 2, apart from the way the new trial individual i' is compared to the current individual i. In this case i' is chosen if and only if it is not worse than i in terms of both the fitness functions, and is better than i for at least one of them. By doing so, a set

Algorithm 1

```
randomly initialize population
evaluate fitness Φ1 and Φ2 for any individual x
while (maximal number of generations g is not reached) do
  begin
    for i = 1 to M do
      begin
        choose a random real number p_sm ∈ [0.0, 1.0]
        if (p_sm < p_m) apply site mutation
        else
          begin
            choose three integers r1, r2 and r3 in [1, M], with r1 ≠ r2 ≠ r3 ≠ i
            choose an integer number s in [1, q]
            for j = 1 to q do
              begin
                choose a random real number ρ ∈ [0.0, 1.0]
                if ( (ρ < CR) OR (j = s) )  x_{i'_j} = x_{r3_j} + F · (x_{r1_j} − x_{r2_j})
                else x_{i'_j} = x_{i_j}
              end
            if ( ((Φ1(x_i') < Φ1(x_i)) AND (Φ2(x_i') ≥ Φ2(x_i))) OR
              ((Φ1(x_i') ≤ Φ1(x_i)) AND (Φ2(x_i') > Φ2(x_i))) )
              insert x_i' in the new population
            else insert x_i in the new population
          end
      end
  end
```

of "optimal" solutions, the so–called *Pareto optimal set*, emerges in that none of them can be considered to be better than any other in the same set with respect to all the single objective functions. The pseudocode of our DE for mapping is shown in the following **Algorithm 1**.

3 Experiments and Results

We assume to have a multisite grid architecture composed of $N = 116$ nodes divided into five sites (Fig. 1). Hereinafter the nodes are indicated by means of the external numbers, so that 37 in Fig. 1 is the fifth of 16 nodes in the site B. Without loss of generality we suppose that all the nodes belonging to the same site have the same power α expressed in terms of millions of instructions per second (MIPS). For example all the nodes belonging to B have $\alpha = 900$.

We have considered for each node three communication bands: the bandwidth β_{ii} available when subtasks are mapped on the same node (*intranode communication*), the bandwidth β_{ij} between the nodes i and j belonging to the same site (*intrasite communication*) and the bandwidth β_{ij} when the nodes i and j belong to different sites (*intersite communication*). For the sake of

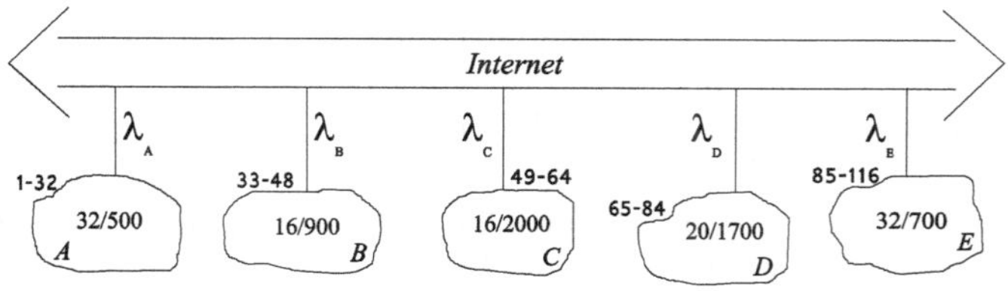

Fig. 1. The grid architecture

simplicity we presume that $\beta_{ij} = \beta_{ji}$ and that all the β_{ii}s have the same very high value (100 Gbit/s) so that the related communication time is negligible with respect to intrasite and intersite communications (Table 1).

Table 1. Intersite and intrasite bandwidths expressed in Mbit/s

	A	B	C	D	E
A	10				
B	2	100			
C	6	3	1000		
D	5	10	7	800	
E	2	5	6	1	100

Moreover we assume to know the local average load $\ell_i(\Delta t) \cdot \alpha_i$ of available grid nodes and, as concerns the reliability, we suppose that $\lambda_w = 0.99$ $\forall w \in \{A, B, \ldots, E\}$, while in any site the nodes have different π values (Table 2).

Table 2. Reliability for the nodes

Sites	A		B		C		D		E	
nodes	1–12	13–32	33–40	41–48	49–58	59–64	65–72	73–84	85–100	101–116
π	0.99	0.96	0.97	0.99	0.97	0.99	0.99	0.97	0.98	0.96

Since a generally accepted set of heterogenous computing benchmarks does not exist, to evaluate the effectiveness of our DE approach we have conceived and explored four scenarios of increasing difficulty: one without communications among subtasks, one in which communications are added, another in which local node loads are also considered and a final in which reliability of a link is decreased. To test the behavior of the mapper also as a function of P, two cases have been dealt with: one in which $P = 20$ and another with $P = 35$. Given the grid architecture, $P = 20$ has been chosen to investigate the ability of the mapper when the number of subtasks is lower than the number of nodes of some of the sites, while $P = 35$ has been considered to evaluate it when this number is greater than the number of nodes in any site.

After a preliminary tuning phase, DE parameters have been set as follows: $\mathcal{M} = 100$, $g = 1000$, $CR = 0.8$, $F = 0.5$, $p_m = 0.2$. All the tests have been made on a 1.5 GHz Pentium 4. For each problem 20 DE runs have been performed. Each execution takes about 1 minute for $P = 20$ and 2 minutes for $P = 35$. Note that these times are negligible since the mapping is related to computationally intensive grid applications which require several hours to be executed.

Henceforth we shall denote by $\boldsymbol{\mu_{\Phi_1}}$ and $\boldsymbol{\mu_{\Phi_2}}$ the best solutions found in terms of lowest maximal resource utilization time and of highest reliability, respectively.

Experiment 1. It has regarded an application of $P = 20$ subtasks with $\gamma_k = 90$ Giga Instructions (GI), $\psi_{km} = 0$ $\forall k, m \in [1, \ldots, P]$, $\ell_i(\Delta t) = 0$ for all the nodes and $\lambda_w = 0.99$ $\forall w \in \{A, B, \ldots, E\}$. Any execution of the DE mapper finds out several solutions, all being non–dominated in the final Pareto front. The solutions at the extremes of the achieved front are:

$$\boldsymbol{\mu_{\Phi_1}} = \{62, 63, 64, 65, 66, 78, 68, 69, 81, 70, 71, 72, 74, 75, 76, 57, 58, 59, 60, 61\}$$

which has fitness values of $\Phi_1 = 52.94$, $\Phi_2 = 0.579$ and uses, as expected, the most powerful nodes of the sites C and D, and

$$\boldsymbol{\mu_{\Phi_2}} = \{47, 43, 63, 63, 64, 59, 60, 61, 62, 59, 64, 61, 66, 62, 68, 69, 70, 71, 48, 46\}$$

which has $\Phi_1 = 100.00, \Phi_2 = 0.668$ and uses the most reliable nodes contained in the sites B, C and D. Furthermore, the system proposes some other non–dominated solutions better balanced in terms of the two goals, like for instance:

$$\boldsymbol{\mu} = \{71, 59, 68, 61, 76, 60, 60, 70, 69, 61, 63, 65, 51, 64, 63, 75, 62, 72, 56, 67\}$$

with $\Phi_1 = 90.0, \Phi_2 = 0.616$. It is up to the user to choose, among the proposed solutions, the one which best suits his/her needs.

Experiment 2. In it all the parameters remain unchanged, but we have added a communication $\psi_{km} = 10$ Mbit $\forall k, m \in [1, \ldots, P]$. In this case we have:

$$\boldsymbol{\mu_{\Phi_1}} = \{72, 73, 74, 75, 66, 77, 78, 69, 80, 81, 82, 83, 84, 65, 76, 67, 68, 79, 70, 71\}$$

with $\Phi_1 = 53.17$ and $\Phi_2 = 0.523$. It is worth noting that the presence of communications yields that all the subtasks are mapped on the same site D, differently from the previous test case. In fact, C is the site with the fastest processors, but if the mapper used the 16 nodes of C only, then on some of them two subtasks should be placed, which would require a computation time on those nodes of $(90000 \cdot 2)/2000 = 90s$, higher than that needed to execute one subtask on a node of D, i.e. $(90000/1700) = 52.94s$. Furthermore, communications between C and D nodes are quite slower than those within the same site due to communication bandwidths. Moreover, we have:

$$\boldsymbol{\mu_{\Phi_2}} = \{70, 71, 64, 65, 62, 67, 71, 69, 70, 66, 72, 68, 72, 65, 66, 67, 68, 69, 60, 61\}$$

with $\Phi_1 = 117.66, \Phi_2 = 0.668$. Also in this case many intermediate solutions are provided, which cannot be shown here for the sake of brevity.

Experiment 3. It is based on the same scenario as in the previous one, but we have taken into account the node loads as well, namely $\ell(\Delta t) = 0.9$ for all the nodes of the sites B and D, while for the site C we assume $\ell_i(\Delta t) = 0.8$ for $i \in [49, \dots, 52]$ and $\ell_i(\Delta t) = 0.6$ for $i \in [53, \dots, 64]$. The best mappings are:

$$\boldsymbol{\mu_{\Phi_1}} = \{85, 86, 87, 88, 89, 90, 91, 92, 93, 94, 95, 96, 97, 98, 99, 100, 101, 102, 103, 104\}$$

with $\Phi_1 = 130.47, \Phi_2 = 0.502$, and

$$\boldsymbol{\mu_{\Phi_2}} = \{63, 1, 5, 4, 60, 5, 1, 7, 3, 12, 2, 6, 3, 6, 2, 8, 9, 10, 8, 11\}$$

with $\Phi_1 = 398.66, \Phi_2 = 0.668$. As concerns $\boldsymbol{\mu_{\Phi_1}}$ it is to remark that all subtasks have been placed on the nodes of E. This might seem strange, since nodes $53 - 64$ in C, being loaded at 60%, are capable of providing an available computation power of 800 MIPS, so they are anyway more powerful than those in E. Nonetheless, since $P = 20$, those nodes would not be sufficient and four less powerful would be needed. The most powerful after those in C are those in E, and, in terms of Φ_1, these two solutions are equivalent, since the slowest resources utilized are in both cases nodes in E, yielding a resource use due to computation of $128.57s$. Moreover, the communications within E are faster than those between C and E, hence the $\boldsymbol{\mu_{\Phi_1}}$ proposed. As regards $\boldsymbol{\mu_{\Phi_2}}$, instead, it correctly chooses among the most reliable nodes in the grid, and strongly uses the nodes $1 - 12$ of A.

Experiment 4. It is as the third, but we have supposed that $\lambda_A = 0.97$. This should cause that, as far as reliability is concerned, solutions containing nodes of A should not be preferred, unlike the previous experiment. The best solutions found are:

$$\boldsymbol{\mu_{\Phi_1}} = \{85, 86, 87, 88, 89, 90, 91, 92, 93, 94, 95, 96, 97, 98, 99, 100, 101, 102, 103, 104\}$$

with $\Phi_1 = 130.47, \Phi_2 = 0.502$ and

$$\boldsymbol{\mu_{\Phi_2}} = \{59, 60, 69, 60, 62, 64, 67, 59, 63, 61, 61, 63, 62, 65, 61, 71, 60, 66, 60, 72\}$$

with $\Phi_1 = 549.47, \Phi_2 = 0.668$. We can see that $\boldsymbol{\mu_{\Phi_1}}$ is the same as before, while $\boldsymbol{\mu_{\Phi_2}}$ has optimally shifted to the most reliable nodes in C and D. This confirms that our mapper correctly deals with low reliability issues.

Experiment 5. It has regarded an application of $P = 35$ subtasks with $\gamma_k = 90$ Giga Instructions (GI), $\psi_{km} = 0 \ \forall k, m \in [1, \dots, P]$, $\ell_i(\Delta t) = 0.0$ for all the nodes and $\lambda_w = 0.99 \ \forall w \in \{A, B \dots, E\}$. The best mappings provided by our tool are:

$$\boldsymbol{\mu_{\Phi_1}} = \{72, 73, 74, 75, 76, 77, 78, 79, 80, 81, 82, 83, 84, 50, 51, 52, 53, 54, 55, 56, 57, 58, 59, 60, 61, 62, 63, 64, 65, 66, 67, 68, 69, 70, 71\}$$

with $\Phi_1 = 52.94s, \Phi_2 = 0.32$ and uses the most powerful nodes in C and D, and

$$\mu_{\Phi_2} = \{66, 59, 60, 56, 59, 41, 61, 62, 69, 71, 65, 63, 62, 72, 67, 68, 53, 72, 44, 61, 62, 66, 71, 47, 74, 66, 63, 65, 63, 68, 71, 70, 43, 61, 65\}$$

with $\Phi_1 = 158.82s, \Phi_2 = 0.465$ and picks the most reliable nodes among the sites B, C and D. Furthermore, the system proposes some other non–dominated solutions which are better balanced in terms of the two goals.

Experiment 6. It is like the fifth but we have added a communication $\psi_{km} = 10$ Mbit $\forall k, m \in [1, \ldots, P]$. The best solutions are:

$$\mu_{\Phi_1} = \{81, 82, 83, 84, 50, 51, 52, 53, 54, 55, 56, 57, 58, 59, 60, 61, 62, 63, 64, 65, 66, 67, 68, 69, 70, 71, 72, 73, 74, 75, 76, 77, 78, 79, 80\}$$

with $\Phi_1 = 74.60s, \Phi_2 = 0.322$ and

$$\mu_{\Phi_2} = \{60, 66, 73, 61, 61, 66, 69, 59, 66, 65, 69, 72, 70, 48, 60, 76, 61, 42, 71, 67, 59, 41, 67, 65, 70, 65, 65, 62, 70, 68, 64, 64, 60, 69, 54\}$$

with $\Phi_1 = 293.14s, \Phi_2 = 0.465$. The mapping μ_{Φ_1} uses only nodes in sites C and D, i.e., the most powerful ones. In fact, in this case the increase in execution time due to communications between sites is lower than that which would be obtained by using just one site and mapping two subtasks on a same node.

Experiment 7. It is like the sixth but we have taken into account the node loads as well, namely $\ell(\Delta t) = 0.9$ for all the nodes of the sites B and D, while for the site C we assume $\ell_i(\Delta t) = 0.8$ for $i \in [49, \ldots, 52]$ and $\ell_i(\Delta t) = 0.6$ for $i \in [53, \ldots, 64]$. This time we have:

$$\mu_{\Phi_1} = \{93, 94, 95, 96, 97, 98, 99, 100, 101, 102, 103, 104, 105, 106, 107, 108, 109, 110, 111, 112, 113, 114, 115, 116, 59, 60, 61, 85, 86, 87, 88, 89, 90, 91, 92\}$$

with $\Phi_1 = 165.85s, \Phi_2 = 0.257$ and

$$\mu_{\Phi_2} = \{1, 6, 63, 8, 9, 2, 5, 1, 2, 6, 7, 5, 6, 7, 8, 9, 10, 24, 12, 1, 2, 3, 9, 57, 4, 5, 6, 7, 8, 9, 10, 11, 64, 65, 66\}$$

with $\Phi_1 = 860.01s, \Phi_2 = 0.47$. In this case μ_{Φ_1} maps 32 subtasks on the site E and 3 on the nodes 59, 60 and 61 of C. This might seem strange, since nodes $53 - 64$ in C, being loaded at 60%, are capable of providing an available computation power of 800 MIPS, so they are anyway more powerful than those in E. Nonetheless, since $P = 35$, those nodes would not be sufficient and 23 less powerful would be needed. The most powerful after those in C are those in E, and, in terms of Φ_1, these two solutions are equivalent since the slowest resources utilized are in both cases nodes in E, yielding a resource use due to computation of $(90000/700) = 128.57s$. Moreover, the intrasite communications of E are faster than the intersite between C and E, hence the μ_{Φ_1} proposed.

It is to note that a number of subtasks on E greater than 32 would imply that more than one subtask should be placed per node. This would heavily increase the computation time. On the contrary, a number of subtasks on C greater than 3 and lower or equal to 12 would leave the computation time unchanged but would increase the amount of intersite communication time.

As regards $\boldsymbol{\mu_{\Phi_2}}$, instead, the suboptimal solution chooses all the nodes with $\pi = 0.99$ apart from node 24 with $\pi = 0.96$ and node 57 with $\pi = 0.97$.

Experiment 8. The scenario is the same as in the previous case, but we have supposed that the reliability $\lambda_A = 0.97$. The best solutions found are:

$$\boldsymbol{\mu_{\Phi_1}} = \{108, 109, 110, 111, 112, 113, 114, 115, 116, 60, 61, 63, 85, 86, 87,\\ 88, 89, 90, 91, 92, 93, 94, 95, 96, 97, 98, 99, 100, 101, 102, 103, 104, 105, 106, 107\}$$

with $\Phi_1 = 165.85s, \Phi_2 = 0.257$ and

$$\boldsymbol{\mu_{\Phi_2}} = \{65, 57, 58, 59, 79, 61, 68, 62, 65, 65, 66, 67, 63, 69, 70, 71, 72, 73,\\ 60, 62, 58, 48, 60, 61, 71, 63, 64, 41, 66, 38, 68, 69, 70, 71, 72\}$$

with $\Phi_1 = 1653.55s, \Phi_2 = 0.437$. Here $\boldsymbol{\mu_{\Phi_1}}$ uses heavily site E and some nodes from C. This solution is slightly different from that in the previous experiment, but has the same fitness value. The solution $\boldsymbol{\mu_{\Phi_2}}$ instead uses the most reliable nodes in sites B, C and D and, as it should be, avoids nodes from A which has $\lambda_A = 0.97$. Moreover, the five non optimal nodes (38, 57, 58, 73 and 79) in the solution are not in E because E, at parity of reliability, has inferior performance due to its minor intrasite bandwidths.

In Fig. 2 we report the final Pareto front of the alternative mapping solutions, equivalent from an evolutionary point of view, found out for one of the runs for this last test. In this case, the solutions range between those with a good reliability Φ_2 and with a high time of use of resources Φ_1 to those with a lower reliability but with a minor use time. As it can be easily perceived from the figure, the solutions which yield better balance in satisfying both goals are those in the intermediate region of the front. In fact such solutions are geometrically closer to the theoretically optimal solution for which $\Phi_1 \rightarrow 0$ and $\Phi_2 \rightarrow 1$.

Table 3 shows for each test (*Exp. no*) for both fitnesses the best final value Φ^b, the average of the final values over the 20 runs $\langle \Phi \rangle$ and the variance σ_Φ. The tests have evidenced a high degree of efficiency of the proposed model in terms of goodness for both resource use and reliability. In fact, efficient solutions have been provided independently of work conditions (heterogenous nodes diverse in terms of number, type and load) and kind of application tasks (computation or communication bound). Moreover, we have that often, especially as regards the problems with $P = 20$ subtasks, $\sigma(\Phi_1) = 0$ which means that in all the 20 runs the same final solution, i.e. the globally best one, has been found.

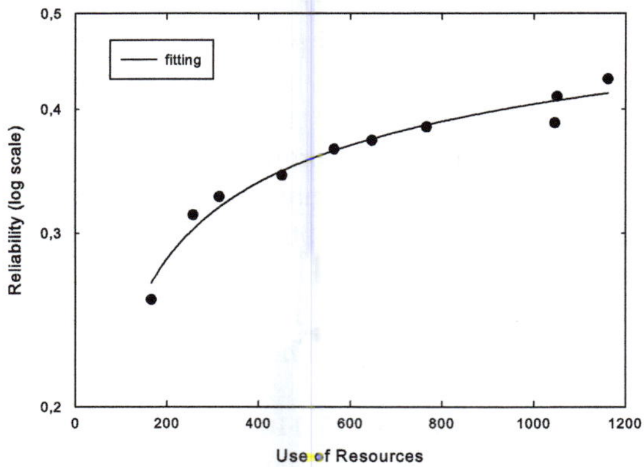

Fig. 2. The solutions on the Pareto front for experiment 8

Table 3. Findings for each experiment

	P=20				*P=35*			
Exp. no.	1	2	3	4	5	6	7	8
Φ_1^b	52.94	53.17	130.47	130.47	52.94	74.60	165.85	165.85
$\langle\Phi_1\rangle$	52.94	53.17	130.47	130.47	93.61	94.93	224.33	165.85
σ_{Φ_1}	0	0	0	0	17.86	15.02	31.44	0
Φ_2^b	0.668	0.668	0.668	0.668	0.465	0.465	0.470	0.437
$\langle\Phi_2\rangle$	0.668	0.668	0.660	0.652	0.443	0.448	0.436	0.419
σ_{Φ_2}	0	0	0.008	0.008	0.009	0.008	0.013	0.011

4 Conclusions and Future Works

This paper faces the grid multisite mapping problem by means of a multiobjective DE considering two goals: the minimization of the use degree of the grid resources and the maximization of the reliability of the proposed mapping. The results show that DE is a viable approach to the important problem of grid resource allocation.

Due to the lack of systems which operate in the same conditions as ours, at the moment, a comparison to ascertain the effectiveness of our multisite mapping approach is difficult. In fact some of these algorithms, such as Min–min, Max–min, XSuffrage [14], are related to independent subtasks and their performance are affected in heterogenous environments. In case of dependent tasks, the classical approaches apply the popular model of Direct Acyclic Graph differently from ours in which no assumptions are made about the communication timing among the processes since we have hypothesized the subtasks co–scheduling.

In addition to reliability, we intend to enrich our tool to manage multiple QoS requirements (performance, bandwidth, cost, response time and so on).

References

1. Buyya, R., Abramson, D., Giddy, J., Stockinger, H.: Economic models for resource management and scheduling in grid computing. Journal of Concurrency and Computation: Practice and Experience 14(13–15), 1507–1542 (2002)
2. Snir, M., Otto, S., Huss-Lederman, S., Walker, D., Dongarra, J.: MPI: The Complete Reference – The MPI Core, vol. 1. The MIT Press, Cambridge, MA (1998)
3. Khokhar, A., Prasanna, V.K., Shaaban, M., Wang, C.L.: Heterogeneous computing: Challenges and opportunities. IEEE Computer 26(6), 18–27 (1993)
4. Foster, I.: Globus toolkit version 4: Software for service–oriented systems. In: Jin, H., Reed, D., Jiang, W. (eds.) NPC 2005. LNCS, vol. 3779, pp. 2–13. Springer, Heidelberg (2005)
5. Mateescu, G.: Quality of service on the grid via metascheduling with resource co-scheduling and co-reservation. International Journal of High Performance Computing Applications 17(3), 209–218 (2003)
6. Fernandez-Baca, D.: Allocating modules to processors in a distributed system. IEEE Transaction on Software Engineering 15(11), 1427–1436 (1989)
7. Wang, L., Siegel, J.S., Roychowdhury, V.P., Maciejewski, A.A.: Task matching and scheduling in heterogeneous computing environments using a genetic–algorithm–based approach. Journal of Parallel and Distributed Computing 47, 8–22 (1997)
8. Kwok, Y.K., Ahmad, I.: Efficient scheduling of arbitrary task graphs to multiprocessors using a parallel genetic algorithm. Journal of Parallel and Distributed Computing 47(1), 58–77 (1997)
9. Braun, T.D., Siegel, H.J., Beck, N., Bölöni, L.L., Maheswaran, M., Reuther, A.I., Robertson, J.P., Theys, M.D., Yao, B.: A comparison of eleven static heuristics for mapping a class of independent tasks onto heterogeneous distributed computing systems. Journal of Parallel and Distributed Computing 61, 810–837 (2001)
10. Kim, S., Weissman, J.B.: A genetic algorithm based approach for scheduling decomposable data grid applications. In: International Conference on Parallel Processing (ICPP 2004), Montreal, Quebec, Canada, pp. 406–413 (2004)
11. Song, S., Kwok, Y.K., Hwang, K.: Security–driven heuristics and a fast genetic algorithm for trusted grid job scheduling. In: IPDP 2005, Denver, Colorado (2005)
12. Price, K., Storn, R.: Differential evolution. Dr. Dobb's Journal 22(4), 18–24 (1997)
13. Fonseca, C.M., Fleming, P.J.: An overview of evolutionary algorithms in multiobjective optimization. Evolutionary Computation 3(1), 1–16 (1995)
14. Dong, F., Akl, S.G.: Scheduling algorithms for grid computing: State of the art and open problems. Technical Report 2006–504, School of Computing, Queen's University Kingston, Ontario, Canada (2006)
15. Fitzgerald, S., Foster, I., Kesselman, C., von Laszewski, G., Smith, W., Tuecke, S.: A directory service for configuring high-performance distributed computations. In: Sixth Symp. on High Performance Distributed Computing, Portland, OR, USA, pp. 365–375. IEEE Computer Society, Los Alamitos (1997)
16. Czajkowski, K., Fitzgerald, S., Foster, I., Kesselman, C.: Grid information services for distributed resource sharing. In: Tenth Symp. on High Performance Distributed Computing, pp. 181–194. IEEE Computer Society, San Francisco, CA, USA (2001)
17. Wolski, R., Spring, N., Hayes, J.: The network weather service: a distributed resource performance forecasting service for metacomputing. Future Generation Computer Systems 15(5–6), 757–768 (1999)
18. Gong, L., Sun, X.H., Waston, E.: Performance modeling and prediction of non–dedicated network computing. IEEE Trans. on Computer 51(9), 1041–1055 (2002)

Latency in Grid over Optical Burst Switching with Heterogeneous Traffic

Yuhua Chen[1], Wenjing Tang[2], and Pramode K. Verma[3]

[1] Department of Electrical and Computer Engineering, University of Houston, Houston, TX, USA
yuhua.chen@mail.uh.edu
[2] Optronic Systems, Inc., Houston, TX, USA
Wenjing.Tang@optronicsys.com
[3] Electrical and Computer Engineering, Telecommunications Systems Program, University of Oklahoma-Tulsa, USA
pverma@ou.edu

Abstract. Optical burst switching (OBS) has been proposed as the next generation optical network for grid computing. In this paper, we envision a heterogeneous Grid served by an Optical Burst Switching framework, where grid traffic co-exists with IP and/or a 10 GE based traffic to achieve economy of scale. This paper addresses the latency that Grid jobs experience in OBS networks. The injection of jumbo size grid jobs can potentially affect the latency experienced by IP/10GE traffic. Simulation results have shown that in Grids served by an optical burst switch, grid jobs consistently have lower latency than co-existing IP/10GE traffic, with a slightly elevated latency of IP/10GE traffic when the size of grid jobs increases. We conclude that given the fact that OBS can efficiently handle the enormous amount of bandwidth made available by DWDM technology, Grid over Optical Burst Switching is a cost effective way to provide grid services, even for latency sensitive grid computing applications.

1 Introduction

Grid computing [1][2] is a potential approach to making the rapidly growing world wide information resources available when and where they are needed, without the need to physically duplicate those resources. Information resources can be characterized as processing and storage resources. The vehicle that makes these resources available remotely is, of course, the communication facility. Design issues associated with communication facility are critically important because the facility must meet the requirements of the specific application; for example, in terms of the needed bandwidth, while at the same time meeting the latency and availability requirements within the price parameters of the intended application.

The global telecommunication infrastructure is rapidly evolving to make grid computing a reality. The *Dense Wavelength Division Multiplexing* (DWDM) technology provides enormous bandwidth at declining transmission costs making it ubiquitous for long haul transmission. Optical switching techniques, especially, *optical wavelength switching* or *lambda switching* techniques are emerging for large

R. Perrott et al. (Eds.): HPCC 2007, LNCS 4782, pp. 334–345, 2007.

bandwidth switching or cross connect applications. Because of the enormous bandwidth (10 gigabits per second or higher) associated with a wavelength, lambda switching by itself is not suitable for most end-user applications. The two complementary switching techniques — *optical burst switching* (OBS) [3][4] and *optical packet switching* — will fill in the needs of lower bandwidth granularity applications. Of these two, optical burst switching, has reached a degree of maturity and is the target technology for complementing lambda switching in the near term.

Optical burst switching has been proposed as the next generation optical network for grid computing [5][6][7]. However, in order to achieve economy of scale, grid jobs will co-exist with IP and/or 10GE (*Gigabit Ethernet*) traffic in the OBS networks. It is important to understand the latency that grid jobs experience in the OBS networks since grid jobs are usually associated with a job completion time, especially for latency sensitive grid computing applications. In addition, the injection of jumbo size grid jobs can potentially affect the latency experienced by IP/10GE traffic.

The rest of the paper is organized as follows. Section 2 provides the background of OBS networks. We present the grid over optical burst switching network architecture in Section 3. In Section 4, the latency of grid jobs and how grid jobs affect the latency of co-existing IP/10GE traffic are discussed in detail. We conclude our work in Section 5.

2 Optical Burst Switching (OBS) Background

We first give a brief introduction to optical burst switching (OBS). Figure 1 illustrates the basic concept for an optical burst switching network. The network consists of a set of OBS routers connected by DWDM links. The transmission links in the system carry tens or hundreds of DWDM channels, any one of which can be dynamically assigned to a user data burst. One (or possibly more than one) channel on each link is used as a control channel to control the dynamic assignment of the remaining channels to data bursts.

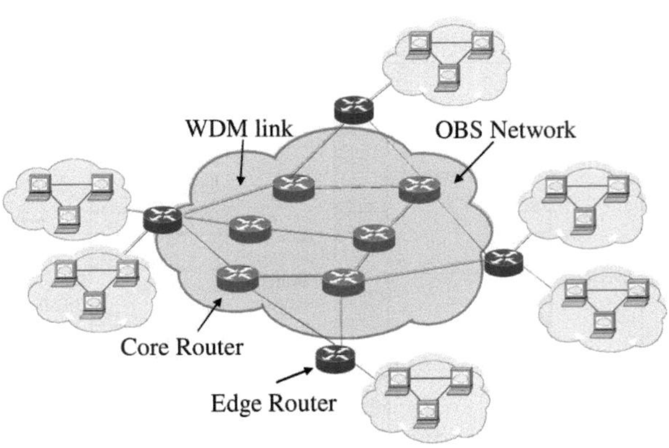

Fig. 1. OBS router architecture

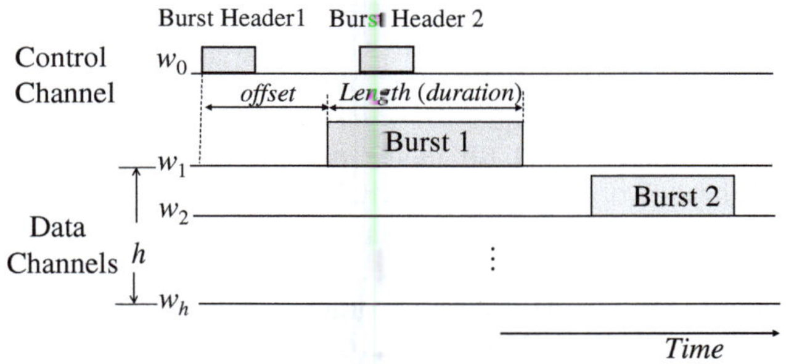

Fig. 2. Bursts and burst headers

An OBS network works as follows. Shortly before the transmission of a data burst on a *data channel*, a *burst header* is sent on the *control channel*, specifying the channel on which the burst is being transmitted and the destination of the burst. The burst header also carries an *offset* field and a *length* field. The offset field defines the time between the transmission of the burst header and the data burst. The length field specifies the time duration of the burst on a DWDM channel. The offset and the length fields are used by the OBS routers to schedule the setup and release of optical data paths *on-the-fly*. Figure 2 shows an example of burst headers sharing the same control channel, while the corresponding data bursts are sent on separate data channels.

An OBS core router, on receiving a burst header, selects an idle channel on the outgoing link leading toward the desired destination. Shortly before the arrival of the data burst, the OBS router establishes a lightpath between the incoming channel that the burst is arriving on and the outgoing channel selected to carry the burst. The data burst can stay in optical domain and flow through the OBS router to the proper outgoing channel. The OBS router forwards the burst header on the control channel of the outgoing link, after modifying the channel field to specify the selected outgoing channel. This process is repeated at every OBS router along the path to the destination.

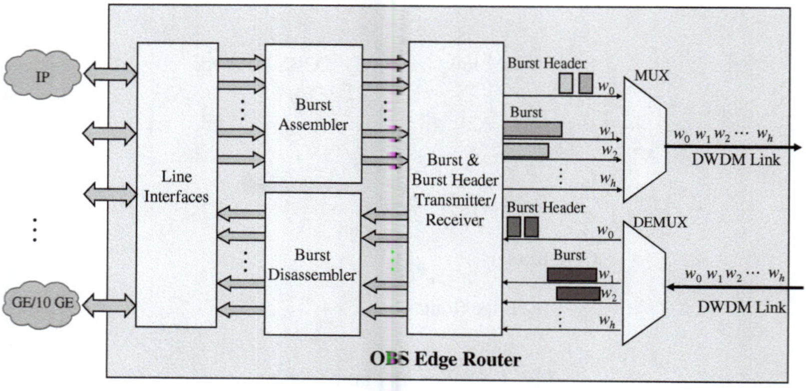

Fig. 3. OBS edge router architecture

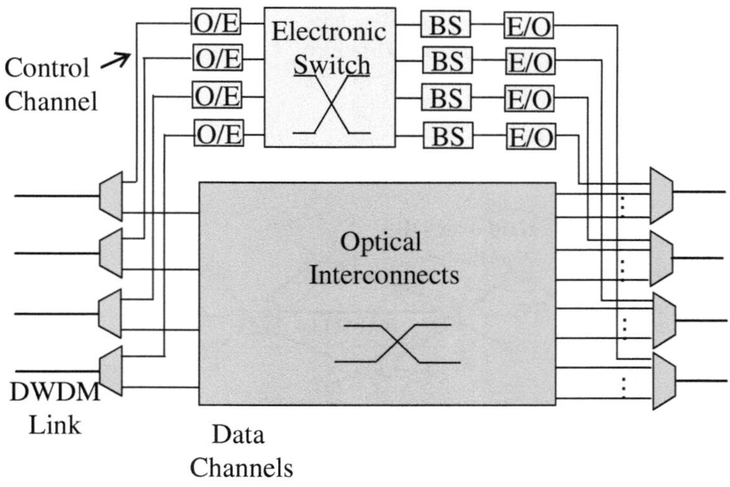

Fig. 4. OBS core router architecture

Figure 3 illustrates the architecture of an OBS edge router. In the ingress direction, packets received on different line interfaces such as IP and *Gigabit Ethernet* (GE)/10 GE are sent to the *Burst Assembler.* The burst assembler classifies the data according to their destinations and QoS levels, and assembles data into different bursts. Once a burst is formed, the burst assembler generates a burst header, which is transmitted on the control channel. After holding the burst for an offset time, the burst assembler releases the data burst to be transmitted on one of the data channels. The control channel and the data channels are combined onto the outgoing DWDM link using a passive optical *multiplexer* (MUX). The outgoing DWDM link is connected to the OBS core router. In the egress direction, the wavelengths on the incoming DWDM link are separated using an optical *demultiplexer* (DEMUX). The burst headers received on the control channel and the data bursts received on data channels are forwarded to the *Burst Disassembler.* The burst disassembler converts bursts back to packets and forwards them to the appropriate line interfaces.

Figure 4 shows the key components of an OBS core router. The architecture consists of two parts, an optical datapath and an electrical control path. The datapath has optical interconnects with/without wavelength conversion capability. The control path includes O/E/O conversion, an electronic switch and a set of *Burst Schedulers* (BSs). Each BS is responsible for making DWDM channel scheduling decisions for a single outgoing link. The electronic switch routes the burst headers received on the control channels of the incoming DWDM links to the corresponding BS according to the destination of the data burst. The BS selects an outgoing channel for the burst and configures the optical switching matrix such that bursts arriving on incoming data channels can pass through to the desired outgoing channels directly without buffering.

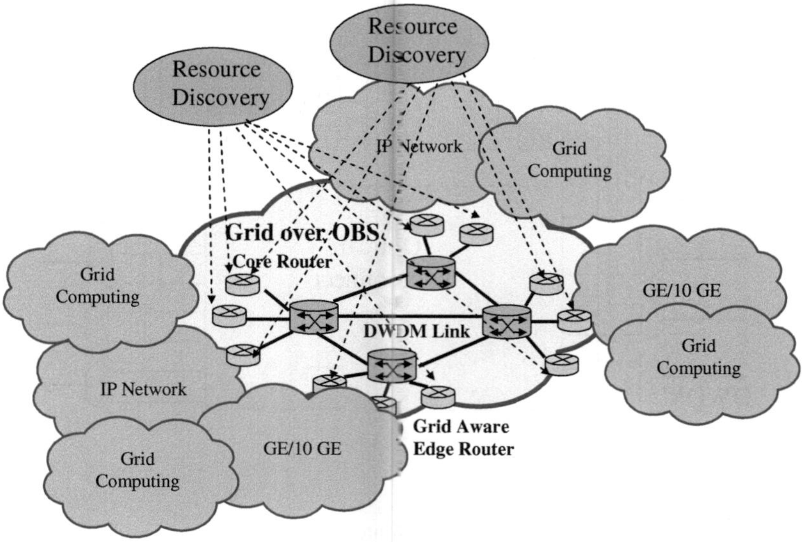

Fig. 5. Grid over optical burst switching architecture

3 Grid over Optical Burst Switching Network Architecture

In this section, we describe a decentralized *Grid over Optical Burst Switching* architecture that consists of grid aware edge routers and transparent core routers as shown in Fig. 5. In the decentralized architecture, grid resources can be connected to the *grid aware edge router* directly, or they can be connected through intermediate networks such as IP network or GE/10GE. Distributed grid resource discovery is achieved as an overlay that can potentially affect the states in all or a subset of the edge routers.

The major rational behind the proposed architecture as oppose to the intelligent OBS router described in [3] is scalability. With the grid computing shifting from scientific computing oriented grids to consumer oriented grids, the grid resources will be owned and managed by many entities, each with their own policy and privilege. If grid resource management is implemented as part of the OBS core router functions, the resource discovery and management can easily become the performance bottleneck. In addition, it is relatively cheap to add grid resources such as hundreds of Terabits of storage, or a server bank, than to gain access to the precious control resources in the OBS network.

The general edge router model shown in Fig. 3 still applies to the grid aware edge router, except that the capability of the *Burst Assembler* block is different, which we call the *Grid Aware Burst Assembler*. Figure 6 shows the functional blocks inside the Grid Aware Burst Assembler. The inputs to the burst assembler are from the line card interfaces.

The *Traffic Classifier* block separates traditional IP/10GE traffic, grid jobs and grid management packets. The traditional IP/10GE traffic is forwarded to the *Router*

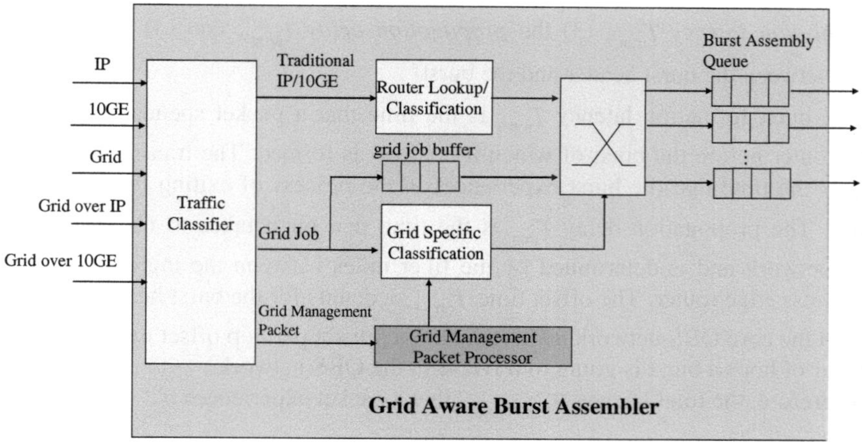

Fig. 6. Grid aware burst assembler

Lookup/Classification block. Grid jobs which arrive in jumbo size, whether or not encapsulated in IP/10GE packets, are sent to the grid job buffer while the grid job control information is processed by the *Grid Specific Classification* Block. The *grid management packets* are directed to the *Grid Management Packet Processor*, which in turn configures the states in the Grid Specific Classification block when necessary.

When the traditional IP/10GE packets determine the destination edge router and possibly the *Quality-of-Service* (QoS) level, they are routed to proper burst assembly queues. The grid specific classification block will determine the destination of the grid job and forwards the jumbo size grid job to the burst assembly queue. When the conditions to form a burst (maximum burst length or maximum timeout) are met, the burst is sent to the next block for burst header generation.

As we can see, grid jobs will merge with traditional IP/10GE traffic into the burst assembly queue. For simplification purpose, from this point on, we use IP traffic with the understanding that IP and 10GE belong to the same category. In the next section, we will investigate the latency performance of the Grid over OBS model under heterogeneous traffic conditions.

4 Latency Analysis with Heterogeneous Traffic

When grid jobs co-exist with traditional IP traffic, the jumbo size grid job can potentially affect the latency of the IP traffic. It is also important to understand the latency characteristics of grid jobs when grid jobs are transported over OBS networks because a time constraint is usually placed on the grid job, especially for latency sensitive grid computing applications.

The latency that packets (both IP packets and grid job jumbo packets) experience in OBS networks is determined by (1) the *burst formation latency* T_{form}; (2) the *transmission latency* T_{tran}; (3) the *propagation delay* T_{prop}; and (4) the *offset* time T_{offset} between the burst header and the burst.

The burst formation latency T_{form} is the time that a packet spends at the ingress edge router before the burst of which it is a part is formed. The transmission latency T_{tran} is the time that the burst experiences in the process of exiting the ingress edge router. The propogation delay T_{tran} is the time that burst takes to traverse the core OBS network and is determined by the fiber miles between the ingress edge router and egress edge router. The offset time T_{offset} accounts for the burst header processing time in the core OBS network and is equal to a pre-set per hop offset multiplied by the number of hops a burst is going to traverse in the OBS network.

Therefore, the total latency $T_{\text{total\_latency}}$ that a packet experiences is

$$T_{\text{total\_latency}} = T_{\text{form}} + T_{\text{tran}} + T_{\text{offset}} + T_{\text{prop}} \,. \tag{1}$$

In this paper, we focus on the factors that can potentially affected by the injection of jumbo size grid jobs. Since the propagation delay T_{prop} and the offset time T_{offset} can be calculated for both IP packets and grid job jumbo packets once the destination edge router is determined, we do not include these two terms in the following discussion. For the rest of the section, we investigate the latency that a packet (IP or grid job) experiences during the burst formation time and the transmission latency that a packet experiences after the burst is launched.

We formulate the problem as follows.

Assume a burst contains m IP packets and n grid jobs before it is launched. Denote the i-th IP packet in the burst p_{ip}^{i} with arrival time t_{ip}^{i} and length $l_{\text{ip}}^{i}, i = 1, \cdots, m$. Denote the j-th grid job jumbo packet in the burst p_{grid}^{j} with arrival time t_{grid}^{j} and length l_{grid}^{j}, $j = 1, \cdots, n$. Without loss of generality, we have

$$t_{\text{ip}}^{1} < t_{\text{ip}}^{2} < \cdots < t_{\text{ip}}^{i} < \cdots < t_{\text{ip}}^{m} < t_{\text{ip}}^{m+1} \,, \; i = 1, \cdots, m \,, \tag{2}$$

and

$$t_{\text{grid}}^{1} < t_{\text{grid}}^{2} < \cdots < t_{\text{grid}}^{j} < \cdots < t_{\text{grid}}^{n} < t_{\text{grid}}^{n+1} \,, \; j = 1, \cdots, n \,. \tag{3}$$

At an OBS edge router, a burst is formed when the maximum burst length $L_{\max}$ is reached, or the maximum timeout $T_{\max}$ is reached [8]. Therefore, based on the above definition, a burst is formed under either of the following conditions:

$$\sum_{i=1}^{m} l_{\text{ip}}^{i} + \sum_{j=1}^{n} l_{\text{grid}}^{j} \ge L_{\max} \,, \tag{4}$$

or

$$\begin{cases} \max(t_{ip}^{m}, t_{grid}^{n}) - \min(t_{ip}^{1}, t_{grid}^{1}) \le T_{max} \\ \min(t_{ip}^{m+1}, t_{grid}^{n+1}) - \min(t_{ip}^{1}, t_{grid}^{1}) > T_{max} \end{cases} . \tag{5}$$

For an IP packet p_{ip}^{i}, the burst formation latency it experiences is

$$T_{form\_ip}^{i} = \max(t_{ip}^{m}, t_{grid}^{n}, T_{max} + \min(t_{ip}^{1}, t_{grid}^{1})) - t_{ip}^{i} . \tag{6}$$

Similarly, the burst formation latency that a grid job p_{grid}^{j} experiences is

$$T_{form\_grid}^{j} = \max(t_{ip}^{m}, t_{grid}^{n}, T_{max} + \min(t_{ip}^{1}, t_{grid}^{1})) - t_{grid}^{j} . \tag{7}$$

Once a burst is launched, a packet (either IP or grid) has to wait for all packets (both IP and grid) in the burst that arrive earlier to be transmitted on the channel.

For an IP packet p_{ip}^{i} that satisfies the following condition

$$t_{grid}^{k-1} < t_{ip}^{i} < t_{grid}^{k} , \tag{8}$$

the transmission latency $T_{tran\_ip}^{i}$ is

$$T_{tran\_ip}^{i} = \sum_{s=0}^{i-1} l_{ip}^{s} + \sum_{t=0}^{k-1} l_{grid}^{t} . \tag{9}$$

Similarly, for a grid job p_{grid}^{j} that satisfies the following condition:

$$t_{ip}^{l-1} < t_{grid}^{j} < t_{ip}^{l} , \tag{10}$$

the transmission latency $T_{tran\_grid}^{j}$ is

$$T_{tran\_grid}^{j} = \sum_{s=0}^{l-1} l_{ip}^{s} + \sum_{t=0}^{j-1} l_{grid}^{t} . \tag{11}$$

The latency $T_{latency\_ip}^{i}$ that an IP packet p_{ip}^{i} experiences is

$$T_{latency\_ip}^{i} = T_{form\_ip}^{i} + T_{tran\_ip}^{i} , \tag{12}$$

and the latency $T^{j}_{\text{latency\_grid}}$ that a grid job p^{j}_{grid} experiences is

$$T^{j}_{\text{latency\_grid}} = T^{j}_{\text{form\_grid}} + T^{j}_{\text{tran\_grid}} \,. \tag{13}$$

The average latency $T_{\text{avg\_latency\_ip}}$ of the IP packets in the burst is

$$T_{\text{avg\_latency\_ip}} = \frac{1}{m}\sum_{i=1}^{m} T^{i}_{\text{latency\_ip}} \,. \tag{14}$$

The average latency $T_{\text{avg\_latency\_grid}}$ of the grid jobs in the burst is

$$T_{\text{avg\_latency\_grid}} = \frac{1}{n}\sum_{j=1}^{n} T^{j}_{\text{latency\_grid}} \,. \tag{15}$$

Based on the above model, we have conducted simulations to study the effect of grid jobs on the latency in grid over optical burst switching under heterogeneous traffic conditions.

Figure 7 illustrates the variation in latency as a function of the grid job load. The grid job load is expressed as a fraction of occupancy of the trunk on which the burst is sent out. The IP packet load is similarly expressed as a fraction of the trunk occupancy. IP packets are assumed to be exponentially distributed with a mean length of 1.5 kBytes. The mean grid job size is 1.5 Mbytes with exponential distribution. As noted earlier, bursts can be formed either due to the maximum timeout interval or due to the maximum allowable burst length. The reduction in latency as the load increases is somewhat anti-intuitive. It is, however, explained by the fact that as the load increase, the maximum allowable burst is quicker to form, thus expediting its exit. The correlation in the latency reduction for both the IP packet and grid job is similarly explained. The average latency is the weighted latency of the two kinds of traffic considered, namely, the packet traffic and the grid job traffic.

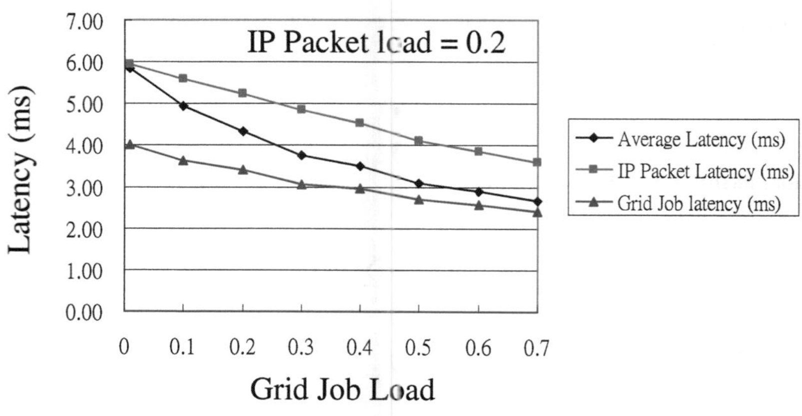

Fig. 7. Latency with fixed IP packet traffic load

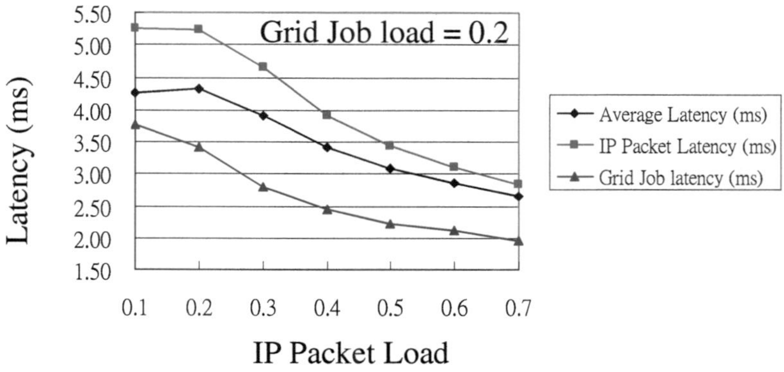

Fig. 8. Latency with fixed grid job traffic load

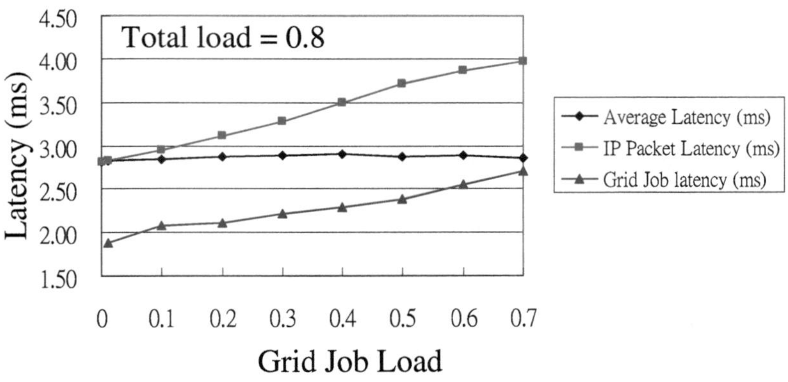

Fig. 9. Latency with fixed total traffic load

Figure 8 presents latency as a function of the grid job load under the condition that the packet load is varied to keep the overall load constant at 0.8. The scenario would imply that as the grid job load increases (as shown on the X-axis), the packet load would correspondingly reduce to keep the overall load constant. The figure clearly indicates the relatively low sensitivity of the average latency to the mixture of traffic *as long as the total traffic is kept constant.*

Figure 9 shows the variation of latency to the IP packet load, when the grid job load is kept constant at 20% of the trunk capacity. As in Fig. 7, the latency falls as the load increases because a higher level of load implies a quicker exit of the burst because of the maximum allowable burst size.

Figure 10 illustrates the impact of the grid job size on latency *when both the packet load and the grid job load are kept constant at occupancies of 0.6 and 0.2 respectively.* As shown in Fig. 8, latency is largely insensitive to the grid job size as long as the overall load is kept constant. However, the packet load latency does increase because

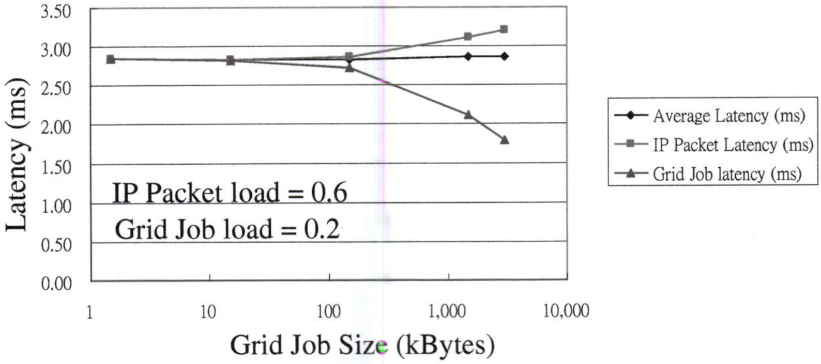

Fig. 10. Latency with varying grid job sizes

packets are crowded out by longer grid jobs. The grid jobs experience lower latency because larger grid jobs utilize the bandwidth more efficiently.

5 Conclusion

In this paper, we have envisioned that grid jobs co-exist with IP/10GE traffic in optical burst switching networks in order to achieve economy of scale. We have investigated the latency of grid jobs when transported over optical burst switching networks. We have also studied the impact of jumbo size grid jobs on the co-existing IP/10GE traffic. Simulation results have shown that in a Grid over OBS, grid jobs consistently have lower latency than co-existing IP/10GE traffic, with a slightly elevated latency for IP/10GE traffic when the size of grid jobs increases. We conclude that given the fact that OBS can efficiently handle an enormous amount of bandwidth made available by DWDM technology, Grid over Optical Burst Switching is a cost effective way to provide grid services, even for latency sensitive grid computing applications.

References

1. Fox, G., Hey, A.J.G., Hey, T., Berman, F.: Grid Computing: Making the Global Infrastructure a Reality. John Wiley & Sons, Chichester (2003)
2. Joshy, J., Craig, F.: Grid Computing. Prentice Hall Ptr (2004)
3. Turner, J.S.: Terabit burst switching. Journal of High Speed Networks 8(1), 3–16 (1999)
4. Qiao, C., Yoo, M.: Optical Burst Switching (OBS) - A new paradigm for an optical internet. Journal of High Speed Networks 8, 69–84 (1999)
5. Leenheer, M.D., Thysebaert, P., Volckaert, B., Turck, F.D., Dhoedt, B., Demeester, P., Simeonidou, D., Nejabati, R., Zervas, G., Klonidis, D., O'Mahony, M.J.: A View On Enabling-Consumer Oriented Grids Through Optical Burst Switching. IEEE Communications Magazine , 124–131 (2006)

6. Simeonidou, D., Nejabati, R., Zervas, G., Klonidis, D., Tzanakaki, A., O'Mahony, M.J.: Dynamic Optical-Network Architectures and Technologies for Existing and Emerging Grid Services. Journal of Lightwave Technology 23(10), 3347–3357 (2005)
7. Simeonidou, D., Nejabati, R.: Grid Optical Burst Switched Networks (GOBS). Global Grid Forum Draft (2005)
8. Xiong, Y., Vandenhoute, M., Cankaya, H.C.: Control architecture in optical burst-switched WDM networks. IEEE Journal on Selected Areas in Communication 18, 1838–1851 (2000)

A Block JRS Algorithm for Highly Parallel Computation of SVDs

Mostafa I. Soliman[1], Sanguthevar Rajasekaran[2], and Reda Ammar[2]

[1] Computer & System Section, Electrical Engineering Department, South Valley University, Aswan, Egypt
soliman@mailer.eun.eg
[2] Department of Computer Science and Engineering, University of Connecticut, Storrs, USA
{rajasek,reda}@engr.uconn.edu

Abstract. This paper presents a new algorithm for computing the singular value decomposition (SVD) on multilevel memory hierarchy architectures. This algorithm is based on one-sided JRS iteration, which enables the computation of all Jacobi rotations of a sweep in parallel. One key point of our proposed block JRS algorithm is reusing the loaded data into cache memory by performing computations on matrix blocks (b rows) instead of on strips of vectors as in JRS iteration algorithms. Another key point is that on a reasonably large number of processors the number of sweeps is less than that of one-sided JRS iteration algorithm and closer to the cyclic Jacobi method even though not all rotations in a block are independent. The relaxation technique helps to calculate and apply all independent rotations per block at the same time. On blocks of size $b \times n$, the block JRS performs $O(b^2n)$ floating-point operations on $O(bn)$ elements, which reuses the loaded data in cache memory by a factor of b. Besides, on P parallel processors, $(2P-1)$ steps based on block computations are needed per sweep.

1 Introduction

Singular value decomposition (SVD) is an extremely powerful and useful tool in Linear Algebra. There are many applications of the SVD in scientific computing and digital signal processing. See [1] for some applications. SVD problem is a very computationally intensive problem that needs to exploit the growing availability of parallel hardware. Thus, many researchers have worked on designing efficient techniques to compute SVDs in parallel to reduce the execution time especially for real time applications. This paper presents our proposed algorithm block JRS, which is based on the one-sided JRS iteration algorithm proposed in [2]. The JRS iteration algorithms enable the computation of all Jacobi rotations of a sweep in parallel. Each sweep of this algorithm consists of several rotations and in each rotation, the value of an off-diagonal element is decreased to a fraction of its current value (instead of to zero).

The well-known sequential bidiagonalization-based Golub-Kahan-Reinsch SVD algorithm [3] takes $O(mn^2)$ time on an $n \times n$ matrix. For large matrices the execution

R. Perrott et al. (Eds.): HPCC 2007, LNCS 4782, pp. 346–357, 2007.

time may be unacceptable. Thus, it is essential to develop efficient parallel algorithms. The bidiagonalization-based SVD algorithm has been found to be difficult to parallelize and hence most works on parallel SVD focus on Jacobi-based techniques. Both two-sided Jacobi and one-sided Jacobi techniques have been studied in this context. Brent and Luk [4] presented a parallel one-sided SVD algorithm using a linear array of n processors, with a run time of $O(mnS)$, where S is the number of sweeps. They also presented an $O(nS)$ time algorithm to compute the singular values of a symmetric matrix using an array of n^2 processors. Zhou and Brent [5] described an efficient parallel ring ordering algorithm for one-sided Jacobi.

Becka and Vajtersic [6] presented a parallel block two-sided Jacobi algorithm on hypercubes and rings with a run time of $O(n^2S)$. They also gave an $O(nS)$ time algorithm on meshes with n^2 processors [7]. Becka et al. [8] proposed a dynamic ordering algorithm for a parallel two-sided block-Jacobi SVD with a run time of $O(n^2S)$. Oksa and Vajtersic [9] designed a systolic two-sided block-Jacobi algorithm with a run time of $O(nS)$. Strumpen et al. [10] presented a stream algorithm for one-sided Jacobi that has a run time of $O(n^3S/p^2)$, where p is the number of processors (p being $O(n^{1/2})$). They created parallelism by computing multiple Jacobi rotations independently and applying all the transformations thereafter. Rajasekaran *et al.* [2] employed the idea of separating rotation computations and transformations. They used a novel scheme to reduce the number of sweeps.

We propose a new algorithm for computing SVDs based on the relaxation technique proposed by Rajasekaran *et al.* [2]. This algorithm is fundamentally different from all the algorithms that have been proposed for SVD. The proposed block JRS algorithm employs the same idea of decreasing (relaxing) the off-diagonal elements proposed by Rajasekaran *et al.* [2]. However, it differs from the JRS algorithms in the partitioning strategy of the input matrix among processors. Besides, it is a highly parallel algorithm designed to exploit the memory hierarchy by processing blocks to reuse the loaded data into cache memories. However, JRS iteration algorithms are working on strips of vectors. On the block JRS, the input matrix $A_{m\times n}$ is divided into blocks of size $b\times n$, where $\lfloor m/2P \rfloor \leq b \leq \lceil m/2P \rceil$ and P is the number of available processors. This means that not all block sizes are necessarily the same, which is more flexible. One of the key points of the block JRS is reusing the loaded data into cache memory by performing the computations on blocks (b rows) instead on strips of vectors. To be specific, it performs $O(b^2n)$ floating-point operations on $O(bn)$ elements, which reuses the loaded data in cache memory by a factor of b. Another key point is the number of sweeps is less than one-sided JRS iteration algorithm and closer to the cyclic Jacobi method even though not all rotations in a block are independent. The relaxation technique helps to calculate and apply multiple rotations per block at the same time.

This paper is organized as follows. The following section reviews the singular value decomposition. Two-sided and one-sided Jacobi as well as one-sided JRS algorithms are explained. Section 3 describes the proposed block JRS iteration algorithms. Our results are shown in Section 4. Finally, Section 5 concludes our paper.

2 Singular Value Decomposition

A singular value decomposition of a real matrix $A_{m\times n}$ is defined as the computation of three matrices $U_{m\times m}$, $\Sigma_{m\times n}$, and $V_{n\times n}$ such that:

$$A_{m\times n} = U_{m\times m}\,\Sigma_{m\times n}\,V^{T}{}_{n\times n}, \tag{1}$$

where $U_{m\times m}$ and $V_{n\times n}$ are orthogonal matrices (i.e. $U^TU = I_m$ and $V^TV = I_n$), and $\Sigma_{m\times n}$ is a diagonal matrix $diag(\sigma_0, \sigma_1, \sigma_2, \ldots\ldots, \sigma_{n-1})$ on top of $(m-n)$ rows of zeros, assuming that $m \geq n$. The σ_i are the singular values of $A_{m\times n}$. Matrix $U_{m\times m}$ contains n left singular vectors, and matrix $V_{n\times n}$ consists of n right singular vectors. The singular values and singular (column) vectors of $U_{m\times m}$ and $V_{n\times n}$ form the relations

$$Av_i = \sigma_i\,u_i \quad \text{and} \quad A^Tu_i = \sigma_i\,v_i. \tag{2}$$

All the existing parallel algorithms use the Jacobi iteration as the basis. The Jacobi iteration algorithm attempts to diagonalize the input matrix $A_{m\times n}$ by a series of Jacobi rotations where each rotation tries to zero-out an off-diagonal element. It needs to perform $n(n-1)/2$ rotations (in the case of a symmetric $n\times n$ matrix) attempting to zero-out all the off-diagonal elements. These $n(n-1)/2$ transformations constitute a sweep. It can be shown that after each sweep the norm of the off-diagonal elements decreases and hence the algorithm converges. It is believed that the number S of sweeps needed for convergence of the sequential Jacobi iteration algorithm is $O(\log n)$ [3]. There are two variants of Jacobi based algorithm, namely, one-sided and two sided.

2.1 Two-Sided Jacobi SVD

The two-sided Jacobi iteration algorithm transforms a symmetric matrix $A_{n\times n}$ into a diagonal matrix $\Sigma_{n\times n}$ by a sequence of Jacobi rotations (J), where each transform attempts to zero-out a given off-diagonal element of $A_{n\times n}$.

$$\Sigma_{n\times n} = (J_n^T \cdots (J_3^T\,(J_2^T\,(J_1^T\,AJ_1)J_2)J_3) \cdots J_n^T) = (J_1J_2J_3 \cdots J_n)^T A (J_1J_2J_3 \cdots J_n) \tag{3}$$

The Jacobi rotation $J(i, j, \theta)$ for an index pair (i, j) and a rotation angle θ is a square matrix that is equal to the identity matrix I plus four additional entries at the intersections of rows and columns i and j:

$$J(i,j,\theta) = \begin{pmatrix} 1 & \cdots & 0 & \cdots & 0 & \cdots & 0 \\ \vdots & \ddots & \vdots & & \vdots & & \vdots \\ 0 & \cdots & c & \cdots & s & \cdots & 0 \\ \vdots & & \vdots & \ddots & \vdots & & \vdots \\ 0 & \cdots & -s & \cdots & c & \cdots & 0 \\ \vdots & & \vdots & & \vdots & \ddots & \vdots \\ 0 & \cdots & 0 & \cdots & 0 & \cdots & 1 \end{pmatrix} \begin{matrix} \\ \\ i \\ \\ j \\ \\ \\ \end{matrix}, \qquad \begin{matrix} \quad & & i & & j & & \end{matrix} \tag{4}$$

where $c = \cos(\theta)$ and $s = \sin(\theta)$. It is clear that the Jacobi rotation is an orthogonal matrix (i.e. $J(i, j, \theta)^T J(i, j, \theta) = I$) from the fact of $\cos^2(\theta) + \sin^2(\theta) = 1$. The c and s are computed as follows. Consider one of the transformations: $B = J^T AJ$, c and s are chosen such that the resultant matrix B is diagonal, *i.e.*, $b_{ij} = b_{ji} = 0$.

$$\begin{pmatrix} b_{ii} & b_{ij} \\ b_{ji} & b_{jj} \end{pmatrix} = \begin{pmatrix} c & s \\ -s & c \end{pmatrix}^T \begin{pmatrix} a_{ii} & a_{ij} \\ a_{ji} & a_{jj} \end{pmatrix} \begin{pmatrix} c & s \\ -s & c \end{pmatrix} \tag{5}$$

By solving this equation and taking the smaller root (see [3] for more detail), c and s are obtained by:

$$c = \frac{1}{\sqrt{1+t^2}} \quad \text{and} \quad s = t*c, \tag{6}$$

where

$$t = \frac{sign(\tau)}{|\tau| + \sqrt{\tau^2 + 1}} \quad \text{and} \quad \tau = \frac{a_{ii} - a_{jj}}{2a_{ij}} \tag{7}$$

Depending on the order of choosing the element to be zeroed, there are classic Jacobi and cyclic Jacobi algorithms. In the classic Jacobi iteration algorithm, each transformation chooses the off-diagonal element of the largest absolute value. However, searching for this element requires expensive computations. Cyclic Jacobi algorithm sacrifices the convergence behavior and steps through all the off-diagonal elements in a row-by-row fashion. For example, if $n = 3$, the sequence of elements is (1, 2), (1, 3), (2, 3), (1, 2), … . The computation is organized in sweeps such that in each sweep every off-diagonal element is zeroed once. Note that when an off-diagonal element is zeroed it may not continue to be zero when another off-diagonal element is zeroed. After each sweep, it can be shown that, the norm of the off-diagonal elements decreases monotonically. Thus the Jacobi algorithms converge.

2.2 One-Sided Jacobi SVD

The two-sided Jacobi algorithms are computationally more expensive than the one-sided algorithms. Moreover, to implement the two-sided Jacobi method, it needs to traverse both row and column of the given matrix; however, matrices are stored either in row-major or column-major format. Thus, one of the two traversals will be less efficient on conventional memory architectures. In other words, one of the two traversals accesses the elements of the input $m \times n$ matrix with unit stride, which is efficient; however, the other traversal performs stride n accesses, which is expensive because of cache miss handling time. On the other hand, the one-sided rotation modifies rows only, which is more suitable for memory hierarchy architectures. Thus to achieve efficient parallel SVD computation the best approach may be to adapt the Hestenes one-sided Jacobi transformation method [11] as advocated in [4, 12].

Hestenes one-sided Jacobi algorithm first produces a matrix $B_{m \times n}$ whose rows are orthogonal by premultiplying $A_{m \times n}$ with an orthogonal matrix $U_{m \times m}$:

$$U_{m\times m} A_{m\times n} = B_{m\times n}, \tag{8}$$

where rows of $B_{m\times n}$ satisfy $b_i^T b_j = 0$ for $i \neq j$. Followed by this $B_{m\times n}$ is normalized by:

$$V_{n\times n} = S^{-1} B_{m\times n}, \tag{9}$$

where $S_{n\times n} = diag(s_1, s_2, \ldots, s_n)$, and $s_i = b_i^T b_i$. It can be easily shown that $A_{m\times n} = U^T{}_{m\times m} S_{m\times n} V_{n\times n}$, which is equivalent to the definition of SVD.

One-sided Jacobi is also realized by a series of Jacobi rotations, but on one side. For a given i and j, rows i and j are orthogonalized by $B_{m\times n} = J^T A_{m\times n}$ where $J = J(i, j, \theta)$ is the same matrix as in the two-sided Jacobi (see equation 4) and:

$$\begin{pmatrix} b_i^T \\ b_j^T \end{pmatrix} = \begin{pmatrix} c & s \\ -s & c \end{pmatrix}^T \begin{pmatrix} a_i^T \\ a_j^T \end{pmatrix}, \tag{10}$$

here c and s are chosen such that $b_i^T b_j = 0$. The solution of them is:

$$c = \frac{1}{\sqrt{1+t^2}} \quad \text{and} \quad s = t^*c, \tag{11}$$

where

$$t = \frac{sign(\tau)}{|\tau| + \sqrt{\tau^2+1}} \quad \text{and} \quad \tau = \frac{a_j^T a_j - a_i^T a_i}{2a_i^T a_j}. \tag{12}$$

As we could see, there is a close similarity between one-sided and two sided versions of the Jacobi algorithm.

2.3 One-Sided JRS Iteration Algorithm

Since any rotation in the one-sided Jacobi algorithm changes only the corresponding (two) rows, there exists inherent parallelism in the Jacobi iteration algorithms. For example, the $n(n - 1)/2$ rotations in any sweep can be grouped into $(n - 1)$ non-conflicting rotation sets each of which contains $n/2$ rotations. The idea of the JRS iteration algorithms is to perform each set of rotations in parallel. One can think of the Jacobi algorithm as consisting of two phases. In the first phase all the rotation matrices are computed. In the second phase the rotation matrices are multiplied to get B, and then S and V are calculated. Consider any rotation operation. JRS performs all the $n(n - 1)/2$ rotations of a sweep in parallel even though not all of these rotations are independent. Followed by this the second phase has to be completed. This involves the multiplication of $O(n^2)$ rotation matrices.

The JRS iteration algorithm also has sweeps and in each sweep we perform rotations. The only difference is that in a given rotation, however, the norm of two rows does not zero-out but rather the value of this norm is decreased by a fraction of its current value. Given two column vectors u_i and u_j, the norm of them is reduced to a fraction of it. In the relaxation technique, c and s are chosen such that $v^T{}_i v_j = \lambda u^T{}_i u_j$ instead of $v^T{}_i v_j = 0$.

From the following transformation

$$\begin{pmatrix} v_i^T \\ v_j^T \end{pmatrix} = \begin{pmatrix} c & s \\ -s & c \end{pmatrix}^T \begin{pmatrix} u_i^T \\ u_j^T \end{pmatrix} \tag{13}$$

it is easily to get

$$v^T_i = c\, u^T_i - s\, u^T_j \text{ and } v^T_j = s\, u^T_i + c\, u^T_j, \tag{14}$$

then

$$v^T_i\, v_j = (c\, u_i - s\, u_j)^T (s\, u^T_i + c\, u^T_j) = u^T_i\, u_j\, (c^2 - s^2) + (u^T_i\, u_i - u^T_i\, u_j)cs = \lambda\, u^T_i\, u_j, \tag{15}$$

where λ is in the interval [0, 1). When $\lambda = 0$, the JRS algorithm gives the same result of the original Jacobi iteration algorithm. The above equation can be solved for s and c as follows:
If $u^T_i\, u_j = 0$, then set $c = 1$ and $s = 0$;
Otherwise

$$\frac{u_i^T u_i - u_j^T u_j}{2u_i^T u_j} = \frac{c}{2s} - \frac{s}{2s} - \frac{\lambda}{2cs}. \tag{16}$$

Let

$$\tau = \frac{u_j^T u_j - u_i^T u_i}{2u_i^T u_j} \text{ and } t = \frac{s}{c}, \tag{17}$$

then

$$(1+\lambda)t^2 + 2\tau t + \lambda - 1 = 0 \tag{18}$$

According to [3], the smaller root should be chosen, so

$$t = \frac{sign(\tau)(1-\lambda)}{|\tau| + \sqrt{\tau^2 + (1-\lambda^2)}} \tag{19}$$

Like in the regular Jacobi rotation, c and s can be computed as:

$$c = \frac{1}{\sqrt{1+t^2}} \quad \text{and} \quad s = t*c. \tag{20}$$

3 Block One-Sided JRS Iteration Algorithm

Any parallel algorithm for computing SVD on $n \times n$ matrix partitions the $n(n - 1)/2$ rotations of a sweep into rotation sets. Each rotation set consists of some number of independent rotations. All the rotations of a rotation set are performed in parallel because they are independent. Most of the parallel SVD algorithms in the literature employ (n - 1) rotation sets, where each rotation set consists of $n/2$ independent rotations. The streaming and JRS iteration algorithms are exceptions, where all the rotations can be calculated in parallel even though they are dependent.

For computers with memory hierarchy, it is often preferable to partition the input data and to perform the computation on the blocks. This approach provides for full reuse of data while the block is held in cache memory. It avoids excessive movement of data to and from memory and gives a surface-to-volume effect for the ratio of arithmetic operations to data movement, i.e., $O(n^3)$ arithmetic operations to $O(n^2)$ data movement. In addition, on architectures that provide for parallel processing, parallelism can be exploited by performing operations on distinct blocks in parallel [13].

The proposed block JRS algorithm divides the rows of the input matrix $A_{n\times n}$ into $2P$ blocks, where P is the number of parallel processors. Not all the blocks should have the same size; the number of rows per block varies from $\lfloor n/2P \rfloor$ to $\lceil n/2P \rceil$. The computations of the block JRS algorithm are done in two main parts. In the first part, each processor calculates all the rotations in the given block even though they are dependent and store them into 1-D arrays $c[]$ and $s[]$. The length of these arrays is $(n/2P)(n/2P\text{-}1)/2$ or $(n^2/8P^2 - n/4P)$, assuming that P divides n. (Each block is of the same size, namely, $n/2P$ rows). For example if $n/2P = 4$, the first processor calculates the values of $c[]$ and $s[]$ for the following sequence of rows (1, 2), (1, 3), (1, 4), (2, 3), (2, 4), and (3, 4), which are six rotations. After calculating the arrays for $c[]$ and $s[]$, each processor applies these rotations to the corresponding block. The same is done for the second block (each processor works on the two blocks). This means that the total number of rotations in the first part of the proposed algorithm equals $2P*(n^2/8P^2 - n/4P)$, which equals $(n^2/4P- n/2)$ rotations.

The second part of our proposed block JRS iteration algorithm calculates then applies the rotations between blocks, where each processor works on two blocks. This part iterates ($2P$ - 1) steps. The well-known round-robin technique is used for generating ($2P$ - 1) orders. Figure 1 shows an example of generating seven orders needed for executing the second part of the block JRS on four processors. In each iteration (step) of the second part, each processor computes arrays of rotation parameters ($c[]$ and $s[]$) for two blocks. Each row of the first block should be orthogonalized with all rows of the second block. A total of $(n/2P)^2$ rotations are computed per processor working on two blocks. For example, assume that $(n/2P) = 4$. For a processor working on blocks 1 and 2, the following sequences are done to calculate $c[1{:}16]$ and $s[1{:}16]$: (1, 5), (1, 6), (1, 7), (1, 8), (2, 5), (2, 6), (2, 7), (2, 8), (3, 5), (3, 6), (3, 7), (3, 8), (4, 5), (4, 6), (4, 7), and (4, 8), which are sixteen rotations. After calculating arrays of rotation parameters ($c[1{:}(n/2P)^2]$ and $s[1{:}(n/2P)^2]$), each processor applies $(n/2P)^2$ rotations on its blocks. Then the round-robin routine is called for generating a new order, as shown in Figure 1. The number of rotations in the second part of the algorithm is $P*(2P - 1)*(n/2P)^2$, which equals $n^2/2 - (n^2/4P)$ rotations. Thus the total number of iterations in the block JRS algorithm equals $n^2/2 - n^2/4P + n^2/4P- n/2 = n(n - 1)/2$.

Listing 1 depicts the proposed block JRS algorithm in detail. The first line calculates δ, which is used as a termination parameter of the repeat-until loop. It equals the norm of the input matrix times the machine epsilon ($\varepsilon = 10^{-15}$). Line 2 of the proposed algorithm initializes up[] and dn[] arrays, which are used by the round-robin routine. Initially up[] array has even numbers (2, 4, 6, …) and dn[] array has odd numbers (1, 3, 5, …). These numbers indicate the block numbers associated with the processors at each step; processor *i* computes and applies the rotations of blocks up[i] and dn[i]. Figure 1 shows how the contents of the up[] and dn[] arrays are changed by calling the round-robin routine. Part one of the block JRS algorithm is depicted in lines 5 to 26, where the size of each block is stored in an array of data structure called SOB[]. SOB[*i*].low and SOB[*i*].high stores the number of the beginning row and last row of the block *i*. Besides, part two is depicted in lines 27 to 51. The last line computes the singular values form the orthogonal vectors.

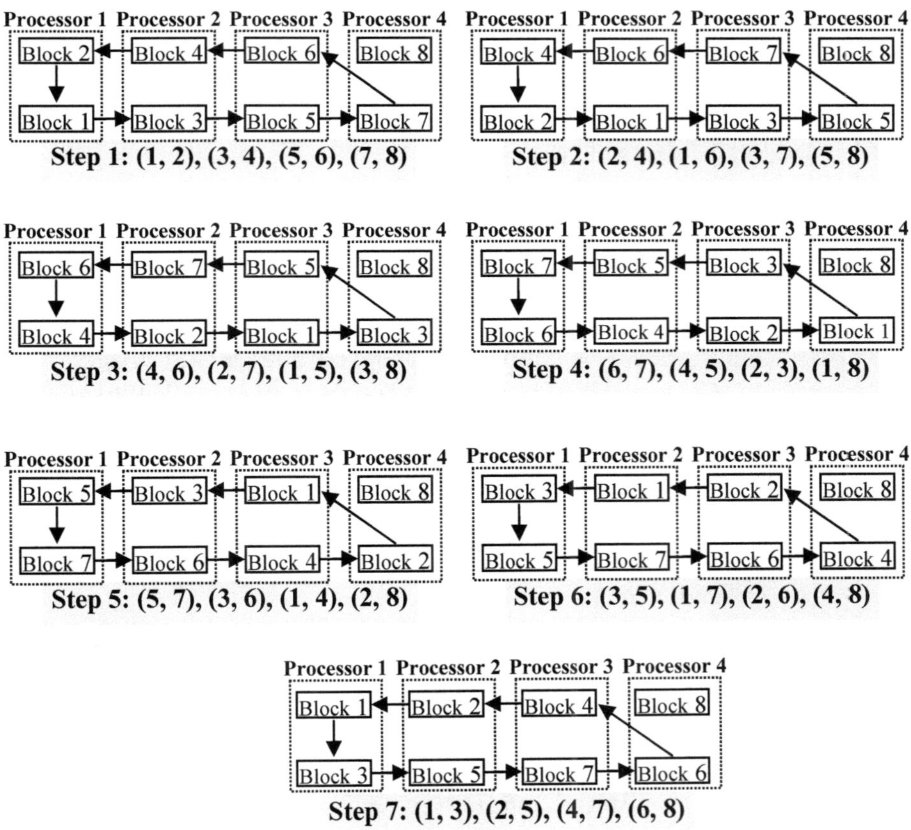

Fig. 1. Round-robin method for generating seven orders on four processors

Listing 1: The proposed block JRS algorithm

```
δ = ε ∑ A[i]^TA[i], 1 ≤ i ≤ n
for i = 1 to P { up[i]=2*i , dn[i]=2*i-1 }
repeat {
    converged = true
    for s = 1 to (2*P) {
        ind = 0
        for i = SOB[s].low to SOB[s].high-1 {
            for j = i+1 to SOB[s].high {
                dii=A[i]^TA[i] , djj=A[j]^TA[j] , dij=A[i]^TA[j]
                if |dij| > δ then converged = false
                if dij ≠ 0 then
                    τ=(djj-dii)/(2*dij),t=sgn(τ)/(|τ|+ sqrt(τ^2+1))
                    c[ind]=1.0/sqrt(t^2+1) , s[ind]=t*c[ind]
                else c[ind] = 1 , s[ind] = 0
                ind = ind + 1
        } }
        ind = 0
        for i = SOB[s].low to SOB[s].high-1 {
            for j = i+1 to SOB[s].high {
                for k = 1 to n {
                    temp = c[ind] * A[i][k] - s[ind] * A[j][k]
                    A[j][k]= s[ind] * A[i][k] + c[ind] * A[j][k]
                    A[i][k] = temp
                }
                ind = ind + 1
    } } }
    for iteration = 1 to (2*P - 1) {
        for s = 1 to P {
            ind = 0
            for i = SOB[up[s]].low to SOB[up[s]].high {
                for j = SOB[dn[s]].low to SOB[dn[s]].high {
                    dii=A[i]^TA[i] , djj=A[j]^TA[j] , dij=A[i]^TA[j]
                    if |dij| > δ then  converged = false
                    if dij ≠ 0 then
                     τ=(djj-dii)/(2*dij),t=sgn(τ)/(|τ|+sqrt(τ^2+1))
                       c[ind]=1.0/sqrt(t^2+1) , s[ind]=t*c[ind]
                    else c[ind] = 1 , s[ind] = 0
                    ind = ind +1
            } }
            ind = 0
            for i = SOB[up[s]].low to SOB[up[s]].high {
                for j = SOB[dn[s]].low to SOB[dn[s]].high {
                    for k = 1 to n {
                        temp=c[ind]*A[i][k] - s[ind]*A[j][k]
                        A[j][k]=s[ind]*A[i][k] + c[ind]*A[j][k]
                        A[i][k]=temp
                    }
                    ind = ind + 1
        } } }
        call round-robin(up[], dn[])
    }
} until converged = true
for i = 1 to n { σ[i] = sqrt(A[i]^TA[i]) }
```

4 Experimental Results

The proposed block JRS algorithm has been implemented and tested on matrices with sizes varying form 50×50 up to 1000×1000 in steps of 50. The contents of these matrices were generated randomly to have a value in the interval [1, 10]. Besides, the number of processors varies form 4 to 20 in steps of 2. We use the norm of the input matrix times 10^{-15} as the convergence parameter.

Figure 2 shows the number of sweeps needed by the bock JRS algorithm. As we can see, the number of sweeps is close to that needed for the cyclic one-sided Jacobi

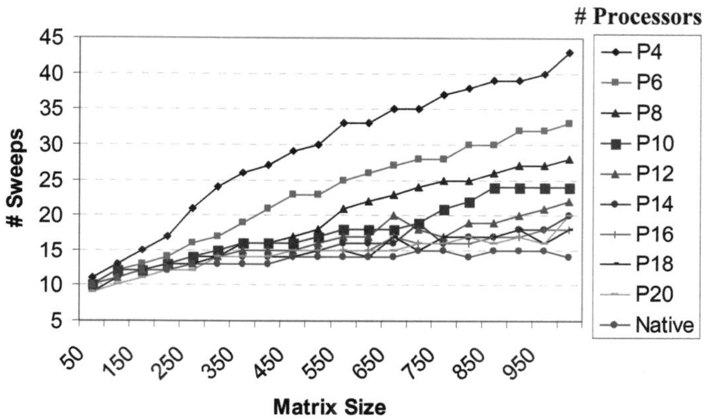

Fig. 2. The number of sweeps for the proposed algorithm

Table 1. The number of sweeps for the block JRS algorithm

Matrix Size		# Processors in Block JRS								
	Native	4	6	8	10	12	14	16	18	20
50	10	11	10	10	10	10	10	10	9	9
100	11	13	12	12	12	11	11	11	11	10
150	12	15	13	12	12	12	12	12	12	11
200	12	17	14	13	13	13	13	13	13	12
250	13	21	16	14	14	13	13	13	13	12
300	13	24	17	14	15	14	14	14	14	14
350	13	26	19	16	16	15	14	14	14	14
400	13	27	21	16	16	15	14	14	14	14
450	14	29	23	17	16	15	15	14	14	15
500	14	30	23	18	17	16	15	15	15	15
550	14	33	25	21	18	17	16	15	15	15
600	14	33	26	22	18	17	16	15	14	15
650	14	35	27	23	18	20	16	17	17	15
700	15	35	28	24	19	18	19	16	15	16
750	15	37	28	25	21	17	16	16	17	16
800	14	38	30	25	22	19	17	16	17	17
850	15	39	30	26	24	19	17	17	17	16
900	15	39	32	27	24	20	18	17	18	17
950	15	40	32	27	24	21	18	18	16	16
1000	14	43	33	28	24	22	20	18	18	20

algorithm when the number of processor is reasonably large. Besides, Table 1 shows the same result.

We have also studied the effect of the relaxation parameter λ on the performance of block JRS. The results are shown in Figure 3. When the number of processors is small the performance of the algorithm is very sensitive to the value of λ. A choice of 0.45 or more for λ seems to be the best the processor range [4:20] tried. When the number of processors is 10 or more even a small value of λ seems to yield good performance. One could identify an optimal value of λ empirically using the ideas given in [2].

Block JRS has also been tested for up to 250 processors and the numbers of sweeps for various values of the relaxation parameter λ are shown in Figure 4. The

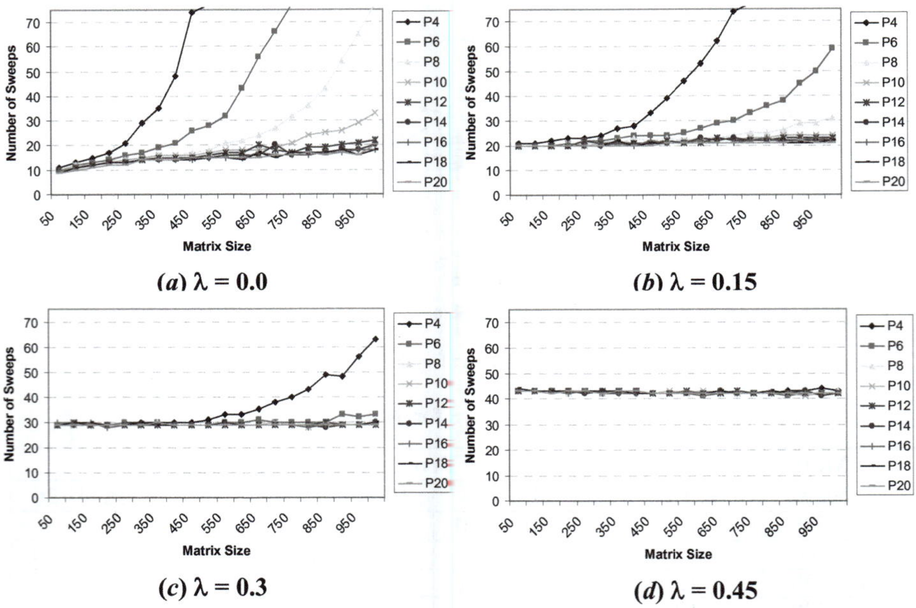

Fig. 3. The effect of λ on the number of sweeps

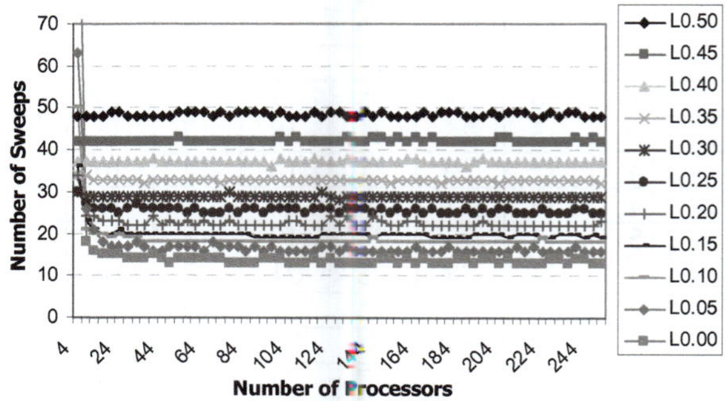

Fig. 4. The number of sweeps required for a matrix 500×500

matrix size used was 500×500. As we could see from this figure, when the number of processors is large, a value of zero or a small value for λ yields the best performance.

5 Conclusion

A new version of JRS iteration algorithm is proposed in this paper for the parallel computation of SVDs, which employs the same idea of decreasing (relaxing) the off-diagonal elements instead of zeroing them out. The block JRS algorithm is a highly parallel algorithm designed to exploit the memory hierarchy by processing blocks to reuse the loaded data into cache memories. It differs form the JRS algorithms in the partitioning strategy of the input matrix among processors. One of the key points of the block JRS is reusing the loaded data into cache memory by performing the computations on blocks. It performs $O(b^2n)$ floating-point operations on $O(bn)$ elements, which reuses the loaded data in cache memory by a factor on b. Another key point of the proposed algorithm is that on reasonably large number of processors, the number of sweeps is close to the cyclic Jacobi method even though not all the rotations in a block are independent. The relaxation technique helps to calculate and apply multiple rotations per block at the same time.

References

1. Klema, V., Laub, A.: The Singular Value Decomposition: Its Computation and Some Applications. IEEE Transactions on Automatic Control AC-25(2) (1980)
2. Rajasekaran, S., Song, M.: A Novel Scheme for the Parallel Computation of SVDs. In: Proc. High Performance Computing and Communications, pp. 129–137 (2006)
3. Golub, G., Van Loan, C.: Matrix Computations, 2nd edn. John Hopkins University Press, Baltimore and London (1993)
4. Brent, R., Luk, F.: The Solution of Singular-Value and Symmetric Eigenvalue Problems on Multiprocessor Arrays. SIAM Journal on Scientific and Statistical Computing 6(1), 69–84 (1985)
5. Zhou, B., Brent, R.: A Parallel Ring Ordering Algorithm for Efficient One-sided SVD Computations. Journal of Parallel and Distributed Computing 42, 1–10 (1997)
6. Becka, M., Vajtersic, M.: Block-Jacobi SVD Algorithms for Distributed Memory Systems I: Hypercubes and Rings. Parallel Algorithms Appl. 13, 265–287 (1999)
7. Becka, M., Vajtersic, M.: Block-Jacobi SVD Algorithms for Distributed Memory Systems II: Meshes. Parallel Algorithms Appl. 14, 37–56 (1999)
8. Becka, M., Oksa, G., Vajtersic, M.: Dynamic Ordering for a Parallel Block-Jacobi SVD Algorithm. Parallel Comp. 28, 243–262 (2002)
9. Oksa, G., Vajtersic, M.: A Systolic Block-Jacobi SVD Solver for Processor Meshes. Parallel Algorithms and Applications 18, 49–70 (2003)
10. Strumpen, V., Hoffmann, H., Agarwal, A.: A Stream Algorithm for the SVD. Technical Memo 641, Computer Science and Artificial Intelligence Laboratory, Massachusetts Institute of Technology (2003)
11. Hestenes, M.: Inversion of Matrices by Biorthogonalization and Related Results. J. SIAP 6(1), 51–90 (1958)
12. Brent, R.: Parallel Algorithms for Digital Signal Processing. Numerical Linear Algebra, Digital Signal Processing and Parallel Algorithms, pp. 93–110. Springer, Heidelberg (1991)
13. Dongarra, J., Foster, I., Fox, G., Kennedy, K., White, A., Torczon, L., Gropp, W.: The Sourcebook of Parallel Computing. Morgan Kaufmann, San Francisco (2002)

Concurrent Number Cruncher: An Efficient Sparse Linear Solver on the GPU

Luc Buatois[1], Guillaume Caumon[2], and Bruno Lévy[3]

[1] Gocad Research Group, INRIA, Nancy Université, France
buatois@gocad.org
[2] ENSG/CRPG, Nancy Université, France
caumon@gocad.org
[3] ALICE - INRIA Lorraine, Nancy, France
levy@loria.fr

Abstract. A wide class of geometry processing and PDE resolution methods needs to solve a linear system, where the non-zero pattern of the matrix is dictated by the connectivity matrix of the mesh. The advent of GPUs with their ever-growing amount of parallel horsepower makes them a tempting resource for such numerical computations. This can be helped by new APIs (CTM from ATI and CUDA from NVIDIA) which give a direct access to the multithreaded computational resources and associated memory bandwidth of GPUs; CUDA even provides a BLAS implementation but only for dense matrices (CuBLAS). However, existing GPU linear solvers are restricted to specific types of matrices, or use non-optimal compressed row storage strategies. By combining recent GPU programming techniques with supercomputing strategies (namely block compressed row storage and register blocking), we implement a sparse general-purpose linear solver which outperforms leading-edge CPU counterparts (MKL / ACML).

1 Introduction

1.1 Motivations

In the last few years, graphics processors have evolved from rendering simple 2D objects to rendering real-time realistic 3D environments. Their raw power and memory bandwidth have grown so quickly that they significantly overwhelm CPU specifications. For instance, a 2GHz CPU has a theoretical peak performance of 8 GFlops (4 floating point operations per cycle using SSE), whereas modern GPUs have an observed peak performance of 350 GFlops [1].

Moreover, graphics card manufacturers have recently introduced new APIs dedicated to *general purpose* computations on *graphics processor units* (GPGPU [2]): CUDA from NVIDIA [3] and CTM from ATI [4]. These APIs provide low-level or direct access to GPUs, exposing them as large arrays of parallel processors.

Numerical solvers play a central role in many optimization problems that can be strongly accelerated by using GPUs. The key point is parallelizing these algorithms in a way that fits the highly parallel architecture of modern GPUs. As shown in Figure 1, our GPU-based Concurrent Number Cruncher (CNC) accelerates optimization algorithms

R. Perrott et al. (Eds.): HPCC 2007, LNCS 4782, pp. 358–371, 2007.

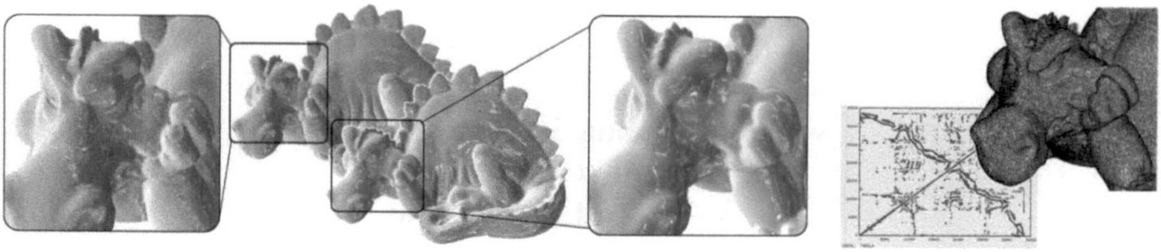

Fig. 1. Our CNC applied to mesh smoothing. Left: initial mesh; Center: smoothed mesh; Right: geometry processing with irregular meshes yield matrices with arbitrary non-zero patterns. Therefore, classic GPGPU techniques cannot be used. Our CNC implements a **general** solver on the GPU that can process these irregular matrices.

and Partial Differential Equations (PDE)s solvers. Our CNC can solve large irregular problems, based on a very general sparse storage format of matrices, and is designed to exploit the computational capabilities of GPUs to their full extent.

As an example, we demonstrate our solver on two different geometry processing algorithms, namely LSCM [5] (mesh parameterization) and DSI [6] (mesh smoothing). As a front-end to our solver, we use the general OpenNL API [7].

In this work, we focus on *iterative* methods since they are easier to parallelize and have a smaller memory footprint than direct solvers, they can be applied to very large sparse matrices, and only a few iterations are needed in interactive contexts. Hence, our CNC efficiently implements on the GPU using the CTM API a Jacobi-preconditioned conjugate gradient solver [8] with a Block Compressed Row Storage (BCRS) of sparse matrices (section 1.2 and 2.1). The BCRS format is much more efficient than the simple Compressed Row Storage format (CRS) by enabling register blocking strategies, which reduce the required memory bandwidth [9].

Moreover, we compare our GPU-CTM implementation of vector operations with the CPU ones from the Intel Math Kernel Library (MKL) [10] and the AMD Core Math Library (ACML) [11], which are highly multithreaded and SSE3 optimized. For sparse matrix operations, we compare our GPU implementation with our SSE2 optimized CPU one since neither the MKL nor the ACML handle the BCRS format. Note that the CUDA BLAS library (CuBLAS) does not provide sparse matrix storage structures.

1.2 The Preconditioned Conjugate Gradient Algorithm

The preconditioned Conjugate Gradient algorithm is a well known method to iteratively solve a symmetric definite positive linear system [8] (extensions exist for non-symmetric systems, see [12]). As it is iterative, it can be used to solve very large sparse linear systems where direct solvers cannot be used due to their memory consumption.

Given the inputs $\mathbf{A}$, $\mathbf{b}$, a starting value $\mathbf{x}$, a preconditioner $\mathbf{M}$, a maximum number of iterations i_{max} and an error tolerance $\varepsilon < 1$, the linear system expressed as $\mathbf{Ax} = \mathbf{b}$ can be solved using the preconditioned conjugate gradient algorithm described as follows:

$i \leftarrow 0$; $\mathbf{r} \leftarrow \mathbf{b} - \mathbf{A}\mathbf{x}$; $\mathbf{d} \leftarrow \mathbf{M}^{-1}\mathbf{r}$;
$\delta_{new} \leftarrow \mathbf{r}^T\mathbf{d}$; $\delta_0 \leftarrow \delta_{new}$;
while $i < i_{max}$ **and** $\delta_{new} > \varepsilon^2\delta_0$ **do**
  $\mathbf{q} \leftarrow \mathbf{A}\mathbf{d}$; $\alpha \leftarrow \frac{\delta_{new}}{\mathbf{d}^T\mathbf{q}}$;
  $\mathbf{x} \leftarrow \mathbf{x} + \alpha\mathbf{d}$; $\mathbf{r} \leftarrow \mathbf{r} - \alpha\mathbf{q}$;
  $\mathbf{s} \leftarrow \mathbf{M}^{-1}\mathbf{r}$; $\delta_{old} \leftarrow \delta_{new}$;
  $\delta_{new} \leftarrow \mathbf{r}^T\mathbf{s}$; $\beta \leftarrow \frac{\delta_{new}}{\delta_{old}}$;
  $\mathbf{d} \leftarrow \mathbf{r} + \beta\mathbf{d}$; $i \leftarrow i + 1$;
end

In this paper, we present an efficient implementation of this algorithm using the Jacobi-preconditioner ($\mathbf{M} = \text{diag}(\mathbf{A})$) for large sparse linear systems using hardware acceleration through the CTM-API [4].

1.3 Previous Works on GPU Solvers

Depending on the discretization of the problem, solving PDEs involves band, dense or general sparse matrices. Naturally, the different types of matrices lead to specific solver implementations. Most of these methods rely on low-level BLAS APIs.

Band matrices
The first type of solver developed on GPUs were band solvers [13]. This is due to the natural and efficient mapping of band matrices into 2D textures. Most of the work done in this field was for solving the pressure-Poisson equation for incompressible fluid flow simulation applied to textures.

Dense matrices
Direct solvers for dense matrices were also implemented on GPUs for a Cholesky decomposition [14], and for both Gauss-Jordan elimination and LU decomposition [15].

General sparse matrices
PDEs discretized on irregular meshes leads to solve irregular problems, hence call for a general representation for sparse matrices. Two authors showed the feasibility of implementing the Compressed Row Storage format (CRS). Bolz et al. [16] use textures to store non-zero coefficients of a matrix and its associated two-level lookup table. The lookup table is used to address the data and to sort the rows of the matrix according to the number of non-zero coefficients in each row. Then, an iteration is performed on the GPU simultaneously over all rows of the same size to complete, for example, a matrix-vector product operation. Another approach was proposed by Krüger and Westerman [13], based on vertex buffers (one vertex is used for each non-zero element). Our CNC method also implements general sparse matrices on the GPU, but uses a more compact representation of sparse matrices, and replaces the CRS format with BCRS (Block Compressed Row Storage) [9] to optimize cache usage and enable register blocking and vector computations.

New APIs, New Possibilities
Previous works on GPGPU used APIs such as DirectX [17] or OpenGL 2.0 [18] to access GPUs and use high-level shading languages such as Brook [19], Sh [20] or Cg [21]

to implement operations. Using such graphics-centric programming model devoted to real-time graphics limits the flexibility, the performance, and the possibilities of modern GPUs in terms of GPGPU.

Both ATI [4] and NVIDIA [3] recently announced or released new APIs, respectively CTM and CUDA, designed for GPGPU. They provide policy-free, low-level hardware access, and direct access to the high-bandwidth latency-masking graphics memory. These new drivers hide useless graphical functionalities to reduce overheads and simplify GPU programming. In addition to our improved data structures, we use the CTM API to improve performance.

1.4 Contributions

The Concurrent Number Cruncher is a high-performance preconditioned conjugate gradient solver on the GPU using the new ATI-CTM API dedicated to GPGPU which reduces overheads and provides fine controls for optimizations. The CNC is based on a general optimized implementation of sparse matrices using BCRS blocking strategies, and optimized BLAS operations through massive parallelization and vectorization of the processing. To our knowledge, this is the first linear solver on the GPU that can be efficiently applied to unstructured optimization problems.

2 The Concurrent Number Cruncher (CNC)

The CNC is based on two components: an OpenGL-like API to iteratively construct a linear system (OpenNL [7]) and a highly efficient implementation of BLAS functions. The following sections present some usual structures in supercomputing used in the CNC, how we optimized them for modern GPUs and important implementation features, which proved critical for efficiency.

2.1 Usual Data Structures in Supercomputing

Compressed Row Storage (CRS)
Compressed row storage [9] is an efficient method to represent general sparse matrices (Figure 2). It makes no assumptions about the matrix sparsity and stores only non-zero elements in a 1D-array, row by row. It uses an indirect addressing based on two lookup-tables to retrieve the data: (1) a row pointer table used to determine the storage bounds of each row of the matrix in the array of non-zero coefficients, and (2) a column index table used to determine in which column the coefficient lies.

The Sparse Matrix-Vector Product Routine (SpMV)
The implementation of a conjugate gradient involves a sparse matrix-vector product (SpMV) that takes most of the solving time [12]. This product $\mathbf{y} \leftarrow \mathbf{Ax}$ is expressed as:

$$\textbf{for } i = 1 \textbf{ to } n,\ y[i] = \sum_j a_{i,j} x_j$$

Since this equation traverses all rows of the matrix sequentially, it can be implemented efficiently by exploiting, for example, the CRS format (code for a matrix of size $n \times n$):

```
for i = 0 to n − 1 do
    y[i] ← 0
    for j = row_pointer[i] to row_pointer[i + 1] − 1 do
        y[i] ← y[i] + values[j] × x[column_index[j]]
    end
end
```

The number of operations involved during a sparse product is two times the number of non-zero elements of **A** [12]. Compared to a dense product that takes $2n^2$ operations, the CRS format significantly reduces processing time.

2.2 Optimizing for the GPU

The massively parallel Single Instruction/Multiple Data (SIMD) architecture of modern GPUs calls for specific optimization and implementation of algorithms. GPUs offer two main levels of parallelism through multiple pipelines and vector processing, and several possibilities of optimizations presented in this subsection.

Multiple Pipelines

Multiple pipelines can be used to process data with parallelizable algorithms. In our implementation, each element of $\mathbf{y}$ is computed by a separate thread when computing the $\mathbf{y} \leftarrow \mathbf{Ax}$ sparse operation (SpMV). This way, each thread iterates through a row of elements of the sparse matrix **A** to compute the product.

To maximize performance and hide memory-latencies, the number of threads used must be higher than the number of pipelines. For example, on an NVIDIA G80 that has 128 pipelines, $\mathbf{y}$ size should be an order of magnitude higher than 128 elements.

Similarly, operations on vectors, as the SAXPY computing the equation $\mathbf{y} \leftarrow \alpha \times \mathbf{x} + \mathbf{y}$, are parallelized by computing a unique element of the result $\mathbf{y}$ per thread.

To parallelize the vector dot product operation, the CNC implements an iterative sum-reduction of the data as in [13]: at each iteration, each thread reads and processes 4 scalars and writes one resulting scalar. The original n-dimensional vector is hence reduced by 4 at each iteration until only one scalar remains, after $\log_4(n)$ iterations.

Vector Processing

Some GPU architectures (e.g. ATI X1k series) process the data inside 4-element vector-processors. For such architectures, it is essential to vectorize the data inside 4-element vectors to maximize efficiency. Since our implementation targets a vector GPU architecture (ATI X1k series), all operations are vectorized. On a scalar architecture like the NVIDIA G80, data do not need to be vectorized. However, for random memory accesses, it is better to read one float4 than four float1 since the G80 architecture is able to read 128bits in one cycle and one instruction. Note that on a CPU it is possible to use SSE instructions to vectorize data processing.

Register Blocking: Block Compressed Row Storage (BCRS)

Unlike previous approaches which investigate, at best, only CRS format [16, 13], our CNC uses an efficient GPU implementation of the BCRS format. Note that, even on the

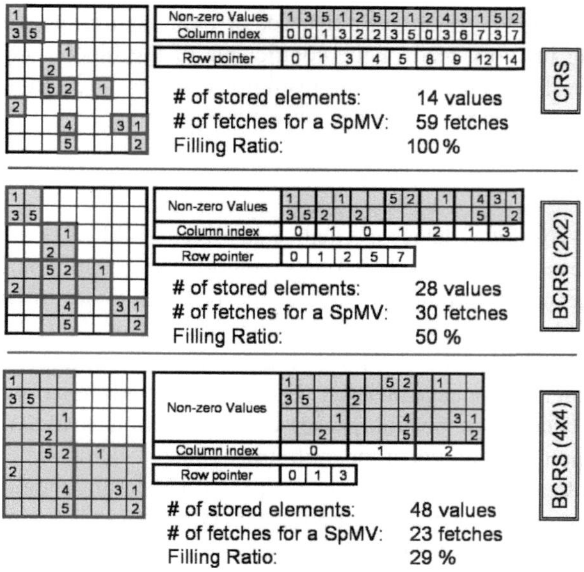

Fig. 2. Compressed Row Storage (CRS) and Block Compressed Row Storage (BCRS) examples for the same matrix. The number of stored elements counts all stored elements, useful or not. The number of fetches required to achieve a sparse matrix-vector product (SpMV) is provided for a 4-vector architecture, less fetches being better for memory bandwidth. The filling ratio indicates the average rate of non-zero data in each block, higher filling ratio being better for the computations.

CPU, BCRS is faster than CRS [9]. The BCRS groups non-zero coefficients in blocks of size $BN \times BM$ to use the maximum memory fetch bandwidth of GPUs, take advantage of registers to avoid redundant fetches, and reduce the number of indirections thanks to the reduced size of the lookup tables.

When computing a product of a block spanned over more than one row, it is possible to read only once the associated values from $\mathbf{x}$ of the block, store these values in registers, and reuse them for each row of the block, saving several texture-fetches. Figure 2 illustrates the influence of the block size on the number of fetches. For 4-component vector architectures, fetching and writing scalars 4 by 4 maximizes both read and write bandwidth. It is therefore useful to process all four elements of $\mathbf{y}$ at the same time, and use blocks of size 4x4. Once the double indirection to locate a block using the lookup-tables is solved, 4 fetches are required to read a 4x4 block, 1 to read the corresponding 4-scalars of $\mathbf{x}$, and 1 write to output the 4-scalars of the product result into $\mathbf{y}$. Values of $\mathbf{x}$ are stored in registers, and reused for each row of the block, which results in fewer fetches than in a classical CRS implementation.

The reduced sizes of the lookup tables reduce memory requirements and, more importantly, the number of indirections to be solved during a matrix operation (each indirection results in dependent memory fetches, which introduces memory latencies).

Although a large block size is optimal for register and bandwidth usage, it is adapted only to compact matrices to avoid sparse blocks (Figure 2). The CNC implements 4x4 blocks, and also 2x2 blocks to handle cases where the 4x4 blocks present a very low filling ratio. In the SpMV operation for a vector architecture, this 2x2 block size is optimal regarding memory read of the coefficients of the block, but not to read the corresponding **x** values and to write the resulting **y** values since only two scalars are read from **x** and written in **y**.

2.3 Technicalities

The CTM (Close-To-Metal) device is a dedicated driver for ATI X1k graphics cards [4]. It provides low-level access to both graphics memory and graphics parallel processors. The memory is exposed through pointers at the base addresses of the PCI-Express-memory (accessible from both the CPU and GPU), and of the GPU-memory (only accessible from the GPU). The CTM does not provide any allocation mechanism, and lets the application fully manage memory. The CTM provides functions to compile and load user assembly codes on the multiprocessors and functions to bind memory pointers indifferently to inputs or outputs of the multiprocessors.

Our CNC implements a high-level layer for the CTM hiding memory management by providing dynamic memory allocation for both GPU and PCI-Express RAM.

As shown in Figure 3, the CNC rolls vectors into 2D memory space to benefit from the 2D memory-cache of GPUs, and hence can store very large vectors in one chunk (on an ATI-X1k, the maximum size of a 2D array of 4-component vectors is 4096^2). The BCRS matrix format uses three tables to store the column indices, the row pointers and the non-zero elements of the sparse matrix. The row pointer and column index tables are rolled as simple vectors, and the non-zero values of the matrix are rolled up depending on the block size of the BCRS format. Particularly, the CNC uses strip mining strategies to efficiently implement various BCRS formats. For the 2x2 block size, data are rolled block by block on one 2D-array. For the 4x4 block size, data are distributed into 4 sub-arrays fulfilled alternatively with 2x2 sub-blocks of the original 4x4 block as in [22]. Retrieving the 16 values of a block can be achieved by four memory fetches from the four sub-arrays at exactly the same 2D index, maximizing the memory fetch capabilities of modern GPUs and limiting address translation computations.

The CNC provides a high-level C++ interface to create and manipulate vectors and matrices on the GPU. Vector or matrix data is automatically uploaded to the GPU memory at their instantiation for later use in BLAS operations. The CNC also pre-compiles and pre-allocates all assembly codes to optimize their execution. To execute a BLAS operation, the CNC binds the inputs and outputs of data on the GPU and calls the corresponding piece of assembly code.

The assembly code performing a BLAS operation is made of very "Close-To-Metal" vector instructions as multiply-and-add or 2D memory fetch. It is finely tweaked using semaphore instructions to asynchronously retrieve data, helping in a better parallelization/masking of the latencies. The assembly code implementing a sparse matrix-vector product counts about a hundred lines of instructions, and uses about thirty registers. The number of used registers closely impacts performance, so we minimized their use.

The SpMV routine executes a loop for each row of blocks of a sparse matrix that can be partially-unrolled to process blocks by pair. This allows to asynchronously pre-load data and better mask the latencies of the graphics memory, especially for consecutive

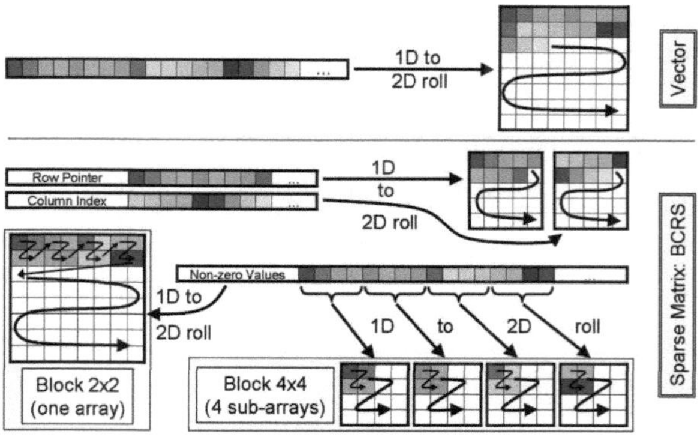

Fig. 3. Vector and BCRS sparse matrix storage representations on the GPU

dependent memory lookups. We increase the SpMV performance of about 30% by processing blocks two by two for both 2x2 and 4x4 block sizes.

As the GPU memory is managed by pointers, outputs of an assembly code can be bound to inputs of the next one, avoiding large overheads or useless data copy. In the conjugate gradient algorithm, the CPU program only calls the execution of assembly codes on the GPU in order, and binds/switches the inputs and outputs pointers. At the end of each iteration, the CPU retrieves only one scalar value (δ) to decide to exit the loop if necessary. Finally, the result vector is copied back to the PCI-Express memory accessible to both the GPU and the CPU. Hence, the whole main loop of the conjugate gradient is fully performed on the GPU.

3 Performance

The preconditioned conjugate gradient algorithm for sparse matrices requires BLAS primitives for which this section provides comparisons in GFlops and percentage of efficiency for CPU and GPU implementations in the CNC. See Table 1 to get the list of implementations used for each operation on each hardware device.

For the SAXPY and SDOT benchmarks on the CPU, we used the Intel Math Kernel Library (MKL) [10] and the AMD Core Math Library (ACML) [11] which are highly multithreaded and optimized using SSE3 instructions. The CPU performance was, on average, close between the MKL and the ACML, hence we choose to only present the results from the former. The SpMV benchmark on the CPU is based on our multithreaded SSE2 optimized implementation, since both the MKL and the ACML do not support the BCRS format.

To be absolutely fair, GPU results take into account all overheads introduced by the graphics cards (cf. section 3.4). All benchmarks were performed on a dual-core AMD Athlon 64 X2 4800+, with 2GB of RAM and an AMD-ATI X1900XTX with 512MB of graphics memory.

Benchmarks were performed at least 500 times to average the results, use synthetic vectors for the SAXPY and SDOT cases, synthetic matrices for the SGEMM case (given for reference), and real matrices built from the set of meshes described in Table 2 for

Table 1. Implementations used according to the operation and computing device: CNC is our, MKL is the Intel one, and CTM-SDK the ATI one

	Device	
Operation	**CPU**	**GPU**
SAXPY	MKL	CNC
SDOT	MKL	CNC
SpMV	CNC	CNC
Pre-CG	MKL+CNC	CNC
SGEMM (for reference)	MKL	CTM-SDK

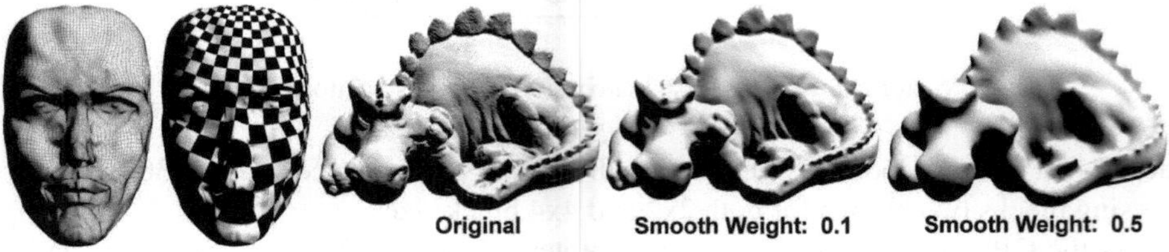

Fig. 4. Parameterization of the Girl Face 3 and smoothing of the Phlegmatic Dragon using our CNC solver on the GPU

Table 2. Meshes used for testing matrix operations. This table provides: the number of unknown variables computed in case of a parameterization or a smoothing of a mesh, denoted by #var, and the number of non-zero elements in the associated sparse-matrix denoted by #non-zero (data courtesy of Eurographics for the Phlegmatic Dragon)

	Parameterization		Smoothing	
Mesh	**#var**	**#non-zero**	**#var**	**#non-zero**
Girl Face 1	1.5K	50.8K	2.3K	52.9K
Girl Face 2	6.5K	246.7K	9.8K	290.5K
Girl Face 3	25.9K	1.0M	38.8K	1.6M
Girl Face 4	103.1K	4.2M	154.7K	6.3M
Phlegmatic Dragon	671.4K	19.5M	1.0M	19.6M

other cases. Figure 4 shows parameterization and smoothing examples of respectively the Girl Face 3 and Phlegmatic Dragon models, both computed using our CNC on the GPU.

3.1 BLAS Vector Benchmarks

Figure 5 presents benchmarks of SAXPY ($\mathbf{y} \leftarrow \alpha \times \mathbf{x} + \mathbf{y}$) and SDOT/sum-reduction operations on both CPU and GPU respectively using the MKL and our CNC. While the performance curve of the CPU stays steady, the GPU performance increases with the vector size. Increasing the size of vectors, hence the number of threads, helps the GPU in masking the latencies when accessing the graphics memory. For the SDOT, performances increase not as fast as for the SAXPY due to its iterative computation that introduces more overheads and potentially more latencies which need to be hidden.

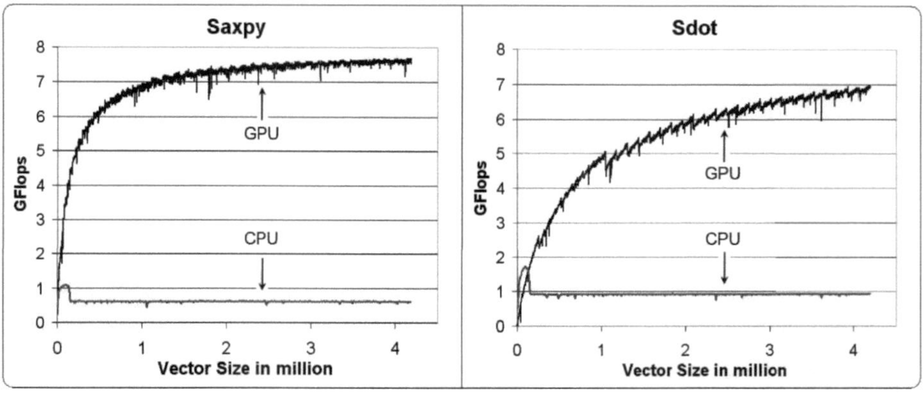

Fig. 5. SAXPY ($\mathbf{y} \leftarrow \alpha \times \mathbf{x} + \mathbf{y}$) and SDOT (sum-reduction of a vector) performances comparison in function of the processed vector size between the CPU MKL implementation and our GPU implementation

At best, our GPU implementation is 12.2 times faster than the CPU implementation of the MKL for the SAXPY and 7.4 times faster for the SDOT.

3.2 SpMV Benchmarks

Testing operations on sparse-matrices is not straightforward since the number and layout of non-zero coefficients strongly govern performance. To test sparse matrix-vector product (SpMV) operations and preconditioned conjugate gradient, we choose five models and two tasks (parameterization and smoothing) which best reflect typical situations in geometry processing (Figure 4 and Table 2).

Figure 6 shows the speed of the SpMV for various implementations and applications. Four things can be noticed: (1) CPU performance stays relatively stable for both problems while the GPU performance increases according to the increasing size of the matrices, (2) thanks to register blocking and vectorization, BCRS 4x4 is faster than 2x2 (which is also faster than CRS) for CPUs and GPUs, (3) GPU is about 3.1 times faster than CPU with SSE2 for significant mesh sizes, and (4) multithreaded-SSE2 implementation enabling vector processing on the CPU is about 2.5 times faster than standard implementation. For very small matrices, the CNC just compares to CPU implementations, but, in that case, a direct solver would be more appropriate.

3.3 Preconditioned Conjugate Gradient Benchmarks

Performance of the main loop of a Jacobi-preconditioned conjugate gradient is provided in Figure 6. Since most of the solving time, about 80%, is spent within the SpMV whatever hardware device is used, the SpMV governs the solver performance, and the comments for the SpMV are also applicable here. GPU solver runs 3.2 times faster than the CPU-SSE2, and CPU-SSE2 solver runs 1.8 times faster than the non-SSE2. As for the SpMV routine, the CNC is inefficient for very small matrices where direct solvers are anyway better.

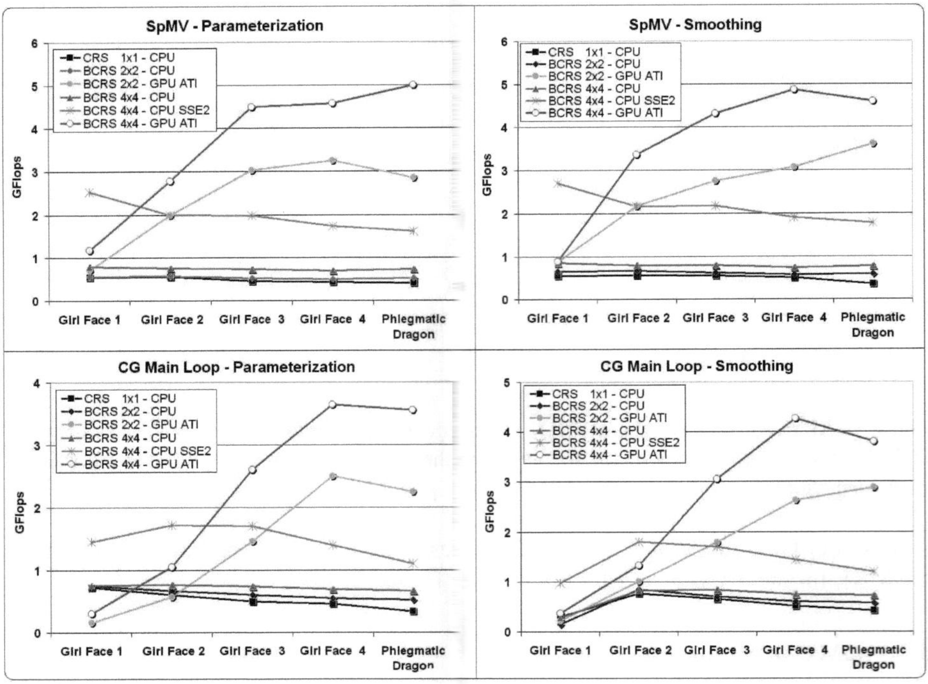

Fig. 6. Performance comparison of the SpMV (sparse matrix-vector product) and the Jacobi-Preconditioned Conjugate Gradient Main Loop between CPU standard, CPU multithreaded-SSE2 optimized, and GPU implementations in case of computing a mesh parameterization or smoothing for CRS (no blocking), BCRS 2x2 and BCRS 4x4

3.4 Overheads and Efficiency

Overheads introduced when copying, binding, retrieving data or executing shader programs on the GPU were the strong bottlenecks of all previous works. New GPGPU APIs like CTM or CUDA can increase performance if well used (cf. section 2.3). During our tests, the total time spent for the processing on the GPU was always higher than 93%, meaning that less than 7% percents were for the overheads, showing a great improvement as compared to previous works.

CPU and GPU architectures are bound by their maximum computational performance and maximum memory bandwidth. Their efficiency mainly depends on the implemented operations (Figure 7). For example, low computation operations like the SAXPY or the SDOT are limited by the memory bandwidth, and not by the theoretical computational peak power. The SAXPY achieves high bandwidth efficiency on GPUs, near 91%, but very low computational efficiency, near 3.1%. Conversely, for reference, a dense matrix-matrix product (SGEMM) on the GPU can achieve a good computational efficiency, near 18.75%, and a very good memory bandwidth efficiency, near 91% (used SGEMM implementation comes from the ATI CTM-SDK). In the case of SpMV operations, both computational and memory bandwidth efficiency are low

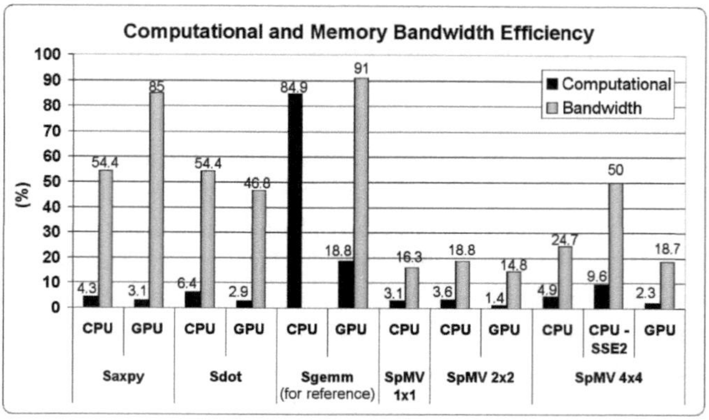

Fig. 7. Percentage of computational and memory bandwidth efficiency of different operations implemented on the CPU and the GPU for 1024^2 vector size, 1024^2 dense matrix size and Girl Face 4 model to test the SpMV in case of a parameterization. See Table 1 to determine used implementations.

on GPUs and CPUs (although SSE2 greatly helps). This is due to the BCRS format that implies strong cache-miss and several dependent memory fetches. Nevertheless, as previously written, our CNC on the GPU performs the SpMV 3 times faster than on the CPU.

4 Conclusions

The Concurrent Number Cruncher aims at providing the fastest possible sparse linear solver through hardware acceleration using new APIs dedicated to GPGPU. It uses *block* compressed row storage, which is faster and more compact than compressed row storage, enabling register blocking and vector processing on GPUs and CPUs (through SSE2 instructions). To our knowledge, this makes our CNC the first implementation of a general symmetric sparse solver that can be efficiently applied to unstructured optimization problems.

Our BLAS operations on the GPU are 12.2 and 8 times faster for respectively SAXPY and SDOT operations than on the CPU using SSE3-optimized Intel MKL library. As compared to our CPU SSE2 implementation, the SpMV is 3.1 times faster, and the Jacobi-preconditioned conjugate gradient 3.2 times. We show that on any device BCRS-4x4 is significantly faster than 2x2, and 2x2 significantly faster than CRS. Note that our benchmarks include all overheads introduced by the computations on the GPU.

While the GPGPU community is waiting for GPU double floating point precision, graphics cards only provide single precision for the moment. Hence, our CNC targets applications that do not require very fine accuracy but very fast performance.

As previously shown, operations on any device are limited either by the memory bandwidth (and its latency), or by the computational power. In most BLAS operations, the limiting factor is the memory bandwidth, which was limited to 49.6 GB/s for our tested GPU. According to ATI announces the memory of their next graphics card - the high-end R600- will handle between 110 to 140 GB/s, which will surely strongly increase the performance of our CNC and the gap with CPU implementations.

We plan to extend the parallelization of the CNC across multi-GPUs within a PC (based on SLI or CrossFire configurations), across PC clusters, or within new visual computing systems like the NVIDIA Quadro Plex containing multiple graphics cards with multiple GPUs inside one dedicated box. Thus, we will use the CUDA API from NVIDIA in the CNC to be compatible with the largest possible panel of GPUs. We will build and release a general framework for solving sparse linear systems, including full BLAS operations on sparse matrices and vectors, accelerated indifferently by an NVIDIA with CUDA, or by an ATI with CTM. Note also that iterative non-symmetric solvers (Bi-CGSTAB and GMRES) can be easily implemented using our framework. We will experiment them in future works.

Acknowledgements

The authors thank the members of the GOCAD research consortium for their support (www.gocad.org), Xavier Cavin, and Bruno Stefanizzi from ATI for providing the CTM API and the associated graphics card.

References

1. Buck, I., Fatahalian, K., Hanrahan, P.: Gpubench: Evaluating gpu performance for numerical and scientific applications. In: Proceedings of the 2004 ACM Workshop on General-Purpose Computing on Graphics Processors, ACM Press, New York (2004)
2. GPGPU (General-Purpose computation on GPUs), http://www.gpgpu.org
3. Keane, A.: CUDA (compute unified device architecture) (2006), http://developer.nvidia.com/object/cuda.html
4. Peercy, M., Segal, M., Gerstmann, D.: A performance-oriented data-parallel virtual machine for gpus. In: ACM SIGGRAPH 2006 (2006)
5. Levy, B., Petitjean, S., Ray, N., Maillot, J.: Least squares conformal maps for automatic texture atlas generation. In: SIGGRAPH 2002, San-Antonio, Texas, USA, ACM Press, New York (2002)
6. Mallet, J.: Discrete Smooth Interpolation. Computer Aided Design 24(4), 263–270 (1992)
7. Levy, B.: Numerical methods for digital geometry processing. In: Israel Korea Bi-National Conference (November 2005)
8. Hestenes, M.R., Stiefel, E.: Methods of Conjugate Gradients for Solving Linear Systems. J. Research Nat. Bur. Standards 49, 409–436 (1952)
9. Barrett, R., Berry, M., Chan, T.F., Demmel, J., Donato, J., Dongarra, J., Eijkhout, V., Pozo, R., Romine, C., der Vorst, H.V.: Templates for the Solution of Linear Systems: Building Blocks for Iterative Methods, 2nd edn. SIAM, Philadelphia, PA (1994)
10. Intel: Math kernel library. www.intel.com/software/products/mkl
11. AMD: Amd core math library. http://developer.amd.com/acml.jsp
12. Shewchuk, J.R.: An introduction to the conjugate gradient method without the agonizing pain. Technical report, CMU School of Computer Science (1994), ftp://warp.cs.cmu.edu/quake-papers/painless-conjugate-gradient.ps.

13. Krüger, J., Westermann, R.: Linear algebra operators for gpu implementation of numerical algorithms. ACM Transactions on Graphics (TOG) 22(3), 908–916 (2003)
14. Jung, J.H., O'Leary, D.P.: Cholesky decomposition and linear programming on a gpu, Scholarly Paper, University of Maryland (2006)
15. Galoppo, N., Govindaraju, N.K., Henson, M., Manocha, D.: LU-GPU: Efficient algorithms for solving dense linear systems on graphics hardware. In: SC 2005, IEEE Computer Society, Washington, DC, USA (2005)
16. Bolz, J., Farmer, I., Grinspun, E., Schröder, P.: Sparse matrix solvers on the GPU: conjugate gradients and multigrid. ACM Trans. Graph. 22(3), 917–924 (2003)
17. Microsoft: Direct3d reference. msdn.microsoft.com (2006)
18. Segal, M., Akeley, K.: The OpenGL graphics system: A specification, version 2.0 (2004), http://www.opengl.org
19. Buck, I., Foley, T., Horn, D., Sugerman, J., Fatahalian, K., Houston, M., Hanrahan, P.: Brook for gpus: stream computing on graphics hardware. ACM Trans. Graph. 23(3), 777–786 (2004)
20. McCool, M., DuToit, S.: Metaprogramming GPUs with Sh. AK Peters (2004)
21. Fernando, R., Kilgard, M.J.: The Cg Tutorial: The Definitive Guide to Programmable Real-Time Graphics. Addison-Wesley Longman Publishing Co. Inc. Boston, MA, USA (2003)
22. Fatahalian, K., Sugerman, J., Hanrahan, P.: Understanding the efficiency of gpu algorithms for matrix-matrix multiplication. In: HWWS 2004: Proceedings of the ACM SIGGRAPH/EUROGRAPHICS conference on Graphics hardware, pp. 133–137. ACM Press, New York (2004)

Adaptive Computation of Self Sorting In-Place FFTs on Hierarchical Memory Architectures

Ayaz Ali, Lennart Johnsson, and Jaspal Subhlok

Department of Computer Science
University of Houston
Houston, TX 77204
{ayaz,johnsson,jaspal}@cs.uh.edu

Abstract. Computing "in-place and in-order" FFT poses a very difficult problem on hierarchical memory architectures where data movement can seriously degrade the performance. In this paper we present recursive formulation of a self sorting in-place FFT algorithm that adapts to the target architecture. For transform sizes where an in-place, in-order execution is not possible, we show how schedules can be constructed that use minimum work-space to perform the computation efficiently. In order to express and construct FFT schedules, we present a context free grammar that generates the FFT Schedule Specification Language. We conclude by comparing the performance of our in-place in-order FFT implementation with that of other well known FFT libraries. We also present a performance comparison between the out-of-place and in-place execution of various FFT sizes.

1 Introduction

The Fast Fourier Transform (FFT) is one of the most widely used algorithms. It is used in many fields of science and engineering, especially in the field of signal processing. Since 1965, various algorithms have been proposed for computing DFTs efficiently. However, the FFT is only a good starting point if an efficient implementation exists for the architecture at hand. Scheduling operations and memory accesses for the FFT for modern platforms, given their complex architectures, is a serious challenge compared to BLAS-like functions. It continues to present serious challenges to compiler writers and high performance library developers for every new generation of computer architectures due to its relatively low ratio of computation per memory access and non-sequential memory access pattern.

FFTW[8,7], SPIRAL[13,6] and UHFFT[12,11,1,2] are three current efforts addressing machine optimization of algorithm selection and code optimization for FFTs. SPIRAL is a code generator system that generates optimized codes for specific size transforms. Both UHFFT and FFTW generate highly optimized straight-line FFT code blocks (micro-kernels) called *codelets,* at installation time. These parametrized code blocks adapt to microprocessor architecture and serve

R. Perrott et al. (Eds.): HPCC 2007, LNCS 4782, pp. 372–383, 2007.

as building blocks in the computation of a larger size transform. The final optimization, specific to a FFT problem, is carried out at run-time by searching the best schedule (plan) among an exponential number of factorizations.

Once the factors are determined, the FFT computation can be carried out in a recursive "cache oblivious" fashion, which derives naturally from the Cooley Tukey Mixed Radix Algorithm [5]. For an input vector $\mathbf{x}$ of size N, where $N = r \times m$, the recursive Cooley Tukey formulation of FFT can be written as:

$$\mathbf{X} = (W_r \otimes I_m) T_m^N (I_r \otimes W_m) \Pi_{N,r} \mathbf{x}$$

where T_m^N is a diagonal "twiddle factors" matrix and $\Pi_{N,r}$ is a mod$-r$ sort permutation matrix. Although these algorithms reduce the complexity of DFT computation from $O(n^2)$ to $O(n \log n)$, the resulting vector is in "digit-reversed" order, requiring non-trivial amount of work to bring the output back in order. Stockham's Autosort framework[10] performs the sorting inside the FFT "butterfly", thereby avoiding an explicit reordering. When an output vector is available (out-of-place FFT), the permutations can be performed in the first rank of butterfly [16] as illustrated in Figure 1(a). This is the most commonly used algorithm for self sorting out-of-place FFT computation in both FFTW and UHFFT. Performing the computation "in-place", i.e., without a separate output vector or temporary array, poses a very difficult problem for self-sorting FFTs. For most transform sizes data reordering is unavoidable in computing in-place and in-order FFTs due to lack of a separate work-space. For sizes that can be factorized in co-prime numbers, Prime Factor Algorithm (PFA) [3,18,17] computes in-place in-order FFT without requiring an additional work-space using special rotated DFT modules (PFA *codelets*). For more prevalent sizes, e.g., powers of 2, [14] and later [19,4] developed the algorithm that was both self sorting and in-place. The algorithm avoids explicit sorting by performing implicit ordering inside the first $\left\lfloor \frac{\log N}{2} \right\rfloor$ ranks as shown in Figure 1(b). However this algorithm can be used only if the factorization of N is a "palindrome", i.e., $N = r \times m \times r$ or $N = r \times r$. The algorithm was later formulated by [9] in a recursive fashion that is suited to parallel and vector architectures. In this paper, we present a recursive formulation of the algorithm that is suited to hierarchical memory architectures because it maintains both temporal as well as spatial locality. We discuss the efficiency of our implementation by comparing it with an alternative factorization strategy. In fact, many different schedules could be constructed to adapt to different architectures. To that end, we have designed a language that is used to construct these schedules dynamically at run-time without requiring to write specific implementations for different FFT sizes.

Expressing the FFT Schedules

Before we discuss the various solutions to the computation of in-place and in-order FFTs, we need to be able to express them in a compact representation. The most commonly used, "Kronecker product" formulation of FFT is an efficient way to express sparse matrix operations. SPIRAL[13], generates optimized

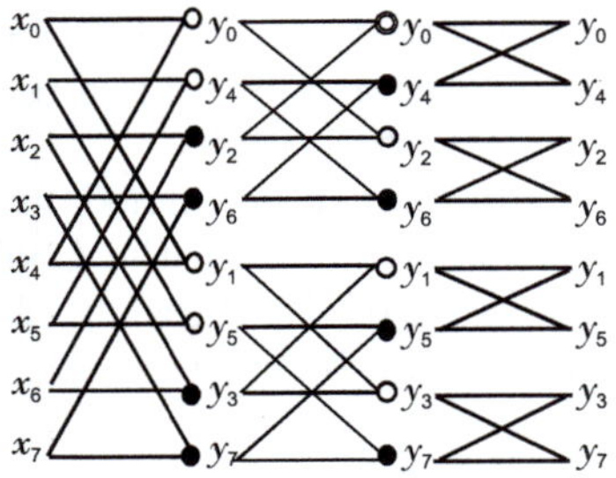

(a) Out-of-place

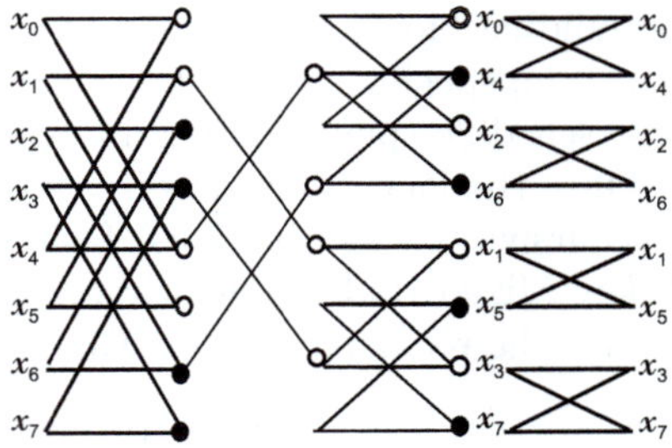

(b) In-place

Fig. 1. Butterflies showing the execution of self sorting FFT of size $N = 8$. (a) The transform is stored to output array in "digit-reversed" order in the first rank. (b) The result is ordered in the first $\lfloor \frac{\log 8}{2} \rfloor = 1$ ranks by performing partial transposed *codelet* transforms on square blocks[19,4].

DSP codes using a pseudo-mathematical representation that is based on the Kronecker product formulation. FFTW [8], uses an internal representation of DFT problems using various data structures, i.e., I/O tensors and I/O dimensions etc. Although they offer an efficient design to solve a complex problem, these representations do not provide sufficient abstraction to the mathematical and implementation level details of FFT algorithms. One of the standard ways to visualize a FFT problem is through a signal flow diagram called the "butterfly" representation. This representation has been adopted across the board by various disciplines of science. However, visualizing large FFT problems through the butterfly is not practical. In this paper we present a language called FFT Schedule Specification Language (FSSL), which is generated from a set of context free grammar productions. The grammar provides a direct and compact translation of the FFT butterfly representation and is easy to understand for computer scientists.

The grammar of FSSL is given in Section 2; where we briefly mention the design overview of UHFFT. We discuss our recursive formulation of self-sorting in-place FFT algorithm in Section 3. And in Section 4, we compare the performance results of our implementation in UHFFT with that of FFTW and Intel's MKL library. Both these libraries are known to perform well on hierarchical memory architectures.

2 UHFFT

UHFFT comprises of two layers, i.e., the code generator (FFTGEN) and the run-time framework. The code generator generates highly optimized small DFT, straight line "C" code blocks called *codelets* at installation time. These *codelets* are combined together by the run-time framework to solve large FFT problems on Real and Complex data. Block diagram of UHFFT2 run-time is given in Figure 2.

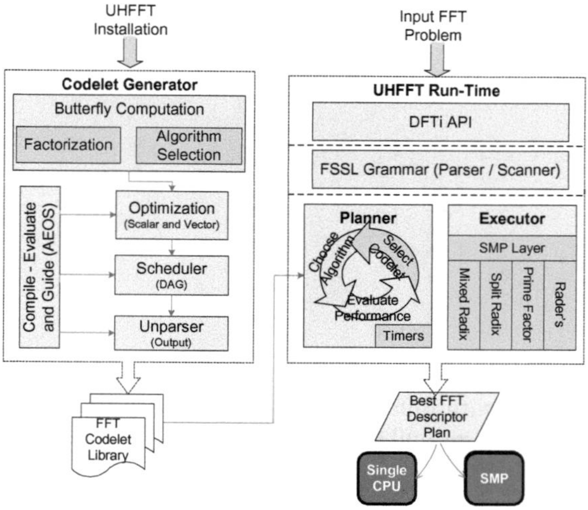

Fig. 2. UHFFT2 block diagram. *Codelets* specified by user are generated at installation time. At run-time the code blocks are used by planner to compose best FFT computation schedules. UHFFT2 supports Intel's DFT API [15] with some extensions.

2.1 Code Generator

UHFFT contains a number of *codelets*, which are generated at installation time. Each *codelet* sequentially computes a part of the transform and overall efficiency of the code depends strongly on the efficiency of these *codelets*. Therefore, it is essential to have a highly optimized set of DFT *codelets* in the library. Apart from the normal DFT *codelets*, FFTGEN generates a few specialized sets of *codelets* that are used in the FFT computation at various stages of the algorithm. Two types of such *codelets* that are relevant to our discussion of self-sorting in-place algorithms are: the "coupled transposed *codelets*" and the "rotated PFA *codelets*". The code generator adapts to the platform, i.e., compiler and hardware architecture by empirically tuning the *codelets* using iterative compilation and feedback [1].

2.2 Planner

A FFT problem is given by a DFTi descriptor [15], which describes various characteristics of the problem, i.e., size, precision, placement, input and output data type, number of threads etc. Once the descriptor is submitted, planner selects the best plan, which may be used repeatedly to compute the transforms of same size. Our current implementation supports two strategies for searching the best plan. Both strategies are based on *dynamic programming* to search the space of possible factorizations and algorithms, given by a tree as shown in Figure 3. The first approach empirically evaluates the sub-plans and avoids the re-evaluation of identical sub-plans by maintaining a lookup table of their performances. This scheme is called *Context Sensitive Empirical Search*. In the second approach, the cost of search is significantly reduced at the expense of

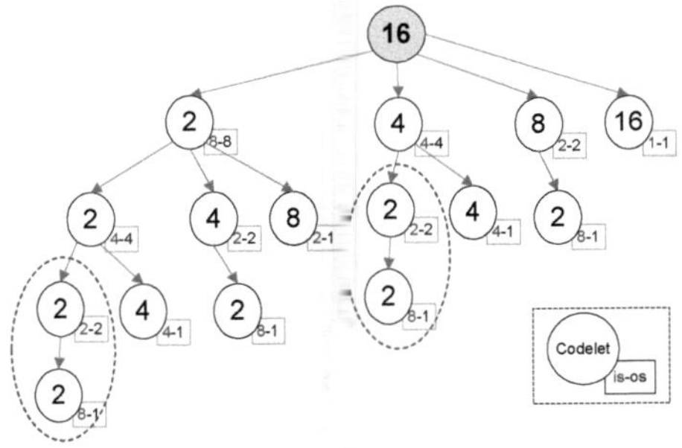

Fig. 3. Search space for size 16 FFT (out-of-place). There are 8 possible plans shown here. *Codelets* are called with different input and output strides at different levels of tree (recursion) also known as ranks.

quality of plan. In this scheme, the cost of a sub-plan is estimated by empirically evaluating only the root of sub-plan (*codelet*) that is encountered in a bottom up traversal. This search strategy is called *Context Free Hybrid Search* as it does not take into account the contextual state of cache due to prior factors of sub-plan. In this paper, we use the first approach because it generates good quality plans at reasonable cost of search.

2.3 FFT Schedule Specification Language (FSSL)

An execution plan specifies the *codelets* that will be used for the FFT computation and also the order (schedule) in which they will be executed. A FFT plan is described in concise language using the grammar given in Table 1. It allows different algorithms to be mixed to generate a high performance execution plan based on properties of the input vector and its factors. Indeed, by implementing a minimal set of rules, adaptive schedules can be constructed dynamically that suit different types of architectures.

3 Self Sorting In-Place FFT Implementation

Our implementation of self-sorting in-place FFTs is based on the algorithms given in [19,4]. These algorithms factorize a given FFT of size N in a palindrome to couple the sorting step with the butterfly computation. However, many alternative factorizations exist for a given transform size that fulfill this requirement. Figure 4 illustrates two factorization schemes for an example of size 48 transform. Even though the two factorizations ultimately use the same factors (*codelets*), their index mapping or data access pattern can be significantly different. An illustration of the memory blocking resulting from the two schedules is given in Figure 5. Notice that the scheme (a) recurses *outward* by first selecting the largest square factors, while scheme (b) recurses *inward* by choosing the largest factor to be in the middle of two small factors. The *outward* recursion generates larger blocks that can be executed independently or in large vectors

Table 1. FFT Schedule Specification Language grammar rules

#	CFG Rules
1-2	ROOT⟶MULTID_FFT | SMPFFT
3-4	MULTID_FFT⟶FFT | [NDFFT , FFT]
5-6	NDFFT⟶NDFFT , FFT | FFT
7	SMPFFT⟶(**mr**$P\mathbb{Z}$ BLOCK , MULTID_FFT) $\mathbb{Z}$
8-9	FFT⟶FFT **mr** MODULE | MODULE
10-11	MODULE⟶CODELET | ORDERED_FFT
12	| (**rader** , FFT) $\mathbb{Z}$
13	ORDERED_FFT⟶(**inplace**$\mathbb{Z}$, FFT)
14	| (**outplace**$\mathbb{Z}$, FFT)
15	| (**pfa**$\mathbb{Z}$, PFAFFT **pfa** ROT_CODELET)
16	CODELET⟶$n \in$ DFT codelets
17	ROT_CODELET⟶$n \in$ Rotated DFT codelets
18	BLOCK⟶**b** $\mathbb{Z}$: $\mathbb{Z}$

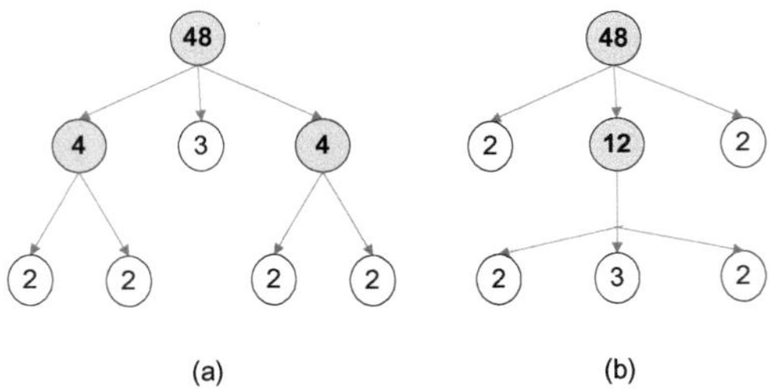

Fig. 4. Two factorization schemes that form a palindrome. Factorization scheme (a) (inplace48, (inplace4, 2mr2)mr3mr(inplace4, 2mr2)) and (b) (inplace48, 2mr(inplace12, 2mr3mr2)mr2). Assuming, only two *codelets* have been generated in the library, namely 2 and 3.

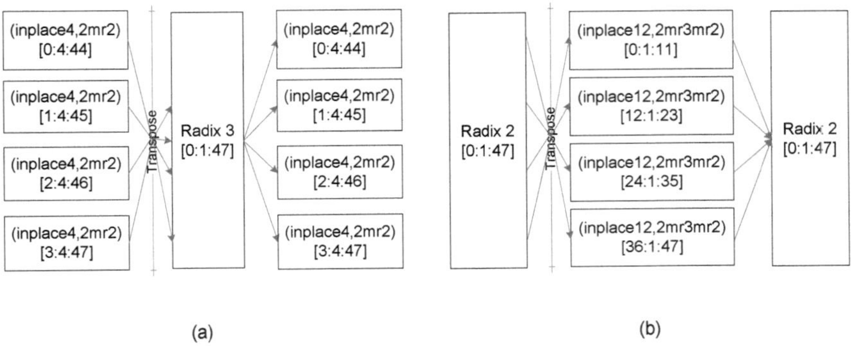

Fig. 5. Data Access pattern using the two factorization schemes shown in Figure 4. Each block shows the vector indices accessed in Matlab vector representation.

at unit strides. However, in order to maintain spatial locality, the blocks will need to be executed in a breadth first fashion thereby reducing the potential for temporal locality between successive ranks. On the other hand, the *inward* recursion allows better blocking that maintains both temporal and spatial locality in a cache oblivious manner. In UHFFT, the planner uses *inward* recursion strategy to factorize and execute the self-sorting in-place algorithm. Following, we give the simple steps that the planner follows to select the best palindrome factorization of size N:

1. Find the factors with odd exponent to construct the middle rank.
   (a) Factorize size N in powers of prime factors, i.e., $N = \prod_{i=0} p_i^{r_i}$, where all the factors p_i are co-primes and r_i is the exponent.
   (b) Based on the above factorization, divide the size $N = q \times s$ in two sets $q = \prod_{i=0} p_i^{e_i}$ and $s = \prod_{i=0} p_i^{o_i}$, where e_i is even and o_i is either 0 or 1.
   (c) Construct the largest possible PFA plan ρ_{mid} for middle rank of size n_{mid}, from all factors of s and optionally only $p_i^{2\times j}$factors of q.
   (d) If all the factors of s were not used (rotated *codelets* are not generated for all prime factors), then the middle rank will be solved using extra buffers.
   (e) If (d) is true then construct the best out-of-place planρ_{mid} using one of the search schemes.
2. Finally, construct the rest of in-place plan of size $\frac{N}{n_{mid}}$ by recursively factorizing it inwards:

$$N = m_0 \times n_{mid} \times m_0$$

$$\Rightarrow N = r_0 \times m_1 \times n_{mid} \times m_1 \times r_0$$

3. The search methods implemented in UHFFT can be used to find the best inplace plan:

$$\rho = (\text{inplace}N, r_0\text{mr}r_1\text{mr}\ldots\text{mr }\rho_{mid}\text{mr}\ldots\text{mr}r_1\text{mr}r_0)$$

In the above algorithm, we start by finding the odd factor to fit in the middle of palindrome. Since PFA algorithm is both self sorting and in-place, we try to construct the middle rank from co-prime factors. In case, there are remaining factors that do not have even exponent, we will need to use a self-sorting out-of-place algorithm that uses a work-space equal to the size of middle rank. Once the middle rank is factorized, rest of the factorization follows recursively in a straightforward manner. To illustrate the algorithm, lets take an example of size $N = 48$. If the library contains rotated *codelet* of size 16, then the largest PFA plan $\rho = (\text{inplace48}, (\text{pfa48}, 3\text{pfa16}))$ of size 48 is constructed which is both self sorting and in-place. Assuming that the rotated *codelet* for size 16 was not generated at installation time. In that case, the plan will use the middle rank of size 12: $\rho = (\text{inplace48}, 2\text{mr}(\text{pfa12}, 3\text{pfa4})\text{mr2})$.

In the above discussion, we omitted the details of "twiddle factors" multiplication, which is performed after each rank. In UHFFT, we have implemented specialized *codelets* that perform the "twiddle factors" multiplication inside the *codelets* thereby avoiding the extra loads and stores. The loads and stores are further reduced by implementing "coupled *codelets*"[19] that perform r small DFTs of size r. These square blocks r^2 are then transposed using registers and written back to the input vector. Notice that this kind of optimization is not possible if *outward* recursion is used because for a large size $N = m \times r \times m$, square matrix $m \times m$ will be too large to fit in the registers.

4 Results

We performed the benchmarking of our self-sorting in-place FFT implementation in UHFFT on Itanium 2 machine and compared its performance against FFTW and Intel's Math Kernel Library (MKL). Specification of the experimental setup is given in Table 2. The performance numbers in this paper, are derived from

Table 2. Experimental Setup

Feature	Description
Architecture	Itanium 2 1.5 Ghz, Cache: 16K/256K/6M, FP Registers=128 CacheLine: 64B/128B/128B, Associativity: 4/8/12
Compiler	Intel C Compiler icc 9.1, Flags=-O3
Libraries	UHFFT-2.0.1beta, FFTW-3.2alpha, MKL-9.1
Data Sets	Powers of 2, Small Co-prime factor sizes, Mixed

the total execution time of a FFT problem. The performance measure, Million Floating Point Operations Per Second (MFLOPS), is calculated from the execution time and commonly used algorithm complexity of FFT, i.e., $5N \log N$. The results were collected for double precision complex FFT of various sizes, ranging from 2 to 16M.

4.1 Performance Comparison of FFT Libraries

UHFFT and FFTW are open source adaptive libraries for computing FFTs efficiently. Intel's MKL is the vendor library that is tuned to Intel's architectures. All the three libraries support both in-place and out-of-place executions of FFT. Figure 6 shows the performance comparison of our implementation of self sorting in-place FFT computation with that of FFTW and MKL for powers of two size FFTs. For very small sizes $(2 - 2^5)$ the FFT is performed within registers using optimized *codelets*, which is why all the three libraries perform equally well. One of the main differences between our strategy in UHFFT and the other two libraries is that FFTW (and quite possibly MKL) would trade memory for performance; extra buffers can be used for improving the performance of code. On the other hand, UHFFT uses buffers only when an in-place in-order plan is not possible. Notice that for sizes larger than Level 2 cache capacity, i.e., 2^{14}, UHFFT performs better than FFTW. Indeed, for data $(N = 2^{18})$ that do not fit

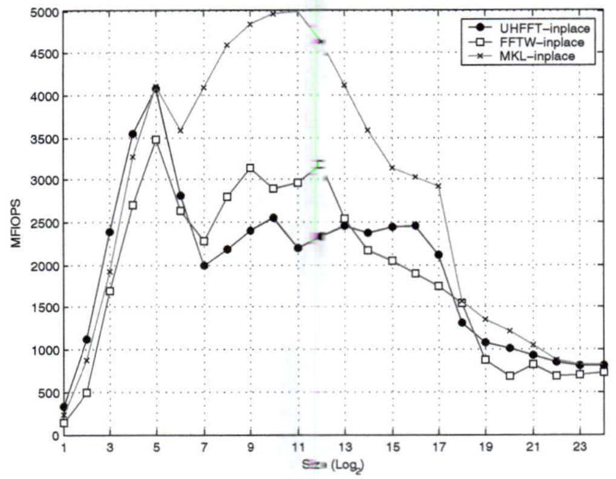

Fig. 6. Powers of 2 Sizes. Complex in-place in-order FFT Computation. UHFFT does not use any extra buffers (except the "twiddle factors" array) in the in-place execution of powers of 2 sizes.

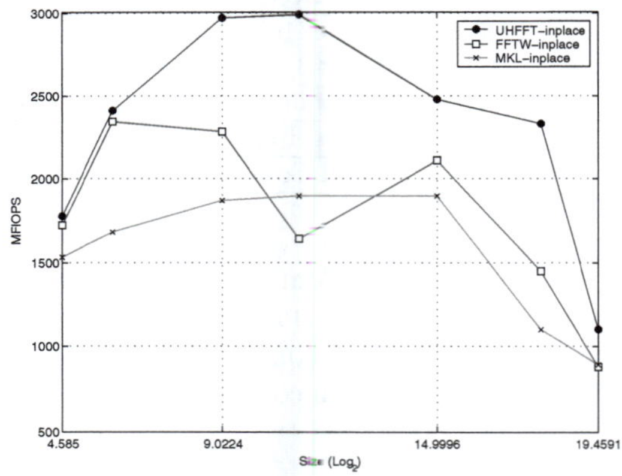

Fig. 7. Sizes with small co-prime factors. Complex in-place in-order FFT. UHFFT uses PFA algorithm, which is known to have lower complexity and produces ordered results. FFTW and MKL are not tuned for the sizes that have co-prime factors.

in any level of cache, our implementation performs consistently better because of superior cache management. Another advantage that UHFFT has over the other two libraries is its support for fast execution of sizes that have co-prime factors. In Figure 7, we computed the in-place FFT of sizes that have co-prime factors. Neither FFTW nor MKL support Prime Factor Algorithm (PFA). Hence, the two libraries have relatively lower performance on sizes that have co-prime factors. In Figure 8, we compared the performance of the three libraries on randomly sampled sizes that have both powers of 2 and prime factors. These results show that UHFFT is equally good at performing in-place self-sorting FFT on non-powers of two sizes, which are less straightforward to factorize in a palindrome.

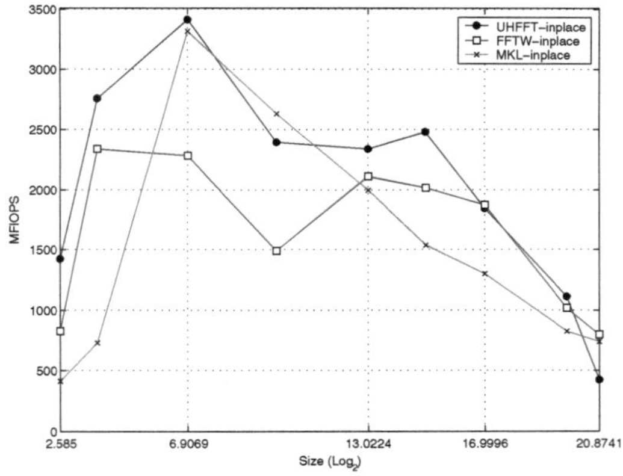

Fig. 8. Sizes with prime as well as powers of two factors. Complex in-place in-order FFT Computation. Not all the sizes presented here result in palindrome of factors, which means that small buffers may be required to compute in-place in-order FFT efficiently.

4.2 Performance Comparison of In-place and Out-of-place FFT

We compared the performance of our in-place implementation of self sorting FFT schedules with that of out-of-place execution. Even though in-place schedules perform many small transposes, overall their performance is quite similar to out-of-place executions as shown in Figures 9(a) & 9(b). The reason that there is no drastic degradation of performance in case of in-place execution is that the transposes are performed inside the *codelets,* i.e., using registers. This avoids the

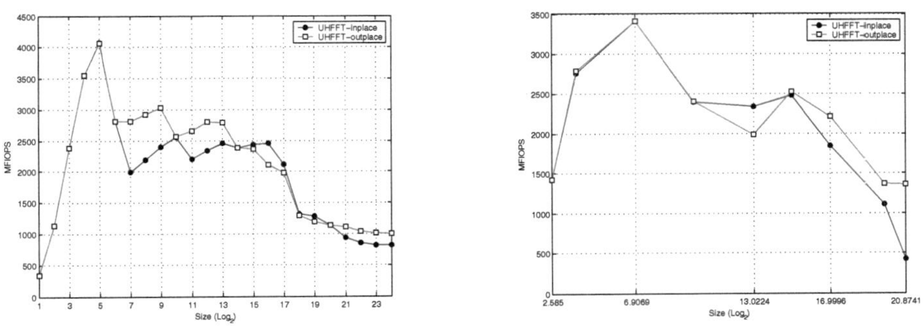

(a) Powers of 2

(b) Non-powers of 2

Fig. 9. Comparison of out-of-place and in-place FFT performance in UHFFT. (a) The performance for small size ($N \leq 2^6$) transforms is identical because the sorting is performed in the registers for both executions. (b) Non Powers of 2 sizes include 6, 14, 120, 968, 8320, 32760, 131040, 915200, 1921920. For sizes 6 and 14, *codelets* are used directly and for sizes 120 and 32760, pfa plans are used for both out-of-place and in-place execution.

extra cost of loads and stores when the re-ordering is performed out of cache. The performance of out-of-place schedules starts dropping slightly earlier than in-place schedules around the L3 cache size mark (2^{18}), because of its extra bandwidth requirement (both input and output vector need to be fetched in the cache).

5 Conclusion

Computing self-sorting in-place FFTs poses a challenge when the execution needs to be carried out without use of additional work-space. The re-ordering performed inside the butterfly can seriously degrade the performance on current hierarchical memory architectures if the problem is not carefully factorized. In this paper we presented a recursive cache oblivious formulation of self-sorting in-place FFT algorithm that is suited to hierarchical memory architectures. Despite requiring the re-orderings to be performed in-place, the performance of self sorting in-place FFT was shown to be at par with out-of-place FFT. This is achieved through careful factorization of the problem such that the partial ordering can be performed within registers when the data has already been loaded for butterfly computation; this avoids extra loads and stores. In order to express FFT algorithms and factorizations, we presented a language, i.e., FFT Schedule Specification Language (FSSL), that defines the schedules in compact representation. To evaluate and validate the efficacy of our recursive strategy, we compared the performance of our implementation with the in-place in-order FFT performance of FFTW and Intel's MKL on Itanium 2. For Powers of two sizes, UHFFT performs competitively against FFTW, without using any extra buffers. For non-powers of two sizes, the performance is much better than that of the other two libraries because UHFFT planner adaptively selects the PFA algorithm for sizes that have co-prime factors.

References

1. Ali, A., Johnsson, L., Mirkovic, D.: Empirical Auto-tuning Code Generator for FFT and Trignometric Transforms. In: ODES: 5th Workshop on Optimizations for DSP and Embedded Systems, in conjunction with International Symposium on Code Generation and Optimization (CGO), San Jose, CA (March 2007)
2. Ali, A., Johnsson, L., Subhlok, J.: Scheduling FFT Computation on SMP and Multicore Systems. In: International Conference on Supercomputing, Seattle, WA (June 2007)
3. Burrus, C.S., Eschenbacher, P.W.: An in-place, in-order prime factor FFT algorithm. IEEE Transactions on Acoustics, Speech, and Signal Processing 29, 806–817 (1981)
4. Burrus, C.S., Johnson, H.W.: An in-order, in-place radix-2 FFT. IEEE Transactions on Acoustics, Speech, and Signal Processing 9, 473–476 (1984)
5. Cooley, J., Tukey, J.: An algorithm for the machine computation of complex fourier series. Mathematics of Computation 19, 297–301 (1965)

6. Franchetti, F., Voronenko, Y., Püschel, M.: FFT program generation for shared memory: SMP and multicore. In: SC 2006, p. 115. ACM Press, New York (2006)
7. Frigo, M.: A fast Fourier transform compiler. In: PLDI 1999: Proceedings of the ACM SIGPLAN 1999 conference on Programming language design and implementation, pp. 169–180. ACM Press, New York (1999)
8. Frigo, M., Johnson, S.G.: The design and implementation of FFTW3. In: Proceedings of the IEEE 1993, vol. 2, pp. 216–231 (2005) special issue on Program Generation, Optimization, and Platform Adaptation
9. Hegland, M.: A self-sorting in-place fast Fourier transform algorithm suitable for vector and parallel processing. Numerische Mathematik 68(4), 507–547 (1994)
10. Loan, C.V.: Computational frameworks for the fast Fourier transform. Society for Industrial and Applied Mathematics, Philadelphia, PA, USA (1992)
11. Mirkovic, D., Johnsson, S.L.: Automatic Performance Tuning in the UHFFT Library. In: Delugach, H.S., Stumme, G. (eds.) ICCS 2001. LNCS (LNAI), vol. 2120, pp. 71–80. Springer, Heidelberg (2001)
12. Mirkovic, D., Mahasoom, R., Johnsson, S.L.: An adaptive software library for fast Fourier transforms. In: International Conference on Supercomputing, pp. 215–224 (2000)
13. Püschel, M., Moura, J.M.F., Johnson, J., Padua, D., Veloso, M., Singer, B.W., Xiong, J., Franchetti, F., Gačić, A., Voronenko, Y., Chen, K., Johnson, R.W., Rizzolo, N.: SPIRAL: Code generation for DSP transforms. In: Proceedings of the IEEE, special issue on Program Generation, Optimization, and Adaptation, vol. 93(2), pp. 232–275 (2005)
14. Singleton, R.C.: An algorithm for computing the mixed radix fast Fourier transform. IEEE Transactions on Audio and Electroacoustics 17, 93–103 (1969)
15. Tang, P.T.P.: DFTI – A New Interface for Fast Fourier Transform Libraries. ACM Transactions on Mathematical Software 31(4), 475–507 (2005)
16. Temperton, C.: Self-Sorting Mixed-Radix Fast Fourier Transforms. Journal of Computational Physics 52, 1–23 (1983)
17. Temperton, C.: Implementation of a Self-Sorting In-Place Prime Factor FFT Algorithm. Journal of Computational Physics 54, 283–299 (1985)
18. Temperton, C.: A new set of minimum-add small-n rotated DFT modules. J. Comput. Phys. 75(1), 190–198 (1988)
19. Temperton, C.: Self-Sorting In-Place Fast Fourier Transforms. SIAM Journal on Scientific and Statistical Computing 12(4), 808–823 (1991)

Parallel Multistage Preconditioners Based on a Hierarchical Graph Decomposition for SMP Cluster Architectures with a Hybrid Parallel Programming Model

Kengo Nakajima

Department of Earth and Planetary Science, The University of Tokyo
7-3-1 Hongo, Bunkyo-ku, Tokyo 112-0002, Japan
nakajima@eps.s.u-tokyo.ac.jp

Abstract. In this work, the Parallel Hierarchical Interface Decomposition Algorithm (PHIDAL) and a hybrid parallel programming model were applied to finite-element based simulations of linear elasticity problems in media with heterogeneous material properties using parallel preconditioned iterative solvers. Reverse Cuthill-McKee reordering with cyclic multicoloring (CM-RCM) was applied for parallelism on each SMP node through OpenMP. The developed code has been tested on the IBM p5-575 and the TSUBAME Grid Cluster using up to 512 cores. Preconditioners based on PHIDAL provide a superior scalable performance and robustness on both architectures in comparison to conventional block Jacobi-type localized preconditioners.

1 Introduction

1.1 Parallel Programming Models on SMP Cluster Architectures

In order to achieve minimal parallelization overheads on SMP (symmetric multiprocessors) clusters, a multi-level *hybrid* parallel programming model is often employed (Fig. 1). In this method, coarse-grained parallelism is achieved through domain decomposition by message passing among SMP nodes, while fine-grained parallelism is obtained via loop-level parallelism inside each SMP node using compiler-based thread parallelization techniques, such as OpenMP [1,2]. Another often-used programming model is the single-level *flat MPI* model (Fig. 1), in which separate single-threaded MPI processes are executed on each processing element (PE) [1,2]. The efficiency of each model depends on hardware performance (CPU speed, communication bandwidth, memory bandwidth, and the balance between these), application features, and problem size [3].

In previous works [1,2], the author developed an efficient parallel iterative solver for finite-element applications on the Earth Simulator (ES) [4] using a three-level hybrid parallel programming model with MPI, OpenMP and vectorization. Multicolor-based reordering methods were applied to distributed data sets on each SMP node in order to achieve an optimum parallel performance of Krylov iterative solvers with ILU/IC (Incomplete LU/Cholesky Factorization) preconditioners. The developed

R. Perrott et al. (Eds.): HPCC 2007, LNCS 4782, pp. 384–395, 2007.

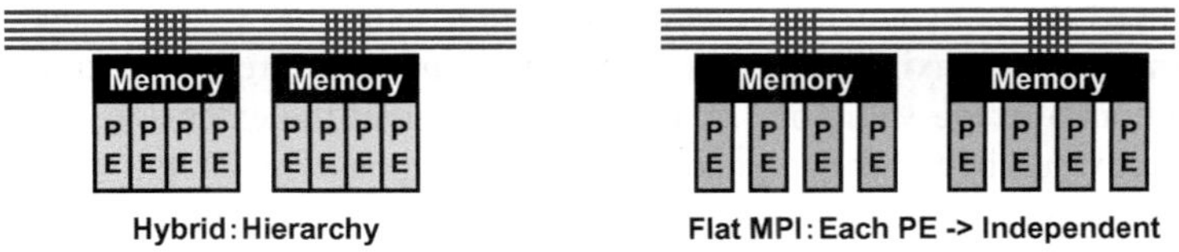

Fig. 1. Parallel programming models on SMP cluster architecture [1,2]

method attained 3.8 TFLOPS with 176 SMP nodes of ES, corresponding to more than 33% of the peak performance (11.3 TFLOPS) [1]. Because of the relatively high sustained memory bandwidth of ES, flat MPI and hybrid parallel programming models are competitive, but the *hybrid* model outperforms *flat MPI*, if the number of processors exceeds 1,000. In general, *hybrid* parallel programming models provide a better performance the larger the number of nodes.

For cases using many colors, fewer numbers of iterations are required for convergence, because there are fewer incompatible nodes [5]. However, performance declines for these cases due to the smaller loop length and greater overhead on vector processors. For the ES, the hybrid parallel programming model was found to be very sensitive to the number of colors.

1.2 Parallel Preconditioned Iterative Solvers

Block Jacobi-type localized preconditioners are widely used for parallel iterative solvers [1,2]. They provide excellent parallel performance for well-defined problems, although the number of iterations required for convergence gradually increases according to the number of processors. However, this preconditioning technique is not robust for ill-conditioned problems with many processors, because it ignores the global effect of external nodes in other domains [1.2]. The generally used remedy is the extension of overlapped elements between domains [6,7]. Usually, this approach reduces the number of iterations required for convergence, but at the expense of requiring additional computations and communications, and so the final elapsed time of the computation is not necessarily reduced.

Table 1 shows the convergence of a parallel iterative solver with block Jacobi-type localized preconditioners for 3D linear elasticity problems in media with heterogeneous material properties on AMD Opteron clusters with 64 cores. In these computations the GPBi-CG (Generalized Product-type methods based on Bi-CG) solver [8]

Table 1. Convergence of parallel iterative solver using a block Jacobi-type localized preconditioner (GPBi-CG solver with localized SGS preconditioner) applied to 3D linear elasticity problems in media with heterogeneous material properties with 1,000,000 elements (3,090,903 DOF). AMD Opteron cluster with 64 cores

Depth of Overlap	Iterations for Convergence	sec.	Average Size per Core	
			Communication Table	Non-Zero Components in Coef. Matrix
0	1004	94.0	3,474	410,009
1	940	99.9	3,474	468,015
2	909	115.5	7,116	496,670

with localized SGS (Symmetric Gauss-Seidel) preconditioner [1,2] has been applied. The minimum and maximum values of Young's modulus are 10^{-3} and 10^{3}, respectively, with an average value of 1.0. It may be seen that the extension of overlapped elements provides better convergence, but at the expense of increasing the elapsed computation time somewhat. *Selective-overlapping* [7] may improve this situation, but the benefits of developing more general preconditioners for scalable and robust computations are clear.

1.3 Overview of PHIDAL

The Parallel Hierarchical Interface Decomposition Algorithm (PHIDAL) algorithm provides robustness and scalability for parallel ILU/IC preconditioners [9]. PHIDAL is based on defining "*hierarchical interface decomposition* (HID)". The HID process starts with a partitioning of the graph, with one layer of overlap. The "stages" are defined from this partitioning, with each stage consisting of a set of vertex groups. Each vertex group of a given stage is a *separator* for vertex groups of a *lower* stage. The incomplete factorization process proceeds by "stage" from lowest to highest. Due to the separation property of the vertex groups at different stages, this process can be carried out in a highly parallel manner. In [9], the concept of *connectors* (small connected sub-graphs) of different *levels* and *keys* are introduced for the purposes of applying this idea to general graphs as follows:

- Connectors of *level-1* (C^1) are the sets of interior points. Each set of interior points is called a *sub-domain*.
- A connector of *level-k* (C^k) ($k>1$) is adjacent to k *sub-domains*
- No C^k is adjacent to any other connector of level-k
- *Key(u)* is the set of sub-domains (connectors of level-1, C^1) connected to vertex *u*.

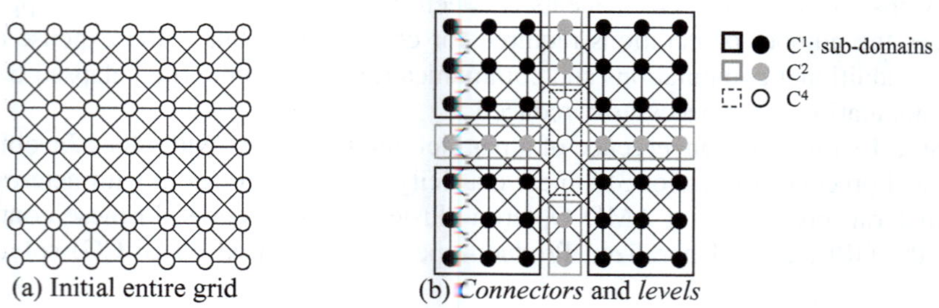

Fig. 2. Partitioning of a 9-point grid into 4 sub-domains

Fig.2 shows the example of the partition of a 9-point grid into 4 domains. In this case, there are 4 connectors of level-1 (C^1, sub-domain), 4 connectors of level-2 (C^2) and 1 connector of level-4 (C^4). Note that different connectors of the same level are not connected directly, but are separated by connectors of higher levels. These properties provide the block structure of the coefficient matrix *A* through reordering the unknowns by this decomposition. If the unknowns are reordered according to their level numbers, from the lowest to highest, the block structure of the reordered matrix would

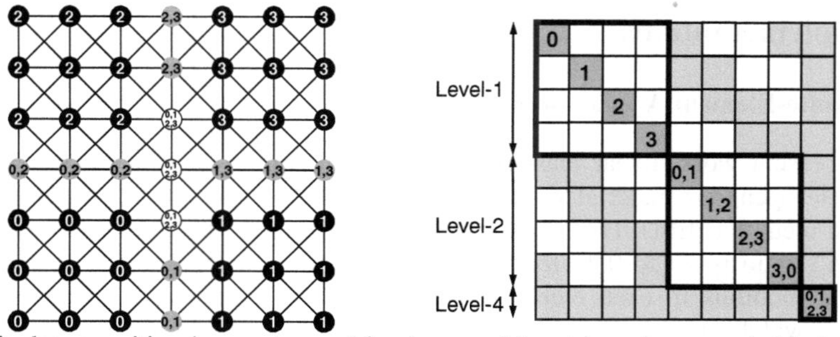

(a) Domain decomposition (*connectors* and *keys*) (b) matrix and non-empty blocks

Fig. 3. Domain/block decomposition of the matrix according to the HID reordering

```
Terms:
• p          number of sub-domains
• |Key(u)|   number of sub-dommains in Key(u)
• L^l        set of vertices whose keys have a
             cardinality equal to l
• V^l(i)     set of vertices belonging to L^l and
             adjacent to vertex i
• Kdeg_u     number of neighbors in V^l(i) that have a
             key different from Key(u)

Initialization:
    for each vertex u ∈ V
        Key(u) := list of subdomains containing vertex u
    end
    for each vertex u ∈ V
        Kdeg_u := |{v ∈ V^l(u) / Key(v) ≠ Key(u)}|
    end
    for l = 1, to p do:
        L^l = {u ∈ V / |Key(u)| = = l}
    end

Main Loop:
    for l = 2 to p do:
        for each vertex u ∈ L^l do:
            Key(u) := Key(u) ∪ ⋃_{k=1,...,l-1} (⋃_{v∈V^k(u)} Key(v))
        end
        while all vertices in L^l have not been processed;
            get u the vertex in L^l such that Kdeg_u is maximum.
            Key(u) := Key(u) ∪ ⋃_{v∈V^k(u)} Key(v)
            m = |Key(u)|
            if m > l then
                L^l := L^l \ {u}
                L^m := L^m ∪ {u}
                for each vertex v ∈ V^l(u)
                    Kdeg_v := Kdeg_v - 1
                end
            endif
        end
    end
```

Fig. 4. Algorithms for HID processes [9]

be as shown in Fig.3. This block structure leads to natural parallelism if ILU/IC decompositions or forward/backward substitution processes are applied. Fig.4 provides algorithms for constructing independent connectors [9]. Thus, PHIDAL-based ILU/IC preconditioners can consider the global effect of external domains in parallel computations, and are expected to be more robust than block Jacobi-type localized ones.

1.4 Overview of the Present Work

In the present work, a hybrid parallel programming model has been applied to finite-element based simulations of 3D linear elasticity problems in media with heterogeneous material properties using parallel preconditioned iterative solvers based on "PHIDAL". Reverse Cuthill-McKee reordering with cyclic multicoloring (CM-RCM) has been applied for parallelism in ill-conditioned problems on each SMP node through OpenMP. The developed code is tested on the IBM p5-575 [10] and the TSUBAME Grid Cluster [11] using up to 512 cores. The rest of this study is organized as follows: In section 2 we outline the details of the present application, and describe the reordering procedures. In section 3 preliminary results of the computations are described, while some final remarks are offered in section 4.

2 Implementations

2.1 Finite-Element Applications

In the present work, linear elasticity problems in simple cube geometries of media with heterogeneous material properties (Fig.5(a)) are solved using a parallel finite-element method (FEM). Tri-linear hexahedral elements are used for the discretization. Poisson's ratio is set to 0.25 for all elements, while a heterogeneous distribution of Young's modulus in each element is calculated by a sequential Gauss algorithm, which is widely used in the area of geostatistics [12]. The minimum and maximum values of Young's modulus are 10^{-3} and 10^{3}, respectively, with an average value of 1.0. The boundary conditions are described in Fig.5(b). The GPBi-CG (Generalized Product-type methods based on Bi-CG) solver with SGS (Symmetric Gauss-Seidel) preconditioner (SGS/GPBi-CG) was applied. Since the present application concerns linear elasticity problems, the coefficient matrices for this application are symmetrical positive definite. Although GPBi-CG is designed for general unsymmetrical matrices, this method is adopted in the present study due to its robustness for ill-conditioned problems [1,2]. In SGS preconditioning, the original matrix *A* is used as a preconditioning matrix, therefore no factorization processes occur.

The code is based on the framework for parallel FEM procedures of GeoFEM [13], and the GeoFEM's local data structure is applied. The local data structures in GeoFEM are node-based with overlapping elements [13]. In FEM-type applications, most communication between processors occurs via information exchange at domain boundaries. The ratio of communication to computation is usually small [1,2]. In the present work, GeoFEM's original partitioner for domain decomposition has been modified so that it can create a distributed hierarchical data structure for PHIDAL. Each *sub-domain* (interior vertices, connectors of level-1) is assigned to an individual SMP node, which corresponds to each MPI process in the *hybrid* parallel programming model [1,2]. Higher-level connectors are distributed to each domain so that load-balancing can be attained and communications can be minimized. Fig.6 shows an example of the final partition of a 9-point grid into 4 domains.

In each local data set, vertices are renumbered according to their level numbers, from the lowest to highest. During forward/backward substitution processes in SGS preconditioning, global communications are required for consistency, when computations at each *level* have been completed. Hierarchical communication tables for these special communications are also created by the modified partitioners.

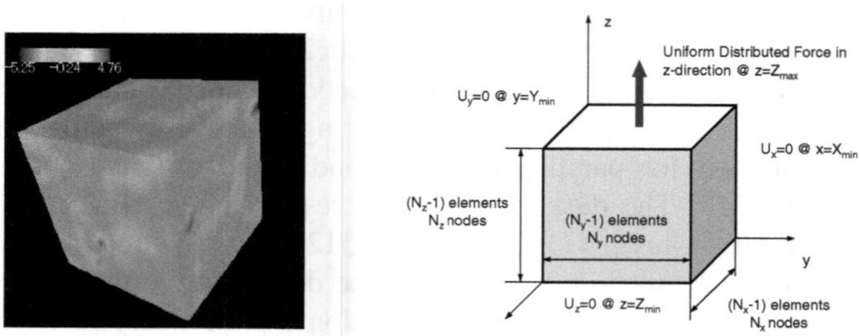

(a) Heterogeneous distribution of material property (b) Boundary conditions

Fig. 5. Simple cube geometries with heterogeneity as domains for 3D linear elasticity problems

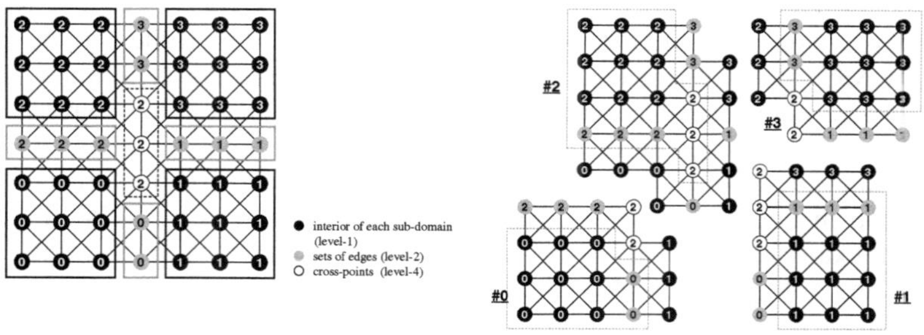

(a) Final partition (number's correspond to ID of partition (0-3))

(b) Distributed local data sets with external vertices

Fig. 6. The final partition of a 9-point grid into 4 domains

2.2 Reordering Procedures

In the *hybrid* parallel programming model, multithreading, such as OpenMP, is applied to each partitioned domain defined in the previous section, where each domain corresponds to a single SMP node. The reordering of vertices in each domain allows the construction of local operations without global dependency but with continuous memory access, in order to achieve optimum parallel performance of Krylov iterative solvers with ILU/IC preconditioners. In previous studies [1,2], the author adopted multicoloring as a reordering method, mainly because a highly parallel performance and load balancing can easily be obtained. In the present work, a Reverse Cuthill-McKee (RCM) type approach is introduced. The RCM method is a typical level set reordering method [1,5,14], which is a *reversing* of the Cuthill-McKee reordering (CMK). In CMK reordering, the vertices adjacent to a visited vertex are traversed from lowest to highest degree, where the *degree* of a vertex refers to the number of vertices connected to it. Multicoloring (MC) is much simpler than RCM. MC is based on an idea in which no two adjacent vertices have the same color. In both methods, vertices of the same color (or level set) are independent. Therefore, parallel operation is possible for the vertices of each color, and the number of vertices of each color should be as large as possible for high granularity in the parallel computation.

RCM (Fig. 7(a)) reordering facilitates a faster convergence of iterative solvers with ILU/IC preconditioners than MC reordering, but leads to irregular numbers of vertices in each level set. For example, in Fig. 7(a), the 1st level set is of size 1, while the 8th level set is of size 8. In contrast, the MC method uses a uniform number of

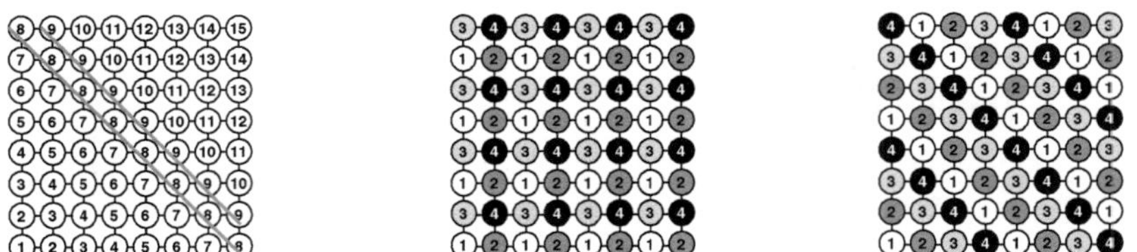

(a) RCM (b) MC: 4 colors (c) CM-RCM: 4 colors

Fig. 7. Example of RCM, MC and CM-RCM coloring for reordering on a 5-point grid [1,2] vertices of each color (Fig. 7(b)). However, it is widely known that the convergence of iterative solvers with ILU/IC preconditioners is rather slow if the vertices are reordered by MC [5]. Convergence can be improved by increasing the number of colors due to the consequent reduction in the number of incompatible local graphs [5], but this reduces the number of vertices of each color. The solution to this trade-off is RCM with cyclic-multicoloring (CM-RCM) [15]. In this method, further renumbering in a cyclic manner is applied to vertices that are reordered by RCM. Figure 7(c) shows an example of CM-RCM reordering. In this case, there are 4 colors: the 1st, 5th, 9th and 13th groups in Fig. 7(a) are classified into the 1st color. There are 16 vertices in each color. In CM-RCM, the number of colors should be large enough to ensure that vertices of the same color are independent.

Figure 8 shows the convergence of the SGS/GPBi-CG solver with MC and CM-RCM reordering on a single AMD Opteron core for small problems (29,791 elements, 98,304 DOF) in elastic media with (a) homogeneous and (b) heterogeneous distributions of material properties. In both cases, CM-RCM yields a faster convergence. In

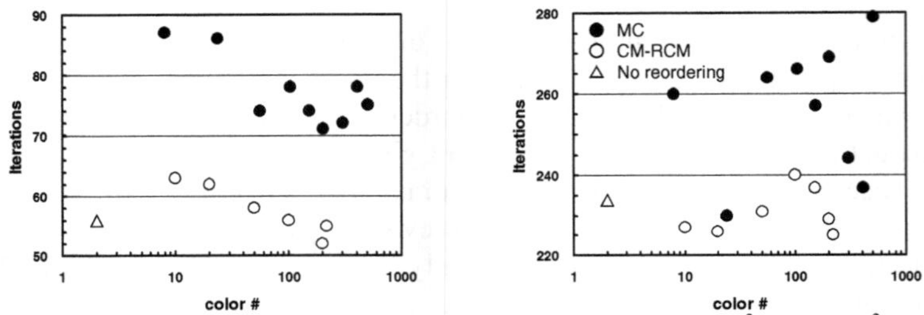

(a)E_{max}= E_{min}= 1.0 (b) E_{max}=10^3, E_{min}=10^{-3}

Fig. 8. Convergence of SGS/GPBi-CG on a single core for a 3D linear elasticity model with a heterogeneous distribution of Young's modulus (E_{max}: maximum, E_{min}: minimum), for 29,791 elements and 98,304 DOF

```
    do lev= 1, LEVELtot
      do ic= 1, COLORtot(lev)
!$omp parallel do private(ip,i,SW1,SW2,SW3,isL,ieL,j,k,X1,X2,X3)
        do ip= 1, PEsmpTOT
        do i = STACKmc(ip-1,ic,lev)+1, STACKmc(ip,ic,lev)
          SW1= WW(3*i-2,R); SW2= WW(3*i-1,R); SW3= WW(3*i  ,R)
          isL= INL(i-1)+1; ieL= INL(i)
          do j= isL, ieL
              k= IAL(j)
             X1= WW(3*k-2,R); X2= WW(3*k-1,R); X3= WW(3*k  ,R)
            SW1= SW1 - AL(9*j-8)*X1 - AL(9*j-7)*X2 - AL(9*j-6)*X3
            SW2= SW2 - AL(9*j-5)*X1 - AL(9*j-4)*X2 - AL(9*j-3)*X3
            SW3= SW3 - AL(9*j-2)*X1 - AL(9*j-1)*X2 - AL(9*j  )*X3
          enddo
          X1= SW1; X2= SW2; X3= SW3
          X2= X2 - ALU(9*i-5)*X1
          X3= X3 - ALU(9*i-2)*X1 - ALU(9*i-1)*X2
          X3= ALU(9*i  )*  X3
          X2= ALU(9*i-4)*( X2 - ALU(9*i-3)*X3 )
          X1= ALU(9*i-8)*( X1 - ALU(9*i-6)*X3 - ALU(9*i-7)*X2)
          WW(3*i-2,R)= X1; WW(3*i-1,R)= X2; WW(3*i  ,R)= X3
        enddo
        enddo
!$omp end parallel do
      enddo

      call SOLVER_SEND_RECV_3_LEV lev,...):   Communications using
                                              Hierarchical Comm. Tables.
    enddo
```

Fig. 9. Forward substitution process of preconditioning written in FORTRAN and MPI with OpenMP directives, global communications using hierarchical communication tables occur at the end of the computation of each level general, convergence is faster the larger the number of colors. The relationship between convergence and number of colors is not monotonic in heterogeneous cases, because only information for connected graphs is considered in the reordering procedures. In the present work, CM-RCM with 10 colors will be applied for performance evaluation. This reordering procedure has to be applied to the vertices at each *level.* The number of vertices in connectors of level 2 and above is usually relatively small. In the present work, the number of colors at each level is specified so that the number of vertices at each level is at least 10. Figure 9 shows *forward substitution* process of preconditioning written in FORTRAN and MPI with OpenMP directives. Global communications using hierarchical communication tables, mentioned in the previous chapter, occur in the end of the computation at each level.

3 Examples

3.1 Hardware Environment

PHIDAL was applied to a parallel FEM code with an SGS/GPBi-CG solver using a *hybrid* parallel programming model. The FEM code solves for simulations of 3D linear elasticity problems in media with heterogeneous material properties, as shown in Fig.5. The developed code has been tested on the IBM p5-575 of the National Energy Research Scientific Computing Center at Lawrence Berkeley National Laboratory (*bassi*) [10], and the TSUBAME Grid Cluster at the Global Scientific Information and Computing Center at Tokyo Institute of Technology [11]. Table 2 presents a summary of the architectural characteristics of the each supercomputer.

The IBM p5-model 575 (IBM p5-575) at NERSC/LBNL [10] is a POWER5-based super scalar system, with 111 SMP nodes, 888 processors, and 3.5 TB memory. Total peak performance is 6.75 TFLOPS. Each node is connected via IBM's "Federation" HPS (High-Performance Switch).

The TSUBAME Grid Cluster at Tokyo Institute of Technology [11] is a scalar system with 655 SunFire X4600 nodes, where each node has 8 sockets (16 cores) of AMD's dual-core-Opteron at 2.4 GHz, connected through Coherent HyperTransport. The overall system has 10,480 cores, and 21.4 TB memory. Total peak performance is 50.4 TFLOPS. SMP nodes are connected through Infiniband 4x/Voltaire ISR 9288 switch. In the present work, only 8 of the cores on each SMP node have been used.

Table 2. Main architectural features of the IBM p5-575 and the TSUBAME Grid Cluster

	IBM p5-575 [10]	TSUBAME [11]
Core#/node	8	16
Clock rate (GHz)	1.90	2.40
Peak performance/PE (GFLOPS)	7.60	4.80
Memory/node (GB)	32	32
Peak Memory BW (GB/sec/node)	100	100
Network BW (GB/sec/node)	4.00	2.50
MPI Latency (μsec)	< 5.0	< 4.0

3.2 Results on the IBM p5-575

Performance of the developed code has been evaluated using between 1 and 8 SMP nodes (8 and 64 cores) of the IBM p5-575. In this evaluation, a strong scaling test has been applied, where the entire problem size is fixed at 1,594,323 DOF (512,000 elements). The hybrid parallel programming model has been applied using 8 threads. Figure 10 shows a comparison between block Jacobi-type localized preconditioner and the present PHIDAL-based one. In the case with 8 cores (= a single SMP node), there are no MPI communications, and "Block Jacobi" and "PHIDAL" provide identical results. As the number of core increases, the number of iterations required for convergence of the "Block Jacobi" increases significantly, but that of the "PHIDAL" stays constant. Both of the methods achieve excellent scalability, but PHIDAL achieves a better performance, because "PHIDAL" can consider the global effect of external domains in parallel computations, and are expected to be more robust. Figure 11 shows the effect of additional communications in SGS/GPBi-CG with PHIDAL. As shown in Fig.9, the PHIDAL algorithm requires additional communications at the end of forward/backward substitution processes at each level in the preconditioning. The ratio of the additional communications to entire computation time for the linear solver increases according to the number of cores. The ratio is 17% for 64 cores.

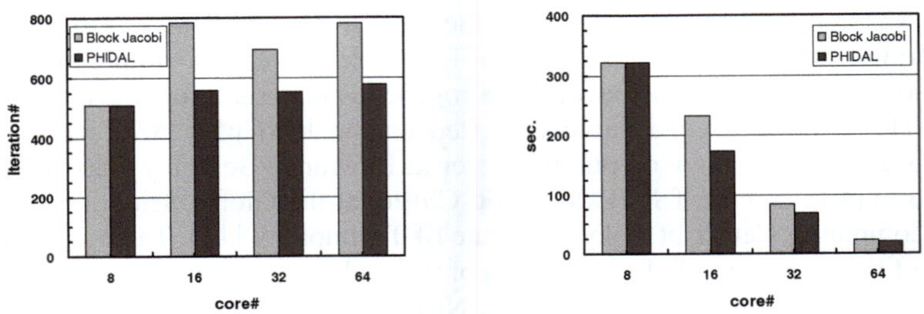

Fig. 10. Convergence (number of iterations required for convergence, and elapsed computation time for the linear solver) of SGS/GPBi-CG on the IBM p5-575 for the 3D linear elasticity model with a heterogeneous distribution of Young's modulus ($E_{max}=10^3$, $E_{min}=10^{-3}$), for 512,000 elements and 1,594,323 DOF, using the *hybrid* parallel programming model

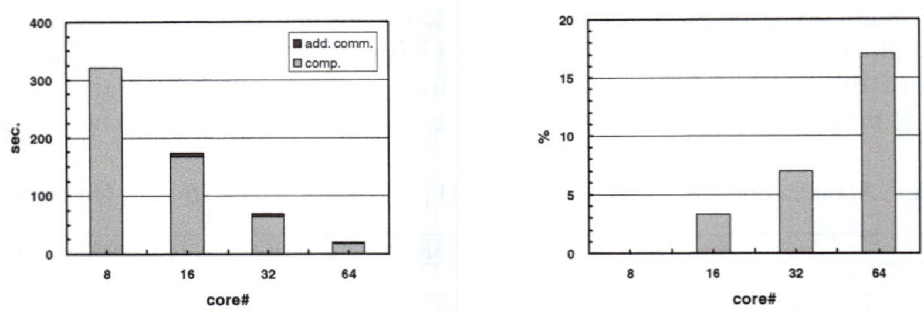

(a) Computation and additional communications (b) Ratio of additional communications

Fig. 11. Convergence of SGS/GPBi-CG with PHIDAL on the IBM p5-575 for the 3D linear elasticity model with a heterogeneous distribution of Young's modulus ($E_{max}=10^3$, $E_{min}=10^{-3}$), for 512,000 elements and 1,594,323 DOF, using the *hybrid* parallel programming model

3.3 Results on TSUBAME

The performance of the developed code has been evaluated using between 8 and 64 SMP nodes (64 and 512 cores) of the TSUBAME Grid Cluster. A strong scaling test has been applied, where the entire problem size is fixed at 6,440,067 DOF (2,097,152 elements). Both a *flat MPI* and a hybrid parallel programming model have been applied, where 8 threads have been used in the latter.

Each plate of Fig. 12 shows the performance of the "Block Jacobi" and "PHIDAL" for the hybrid parallel programming model. Although both of the methods achieve excellent scalability, PHIDAL achieves the superior performance. Figure 13 shows effect of additional communications in SGS/GPBi-CG with PHIDAL. The ratio of the additional communications to entire computation time for the linear solver increases according to number of cores; the ratio is 19% for 512 cores.

Figure 14 shows a comparison between *flat MPI* and the *hybrid* parallel programming models for PHIDAL. In general, flat MPI outperforms the hybrid parallel programming model, mainly because of the efficiency of memory. However, as the number of cores increases, the problem size for each core (or SMP node) decreases, and the performance of hybrid parallel programming model improves markedly. The effect of additional communications is significant in *flat MPI*, and the ratio of these communications to the entire solver process is more than 40% at 512 cores.

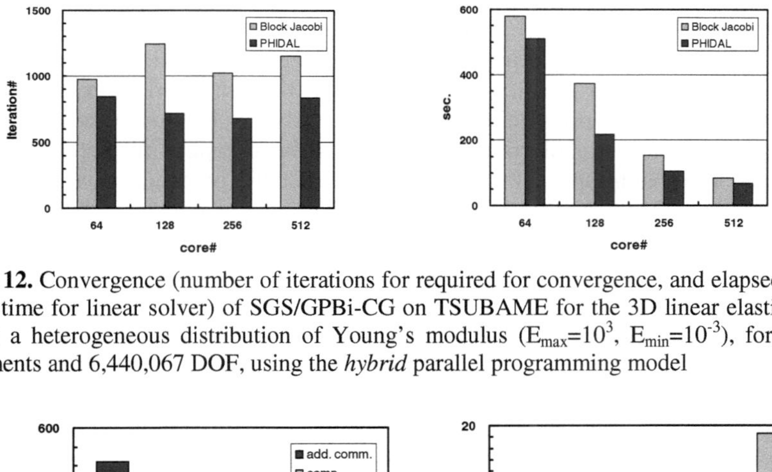

Fig. 12. Convergence (number of iterations for required for convergence, and elapsed computation time for linear solver) of SGS/GPBi-CG on TSUBAME for the 3D linear elasticity model with a heterogeneous distribution of Young's modulus ($E_{max}=10^3$, $E_{min}=10^{-3}$), for 2,097,152 elements and 6,440,067 DOF, using the *hybrid* parallel programming model

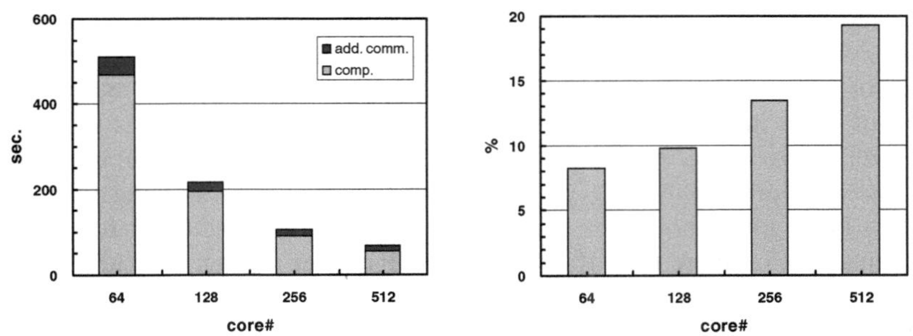

(a) Computation and additional communications (b) Ratio of additional communications

Fig. 13. Convergence of SGS/GPBi-CG with PHIDAL on TSUBAME for the 3D linear elasticity model with a heterogeneous distribution of Young's modulus ($E_{max}=10^3$, $E_{min}=10^{-3}$), for 2,097,152 elements and 6,440,067 DOF, using the *hybrid* parallel programming model

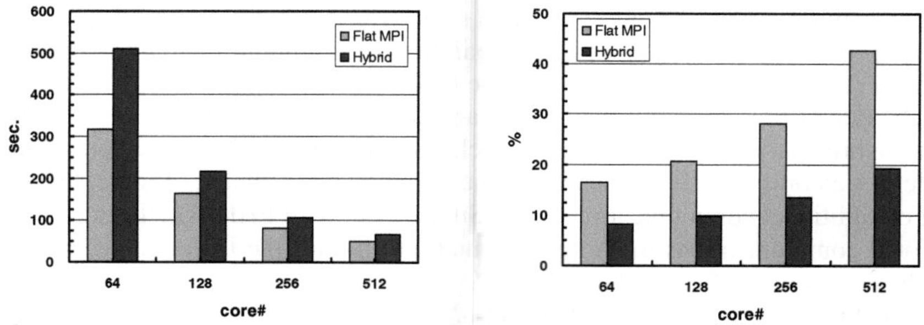

(a) Elapsed computation time for linear solvers (b) Ratio of additional communications

Fig. 14. Convergence of SGS/GPBi-CG with PHIDAL on TSUBAME for the 3D linear elasticity model with a heterogeneous distribution of Young's modulus ($E_{max}=10^3$, $E_{min}=10^{-3}$), for 2,097,152 elements and 6,440,067 DOF

4 Summary and Future Works

In this work, "PHIDAL (Parallel Hierarchical Interface Decomposition Algorithm)" using a hybrid parallel programming model has been applied to a finite-element solution of simulations of linear elasticity problems in media with heterogeneous material properties using parallel preconditioned iterative solvers. Reverse Cuthill-McKee reordering with cyclic multicoloring (CM-RCM) has been applied for parallelism on each SMP node through OpenMP. PHIDAL-based ILU/IC preconditioners can consider the global effect of external domains in parallel computations, and are expected to be more robust than block Jacobi-type localized ones.

The developed code has been tested on the IBM p5-575 and the TSUBAME Grid Cluster using up to 512 cores. A preconditioner based on PHIDAL provides a superior scalable performance on both architectures than a conventional block Jacobi-type localized preconditioner, although the PHIDAL-based approach requires additional communications at each level.

A comparison between flat MPI and hybrid parallel programming models for PHIDAL on TSUBAME for strong scaling shows that flat MPI is much better if the number of cores is small, but these two methods are more closely matched as the number of cores increases. The effect of additional communications is significant in cases with many cores, especially for the flat MPI parallel programming model. PHIDAL-based preconditioning with the hybrid parallel programming model is expected to be a good choice for excellent scalable performance and robustness for implementations on more than 10^4 cores on SMP cluster architectures and clusters of multi-core processors. A CM-RCM reordering provides a more robust convergence than MC reordering, but the relationship between number of colors and convergence is not straightforward in ill-conditioned cases with heterogeneity. Further investigation of effective reordering techniques for ill-conditioned problems is important.

In the present work, no fill-in processes have been considered in PHIDAL procedures. The results of using PHIDAL in comparison with conventional block Jacobi-type localized preconditioning have therefore not been so significant. More robust preconditioning methods based on PHIDAL may be developed by considering fill-ins inside and between connectors. Moreover, the developed methods may further be evaluated on various types of SMP cluster architectures and clusters of multi-core processors.

Acknowledgements

This work is supported by the 21st Century Earth Science COE Program at the University of Tokyo, and CREST/Japan Science and Technology Agency. The author would like to thank the Global Scientific Information and Computing Center at Tokyo Institute of Technology, and the National Energy Research Scientific Computing Center at Lawrence Berkeley National Laboratory, for use of their computer resources.

References

1. Nakajima, K.: Parallel Iterative Solvers of GeoFEM with Selective Blocking Preconditioning for Nonlinear Contact Problems on the Earth Simulator. In: ACM/IEEE Proceedings of SC2003 (2003)
2. Nakajima, K.: The Impact of Parallel Programming Models on the Linear Algebra Performance for Finite Element Simulations. Lecture Notes in Computer Science 4395, 334–348 (2007)
3. Rabenseifner, R.: Communication Bandwidth of Parallel Programming Models on Hybrid Architectures. In: Zima, H.P., Joe, K., Sato, M., Seo, Y., Shimasaki, M. (eds.) ISHPC 2002. LNCS, vol. 2327, pp. 437–448. Springer, Heidelberg (2002)
4. Earth Simulator Center: http://www.es.jamstec.go.jp/
5. Doi, S., Washio, T.: Using Multicolor Ordering with Many Colors to Strike a Better Balance between Parallelism and Convergence. In: Proceedings of RIKEN Symposium on Linear Algebra and its Applications, pp. 19–26 (1999)
6. Washio, T., Hisada, T., Watanabe, H., Tezduyar, T.E.: A Robust and Efficient Iterative Linear Solver for Strongly Coupled Fluid-Structure Interaction Problems. Computer Methods in Applied Mechanics and Engineering 194, 4027–4047 (2005)
7. Nakajima, K.: Parallel Preconditioning Methods with Selective Fill-Ins and Selective Overlapping for Ill-Conditioned Problems in Finite-Element Methods. Lecture Notes in Computer Science 4489, 1085–1092 (2007)
8. Zhang, S.L.: GPBi-CG: Generalized Product-type methods based on Bi-CG for solving nonsymmetric linear systems. SIAM Journal of Scientific Computing 18, 537–551 (1997)
9. Henon, P., Saad, Y.: A Parallel Multistage ILU Factorization based on a Hierarchical Graph Decomposition. SIAM Journal for Scientific Computing 28, 2266–2293 (2007)
10. National Energy Research Scientific Computing Center, Lawrence Berkeley National Laboratory, http://www.nersc.gov/
11. TSUBAME Grid Clusterm, http://www.gsic.titech.ac.jp/
12. Deutsch, C.V., Journel, A.G., GSLIB,: Geostatistical Software Library and User's Guide, 2nd edn. Oxford University Press, Oxford (1998)
13. GeoFEM, http://geofem.tokyo.rist.or.jp/
14. Saad, Y.: Iterative Methods for Sparse Linear Systems. 2nd edn. SIAM (2003)
15. Washio, T., Maruyama, K., Osoda, T., Shimizu, F., Doi, S.: Efficient implementations of block sparse matrix operations on shared memory vector machines. In: Proceedings of The 4th International Conference on Supercomputing in Nuclear Applications (SNA2000) (2000)

High Performance FFT on SGI Altix 3700

Akira Nukada[1,2], Daisuke Takahashi[3,1], Reiji Suda[2,1], and Akira Nishida[2,1]

[1] Japan Science and Technology Agency, Saitama 332-0012, Japan
[2] The University of Tokyo, Tokyo 113-8685, Japan
{nukada,reiji,nishida}@is.s.u-tokyo.ac.jp
[3] University of Tsukuba, Ibaraki 305-8573, Japan
daisuke@cs.tsukuba.ac.jp

Abstract. We have developed a high-performance FFT on SGI Altix 3700, improving the efficiency of the floating-point operations required to compute FFT by using a kind of loop fusion technique. As a result, we achieved a performance of 4.94 Gflops at 1-D FFT of length 4096 with an Itanium 2 1.3 GHz (95% of peak), and a performance of 28 Gflops at 2-D FFT of 4096^2 with 32 processors. Our FFT kernel outperformed the other existing libraries.

1 Introduction

The Fast Fourier Transform (FFT) [1] algorithm is now used in many fields, and its fast adaptation to computer systems continues to be expected. We developed a high-performance FFT on SGI Altix 3700 [2], which is a hardware distributed shared memory parallel computer with Intel Itanium 2 processors.

The performance of floating-point operations in modern microprocessors has significantly matured. The next issue is how to use the FPUs efficiently. In this paper, we focus on the efficiency of executing instructions. Several optimization techniques for nested loops have been researched [3]. But they are not adequate for archieving perfect performance of processors such as the Itanium 2, because the overheads of the loop initializations are too large. Various kinds of loop fusion methods for FFTs on vector processors have been proposed [4,5]. Those methods require index arrays, which are usually larger than the capacity of the cache memory. For this reason, conventional loop fusion methods are not suitable for non-vector processors like Itanium 2. Therefore, we propose a new loop fusion method without index arrays, and implemented highly optimized FFT kernels for the Itanium 2.

2 IA-64

Intel Architecture 64 (IA-64) [6] is a new 64-bit processor architecture developed by Intel and Hewlett Packard. As a major step forward from the IA-32, Intel's 32-bit architecture, the IA-64 is equipped with many advanced and challenging technologies.

R. Perrott et al. (Eds.): HPCC 2007, LNCS 4782, pp. 396–407, 2007.

Table 1. The capacities of the cache memories

	Itanium	Itanium 2	
	Merced	McKinley	Madison
Level-1 Inst.	16kB	16kB	16kB
Level-1 Data	16kB	16kB	16kB
Level-2 Unified	96kB	256kB	256kB
Level-3 Unified	2MB	1.5/3MB	3/6MB

Itanium is the first product of the IA-64 architecture, codenamed 'Merced', and was released in 2001. The second product, codenamed 'McKinley', was released as Itanium 2 in 2002, and the third product, codenamed 'Madison' was released in 2003.

2.1 EPIC

The IA-64 uses a kind of VLIW (Very Long Instruction Word) [7] architecture. A 16-byte 'bundle' usually contains three instructions, and the Itanium and the Itanium 2 processor can execute two bundles, i.e., up to six instructions, in each cycle. This number is large compared with other processors. EPIC (Explicitly Parallel Instruction Computing) [8] technology plays an important role in exploiting its performance.

Checking the dependencies between instructions in order to execute many instructions in one cycle requires a great deal of hardware resources. For this reason, the number of instructions to be executed in a single cycle is limited. In the case of EPIC, the dependencies between the instructions are checked at the compilation time, and independent instructions are explicitly grouped to avoid dependency. Thus, the grouped instructions can be subsequently executed without checking dependencies.

Itanium 2 processors have four memory units, two integer units, two floating-point units, and three branch units. Up to six instructions are sent to those execution units.

2.2 Memory Hierarchy

The IA-64 processors have a level-1 instruction cache, a level-1 data cache, a level-2 unified cache, and a level-3 unified cache. Their capacities depend on the models as listed in Table 1.

In the IA-64 architecture, the floating-point data do not pass the level-1 cache, but are directly transferred from the level-2 cache. The throughput of the level-2 cache is large, but the latency is long.

2.3 Predicate Register

In case of the IA-64 instruction set architecture, predicate registers can be specified to almost all instructions. The predicate register is a single-bit register, and 64 registers in total are available. If a predicate register is specified to an instruction, the instruction will be executed only if the value of the predicate register is 1. The values of the predicate registers are changed by compare instructions, and logical operations between the predicate registers are also available.

Implementations with branch instructions always have the risk of penalties for the branch prediction misses. But implementations with predicate registers have no such risk; therefore, even more complex operations including conditional statements can be executed efficiently. Although an instruction is not executed if the value of the specified predicate register is 0, the instruction nonetheless occupies an instruction slot. In practice, it is a rare case that this matters, because the IA-64 processors can execute six instructions in each cycle.

2.4 Register Rotation

The IA-64 architecture uses the mechanism of register rotation. The relations between the logical register numbers and the physical register numbers can be rotated mainly by the branch instructions for loops described later. The IA-64 architecture has 128 floating-point registers, and 96 registers from the number 32 can be rotated. In general, a long sequence of instructions is required to utilize such a large number of registers. But the mechanism of the register rotation enables a short compact sequence of the instructions to use many physical registers. There are 64 predicate registers, where 48 registers from the number 16 can be rotated. The general purpose (integer) registers also can be rotated partially.

2.5 Branch Instructions for Loop

The IA-64 instruction set defines several kinds of branch instructions, including a normal branch instruction, a branch instruction for calling functions, a branch instruction for returning from functions, and branch instructions for loops. The conditional branches are implemented by the normal branch instruction with the predicate register.

The branch instructions for loops reference the Loop Counter register (LC) and the Epilogue Counter register (EC) to control the iteration. The LC is set to (iteration count - 1) of the original loop. The EC is set to the number of extra iterations required by the software pipelining. The '**br.cloop**' instruction only references the LC. On the other hand, the branch instructions for the modulo-scheduled loops [9] such as '**br.ctop**' reference both the LC and EC, and the registers are rotated.

Those branch instructions reference the value of the LC, and jump to a specified address as long as the value is positive, and the LC is decreased by one. When the LC becomes zero, the branch instructions for the modulo-scheduled loops also reference the value of the EC and decrease it. In case of the modulo-scheduled loop, the predicate registers are rotated by the branch instructions. After rotating, the predicate register number 16 (p16) is set to 1 as long as the value of the LC is positive. When the LC becomes 0, p16 is set to 0. Using the features of the predicate registers, the register rotation, and the branch instruction for loops, we can describe an efficient modulo-scheduled loop very simply.

3 Conventional Optimizations of Nested Loop of FFT

Many of the traditional implementations of the FFT algorithm have nested loops. An example of the radix-2 FFT kernel [5] is described below:

```
double complex vin[n1][2][n2], vout[2][n1][n2];
double complex w, table[n1][n2];
for (i = 0; i < n1; i++) {
  w = table[i][0];
  for (j = 0; j < n2; j++) {
    t = w * vin[i][1][j];
    s = vin[i][0][j];
    vout[0][i][j] = s + t;
    vout[1][i][j] = s - t;
  }
}
```

The array '**table**' is a trigonometric table called the twiddle factor table. Stockham's auto-sort algorithm [4,5] is used in this example. For this reason, the input array '**vin**' can not be overwritten. The results are written to another output array '**vout**'.

The iteration counts of the nested loops are n1 and n2, respectively, and they change during the computation of the FFT. For example, using the radix-2 FFT kernel for the input size $N(= n1 * n2 * 2) = 64$, $(n1, n2)$ becomes $\{(1, 32), (2, 16), (4, 8), (8, 4), (16, 2), (32, 1)\}$. When the iteration count of the inner loop n2 is too small, the efficiency of executing instructions is generally degraded because of the overhead around the loop. There are several loop optimization methods [3] proposed for such a case.

3.1 Loop Interchange

If one of the iteration counts of the nested loops is large enough, the performance may be improved by interchanging the inner loop and the outer loop [4,5].

```
if (n1 > n2) {
  for (j = 0; j < n2; j++) {
    for (i = 0; i < n1; i++) {
      w = table[i][0];
      t = w * vin[i][1][j];
      s = vin[i][0][j];
      vout[0][i][j] = s + t;
      vout[1][i][j] = s - t;
    }
  }
} else {
  ....
}
```

In this example, the loop interchange is performed only if the iteration count of the outer loop is larger than that of the inner loop. But the loop interchange does not always improve the performance in the case of the FFT. The loop interchange increases the number of load operations for the twiddle factor table, and the memory accesses to the input and output buffers are no longer contiguous.

3.2 Unrolling of Inner Loop

The efficiency of executing instructions decreases when the iteration count of the inner loop is small. The rate of the performance degradation depends on the overhead of the loops such as the penalty for the branch prediction misses and so on. Assume that branch prediction misses occur once every n2 iterations. A drastic performance degradation results only if n2 is too small. Thus, we should unroll the inner loop only in those cases.

```
switch (n2) {
case 1:
  for (i = 0; i < n1; i++) {
    w = table[i][0];
    t = w * vin[i][1][0];
    s = vin[i][0][0];
    vout[0][i][0] = s + t;
    vout[1][i][0] = s - t;
  }
  break;
case 2:
  for (i = 0; i < n1; i++) {
    w = table[i][0];
    t = w * vin[i][1][0];
    s = vin[i][0][0];
    vout[0][i][0] = s + t;
    vout[1][i][0] = s - t;
    t = w * vin[i][1][1];
    s = vin[i][0][1];
    vout[0][i][1] = s + t;
    vout[1][i][1] = s - t;
  }
  break;

  ....
}
```

The improvement in performance achieved by unrolling is especially large in the case of smaller n2 values. As n2 becomes larger, the ratio of improvement decreases. The exact threshold at which n2 should be unrolled should be determined in consideration of the size of the instruction cache memory and so on. In general, the use of unrolling for large n2 values is not recommended.

3.3 Loop Fusion Using Index Array

When computing on vector computers, the iteration count of the inner loop is a very important factor. To achieve high performance with vector processing, the iteration count must be sufficiently large. Using the loop interchange, the iteration count may become $\sqrt{N/2}$ in the worst case, and this is not sufficiently large for small N values. Pease proposed a method to combine nested loops into a single loop [10]. The method prepares index arrays to realize non-contiguous access in nested loops and performs gather operations using them. The example of n2 = 4 is described below.

```
double complex vin[n1 * n2 * 2], vout[n1 * n2 * 2];
double complex w, table[n1 * n2];
long index1[] = {0, 1, 2, 3, 8, 9, 10, 11, 16, ...};
long index2[] = {0, 0, 0, 0, 1, 1, 1, 1, 2, ...};
for (i = 0; i < n1 * n2; i++) {
  w = table[index2[i] * n2];
  t = w * vin[index1[i] + n2];
  s = vin[index1[i]];
  vout[i] = s + t;
  vout[i + n1 * n2] = s - t;
}
```

Although the iteration count of the loop becomes large, the method requires $O(N \log N)$ memory space for the index arrays, and additional memory access to the index arrays is required.

4 Loop Fusion Without the Index Arrays

In the case of the modulo scheduled loop of Itanium 2, the throughput of executing instructions is very high, whereas the overheads of the loops are large. Therefore, the technique of loop interchange is not sufficient to exploit the potential. The conventional loop fusion method requires large index arrays. This causes many cache misses; therefore, those methods are not suitable for non-vector processors including Itanium 2. Thus, we propose a new loop fusion method without index arrays.

At first, we observed the memory access patterns for '**vin**', '**vout**', and '**table**'. The addresses for '**vin**' move 16 bytes forward at each iteration; additionally, they move (n2 * 16) bytes forward once every n2 iterations. The addresses for '**table**' move (n2 * 16) bytes forward once every n2 iterations. The addresses for '**vout**' move 16 bytes forward at each iteration. Using the operations of pointers, they can be described as follows:

```
double complex *vin0 = vin, *vin1 = vin + n2;
double complex *vout0 = vout, *vout1 = vout + n1 * n2;
double complex *ptable = table;
count = n2;
```

```
for (i = 0; i < n1 * n2; i++) {
  w = *ptable;
  t = w * *vin1; s = *vin0;
  *vout0 = s + t; *vout1 = s - t;
  vin0 += 1; vin1 += 1;
  vout0 += 1; vout1 += 1;
  count -= 1;
  if (count == 0) {
    vin0 += n2; vin1 += n2;
    ptable += n2;
    count = n2;
  }
}
```

The variable '**count**' is used to count down from n2, and the adjustments of the pointers are performed when the count becomes zero. Although the number of address operations is increased, the nested loops can be combined into a single loop. The latest super-scalar processors can execute many instructions in each cycle. In particular, IA-64 processors can execute six instructions, and the post increment feature of the load and store instructions is usable to increase the addresses by 16 bytes.

4.1 Implementation on Itanium 2

Combining the nested loops of the FFT into a single loop, the iteration count is always sufficiently large. This may improve the efficiency of executing instructions, while the '**if**' statement appears in the loop. If the statement is implemented with branch instructions, the branch prediction misses degrade the performance. For this reason, this kind of transform is usually not performed. But in the case of implementation with the predicate register of IA-64, the transform is effective.

The memory accesses for the twiddle factor table are performed at each iteration. Since the accesses to the same addresses are repeated in n2 contiguous iterations, the lines are probably stored in the cache memory. The number of load operations actually increases; therefore, the high-performance (cache) memory access such as that of IA-64 is required.

4.2 Implementation on Other Architecture

The loop fusion method can be implemented, not with predicate registers but with generic integer operations. If n2 is a power of two, it is clearly possible to implement with logical operations as follows:

```
count = 0;
for (i = 0; i < n1 * n2; i++) {
```

```
    ....
    count += 1;
    b = count & n2;
    count -= b;
    vin0 += b; vin1 += b;
    ptable += b;
}
```

This is because we can use n2 as the bit mask and create the counter easily if n2 is a power of two. Even if not, a similar operation is possible. The following example uses the arithmetic shift right operation.

```
count = n2 - 1;
for (i = 0; i < n1 * n2; i++) {
    ....
    count -= 1;
    b = count >> 63; shift right arithmetic.
    b &= n2;
    count += b;
    vin0 += b; vin1 += b;
    ptable += b;
}
```

The size of the integers is assumed to be 64 bits. As the result of the 63-bit arithmetic shift right operation, the significant bit will be copied to all the other bits. For negative numbers, all bits become one. Otherwise, all bits become zero. Using this procedure for the mask operation, the variable '**b**' becomes n2 once every n2 iterations.

Almost all the processors have the arithmetic shift operation; therefore, the loop fusion method is usable not only for IA-64 but also for other processors.

5 Implementation

The floating-point units of the IA-64 processors execute Fused Multiply-Add (FMA) operations. Several FFT kernels for FMA have been proposed [11,12,13]. We used the FFT kernels proposed by Goedecker and Linzer, which require a minimum number of FMA operations. To compute the FFT of the length of powers of two, we can use the radix-2, radix-4, and radix-8 [14] FFT kernel and so on. Table 2 shows the numbers of FMAs, and the load and store operations required in those kernels. Using the radix-4 and radix-8 kernels, the number of the FMA operations becomes smaller than with the radix-2 kernel. We used the radix-4 mainly, and also used the radix-8 if required. Although the sizes of transforms are limited to powers of two in this paper, our loop fusion method can be used for kernels of radix 3, 5, or any other radix.

An optimized assembly code of the radix-4 kernel is used in later experiments. Using the radix-4 kernel for the FMA, the nested loops are combined into a single

Table 2. The number of FMA and load/store operations of the FFT kernels for FMA

	FMA	load	store
Radix-2	6	6	4
Radix-4	22	14	8
Radix-8	66	30	16

loop, and then the software-pipelined modulo-scheduled loop is generated. The instructions are well scheduled in consideration of the dependencies and latencies. In case of the Itanium 2 (Madison), the latency of the FMA is 4 cycles. The latencies of the floating-point load operations from the level-2 cache and the level-3 cache are 6 cycles and at least 13 cycles, respectively. In our optimized kernel, the load operations are scheduled for 22–33 cycles before the data is used. Each iteration takes 11 cycles, and 4 iterations are added for the epilogue. Thus, $11(N/4+4)$ cycles in total are required.

When the twiddle factor is 1.0, the multiplication by it can be skipped. This optimization is often used to reduce the number of floating-point operations. In case of the sample code in Section 4, the twiddle factors become 1.0 in the first n2 iterations. The radix-4 kernel has three complex multiplications by the twiddle factors. If they are skipped, the number of the FMA operations is reduced to 16. We also generated an optimized modulo-scheduled loop for them. Its epilogue count is 5, and $8(n2+5)$ cycles in total are required. The other part, that is, multiplications for which the twiddle factor is not 1.0, is computed normally and requires $11(N/4-n2+4)$ cycles. As the result of skipping the multiplication, $(3*n2-40)$ cycles are eliminated. Therefore, it is effective only if n2 is larger than 13. Otherwise, the overhead of splitting the loop is larger. In our experiments, N is a power of two. We used this optimization only if n2 was not less than 16.

For the 2-D FFT, the optimized 1-D FFT kernel is used. And it is parallelized using OpenMP as follows.

```
double complex v[N][N+1];
double complex work1[N], work2[N];
double complex table[N];

#pragma omp parallel for private(work1)
for (i = 0; i < N; i++)
  FFT1D(N, v[i], work1, table);

#pragma omp parallel for private(j,work1,work2)
for (i = 0; i < N; i++) {
  for (j = 0; j < N; j++) work1[j] = v[j][i];
  FFT1D(N, work1, work2, table);
  for (j = 0; j < N; j++) v[j][i] = work1[j];
}
```

In the transform along the second axis, the data in non-contiguous addresses are copied into the work space, and they are written back after the transform for the efficient memory access. The 2-D array of $N \times (N+1)$ is allocated to store $N \times N$ data. This is important to improve the performance of the memory access because N is a power of two in our experiments.

In case of the Altix 3700, the physical memory pages are allocated by the first-touch policy. The 2-D array is initialized in parallel to allocate the physical memory pages of all nodes equally.

6 Performance Evaluations

Using the optimized FFT kernels, the performance on the SGI Altix 3700 was evaluated. Table 3 shows the specifications of the system.

6.1 Performance of 1-D FFT

The performance of the 1-D FFT is shown in Table 4. Only one processor (single thread) was used for computation. The elapsed time required to repeat the forward and backward transforms $\lfloor 300/\log_2 N \rfloor$ times was measured. The performance (Gflops) was calculated from this value. Scaling of the results by $(1/N)$ was not performed. The number of floating-point operations in a 1-D FFT of length N was assumed to be $5N \log_2 N$. To compare the performance of the optimized FFT kernel with that of other existing libraries, the performances of FFTW 3.1 [15], Intel Performance Primitive Signal Processing 5.1.1 (IPPS), Intel Math Kernel Library 8.0 (MKL), and SGI Scientific Computing Software Library 1.5 (SCSL) were also measured. To compile the FFTW library, Intel C/C++ Compiler 9.1.042 and compiler option '-O3' were used.

Table 3. The specification of SGI Altix 3700

Processor	Intel Itanium 2 (Madison)
Clock Frequency/Peak Gflops	1.3 GHz/5.2 Gflops
# of Proc.	32
Level-2/3 cache	256kB/3MB
Memory	32GB
Operating System	SGI ProPack 2.4 for Linux

Table 4. The performance (Gflops) of 1-D FFTs

N	256	512	1024	2048	4096
Optimized	4.18	4.56	4.80	4.86	4.94
FFTW 3.1	2.94	3.11	3.14	2.92	3.01
IPPS 5.1.1	4.10	4.25	4.33	4.37	4.03
MKL 8.0	3.27	3.66	3.72	3.86	2.10
SCSL 1.5	2.87	3.13	3.32	3.44	3.29

Table 5. The use rate of the FPUs

N	256	512	1024	2048	4096
# of FMAs	5152	11712	26144	57920	127008
Time	2.43us	5.04us	10.6us	23.1us	49.7us
Use rate	81.4%	89.2%	94.2%	96.2%	98.2%

Table 6. The performance (Gflops) of parallel 2-D FFT of 4096^2

# of PEs	1	2	4	8	16	32
Optimized	1.35	2.52	4.37	8.37	15.8	28.2
SCSL 1.5	0.82	1.44	2.54	4.36	6.91	10.6
MKL 8.0	0.62	0.95	1.51	2.83	4.92	7.99
FFTW 3.1	0.69	0.61	0.91	1.25	2.02	3.20

The optimized kernel (Optimized) we developed highly outperformed other libraries. As the result of the loop fusion, the iteration count became $O(N)$ and the performance increased as N became larger.

Table 5 shows the number of floating-point FMA operations and the use rate of the FPUs. The number of FMA operations only counts the operations which are actually completed; therefore, it does not include the operations retired in the prologue and epilogue parts of the modulo-scheduled loop. Despite that, the use rate of the FPUs is very high, reaching up to 98.2%. The rate was improved by loop fusion.

For FFTs of larger sizes, the block FFT algorithm based on six-step FFT is suitable. In the block FFT algorithm, multicolumn FFTs of the sizes evaluated in the Table 4 are computed in the cache memory.

6.2 Performance of 2-D FFT

Table 6 shows the performance of the parallel 2-D FFT of 4096^2. The performances with 1,2,4,8,16, and 32 threads were measured. The speed-up of the optimized kernel was about 20.8 with 32 threads. In the case of a single thread, all the data were available in the memory of the local node; therefore, no data transfer between nodes was required. In consideration of that, the speed-up was sufficiently large.

7 Concluding Remarks

We have developed a high-performance FFT on the SGI Altix 3700, which is a hardware distributed shared memory parallel computer with Itanium 2 processors. To exploit the improved performance of the modulo-scheduled loop of the IA-64 architecture, we proposed a loop fusion method for nested loops of the FFT. In contrast to conventinal loop fusion methods, our method requires no index arrays, and also can be used for other processor architectures because it can be implemented with a series of generic instructions. This improved the efficiency of executing instructions, and the use rate of the FPUs reached 98.2%. Using the Itanium 2 1.3 GHz, we achieved a performance of 4.94 Gflops (95% of peak) in the 1-D FFT, and 28 Gflops in the 2-D FFT with 32 processors. The optimized FFT kernel outperformed the other existing libraries.

Acknowledgments. This work was partially supported by CREST of JST.

References

1. Cooley, J.W., Tukey, J.W.: An Algorithm for the Machine Calculation of Complex Fourier Series. Math. Comput. 19, 297–301 (1965)
2. Dunigan, T.H., Vetter, J.S., Worley, P.H.: Performance evaluation of the SGI Altix 3700. In: ICPP, pp. 231–240 (2005)
3. Wolf, M.E., Lam, M.S.: A loop transformation theory and an algorithm to maximize parallelism. IEEE Trans. Parallel Distrib. Syst. 2, 452–471 (1991)
4. Swarztrauber, P.N.: FFT algorithms for vector computers. Parallel Computing 1, 45–63 (1984)
5. Van Loan, C.: Computational Frameworks for the Fast Fourier Transform. SIAM Press, Philadelphia, PA (1992)
6. Intel Coporation: Itanium Architecture Software Developer's Manual Revision 2.1 (2002)
7. Colwell, R.P., et al.: A VLIW architecture for a trace scheduling compiler. IEEE Trans. on Computers 37, 967–979 (1988)
8. Gwennap, L.: Intel, HP make EPIC disclosure. Microprocessor Report 11, 1–9 (1997)
9. Rau, B.R.: Iterative modulo scheduling: An algorithm for software pipelining loops. In: Proc. 27th Annual International Symposium on Microarchitecture, San Jose, CA, pp. 63–74 (1994)
10. Pease, M.C.: An adaptation of the fast Fourier transform for parallel processing. J. ACM 15, 252–264 (1968)
11. Linzer, E.N., Feig, E.: Implementation of efficient FFT algorithms on fused multiply-add architectures. IEEE Trans. Signal Processing 41, 93–107 (1993)
12. Goedecker, S.: Fast radix 2,3,4 and 5 kernels for fast Fourier transformations on computers with overlapping multiply-add instructions. SIAM J. Sci. Comput. 18, 1605–1611 (1997)
13. Karner, H., et al.: Multiply-Add Optimized FFT Kernels. Math. Models and Methods in Appl. Sci. 11, 105–117 (2001)
14. Bergland, G.D.: A fast Fourier transform algorithm using base 8 iterations. Math. Comp. 22, 275–279 (1968)
15. Frigo, M., Johnson, S.G.: The design and implementation of FFTW3. Proceedings of the IEEE 93, 216–231 (2005) (special issue on "Program Generation, Optimization, and Platform Adaptation)

Security Enhancement and Performance Evaluation of an Object-Based Storage System

Po-Chun Liu, Sheng-Kai Hong, and Yarsun Hsu

Department of Electrical Engineering
National Tsing Hua University Hsinchu, 30013, Taiwan
{pcliu,phinex}@hpcc.ee.nthu.edu.tw, yshsu@ee.nthu.edu.tw

Abstract. Object-based storage offloads some works of file systems to storage devices to improve security, scalability, and performance. Security is a main concern when sharing data over network. We examine the security model of object-based storage and find that there is some problem in the model. It can be disabled by modifying specific field in the command. We propose a solution to this problem by encryption that makes unauthenticated clients impossible to alter the field. The overhead of this encryption is quite low. Thus the performance of our enhanced object-based storage system is comparable to that of the original one while offering an enhanced security. In addition, we have compared the performance of OSD systems with that of iSCSI and NFS. The write performance of an object-based storage system is much better because it can offload some tasks to storage devices, and the CPU usage at client side is also largely reduced.

1 Introduction

The design of storage subsystems becomes an important issue with the improvement of computer system. Although direct attached storage (DAS) devices have much better performance compared to network attached storage (NAS), the advantages of NAS, such as scalability and file sharing, make it become important in large scale computer systems. However, when a storage subsystem is connected to internet, security is always a main concern for system designers. Object-based storage systems [1][2] provide a security model which protects the shared data in storage devices and eases the storage subsystem design in a very large scale computer system.

The main concept of object-based storage is to offload the space management component of existing file system to the storage device itself. Application clients thus request for an object (or file) instead of many disk blocks. The new interface can reduce traffics between application clients and storage devices. And the application clients only need to manipulate hierarchy management, naming, and user access control. The work data structure mapped to physical organization and disk space management is left to be done in storage device.

The object-based interface is defined in details by OSD T10 standard [3] and the aim of our work is to build an OSD (Object-based Storage Device)

R. Perrott et al. (Eds.): HPCC 2007, LNCS 4782, pp. 408–419, 2007.

system compliant to this standard. Our implementation is based on IBM Object Store Initiator [4][5] and Intel iSCSI/OSD reference implementation [6].However, we find that there is a potential problem in their security model. The security method can be altered by unauthenticated application clients to disable the security model. In this paper, we propose an enhanced security model to resolve the potential problem. The main idea is to encrypt the security method field and make others impossible to alter the security method. The encryption does not add much computing overhead to the system. We will illustrate the architecture and design issues of our enhanced OSD system.

Section 2 gives the system architecture of our OSD system and briefly explains the design of the system. Section 3 discusses the security model of OSD T10 standard and points out the potential problem. In section 4 we discuss the possible solutions for the potential problem of the security model and give the performance result in section 5. We give the conclusion and what to do in the future in section 6.

2 OSD Environment

Currently the object-based hard disk drives compliant to the T10 standard are still not available and our OSD storage system is based on Intel iSCSI/OSD reference implementation, which consists of initiators and object-based targets, to emulate the underlying Object-based Storage Device (OSD). Initiators and targets are connected via iSCSI protocol over IP network [7][8]. The initiator prepares the OSD SCSI commands, requests for credential from security manager and packs it into OSD Command Descriptor Block (CDB) format which is a 200 bytes variable length CDB [9], and then issues it. The object based target receives the OSD CDB from the initiators, checks whether the command is illegal and then executes the command if it is a legal request. The architecture of the system is shown in Fig. 1, which consists of initiator, security manager, policy/storage manager, and storage device server.

2.1 Initiator

The initiator acts as an application client, which issues the OSD SCSI commands to access the storage device server. The difference between the OSD system and the traditional storage subsystem is that an application client of traditional storage systems needs to allocate and manage disk space for specific files while an OSD system does not care about it. An application client of OSD systems only needs to specify which object it wants to access and the mapping between the file name and the object ID. The storage space allocation and management are delegated to storage device server. The only thing an application client needs to do is specifying which object and its offset to be accessed. The application client does not concern about the disk space and thus its loading is eased compared to traditional storage system.

When an application client wants to issue an OSD command, it prepares parameters such as partition ID, object ID, the length and the starting address of

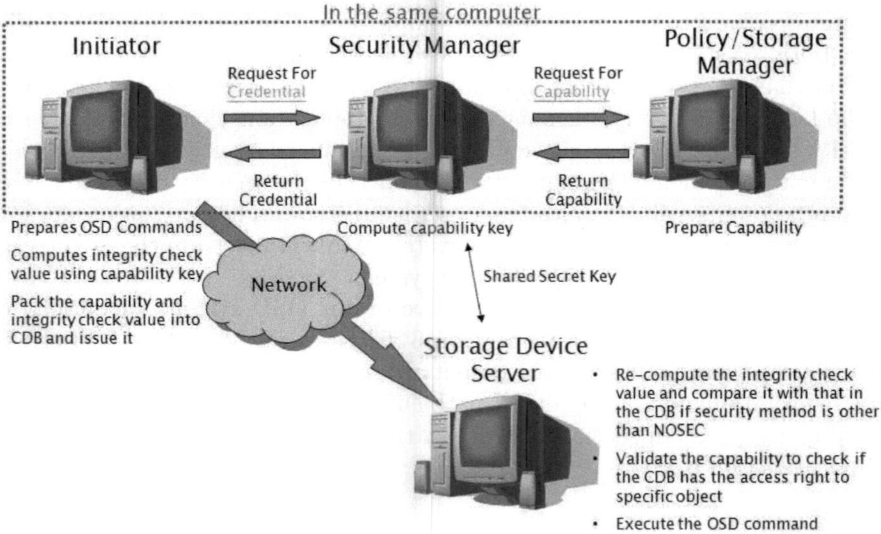

Fig. 1. The architecture of our OSD system

the object it wants to access. Before it issues the command, it needs to request a credential, which contains capability and capability key, from security manager. The capability is generated by policy/storage manager which coordinates requests from different application clients and determines their access rights. Once the application client gets the credential, it packs the capability contained in the credential into OSD CDB and sends the CDB via iSCSI initiator over network. The command it issues can be verified by storage device server and executed correctly with appropriate access right.

There are four supported security methods, NOSEC, CAPKEY, CMDRSP, and ALLDATA [3][10]. The NOSEC method means security model is disabled and all data is unprotected. The CAPKEY method protects the capability of the CDB and the CMDRSP protects entire CDB and responses from targets. The ALLDATA method makes sure all of the data between initiators and targets are under protection. If the security model is enabled, the initiator needs to compute an integrity check value before sending the command. In our system we use HMAC (Hash Message Authentication Code) [11] algorithm and use the secret key generated by security manager to compute integrity check value of specific data depending on the security method. The integrity check value is then packed into OSD CDB and the command is issued.

2.2 Security Manager

The main purpose of the security manager is to generate a key for initiators to compute the integrity check value if the security method is not NOSEC. Once the security manager receives a request from an application client, it hands over the request to policy/storage manager and requests for a capability. After the security manager receives a capability back from the policy/storage manager, it packs the capability and OSD system ID of storage device server into credential. Then the security manager uses HMAC algorithm with the secret key shared with the storage device server to compute a check value of this credential. The check value, also called capability key, is the key for initiators to compute integrity check values. The security manger packs the check value into the credential and the credential is then sent back to the application client.

2.3 Policy/Storage Manager

The policy/storage manager receives requests from security manger and prepares capabilities depending on different requests. The capability contains parameters such as security method, integrity check value algorithm, read/write permissions, allowed object ID for specific command, and so forth. Without the appropriate capability an application client's access to the storage device server will be rejected. With an appropriate capability, the policy/storage manager can make application clients access only specific objects of the storage device server but not the entire disk space.

2.4 Storage Device Server

The storage device server receives OSD commands via iSCSI target driver from application clients and handles the space allocation and management according to the requests. Once it receives a request, it will verify whether the command had been tampered with by checking the integrity check value if the security method is other than NOSEC. The storage device server will first reconstruct the credential by copying both the capability from the CDB and the OSD system ID from the Root Information attributes page. Then it uses the key shared with security manager and HMAC algorithm to compute a check value of this reconstructed credential. Finally, the storage device server computes an integrity check value of the data depending on different security method by use of this check value. It then compares the reconstructed integrity check value with that in the CDB computed by application clients. If the two integrity check values are identical, we can assure that the protected data has not been tampered with. And if the two values mismatch, the protected data has been altered unintentionally or maliciously and thus the storage server will reject the command.

If the security method is CAPKEY, we can assure that the capability is the one that is prepared by the policy/storage manager. If the security method is CMDRSP, we are confident that the OSD CDB is issued by authenticated application clients. The ALLDATA security method makes sure that all the data transmitted between authenticated application clients and storage device server are not be altered.

If the command is not tampered with, the storage device server then checks whether the accessed object is allowed and the access right of the command depending on the capability in the CDB. The storage device server will also check the parameters such as object created time. Moreover, the specific permission bits must be set according to the service action and object type. Only when all the parameters of capability are verified can the storage device server execute the OSD command.

The storage device server translates the OSD commands to VFS (Virtual Filesystem Switch) system calls and exploits existing Linux file system such as ext3 to do the space allocation and management. For example, OSD-CREATE-PARTITION can be translated to mkdir() of Linux file system to create a directory to be treated as a partition of the object-based storage. We treat directories of Linux file system as partitions and ordinary files as user objects.

2.5 Potential Problem

The security model of OSD systems provide some level of protection to avoid malicious alteration of data over network. The credential integrity check value, also called as capability key, is generated by security manager to compute the integrity check value of protected data. The unauthenticated application clients have no idea how to reconstruct the capability key because the key used to compute the credential integrity check value is only known by the security manger and the storage device server. Thus any unauthenticated application client cannot generate a valid integrity check value without knowing capability key and this is why the security model can avoid any malicious alteration by unauthenticated application clients.

Even though the integrity check value based security model protects the data from being altered, the CDB still can be seen by others. Unauthenticated application clients can capture the capability by monitoring the CDB sent by authenticated ones. If they change the security method field in the capability to NOSEC and set integrity check value of CDB to 0s, then they can still access the objects that the captured capability allows. All unauthenticated clients have to do is to capture a valid capability, modify the security method, and send the command using the modified capability to disable the security model. That is to say, unauthenticated application clients can force storage device server not checking the integrity check value by changing the security method field to NOSEC.

Because the security method of certain command fully depends on the capability, the storage device server will check the integrity check value only if the security method field of capability is not NOSEC. The security model will not work if unauthenticated application clients can tamper the security method field. In this study, we propose a solution to resolve this potential problem and assure that the security model always works.

3 Enhanced Security Model

The most intuitive solution to prevent unauthenticated application clients from tampering the security method field is to make it impossible for them to alter the capability. Encrypting the transmitted data is one way to hide the contents of the data. Even if unauthenticated application clients can capture the CDBs sent by authenticated ones, they still do not know how to decrypt the data without the key, which is only known by storage device server, and thus they cannot tamper the capability.

The public key crypto-system is a well known method to encrypt and decrypt transmitted data. The receiver has its own private key and publishes a public key, which is usually generated by the private key, to the one which wants to transmit data to it. The transmitter encrypts the data with the public key of receiver and the receiver decrypts the data by the private key. Others can't decrypt the cipher text without the private key thus they don't know what the data is and cannot tamper it.

A well known RSA algorithm uses the public key crypto-system with a set of public and private keys to encrypt and decrypt. But the computing complexity of RSA is very high because the encryption and decryption of RSA need to compute the power of some integer. Although there are some methods to speed up the computation, it still needs lots of computing resources. We combine the concept of public key crypto-system and error correction coding, which is first proposed by McEliece [12], to encrypt the transmitted data and decrypt the cipher text. The encryption and decryption complexity is very low because all operations are in binary field and addition can be done by logic exclusive OR and multiplication can be done by logic AND. This largely decreases the need of computing power in encryption and decryption.

We design an encryption procedure in the initiator. It encrypts the specific field of the capability including security method with the public key of the OSD storage server before issuing the CDB. To encrypt specific data, just like the encoding process of error correction code encoder, it just multiplies the specific data by the public key G' of the OSD storage server, where G' is generated by the three private keys. Then it randomly flips t bits of encrypted data, which stands for t errors during transmission. The encrypted data can be described by equation $c = mG \oplus e$, where m is K bits data and e is error pattern with weight t. The public key G' is a $K \times N$ matrix, and K is the number of bits of data to be encrypted and N is the number of bits after encryption. We pack the redundant bits N - K into the reserved fields of OSD CDBs and thus there is no transmission overhead.

There are three private keys, G, S, and P generated by OSD storage server. G is a generating matrix of a (N, K) code, P is an N by N permutation matrix and S is a nonsingular matrix with dimension K. Currently G is a generating matrix of a Hamming Code with easily generated generating matrix, thus how to design a random non-singular matrix is our main concern. We first generate a matrix with entries of random 0s or 1s and examine whether it is invertible. If the inverse of the randomly generated matrix exists, then we have found the non-singular matrix; if not, we regenerate another random matrix and examine it again. We can find the non-singular within four tries because the probability of a randomly generated matrix being non-singular is more than 25% [13][14]. Then the public key is generated by multiplying S, G, and P, $G' = SGP$.

The OSD storage server must decrypt the received CDB before computing the integrity check value. Because $G' = SGP$, we can rewrite the encrypt data

$c = mSGP \oplus e$. The server first multiplies c with the inverse of P and gets the equation: $c' = (mSGP \oplus e)P^{-1} = mSG \oplus eP^{-1} = m'G \oplus e'$. Here e' has the same weight as e and then the server can decode c' with error pattern e' to m' using generating matrix G. We compute the syndrome of c' and find the least weight error pattern and then decode c' to m'. Once m' is found, the original message can be decrypted by multiplying m' by the inverse of S because $m' = mS$ and then $m = m'S^{-1}$.

Both the encryption and the decryption process are very simple because the matrix multiplications can be replaced with logic AND in binary field. With this insignificant computing overhead we can provide an enhanced security model without adding much cost. Unauthenticated clients now cannot capture a valid capability to generate an unauthenticated CDB by just changing the security method in the capability.

4 Performance Evaluation

In this section we will show the performance of our system and compare it with other file systems. We use two identical machines with configuration listed in Table 1 and connect them with a 3COM gigabits switch. We use the program bonnie++ version 1.03a [15] with file size set to 1 GB and block size 8 KB for all of the performance evaluations. We evaluate the performance under five environments:

OSD-NOSEC: In this environment we use the OSD system according to OSD T10 standard with security manager disabled. The keys and integrity check values are all set to zero. In the initiator we implement an OSD file system based on the Intel iSCSI/OSD reference implementation, and in the target we exploit existing ext3 file system to manage the disk space. The initiator and target are connected through a gigabits network.

OSD-CAPKEY: In this environment we use the OSD system with security method set to CAPKEY. The security manager needs to generate a key for initiators to compute integrity check value. The initiator and target are implemented in the same way as OSD-NOSEC.

ENHANCED: This uses the enhanced security model we have proposed. The architecture is the same as an OSD system except an additional encryption process.

iSCSI: We use a storage subsystem which supports dual 1 Gbits iSCSI ports as the underlying storage medium. In the initiator we use Linux-iSCSI driver [16] to connect to the iSCSI storage through a gigabits network and mount it as a local disk.

NFS: A well known Network File System. Clients and file server are also connected through the same network as above.

In order to reduce variations between different machines, in all of the tests we use the same initiator and target.

Fig. 2(a) shows the performance when writing a character at a time of a total 1 GB file. Our enhanced security model has comparable performance to that of

Table 1. Configurations for initiator and target

CPU	AMD Sempron Processor 2500+
Memory	512MB DDR 400 RAM
HDD	WD IDE 7200rpm 160GB with 8MB buffer
NIC	3COM gigabits NIC 3C200-T

OSD-CAPKEY. The figure shows that our enhanced model adds almost zero overhead compared with the OSD T10 standard, but our proposed OSD system offers stronger security. This is because the matrix multiplication process can be replaced with logic AND in binary field. The overhead of security method CAPKEY over NOSEC is about 0.15 % . The overhead is insignificant because the HMAC process is quite simple and thus requires little computing power. Compared with iSCSI, our system is about 20 % faster when writing a character at a time. This is because that the iSCSI storage is block based, and the local file system packs a character into a block at a time, thus there is a huge overhead to write only a character at a time. When the OSD file system issues a WRITE command, it can issue another one as soon as it receives an acknowledgement response from the target without waiting for the target server to process the command. That is why the writing performance of an OSD system is better than iSCSI storage. The same reason accounts for the poor performance of NFS. Moreover, NFS is based on RPC (Remote Procedure Call) a client, after issuing a write command, needs to wait until receiving an acknowledgement from NFS server and this is why NFS is about 35.95 % worse than an OSD system.

Fig. 2(b) shows the performance when writing a block at a time for a 1 GB file. The performance improvement for both OSD systems and iSCSI storage are significant because a block now contains 8 KB data to be written instead of 1 byte in the per character write, thus the writing overhead is largely reduced. We can see that our enhanced model still offers stronger security with little performance degradation. Again we can see that the overhead of CAPKEY is still very small over NOSEC, and our OSD system performs better than the iSCSI storage because application clients do not have to manage the disk space. We find that the CPU usage for iSCSI storage in such a test is about 3 times of

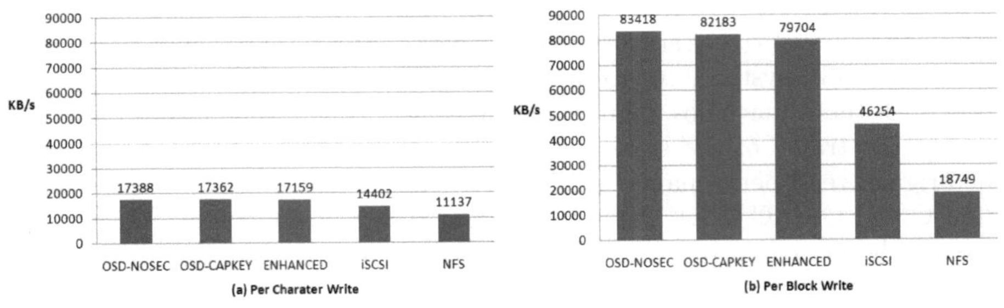

Fig. 2. (a)Performance when writing a character at a time (b)Performance when writing a block at a time

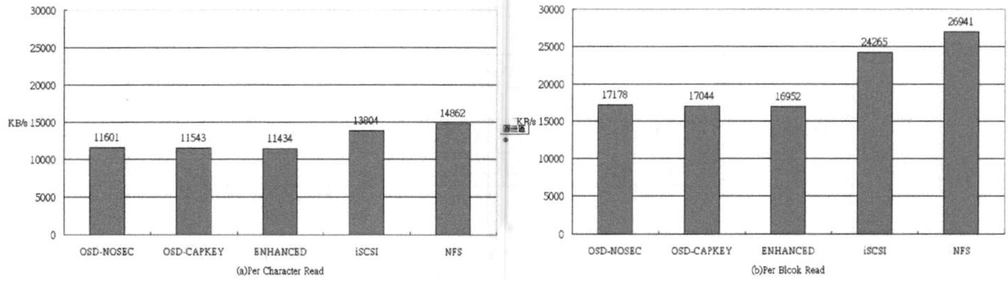

Fig. 3. (a)Performance when reading a character at a time (b)Performance when reading a block at a time

that for an OSD system (9.7% for an OSD system and 31.7% for iSCSI storage). This is because clients need to allocate and manage the disk space before issuing a write command but nothing of this sort has to be done in the OSD system. The performance difference between the OSD system and iSCSI storage in this test is due to disk performance, where performance of iSCSI storage is limited by disk. We can improve the performance by about 65 % if we configure the iSCSI storage to a RAID 5 storage subsystem. The performance improvement for NFS is smaller compared with OSD system and iSCSI storage due to RPC procedure because every command is limited by the RPC response time. The write performance of our OSD system is much better in this test.

Fig. 3(a) shows the performance of reading a character at a time. The poor performance in all environments is now again due to the huge overhead if it only reads a character at a time because the entire block needs to be transferred while only one character is needed. Now the OSD system is about 17 % slower than the iSCSI storage and this is due to the overhead of OSD device server when processing a READ command. Unlike in the writing case, where the OSD file system does not need to wait for the target server to process the command, it now needs to wait for the target server to process the command and return accessed data. The overhead of the target server processing the OSD command includes security check, capability validation and command execution. Thus the overhead causes the performance degradation in reading data compared to iSCSI storage. But the CPU usage of initiator is reduced by about 18 % (78.2 % in iSCSI storage and 63.7% in the OSD system with NOSEC) in the OSD system due to the offloading of disk space management to the target server. The NFS performs better than both iSCSI storage and OSD systems due to the use of a client cache.

Because these machines share the switch with other computers in our laboratory, the traffic of the switch can impact the performance of this system. The original read performance of ENHANCED is somewhat better than OSD-NOSEC and OSD-CAPKEY due to different traffic of the switch. Thus we re-measure the performance of these three systems by disconnecting all the other computers and get the result shown in Fig. 3(a).

Fig. 3(b) shows the performance of reading a block at a time. The overall performance is better than per character read performance as shown in Fig. 3(a). The iSCSI storage and NFS improve a lot due to lower overhead when reading

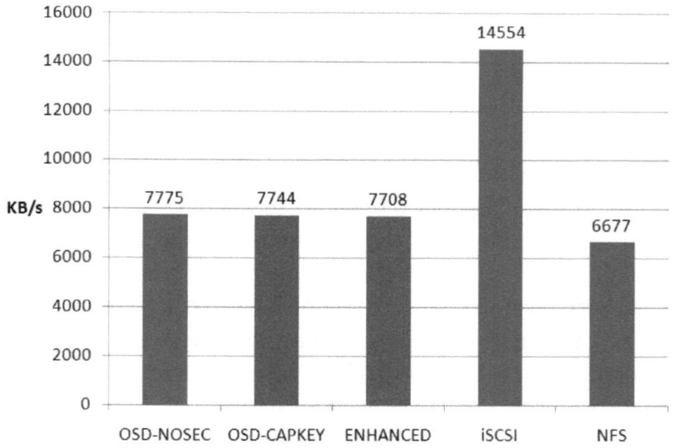

Fig. 4. Performance when rewriting an existing file

a block at a time. An entire block now contains valid data and traffics between client and server are reduced. The improvement in OSD systems is limited by the overhead to process an OSD command.

Fig. 4 shows the performance result of rewriting an existing file. The performance of our proposed system is still comparable to that of the OSD T10 standard. The rewriting performance of the OSD system is not as good as that of iSCSI storage because it needs to copy the data from storage server to a client's local buffer and then send back the rewritten data. The performance degradation is due to poorer read performance of the OSD system. Another reason is that there are two file systems in the OSD systems, OSD file system for initiator and ext3 file system for target. We have to use ext3 file system in the target to emulate an OSD storage device because currently no object-based disks are manufactured and available. We expect the performance of rewrite to be better if the underlying storage medium is real object-based disks. Another method to improve rewrite performance is to design a REWRITE command, which is a new command not defined in OSD T10 standard. Initiator can issue this new command with appropriate data and let the storage server process the rewriting process. In this way, the data needs not to be copied to application clients and thus the rewriting performance can be improved. This is left as our future implementation.

5 Conclusion and Future Work

The OSD system is a new architecture for storage subsystem. It offloads some tasks to storage devices to improve security management, scalability, and even performance. However, the security model of OSD T10 standard can be disabled by altering specific field in the capability. In this paper we point out a problematic security issue and propose to resolve the problem with encryption. The encryption process can be simplified in the binary field by using McEliece based cryptosystem. The overhead of our OSD system is almost zero while offering an enhanced security.

We have also studied storage subsystem performance under five different environments using Bonnie++ benchmark suit. In general the CPU usage in the client side of OSD systems can be reduced significantly. We also find that our OSD system performs much better than the others when writing a file because a client's OSD file system dose not need to manage disk space. An application client can issue another command right after receiving an acknowledgement response from storage device server without waiting for it to execute the command. The read and rewrite performances are not as good as write performance due to the overhead of the file system used to emulate object-based disk in the storage server.The degradation is expected to be improved significantly when real object-based disk becomes available in the future.

The first task to do in the future is to improve the rewrite performance. We want to build an OSD system with performance comparable to iSCSI storage, thus we first need to improve the performance when rewriting a file. The reason that rewrite performance is not as good as write performance is that data needs to be read from the server to the client side, and the data is then modified and sent back to the server. We propose to implement a new command REWRITE to be used to eliminate this unnecessary data read from a storage server to an application client. The new data is sent to the server and the modification is done in the server side directly.

The ultimate goal is to design a real object-based disk. Currently the OSD server emulates object-based storage disk by utilizing the Linux ext3 file system to manage the disk space of the server and its overhead is very large. If we can design an intelligent disk which can manage disk space by itself, the read performance improvement will be significant. And this will take the storage technique to another era.

References

1. Mesnier, M., Ganger, G.R., Reidel, E.: Object-Based Storage. IEEE Communications Magazine 41(8), 84–90 (2003)
2. Du, D., He, D., Hong, C., Jeong, J., Kher, V., Kim, Y., Lu, Y., Raghuveer, A., Sharafkandi, S.: Experiences Building an Object-Based Storage System based on the OSD T-10 Standard. In: MSST 2006 (2006)
3. SCSI Object-Based Storage Device Commands - 2(OSD-2), Project T10/1731-D, revision 0 (2004)
4. IBM Object Store `http://www.haifa.il.ibm.com/projects/storage/objectstore/index.html/`
5. Azagury, A., et al.: Towards an Object Store. In: MSS 2003 (2003)
6. Intel iSCSI/OSD reference implementation `http://sourceforge.net/projects/intel-iscsi/`
7. Satran, J., et al.: Draft-ietf-ips-iscsi-20 (2003)
8. Meth, K.Z., Santran, J.: Design of the iSCSI Protocol. In: MSS 2003 (2003)
9. SCSI Architecture Model - 4(SAM-4), Project T10/1683-D, revision 5(2006)

10. Factor, M., Nagle, D., Naor, D., Reidel, E., Satran, J.: The OSD security protocol. In: Proceedings of 3rd International IEEE Security in Storage Workshop (2005)
11. The Keyed-Hash Message Authentication Code. Federal Information Processing Standards Publication 198 (2006)
12. McEliece, R.J.: A public-key cryptosystem based on algebraic coding theory. JPL DSN Progress Report, 42-44, pp. 114–116 (1978)
13. Li, Y.-X., Li, D.-X., Wu, C.-K.: How to Generate a Random Nonsingular Matrix in McElieces Public-Key Cryptosystem. Singapore ICCS/ISITA, vol.1, pp. 268–269 (1992)
14. Preneel, B., Bosselaers, A., Govaerts, R., Vandewalle, J.: A Software Implementation of the Mceliece Public-Key Cryptosystem. In: Proceedings of the 13th Symposium on Information Theory in the Benelux, Werkgemeenschap voor Informatie- en Communicatietheorie, pp. 119–126 (1992)
15. Bonnie++ benchmark suit, `http://www.coker.com.au/bonnie++/`
16. Linux-iSCSI Project, `http://linux-iscsi.soureforge.net`

Strategies and Implementation for Translating OpenMP Code for Clusters

Deepak Eachempati, Lei Huang, and Barbara Chapman

Department of Computer Science
University of Houston
Houston, TX 77204-3475
{dreachem,chapman}@cs.uh.edu, lhuang5@mail.uh.edu

Abstract. OpenMP is a portable shared memory programming interface that promises high programmer productivity for multithreaded applications. It is designed for small and middle sized shared memory systems. We have developed strategies to extend OpenMP to clusters via compiler translation to a Global Arrays program. In this paper, we describe our implementation of the translation in the Open64 compiler, and we focus on the strategies to improve sequential region translations. Our work is based upon the open source Open64 compiler suite for C, C++, and Fortran90/95.

1 Introduction

MPI is still the most popular and successful programming model for clusters. It, however, is error-prone and too complex for most non-experts. OpenMP is a programming model designed for shared memory systems that provides simple syntax to achieve easy-to-use, incremental parallelism, and portability. However, it is not available for distributed memory systems including widely deployed clusters. We believe that it is feasible to use compiler technologies to extend OpenMP to Clusters to alleviate the programming efforts on cluster. We have developed strategies[7][8][12] to implement it via Global Arrays (GA)[15].

GA is a library that provides an asynchronous one-sided, virtual shared memory programming environment for clusters. A GA program consists of a collection of independently executing processes, each of which is able to access data declared to be shared without interfering with other processes. GA enables us to retain the shared memory abstraction, but at the same time makes all communications explicit, thus enabling the compiler to control their location and content. Considerable effort has been put into the efficient implementation of GA's one-sided contiguous and strided communications. Therefore, we can potentially generate GA codes that will execute with high efficiency using our translation strategy. On the other hand, our strategy shares some of the problems associated with the traditional SDSM approach to translating OpenMP for clusters: in particular, the high cost of global synchronization, and the difficulty of translating sequential parts of a program.

The traditional approach to implementing OpenMP on clusters is based upon translating it to software Distributed Shared Memory systems (DSMs), notably TreadMarks[6] and Omni/SCASH[17]. The strategy underlying such systems is to manage

R. Perrott et al. (Eds.): HPCC 2007, LNCS 4782, pp. 420–431, 2007.

shared memory by migrating pages of data, which unfortunately incurs high overheads. Software DSMs perform expensive data transfers at explicit and implicit barriers of a program, and suffer from false sharing of data at page granularity. They typically impose constraints on the amount of shared memory that can be allocated. But this effectively prevents their applications to large problems. [4] translates OpenMP to a hybrid MPI+software DSM in order to overcome some of the associated performance problems. This is a difficult task, and the software DSM could still be a performance bottleneck. [1] translates OpenMP directly to MPI programs.

In this paper, we discuss our approach and implementation for translating OpenMP programs to Global Arrays, especially we present our strategies for handling sequential regions of OpenMP efficiently. The remainder of this paper is organized as follows. We first describe our approach that translates OpenMP to GA in Section 2, and follow by three strategies for handling the sequential parts of OpenMP programs. We will then present a detailed implementation for the work based on the OpenUH compiler. Related work and conclusions are described in the subsequent sections.

2 Our Translation Approach

GA [15] was designed to simplify the programming methodology on distributed memory systems by providing a shared memory abstraction. It does so by providing routines that enable the user to specify and manage access to shared data structures, called global arrays, in a FORTRAN, C, C++ or Python program. GA permits the user to specify block-based data distributions for global arrays, corresponding to HPF BLOCK and GEN_BLOCK distributions which map the identical length chunk and arbitrary length chunks of data to processes respectively. Global arrays are accordingly mapped to the processors executing the code. Each GA process is able to independently and asynchronously access these distributed data structures via get or put routines.

In principle, translating OpenMP programs into GA programs is not difficult since both of them have the concept of shared data and the GA library features match most OpenMP constructs. Almost all OpenMP directives can be translated into GA or MPI library calls at source level. (We may use these together if needed, since GA was designed to work in concert with the message passing environment.) Most OpenMP library routines and environment variables can be also be translated to GA routines. Exceptions are those that dynamically set/change the number of threads, such as OMP_SET_DYNAMIC, OMP_SET_NUM_THREADS.

Our translation strategy follows OpenMP semantics and translates an OpenMP code to a SPMD style GA program. OpenMP threads correspond to GA processes: a fixed number of processes are generated and terminated at the beginning and end of the corresponding GA program. Instead of flexible forking and joining threads in OpenMP, the GA programs have to keep the fixed number of processes and manage them from the beginning. OpenMP shared data are translated to global arrays that are distributed among the GA processes; all variables in the GA code that are not global arrays are called private variables. OpenMP private variables are replicated to each GA process. All shared OpenMP arrays are given a data distribution and translated to global arrays.

2.1 Strategies for Translating Sequential Regions

All strategies for implementing OpenMP on cluster have a problem with sequential regions. A process executing sequential code may need to read or write to shared data which potentially could be distributed across multiple cluster nodes. A major challenge is to devise an effective compiler/runtime strategy to minimize this communication overhead. A simple approach for handling sequential regions would be to restrict execution of enclosed statements to the master process. However, this presents another problem. If the executing process encounters a parallel region, other processes would need to be available to join in executing it. We refer to this as the control flow problem for translating OpenMP sequential region on clusters. We have identified 3 general translation strategies for resolving this control flow problem, which we discuss below.

Parallel Control Flow. For this approach, we translate sequential regions such that only one process (typically the master process) executes assignment statements or I/O operations. However, to ensure every process is available for a parallel region, control flow must be executed by all processes if it encloses a parallel region. While sometimes this can be determined at compile time, procedure calls can make it difficult to statically determine if a control flow construct encloses a parallel region. In the cases a control flow construct is free of enclosed parallel regions or calls to procedures with parallel regions, it need only be executed by a single process. Otherwise, all processes must execute the control flow statement. In a similar way, a call to a procedure must be executed by all processes if there is the potential of encountering a parallel region down that call path.

Another issue is that all processes must identically evaluate the control flow condition expression. This means that any input values required for the condition expression must be consistent across all processes. To handle this, the condition expression is evaluated and written to a temporary variable by the executing single process. Then, all processes execute a collective broadcast operation, which will send the resulting value of the expression to all processes. All processes then use this value to evaluate the control flow condition.

Redundant Execution Approach. Another way of tackling the control flow problem in sequential regions is to simply allow every process to redundantly execute the entire sequential region, except for certain I/O operations (e.g. writing to a file). There are multiple potential advantages with this approach. Firstly, it can be implemented with a more straightforward translation then parallel control flow. And in theory, if all shared data is replicated for every process, then a sequential region can be redundantly executed by every process without requiring communication to ensure shared data consistency. And, of course, it would not be necessary to broadcast control flow conditions either.

There are, however, some significant drawbacks to this approach as well. It could be impractical to replicate shared data for every process if the data consisted of very large shared arrays. Often, we would prefer to distribute such arrays across the cluster processes. But if we do this then redundant accesses of such data in sequential regions would require communication to/from every process, a considerable drain on the cluster's network bandwidth. Also, for certain applications with resource-intensive sequential regions (e.g. graphical interfaces), replicating all sequential code is impractical. For these reasons, we believe parallel control flow is a more suitable translation strategy.

Idle Process Invocation. The final strategy we have considered for dealing with this problem is for the master process to invoke idle helper processes to work on parallel regions as they are encountered. In this approach, only the master process executes sequential regions while the rest of the processes remain idle. Furthermore, the compiler outlines all parallel regions in the program to separate procedures, and inserts calls to these procedures where the regions used to be. When the master process encounters a call to an outlined parallel region, it will broadcast an index and any shared data references for the procedure to the rest of the processes. All processes are synchronized just before beginning and exiting execution of the outlined parallel region.

2.2 Translation of Parallel Region

Before delving into implementation details, we present a short example to illustrate how we translate an OpenMP parallel region into an SPMD-style code for clusters.

```
!$omp parallel shared(a)
   do k = 1, MAX
!$omp do
     do j = 1, SIZE_Y
       do i = 1, SIZE_X-1
          a(i,j) = a(i+1, j) + ...
       end do
     end do
!$omp end do
   end do
!$omp end parallel
```

(a) Original OpenMP

```
! tell runtime to register 'a' as a shared variable (creates
! corresponding GA)
call clustermp_register_var('a', varid, TYP_DBL, SIZE_X, SIZE_Y, 2, A)
! signals to runtime we are entering a parallel region
call clustermp_entering_parallel()
do k0 = 1, MAX
! runtime calculates loop bounds based on process id
   jlo = clustermp_getlowerbound(1, SIZE_Y)
   jhi = clustermp_getupperbound(1, SIZE_Y)
! compiler determines region of array that is read in parallel loop
   call clustermp_read(IS_LOCAL, a, 'a', varid,2,SIZE_X,jlo,jhi)
   do j0 = jlo, jhi
     do i0 = 1, SIZE_X-1
       if ( clustermp_cond_exec() .eq. 1 ) then
         a(i0,j0) = a(i0+1,j0) + ...
       end if
     end do
   end do
! compiler determines regiong of array that is written in parallel loop
   call clustermp_write(IS_LOCAL, a, 'a', varid,1,SIZE_X-1,jlo,jhi)
   call clustermp_barrier()
end do
call clustermp_leaving_parallel()
```

(b) Resulting Parallel Code for Cluster

Fig. 1. OpenMP code is translated into an SPMD-style parallel program by our compiler. Calls to the ClusterMP runtime are inserted to manage shared data and coordinate work in sequential and parallel regions.

Fig.1 (a) depicts a simple OpenMP parallel region where work is distributed across the array *a*'s second dimension. Suppose the size of *a*'s first dimension is *SIZE_X*, and the size of *a*'s second dimension is *SIZE_Y*. We determine that *a* is a shared variable with work distributed across its second dimension, and we register it as such with our runtime Fig. 1 (b), line 3. This registration will create a corresponding GA for the array *a* that is block distributed across each process executing on the cluster. A handle for the GA is managed internally by the runtime, which allows potential runtime optimizations such as redistribution based on dynamic access patterns. Next, we signal the runtime to enter a parallel state, and we commence with the *MAX* iterations.

On lines 8 and 9 in Fig. 1 (b), lower and upper bounds for the second dimension are calculated by each process at runtime. On line 11, each process performs a blockwise "get" for array elements it will be reading in the loop. We apply a condition for the assignment on line 14, which restricts execution of the enclosed block of statements depending based on the process id and the OpenMP state (i.e. sequential or parallel region). At the end of the iteration, each process will "commit" the local writes it performed via the *clustermp_write* call and then synchronize with the other processes before starting the next iteration. Finally, after all iterations have completed, we tell the runtime that the parallel region has completed.

3 Implementation

We have implemented the parallel control flow translation strategy for sequential regions, described in the previous section. Our implementation consists of two phases: (1) we use the OpenUH compiler to translate OpenMP programs into a corresponding cluster-enabled parallel program, and (2) we created a runtime library, *ClusterMP*, which wraps calls to the Global Arrays toolkit for creating, reading, and writing shared data in a cluster environment. We present implementation details for the compilation and runtime phases below.

3.1 OpenUH Compiler

The OpenUH compiler is based on Open64, an open source compiler for C, C++, and Fortran 90. Fig. 2 shows a component diagram for the OpenUH backend, where we have implemented our OpenMP translation for execution on clusters.

Our translation approach is to leverage as much as possible the ClusterMP runtime in order to reduce complexity in our OpenMP translation. Consequentially, our translation scheme is fairly straightforward and does not assume an underlying use of global arrays by the runtime. We also feel there is considerable scope for optimizations to be carried out both at the compiler and runtime level. In this section, we describe the translation scheme employed by OpenUH to translate OpenMP programs for execution on a cluster in conjunction with the ClusterMP runtime.

In our translation, each procedure is handled separately. Specifically, if it is the main procedure the compiler will insert a call to initialize the ClusterMP runtime (*clustermp_init*) at the beginning of the procedure's body and a call to shutdown the ClusterMP runtime (*clustermp_finalize*) at the very end of the procedure's body. For all procedures, a runtime call (*clustermp_create_locals*) is inserted to signal to the

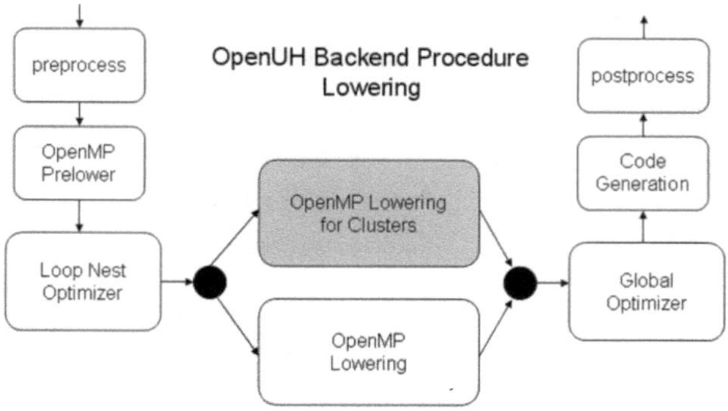

Fig. 2. We use the OpenUH compiler to implement the OpenMP translation for clusters

runtime that a new procedure has been entered. This will cause the runtime to push a new shared data frame for the current procedure onto the runtime's shared data frame stack. Then, a runtime call (*clustermp_check_args*) is inserted to communicate the procedure's parameters to the runtime.

The compiler first traverses the IR tree for the procedure and adds any identified shared variables to a shared data list. Calls (*clustermp_register_var*) are inserted for each shared variable to register it with the runtime. Information passed includes array bounds information (all dimension sizes are set to 1 if it is a scalar), the variable name, a unique identification number, and the variable data type. Next, the compiler traverses the IR a second time and restructures the code such that computations in sequential regions are handled by the master process, and all processes participate in the parallel regions as specified by the OpenMP directives. Runtime calls are inserted to read/write shared data, check or modify the dynamic OpenMP state (e.g. in a sequential or parallel region?), and to receive the assigned iterations for OpenMP DO loops. Finally a runtime call (*clustermp_pop_locals*) is inserted at the end of the procedure to signal the runtime to pop the current shared data frame from its shared data frame stack.

3.2 ClusterMP Runtime

Considerable research has been carried out for executing OpenMP programs on cluster architectures. We believe the GA toolkit is a suitable target for translating OpenMP programs. It allows for a relatively straightforward translation due to its shared memory abstraction. But unlike software distributed memory systems (SDSMs), which are traditionally used for implementing OpenMP on clusters, GA makes all communication explicit which is useful for controlling and optimizing data communication. For our implementation of OpenMP on clusters we developed a runtime library, ClusterMP, which wraps functionality provided by the GA toolkit and provides management of shared data. In this section, we present an overview of the ClusterMP Runtime implementation.

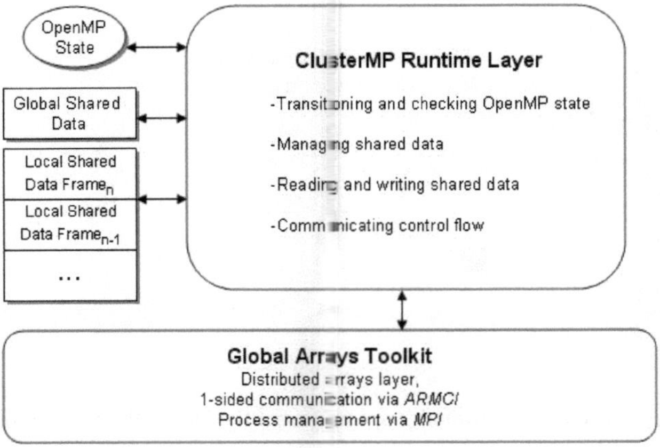

Fig. 3. The runtime core functionality can be split into four categories: (1) transitioning and checking OpenMP state, (2) managing shared data, (3) reading and writing to shared data, and (4) communicating control flow information to all processes

OpenMP State Transition. OpenMP allows parallel regions to extend past the lexical extent of the OMP parallel directive and into called procedures. Consequentially, we must maintain dynamic OpenMP state information which indicates whether the executing process is currently in a sequential or parallel region. We add other states to distinguish OpenMP single and critical regions as well. The compiler inserts runtime calls into the translated code to initiate state transition as appropriate. All processes, by default, start in the sequential state. When transitioning from sequential to parallel (via *clustermp_entering_parallel*) or from parallel to sequential (via *clustermp_leaving_parallel*), the runtime will implicitly invoke a barrier operation to synchronize all the processes, unless a *nowait* clause was specified.

The value of retaining OpenMP state for the purposes of our translation strategy is to identify whether a process may execute a conditional block. As explained in our description of the compiler translation, we place all assignment statements (i.e. computation) in a conditional block. The condition is evaluated to true at runtime simply if the process ID is 0 (it is the master process) or the OpenMP state is set to parallel or critical. This will restrict other processes from performing computation in sequential regions or OpenMP single blocks.

Managing Shared Data. The runtime uses two data structures for maintaining metadata corresponding to all shared variables identified for the running process. The *Global Shared Data* table holds metadata for shared global variables. The *Local Shared Data Frame* (*LSDF*) stack holds metadata for all shared variables local to each procedure in the call stack. The shared data we support may be either a scalar or multi-dimensional array with a base data type of integer, floating point, or complex number. For each shared variable, we track the variable name, a unique variable id, a global array handle, and the number of dimensions and the size of each dimension (1 and 1 for scalars).

We use global arrays to represent shared data. Whenever clustermp_register_var is invoked for a shared variable, the ClusterMP runtime will attempt to create a corresponding global array for the variable with an appropriate data distribution. Information for this global array is then added to the appropriate Shared Data table. When a process enters a called procedure, *clustermp_create_locals* is called to create a new *LSDF* that records information for shared variables encountered in the procedure. This frame is pushed onto a stack which corresponds to the call stack for the process. Global array information for parameters passed into a procedure can be copied from the previous *LSDF* into the new frame, using the calls *clustermp_set_args* and *clustermp_check_args*.

Reading and Writing to Shared Data. The compiler inserts a call to *clustermp_read* whenever data that is not explicitly scoped as private is read, and it inserts a call to clustermp_write whenever such data is written to. The *clustermp_read* operation will perform a one-sided get on the corresponding global array for the variable being read, and copy the retrieved value into the variable address (which is essentially treated as a "local buffer" for the global array). Similarly, *clustermp_write* will perform a one-sided put on the corresponding global array using the value provided in the variable. If either function is invoked for a variable that hasn't been registered, then it immediately returns. The steps taken for reading and writing shared data is identical for sequential and parallel regions.

Clearly, this approach incurs very high overhead, simply due to the fact that a runtime call is made for each variable access. This means that it is the job of the runtime to minimize communication overhead as much as possible. We have investigated various strategies for minimizing communication – including prefetching read data and delaying writes to shared data until a synchronizing barrier is reached. We believe there is considerable scope for optimizing the overhead of reading or writing shared data, both in the compiler and the runtime.

Communicating Control Flow. In our translation, the master process must communicate the resulting value of the Boolean expression for control flow constructs. The routine *clustermp_comm_cf* is used to communicate this information. If the OpenMP state is not sequential or single, then this routine does nothing (all processes already have the necessary information to correctly execute the control flow). Otherwise, we broadcast the value specified from the master process to all other processes.

4 Evaluation

In this section, we discuss the results of a set of experiments we ran to evaluate the three strategies for sequential translation described in Section 2.1. We look at execution time and frequency of various GA routine calls. Experiments were conduceted on *Atlantis*, a cluster of dual-Itanium2 machines connected by a high performance myrinet interconnect. We used OpenUH 1.0 to implement the parallel control flow (PCF) translation and we create *ClusterMP* programs by hand to test redundant execution (RE) and idle process invocation (IPI). These experiments also used the GA 4.0.5 and Intel OpenMPI 1.2.3 libraries.

4.1 Benchmark

To test our implementation, we devised a benchmark which would contain a configurable mixture of sequential regions, with read and write I/O, and parallel regions. Within the main iteration loop of the benchmark: (1) the contents of a large file are read into a buffer, (2) an arithmetic expression is evaluated based on the interation count and one of multiple procedures are called accordingly. (3) a parallel routine (*Jacobi*) may or may not be called (likelihood increases proportionally with a specifed *JacobiFrequency* parameter), and (4) contents of file are written back out to an output file. We ran this benchmark, translated using the above three strategies, on 4 cluster nodes.

4.2 Performance Results

The results in Fig. 4 (a) show that the PCF approach yielded the best total execution time as the frequency of Jacobi interjection increases. While from Fig. 4 (b) it is clear that sequential execution time was much greater for PCF than either RE or IPI, we see also that each invocation of Jacobi is executed faster (Fig. 4 (c)). This is an interesting result, since all three approach should yield identical translations for *parallel* regions, and indeed we verified that all three translations of Jacobi had an identical number of calls to *ga_create, ga_get, ga_put, ga_sync*, and *ga_destroy.* When worker processes are expected to replicate execution (as in RE) in sequential regions, this can cause them to arrive late at the parallel region which in turn results in extra wait time to complete necessary collective operations. Thus, even though the RE

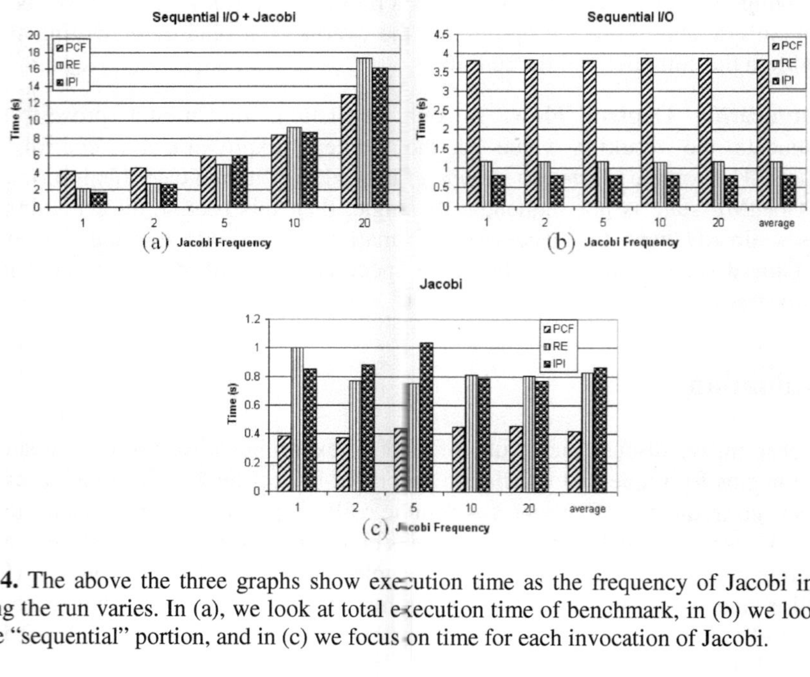

Fig. 4. The above the three graphs show execution time as the frequency of Jacobi invoked during the run varies. In (a), we look at total execution time of benchmark, in (b) we look only at the "sequential" portion, and in (c) we focus on time for each invocation of Jacobi.

approach can yield superior sequential execution time (*for the master process*), overall performance can be potentially be degraded.

We are particularly intrigued by the results of the IPI approach, which in effect mimics the fork-join model of traditional shared memory programs. Sequential execution times are very small, as expected, since only the master process executes it. As with RE though, there are unexpected overheads when executing the parallel regions that are currently being investigated.

5 Related Work

Our approach to implementing OpenMP for distributed memory systems has a number of features in common with approaches that translate OpenMP to Software DSMs [17][2][6][5] or Software DSM plus MPI [4], or directly to MPI [1] for cluster execution. All of these methods need to determine the data that has to be distributed across the system, and must adopt a strategy for doing so. Also, the work distribution for parallel and sequential regions has to be implemented, and it is typically the latter that leads to problems. Note that it is particularly helpful to perform an SPMD-style, global privatization of OpenMP shared arrays before translating codes via any strategy for cluster execution, due to the inherent benefits of reducing the size and number of shared data structures and of obtaining a large fraction of references to (local) private variables.

On the other hand, our translation to GA is distinct from other approaches in that ours promises higher levels of efficiency via the construction of precise communication sets. The difficulty of the translation itself lies somewhere between the translation to MPI and the translation to Software DSMs. First, the shared memory abstraction is supported by GA and Software DSMs, but is not present in MPI. It enables a consistent view of variables and a non-local datum is accessible if given a global index. In contrast, only the local portion of the original data can be seen by each process in MPI. Therefore manipulating non-local variables in MPI is inefficient since the owner process and the local indices of arrays have to be calculated. Furthermore, our GA approach is portable, scalable and does not impose limitations on the shared memory space. The everything-shared SDSM as presented in [3] attempts to overcome the relaxation of the coherence semantics and the limitation of the shared areas in other SDSMs. It does solve commonly existing portability problems in SDSMs by using an OpenMP run-time approach, but it is hard for such a SDSM to scale with sequential consistency. Second, the non-blocking and blocking one-sided communication mechanisms offered in GA allow for flexible and efficient programming. In MPI-1, both sender process and receiver process must be involved in the communication. Care must be taken with the ordering of communications in order to avoid deadlocks. Instead, get and/or put operations can be handled within a single process in GA.

6 Conclusion

Despite the widespread use multicores, clusters are increasingly used as compute platforms for a growing variety of applications. Extending the simple programming model OpenMP to clusters will provide an efficient program paradigm to allow many non-experts to exploit the capacity of clusters. OpenMP is designed for shared memory systems, and it seems inappropriate to add new language features for cluster support into it. A joined compiler and runtime system is a more practical and promising approach for bridging the gap between the language and hardware.

This paper describes our strategies in implementing OpenMP on clusters by translating OpenMP codes to equivalent GA ones. This approach has the benefit of being relatively straightforward. This strategy has the advantage of relative simplicity together with reasonable performance, without adding complexity to OpenMP for both SMP and non-SMP systems. We have presented the strategies for efficiently handling sequential regions in an OpenMP program. We describe our implementation for the translation from OpenMP to GA in the Open64 compiler [16], an open source compiler that supports OpenMP and which we have enhanced in a number of ways already. We have developed a runtime system on top of the GA library for facilitating the compiler implementation. The rich set of analyses and optimizations in Open64 may help us create efficient GA codes. We will continue working on optimizing our translation by utilizing existing compiler analyses. We plan to build a framework to conduct parallel data flow analysis for OpenMP to improve the performance.

References

1. Basumallik, A., Eigenmann, R.: Towards Automatic Translation of OpenMP to MPI. In: ICS 2005: Proceedings of the 19th Annual International Conference on Supercomputing, pp. 189–198. ACM Press, New York (2005)
2. Basumallik, A., Min, S.-J., Eigenmann, R.: Towards OpenMP Execution on Software Distributed Shared Memory Systems. In: Proceedings of the 4th International Symposium on High Performance Computing, pp. 457–468. Springer, Heidelberg (2002)
3. Costa, J.J., Cortes, T., Martorell, X., Ayguade, E., Labarta, J.: Running OpenMP Applications Efficiently on an Everything-Shared SDSM. In: IPDPS 2004, IEEE Computer Society Press, Los Alamitos (2004)
4. Eigenmann, R., Hoeflinger, J., Kuhn, R.H., Padua, D., Basumallik, A., Min, S.-J., Zhu, J.: Is OpenMP for Grids? In: IPDPS, Fort Lauderdale, FL (April 2002)
5. Hoeflinger, J.P.: Extending OpenMP to Clusters. Technical report. Intel Corporation (2006)
6. Hu, Y.C., Lu, H., Cox, A.L., Zwaenepoel, W.: OpenMP for Networks of SMPs. Journal of Parallel Distributed Computing 60, 1512–1530 (2000)
7. Huang, L., Chapman, B., Liu, Z.: Towards a More Efficient Implementation of OpenMP for Clusters via Translation to Global Arrays. Parallel Computing 31(10-12) (2005)
8. Huang, L., Chapman, B., Kendall, R.: OpenMP for Clusters. In: The Fifth European Workshop on OpenMP, EWOMP 2003. Aachen, Germany (2003)
9. Kee, Y.-S., Kim, J.-S., Ha, S.: ParADE: An OpenMP Programming Environment for SMP Cluster Systems. In: Proceedings of the ACM/IEEE Supercomputing 2003 Conference. Phoenix (2003)
10. Liao, C., Hernandez, O., Chapman, B., Chen, W., Zheng, W.: OpenUH: An Optimizing, Portable OpenMP Compiler. Concurrency and Computation: Practice and Experience, Special Issue on CPC selected papers (2006)

11. Liu, Z., Chapman, B., Wen, Y., Huang, L., Hernandez, O.: Analyses for the Translation of OpenMP Codes into SPMD Style with Array Privatization. In: Voss, M.J. (ed.) WOMPAT 2003. LNCS, vol. 2716, pp. 26–41. Springer, Heidelberg (2003)
12. Liu, Z., Huang, L., Chapman, B., Weng, T.-H.: Efficient Implementation of OpenMP for Clusters with Implicit Data Distribution. In: Chapman, B.M. (ed.) WOMPAT 2004. LNCS, vol. 3349, pp. 121–136. Springer, Heidelberg (2005)
13. Matsuba, H., Ishikawa, Y.: OpenMP on the FDSM Software Distributed Shared Memory. In: The Fifth European Workshop on OpenMP, EWOMP 2003. Aachen, Germany (September 2003)
14. Nieplocha, J., Carpenter, B., ARMCI,: A Portable Remote Memory Copy Library for Distributed Array Libraries and Compiler Run-Time Systems. In: Proceedings of the 11 IPPS/SPDP'99 Workshops Held in Conjunction with the 13th International Parallel Processing Symposium and 10th Symposium on Parallel and Distributed Processing, pp. 533–546. Springer, Heidelberg (1999)
15. Nieplocha, J., Harrison, R.J., Littlefield, R.J.: Global Arrays: A Nonuniform Memory Access Programming Model for High-Performance Computers. Journal of Supercomputing 10, 169–189 (1997)
16. Open64 Compiler and Tools (November 2003), http://open64.sourceforge.net/
17. Sato, M., Harada, H., Hasegawa, A., Ishikawa, Y.: Cluster-Enabled OpenMP: An OpenMP Compiler for SCASH Software Distributed Share Memory System. Scientific Programming, Special Issue: OpenMP 9(2,3), 123–130 (2001)

Optimizing Array Accesses in High Productivity Languages

Mackale Joyner, Zoran Budimlić, and Vivek Sarkar

Rice University, Houston TX
{mjoyner,zoran,vsarkar}@cs.rice.edu

Abstract. One of the outcomes of DARPA's HPCS program has been the creation of three new high productivity languages: Chapel, Fortress, and X10. While these languages have introduced improvements in language expressiveness and programmer productivity, several technical challenges still remain in delivering high performance with these languages. In the absence of optimization, the high-level language constructs that improve productivity can result in order-of-magnitude runtime performance degradations.

This paper addresses the problem of efficient code generation for high level array accesses in the X10 language. Two aspects of high level array accesses in X10 are important for productivity but also pose significant performance challenges: the high level accesses are performed through Point objects rather than integer indices, and variables containing references to arrays are rank-independent. Our solution to the first challenge is to extend the X10 compiler with *automatic inlining and scalar replacement of Point objects*. Our partial solution to the second challenge is to use X10's *dependent type system* to enable the programmer to annotate array variable declarations with additional information for the rank and region of the variable, and to allow the compiler to generate efficient code in cases where the dependent type information is available. Although this paper focuses on high level array accesses in X10, our approach is applicable to similar constructs in other languages.

Our experimental results for single-thread performance demonstrate that these compiler optimizations can enable high-level X10 array accesses with implicit ranks and Points to improve performance by up to a factor of 5.4× over unoptimized X10 code, and to also achieve performance comparable (from 48% to 100%) to that of lower-level Java programs. These results underscore the importance of the optimization techniques presented in this paper for achieving high performance with high productivity.

1 Introduction

The Defense Advanced Research Projects Agency (DARPA) has challenged supercomputer vendors to increase development productivity in high-performance scientific computing by a factor of 10 by the year 2010. DARPA has recognized that constructing new languages designed for scientific computing is important to meeting this productivity goal. Cray (Chapel), IBM (X10), and Sun (Fortress) have developed new high productivity languages in response to this challenge.

R. Perrott et al. (Eds.): HPCC 2007, LNCS 4782, pp. 432–445, 2007.

While these languages' abstractions suitably provide the mechanisms necessary to improve productivity in high-performance scientific computing [10], compiler optimizations are crucial to minimizing performance penalties resulting from the abstractions.

This paper addresses the problem of efficient code generation for high level array accesses in the X10 language. There are two aspects of high level array accesses in X10 that are important for productivity but that also pose significant performance challenges. First, the high level accesses are performed through Point objects rather than integer indices. Points support an object-oriented approach to specifying sequential and parallel iterations over general array regions and distributions in X10. As a result, the Point object encourages programmers to implement reusable high-level iteration abstractions to efficiently develop array computations for scientific applications without having to manage many of the details typical for low level scientific programming. However, the creation and use of new Point objects in each iteration of a loop can be a significant source of overhead. Second, variables containing references to arrays are rank-independent *i.e.,* by default, the declaration of an array reference variable in X10 does not specify the rank (or dimension sizes) of its underlying array. This makes it possible to write rank-independent code in X10, but poses a challenge for the compiler to generate efficient rank-specific code. Our solution to the first challenge is to extend the X10 compiler so as to perform *automatic inlining and scalar replacement of Point objects.* We have a partial solution to the second challenge that uses X10's *dependent type system* to enable the programmer to annotate selected array variable declarations with additional information for the rank and region of the variable, and to extend the compiler so as to generate efficient code in cases where the dependent type information is available. In the future, we plan to evaluate existing algorithms in the literature for rank and region analysis to test their effectiveness for X10 arrays, so as to reduce the need for the programmer to provide the dependent type annotations.

Our experimental results for single-thread performance demonstrate that these compiler optimizations can enable high-level X10 array accesses with implicit ranks and Points to improve performance by up to a factor of 5.4× over unoptimized X10 code, and to also achieve performance comparable (from 48% to 100%) to that of lower-level Java programs. Even though the current prototype X10 implementation [18] targets Java as its execution platform, we expect the code optimizations presented here to be applicable to other source languages (including Chapel and Fortress) and other target languages (including C and FORTRAN). Further, recent improvements in Java optimization and implementation technologies show that Java performance can also approach that of native FORTRAN and C for some high-performance scientific applications [15]. Thus, we believe that the experimental results in this paper are also indicative of the impact that the optimizations will have on future production-strength implementations of the new high-productivity languages.

Section 2 discusses X10 language constructs related to arrays, points, regions, and point-wise loops. Section 3 describes the optimizations we utilize to enhance the performance of applications employing these specific language constructs. Finally, section 4 presents the experimental results obtained from these compiler optimizations.

2 X10 Language Overview

In this section, we summarize X10 features related to *arrays*, *points*, *regions* and *loops* [7], and discuss how they contribute to improved productivity in high performance computing. Since the introduction of arrays in the FORTRAN language, the prevailing model for arrays in high performance computing has been as a contiguous sequence of elements that are addressable via a Cartesian index space. Further, the actual layout of the array elements in memory is typically dictated by the underlying language *e.g.,* column major for FORTRAN and row major for C. Though this low-level array abstraction has served us well for several decades, it also limits productivity due to the following reasons:

1. *Iteration.* It is the programmer's responsibility to write loops that iterate over the correct index space for the array. Productivity losses can occur when the programmer inadvertently misses some array elements in the iteration or introduces accesses to non-existent array elements (when array indices are out of bounds).
2. *Sparse Array accesses.* Iteration is further complicated when the programmer is working with a logical model of sparse arrays, while the low level abstraction supported in the language is that of dense arrays. Productivity losses can occur when the programmer introduces errors in managing the mapping from sparse to dense indices.
3. *Language Portability.* The fact that the array storage layout depends on the underlying language (*e.g.,* C vs. FORTRAN) introduces losses in productivity when translating algorithms and code from one language to another.
4. *Limitations on Compiler Optimizations.* Finally, while the low-level array abstraction can provide programmers with more control over performance, there is a productivity loss incurred due to its interference with the compiler's ability to perform data transformations for improved performance (such as array dimension padding and automatic selection of hierarchical storage layouts).

The X10 language addresses these productivity limitations by providing higher-level abstractions for arrays and loops that build on the concepts of *points* and *regions* (which were in turn inspired by similar constructs in languages such as ZPL). A *point* is an element of an n-dimensional Cartesian space ($n \geq 1$) with integer-valued coordinates, where n is the *rank* of the point. A *region* is a set of points, and can be used to specify an array allocation or an iteration construct such as the point-wise for loop. The benefits of using points inside of *for* loops include: potential reuse of common iteration patterns via storage inside of regions and simple point references replacing multiple loop index variables to access array elements. We use the term, *compact region*, to refer to a region for

Region operations:

```
R.rank ::= # dimensions in region;
R.size() ::= # points in region
R.contains(P) ::= predicate if region R contains point P
R.contains(S) ::= predicate if region R contains region S
R.equal(S) ::= true if region R and S contain same set of points
R.rank(i) ::= projection of region R on dimension i (a one-dimensional region)
R.rank(i).low() ::= lower bound of i-th dimension of region R
R.rank(i).high() ::= upper bound of i-th dimension of region R
R.ordinal(P) ::= ordinal value of point P in region R
R.coord(N) ::= point in region R with ordinal value = N
R1 && R2 ::= region intersection (will be rectangular if R1 and R2 are rectangular)
R1 || R2 ::= union of regions R1 and R2 (may or may not be rectangular,compact)
R1 - R2 ::= region difference (may or may not be rectangular,compact)
```

Array operations:

```
A.rank    ::= # dimensions in array
A.region ::= index region (domain) of array
A.distribution ::= distribution of array A
A[P] ::= element at point P, where P belongs to A.region
A | R ::= restriction of array onto region R (returns copy of subarray)
A.sum(), A.max() ::= sum/max of elements in array
A1 <op> A2 ::= returns result of applying a point-wise op on array elements,
               when A1.region = A2. region
               (<op> can include +, -, *,  and / )
A1 || A2 ::= disjoint union of arrays A1 and A2
             (A1.region and A2.region must be disjoint)
A1.overlay(A2) ::= array with region, A1.region || A2.region,
                   with element value A2[P] for all points P in A2.region
                   and A1[P] otherwise.
```

Fig. 1. Region operations in X10

which the set of points can be specified in bounded space[1], independent of the number of points in the region. Rectangular, triangular, and banded diagonal regions are all examples of compact regions. In contrast, sparse array formats such as compressed row/column storage are examples of non-compact regions.

Points and regions are first-class value types [1] in X10 — a programmer can declare variables and create expressions of these types using the operations listed in Figure 1. In addition, X10 supports a special syntax for point construction — the expression, "`[a,b,c]`", is implicit syntax for a call to a three-dimensional point constructor, "`point.factory(a,b,c)`", and also for variable declarations — the declaration, "`point p[i,j]`" is exploded syntax for declaring a two-dimensional point variable `p` along with integer variables `i` and `j` which correspond to the first and second elements of `p`. Further, by requiring that points and regions be value types, the X10 language ensures that individual elements of a point or a region cannot be modified after construction.

A summary of array operations in X10 can be found in Figure 1. A new array can be created by restricting an existing array to a sub-distribution, by combining multiple arrays, and by performing point-wise operations on arrays with the

[1] For this purpose, we assume that the rank of a region can be assumed to be bounded.

Java version:

```
double G[][] = new double[M][N];
. . .
int Mm1 = M-1; int Nm1 = N-1;
for (int p=0; p<num_iterations; p++) {
   for (int i=1; i<Mm1; i++) {
      double[] Gi = G[i]; double[] Gim1 = G[i-1]; double [] Gip1 = G[i+1];
      for (int j=1; j<Nm1; j++)
         Gi[j] = omega_over_four * (Gim1[j] + Gip1[j] + Gi[j-1] + Gi[j+1])
                 + one_minus_omega * Gi[j];
   } // for i
} // for p
```

X10 version (rank-specific):

```
region R = [0:M-1,0:N-1]; double[.] G = new double[R];
. . .
region R_inner = [1:M-2,1:N-2]; // R_inner is a subregion of R
for (int p=0; p<num_iterations; p++) {
   for (point t : R_inner) {
         G[t] = omega_over_four * (G[t+[-1,0]] + G[t+[1,0]]
                + G[t+[0,-1]] + G[t+[0,1]]) + one_minus_omega * G[t];
   } // for t
} // for p
```

X10 version (rank-independent):

```
. . .
region R_inner = ... ; // Inner region as before
region stencil = ... ; // Set of points in stencil
double omega_factor = ... ; // Weight used for stencil points
for (int p=0; p<num_iterations; p++) {
   for (point t : R_inner) {
      double sum = one_minus_omega * G[t];
      for (point s : stencil) sum += omega_factor * G[t+s];
      G[t] = sum;
   } // for t
} // for p
```

Fig. 2. Java Grande SOR benchmark

same region. Note that the X10 array allocation expression, "`new double[R]`", directly allocates a multi-dimensional array specified by region `R`. In its full generality, an array allocation expression in X10 takes a *distribution* instead of region. However, we will ignore distributions in this paper, since we limit our attention to single-place executions.

As an example, consider the Java and code fragments shown in Figure 2 for the Java Grande Forum [12] SOR benchmark[2]. Note that the Java version involves a

[2] For convenience, we use the same name, `G`, for the allocated array as well as the array used inside the SOR computation, even though the actual benchmark uses distinct names for both.

lot of manipulation of explicit array indices and loops bounds that can be error prone. In contrast, the rank-specific X10 version uses a single `for` loop to iterate over all the points in the inner region (`R_inner`), and also uses point expressions of the form "`t+[-1,0]`" to access individual array elements. One drawback of the *point-wise* `for` loop in the X10 version is that (by default) it leads to an allocation of a new point object in every iteration for the index and for each subscript expression, thereby significantly degrading performance. Fortunately, the optimization techniques presented in this paper enable the use of point-wise loops as in the bottom of Figure 2, while still delivering the same performance as manually indexed loops as in the top of Figure 2.

Figure 2 also contains a *rank-independent* X10 version. In this case, an additional loop is introduced to compute the weighted sum using all elements in the stencil. Note that the computation performed by the nested `t` and `s` `for` loops in this version can be reused unchanged for different values of `R_inner` and `stencil`.

3 Improving Performance of Applications with X10 Language Abstractions

This section has two areas of focus. First, we discuss a compiler optimization we employ to reduce the overhead of using *points* in X10. Second, we use X10's *dependent type system* to further improve code generation. As an example, Figure 3 contains a simple code fragment illustrating how X10 arrays may be indexed with points in lieu of loop indices. Figure 4 shows the unoptimized Java output generated by the reference X10 compiler [18] from the input source code in Figure 3. The *get* and *set* operations inside the *for* loops are expensive, and this is further exacerbated by the fact that they occur within innermost loops.

To address this issue, we have a developed an optimization that is a form of *object inlining*, specifically tailored for value-type objects. Object inlining [2,4,8,9] is a compiler optimization for object-oriented languages that transforms objects into primitive data, and the code that operates on objects into code that operates on inlined data. Budimlić [2] and Dolby [8] introduced object inlining as an optimization for Java and C++. General object inlining requires complex escape analysis and concrete type inference, and the transformation is irreversible (once unboxed, objects in general cannot be "reboxed").

However, because points in X10 are value types, we can safely optimize all array accesses utilizing point objects by replacing them with an object inlined point array access version. A value object has the property that once the program initializes the object, it cannot subsequently modify any of the object's fields. This prevents the possibility of the code modifying *point p* in Figure 3 in between the assignments – a situation that would prevent the inlining of the point. As a result, we can inline the point object declared in the *for* loop header. Figure 5 shows the results of applying this point optimization to the loop we introduce in Figure 3, and Figure 6 shows the resulting Java code.

```
region arrayRegion1 = [0:datasizes_nz[size]-1];
...
//X10 for loop
for (point p : arrayRegion1) {
   row[p] = rowt[p];...
   col[p] = colt[p];...
   val[p] = valt[p];...
}
```

Fig. 3. X10 source code of loop example taken from the Java Grande sparsematmult benchmark

```
//X10 for loop body translated to Java
for  ... {
   ... // Includes code to allocate a new point object for p
  (row).set(((rowt).get(p)),p);...
  (col).set(((colt).get(p)),p);...
  (val).set(((valt).get(p)),p);...
}
```

Fig. 4. Java source code of loop following translation from X10 to Java by X10 compiler

3.1 Point Inlining Algorithm

We perform a specialized version of object inlining [2] to inline *points*. There are two main differences between points and the objects traditionally considered as candidates for object inlining. First, a point variable can have an arbitrary number of fields because a programmer may use points to access arrays of different rank. Second, a point variable may appear in an X10 loop header. Consequently, the specialized object inlining algorithm must transform the X10 loop header by using the inlined point fields as loop index variables. As a result, this may lead to nested *for* loops if the point variable is a multi-dimensional point.

Figure 7 shows the point inlining algorithm. The first step in the algorithm is to use type analysis to discover the rank of all X10 points in the program. Recall, developers may omit rank information when declaring X10 points. However, we need to infer rank information to inline the point. We obtain rank information for points from both point assignments and array domain information found in X10 loop headers. Because points have the value type property, we inline/unbox every point with an inferred rank. When encountering method calls passed point arguments, we reconstruct the inlined point by creating a new point instance, but ensure that this overhead is only incurred on paths leading to the method calls by allowing the code to work with both original and unboxed versions of the point. Finally, when possible, we convert a point-wise X10 loop into a set of nested *for* loops using the X10 loop's range information for each dimension in the region.

```
//X10 optimized for loop
for (int i = 0;  i <= datasizes_nz[size] -1; i +=1) {
  // No point allocation is needed here
  row[i] = rowt[i];...
  col[i] = colt[i];...
  val[i] = valt[i];...
}
```

Fig. 5. X10 source code following optimization of X10 loop body

```
//X10 optimized for loop translated to Java
for (int i = 0;  i <= datasizes_nz[size] -1; i +=1) {
  (row).set(((rowt).get(i)),i);...
  (col).set(((colt).get(i)),i);...
  (val).set(((valt).get(i)),i);...
}
```

Fig. 6. Java source code of loop following translation of optimized X10 to Java by X10 compiler

3.2 Use of Dependent Type Information for Improved Code Generation

When examining the Java code generated for the optimization example discussed in the previous section (Figure 6) we see that even though the point object has been inlined, significant overheads still remain due to the calls to the get/set methods. These calls are present because by default, the declaration of an array reference variable in X10 does not specify the rank (or dimension sizes) of its underlying array. This makes it possible to write rank-independent code in X10, but poses a challenge for the compiler to generate efficient rank-specific code. In this example, all regions and array accesses are one-dimensional, so it should be possible for the compiler to generate code with direct array accesses instead of method calls. Ideally, this information should be deduced automatically by the compiler (*e.g.*, by propagating rank information from the array's allocation site to all its uses), but in general it requires intra- and inter-procedural rank and region analysis of X10 programs which is beyond the scope of this paper and a subject for future work. Instead, the partial solution in this paper is to use the *dependent type system* [11] available in version 1.01 of the X10 language [16] to enable the programmer to annotate selected array variable declarations with additional information for the rank and region of the variable, and to extend the X10 compiler so as to generate efficient code in cases where the dependent type information is available. A key advantage of dependent types over pragmas is that type soundness is guaranteed statically with dependent types, and dynamic casts can be used to limit the use of dependent types to performance-critical code regions.

```
//flow-insensitive point inlining algorithm

//init pass
for each region r
  r's rank = TOP
for each point p
  p's rank = TOP

//gather rank information
for each AST node n
  case(assignment)
    if (n.lhs == point OR region)
      n.lhs rank = merge(n.lhs's rank, n.rhs's rank)
  case(x10 loop)
    point p = s.formal();
    region r = s.domain();
    p's rank = merge(p's rank, r's rank);

//merge rank using lattice
merge(rank l, rank r) {
  return l ^ r where :
  TOP ^ r = r;
  BOTTOM ^ r = BOTTOM;
  c1 ^ c2 = c1, if c1 equals c2 else BOTTOM;
}

//inline points
for each AST node n
  if (get_rank(n) == CONSTANT) //inlineable point found
    switch(n)
      case(point declaration)
        inline(n);
      case(point use)
        inline(n);
      case(method call argument)
        reconstruct_point(n);
      case(loop with formal point)
        convert_loop(loop)
```

Fig. 7. Algorithm for X10 point inlining

To illustrate this approach, Figure 8 contains an extended version of the original X10 code fragment in Figure 3 with a dependent type declaration shown for array `row`. Similar declarations need to be provided for the other arrays as well. The X10 compiler ensures the soundness of this type declaration i.e., it does not permit the assignment of any array reference to `row` that is not guaranteed to satisfy the properties. For the one-dimensional case, we extended the code generation performed by the reference X10 compiler [18] to generate the optimized code shown in Figure 9 for array references with the appropriate dependent type declaration. Performing this optimized code generation for multi-dimensional arrays with dependent types is a subject for future work.

4 Performance Results

We ran all experiments on a 1.25 GHz PowerPC G4 with 1.5 GB of memory using the Sun Java Hotspot VM (build 1.5.0_07-87) for Java 5 with the -Xms2000M

```
// X10 array declarations with dependent type information
//   rank==1    ==> array is one-dimensional
//   rect       ==> array's region is dense (rectangular)
//   zeroBased ==> lower bound of array's region is zero
double[: rank==1 && rect && zeroBased ] row = ... ;

. . .

region arrayRegion1 = [0:datasizes_nz[size]-1];
//X10 for loop
for (point p : arrayRegion1) {
   row[p] = rowt[p];...
   col[p] = colt[p];...
   val[p] = valt[p];...
}
```

Fig. 8. X10 for loop example from Figure 3, extended with dependent type declarations

```
//X10 optimized for loop translated to Java
for (int i = 0;  i <= datasizes_nz[size] -1; i +=1) {
  ((DoubleArray_c) row).arr_[i] = ((DoubleArray_c) rowt).arr_[i] ;...
  ((DoubleArray_c) col).arr_[i] = ((DoubleArray_c) colt).arr_[i] ;...
  ((DoubleArray_c) val).arr_[i] = ((DoubleArray_c) valt).arr_[i] ;...
}
```

Fig. 9. X10 for loop body translated from X10 to Java by X10 compiler

-Xmx2000M options to set the heap size to 2 GB (we used < 1.5 GB in practice). We measured performance results on the Java Grande benchmarks. All benchmark results are obtained using the class A versions of the benchmark. We report results for 3 different versions of the benchmark suite. Version 1 is essentially the original Java version obtained from the Java Grande Forum web site [12] renamed with the *.x10* extension – we use this version as the baseline since the X10 compiler currently translates X10 code into Java. Version 2 is an unoptimized direct translation of the Java version into X10, with all Java arrays converted into X10 arrays and integer subscripts replaced by *points*. Version 3 uses the same input X10 program as in Version 2 but turns on the optimizations described in this paper. All results include runtime array bounds checks, null pointer checks and other checks associated with a Java runtime environment.

Table 1 shows the impact of the optimizations by comparing the performance of Versions 2 and 3. Performance improvements in the range of 1.6× to 5.4× were observed for 7 of 8 benchmarks in Table 1. We observed no improvement in the *series* benchmark because its performance is dominated by scalar (rather than array) operations.

Table 2 compares the performance of the Java baseline (Version 1) with the optimized X10 (Version 3) by reporting the execution time ratio for Version 1 relative to version 3. For 6 of 8 benchmarks the ratio is in the range of 0.48

Table 1. Results from optimizing points in X10 version of Java Grande benchmarks

Benchmarks	*Runtime Performance in seconds*		*Speedup Factor*
	Unopt. X10 (Version 2)	*Opt. X10 (Version 3)*	*(Version 2)/(Version 3)*
sparsematmult	57.97	13.83	4.1×
crypt	8.14	4.79	1.7×
lufact	52.87	18.86	2.8×
sor	508.49	93.41	5.4×
series	19.01	18.95	1.0×
moldyn	2.39	1.19	2.0×
montecarlo	7.59	3.49	2.2×
raytracer	2.27	1.43	1.6×

Table 2. Comparison of applied compiler optimizations to X10 array point accesses versus the original version with Java arrays

Benchmarks	*Runtime Performance in seconds*		*Performance Ratio*
	Orig. Java (Version 1)	*Opt. X10 (Version 3)*	*(Version 1)/(Version 3)*
sparsematmult	9.75	13.83	0.71
crypt	4.60	4.79	0.96
lufact	1.38	18.86	0.07
sor	6.06	93.41	0.07
series	19.01	18.95	1.00
moldyn	0.57	1.19	0.48
montecarlo	3.00	3.49	0.86
raytracer	1.28	1.43	0.90

to 1.00, showing that the performance gap is at most a factor of 2 for these benchmarks. For the two remaining benchmarks, *lufact* and *sor*, the ratio is 0.07 indicating that the Java version is 14.3× faster than the X10 version in these two cases. This gap is primarily due to the multi-dimensional array computations in the two benchmarks, and the fact that the efficient code generation discussed in Section 3.2 currently does not support arrays with rank > 1. Enabling efficient code generation for multi-dimensional X10 arrays and comparison to C/Fortran benchmark versions is a subject for future work.

5 Related Work

Object Inlining [2,4,8,9] is a compiler optimization for object-oriented languages that transforms objects into primitive data, and conversely the rest of the program code that operates on objects into code that operates on inlined data. It is closely related to "unboxing" [14] for functional languages. Budimlić [2] and Dolby [8] introduced object inlining as an optimization for object-oriented languages, particularly for Java and C++. General object inlining requires complex escape analysis and concrete type inference, and the transformation is irreversible (once unboxed, objects cannot always be reboxed). Joyner [6,13] extended the analysis to allow more objects and arrays of objects to be inlined in scientific, high performance Java programs. This paper presents object inlining for *points* and other *value* objects in X10, which is a less general, but more effective and more applicable (*all* value objects can be boxed and unboxed freely) form of object inlining.

Wu et al. [17] presented *Semantic Inlining* for *Complex* numbers in Java, an optimization closely related to object inlining. Their optimization incorporates the knowledge about the semantics of a standard library (Complex numbers) into the compiler, and converting all the Complex numbers into data structures containing the real and imaginary part. Although this optimization achieves the same effect as object inlining for Complex numbers, it is less general since it requires compiler modifications for any and all types of objects for which one desires to apply this optimization.

The point-wise *for* loop language abstraction is not unique to the X10 language. Titanium [19], a Java dialect, also has *for* loops which iterate over points in a given domain. The Titanium compiler also performs an optimization to remove points appearing inside *for* loops. However, there are a couple of differences between our approach and the one applied in Titanium. First, because in X10 the rank specification of both points and arrays is not required at the declaration site, we employ a type analysis algorithm to determine the rank for all X10 arrays. Second, object inlining in X10 is directly applicable to all *value* objects, not just points, and thus is a more general optimization.

6 Conclusions and Future Work

In this paper, we discussed the Point abstraction in high-productivity languages, and described compiler optimizations that reduce their performance overhead. We conducted experiments that validate the effectiveness of our optimizations and demonstrate that these optimizations can enable high-level X10 array accesses written with implicit ranks and Points to achieve performance comparable to that of low-level programs written with explicit ranks and integer indices. The experimental results showed performance improvements in the range of 1.6× to 5.4× for 7 of 8 Java Grande benchmark programs written in X10, as a result of these optimizations. Further, for 6 of 8 benchmarks, the performance ratio of the optimized X10 versions relative to the low-level Java versions was in the range of 0.48 to 1.00, showing that the performance gap is at most a factor of 2 for these benchmarks. These results emphasize the importance of the optimizations we have presented in this paper as a step towards achieving high performance for high productivity languages.

We plan to investigate possible optimizations to the X10 array implementation that brings it closer to Java array performance. We will be exploring the ways to communicate static compiler analysis information to the run-time environment to further speed up array accesses, for example by eliminating array bounds checks whenever possible.

We will examine the achievability of an object inlining framework that would expand inlining to more general types of objects. This framework will require a sophisticated concrete type analysis for high-productivity languages, which is an exciting problem in its own right.

Acknowledgments

We would like to thank the anonymous reviewers for their detailed feedback on the paper.

We are grateful to all X10 team members for their contributions to the X10 software used in this paper. We would like to especially acknowledge Vijay Saraswat's work on the design and implementation of dependent types in X10, and Rajkishore Barik's work on optimized code generation for rectangular loops in X10.

We would also like to acknowledge the contributions of the late Ken Kennedy, who for a long time led a multi-institutional effort of bringing high-productivity and high-performance together and who was particularly enthusiastic about this project and participated in its early stages.

Mackale Joyner and Zoran Budimlić are supported in part by an IBM University Relations Faculty Award. While at IBM, Vivek Sarkar's work on X10 was supported in part by the Defense Advanced Research Projects Agency (DARPA) under its Agreement No. HR0011-07-9-0002.

References

1. Bacon, D.F.: Kava: a Java dialect with a uniform object model for lightweight classes. In: Proceedings of the 2001 joint ACM-ISCOPE conference on Java Grande, Palo Alto, California, pp. 68–77 (2001)
2. Budimlić, Z.: Compiling Java for High Performance and the Internet. PhD thesis. Rice University (2001)
3. Budimlić, Z., Kennedy, K.: JaMake: A Java Compiler Environment. In: 3rd International Conference on Large Scale Scientific Computing, pp. 201–209 (2001)
4. Budimlić, Z., Kennedy, K.: Optimizing Java: Theory and practice. Concurrency: Practice and Experience 9(6), 445–463 (1997)
5. Budimlić, Z., Kennedy, K.: Prospects for Scientific Computing in Polymorphic, Object-Oriented Style. In: the Proceedings of the 9th SIAM Conference on Parallel Processing for Scientific Computing, San Antonio, Texas (1999)
6. Budimlić, Z., Joyner, M., Kennedy, K.: Improving Compilation of Java Scientific Applications. The International Journal of High Performance Computing Applications (2006)
7. Charles, P., Donawa, C., Ebcioglu, K., Grothoff, C., Kielstra, A., Praun, C.: X10: An object-oriented approach to non-uniform cluster computing. In: OOPSLA 2005 Onward! Track (2005)
8. Dolby, J.: Automatic Inline Allocation of Objects. In: Proceedings of ACM SIGPLAN conference on POPL, Las Vegas, Nevada (1997)

9. Dolby, J., Chien, A.: An Automatic Object Inlining Optimization and its Evaluation. In: Proceedings of the 2000 ACM Sigplan Conference on Programming Language Design and Implementation, pp. 345–357. ACM Press, New York (2000)
10. Ebcioglu, K., Sarkar, V., El-Ghazawi, T., Urbanic, J.: An Experiment in Measuring the Productivity of Three Parallel Programming Languages. In: P-PHEC 2006 (2006)
11. Harper, R., Mitchell, J.C., Moggi, E.: Higher-order modules and the phase distinction. In: POPL 1990: Proceedings of the 17th ACM SIGPLAN-SIGACT symposium on Principles of programming languages, pp. 341–354. ACM, New York (1990)
12. The Java Grande Forum benchmark suite, `http://www.epcc.ed.ac.uk/javagrande`
13. Joyner, M.: Improving Object Inlining for High Performance Java Scientific Applications. Master's Thesis. Rice University (2005)
14. Leroy, X.: Unboxed objects and polymorphic typing. In: Proceedings of the 19th Symposium on the Principles of Programming Languages, pp. 177–188 (1992)
15. Markidis, S., Lapenta, G., VanderHeyden, W.B., Budimlić, Z.: Implementation and Performance of a Particle-in-cell code Written in Java. Concurrency and Computation: Practice and Experience 17, 821–837 (2005)
16. Saraswat, V.: Report on the experimental language x10 version 1.01. `http://x10.sourceforge.net/docs/x10-101.pdf`
17. Wu, P., Midkif, S., Moreira, J., Gupta, M.: Efficient support for complex numbers in Java. In: Proceedings of the ACM 1999 conference on Java Grande, pp. 109–118 (1999)
18. X10 Prototype Implementation, `http://x10.sf.net`
19. Yelick, K., Semenzato, L., Pike, G., Miyamoto, C., Liblit, B., Krishnamurthy, A., Hilfinger, P., Graham, S., Gay, D., Colella, P., Aiken, A.: Titanium: a high-performance Java dialect. Concurrency: Practice and Experience 10(11-13), 825–836 (1998)

Software Pipelining for Packet Filters*

Yoshiyuki Yamashita[1] and Masato Tsuru[2]

[1] Saga University, Honjyo 1, Saga, 840-8502 Japan
yaman@is.saga-u.ac.jp
[2] Kyushu Institute of Technology, Kawazu 680-4, Iizuka, 820-8502 Japan
tsuru@ndrc.kyutech.ac.jp

Abstract. Packet filters play an essential role in traffic management and security management on the Internet. In order to create software-based packet filters that are fast enough to work even under a DOS attack, it is vital to effectively combine both the higher-level optimization related to algorithmic structure and the lower-level optimization related to acceleration techniques in compiler study. In the present paper, we focus on the lower-level (machine code) optimization using software-pipelining, and report experimental results that indicate the potential of our approach for accelerating packet filter performance. The technical difficulty is that the packet filter is a lump of conditional branches, so that standard optimization techniques usually applied to basic blocks is not directly applicable to this problem. Using predicated execution and enhanced modulo scheduling, we solve this problem and achieve 20 times higher performance compared with a conventional interpreter-based packet filter. We also compare the proposed filters and compiler-based packet filters, and obtain a better than two-fold increase in performance.

1 Introduction

Packet filters, which basically inspect the header and/or payload of each incoming packet and perform appropriate actions on each packet, are essential for traffic management and security management on the Internet and so are implemented in a variety of systems/devices, such as IP routers and firewalls. Software-based packet filters are cost-effective and flexible but are general relatively slow, whereas hardware-based packet filters (e.g., those using ASIC or FPGA[8,11,6] for pattern matching) are fast but expensive and less flexible.

Recently, the rapid growth of network bandwidth has led to the requirement for high-speed packet filters. On the other hand, emerging applications of packet filters require increases scalability and flexibility of filter rules. In particular, in the case of DOS attacks, it is important to drop an enormous number of useless packets efficiently, while valid packets should be handled properly. In addition, these filters should be easily modified in response to new circumstances or new

* This work was supported in part by Hitachi, Ltd, the Ministry of Internal Affairs Communications, Japan, and JSPS, Grant-in-Aid for Scientific Research (C) (19500057).

R. Perrott et al. (Eds.): HPCC 2007, LNCS 4782, pp. 446–459, 2007.

requirements. In order to realize a packet filter that manages both flexibility and high-speed in a cost-effective manner, it is of practical importance to establish an approach by which to create software-based packet filters that are fast enough to work even when under DOS attack. This requires the effective combination of both higher-level optimization related to algorithmic structure adapted to the input packet sequence and lower-level (machine code) optimization related to acceleration techniques in a compiler study.

We will focus on the lower-level optimization. Several studies have attempted to produce a native machine code from a packet filter rule and to make the filter faster compared with the conventional interpreter-based packet filter[2,3,7]. However, to the best of our knowledge, none of these studies attempted to apply the state-of-the-arts optimization techniques based on software-pipelining. Although software-pipelining is a common technique in compiler construction, it is technically difficult to apply software-pipelining to a packet filter because the filter consists of several (more than 20 in some cases) conditional branches, but the standard software-pipelining algorithm is not applicable to programs with conditional branches. We solve this program with predicated execution[4] and enhanced modulo scheduling[9].

The present paper is an extension of our previous research [10], which showed that we achieved from 10 to 20 times higher performance. In the present paper, we will show that further improvement of the algorithms and more appropriate choices of scheduling parameters result in 12 to 28 times higher performance and an average of 20 times higher performance. In addition, in order to explain our optimization techniques in detail, we present concrete code examples in the C language and the assembly language using an Intel IA-64 Itanium 2 processor.

2 Framework

For the purpose of simplification, we consider a packet filtering procedure to consist of two serial procedures. The first procedure classifies each received packet into one of three categories: usually *accept*, *drop*, or *forward*, according to predefined filter rules and the context of the packet. The second procedure is to process the packet according to the above classification. These two procedures are usually invoked serially within a single loop, such as

```
    for(;;) {
        n_packets = receive();
@       for(i = 0; i < n_packets; i++) {
            result = filter(i);
            action(result);
@       }
    }
```

where we assume that when a packet stream is incoming at a very high rate (e.g., due to a DOS attack), two or more consecutive packets are received by each `receive()` and are stored in the receiving buffer, after which they are continuously processed in the internal loop.

Simply speaking, making a packet filter faster is approximately equivalent to executing this loop in a shorter time. Hence, in the present paper, we apply code optimization techniques to this loop. The function `filter()` in the above loop usually consists of only logical and arithmetic operations. Therefore, we expect that it is rather easy to optimize this function with various code optimization techniques. On the other hand, since the function `action()` includes complex tasks, the function must perform tasks such as hoist a packet to an upper-layer and send a packet to other network ports, it is almost impossible to apply optimization techniques to this function. Then, in order to obtain the maximum effects of optimization, in other words, to enable the application of loop optimization techniques, we divide the above loop into two loops as follows:

```
for(;;) {
    n_packets = receive();
    for(i = 0; i < n_packets; i++){     // Loop1
        result[i] = filter(i);
    }
    for(i = 0; i < n_packets; i++){     // Loop2
        action(result[i]);
    }
}
```

We therefore concentrate on how to optimize `Loop1` efficiently.

The main idea is to process multiple consecutive packets together using a software-pipelined procedure. This approach is especially effective in suffering from DOS attacks, where a major problem is how to prevent the unprocessed packets from being incessantly lost due to buffer overflow, that is, how to increase the maximum acceptable input packet rate. Consider a very simple analytical model. Let T_1 and T_2 be the average times of executing (non-optimized) `filter()` for one packet and `action()` in cases of *accept* and *forward* for one packet, respectively. We also assume the execution time of `action()` in case of *drop* is negligible. Then, let α denote the ratio of non-dropped (valid) packets to all of the incoming packets. In addition, let A denote the acceleration ratio of the execution time in optimizing `Loop1`, when many consecutive packets are processed together. The average processing time per packet is then $T_1/A + \alpha T_2$. In a DOS attack situation, we can assume that $\alpha \ll 1$ because the number of received packets is huge but the number of valid packets does not vary from a non-DOS attack situation. Therefore, the ratio of the optimized processing time to the non-optimized processing time is approximately given as $(T_1/A + \alpha T_2)/(T_1 + \alpha T_2) \approx 1/A$. It shows that the code optimization in `Loop1` directly affects the total performance of the packet filter.

3 Loop Optimization

In this section, we briefly explain the three steps of the optimizations compared herein. Our proof-of-concept prototype of the packet filter is based on `tcpdump`[5] (or `bpf` as its basis). Actually, the compilers for packet filters in our experiments are implemented by rewriting the source program of `tcpdump`.

List 1: `bpf`'s filter machine code for rule, `ip and tcp`

```
(000)   ldh     [12]
(001)   jeq     #0x800    jt 2    jf 5
(002)   ldb     [23]
(003)   jeq     #0x6      jt 4    jf 5
(004)   ret     #96
(005)   ret     #0
```

3.1 Interpreter

`tcpdump` translates a filter rule into a code for the pre-defined virtual machine, called a *filter machine*, which consists of an accumulator, an index register, and an array of scratch memory. `tcpdump` then executes the code on the interpreter that simulates the filter machine. The code is a sequence of `bpf` instructions. `bpf` instructions can load a part of a packet contents into a virtual register, perform arithmetic on virtual registers, and return an integer value indicating whether a packet is accepted.

For example, let us consider the `tcpdump` rule, `ip and tcp`, which accepts only IPv4 tcp packets. List 1 shows the corresponding filter machine code of `bpf` instructions. Here, the instructions `ldh`, `ldb`, `jeq`, and `ret` denote load a half word, load a byte, jump if equal, and return a constant value, respectively. The tags `jt` m and `jf` m denote that the execution jumps to location m when the comparison yields *true* and *false*, respectively. This code corresponds to the function `filter(i)` in the previous section.

3.2 Compiling into a C Program

Since `tcpdump` translates a filter rule into a filter machine code and executes the code on an interpreter, the filtering process is very slow. The process generally becomes faster when we compile the rule into a native code. Several studies???[3,7] have already reported that the filter becomes approximately five times faster. As the first step of our research, we compile the rule into a native code. Specifically, we translate a filter rule into a C program and compile the C program into a native code using an existing C compiler.

In the present paper, we examine two types of C programs because we expect that compilers treat them differently. The first type is a C program in which each `bpf` instruction in a filter machine code is straightforwardly replaced with equivalent C statements, including labels and `goto` statements. For example, List 2 is the C program translated from List 1. The other type of C program is a structured program without `goto` statements, like List 3. Structured programs are sometimes more redundant than the unstructured programs but are expected to be compiled into more efficient codes by highly-optimizing C compilers.

Rewriting the source program of the `bpf` interpreter, we can easily build a translator that translates a given filter machine code into both unstructured and structured C programs. The C compilers the authors used in this research are `gcc`, the Gnu C compiler, and `icc`, Intel's C compiler.

List 2: unstructured C program for List 1

```
for(i = 0; i < n_packets; i++) {
    byte* p = packets[i];
    L0: A = EXTRACT_SHORT(&p[12]);
    L1: if(A == 0x800) goto L2; else goto L5;
    L2: A = p[23];
    L3: if(A == 6) goto L4; else goto L5;
    L4: A = 96; goto L6;
    L5: A = 0; goto L6;
    L6: result[i] = A;
}
```

List 3: structured C program for List 1

```
for(i = 0; i < n_packets; i++) {
    byte* p = packets[i];
    short tmps = EXTRACT_SHORT(&p[12]);
    if(tmps == 0x800) {
        byte tmpb = p[23];
        if(tmpb == 6) A = 96;
        else A = 0;
    } else A = 0;
    result[i] = A;
}
```

3.3 Simple List Scheduling

A C compiler sometimes generates slow code because the compiler has little information specific to a given C program so that it is likely to select a conservative but inefficient instruction pattern. As the second step of our research, we build a translator that directly translates a filter machine code into a native assembly code using elaborately selected instruction patterns. A simple instruction scheduling, so-called *list scheduling*, is applied to individual basic blocks.

List 4 is an example of such simply list-scheduled code translated from List 3, based on the Intel IA-64 architecture. Here, we assume that the registers `r100` and `r110` hold the addresses of arrays `packets` and `results` (see List 2), respectively, and that `r10` and `r20` hold the constant values `0x800` and `96`, respectively. For ease of reading, in the present paper, every register used in the codes is singly assigned, although register allocation algorithms are applied to actual codes. Double semicolons "`;;`" indicate a boundary of adjacent instruction groups, in each group of which all of the instructions can be executed simultaneously.

3.4 Software Pipelining

As the third step of our optimization, we apply software-pipelining techniques[1], which are highly sophisticated aggressive instruction scheduling techniques for

List 4: naive code for List 3

```
L0:  ld8 r32 = [r100],8          ;; // r100 == &packets[i]
     adds r40 = 12,r32           ;;
     ld2 r50 = [r40]             ;;
     cmp.eq p6,p7 = r10,r50         // r10 == 0x800
     (p7)br L2                   ;;
     adds r60 = 23,r32           ;;
     ld1 r70 = [r60]             ;;
     cmp.eq p6,p7 = 6,r70
     (p7)br L1                   ;;
     mov r80 = r20                  // r20 == 96
     br L3                       ;;
L1:  mov r80 = r0
     br L3                       ;;
L2:  mov r80 = r0                ;;
L3:  st4 [r110] = r80,4             // r110 == &results[i]
     br.ctop.dptk L0
```

loops. Although software pipelining techniques are popular and have been implemented for several existing compilers, including `gcc`, most of the compilers can only apply the techniques to loops whose bodies are basic blocks, but cannot apply the techniques to loops with conditional branches like List 2 or List 3.

In order to obtain software-pipelined codes for loops with conditional branches, we adopt two special techniques. The first is called *predicated execution*[4] (PE for short), which can be performed by processors with the facility of predicate registers, such as the Intel Itanium 2 [4]. The second is called *enhanced modulo scheduling* (EMS for short) [9], which can be performed by any RISC processor, but generates code of larger size. Since these techniques are the primary theme in the present paper, we will explain them in detail in the following two sections.

4 Predicated Execution

In this section, we briefly explain the basic concepts of PE and the software-pipelining technique based on PE.

4.1 Predication

Predicate registers are used to eliminate branch instructions, which may seriously slow program execution. List 5, translated from List 4, is an example of code that uses predicate registers. For example, the instruction `cmp.eq p10,p20 = r10,r50` assigns true and false values to predicate registers `p10` and `p20`, respectively, if the values of `r10` and `r50` are equal. Otherwise, this instruction assigns false and true values to predicate registers `p10` and `p20`, respectively. The instruction `(p10)adds r60 = 23,r32` performs addition if `p10` is true. Otherwise, this instruction works as a `nop` instruction, and the instruction is said to be *nullified*. The compare instruction `(p10)cmp.eq.unc p30,p40 = 6,r70`, which is qualified with the predicate register `p10`, assigns the appropriate truth values

List 5: naive PE code for List 3

```
L0:  ld8 r32 = [r100],8                  ;;
     adds r40 = 12,r32                   ;;
     ld2 r50 = [r40]                     ;;
     cmp.eq p10,p20 = r10,r50            ;;
     (p10)adds r60 = 23,r32
     (p20)mov r80 = r0                   ;;
     (p10)ld1 r70 = [r60]                ;;
     (p10)cmp.eq.unc p30,p40 = 6,r70     ;;
     (p30)mov r80 = r20
     (p40)mov r80 = r0                   ;;
     st4 [r110] = r80,4
     br.ctop.dptk L0
```

List 6: software-pipelined PE code (kernel) for List 3

```
                                               //stage
L0:  ld8 r32 = [r100],8                          //1
     (p12)cmp.eq.unc p30,p40 = 6,r71             //4
     (p11)adds r60 = 23,r34                      //3
     ld2 r50 = [r41]                             //2
     (p41)mov r82 = r0                           //5
     (p21)mov r80 = r0                   ;;      //3
     (p11)ld1 r70 = [r60]                        //3
     cmp.eq p10,p20 = r10,r50                    //2
     adds r40 = 12,r32                           //1
     st4 [r110] = r82,4                          //5
     (p30)mov r81 = r20                          //4
     br.ctop.dptk L0
```

to p30 and p40 according to the results of comparison if p10 is true. Otherwise, the instruction assigns false values to both p30 and p40.

4.2 Software pipelining for Predicated Code

Since a loop body without branches is a straight-line code, we can optimize the loop by applying a standard method of software-pipelining[1] (or modulo scheduling as a concrete algorithms). List 6 is such a software-pipelined predicated code (hereinafter PE code) optimized from List 5 under the assumption that every memory access latency is one machine cycle (MC). In order to minimize the initiation interval of the loop iterations (hereinafter I.I.), we use register rotation[4] for both predicate and integer registers. In List 6, for example, just after the loop-back branch instruction `br.ctop` is executed, the predicate registers p10, ... , p40 are renamed p11, ..., p41, respectively, and the integer registers r32, ..., r81 are renamed r33, ..., r82, respectively. The code in List 6 consists of five software pipeline stages, denoted 1 through 5 following the corresponding instruction at each line of List 6.

The idealistic I.I. of the loop in List 6 is 2 MC because the first six instructions of the loop can be executed simultaneously and so the last six instructions can also be executed, although it is almost impossible to attain I.I. because the memory access latency is longer than 1 MC and often varies greatly.

A major benefit of using PE is that PE frees us from branches (and branch penalties) and can facilitate the application of software-pipelining. Conversely, a major disadvantage is that the instructions that would be nullified do not contribute any effective computation and waste processor resources. In general, a PE code becomes slower when the code size (precisely speaking, the number of nullified instructions) increases.

5 Enhanced Modulo Scheduling

Since it is difficult to explain the EMS algorithm[9] in detail in the limited space available here, we only present the concept of how the processor runs effectively on the code generated by the EMS algorithm. The EMS algorithm generates

List 7: EMS code (kernel) for List 3

```
Lxx_tx:                        //stage
    ld8 r32 = [r100],8         //1
    st4 [r110] = r82,4         //4
    cmp.eq p6,p7 = r10,r51     //2
    (p7)br Lfx_tx          ;;
Ltx_tx:
    adds r40 = 12,r32          //1
    adds r60 = 23,r33          //2
    cmp.eq p6,p7 = 6,r71       //3
    (p7)br Ltx_tf          ;;
Ltx_tt:
    ld2 r50 = [r40]            //1
    mov r81 = r20              //3
    ld1 r70 = [r60]            //2
    br.ctop.dptk Lxx_tx
    br Lexit               ;;
Ltx_tf:
    ld2 r50 = [r40]            //1
    mov r81 = r0               //3
    ld1 r70 = [r60]            //2
    br.ctop.dptk Lxx_tx
    br Lexit               ;;
Lfx_tx:
    adds r40 = 12,r32          //1
    mov r80 = r0               //2
    cmp.eq p6,p7 = 6,r71       //3
    (p7)br Lfx_tf          ;;
```

continues to the right column

```
Lfx_tt:                        //stage
    ld2 r50 = [r40]            //1
    mov r81 = r20              //3
    br.ctop.dptk Lxx_xx
    br Lexit               ;;
Lfx_tf:
    ld2 r50 = [r40]            //1
    mov r81 = r0               //3
    br.ctop.dptk Lxx_xx
    br Lexit               ;;
Lxx_xx:
    ld8 r32 = [r100],8         //1
    st4 [r110] = r82,4         //4
    cmp.eq p6,p7 = r10,r51     //2
    (p7)br Lfx_xx          ;;
Ltx_xx:
    adds r40 = 12,r32          //1
    adds r60 = 23,r33      ;;  //2
    ld2 r50 = [r40]            //1
    ld1 r70 = [r60]            //2
    br.ctop.dptk Lxx_tx
    br Lexit               ;;
Lfx_xx:
    adds r40 = 12,r32          //1
    mov r80 = r0           ;;  //2
    ld2 r50 = [r40]            //1
    br.ctop.dptk Lxx_xx    ;;
Lexit:
```

such a code. Strictly speaking, the EMS algorithm originally given in [9] treats a loop having a few unnested conditional branches and is not sufficient to optimize the loops with *many* (more than 20 in the present cases, as shown in the next section) deeply-nested (maximum depth 10) conditional branches. The authors are now preparing another paper that explains how to treat such complex loops.

5.1 Difference from Predicated Execution

A software-pipelined PE code uses a set of predicate registers to memorize trace information of the evaluation results of compare instructions. Review the code in List 6. The rotating predicate register `p10` is defined at the second stage and is referred at the third and fourth stages as `p11` and `p12`, respectively. `p20` is defined as the second stage and referred at the third stage as `p21`. `p30` is defined and referred at the fourth stage. `p40` is defined at the fourth stage and referred at the fifth stage as `p41`. Thus, at the beginning of every new iteration, the predicate information that must live over iterations is held in the set of the registers, (`p11`, `p12`, `p21`, `p41`), so that there are at least 16 (=2^4) patterns of code executions for the one code in List 6.

The EMS algorithm does not use predicate register, but instead generates two or more code patterns corresponding to the set of the truth values the predicate registers possess in the PE code.

5.2 Code Example

List 7 is the kernel part of the software-pipelined code generated with the EMS algorithm (hereinafter EMS code). Each basic block in the code has a label of form La_1b_1_a_2b_2, where the letters a_i and b_i represent the truth values of the results of the comparisons `cmp.eq p6,p7 = r10,r51` and `cmp.eq p6,p7 = 6,r71`, respectively, for the i-th past iteration. The letters '`t`', '`f`', and '`x`' denote the true, false, and undefined (or useless) values, respectively. For example, the label `Lxx_tx` at the first line in List 7 represents the situation in which no comparison has yet been evaluated for the most recent iteration but the comparison `cmp.eq p6,p7 = r10,r51` has been evaluated affirmatively for the second most recent iteration. If a series of adjacent packets are all IPv4 tcp packets, the program traverses only the three basic blocks labeled `Lxx_tx`, `Ltx_tx`, and `Ltx_tt`. If none of the packets is a IPv4 tcp packet, the program traverses the basic blocks labeled `Lxx_xx` and `Lfx_xx`. Otherwise the trace becomes more complicated. Under the assumption that there is no branch penalty, the ideal I.I. of the code in List 7 is 3 MC, even if the processor runs on any path.

In contrast to the PE code, the EMS code is not free from branch penalty in general but rather contains no nullified instructions because the EMS code only executes the instructions required for the computation. Therefore, the EMS code is faster than the PE code if the penalty of the nullified instructions is greater than the branch penalty.

Table 1. Filter Rule in Our Experiments

No	Rule	depth	#ifs	#insts	matching ratio
1	`ip`	1	1	7	100.0%
2	`ip and tcp`	2	2	11	99.8%
3	`tcp`	3	3	17	99.8%
4	`ip and udp`	2	2	11	0.2%
5	`udp`	3	3	17	0.2%
6	`ip and dst net` ⟨ip_address⟩	2	2	15	4.9%
7	`dst net` ⟨ip_address⟩	4	6	36	4.9%
8	`ip and net` ⟨ip_address⟩	3	3	22	10.9%
9	`net` ⟨ip_address⟩	5	9	59	10.9%
10	`ip and dst port` ⟨number⟩	7	19	66	9.3%
11	`dst port` ⟨number⟩	8	29	88	9.3%
12	`ip and port` ⟨number⟩	9	19	90	21.41%
13	`port` ⟨number⟩	10	29	125	21.41%

6 Experiments

This section reports the experimental results obtained using the code optimization methods explained in Sections 3.1, 3.2, 3.3, 4, and 5.

6.1 Environment

Table 1 shows 13 examples of the filter rules of `tcpdump` examined herein. Each rule is simple, short, and uni-functional, although, in reality, a longer rule that combines these simple rules is likely to be employed. This is because examining each simple rule separately is suitable for analyzing the effect of our optimizations in depth. In order to clarify how the rule is complex to optimize, the columns *depth*, *#ifs*, and *#insts* in the table show three characteristics of the structured C program in Section 3.2 and the native assembly code in Section 3.3 corresponding to each rule. The depth is the maximum depth of nests of `if` constructs that the program contains, #ifs is the number of `if` constructs that the program contains, and #insts is the number of assembly instructions not including branch instructions.

The sample packets used in the experiments were captured from the network of the authors' lab using `tcpdump -w`. The total number of packets was 10,000. The column *matching ratio* in Table 1 is the ratio of the packets accepted by each rule. In our experiments, the 10,000 packets are loaded on a large buffer (virtual network). Then, every `n_packets` packets are copied from the virtual network buffer to the receiving buffer and are processed by `Loop1` 10,000/`n_packets` times (recall Section 2). The total execution time is then divided by 10,000 to obtain the execution time per packet for the given `n_packets`. For performance evaluation purposes, we use the execution time for the optimal `n_packets`, which varies depending on the filter rule. Experiments were performed on an Intel IA-64 Itanium 2 processor (900 MHz, revision 7), Linux version 2.4.18-1.

Table 2. Execution Time per Packet (MC)

No	T_{ip}	T^u_{gcc}	T^s_{gcc}	T^u_{icc}	T^s_{icc}	T_{sls}	T_{pe}	T_{ems}	T_{hyb}
1	80.6	17.5	17.6	13.4	7.6	11.2	4.3	4.3	4.3
2	126.7	19.4	19.5	15.5	[illegible].6	14.1	4.5	5.8	4.5
3	143.6	20.6	21.7	16.2	[illegible].6	15.3	5.5	7.5	5.5
4	126.5	19.5	19.6	14.6	[illegible].6	14.1	4.5	7.7	4.5
5	143.6	20.6	21.7	16.2	[illegible].6	15.3	5.5	7.6	5.5
6	132.5	188.1	188.8	184.4	1[illegible]9.5	18.7	6.9	10.4	6.9
7	132.7	190.6	189.8	187.1	1[illegible]4.9	19.0	10.7	9.9	10.7
8	181.8	332.0	332.0	330.1	3[illegible]9.2	24.9	9.1	13.5	9.1
9	181.6	333.7	333.5	331.4	3[illegible]0.5	25.2	18.2	14.4	14.4
10	266.2	54.3	53.6	32.5	[illegible]4.5	30.0	18.6	15.1	15.1
11	283.2	55.1	55.4	30.7	[illegible]5.4	31.1	21.2	16.9	16.9
12	309.5	64.6	67.3	37.1	[illegible]0.6	37.2	22.2	20.5	20.5
13	326.5	66.7	69.4	37.6	[illegible]9.5	38.0	28.8	22.2	22.2

6.2 Results

Table 2 shows the average execution time per packet for `Loop1` (in units of machine cycle time). Here, T_{ip} is the execution time of the original filter function of `tcpdump`, T^u_{gcc} and T^s_{gcc} are the execution times of unstructured and structured C programs, such as List 2 and List 3, respectively, compiled by `gcc` 2.96 with a `-O3` option, T^u_{icc} and T^s_{icc} are the execution times of unstructured and structured C programs compiled by Intel `icc` 9.1 with a `-O3` option, T_{sls} is the execution time of the simply list-scheduled code, such as List 4, T_{pe} is the execution time of the software-pipelined PE code, such as List 6, and T_{ems} is the execution time of the EMS code, such as List 7. Since a software-pipelined code and its execution time can vary greatly depending on the memory access latency, we profiled the possible execution times in terms of the memory access latency and selected the optimal execution time as T_{pe} (or T_{ems}). Table 3 shows the acceleration ratios $A_x = T_{ip}/T_x$, and Figure 1 shows this information in the form of a bar graph. From the tables, we observe the following:

The ratio A^u_{gcc} is approximately 5, except for cases No. 5 through No. 8, which agrees with the results reported in [7]. The ratio A^s_{gcc} is approximately A^u_{gcc} for all cases, because `gcc` applies no special optimization to loops with conditional branches, regardless of their program structures. A^s_{icc} is approximately twice as large as A^u_{icc} for cases No. 1 through No. 5, because `icc` optimizes the loops of which the bodies are small and well-structured by applying software pipelining with predicated execution but sacrifices optimization when the program size is sufficiently large.

The ratio A_{pe} shows that the PE code is at least 10 times faster and approximately 18 times faster on average than the interpreter-based execution. This ratio decreases as the number #insts increases. The ratio A_{ems} shows that

Table 3. Acceleration Ratio ($A_x = T_{ip}/T_x$)

No	A_{ip}	A^u_{gcc}	A^s_{gcc}	A^u_{icc}	A^s_{icc}	A_{sls}	A_{pe}	A_{ems}	A_{hyb}
1	1.0	4.6	4.6	6.0	10.6	7.2	18.7	18.7	18.7
2	1.0	6.5	6.5	8.2	14.7	8.9	28.2	21.8	28.2
3	1.0	7.0	6.6	8.9	15.0	9.4	26.1	19.1	26.1
4	1.0	6.5	6.5	8.7	14.7	9.0	28.1	16.4	28.1
5	1.0	7.0	6.6	8.9	15.0	9.4	26.1	18.9	26.1
6	1.0	0.7	0.7	0.7	0.7	7.1	19.2	12.7	19.2
7	1.0	0.7	0.7	0.7	0.7	7.0	12.4	13.4	12.4
8	1.0	0.5	0.5	0.6	0.6	7.3	20.0	13.5	20.0
9	1.0	0.5	0.5	0.5	0.5	7.2	10.0	12.6	12.6
10	1.0	4.9	5.0	8.2	7.7	8.9	14.3	17.6	17.6
11	1.0	5.1	5.1	9.2	8	9.1	13.4	16.8	16.8
12	1.0	4.8	4.9	8.3	7.6	8.3	13.9	15.1	15.1
13	1.0	4.9	4.7	8.7	8.3	8.6	11.3	14.7	14.7
ave.	1.0	5.7*	5.6*	8.3*	11.3*	8.3	18.6	16.3	19.7

* Calculated without No. 6 to No. 9.

the EMS code is at least 12 times faster and 16 times faster on average than the interpreter-based execution. Since the declining speed of A_{ems} is lower than that of A_{pe}, A_{ems} is larger than A_{pe} when #insts is greater than 30. This will be discussed further in the following subsection.

T_{sls} can be regarded as the minimum execution time (or, conversely, as the maximum execution speed) of a program to which no software-pipelining method has been applied, because in all cases except cases No. 6 through No. 9, T_{sls} is approximately T^u_{icc}, which is the execution time of the program compiled by the Intel compiler optimized for the Itanium 2 processor. Then, A_{ems} and A_{pe} are approximately twice as large as A_{sls} on average. Roughly speaking, this ratio with respect to A_{sls} indicates the effect of software-pipelining.

The values in entries $\{\text{No. } 6, ..., \text{No. } 9\} \times \{T^u_{gcc}, T^s_{gcc}, T^u_{icc}, T^s_{icc}\}$ are extraordinarily large compared to the corresponding values for T_{ip}. This is because memory access conflicts occur on the Itanium 2 machine used in this study, and the compilers `gcc` and `icc` can never recognize this symptom. The algorithms associated with T_{sls}, T_{pe}, and T_{ems} manage to avoid this conflict by using a special load/store pattern.

6.3 Comparison of PE and EMS Algorithms

In general, T_{pe} is smaller than T_{ems} in simple rules, (cases No. 1 through No. 5) and conversely T_{ems} is smaller than T_{pe} in complex rules. (cases No. 10 through No. 13). This is because the PE code is affected by the number of nullified instructions, which is usually proportional to the code size (or the complexity of the rule). On the other hand, the EMS code contains no nullified instruction, but rather branch instructions that cause branch penalties. In a simple rule, the

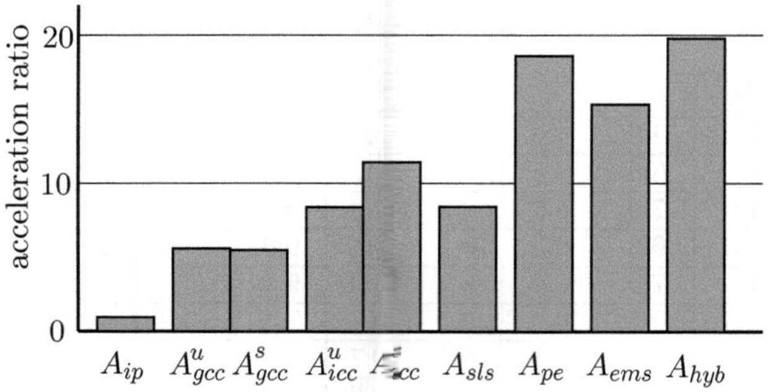

Fig. 1. Average Acceleration Ratios

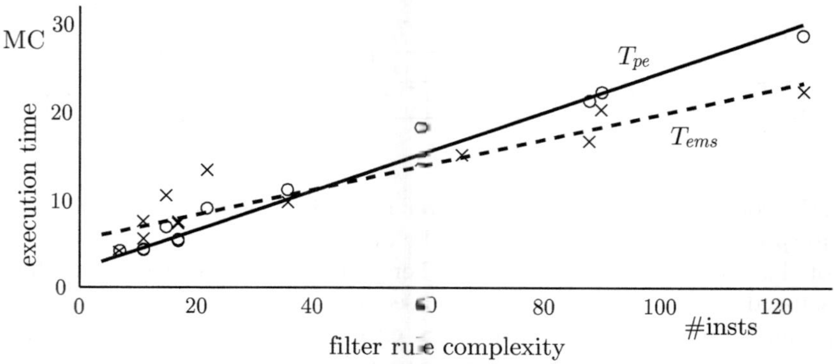

Fig. 2. Linear Approximation of Execution Times

branch penalties dominate the slowdown of the execution but, in a complex rule, the nullified instructions dominate the slowdown.

Figure 2 shows the plots of all T_{pe} (shown as circles in the graph) and T_{ems} (shown as cross-hairs in the graph) in terms of #insts. Using the least mean square method, we can approximate T_{pe} and T_{ems} as the following liner equations:

$$T_{pe} = 0.22\#\text{insts} + 2.93, \qquad T_{ems} = 0.14\#\text{insts} + 6.01$$

as in Figure 2 with considerably small errors. The root mean square errors of these equations are 1.13 and 1.86, respectively. The intersection of these two lines is located at #insts = 39.3. Therefore, we can alternatively select the better software-pipelining algorithm according to the size of #insts. The execution time T_{hyb} in Table 2 is a value obtained alternatively from T_{pe} and T_{ems} and so is the acceleration ratio A_{hyb} in Table 3.

Consequently, this hybrid algorithm is approximately 20 (=19.7) times faster than the interpreter-based execution on average, 3.5 times faster than `gcc`-based execution, and 1.7 times faster than `icc`-based execution.

References

1. Appel, A.W.: Modern Compiler Implementation in C. Cambridge University Press, Cambridge (1997)
2. Begel, A., McCanne, S., Graham, S.: BPF+: Exploiting Global Data-Flow Optimization in a Generalized Packet Filter Architecture. In: ACM SIGCOMM 1999 (1999)
3. Cristea, M.L., Bos, H.: A Compiler for Packet Filters. In: Proceedings of ASCI 2004 (2004)
4. Intel Itanium Architecture Software Developer's Manual, `http://www.intel.com/design/itanium2/documentation.htm`
5. Jacobson, V., et al.: tcpdump(1), bpf.., Unix Manual Page (1990)
6. Kumar, S., Dharmapurikar, S., Yu, F., Crowly, P., Turner, J.: Algorithms to Accelerate Multiple Regular Expressions Matching for Deep Packet Inspection In: ACM SIGCOMM 2006 (2006)
7. Okumura, T., Mossé, D., et al.: Network QoS Management Framework for Server Clusters An End-Host Retrofitting Event-Handler Approachusin Netnice. In: 3rd Int. Symp. on Cluster Computing and the Grid (2003)
8. Singh, S., Baboescu, F., Varghese, G., Wang, J.: Packet Classification Using Multidimensional Cutting. In: ACM SIGCOMM 2003 (2003)
9. Warter, N.J., Haab, G.E., Bockhaus, J.W.: Enhanced Modulo Scheduling for Loops with Conditional Branches. In: IEEE MICRO-25 (1992)
10. Yamashita, Y., Tsuru, M.: Code Optimization for Packet Filters. In: SAINT2007, Workshop on Internet Measurement Technology and its Applications to Building Next Generation Internet (2007)
11. Yusuf, S., Luk, W.: Bitwise Optimised CAM for Network Intrusion Detection Systems. In: Int. Conf. Field Programmable Logic Appl. (2005)

Speculative Parallelization – Eliminating the Overhead of Failure

Mikel Luján[1], Phyllis Gustafson[2], Michael Paleczny[2], and Christopher A. Vick[2,*]

[1] The University of Manchester
[2] Sun Microsystems Laboratories
mikel.lujan@manchester.ac.uk,
{phyllis.gustafson,michael.paleczny,christopher.vick}@sun.com

Abstract. Existing runtime parallelization techniques impose severe performance penalties when a speculative parallelization is attempted and fails. Some techniques require a sequential restart of the speculative execution while others only disregard the work after the first point of failure. This paper introduces a new technique that reduces the performance overhead of failure to less than 1% on standard processors through a combination of hoisting the failure path and partitioning work to a *Coinspector Thread.*

1 Introduction

Runtime compilation techniques have become mainstream for commercial applications since the introduction of Java™ in 1995. However, runtime compilation has not been adopted by industrial strength High Performance Computing (HPC) compilers, although, for example, runtime parallelization has been investigated since the late 1980s. The motivation for runtime parallelization are applications for which the data dependences can only be analyzed at runtime. Charm, Gaussian, Dyna-3D, Spice and Chaos are examples of such HPC applications, but runtime parallelization has also been tested with SPEC® CPU and SPECjvm®.

An example loop construct:

```
for (i = lb; i< up; i = i + s) {
     A[indx[i]] = A[jndx[i]] + expression;
}
```

where `indx` and `jndx` are indirection arrays, and `lb`, `up` and `s` are integers, illustrates the need for runtime parallelization. In general the contents of these indirection arrays (`indx` and `jndx`) cannot be known until runtime. Thus, a static

* This material is based upon work supported by DARPA under Contract No. NBCH3039002. Any opinions, findings and conclusions or recommendations expressed in this material are those of the author(s) and do not necessarily reflect the views of DARPA or DOI/NBC.

R. Perrott et al. (Eds.): HPCC 2007, LNCS 4782, pp. 460–471, 2007.

parallelizing compiler can only identify that elements of array A are read from and written to by each iteration. Consider that the iterations of the loop are divided, for example, into two equal size partitions and that each partition is scheduled to run in parallel on a different processor. When there are no common values stored in `indx` and `jndx`, and `indx` contains no repetitions, the reads and writes can be performed in parallel maintaining sequential semantics. But these properties that guarantee safe parallelization cannot be determined at compilation time.

Several approaches have been proposed for runtime parallelization. One common problem for these approaches is the cost of failures. In the context of this paper, a *failure* is considered either an attempt to parallelize at runtime for which sequential semantics could not be maintained or an attempt to parallelize at runtime for which sequential semantics are maintained, but the execution time takes longer than a sequential execution.

The contribution of this paper is a technique to ensure that despite runtime parallelization failures, the associated execution time overhead is minimal. The technique targets speculative parallelization and it is inspired by the inspector-executor model. The core idea is to preserve the sequential execution without adding anything extra, but at the same time to enable the speculative execution including the required extra tracking of memory accesses and data dependence tests. The technique does not assume any computer architecture support to facilitate speculative parallelization.

The paper is organized as follows. Section 2 provides an overview of the inspector-executor model and speculative techniques for runtime parallelization. Section 3 introduces the new technique to minimize the overhead of failure in speculative parallelization — the *Coinspector Thread* — and Section 4 shows the performance evaluation. Section 5 presents computational patterns that can require additional overhead and an analysis of their effect on how much the new technique can eliminate the overhead of failure. The summary of the paper is presented in Section 6.

2 Background and Related Work

In a nutshell, current industrial parallelizing compilers analyze code regions and try to prove that no data dependences will occur in any parallel execution. To achieve this, the analysis sets up equations describing the read and write memory accesses within a given code region. When there are intersection points between these equations, if parallelized, the sequential semantics cannot be guaranteed. A kind of data dependence captured in this case is known as *read-after-write* (RAW) — a read access to a given variable `C` occurs after a write access to the same variable `C` (e.g. `C = B + 5;... A = C - H;`).

When the equations describing the write accesses intersect, sequential semantics cannot be guaranteed if parallelized without other code transformations. This kind of data dependence is known as *write-after-write* (WAW) — a write access to a given variable D occurs after another write access to the same variable D (e.g. `D = B + 5; ... D = A / C;`).

Parallelizing compilers work around WAW dependences by creating a privatized copy of the offending variables for each parallel thread — *variable privatization*. Each parallel thread writes directly to the private copy of the variable instead of the shared original variable. After the parallelized code region, parallelizing compilers place a new section of code whose sole purpose is to copy out from the private variables, and if necessary accumulate the private results, to the original shared variable.

The variable privatization transformation also takes care of the remaining data dependence type — *write-after-read*. For brevity hereafter this last data dependence type is omitted as it is solved by the treatment for WAW dependences.

Complementary to the a priori approach based on proving mathematically the absence of data dependences, runtime parallelization has been investigated for two decades [1,2]. Rauchwerger [3] presents a survey of the main developments covering the first decade. Almost all of those developments fall under the inspector/executor model and target array-based partially parallel loops. Accordingly, the basis of this model is to establish at runtime the different sets of iterations — *wavefronts* — without data dependences and then execute these wavefronts in parallel. At its simplest, the **inspector** is a single thread chartered with identifying the wavefronts. To achieve this, the inspector only needs to track the memory accesses and thus it executes all the iterations of the loop, but not all the instructions of the loop body. The compiler generates a stripped down version of the loop with the minimum set of instructions needed to calculate the memory addresses accessed. With the wavefronts in hand, *executor* threads run independent wavefronts in parallel knowing that the sequential semantics are intact. The scalability of this model is limited by how long it takes to run the inspector phase. In the worst case scenario the stripped down loop is exactly the same as the original loop and only one inspector thread can be used.

Rather than establishing safety before executing the code region in parallel, it is possible to establish safety a posteriori — *speculative parallelization*. An implementation of speculative parallelization requires a means for restoring the program state to that before the speculation started.

One recovery mechanism is to checkpoint the variables that may be modified during the speculation. The speculative threads then execute in parallel the code region plus some extra instructions to log the memory accesses (reads and writes) in new data structures generated by the compiler. Hereafter, these data structures are referred to as *shadow structures*. Once the speculative threads have finished the execution of the code region, speculative parallelization needs to test for data dependence violations relying on the information stored in the shadow structures. When the test is passed, the sequential semantic equivalence has been established and the program state contains the correct values. However, when the test fails, the program state contains incorrect values and it is necessary to recover from the checkpoint. With this recovery mechanism, suggested by [4], the execution time of a given sequential region of code for which one speculative parallelization is attempted and failed can be modeled as

$$T_f = T_s + \frac{T_s}{p} + \max_{i=1..p} T_l^i + T_c + T_t + T_r \ ,$$

where T_f is the total execution of the failed speculative parallelization, T_s is the sequential execution time, p is the number of speculative threads executing in parallel, T_l is the time spent logging memory accesses by a speculative thread, T_c is the time to take the checkpoint, T_t is time spent applying the test, and T_r is time to recover the program state from the checkpoint.

The checkpoint recovery mechanism can be modified by making the shadow structures more complex. Instead of taking the checkpoint, the shadow structures in addition to tracking the memory accesses now store information for *undo logs*. This enables an implementation of speculative parallelization to eliminate T_c (checkpoint time) and to save part of the speculative computation by undoing the modifications up to the first point of failure as identified during the test. Garzarán *et al.* [5] study various mechanisms to store the data generated during speculative parallelization and one of those studied covers the undo logs.

In these two recovery mechanisms, WAW dependences can occur since privatized data structures are not considered. Using privatization the results of speculative parallelization do not modify the program state directly. Only after the test has passed are the results from the privatized variables copied out to the program state. When the test fails it is enough to disregard the privatized variables of those speculative threads that have computed results that correspond to instructions after the first point of failure according to the sequential order of execution. With this mechanism, the cost of failure can be modeled as

$$T_f = (1 - \alpha_f T_s) + \alpha_f \frac{T_s}{p} + \max_{i=1..p} T_{lp}^i + T_t + \alpha_f T_{com} \ ,$$

where T_f is the total execution time including a failed speculative parallelization, T_s is the sequential execution time, α_f is the ratio of speculative execution which can be saved, p is the number of speculative threads executing in parallel, T_{lp} is the time spent by a speculative thread logging memory accesses and writing to privatized variables, T_t is the time spent applying the test, and T_{com} is the time to copy-out the results from the privatized variables. Dang *et al.* [6], Cintra and Llanos [7], Rundberg and Stenström [8] and Garzarán *et al.* [5] follow this mechanism. Note that Garzarán *et al.* [5] not only consider this mechanism, but also studied several options, as mentioned above.

Dang *et al.* [6], and Cintra and Llanos [7] proposed to reduce the resource requirements and the overhead of failure by using a *sliding window* mechanism. Rather than applying speculative parallelization to an entire iteration space in one go, the iteration space is partitioned into blocks of contiguous iterations. The block containing the first iterations of the loop constitutes the first window of iterations. Speculative parallelization is then applied to this window. That is, each speculative thread is assigned to execute different subsets of iterations contained within the window. When the speculation succeeds the window slides to the immediately following block of iterations and the window only contains the iterations of this block. These steps are applied repeatedly until the window has slid across the entire iteration space and thereby the loop has been completely executed. If the speculation fails for any given window of iterations, only the iterations of the window for which speculation failed are executed sequentially. Alternatively, it is also possible to commit the results of those speculative threads that have executed instructions corresponding to iterations situated before the first point of failure according to the sequential order of execution. In this case, the window slides up to the iterations of the speculative thread identified as containing the first point of failure. The sliding window advances without leaving behind any unexecuted and uncommitted iteration. The window may increase its size to include all the iterations in the immediately following block or it may keep the same window size, effectively repartitioning the remaining iterations in the iteration space. The last mathematical model describing T_f still holds, but instead of covering the whole iteration space of a loop, it models a subset of that iteration space, a window. The overhead of failure has not been reduced as such, but probabilistically the chances of encountering failures are reduced due to fewer iterations being speculated upon. As long as several windows can successfully execute in parallel, the accumulated overhead of failure may be reduced.

The contribution of this paper is how to make $|T_f - T_s|$ as close to 0 as possible without assuming any special functionality to facilitate speculative parallelization from the computer architecture.

3 The Coinspector Thread

Practical application of speculative parallelization requires not only the possibility of a performance improvement as demonstrated by previous work (software [9,4,10,8,6,7] and hardware [11,12,13,14,15,16,5,17]), but also assurances that users will not be confronted with a "double-edged" sword due to this optimization. Otherwise, the question of when to apply speculative parallelization becomes complex and critical.

The existing techniques for speculative parallelization, at best, offer not to waste all the results generated by the speculation and re-execute from the first point of a data dependence occurrence. The Coinspector Thread technique targets software speculative parallelization using privatization and borrows concepts from the inspector-executor model. The core idea is to preserve the sequential execution without adding anything extra, while at the same time enabling the speculative execution to proceed including the required extra tracking of memory accesses and data dependence tests.

Figure 1 shows a graphical representation of a speculative parallelization execution. Privatization has been applied and no computer architecture support for speculative parallelization is required. The vertical axis represents the iteration space under consideration. Without loss of generality, these iterations can span the entire set of loop iterations or a windowed subset as described in Section 2.

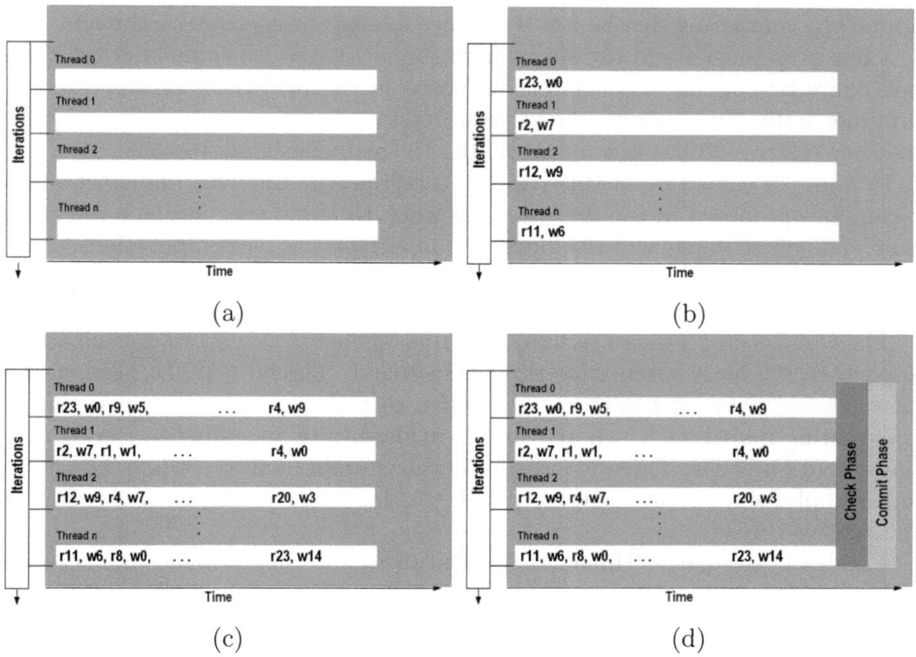

Fig. 1. Speculative Parallelization with Privatized Variables

Neither speculative parallelization nor the Coinspector Thread technique are limited to loop constructs. An iteration space is selected for easier presentation and comprehension. On the horizontal axis, execution time progresses from left to right. Figure 1 (a)–(d) are snapshots at different points during the speculation process. Figure 1 (a) presents the snapshot just before the speculative threads start execution. For example, Thread 0 has been assigned iterations 0-to-9, Thread 1 has been assigned 10-to-19 and so on. Each thread will execute the body of the loop augmented with extra instructions to track memory accesses using the shadow structures. Figure 1 (b) advances in time and the speculative threads have completed one iteration. For example, Thread 0 has completed iteration number 0 while Thread 1 has completed iteration number 10. In this example there are one read and one write memory operation tracked per iteration. Each speculative thread has executed the instructions to record the memory accesses and it has written the result of the write in a private variable. For example, Thread 2 has read from memory address 12 and it has written to memory address 9. However, the write operation has not been performed directly at memory address 9. Instead the value has been stored in a private variable of Thread 2. Figure 1 (c) fast-forwards the speculative threads to a point in time where all of them have finished executing their assigned iterations. For example, Thread n has written to memory addresses 6, 0, . . . 14 and read from memory addresses 11, 8, . . . 23. The last snapshot of the sequence shows that the speculative threads have taken part in the check phase where the test has been applied by comparing the shadow structures among the speculative threads. The test has been passed and the speculative threads have also completed the commit phase by copying out the results stored in the privatized variables to the program state. For example, the value stored in private memory of Thread n for memory address 23 has now been written to the memory address 23.

It is out of the scope of this paper to describe the different implementation techniques proposed in the literature for when to apply speculative parallelization, and how to implement the different phases (speculative computations, check and commit phases). The technique described here has been implemented with several variants to confirm its applicability.

The Coinspector Thread technique can be explained as a set of optimizations applied to the least speculative thread. In Fig. 1, Thread 0 is the least speculative thread. Thread 1 is more speculative than Thread 0 in the sense that it is executing iterations which are further in the future according to the sequential execution order. Thread 0 was executing iterations 0-to-9 while Thread 1, for example, was executing iterations 10-to-19. Similarly, Thread n is the most speculative thread.

The least speculative thread is referred to as the *Base Thread* and receives the following optimizations:

1. The Base Thread executes the original loop body without requiring writing to private memory or logging of read memory accesses. As noted by Gupta and Nim [9], no preceding speculative writes can invalidate the reads which renders unnecessary the logging of read memory accesses, no RAW dependences.
2. The Coinspector Thread offloads from the Base Thread the logging of write memory accesses by executing (as the inspector thread does in the inspector/executor model) a stripped down version of the loop body for the set of iterations assigned to the Base Thread.
3. The Base Thread, although it has been assigned a set of iterations, continues executing beyond that set until the whole loop is completed or the speculation has succeeded.

The Coinspector Thread is the key mechanism that enables the Base Thread to be performing a sequential execution completely unaware of the speculative threads, of logging memory accesses, or of applying tests to establish the sequential semantic equivalence. Unlike the Base Thread, the Coinspector will never execute more than the assigned set of iterations. This is sufficient to check for successful parallelization.

Figure 2 presents a graphical representation of a speculative parallelization using the Coinspector Thread. Figure 1 can be compared against this figure to observe the differences. Thread 0 has become the pair Base and Coinspector Threads and both threads are assigned the same set of iterations. The rest of the threads and iteration assigments remain identical. Figures 2 (a)–(c) illustrate that the Base Thread does not take part in any logging of memory accesses. The

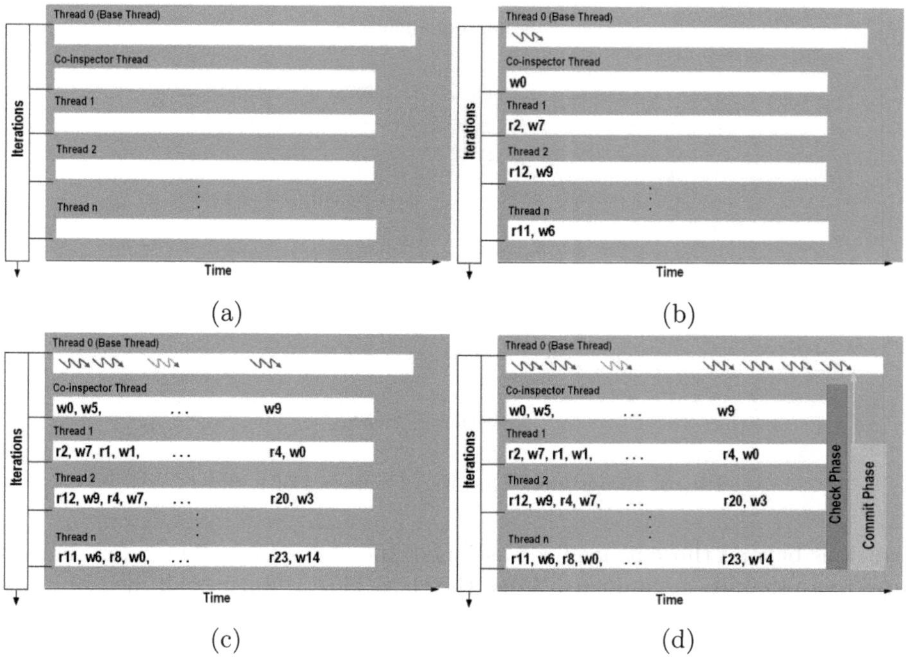

Fig. 2. Speculative Parallelization with the Coinspector Thread

Base Thread is running the original loop body writing directly to memory; no privatization. The Coinspector and Threads 1–n progress through the iterations logging the memory accesses. The Coinspector is duplicating some of the computations performed by the Base Thread to be able to keep a record of the write memory accesses made by the Base Thread. The Coinspector Thread is not actually performing the write operations or storing the values unless needed for logging the write memory accesses. In such cases the values are stored in privatized variables since a result generated by the Coinspector will never need to be copied out to the program state. As in Fig. 1, Threads 1–n advance by executing the original loop body augmented with the extra logging instructions and performing the write operations in private variables. Figure 2 (d) illustrates how the Base Thread continues executing without being involved in the check phase. Only once the Coinspector Thread together with Threads 1–n have passed the test in the check phase, the Base Thread gets a notification to stop because the speculation was successful. In such case, the commit phase will need only those results generated by speculative threads that have not been passed by the Base Thread. If the Base Thread has passed a speculative thread, the program state already contains the write values that would need to be copied out from the passed speculative thread. This situation would occur, for example, if the Base Thread has executediterations 0-to-9 and also 10-to-19, and iterations 10-to-19 were assigned to Thread 1. Note that although the Base Thread can execute

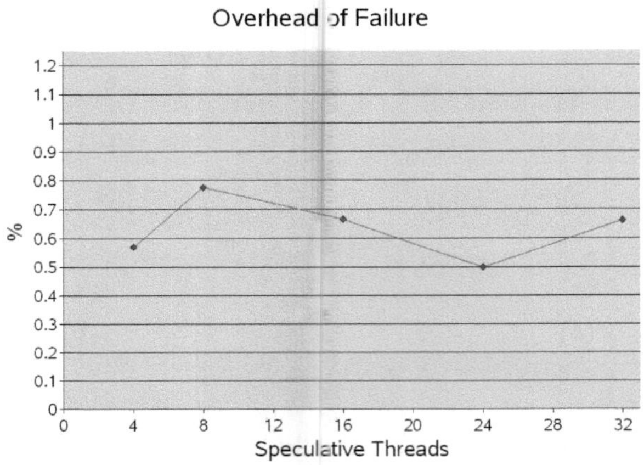

Fig. 3. Experiment Results - Overhead of Failure

iterations beyond those it has been assigned, the Coinspector Thread never executes more than its assignment. When the Base Thread finishes all the iterations in the loop before getting notified, then the Base Thread requests the speculation be discarded. In the check phase when the test fails, the Coinspector Thread as such does not prevent exploiting those correct results (i.e. those speculative threads assigned to iterations prior to the first point of failure). The Base Thread is notified of the partial success and will restart execution once those speculative threads that passed the test have completed the commit phase.

4 Performance Evaluation

To test the value of the Coinspector Thread in eliminating the overhead of speculative failure, a code of interest in the DARPA High Productivity Computing Systems program is considered, the Gyrokinetic Toroidal Code (GTC). GTC is a 3-D particle-in-cell code developed at the Princeton Plasma Physics Lab [18]. Two sections of the code are of interest for speculative parallelization: the `compute_mapping` and `deposit` routines. These routines map particles to a 3-D grid. The benchmark code is a synthesized loop from these two routines. The data set was engineered to ensure that one failure would occur. When a failure occurs, all of the speculative results, including correct ones sequentially earlier than the failure, are discarded. The benchmark experiments map 1 billion particles into a 1 million element 3-D array.

The experiments run on a Sun Fire™ E6900 server running Solaris™ 10 with 24 1.35GHz UltraSPARC® IV processors. Each UltraSPARC® IV processor is dual core adding up to a total of 48 cores in the server. Sun Fire™ E6900 servers offer a cache coherent shared memory system and this server has 192GB of physical memory installed.

Figure 3 presents the overhead of failure as a percentage of T_s (sequential execution time without any instrumentation due to speculation). Although the overhead fluctuates depending on the number of speculative threads taking part, the maximum overhead recorded is 0.77%.

The Coinspector Thread takes a thread slot that, if speculation were successful, could have been exploited by executing an extra speculative thread. Due to space constraints such an evaluation is out of scope, but note that the actual speedup thanks to that extra speculative thread is completely dependent on the computer architecture and application. It is well established that speedup curves flatten and decrease beyond a given number of threads. For example, the best speedup on 16 processors published in [6] is around 9 with most values being between $5 - 6$. Given the current industry trend towards multi-core and many-core processors (such as Sun's Niaraga processor with 32 hardware threads), it seems more challenging to develop scalable applications than to find a hardware slot for the Coinspector Thread.

5 Limits of the Coinspector Thread

The characteristics of the original code from which a compiler extracts the instructions for the Coinspector Thread has a direct impact on how much overhead of failure can be eliminated. Consider the example loop:

```
for (i = 0; i < 100; i++) {
     A[indx[i]]] = A[i] + 3.14159; // statement S1
     indx[i] = expression; // statement S2
}
```

where `indx` is an indirection array and `A` is an array of doubles. The Coinspector Thread executes a version of this loop that does not interfere with the Base Thread. Since all changes done by the Base Thread are correct and are safe if seen by later speculative threads, these are done directly to the original data structures. The problem is that a variable used to determine the memory addresses of write operations to array `A` can be modified by statements inside the loop body. For example, consider that the Base and Coinspector Threads have been assigned to iterations 0-to-9 and both execute the loop body as presented. In the first iteration (`i = 0`), the Base Thread can execute Statement S2 before the Coinspector Thread has been able to read the contents of `indx[0]` for Statement S1. In specific, the Base Thread could replace the value 7 with the value 8 in `indx[0]`. Thus, the Coinspector Thread would believe that the write to array `A` had been done at index position 8 and would incorrectly generate the corresponding log entry. This is one illustration of the general case in which the memory access patterns that need to be logged are not reproducible. However, even in these cases it is possible to use a Base and Coinspector Thread whenever speculative parallelization is applicable.

There are several alternatives available when speculating on this type of code section, however all of them introduce some additional overhead. The choices range from first, using prior techniques with logging overhead in the base thread, second, privatizing a portion of the index array for use by the Coinspector and speculative threads, and third, delaying the progress of the base thread so that it does not interfere with correct logging by the Coinspector or correct execution by later speculative threads. In the second case, the Base Thread does not start execution until the privatization is completed. This delay means that the sequential execution has been halted and, thereby, the worst case analysis for the execution time in case of failure T_f is the sequential time T_s plus the time for privatizing data in the Coinspector Thread T_{pct}.

It is worth noting that when the operations performed by the `expression` in Statement S2 are reversible, for example an increment `indx[i]++`, the Coinspector Thread has the option of reversing the computation to recover the original array index for its log. However, even in these situations the Coinspector Thread must know whether it is ahead or behind the Base Thread. One way to accomplish this is to force the Coinspector Thread to lag the Base Thread so that the Base Thread can run at full speed.

An example which introduces a similar difficulty for both previous speculative parallelization techniques and the Coinspector Thread is when the access pattern is not predictable.

```
for (i = 0; i < 100; i++) {
     A[random()] = A[i] + 3.14159; // statement S3
}
```

This lack of predictability can be resolved by applying a straightforward inspector loop to generate the access pattern in advance. This access pattern can then be stored for use by all threads.

6 Conclusions

Practical application of speculative parallelization requires not only the possibility of a performance improvement as demonstrated by previous work, but also assurances that users will not be confronted with a severe performance degradation. Such risk may prevent wide spread adoption of speculative parallelization by industrial strength compilers.

The contribution of this paper is a technique to preserve the sequential execution without adding anything extra (Base Thread), but at the same time to enable the speculative execution including the required extra tracking of memory accesses and data dependence tests (Coinspector Thread).

The performance evaluation has tested the technique under various numbers of speculative threads (from 4 up to 32). The performance results show that the Coinspector Thread enables the overhead to execute regions of code including failed speculative parallelization to be less than 1%. To the best of the authors' knowledge, this is the first technique delivering such a low overhead.

References

1. Zhu, C.Q., Yew, P.C.: A scheme to enforce data dependence on large multiprocessor systems. IEEE Transactions on Software Engineering 13(6), 726–739 (1987)
2. Midkiff, S.P., Padua, D.A.: Compiler algorithms for synchronization. IEEE Transactions on Computers 36(12), 1485–1495 (1987)
3. Rauchwerger, L.: Run-time parallelization: Its time has come. Parallel Computing 24(3-4), 527–556 (1998)
4. Rauchwerger, L., Padua, D.A.: The LRPD Test: Speculative run-time parallelization of loops with privatization and reduction parallelization. IEEE Transactions of Parallel and Distributed Systems 10(2), 160–180 (1999)
5. Garzarán, M.J., Prvulovic, M., Llabería, J.M., Viñals, V., Rauchwerger, L., Torrellas, J.: Tradeoffs in buffering speculative memory state for thread-level speculation in multiprocessors. ACM Transactions on Architecture and Code Optimization 2(3), 247–279 (2005)
6. Dang, F.H., Yu, H., Rauchwerger, L.: The R-LRPD Test: Speculative parallelization of partially parallel loops. In: IPDPS 2002, pp. 20–29 (2002)
7. Cintra, M., Llanos, D.R.: Design space exploration of a software speculative parallelization scheme. IEEE Transactions of Parallel and Distributed Systems 16(5), 1–15 (2005)
8. Rundberg, P., Stenström, P.: An all-software thread-level data dependence speculation system for multiprocessors. Journal of Instruction Level Parallelism 3 (2001)
9. Gupta, M., Nim, R.: Techniques for speculative run-time parallelization of loops. In: Proceedings of the 1998 ACM/IEEE Conference on Supercomputing – SC 1998 (1998)
10. Bruening, D., Devabhaktuni, S., Amarasinghe, S.: Softspec: Software-based speculative parallelism. In: Third ACM Workshop on Feedback-Directed and Dynamic Optimization – FDDO-3 (2000)
11. Sohi, G.S., Breach, S.E., Vijaykumar, T.N.: Multiscalar processors. In: ISCA 1995, pp. 414–425 (1995)
12. Marcuello, P., González, A.: Clustered speculative multithreaded processors. In: ICS 1999, pp. 365–372 (1999)
13. Tsai, J.Y., Huang, J., Amlo, C., Lilja, D.J., Yew, P.C.: The Superthreaded processor architecture. IEEE Transactions on Computers 48(9), 881–902 (1999)
14. Oplinger, J.T., Heine, D.L., Lam, M.S.: In search of speculative thread-level parallelism. PACT 1999, 303–313 (1999)
15. Prvulovic, M., Garzarán, M.J., Rauchwerger, L., Torrellas, J.: Removing architectural bottlenecks to the scalability of speculative parallelization. ISCA 2001, 204–215 (2001)
16. Chaudhry, S., Tremblay, M.: Space-time dimensional computing for Java^{tm} programs on the MAJC architecture. In: Java Microarchitectures (2002)
17. Sarangi, S.R., Wei Liu, J.T., Zhou, Y.: Reslice: Selective re-execution of long-retired misspeculated instructions using forward slicing. In: Proceedings of the 38th annual IEEE/ACM International Symposium on Microarchitecture – MICRO 38, pp. 257–270 (2005)
18. Oliker, L., Canning, A., Carter, J., Shalf, J., Ethier, S.: Scientifc computations on modern parallel vector systems. In: Proceedings of the ACM/IEEE SC2004 Conference on Supercomputing (2004)

Power-Aware Fat-Tree Networks Using On/Off Links

Marina Alonso[1], Salvador Coll[2], Vicente Santonja[1],
Juan-Miguel Martínez[1], Pedro López[1], and José Duato[1]

[1]Dept. of Computer Engineering
[2]Dept. of Electronic Engineering
Universidad Politécnica de Valencia
Camino de Vera s/n
46022 Valencia, Spain
{malonso,scoll,visan,jmmr,plopez,jduato}@upvnet.upv.es

Abstract. Nowadays, power consumption reduction techniques are being increasingly used in computer systems, and high-performance computing systems are not an exception. In particular, the power consumed by the interconnect circuitry has a non-negligible contribution to the total system budget. In this scenario, fat-tree interconnection networks are one of the most popular topologies. This topology is particularly well-suited for applying power consumption reduction techniques since it provides multiple alternative paths for each source/destination pair. In this paper, we present a mechanism that dynamically adjusts the available network bandwidth by switching links on and off, according to the traffic requirements. This mechanism provides significant reduction in power consumption while maintaining the original underlying routing algorithm, at the expense of slight latency increase for low loads.

1 Introduction

Nowadays power consumption reduction techniques are being increasingly used in computer systems, and high-performance computing systems are not an exception. Most of these systems are clusters (this is the architecture of 72.20% of the 500 systems listed in the November 2006 edition of the Top500 Supercomputers sites [1]). In particular, the power consumed by the interconnection network circuitry has a significant contribution to the total system budget. For example, the routers and links in a Mellanox server blade, consume about 37% of the total power budget [2]. In this scenario, fat-tree interconnection networks are one of the most popular topologies due to their high bisection bandwidth and ease of application mapping for arbitrary communication topologies [3]. But most applications have communication topology requirements that are far less than the total connectivity provided by fat-trees. Vetter and Mueller show that applications that scale most efficiently to large numbers of processors use point-to-point communications patterns where the average number of distinct destinations is relatively small [4]. This provides strong evidence that

R. Perrott et al. (Eds.): HPCC 2007, LNCS 4782, pp. 472–483, 2007.

many application communication topologies exercise a small fraction of the resources provided by fat-trees [5]. Moreover, traffic in an interconnection network exhibits large spatial and temporal variance, leading to inactivity periods at several links in the network [6]. On the other hand, fat-trees are particularly well-suited for applying power consumption reduction techniques since they provide multiple alternative paths for each source/destination pair. This paper shows that there is a chance to reduce power consumption by dynamically switching on/off links based on the traffic they support while running a set of applications.

Several power reduction techniques for interconnection networks have been proposed. Most of them are based on Dynamic Voltage Scaling (DVS). DVS was originally proposed, and now is widely deployed, for microprocessors. When applied to networks, this approach allows DVS links to work in a discrete range of frequencies and supply voltages, which leads to different levels of power consumption in response to their traffic utilization. The history-based DVS policy proposes to use past network utilization to predict future traffic, therefore tuning dynamically link frequency and voltage to reduce network power consumption [7]. Stine and Carter compare DVS with the use of adaptive routing in non DVS links, showing that, as long as the network provides enough bandwidth to meet the needs of the application, an adaptively-routed network can improve latency with the same power consumption [8]. DVS has significant drawbacks: it requires a sophisticated hardware mechanism to ensure correct link operation during scaling, it consumes significant CMOS area, and DVS links continue to consume power even while idle.

Other techniques are based on the use of on/off links that are selectively switched on and off according to their utilization [2,9,6]. In order to avoid deadlocks, adaptive routing algorithms must be used. Kim et al. also investigate hybrid techniques based on both DVS and on/off links. The idea is to shut down DVS links when traffic drops to very low levels [2].

In this paper, we present a new method to reduce power consumption in fat-trees based on the use of on/off links. The rest of the paper is organized as follows. Section 2 formalizes fat-tree network topology considering it a particular class of k-ary n-tree, including a description of packet routing. Section 3 describes the proposed power saving mechanism. Our proposal is evaluated using simulation in Section 4 and, finally, some conclusions are drawn in Section 5.

2 Fat-Trees

A k-ary n-tree is composed of $N = k^n$ processing nodes and $S = nk^{n-1}$ k-ary switches. Every switch has $2k$ ports, k "up" links and k "down" links. Processing nodes are identified by $(p_0, p_1, \ldots p_{n-1})$ where $p_i \in \{0, 1, \ldots, k-1\}$ for $0 \leq i \leq n-1$, and each switch is identified by $(w_0, w_1, \ldots w_{n-2}, l)$, where $w_i \in \{0, 1, \ldots, k-1\}$ for $0 \leq i \leq n-2$ and $l \in b\{0, 1, \ldots, n-1\}$ is the level of the switch (0 is the root level).

- Two given switches, $(w_0, w_1, \ldots w_{n-2}, l)$ and $(w'_0, w'_1, \ldots w'_{n-2}, l')$ are connected if and only if $l' = l + 1$ and $w_i = w'_i$ for all $i \neq l$. The link connecting both switches is labeled with w'_l on the level l switch and with $k + w_l$ on the l' switch.
- There is a link between the switch $(w_0, w_1, \ldots w_{n-2}, l)$ and the processing node $(p_0, p_1, \ldots p_{n-1})$ if and only if $w_i = p_i \; \forall i \in \{0, \ldots, n-2\}$.

The labeling scheme shown in the previous definitions makes the k-ary n-tree a delta network: any path starting from a level 0 switch and leading to a given node $p_0, p_1, \ldots, p_{n-1}$ traverses the same sequence of links (p_0 at level 0,p_1 at level 1,..., p_{n-1} at level $n-1$) [10]. An example of such labeling is shown in Figure 1, for a quaternary fat-tree of dimension 3 (64-node network), that is a 4-ary 3-tree.

Given a k-ary n-tree, the Minimal Tree (MT) is the subset of the tree composed of all the processing nodes, a subset of the communication switches, and the edges between them. A switch $(w_0, w_1, \ldots w_{n-2}, l)$ belongs to the MT if one of the following properties holds:

1. $l < n - 1$ and $w_i = 0 \;\; \forall i \in \{l, \ldots, n-2\}$.
2. $l = n - 1$ (all the switches in the level $n - 1$ belong to the MT).

Within these switches, all the "down" links and the "up" links with index k also belong to the Minimal Tree. The Minimal Tree of a quaternary fat tree of dimension 3 (4-ary 3-tree) is shown in Figure 1 using thicker lines.

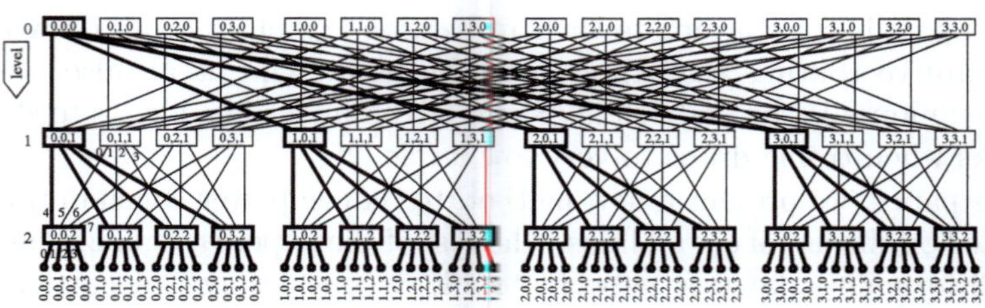

Fig. 1. 4-ary 3-tree node, switch and edge labels. The Minimal Tree is highlighted.

Minimal routing between any pair of processing nodes can be accomplished by sending the message to one of the nearest common ancestor switches and from there to the destination. Hence, each message experiences two routing phases: an ascending phase, from the processing node to a nearest common ancestor, followed by a descending phase. While the descending phase is necessarily deterministic, since there is a single path from a nearest common ancestor switch to the destination, there could be alternative routes to reach a nearest common ancestor. The availability of alternative routes makes it possible to randomly choose ascending links or even implementing an adaptive algorithm that makes a decision according to the local state of the switch, avoiding congested links.

3 Description of the On/Off Power Saving Mechanism

The proposed power saving mechanism is based on dynamically switching links on/off as a function of the required network throughput, and improves a preliminary version of this mechanism [11]. We consider bidirectional links that can be turned on/off in a given direction, either ascending or descending. Every switch in the network periodically measures outgoing traffic and controls the number of operating outgoing links depending on traffic variations. A subset of network links, which is defined as the Minimal Tree, fixing the maximum level of power saving, cannot be switched off in order to maintain the network connectivity.

In order to dynamically turn links on and off according to their utilization, the average utilization of all the "up" links in a switch, u_{up}, is periodically obtained. Two thresholds are defined to control the mechanism behavior: U_{off} is the turn off threshold, and U_{on} is the turn on threshold. If $u_{\text{up}} < U_{\text{off}}$ one of the "up" links is turned off, while when $u_{\text{up}} > U_{\text{on}}$, an inactive "up" link is turned on.

The power saving mechanism controls the network links state according to the following general rules:

- Up links: the utilization of the up links of a given switch is used to decide whether to turn on or turn off up links. This decision propagates upward to guarantee at least one path to level 0 switches (this is required to provide a path to every destination), and downward to guarantee descending routes to processors (for the same reason).
- Down links: these links are turned on/off all at once. Down links at a given switch are turned off when that switch cannot receive descending traffic, that is when all its input links (ascending and descending) have been turned off. Those links will be turned on again when some input link is turned on.
- The underlying routing algorithm need no changes since the mechanism is limited to those switches and links that do not belong to the MT (see Section 2), and the MT provides the minimum paths needed to maintain all the processing nodes connected.

A detailed description of the mechanism can be found in previous work [11].

The mechanism performance can be tuned by setting the values of the thresholds (U_{on} and U_{off}) used to control link on/off switching. Their effect can be analyzed by considering that threshold average indicates the mechanism aggressiveness while threshold difference (or hysteresis band) indicates mechanism responsiveness. Aggressiveness is related to the maximum power saving requested to the mechanism. Responsiveness refers to its ability to follow load changes.

The range of possible threshold values is conditioned by several limiting factors [12] that are summarized in the map of possible thresholds shown in Figure 2. This diagram is represented as a function of threshold difference and average. Any point inside the shaded region provides a valid configuration, with different responsiveness and aggressiveness as indicated with the bars depicted together with the axis in the graph.

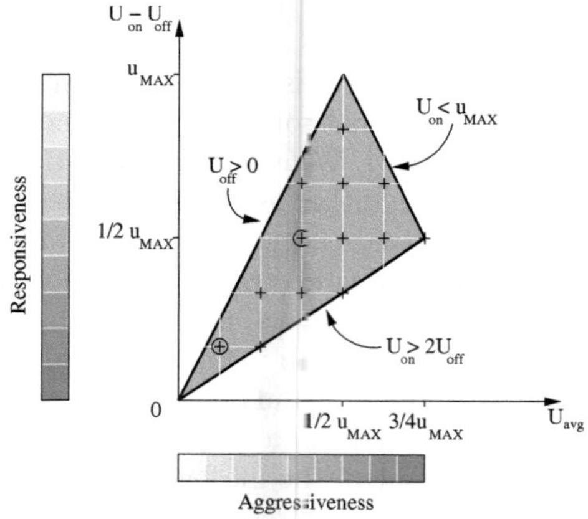

Fig. 2. Map of possible thresholds

3.1 Static Thresholds

Our mechanism uses static thresholds, as their value is constant regardless of the number of links that are on or off. Nevertheless, the mechanism generates an effect of increase (decrease) in the utilization of the available links as long as some of them are turned off (on). This effect could also be analyzed as whether the thresholds were reduced proportionally to the available throughput and the load was calculated relative to the full throughput. We refer those thresholds to as the effective thresholds. The effective threshold values as a function of the number of outgoing links in ascending direction for a 4-ary n-tree are indicated in Table 1. Factor column shows the fraction of available outgoing throughput used to calculate the effective thresholds. According to that, $\frac{3}{4}U_{\text{on}}$ can be considered as the minimum utilization for having all links active, while $\frac{2}{4}U_{\text{off}}$ is the maximum utilization that guarantees maximum power saving for a particular switch.

Table 1. Static threshold values according to the available links for a 4-ary switch

	Static thresholds			Effective thresholds	
On Up Links	On	Off	Factor	On	Off
4	U_{on}	U_{off}	1	U_{on}	U_{off}
3	U_{on}	U_{off}	3/4	$\frac{3}{4}U_{\text{on}}$	$\frac{3}{4}U_{\text{off}}$
2	U_{on}	U_{off}	2/4	$\frac{2}{4}U_{\text{on}}$	$\frac{2}{4}U_{\text{off}}$
1	U_{on}	U_{off}	1/4	$\frac{1}{4}U_{\text{on}}$	$\frac{1}{4}U_{\text{off}}$

Figure 3(a) shows the switch state (given by the number of ascending active links) versus the load traversing the switch in ascending direction for a 4-ary fat-tree. The state with all the links in the off state is not shown, since the last

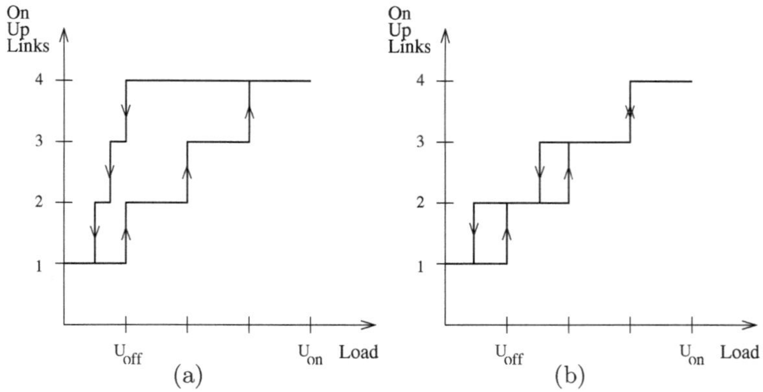

Fig. 3. Switch state as a function of traffic with (a) static and (b) dynamic thresholds

ascending link is turned off (except for the MT) when all the incoming links in the ascending direction have been already turned off.

The reduction in available throughput when reducing the active outgoing links generates an effect that can be viewed as a reduction in the hysteresis band (increase on responsiveness) of the mechanism. This is a positive effect since the network is more sensitive to congestion when the fraction of active links is reduced. In that situations, having a higher responsiveness will increase the mechanism agility to react against small changes in traffic, providing additional active links if needed. Otherwise a low responsiveness could lead to network congestion during limited periods of time, with a significant increase in latency.

An important limitation of the static thresholds is that U_{on} must be higher than $2 \cdot U_{\text{off}}$. This makes the threshold average to be low for situations with several active links, and hence making the mechanism less aggressive, which means reducing the margin for power saving. Moreover, this condition precludes the use of a mechanism that is both very aggressive and very responsive (see Figure 2).

3.2 Dynamic Thresholds

As an alternative, we have devised a version of the mechanism which is based on dynamic thresholds. The main objective is trying to divide the full range of network utilization in as many slots as indicated by the network arity. On average every switch distributes the ascending traffic among k links, when the network is fully active. If the traffic decreases $1/k$ of the nominal traffic requiring the full switch throughput it seems reasonable to reduce the available throughput by exactly the same amount, thus turning off one link.

The dynamic implementation is based on a fixed "on" threshold, U_{on}, and a dynamic version of the "off" threshold, U_{off}, that depends on the number of active outgoing links according to the following expression:

$$U_{\mathrm{off}_i} = \frac{U_{\mathrm{on}} \cdot (i-1)}{k}$$

i being the number of active outgoing links in the ascending direction.

Considering a 4-ary n-tree, the set of dynamic thresholds together with the effective thresholds is indicated in Table 2, with the corresponding state transition diagram shown in Figure 3(b).

The dynamic implementation of thresholds provides significant power saving improvements, since the fraction of switch utilization where some links are turned off increases with respect to the static version. This result is validated in Section 4. The overlap among the transitions between 4 and 3 on links is not a problem (multiple alternative transitions due to traffic oscillations) since the mechanism operation is based on the average switch utilization during fixed length periods (Section 4.2) which has the effect of filtering quick traffic changes.

Table 2. Dynamic threshold values according to the available links for a 4-ary switch

	Dynamic thresholds			Effective thresholds	
On Up Links	On	Off	Factor	On	Off
4	n.a.	$\frac{3}{4}U_{\mathrm{on}}$	1	n.a.	$\frac{3}{4}U_{\mathrm{on}}$
3	U_{on}	$\frac{2}{4}U_{\mathrm{on}}$	3/4	$\frac{3}{4}U_{\mathrm{on}}$	$\frac{6}{16}U_{\mathrm{on}}$
2	U_{on}	$\frac{1}{4}U_{\mathrm{on}}$	2/4	$\frac{2}{4}U_{\mathrm{on}}$	$\frac{2}{16}U_{\mathrm{on}}$
1	U_{on}	0	1/4	$\frac{2}{4}U_{\mathrm{on}}$	0

4 Performance Evaluation

In this section, we study, using simulation, the impact of the power reduction mechanism on latency. The metrics used in this study are the average latency of a message (measured from generation to delivery time) and the relative power consumption of the links as compared with the default system.

Two types of graphs are presented: the first shows the relative power consumption of the network links as a function of the injected traffic. These graphs includes two separate curves corresponding to the network behavior when an increasing or decreasing load is applied Assuming that uniform traffic is used, this representation provides a view of the power consumption hysteresis band for the whole network. Note that the graph corresponding to the increasing workload should be read from left to right, while the one corresponding to decreasing traffic should be read from right to left. The second graph type shows the latency evolution for the same range of injected traffic, including results with the network links fully operational.

4.1 Network and Traffic Model

Our simulator models a wormhole switching network at the flit level [13]. The network is composed of switch nodes and processor nodes. The switches contain a routing control unit, a crossbar and as many physical links as indicated by the network arity. Physical links are split into three virtual channels, with capacity for four flits. The results have been obtained for quaternary fat-trees of dimension 4 (256 nodes).

The network load is defined by the message generation rate at each node, the message size, and the destination of each message. Initially only the links in the minimal tree are active. For each test, the generation rate is kept constant at 0.01 flits/cycle/node during 60000 cycles, increased to its maximum value (0.60 flits/cycle/node) with constant slope during 120000 cycles, maintained at the maximum rate during 120000 cycles and decreased to the minimum load, again in 120000 cycles. A synthetic workload based on the uniform distribution is used. All the nodes in the network have the same behavior: the message inter arrival time is generated according to the workload type, the message length is fixed to 16 flits and, the destination node for each message is chosen among all the nodes in the network (except the source node) with the same probability.

We use a uniform distribution for the destinationss because it corresponds to the worst case, as this workload produces a sustained traffic all over the network. If we had selected a workload that exhibits locality, some parts of the network would not receive any traffic. In that case, our mechanism would obtain much better results by permanently keeping the links in this area switched off.

4.2 Parameters of the Proposed Mechanism

As explained in Section 3, at a given time the operational level of a link depends on its utilization. The dynamics of the model is driven by the off threshold U_{off} and the on threshold U_{on}. In the following figures, we explore part of the design space by selecting different values of U_{off} and U_{on} in order to achieve different goals of responsiveness and aggressiveness for the power saving mechanism. The complete set of tested configurations is given by the cross (+) signs on the map of possible thresholds shown on Figure 2.

On the other hand, a link cannot be instantaneously turned on, but it requires a time T_{on}. Turning off a link also needs some time T_{off} to decrease the circuit voltage level to zero. When a link is turned off, we assume that it becomes immediately unavailable but it continues consuming power until T_{off} cycles have elapsed. Similarly, when a link is turned on, the new link is available to messages after T_{on} cycles, but power consumption increases at once. Based on the values reported by Kim et al., we have used $T_{\text{on}} = T_{\text{off}} = 1000$ clock cycles [14,9]. The state of the network is periodically checked to decide if it is necessary to turn on or off any link. We use a period greater than T_{on} and T_{off} in order to allow network stabilization after the changes. Specifically, the check period used is 2000 clock cycles.

4.3 Results with Static Thresholds

The power consumption and the latency results for three selected points from Figure 2 (highlighted with a circle) are presented below.

Figures 4(a) and 4(b) show results for the static thresholds $[U_{on}, U_{off}] = [0.030, 0.150]$. The power curves are very similar to those predicted by Figure 3, since they are the result of the overlapped effect of all network switches. These curves clearly show four different and stable network states. Each one of them is given by the average network performance of network states where the switches directly connected to processors have one, two, three or four active ascending links. This is also due in part to the fact that the reported experiments have been performed with uniform traffic to show the average network behavior. As expected, the network is at 100% power consumption with ascending traffic when the injected traffic surpasses $\frac{3}{4}U_{\text{on}}$ ($\frac{3}{4}U_{\text{on}} = 0.11$ for this test) for the 4-ary 4-tree topology.

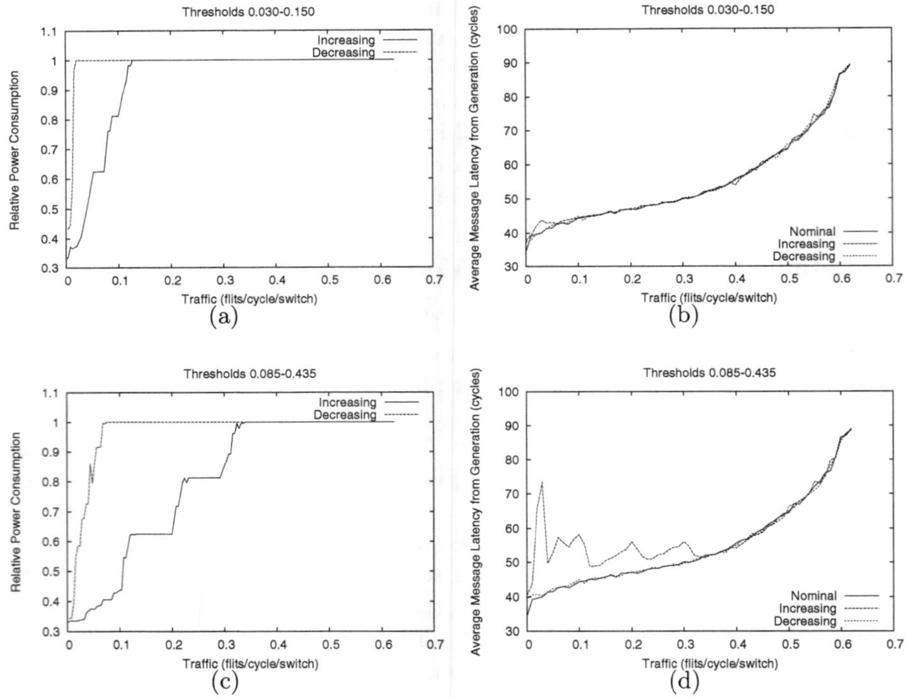

Fig. 4. Power and latency versus traffic with static thresholds

Figures 4(c) and 4(d) show results for the static thresholds $[U_{on}, U_{off}] = [0.085, 0.435]$, that provide a more aggressive power reduction mechanism. For this reason, the switches remain partially disconnected for higher loads and, as a consequence, the latency penalty increases. As can be seen the latency obtained for increasing traffic experiences several peaks for low traffics, which are due to the availability of a fraction of the total network throughput until the turning on of additional link allows higher loads. The latency curve with increasing traffic denotes that the effective U_{on} thresholds are very close to the maximum traffic each one of the stable network states can deliver.

4.4 Results with Dynamic Thresholds

In this section, we report the results obtained for the selected points when using the dynamic thresholds version of the mechanism. As can be seen in Figure 5, the hysteresis band width reduction provides additional power reductions at no additional latency cost.

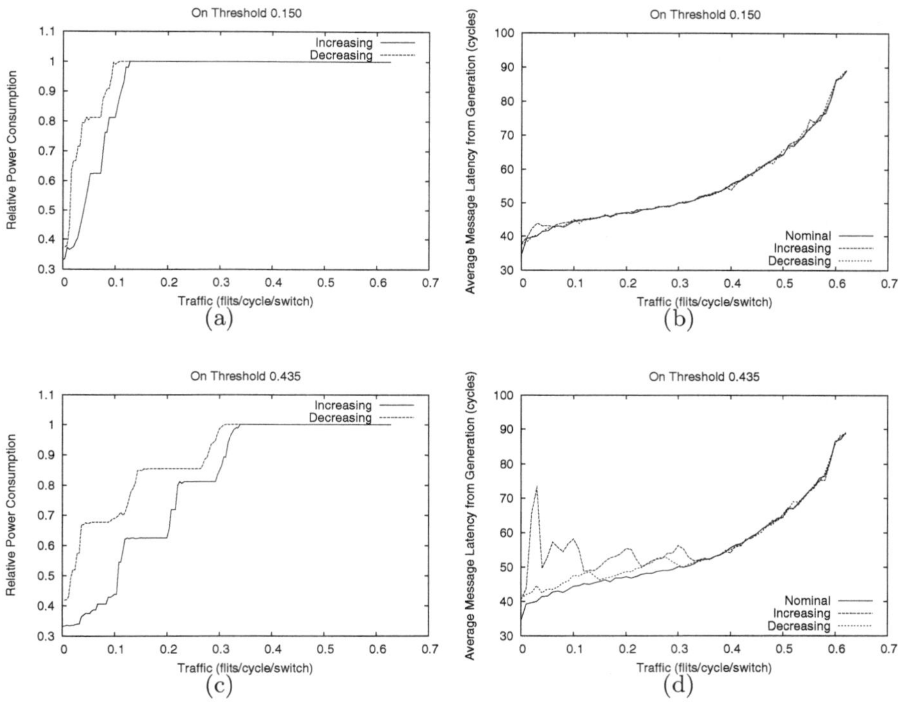

Fig. 5. Power and latency versus traffic with dynamic thresholds

The main benefits of the dynamic thresholds performance are based on the displacement of the decreasing power consumption traffic curve closer to the increasing one. The relative power saving increases dynamic thresholds provide range between 50% and 30% for the tests reported in this paper. It is important to note that the increasing traffic power consumption curves make state transitions for higher loads; hence the network experiences higher latency penalties than for decreasing traffic, as shown in the latency graphs.

The rest of the experiments corresponding to the points in the area shown in Figure 2 confirms the results reported in this paper. In particular, more aggressive thresholds introduce significant latency penalties. In some cases, latency for very low loads becomes similar to the nominal latency obtained with the maximum traffic rate. Therefore, these very aggressive thresholds may not be interesting.

5 Conclusions

In this paper, we have presented a novel technique to reduce power consumption in fat-tree interconnection networks. Two important contributions of our mechanism are its simple implementation and the fact that the underlying routing algorithm does not need to be modified. The mechanism can be set to provide different levels of sensitivity to traffic variations. Moreover, power reduction policies with different levels of aggressiveness can be set, too. Hence, different ratios of power saving versus performance penalty can be obtained. Another significant contribution is the improvement of our original mechanism implementation by defining a dynamic behavior of the thresholds that control the mechanism operation. The dynamic version of the mechanism significantly outperforms the static approach at no additional performance cost.

Our results, obtained by simulation on 4-ary 4-tree networks, show that significant power savings can be obtained with moderate latency penalty, by selecting a conservative power reduction policy. Additional power savings can be obtained as well by further stressing the network at the cost of increasing latency. As future work we will further explore our proposal with realistic communication traffic patterns and we will analyze the eventual local congestion that may arise when many links are in the off state in order to improve network power-performance.

References

1. TOP500 Supercomputer Sites (2007), `http://www.top500.org`
2. Kim, E.J., et al.: Energy optimization techniques in cluster interconnects. In: ISLPED 2003, pp. 459–464 (2003)
3. Petrini, F., Vanneschi, M.: Performance analysis of wormhole routed k-ary n-trees. Int. Journal on Foundations of Computer Science 9(2), 157–177 (1998)
4. Vetter, J., Mueller, F.: Communication characteristics of large-scale scientific applications for contemporary cluster architectures. In: IPDPS (2002)
5. Shalf, J., Kamil, S., Oliker, L., Skinner, D.: Analyzing ultrascale application communication requirements for a reconfigurable hybrid interconnect. In: Proceedings of the ACM/IEEE SC 2005 Conference, SuperComputing 2005 (2005)
6. Soteriou, V., Peh, L.S.: Design-space exploration of power-aware on/off interconnection networks. In: ICCD 2004, San Jose, pp. 510–517 (2004)
7. Shang, L., Peh, L.S., Jha, N.K.: Dynamic voltage scaling with links for power optimization of interconnection networks. In: Proceedings of the 9th Int. Symposium on High-Performance Computer Architecture (HPCA-9), Anaheim, CA, pp. 79–90 (2003)
8. Stine, J.M., Carter, N.P.: Comparing adaptive routing and dynamic voltage scaling for link power reduction. Computer Architecture Letters (2004)

9. Soteriou, V., Peh, L.S.: Dynamic Power Management for Power Optimization of Interconnection Networks Using On/Off Links. In: Hot Interconnects 11, Stanford University, Palo Alto CA (2003)
10. Duato, J., Yalamanchili, S., Ni, L.: Interconnection Networks: an Engineering Approach. Morgan Kaufmann, San Francisco (2002)
11. Alonso, M., Coll, S., Martínez, J.M., Santonja, V., López, P.: Dynamic power saving in fat-tree interconnection networks using on/off links. In: HPPAC 2006, Rhodes Island, Greece, IEEE Computer Society, Los Alamitos (2006)
12. Alonso, M., Martínez, J.M., Santonja, V., López, P.: Reducing Power Consumption in Interconnection Networks by Dynamically Adjusting Link Width. In: Danelutto, M., Vanneschi, M., Laforenza, D. (eds.) Euro-Par 2004. LNCS, vol. 3149, pp. 882–890. Springer, Heidelberg (2004)
13. Duato, J.: A New Theory of Deadlock-Free Adaptive Routing in Wormhole Networks. IEEE Transactions on Parallel and Distributed Systems 4(12), 1320–1331 (1993)
14. Kim, J., Horowitz, M.A.: Adaptive Supply Serial Links with sub-1V Operation and Per-pin Clock Recovery. IEEE Journal of Solid State Circuits 1403–1413 (2002)

Efficient Broadcasting in Multi-radio Multi-channel and Multi-hop Wireless Networks Based on Self-pruning

Li Li, Bin Qin, Chunyuan Zhang, and Haiyan Li

School of Computer Science
National University of Defense Technology
Changsha, Hunan province, 410073, China
{liligfkd,qbsn,cyzhang,hyli}@163.com

Abstract. An important question in multi-radio multi-channel and multi-hop networks is how to perform efficient network-wide broadcast. Currently almost all broadcasting protocols assume a single-radio single-channel network model. Simply using them in multi-channel environment without careful enhancement will result in unnecessary redundancy. In this paper, we focus on reducing the amount of redundant traffic of broadcasting under multi-channel environment. We propose a general model for broadcasting and reduce the efficient broadcast problem into the minimal strong connected dominating set problem of the *interface-extend* graph which extends the original network topology across interfaces. Using interface-extend graph, we describe our Multi-Channel Self-Pruning broadcast protocol and simulation shows that our protocol can significantly reduce the transmission cost. To the best of our knowledge, our work is the first self-pruning broadcast scheme in this area.

1 Introduction

Multi-hop wireless networks such as mobile ad hoc networks, sensor networks, and mesh networks have garnered increasing attention over the last few years. However, a fundamental obstacle to building large scale multi-hop networks is the insufficient network capacity when route lengths and network density increase due to the limited spectrum shared in the neighborhood [1]. Wireless technologies, such as IEEE 802.11, provide multiple non-overlapping channels and recent advancements in wireless technology render the usage of multiple radios affordable. So the use of multiple radios which tuned to orthogonal channels becomes very promising because it can significantly improve the capacity of the network by employing concurrent transmissions in different channels, and that motivates the development of new protocols designed for multi-channel operation.

An important question in multi-radio multi-channel multi-hop networks which we attempt to address in this paper is how to perform efficient network-wide broadcast in such networks. Broadcasting is frequently used in multi-hop networks not only for data dissemination, but also for route discovery in reactive unicast routing protocols [2]. The presence of several multi-party applications—such as natural disaster warning, terrorist threat alert, local content distribution and multimedia gaming—also

R. Perrott et al. (Eds.): HPCC 2007, LNCS 4782, pp. 484–495, 2007.

imposes more capacity requirements to the broadcast protocol. However, naive broadcast scheme will generate an excessive amount of redundant traffic and exaggerates interference in the shared medium among neighboring nodes, which is called the broadcast storm problem [3]. A vast amount of broadcasting protocols such as probability-based methods, area-based methods, and neighbor-knowledge-based methods [4] have been proposed to mitigate the broadcast storm problem. However, all of the above protocols assume a single-radio single-channel (SR-SC) model.

There exist large amount of research on channel assignment and protocol design for MR-MC networks, but the study on broadcasting is very limited. The use of multiple radios and multiple channels proposes new challenges to broadcast protocol design. In SR-SC networks with omni-directional antenna, a transmission by a node can be received by all neighboring nodes that lie within its communication range, and this is called 'wireless broadcast advantage' (WBA). However, when multiple channels are being used, a packet broadcast on a channel is received only by those nodes listening to that channel. Simply using the SR-SC broadcast protocols without careful enhancement will result in unnecessary redundancy. For example, in figure 1, node *a* initials a network-wide broadcast process. Under the SR-SC broadcast protocols which will only choose node *b* as the forward node, totally 4 transmissions are needed to cover all nodes. However, if we choose node *b* (use channel 1) and node *f* (use channel 4) to forward packets, only 3 transmissions are sufficient to complete the broadcast.

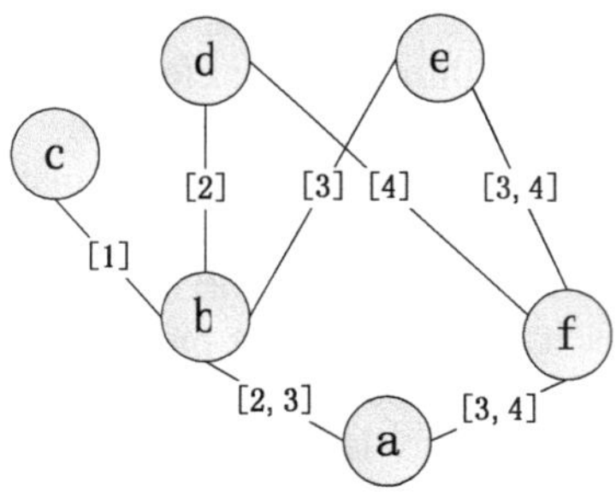

Fig. 1. A 6-node network with 4 available channels. The number in [] represents the assigned channel of the link.

In this paper, we consider to mitigate the broadcast storm problem in MR-MC networks. The objective is to achieve full coverage, and at the same time reduce the amount of redundant traffic. We show that the efficient broadcast problem in MR-MC environment can be reduced into the minimal strong connected dominating set problem of the *interface-extend* graph which extends the original network topology across interfaces. Using interface-extend graph, we describe our protocols called Multi-Channel Self-Pruning (MCSP) broadcast protocols, both in static (virtual backbone) and dynamic approach, extending the localized neighbor-knowledge-based broadcast protocols called self-pruning [6][7][8] in SR-SC environment. Our simulation results show that our MCSP protocol can significantly minimize redundant traffic. To the best our knowledge, our work is the first neighbor-knowledge-based broadcast scheme in MR-MC environment.

The rest of the paper is organized as follows. Section 2 reviews the existing broadcast schemes. Section 3 presents the network model and defines the efficient broadcasting problem in MR-MC wireless networks. In Section 4, we propose the *interface-extend* graph and describe the MCSP broadcast protocol and its properties. Simulation results are presented in Section 5, and Section 6 concludes this paper.

2 Relate Works

Williams and Camp [4] divided broadcast techniques into four categories: simple flooding, probability-based methods, area-based methods, and neighbor-knowledge-based methods. Blind flooding may be the simplest form of broadcasting. In blind flooding, upon receipt of a new broadcast packet, a node simply sends it to all its neighbors. This, however, causes serious network congestion and collision. In probability-based and area-based methods, each node estimates its potential contribution to the overall broadcasting to make a decision whether or not to forward the packet. Though smaller forward node sets can be generated, they cannot ensure the full coverage. Neighbor-knowledge-based methods are based on the following idea: select a small set of nodes to form a connected dominating set (CDS) as virtual backbone to forward packet. A node set is a dominating set if every node in the network is either in the set or the neighbor of a node in the set. In [10], it is demonstrated that broadcast scheme based on a backbone of size proportional to the minimum connected dominating set guarantees a throughput within a constant factor of the broadcast capacity.

Neighbor-knowledge-based algorithms can be divided into neighbor-designating methods and self-pruning methods. In neighbor-designating methods [11][12][13], each forward node uses a greedy algorithm to selects a few 1-hop neighbors as new forward nodes to cover its 2-hop neighbors. The forward node list is piggybacked in the broadcast packet and each forward node in turn designates its own forward node list. In self-pruning methods, each node determines it own status (forward or non-forward) according the local topology information and broadcast routing history information. Wu and Li [7] proposed a marking process and *Rule k* which can make use of local topology and priority among nodes to determine a small CDS. Peng and Lu's SBA [6] uses a random backoff delay to discover more forwarded nodes, and then uses a neighbor elimination scheme to determine the forward status for each node. A generic self-pruning scheme was proposed by Wu and Dai [8] to unify all the above self pruning protocols. In [14][15][16], some schemes are proposed for broadcasting using directional antennas.

All of the aforementioned protocols assume a SR-SC model. Broadcasting in MR-MC networks is very limited in literature. Kyasanur and Vaidya [5] simply propose to transmit a copy of the broadcast packet on every channel or use a separate broadcast channel at the expense of a dedicated interface. Qadir and Chou [17] design a set of centralized algorithms to achieve minimum broadcasting latency in multi-radio multi-channel and multi-rate mesh networks. However, the centralized approach results in a nontrivial overhead to construct and maintain the broadcast tree.

3 Network Model and Problem Formulation

3.1 Network Model

We consider a multi-radio multi-channel multi-hop network in which all nodes communicate with one another based on the IEEE 802.11 MAC protocol. It's assumed that there are totally C non-overlapping orthogonal frequency channels in the system and each node v is equipped with $I(v)$ omni-directional radio interfaces, $I(v) \leq C$. The unit disk graph model is used to model the transmission. A channel assignment scheme A assigns each node v, $I(v)$ different channels denoted by the set: $A(v) = \{a_1(v), ..., a_{I(v)}(v) \mid \forall i, 1 \leq a_i(v) \leq C; \forall i \neq j, a_i(v) \neq a_j(v)\}$, where $a_i(v)$ represents the channel assigned to ith radio interface of node v. Generally speaking, the channel assignment scheme can be classified into static, dynamic, and hybrid approach [5]. However, in our work, we currently assume that the channel assignment is given independently from our broadcasting because the channel assignment strategy is influenced by many factors, such as the unicast traffic. We further assume the channel assignment is static during the process of broadcasting and can keep the networks connected. Recognizing that channel assignment in MR-MC networks plays an important part in the actual performance, we will jointly consider channel assignment and broadcasting in our future work.

Given a channel assignment scheme A, we can use an undirected graph $G = (V, E)$ to model the MR-MC network topology, where V is the set of vertices and E is the set of edges. A vertex in V corresponds to a wireless node in the network. An edge $e = (u, v, k)$, corresponding to a communication link between nodes u and v under channel k, is in the set E if and only if $k \in A(u) \cap A(v)$ and $d(u, v) \leq r$, where $d(u, v)$ is the Euclidean distance between u and v, and r is the communication range of the transmission. Note that G may be a multi-graph, with multiple edges between the same pair of nodes, when the node pair shares two or more channels.

For each node v, $N_k(v)$ denotes the set of neighbors of v that are using channel k, and $N(v) = N_1(v) \cup ... \cup N_c(v)$ is v's neighbor set. Note that a neighbor may appear in several $N_i(v)$.

3.2 Problem Formulation

In SR-SC networks, some nodes (called forward nodes) are selected to form connected dominating set (CDS) to relay the packet. There're two approaches that can be adopted: one is the static approach, i.e. the virtual backbone method, where the CDS is constructed based on the network topology, but irrelative to any broadcasting; another is the dynamic approach, where the CDS is constructed for a particular broadcast request, and dependent on the progress of the broadcast process. In MR-MC environment, we will consider both approaches.

First, we define the forward scheme, F, as a function on V, where $F(v)$ is the set of node v's forward channels, i.e. the channels that node v uses to relay broadcast packets, $F(v) \subseteq A(v)$. We use $B = \{v \mid v \in V, F(v) \neq \varnothing\}$ to denote the forward node set. For two nodes $u \in V$ and $v \in B$, we say u is reachable from v under forward scheme F, if $u = v$ or there exists a path P: $(v_1 = v, \ldots, v_l = u)$, satisfying $v_i \in B$ and $F(v_i) \cap A(v_{i+1}) \neq \varnothing$, $i = 1, \ldots, l-1$. For example, in figure 2, under forward scheme $F = \{a[3], b, c[2,3], d[1], e, f\}$ (which means node a uses channel 3, node c use channel 2 and 3, node d uses channel 1 as forward channels, and other nodes don't forward packets), node d can reach node f and b.

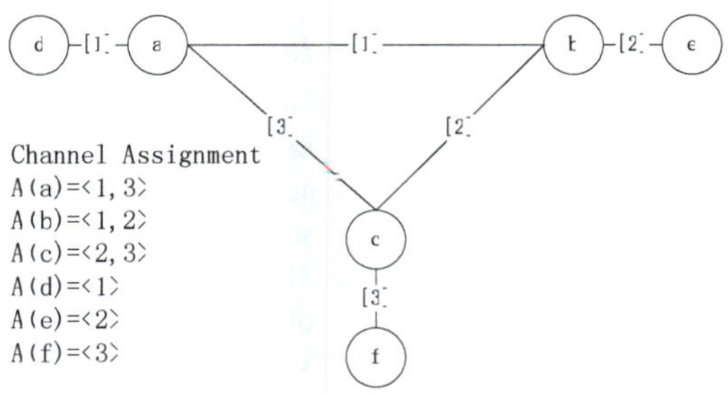

Fig. 2. An example for illustration the problem formulation

In the static approach, we say a forward scheme F can form a virtual backbone if $\forall u \in V, \forall v \in B$, u is reachable from v under F. Obviously, for any broadcast process with source node $s \in B$, every other node can receive s' packets. For broadcast process with source node $s \in V - B$, there must exist a node $u \in B$ and $F(u) \cap A(s) \neq \varnothing$, so s can send data to node u, then to other nodes. Compared with the broadcasting with source node in B, only one more transmission is needed. For example, in figure 2, both forward schemes $F_1 = \{a[1], b[2], c[3], d, e, f\}$ and $F_2 = \{a[1,3], b[1,2], c[2,3], d, e, f\}$ can form virtual backbone.

In the dynamic approach for a particular broadcast with node s as source node, we say a forward scheme F achieves full delivery if $\forall u \in V$, u is reachable from node s. For example, in figure 2, forward scheme $F_1 = \{a[1], b[2], c[2,3], d, e, f\}$ can achieve full delivery for the broadcast process with source node f.

Our aim is to ensure cover every node, and at the same time reduce the amount of redundant traffic. Next we define the transmission cost of a forward scheme F as $|F| = \sum_{v \in B} |F(v)|$, where $|F(v)|$ is the number of forward channels of node v. So our efficient broadcasting problem in MR-MC networks can be defined as follows: given networks G under channel assignment scheme A, find the forward scheme F with minimum transmission cost $|F|$ that can form a virtual backbone in the static approach or achieve full delivery in the dynamic approach. Obviously, forward scheme $F = \{a[1], b[2], c[3], d, e, f\}$ can form a virtual backbone with minimum transmission cost in figure 2.

Efficient broadcasting in SR-SC networks is a special case of the above problem with $C = 1$. It is equal to find the minimal connected dominating set (MCDS) which is proved to be NP-complete [18]. So we can get the following theorem.

Theorem 1. Given a multi-radio multi-channel network G under channel assignment scheme A, it's NP-hard to find an efficient broadcasting scheme for the networks.

4 Proposed Scheme

We first review the self-pruning protocol [8] under omni-directional SR-SC model as a trivial example solution to the above problem. In [8], each node computes the coverage of its neighborhood. A neighbor node v is covered from the view of node u if $\forall w \in N(u), w \neq v$, there exists a replace path that connects w and v via several intermediate nodes (if any) with higher priority values than the priority value of u. If all neighbor nodes are covered from the view of v, v has a non-forward node status, otherwise, it will forward the packet.

We use figure 2 to illustrate the problem of broadcasting under MR-MC model using the above scheme. From the view of node a, all its neighbors are not covered, and it will select channel 1 and 3 as forward channels to cover all neighbors. So simply using the scheme of [8] will result in the forward scheme $F = \{a[1,3], b[1,2], c[2,3], d, e, f\}$. Though forming a virtual backbone, apparently F is not the most efficient scheme.

In SR-SC networks with omni-directional antennas, a transmission by a node can be received by all neighboring nodes within its communication range. The 'wireless broadcast advantage' (WBA) makes broadcasting in SR-SC wireless networks fundamentally different from broadcasting in wired networks where the cost to reach two neighbors is generally the sum of the costs to reach them individually. This arise the shift in paradigm from the 'link-centric' nature of wired networks to the 'node centric' nature of wireless communications. However, when multiple radios and multiple channels are used, a packet broadcast on a channel is received only by those nodes listening to that channel. Motivated by the above example, we argue that we should shift the paradigm from the 'node centric' to 'channel/interface centric' in MR-MC environment. In this section, we will reduce the efficient broadcast problem in MR-MC environment into the minimal strong connected dominating set problem of the *interface-extend* graph which extends the original network topology across interfaces. Then we propose our Multi-Channel Self-Pruning (MCSP) broadcast protocol, both in static and dynamic approach and describe its property.

4.1 Extended Graph G Across Interface

In this subsection, we extend the original graph G into *interface-extend* graph G'. The basic idea here is to treat every interface of every node in MR-MC networks as a vertex of a directed graph. Using interface-extend graph, we will show the efficient broadcast problem in MR-MC environment can be reduced into the minimal strong connected dominating set problem of G'.

Definition 1. ***Interface-Extend Graph***
For an undirected connected graph $G = (V, E)$, we construct a directed graph $G' = (V', E')$, where $V' = \{v_i \mid v \in V, i = 1, ..., I(v)\}$ is the set of vertices and

$E' = \{ <v_i, u_j> | v = u \& i \neq j \text{ or } v \neq u \& (v, u, a_i(v)) \in E \}$ is the set of directed edges. We call G' the interface-extend graph of G.

Figure 3 shows the interface-extend graph of figure 2. As we can see, the original multi-edge graph is changed into a directed simple graph.

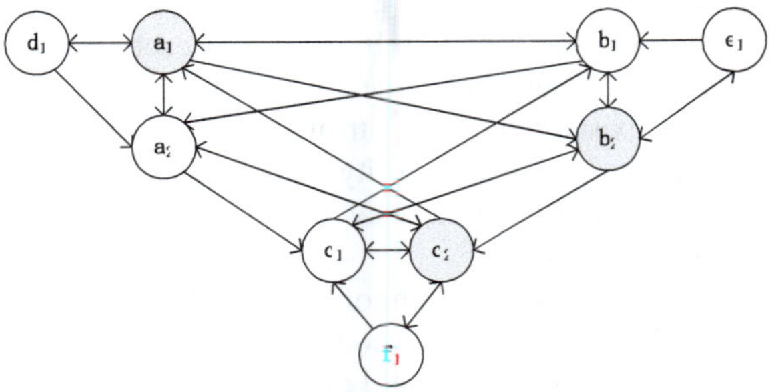

Fig. 3. Interface-extend graph of figure 2

Definition 2. ***Strong connected dominating set***
In a strong connected directed graph $G' = (V', E')$, a set $S' \subseteq V'$ is a strong connected dominating set of G' if every vertex in $V' - S'$ is dominated by at least one vertex in S' (i.e. $\forall v' \in V' - S'$, $\exists u' \in S'$ satisfying $<u', v'> \in E'$) and the deduced graph $G'[S']$ is strong connected (i.e. $\forall u' \in S', \forall v' \in S'$, there exists a directed path in $G'[S']$ from u' to v'). For example, in figure 3, vertex a_1, b_2 and c_2 form a strong connected dominating set.

Next, we use the following theorem to reduce the broadcast problem in MR-MC networks G into the minimal strong connected dominating set problem.

Theorem 2. To find the virtual backbone with minimum transmission cost for a MR-MC network G is equivalent to find the minimal strong connected dominating set of the interface-extend graph G'.

Proof. Let $F^* = \{ F \mid \textit{forward scheme } F \textit{ that can form a virtual backbone in } G \}$, $D^* = \{ D \mid D \textit{ is a strong connected dominating set of } G' \}$, and $g : F^* \rightarrow 2^{V'}$, $g(F) = \{ v'_{i,m} \mid i \in B, a_m(i) \in F(i) \}$. From definition 2, we can prove $g(F) \in D^*$ and $|F| = |g(F)|$.

For $\forall D \in D^*$, we can construct a forward scheme F, $F(v_i) = \{ m \mid v'_{i,m} \in D \}$, $\forall v_i \in V$. It's easy to verify that F forms a virtual backbone of G and $g(F) = D, |F| = |D|$.

We can also prove $\forall F_1, F_2, F_1 \neq F_2, g(F_1) \neq g(F_2)$. So g is a bijective mapping from F^* to D^* and $|F| = |g(F)|$. And the virtual backbone with minimum transmission cost in MR-MC networks G can be reduced into the minimal strong connected dominating set problem in directed graph G'. □

4.2 Multi-channel Self-pruning (MCSP) Broadcast Protocol

Theorem 2 implies that we can get the best forward scheme through seeking for the minimal strong connected dominating set of correspond interface-extend graph. The localized approximation algorithms for minimal strong connected dominating set have been studied in [9]. In this subsection, using the localized interface-extend graph, we propose our MCSP broadcast protocol, extending the self-pruning protocol in [8][9]. We redefine new priority among all interfaces using a combination of node ID and the interface/channel properties (such as channel degree $|N_k(v)|$, interface ID and so on). For other self-pruning protocols, they also can be adapted with some modification.

In MCSP, neighborhood information can be collected via exchanging "Hello" messages among neighbors. Periodically, each node broadcasts "Hello" packets on each channel. In the *kth* round of information exchange, the hello packet contains $(k-1)$-hop neighbor's channel assignment and priority information. After m round of information exchange, where generally $m \le 3$, each node can build its local $(m-1)$-hop interface-extend graph.

Figure 4 shows the MCSP algorithm for virtual backbone construction in the static approach.

Algorithms MCSP (the static virtual backbone approach)
For each node v

1. *Calculate the uncovered interface set Uncovered_Set$_i$, for every interface v_i, $i=1,\ldots,I(v)$ in the interface-extend graph G'*
2. *Uncovered_Set* = $\sum_{1\le i\le I(v)}$ *Uncovered_Set$_i$*
3. *Calculate the uncovered node set from Uncovered_Set*
4. *If Uncovered_Set =∅, node v has a non-forward status, otherwise use greedy algorithm to compute the forward channels $F(v)$ that cover all the uncovered neighbors.*

Fig. 4. MCSP algorithm in virtual backbone approach

Similar to the coverage condition of [8], we say an interface w' is covered from the view of u' if w' is an out-neighbor of u', and for any u''s in-neighbor v', there exist a replace path that connects from v' to w' via several intermediate nodes (if any) with higher priority values than that of u'. Note that here u', v' and w' are all interfaces of the interface-extend graph.

Every interface can make decision independently. However, interfaces on the same node can interact with each other without extra communication cost. So in the above algorithm, every interface first calculates its own uncovered neighbors, then we combine them and use greedy algorithms to reselect forward channels in order to save extra transmission. In the example of figure 2, if we use channel degree $N_k(v)$ as interface's priority, MCSP will mark the interface a_1, b_2 and c_2 to forward, which form a minimal strong connected dominating set of figure 3.

We also present the MCSP algorithm in the dynamic approach in Figure 5.

In the dynamic approach, every node can use broadcast routing history to further eliminate the uncovered neighbors. Broadcast routing history can be piggyback in the packet. In the example of figure 1, when node *a* initials a broadcast process, it will use channel 3 to cover all its neighbors, and then node *b* will use channel 1, node *f* will use channel 4 to forward packets. Please note that the broadcast scheme of the dynamic approach sometimes can't form a virtual backbone. In figure 1, when node *c* initials a broadcast, the above forward scheme can't achieve full delivery.

Algorithms MCSP (the dynamic approach)

1. *For source node s, use greedy algorithm to compute the forward channels $F(s)$ that cover all neighbors*
2. *For other node v, when v fist receives a new packet*

   2.1. *For each interface v_i of node v, compute the uncovered interface set*

   2.1.1. *Forward_Set$_i$ = all known forwarded interface*

   2.1.2. *Covered_Set$_i$ = Forward_Set$_i$ + { $N_{out}(v')$ | $v' \in$ Forward_Set }*

   2.1.3. *While there exists an interface $w' \in N_{out}(u')$, $u' \in$ Covered_Set and* Priority *(u') >* Priority *(v_i)*
   Covered_Set$_i$ = Coverws_Set$_i$ +{ w' },

   2.1.4. *Uncovered_Set$_i$= $N_{out}(v_i)$ - Cover_Set$_i$*

   2.2. *Compute the forward channels*

   2.2.1. *Uncovered_Set =* $\sum_{1 \le i \le I(v)}$ *Uncovered_Set$_i$*

   2.2.2. *If Uncovered_Set =∅, node v has a non-forward status, otherwise use greedy algorithm to compute the forward channels $F(v)$ that can cover all the uncovered neighbors.*

Fig. 5. MCSP algorithm in the dynamic approach

Following theorem guarantees that the MCSP protocol in the dynamic approach can assure every node eventually receives the broadcast packet of the source node s.

Theorem 3. The forward scheme determined by MCSP in the dynamic approach achieves full delivery.

Proof. We use contradiction to conclude the theorem. Suppose there exists a non-empty node set $M \subseteq V$ that every node in M is not reachable from the source node s. Let $M' = \{v_i \mid v \in M, i = 1,...,I(v)\}$. So there must exist a non-empty interface set $U' \subseteq N_{in}(M') - M'$ in which every interface in U' is reachable from one interface of the source node s. Let $u_k = \max_{v_i \in U'}\{priority(v_i)\}$. Let $v' \in N_{out}(u_k) \cap M'$. u_k doesn't forward packet, so v' is covered from local view of u_k. However, according to 2.1.1-2.1.3, v' cannot be covered, because:

1. If v' is a known forwarded interface (2.1.1) or v' is a neighbor of a known forwarded interface (2.1.2), v' is reachable from one interface of the source node s, which contradicts the assumption that $v' \in M'$.

2. If v' is an out-neighbor of a covered interface w' and Priority (w') > Priority (u_k) (2.1.3), according to the loop of 2.1.3, there exists a path *P*: $(x', y_1', y_2', ..., y_l', v')$ from a known forwarded interface x' to interface v', where Priority (y_i') > Priority (u_k), i=1,2,...,l. Because $v' \in M'$, there is at least one interface y_j' in *P* that is reachable from one interface of s but $y_{j+1}' \in M'$, $1 \leq j \leq l$, so, $y_j' \in U'$, but Priority (y_j') > Priority(u_k), which contradicts the assumption that u_k is the interface in U' that has the highest priority. □

5 Simulation

The proposed MCSP protocols have been implemented in ns-2. For comparison purpose, we implement a centralized broadcast algorithm (CBA) which is similar to what Das et al. [19] proposed. The centralized broadcast algorithm finds the forward scheme by growing a directed tree T in the interface-extend graph starting from an interface with the maximum channel degree, and adding new interface to T according to its effective channel degree (number of neighbors that are not covered). Then we can translate T to the resulting forward scheme. The centralized style makes the algorithm unpractical since it requires global information to compute the forward scheme. However, it can produce a near-optimal result. Here we use it as a substitution of the "perfect" algorithm that produces the optimal forward scheme. The original self-pruning protocols (OSP) [8] are also implemented for comparison. We evaluate the above 3 algorithms both in static and dynamic approach in terms of efficiency and reliability.

The simulated MR-MC network is deployed in a 1000m×1000m area with 20-110 nodes. Each node is equipped with four radios and twelve 2Mb/s channels are available in the system. The communication range for all nodes is 250m and the interference range is 500m. All nodes are randomly deployed and interfaces are randomly assigned with a constraint of full network connectivity. The "Hello" message interval is *1*s and every node gathers *2*-hop local topology and channel assignment information. 1-hop broadcasting routing history information is piggybacked in the broadcast packet. We use channel degree $|N_k(v)|$ followed by interface ID and node ID to break tie as interface's priority value.

Figure 6 and 7 present the comparison of CBA, OSP and MCSP in generated number of forward nodes and forward channels. Generally speaking, the static approach has a larger set of forward nodes and forward channels than dynamic approach. The centralized broadcast algorithm (CBA) has the smallest set of forward node set and forward channels. The OSP protocol have a little smaller forward nodes set than MCSP protocol, but has much more forward channels thus more transmission cost than MCSP protocol, especially when node number is large. When node number is larger than 60, MCSP can save 25-30% of OSP's transmission cost.

MCSP and other neighbor-knowledge-based broadcast protocol can cover all nodes. But because of the deficiency of the contention-based 802.11 MAC mechanism, collisions are likely to occur and cause some damage. Fig. 8 compares reliability in terms of delivery ratio. Flooding achieves almost 100 percent delivery in

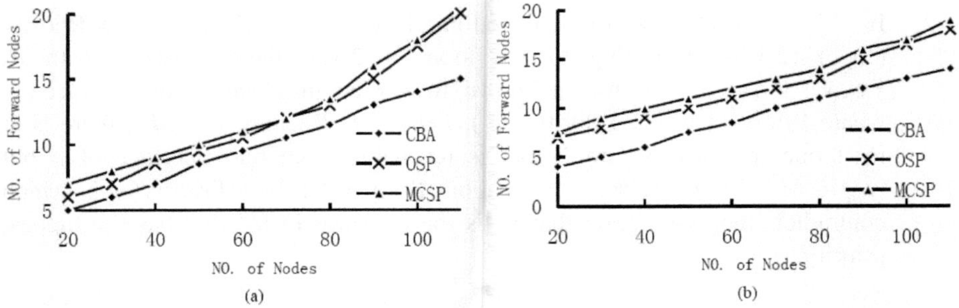

Fig. 6. Forward node number. (a) in static approach (b) in dynamic approach.

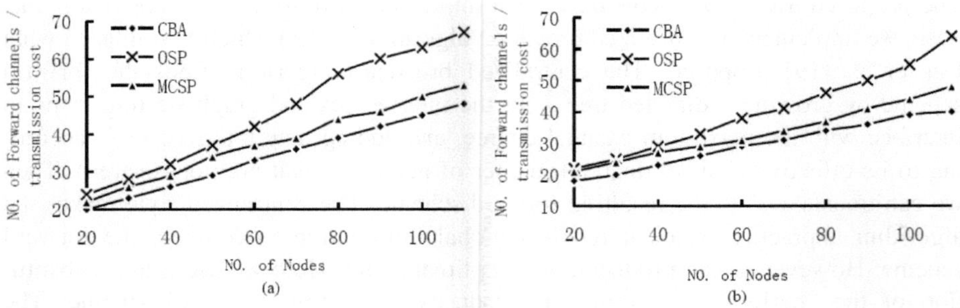

Fig. 7. Transmission cost. (a) in dynamic approach, (b) in static approach.

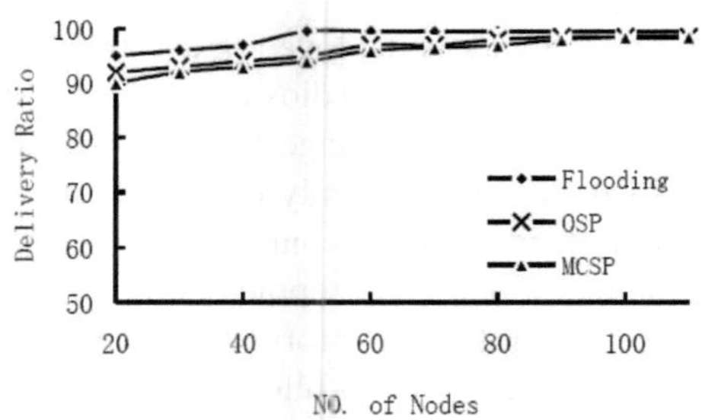

Fig. 8. Delivery ratio versus network size

networks with more than 60 nodes. The delivery ratios of MCSP and OSP are less than that of flooding, but when the node number is larger than 100, they achieve almost the same level.

6 Conclusion

This paper aims to provide a general model for broadcasting in multi-radio multi-channel multi-hop networks that uses self-pruning techniques to reduce the transmission cost. We reduce the efficient broadcast problem into the minimal strong connected dominating set of the interface-extend graph. We propose our MCSP protocol, both in virtual backbone approach and dynamic approach. The simulation result shows that our protocol can significantly reduce the transmission cost. In our future work, we will jointly consider the channel assignment and broadcasting problem that will take into account the impact of interference.

References

1. Gupta, P., Kumar, P.R.: The Capacity of Wireless Networks. IEEE Trans. on Information Theory 46(2), 388–404 (2000)
2. Perkins, C.E., Moyer, E.M., Das, S.R.: Ad hoc on-demand distance vector (AODV) routing", IETF Internet draft, draft-ietf-manet-aodv-05.txt (2000)
3. Tseng, Y.-C., Ni, S.-Y., Chen, Y.-S., Sheu, J.-P.: The broadcast storm problem in a mobile ad hoc network. Wireless Networks 8(2/3), 153–167 (2002)
4. Williams, B., Camp, T.: Comparison of broadcasting techniques for mobile ad hoc networks. In: Proceedings of MobiHoc, pp. 194–205 (2002)
5. Kyasanur, P., Vaidya, N.H.: Routing and interface assignment in multichannel multi-interface wireless networks. In: Wireless Communications and Networking Conference (2005)
6. Peng, W., Lu, X.: On the reduction of broadcast redundancy in mobile ad hoc networks. In: Proceedings MobiHoc, pp. 129–130 (2002)
7. Dai, F., Wu, J.: Distributed dominant pruning in ad hoc wireless networks. Florida Atlantic University, Technical Report TR-CSE-FAU-02-02 (2002)
8. Wu, J., Dai, F.: A generic distributed broadcast scheme in ad hoc wireless networks. IEEE Transactions on Computers (10), 1343–1354 (2004)
9. Wu, J.: Extended dominating-set-based routing in ad hoc wireless networks with unidirectional links. IEEE Transactions on Parallel and Distributed Computing (1-4), 327–340 (2002)
10. Keshavarz-Haddad, A., Ribeiro, V., Riedi, R.: Broadcast capacity in multi-hop wireless networks. In: Proc. of ACM MobiCom (2006)
11. Peng, W., Lu, X.: AHBP: An efficient broadcast protocol for mobile ad hoc networks. Journal of Science and Technology, Beijing, China (2002)
12. Lim, H., Kim, C.: Multicast tree construction and flooding in wireless ad hoc networks. In: Proceedings of MSWiM (2000)
13. Qayyum, A., Viennot, L., Laouiti, L.: Multipoint relaying: An efficient technique for flooding in mobile wireless networks. INRIARapport de recherche, Tech. Rep. 3898 (2000)
14. Hu, C., Hong, Y., Hou, J.: On mitigating the broadcast storm problem with directional antennas. In: Proc. of IEEE ICC (2003)
15. Dai, F., Wu, J.: Efficient broadcasting in ad hoc wireless networks using directional antennas. IEEE Transactions on Parallel and Distributed Systems (4), 1–13 (2006)
16. Roy, S., Hu, Y.C., Peroulis, D., Li, X.-Y.: Minimum-Energy Broadcast Using Practical Directional Antennas in All-Wireless Networks. In: Proc. of IEEE InfoCom, IEEE Computer Society Press, Los Alamitos (2006)
17. Qadir, J., Chou, C.T., Misra, A.: Minimum Latency Broadcasting in Multi-Radio Multi-Channel Multi-Rate Wireless Mesh Networks. In: Proc. of IEEE SECON, IEEE Computer Society Press, Los Alamitos (2006)
18. Garey, M.L., Johnson, D.S.: Computers and Intractability: A Guide to the Theory of NP-Completeness. W H Freeman, San Francisco (1979)
19. Das, B., Sivakumar, R., Bharghavan, V.: Routing in ad-hoc networks using a spine. In: Proc. of IC3N (1997)

Open Box Protocol (OBP)

Paulo Loureiro[1], Saverio Mascolo[2], and Edmundo Monteiro[3]

[1] Polytechnic Institute of Leiria, Leiria, Portugal
loureiro.pjg@gmail.pt
[2] Politecnico di Bari, Bari, Italy
saverio.mascolo@gmail.com
[3] University of Coimbra, Coimbra, Portugal
edmundo@dei.uc.pt

Abstract. In this paper we propose a new explicit congestion control approach, Open Box Protocol (OBP). The OBP gives sources the capacity to look inside the network and to make their congestion control decisions, based, not only on packets loss or packets delay, but also based on another kind of information. For example, the most restricted interface capacity, the available bandwidth, the RTT variations or the presence of heterogeneous transmission means. In the paper we describe the OBP and discuss its evaluation results based on ns-2 simulations. The results show the OBP´s capacity to provide traffic sources with the required information, in static or dynamic network scenarios, and to make correct and quick congestion control decisions. Also, it is visible that the OBP avoids the full queue problem and tries to keep the queues near zero occupation.

1 Introduction

Congestion control algorithms have many objectives to address. They must maintain the performance independently of the flows size distribution; they must guarantee short flow completion times in presence of a mix of flows; they have to ensure all flows with a fair share of available bandwidth; they have to make an efficient use of high bandwidth-delay links; they must have the capacity to react to sudden changes in network paths (wireless links); algorithms must be stable, robust [1], [2] and deployable in the Internet. On the other hand, implicit congestion control algorithms [3], [4], [5], [6], the most used in the Internet [7], give traffic sources poor information about network state, typically information about packets loss [3] and / or round trip time (RTT) variations [5], [6], [8]. With these two pieces of information, congestion control algorithms have to control the congestion inside the network. The great number of proposals about congestion control algorithms shows that any algorithm that only uses these two pieces of information will have limitations in making a better use of network capacity.

Internet traffic is of dynamic nature and the characteristics of this traffic are changing. Beyond the traditional applications, Internet is used by a set of new applications, for example the multimedia applications VoIP and IPTV. This kind of applications has changed the characteristics of the internet traffic and therefore the

R. Perrott et al. (Eds.): HPCC 2007, LNCS 4782, pp. 496–507, 2007.

congestion control actions must be adjusted. Besides that, the wireless networks are in expansion. In this case the congestion control decisions may not be the best when using the detection of packets loss as the only criterion to identify the congestion situations. In wireless networks the corruption in packets is more frequent than in wire networks. Moreover, the current network capacity is larger than in the past. This new Internet has new congestion control challenges and the transport protocols have to control the congestion situations and efficiently use the network resources. It is important to bear in mind that short capacity networks still exist and need to coexist with high capacity networks.

Congestion control solutions based on router collaboration have good potential because routers are the place where the congestion normally occurs. Therefore, they can quickly detect the congestion situations. Using routers, the congestion symptoms are identified in a direct way because routers detect that the amount of packets that arrives at interfaces is bigger than the capacity of the interface to process all the packets. Router based congestion control mechanisms allow quick congestion detection and the actions taken to control the congestion can be more effective. The actions to control the congestion are taken with delay in reference to the instant when the congestion began. This factor is important because the congestion duration can be shortened, the packet loss reduced and the network resources better used.

Traditional TCP's congestion control and avoidance algorithms [14] are powerful but not enough to provide good service in a lot of network conditions since they handle the network as if it were a black box [9]. The goal of our work is to create a new congestion control model, which we call Open Box Protocol (OBP), using router collaboration to identify the network resources along the path and to provide this information to end systems. Additionally, the congestion control decisions are made by the end systems by using the information received from routers. Moreover, the model must have the capacity to efficiently use the network resources, avoid congestion, reacting well to sudden changes in network paths, being easily deployed in the Internet and coexisting with other congestion control protocols.

2 Background and Related Work

Over the past few years, several solutions have been proposed to give TCP better and more network feedback, beyond packets loss information and RTT variations. In addition, the research community has been specifying alternative solutions to the TCP architecture. As the OBP, some of these models are classified in category of “modification of the network infrastructure”. They are briefly explained as follows.

The ECN [9], [10] and the Quick-Start [11] are two solutions that use the router collaboration to address the congestion control problem. With the ECN, the router collaboration is done by detecting congestion situations and by informing the end systems about this situation.

With the Quick-Start, the collaboration of routers is used to decide the value of the initial congestion window. The algorithms slow start, congestion avoidance, fast retransmission and fast recovery [14] are still used and therefore the problems associated to theses mechanisms remain.

Explicit Control Protocol (XCP) [12], [13] is designed to work well in networks with large bandwidth-delay products. The XCP generalizes the Explicit Congestion Notification proposal (ECN). Routers provide feedback, in terms of incremental window changes, to the sources in multiple round-trip times, which works well when all flows are long-lived.

Rate Control Protocol (RCP) [1] is designed to efficiently use high bandwidth-delay product networks, such as the long optical links; to be stable independently of link-capacity, round-trip times and number of active flows and to try to emulate processor sharing. Each router maintains a single fair-share rate per link. Each packet carries the rate of the bottleneck link. The RCP gives the same transmission rate to all flows. Another relevant point, at routers, the RCP uses information presents in packets, and received from end systems, to decide the fair-share rate. This means that routers cannot simultaneously receive RCP packets and packets from other transport protocols. The implementation of this model in Internet is conditioned by this factor.

The differences between the XCP, the RCP and the OBP are the kind of feedback that end systems receive and the entities that have to make decisions about congestion control. Opposite to the ECN and the Quick-Star, the OBP makes all congestion control decisions based on information received by routers. The OBP is computationally simpler than the XCP and the RCP, since routers do not have to make decisions about congestion control and only need to provide feedback information about the network state.

3 Open Box Protocol (OBP)

We will answer the following question: Is it possible for end systems to constantly see the network state between the source and the receiver? The answer is yes if we can represent the network path through a small set of variables and if we can continually put this information at sources. With this information, the sources can quickly make decisions about the efficient use of network capacity and quickly adapt to sudden network changes.

Any flow begins with an SYN packet that will be acked to give the initial values of the OBP protocol parameters. When this packet arrives at first router, the variables that represent the network state are updated. At second router the packet is evaluated and if any variable needs to be updated the router performs this exchange. Information about the network state will arrive at the end system inside the ACK packets. The end system, with this information, can make congestion control decisions for efficiently use the network resources and avoid the congestion.

3.1 State of Network Path

To represent the network path we only need the information that is used or important to make decisions about congestion control. It is not relevant to represent the network path at other levels. Fig.1 shows a network path with four routers. Each box represents the router output interface and has two kinds of information: the output interface capacity and the available bandwidth. In this example the *narrow link* is 45 Mbps (the most restricted interface capacity) and the available bandwidth is obtained in *tight link* (the most restricted available bandwidth).

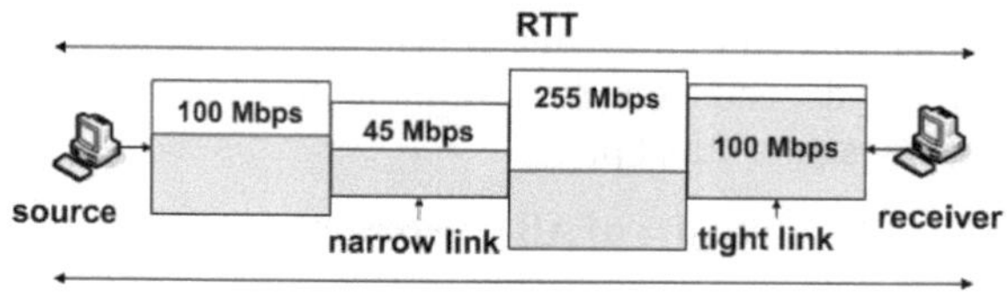

Fig. 1. Network state

To represent the network path from one end system to another one needs the following variables: narrow link (NL) - along the path, this is the link with the most limited capacity (Mbps); tight link (TL) - along the path, this is the link with the least available bandwidth (Mbps); round trip time (RTT) - time needed to send a data packet and to receive the associated ACK packet (seconds); heterogeneous path (HP) - having or not heterogeneous means along path (boolean), for example wireless links. With the exception of RTT, which is obtained by using information from TCP header, all the other variables are carried inside data packets in IP header.

3.2 Router Processing

The OBP considers that, when a new packet is inside the first router of the network path, and before the packet leaves, the router updates three variables: NL, TL and HP. When this packet is at the second router these variables are changed or not. The value of the NL is changed if the output interface capacity is less than the previous one. The value of the TL is changed if the available bandwidth is less than the previous one. The value of the HP is changed if this router has a wireless link. The available bandwidth at the TL is obtained through the following steps: at the output interface, and for short periods of time, all packets that get in are counted. Then, all of the packets that got in are divided by the period of time. This procedure gives us the used bandwidth.

3.3 Source and Destination Processing

Unlike other explicit congestion control protocols, the OBP congestion control decisions are made in sources. However, the OBP makes decisions using explicit information received from the network.

New flows begin with a high transmission rate allowed by the network. The initial transmission rate depends on the available bandwidth at TL and the interface capacity at NL, and is calculated after the sources have received the SYN-ACK packet. This method assures short completion times for short flows. Every time the sources receive an ACK packet the transmission rate is tuned. This is done using the feedback information received from the network. These transmission rate adjustments are done to come near to zero the available bandwidth and the RTTs near the minimum. The RTT near minimum means that the queue occupation is near zero. The OBP model tries to efficiently use the network path capacity and tries avoiding congestion. This means that the available bandwidth must always be near zero;

The following equations show how the transmission rate is adjusted. The initial transmission rate $W(_{t0})$ depends on the available bandwidth $AB(_{t0})$, the capacity $CN(_{t0})$ at NL and the constants α and β.

$$W(_{t0}) = \alpha * AB(_{t0}) + \beta * CN(_{t0})$$

Every time a new ACK packet is received, the feedback information inside the packet is used to make adjustments in transmission rate. These adjustments are done based on feedback information and based on an equilibrium point. The equilibrium point is updated in multiples of RTT and is calculated based on the mean of the transmission rate during the previous period (this period is equal to an RTT average). The transmission rate is updated whenever an ACK packet is received and is always around the equilibrium point. Fig. 2 shows, for a flow, an example of the behavior of these two variables: the transmission rate and the equilibrium point. In this example, at the beginning, we can see that the transmission rate is always higher than the equilibrium point. This means that there is available bandwidth to be used. After the first RTTs the source finds the equilibrium point that enables filling the entire network path between the source and the destination.

The network warnings can include negative available bandwidth - this case corresponds to a transmission rate below the equilibrium point; or positive available bandwidth – in this case corresponds to a transmission rate above the equilibrium point. Although this image is obtained from a test with 100 flows, it is visible, for this flow, that the OBP quickly finds the maximum equilibrium point and stabilizes in this value. This situation corresponds to the efficient use of network resources.

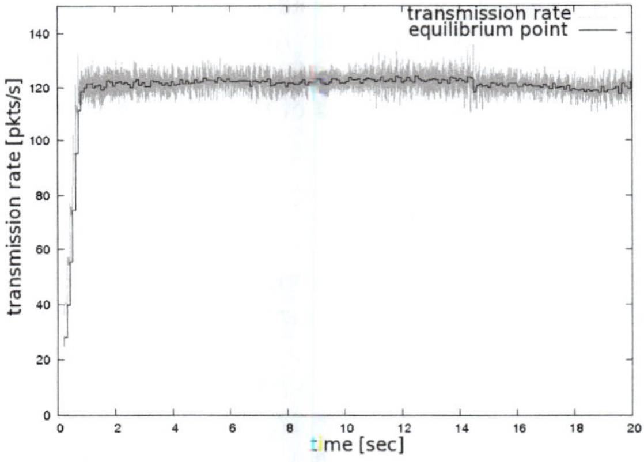

Fig. 2. Transmission rate vs Equilibrium point of a flow. This test had 100 flows, a bottleneck link equal to 1 Gbps and the RTT equal to 0.1 ms.

The transmission rate is updated every time an ACK packet is received. The transmission rate $W(_t)$ depends on the current equilibrium point $EP(_k)$, the available bandwidth $AB(_t)$, the capacity $CN(_t)$ at NL and a constant δ. Also, it is affected by the RTT if this value is different from the minimum RTT, affected by a constant μ.

$$W(_t) = EP(_k) + EP(_k)*[(\delta * AB(_t))/(AB(_t)+CN(_t))] + EP(_k)*\mu*[RTT_{min} - RTT]$$

This formula allows obtaining transmission rates around the equilibrium point. If the AB(t) is near zero the W(t) obtained is the EP(k). However, if the AB(t) received is negative or if the RTT is large, the value obtained of W(t) is less than the EP(k). In an extreme case, the W(t) obtained may be near zero, for example if the AB(t) has a high negative value or if the RTT is very high. This solution protects the network against congestion collapse, because it can instantly reduce the transmission rate to few packets [2]. This formula also enables a quick adaptation to sudden or transient events [2] as it admits changes in the transmission rate whenever ACK packet is received.

The equilibrium point is updated one time per RTT. When this occurs, the equilibrium point is updated with the mean of all transmission rates calculated whenever an ACK packet is received during the previous period.

$$EP(k) = mean(\sum W_{(t)}); \text{ during last RTT}$$

The formulas used by OBP assure that the increase in transmission rate is always decided by the feedback received from the routers. Also, the transmission rate can be updated every time an ACK packet is received. Opposite to this behavior the traditional congestion control algorithms allow the sources to increase the transmission rate without knowing if the network is near congestion.

At destination end systems, and at transport level, the ACK has three new variables, NL, TL and HP. These variables are filled with information received from data packet. Information present in ACK packet arrives at the source.

3.4 Deployable in Internet

To be deployable in internet any new congestion control approach should consider the overhead in terms of packet header size, the added complexity at end systems or routers and the interaction with other transport protocols [16]. In terms of packet size overhead this model has three new variables: NL, TL and HP. The NL and the TL are represented in units of KBps and use 3 bytes each. The HP is a boolean (1 bit).

In terms of complexity, the OBP implementation puts the load processing in the sources side. Moreover, we can have the network being used by flows that are controlled by different congestion control algorithms, among which the OBP.

4 Evaluation

To evaluate the OBP model, we have created simulations on the ns-2 simulator [15] (version 2.30) with the OBP implementation. Fig. 3 shows the network topology used in the simulations.

To evaluate the performance of OBP we did tests with long flows. We also did with flows created by Poison distribution and the flows size was defined by Pareto distribution. We equally tested the OBP and compared it with the TCP Reno, the XCP, the RCP and the TCP Reno with Quick-Star.

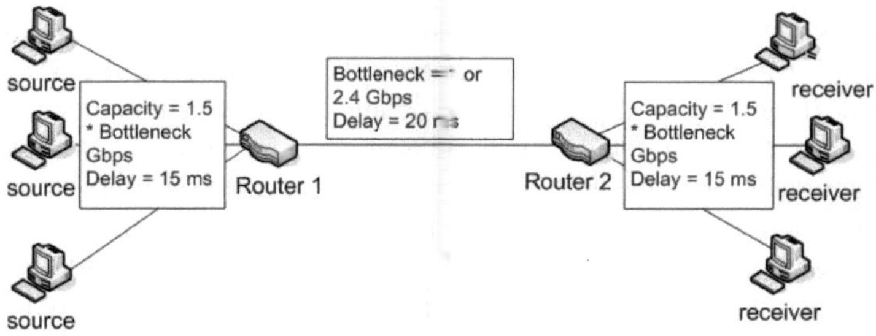

Fig. 3. Network topology

4.1 Behavior with Long Flows

In these two tests we generated 100 flows at instant 0 seconds. The bottleneck link was equal to 1 Gbps or 2.4 Gbps. The RTT was 100 ms, and the size was 1000 bytes per packet. The transmission rate used the following configurations: $\alpha = 0.0$, $\beta = 0.002$, $\mu = 1.0$, $\delta = 1.0$. The objective of these tests was to verify if the available bandwidth, at bottleneck link, was near zero; which meant that sources generated enough traffic to fill the path, and if the RTT had no oscillations, which meant that the routers' queues were near zero occupation.

Convergence and stability. We can see in Fig. 4 that, in few RTTs, the sources generated traffic to fill the path and the feedback received about the available bandwidth was near zero. The available bandwidth can be less than zero. This means that, at those instants the amount of traffic generated is greater than the capacity of output interface. In this case, packets were momentarily stored in the routers' queues.

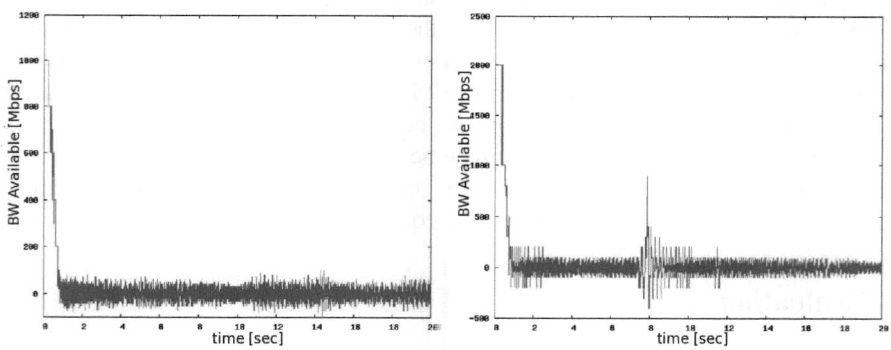

Fig. 4. Available bandwidth sent to sources by the bottleneck router, with the bottleneck interface equal to 1 Gbps and 2.4 Gbps

The state of routers' queues. The OBP also uses the time that a packet spends inside the routers' queues to make congestion control decisions. This situation allows maintaining the queues' occupation near zero, or with tendency to near zero. Fig. 5 shows that the RTT is always near 0.1 ms. By these results we can conclude that the OBP has capacity to generate traffic that tends to use the path capacity, without filling the routers' queues.

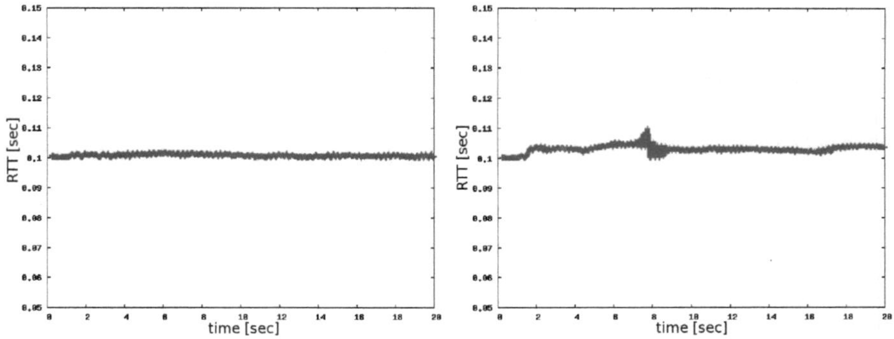

Fig. 5. Packets' RTT received at sources. The bottleneck link is 1 Gbps and 2.4 Gbps.

4.2 OBP Behavior with Variation of Traffic Characteristics

In this section our goal is to find out the OBP performance in presence of traffic with dynamic characteristics. The simulations used the Pareto distribution to define the flows' size, with the mean of 125 packets (bottleneck link 1 Gbps) and 300 packets (bottleneck link 2.4 Gbps), and the shape was 1.8. The arrival of flows was defined by the Poisson distribution with the lambda equal to 950 packets. The duration of the tests was 20 seconds and the packets had 1000 bytes.

The results of these tests are analyzed through the available, the average completion time per flow and the RTT. The average completion time represents the difference between two instants, SYN packet departure and arrival of the last data packet.

Convergence and stability. Fig. 6 shows the available bandwidth. Although these tests generated 950 new flows per second and the flows' size varied between 56 and 16340 packets (1 Gbps) or between 134 and 39216 packets (2.4 Gbps), the results show the capacity of OBP model to manage the transmission rate and to fill the network path without congesting the bottleneck link. The Poisson distribution is used to define the instant of flows creation, so there are some variations in the available bandwidth along the time. The stability of the OBP is equally confirmed through these results and the available bandwidth tends to zero.

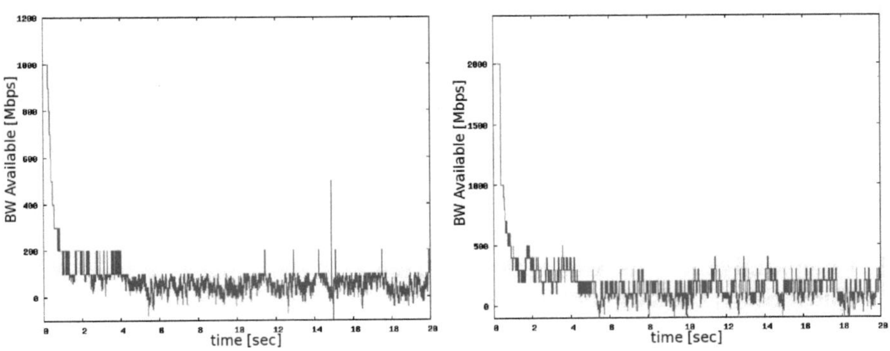

Fig. 6. Available bandwidth. The bottleneck is 1 Gbps and 2.4 Gbps.

The state of routers' queues. The RTT calculated at the sources is always near the minimum RTT. The OBP tries to use the available bandwidth, without filling the routers' queues. This is visible in Fig. 7 which shows the RTT that includes the propagation delay (0.1 ms) and the extra time spent inside routers' queues. This extra time is near zero.

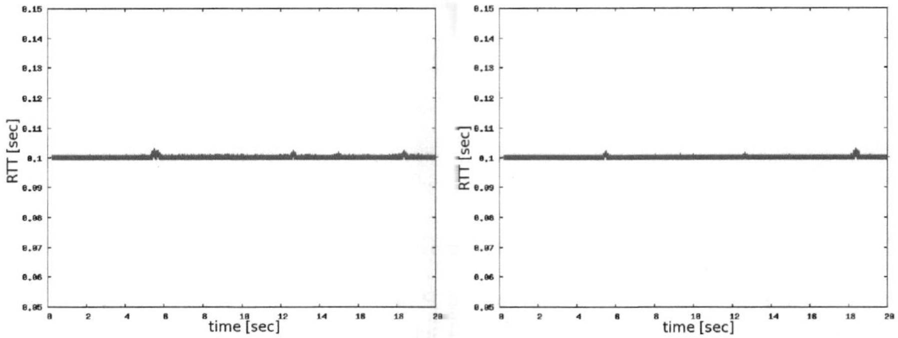

Fig. 7. Packets' RTT. The bottleneck link is 1 Gbps and 2.4 Gbps.

Average completion time. The average completion time is a good metric for elastic flows and for short flows because it gives the necessary time to transfer all data of the flow. Fig. 8 shows that there are not great variations in completion time of flows with identical size. The exceptions are related to the instants of flows' creation and, also, to the network load at those instants. The behavior of the OBP is good with the bottleneck link of 1 Gbps likewise 2.4 Gbps. These results are again analyzed in the next section where they will be compared with the results of other transport protocols.

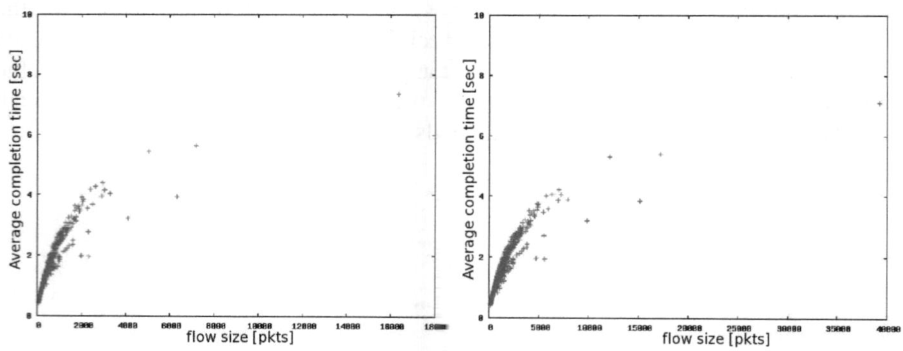

Fig. 8. Average completion time per flow size. The bottleneck link is 1 Gbps and 2.4 Gbps.

4.3 Comparative Results Among the OBP, the TCP Reno, the TCP Reno with Quick-Start, the XCP and the RCP

In this section our goal is to compare the OBP performance with other transport protocols. In these tests we compare the OBP with the TCP Reno, the XCP, the RCP and the TCP Reno with Quick-Start with the request rate equal to 100 KBps. These tests were done using the Pareto distribution to define the flows' size, with mean of 1250 packets and shape 1.8. The bottleneck link was 1 Gbps and the queue' size was 2000 packets. The arrival of flows was by Poisson distribution with lambda 95. The duration of tests was 20 seconds and packets had 1000 bytes. The results of these tests are analyzed through the average completion time per flow, the average throughput and the RTT of each packet.

Completion time and Throughput. From Fig. 9 and Fig. 10, the average completion time for the OBP is more stable than for the TCP Reno, the TCP Reno with Quick-Start and the XCP. This is visible by the smallest variations between flows with near sizes. The longer flows are also concluded more quickly by the OBP than by the TCP Reno, the TCP Reno with Quick-Start and the XCP.

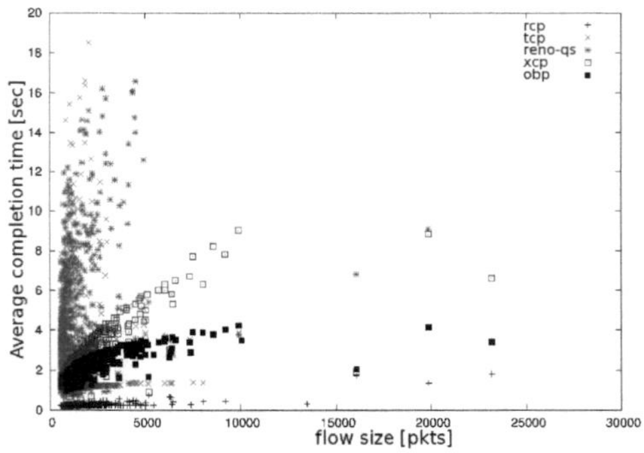

Fig. 9. Average completion time per flow

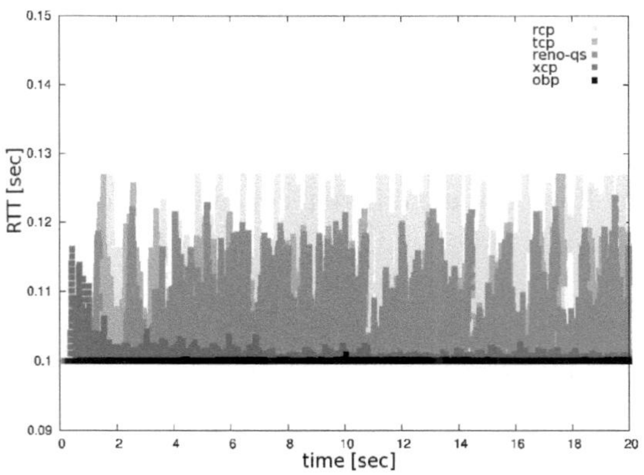

Fig. 10. Packets' RTT

The results of the RCP have to be analyzed together with the packets' RTT results. The RCP is very aggressive and defines high transmission rates. This is visible in Fig. 10 where the packets' RTT is always higher than the propagation delay value. The consequence of this aggressiveness is the packets loss, showed by Fig. 11, where the RCP' throughput is less than the OBP or the XCP.

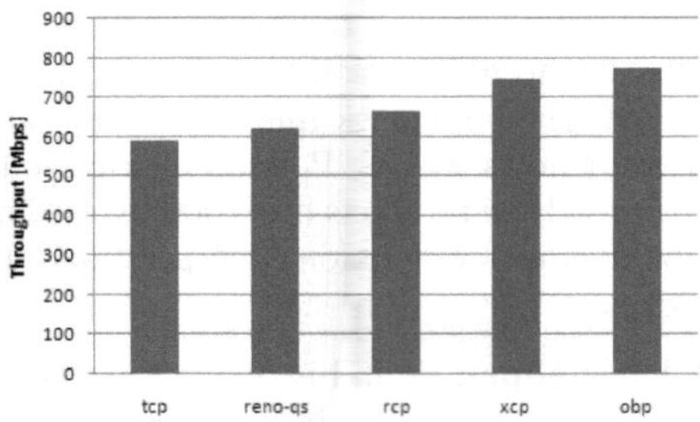

Fig. 11. Average throughput reached in 20 second

5 Conclusions and Future Work

In this paper we present Open Box Protocol (OBP). This solution enables the efficient use of the network capacity because the sources make congestion control decisions based on information about the network state. The OBP gives sources capacities to look inside the network and to make their congestion control decisions, based on the most restricted interface capacity, the available bandwidth, the RTT variations and the heterogeneous transmission means.

We have shown through analysis and experimental evaluation that the OBP has capacity to put information about the network state in the sources. Also, it is visible that the OBP can efficiently use the network bandwidth, keeping the routers' queues near zero occupation. We have equally shown that OBP can have better performance than others congestion control solutions. Moreover, the OBP implementation puts the load processing on the sources side, in opposition to other congestion control approaches, which make congestion control decisions for all flows by the same routers, as the case of the XCP and the RCP. The OBP can be used in networks where there are flows that use other congestion control protocols, because the OBP only needs to receive from routers congestion control information.

As part of our future work, we plan to test the OBP in wireless networks as in networks shared by flows that use different congestion control approaches.

References

1. Dukkipati, N., Kobayashi, M., Zhang-Shen, R., McKeown, N.: sa: Processor Sharing Flows in the Internet. In: IWQoS, Passau, Germany(2005)
2. Floyd, S.: Metrics for the Evaluation of Congestion Control Mechanisms. Internet draft - IETF, Expires (August 2007)
3. Jacobson, V.: Congestion Avoidance and Control. In: ACM SIGCOMM (1988)
4. Floyd, S., Henderson, T.: New Reno Modification to TCP's Fast Recovery. RFC 2582 (1999)
5. Jin, C., Wei, D.X., Low, S.H.: FAST TCP: Motivation, Architecture, Algorithms, Performance. Hong Kong. In: Proceedings of IEEE Infocom (2004)
6. Floyd, S.: HighSpeed TCP for Large Congestion Windows. RFC 3649 (December 2003)

7. Medina, A., Allman, M., Floyd, S.: Measuring the Evolution of Transport Protocols in the Internet. Computer Communications Review (April 2005)
8. Mascolo, S., Casetti, C., Gerla, M., Sanadidi, M., Wang, R.: TCP Westwood: End-to-End Bandwidth Estimation for Efficient Transport over Wired and Wireless Networks. In: Proceedings of ACM Mobicom, Rome, Italy (2001)
9. Salim, J.H., Ahmed, U.: Performance Evaluation of Explicit Congestion Notification (ECN) in IP Networks. RFC 2884 (2000)
10. Ramakrishnan, K., Floyd, S., Black, D.: The Addition of Explicit Congestion Notification (ECN) to IP. RFC 3168 (2001)
11. Floyd, S., Allman, M., Jain, A., Sarolahti, P.: Quick-Start for TCP and IP. RFC 4782 (2007)
12. Falk, A., Pryadkin, Y., Katabi, D.: Specification for the Explicit Control Protocol (XCP). Internet-Draft, Expires (May 9, 2007)
13. Katabi, D., Handley, M., Rohrs, C.: Internet Congestion Control for High Bandwidth-Delay Product Networks. In: Proceedings of ACM Sigcomm, Pittsburgh (August 2002)
14. Stevens, W.: TCP Slow Start, Congestion Avoidance, Fast Retransmit, and Fast Recovery Algorithms. RFC 2001 (1997)
15. The Network Simulator, http://www.isi.edu/nsnam/ns/
16. Floyd, S., Allman, M.: Specifying New Congestion Control Algorithms. Internet-Draft (December 2006)

Stability Aware Routing: Exploiting Transient Route Availability in MANETs

Pramita Mitra[1], Christian Poellabauer[1], and Shivajit Mohapatra[2]

[1] Department of Computer Science and Engineering, University of Notre Dame, Notre Dame, IN 46556
{pmitra,cpoellab}@cse.nd.edu

[2] Applications Research Center, Motorola Labs, 1295 E. Algonquin Road, MD: 2nd floor, Schaumburg, IL 60196
mopy@labs.mot.com

Abstract. The highly dynamic character of a mobile ad-hoc network (MANET) poses significant challenges on network communications and resource management. Previous work on routing in MANETs has resulted in numerous routing protocols that aim at satisfying constraints such as minimum hop/distance or low energy. Existing routing protocols often fail to discover stable routes between source and sink when route availability is transient, e.g., due to mobile devices switching their network cards into low-power sleep modes whenever no communication is taking place. In this paper, we introduce a stability aware dynamic source routing protocol (SA-DSR) that is capable of predicting the stability (i.e., expiration time) of multiple routes. SA-DSR then selects the route that minimizes hop count while staying available for the expected duration of packet transmission. Comparisons of SA-DSR to the original DSR (Dynamic Source Routing) protocol indicate a significant (up to 31%) increase in successful packet transmissions with comparable route establishment and maintenance overheads.

1 Introduction

Conventional MANET routing protocols do not consider power as a design constraint, instead, they tend to search for optimal routes in terms of delay. In such algorithms, connection between two nodes is established through nodes on the shortest path routes, which may however result in quick depletion of the battery of the nodes along the most heavily used routes in the network and eventual network partition and low connectivity.

Dynamic Power Management (DPM) in MANETs has gained huge popularity over the last decade. Mobile nodes in wireless ad-hoc networks often put their wireless network cards to sleep when they are not transmitting or receiving data. In most MANETs, wireless traffic is infrequent and recent work [1] shows that wireless network cards should be turned inactive for 50% or less of the entire lifetime to obtain a balance between optimal power-saving and sustained network connectivity. But sleep modes can lead to loss of network connectivity

R. Perrott et al. (Eds.): HPCC 2007, LNCS 4782, pp. 508–520, 2007.

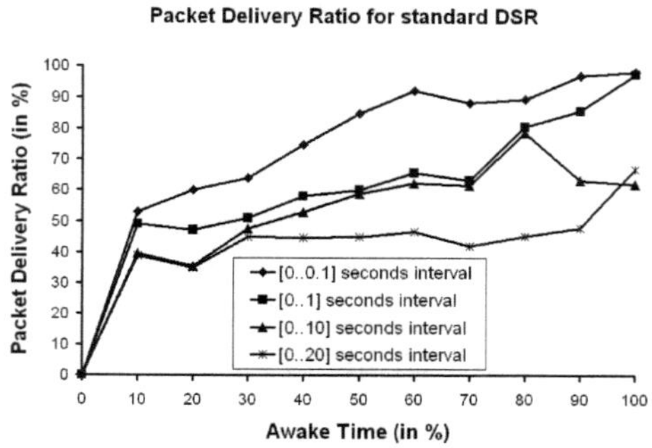

Fig. 1. DSR Packet Delivery Ratio with varying Sleep-Awake Intervals

and hence lower the packet delivery ratio. Figure 1 shows DSR's rate of successful packet transmissions in a MANET of 100 nodes with arbitrarily chosen routes and periods of network activity ranging from 0% to 100% in repeated random interval in the ranges of [0..0.1] seconds, [0..1] seconds, [0..10] seconds, and [0..20] seconds, respectively.

This paper[1] focuses on routing in MANETs with transient route availabilities, i.e., route establishment takes into consideration the expiration time and therefore the stability of a potential route. This approach, called *Stability Aware Dynamic Source Routing (SA-DSR)*, is based on the prediction of future sleep times of mobile nodes (i.e., the times when mobile nodes' DPM techniques will turn off their network cards). The goal of this approach is to introduce DPM-awareness in routing decisions and thereby to increase the number of successful packet transmissions. While SA-DSR is an extension to the well-known on-demand Dynamic Source Routing (DSR) [10], the concept of stability awareness can be added to any routing protocol. A variety of *stability predictors* can be used, including *hints* given by applications and/or the DPM mechanism. In this paper, SA-DSR monitors the *packet scheduler queue* for the network driver and predicts the next down-time of the network card. That is, SA-DSR conservatively predicts that the network card will enter its next sleep mode after all packets in the queue have been transmitted plus a driver-specific timeout value.

The remainder of the paper is organized as follows: Section 2 compares SA-DSR to previous work. Section 3 discusses the system model for the proposed routing protocol. In Section 4, we describe the details of SA-DSR, followed by simulation results in Section 5. Finally, Section 6 concludes the paper.

2 Related Work

In [2], Chin et al. present their experience of implementing two popular MANET protocols, AODV [11] and DSDV [12], and report that due to fading and transient network links both of these protocols fail to offer a stable route over any

[1] This work was made possible by NSF grant 0545899.

multi-hop network connection. They also suggest the need for using a route stability metric during the route discovery phase of MANET routing protocols. This observation has given rise to a number of algorithms that address energy management and routing in wireless ad-hoc networks. We broadly classify these algorithms into three categories: *Transmit Power Aware Routing*, *Residual Energy Aware Routing* and *Network Sleep Time-Aware Routing*. This section compares our SA-DSR protocol to these related categories.

2.1 Transmit Power Aware Routing

Such algorithms minimize the transmit power required for packet transmission or adjust the transmit power of nodes with varying network traffic and remaining node energy. In [3], Toh et al. propose the conditional max-min battery capacity routing algorithm which chooses the route with minimal total transmission power if all nodes in the route have remaining battery capacities higher than a threshold; otherwise, routes that consist of nodes with the lowest remaining battery capacities are avoided. In [4], Tarique et al. integrate two common energy management approaches: they use a load sharing approach for routing decisions and a transmit power control approach for link by link power adjustments. They employ their approach to enhance DSR [10].

Such algorithms in general select the minimum transmit power cost routes. Though some of them take the node residual energy into account, but mostly nodes along the least transmit power cost routes tend to die soon since these routes now become the most heavily used ones instead of the min-hop ones. This is harmful since the nodes which die early are precisely the ones that are needed most to maintain network connectivity. The SA-DSR protocol does find the optimal route not only based on a metric like min-hop, but also a second metric (reliability). It finds multiple stable routes for a particular pair of source and sink nodes and thus maintains the network connectivity.

2.2 Residual Energy Aware Routing

Such algorithms minimize the residual energy of the nodes and select the most residual energy or least battery cost routes. In [5], Marbukh et al. aim at preserving network connectivity by choosing routes according to the remaining battery life of nodes along the route. They use a power draining factor to accurately predict the residual battery life time. In [6], Venugopal et al. study various ad-hoc network protocols in terms of robustness and conclude that the robustness of a routing protocol is restricted by its remaining energy. Further, they present a Max-Min Energy DSR (MME-DSR) route selection algorithm to select the optimal energy route. In [7], Maleki et al. propose a lifetime prediction routing protocol for MANETs that maximizes the network lifetime by selecting routes that minimize the variance of the residual energies of the nodes in the network and include the rate of energy discharge into the cost function to improve network lifetime. They argue that mobility of nodes can affect the traffic pattern through the nodes and the recent history is a good indicator of this traffic.

These works assume that it is better to use a higher transmit-power cost route if the least transmit-power cost route consists of nodes with small amount of residual energy. Nodes usually do idle listening when there is no significant traffic. Such algorithms never completely turn off the nodes in absence of traffic. SA-DSR realizes a better approach for power-saving by utilizing a Dynamic Power Management (DPM) module that puts wireless nodes into a sleep mode when the node is not transmitting or receiving data. It finds stable routes as predicted by a DPM-aware *Route Stability Prediction Algorithm* and thus ensures acceptable network connectivity.

2.3 Network Sleep Time Aware Routing

In [8], Chia et al. propose that devices which are not currently active in any data communication may enter a sleep state, but can be powered up remotely through a signal using a simple circuit based on RF technology. Radio devices select different time-out values (*sleep patterns*), to enter various sleep states depending on their battery status and quality of service. In [9], Singh et al. employ a MAC layer protocol for PAMAS (Power Aware Multiple Access protocol with Signaling) in which nodes overhearing transmissions between two other nodes turn themselves off and wake up after an interval of time equal to the total transmission time as indicated in the RTS/CTS message exchange between the sender-receiver pairs. They deploy metrics such as *minimize energy consumed per packet* or *minimize time to network partition*, and verify these metrics with their proposed MAC layer protocol.

In our SA-DSR protocol, the sleep and awake schedule is determined from prediction of link expiration based on the queue contents of the packet scheduler and the network interface device timeout value. The DPM schedule is somewhat conservative since it ignores the possibility of more packets being added before the timeout expires.

3 System Model

DPM supports energy conservation by making mobile nodes put their wireless network cards to sleep when no data communication is taking place. A consequence of this technique is that mobile nodes will be unreachable for large periods of time. Therefore, we need to know the accurate network 'up' and 'down' times (DPM schedule) in order to introduce DPM awareness in routing decisions. Currently SA-DSR predicts the DPM schedule for mobile nodes from the queue contents of the packet scheduler and the network device timeout value. Toward that end, the SA-DSR protocol computes the minimum time to transmit all packets currently residing in the packet scheduling queue and adds the device-specific timeout value, i.e.: $t_{exp} = n * t_{send} + t_{timeout}$, where n is the number of bytes to be sent, t_{send} is the minimum time needed to transmit one byte over the medium, and $t_{timeout}$ is the amount of time the network card stays in active mode after the last transmission.

We define the *route uptime factor* (RUF) as a metric which indicates the *next earliest time* when the link between any pair of adjacent nodes on a route is going to be interrupted due to one (or both) of the nodes being put to sleep. Now we derive RUF as follows:

If we assume nodes as vertices and the links between the nodes as edges connecting them, then let G(V,E) be the graph representing the network topology where V is the set of vertices and E is the set of edges. Let

$$R_{ij} = (V_i, V_{i+1}, V_{i+2}, \ldots, V_K, V_{K+1}, \ldots, V_{j-1}, V_j) \tag{1}$$

be the route from source node V_i to V_j through intermediate nodes V_k,V_{k+1}, etc. Let Ω_{ij} be the set of all possible alive routes between V_i and V_j. The DPM sleeping schedule S_{ij} for the route R_{ij} is defined as

$$S_{ij} = ({t_i}^{off}, {t_{i+1}}^{off}, {t_{i+2}}^{off}, \ldots, {t_K}^{off}, {t_{K+1}}^{off}, \ldots, {t_{j-1}}^{off}, {t_j}^{off}) \tag{2}$$

where ${t_i}^{off}$ is the next earliest time to sleep for node V_i.We define the *link uptime vector* or L_{ij} for the route R_{ij} as

$$L_{ij} = ({t_i}^{uptime}, {t_{i+2}}^{uptime}, {t_{i+2}}^{uptime}, \ldots, {t_K}^{uptime}, {t_{K+1}}^{uptime}, \ldots, {t_{j-1}}^{uptime}) \tag{3}$$

where ${t_i}^{uptime}$ is the uptime of the link $E_{i,i+1}$ connecting nodes V_i and V_{i+1} and is defined by $\min({t_i}^{off}, {t_{i+1}}^{off})$, since uptime of a link is determined by how long the link will be alive before breaking down due to one of its end nodes going to sleep and thus essentially is expressed by the minimum of the DPM sleeping schedule of the end nodes. The *route uptime factor* RUF_{ij} for route R_{ij} can be expressed as the minimum of the link uptime vector L_{ij} along the route since it will indicate how long the route will be alive before breaking down due to the break in any of its constituent links:

$$RUF_{ij} = min({t_i}^{uptime}\ V_i \in V, {t_i}^{uptime} \in L_{ij}) \tag{4}$$

Therefore, we can summarize the problem as obtained in Problem 1.

Problem 1. *Given the next earliest time to sleep ${t_i}^{off}$ for each node $V_i \in V$ in the graph G(V,E), accumulate the set of all possible routes Ω_{ij} between nodes V_i and V_j with the corresponding route uptime factors RUF_{ij} for each $R_{ij} \in \Omega_{ij}$ and find the min-hop route R_{ij} from the set of all stable routes Ω_{ij}. If there are more than one routes with the same min-hop length, then the one with the maximum route uptime factor value is selected.*

4 Stability Aware Dynamic Source Routing

There are two key differences between SA-DSR and standard DSR. Firstly, SA-DSR introduces *DPM Awareness in routing decisions* at intermediate nodes. The discovered routes are always stable since the routing module employs a *Route Stability Prediction (RSP) Algorithm* (Algorithm 1) and proactively discards the forwarding of RREQ packets for predicted unstable routes, on the basis of the Link Uptime Vector L_{ij} contained in the RREQ packet. L_{ij} is dynamically formed by intermediate nodes during the route discovery phase. Secondly, unlike DSR, SA-DSR finds *multiple stable routes* to the target node by allowing the intermediate nodes to rebroadcast multiple RREQ packets with the same <source address, request id> pair containing distinct source routes. Although having more route discovery can contribute to an increase in the total network traffic, this increase in traffic is expected to be compensated by proactively avoiding the forwarding of RREQ packets for predicted unstable routes by intermediate nodes.

SA-DSR consists of three steps which will be described in the remainder of this section; (1) Route discovery phase, (2) Route reply phase, and (3) Route selection phase in the source node.

4.1 Route Discovery Phase

When a source node needs to send a data packet to a target node, it first searches its routing table for any entry using the target node address as the key. An entry in the routing cache contains a list of stable routes to the target node. If a routing cache entry is found, then the source node picks a route based on the *Route Selection (RS) Algorithm* (Algorithm 3). If no entry is found for the target node, then the source node initiates a route discovery for the target.

SA-DSR adds five new entry types to the RREQ packet format of standard DSR.

- Link Uptime Vector $(t_i{}^{uptime}, i \in (1, \ldots, N-1))$ for the route,
- DPM Sleeping Schedule for the last upstream node $(t_{upstrm}{}^{off})$,
- Timestamp at source node $(T_{stamp}{}^{src})$,
- Timestamp at last upstream node $(T_{stamp}{}^{upstrm})$ and,
- Per Hop Transmission Time $(T_{trans}{}^{perhop})$.

SA-DSR allows intermediate nodes to forward multiple RREQ packets with the same <source address, request id> pair if the packets contain distinct source routes. During the RREQ lookup at intermediate nodes, the 4-tuple <source address, request id, last upstream node address, partial route length> is checked with each entry in the recently seen requests list for possible match. If no match is found, then the RREQ contains a distinct source route and is eligible to be forwarded if the contained source route is predicted to be stable.

Route Stability Prediction (RSP) Algorithm: SA-DSR predicts the route stability using a link by link stability prediction. Each intermediate node executes the RSP algorithm for each received RREQ and predicts the stability of the link between itself and the last upstream node. All previous links in the source route are assumed to be stable, otherwise the previous upstream nodes would not have forwarded the RREQ packet. Thus the stability of the current link ensures the stability of the entire source route. For each received RREQ, the RSP algorithm in the $K+1$-th intermediate node V_{K+1} calculates the uptime of the link between itself and the last upstream node recorded in the RREQ and appends it to the Link Uptime Vector in the RREQ. It also calculates a running average of the per hop transmission time from the source node to itself $T_{trans}{}^{perhop}$, which takes into account varying per hop latency in case of the mobile nodes having various types of wireless interface devices (i.e., 802.11 or Bluetooth). If the uptime is less than $T_{stamp}{}^{src} + (3 * K * T_{trans}{}^{perhop})$, then the link will not be stable for the entire period of 3-way exchanges of the RREQ, the following RREP and then the data packet. Hence the intermediate node discards the RREQ. The detailed RSP algorithm is described in Algorithm 1.

Algorithm 1. Route Stability Prediction (RSP) Algorithm at the local intermediate node V_{K+1}

Input: received RREQ packet from last upstream node,
next earliest time to sleep $t_{K+1}{}^{off}$ for itself.
Output: boolean true is route predicted alive, or false

boolean alive:= true;
/*Compute Uptime $t_K{}^{uptime}$ of the link $E_{K,K+1}$*/
$t_K{}^{off}$:= packet.getSleepingScheduleUpstreamNode();
$t_K{}^{uptime}$:= min($t_{K+1}{}^{off}$,$t_K{}^{off}$);
/*Compute Trans Time from Last Upstream Node to itself*/
$T_{stamp}{}^{upstrm}$:= packet.getTimeStampAtUpstreamNode();
$T_{trans}{}^{perhop}$:= packet.getPerHopTransTime();
$T_{trans}{}^{upstrm}$:= Current System Time - $T_{stamp}{}^{upstrm}$;
/*Compute Per Hop Trans Time from Source Node to itself*/
K:= packet.getNumLinkUptimeVector();
$T_{trans}{}^{perhop}$:= $((K * T_{trans}{}^{perhop}) + T_{trans}{}^{upstrm})/(K+1)$;
/*Predict if the route will be stable*/
$T_{stamp}{}^{src}$:= packet.getTimestampAtSourceNode();
$T_{transTot}{}^{src}$:= $T_{stamp}{}^{src} + (3 * K * T_{trans}{}^{perhop})$;
if ($T_{transTot}{}^{src} > t_K{}^{uptime}$)
  alive:= false;
return alive;

If the partial source route $R_{src,K+1}$ is predicted stable, then the intermediate node V_{K+1} rebroadcasts the augmented RREQ which has the following fields as modified:

- The route record list appended by its own address.
- The Link UptimeVector augmented by the uptime of the link ($E_{K,K+1}$)
- Its own next earliest time to sleep as the DPM Sleeping Schedule of the last upstream node.
- The calculated running average Per Hop Transmission Time from the source node to itself as the Per Hop Transmission Time.
- Its current time as the new TimeStamp at the last upstream node.

The intermediate node V_{K+1} stores the <source address, request id, last upstream node address, partial route length> 4-tuple in its recently seen requests list.

4.2 Route Reply Phase

When the RREQ reaches the target node, it executes the RSP algorithm in the same way as the intermediate nodes. If the source route is predicted to be stable, the target node sends an RREP packet back to the source along the reverse path recorded in the RREQ. SA-DSR adds three new entry types to the standard DSR RREP packet format:

1. *Link Uptime Vector:* The L_{ij} for the entire source route.
2. *Earliest Down Time:* ${T_{down}}^{route}$ The minimum of all the Link Uptime Vector elements.
3. *Estimated Transmission Time:* ${T_{transTot}}^{src}$, as calculated in the route stability prediction phase.

The RSP algorithm during the route discovery phase did not take into account the length of the entire source route to the target node and hence made a prediction based on the stability of the partial route from the source node to itself. This might lead to some false predictions and hence needs to be corrected at the intermediate nodes during forwarding of the RREP packet. The intermediate nodes execute the *Modified Route Stability Prediction (MRSP) Algorithm* (Algorithm 2). Suppose there are total L links recorded in the source route of the RREQ and it arrives at the K-th intermediate node V_K (counting from the source node). V_K computes the time ${T_K}^{more}$ at which the data packet will arrive at the target node after a two-way travel of total $(L + K - 1)$ number of links, since the RREP travels another $K - 1$ links to arrive at the source node and the data packet travels L links to arrive at the target node. If this ${T_K}^{more}$ is less than ${T_{down}}^{route}$ of the route, then V_K discards the RREP.

Algorithm 2. Modified Route Stability Prediction (MRSP) Algorithm

Input: received RREP packet from last upstream node.
Output: boolean true is route predicted alive, or false

L:= packet.getNumLinkUptimeVector();
K:= packet.retNodeNum(localAddr);
/*Given the address of the current node, the above function
returns its node number in the source route*/
${T_{transTot}}^{src}$:= packet.getTransTime();
$T_{transPerHop}$:= ${T_{transTot}}^{src}/L$;
${T_K}^{more}$:= current System Time + $(L + K - 1)$*$T_{transPerHop}$;
${T_{down}}^{route}$:= packet.getEarliestDownTime();
if (${T_K}^{more} <= {T_{down}}^{route}$)
  return true;
else
  return false;

4.3 Route Selection Phase at the Source Node

A successfully received RREP in SA-DSR always ensures that the route is predicted to be stable for the estimated data transmission period (because of the conservative stability prediction used in our approach). A source node on having a successful RREP packet back, caches the learned route, the RUF for the route and estimated transmission time along the route in its routing table (the last two items are obtained from the *Earliest Down Time* and *Estimated Transmission Time* fields of the RREP respectively). The routes are stored in increasing order of their length in the routing table. When the source node sends a data packet, it exhaustively searches its routing table according to the *Route Selection (RS) Algorithm*, (Algorithm 3). The RS algorithm selects the min-hop stable route (the first entry in the routing table) for the target node. If there are more than one entry with the same min-hop length, then the route with the maximum value of RUF is chosen, since it has the highest predicted lifetime.

Algorithm 3. Route Selection (RS) Algorithm

```
Input: routing table and the target node address
Output: the min-hop route from the set of stable routes.
/*initialize pointer to the first entry of routing table*/
rt_ptr:= routingTable.returnEntry(targetAddr);
min_hop:= MAX_HOPCOUNT;
route:= rt_ptr->entry;
/*extract the min_hop stable route from the set of stable routes and
if more than one entries with same length then extract the max-RUF one*/
while (rt_ptr){
  if (rt_ptr->entry.hopCount == min_hop){
    minRUFvalue:= min(route.RUFvalue , rt_ptr->entry.RUFvalue);
    if (minRUFvalue == rt_ptr->entry.RUFvalue)
      route:= rt_ptr->entry;
  }
  if (rt_ptr->entry.hopCount < min_hop){
    min_hop:= entry.hopCount;
    route:= rt_ptr->entry;
  }
  rt_ptr:= rt_ptr->next;
}
currentRUFvalue:= route.RUFvalue;
totalTransTime:= current System Time + route.TransTime;
if (totalTransTime < currentRUFvalue)
  return route;
else
  return NULL;
```

5 Experimental Results

In order to evaluate the performance of SA-DSR, we performed simulations to study large and complex network topologies. We chose the JiST/SWANS simulation environment. JiST[2], Java in Simulation Time, is a discrete event simulator designed to run over a standard Java virtual machine. SWANS, a Scalable Wireless Ad hoc Network Simulator, is built on top of the JiST platform to provide the tools needed to construct a wireless mobile ad hoc network.

5.1 Simulation Setup

- The simulation driver gives us one sink node and a variable number of transmitting source nodes. SA-DSR will be implemented in a distributed event-based publish subscribe system where there will be only one event broker to match and deliver the events to the corresponding customers. The reason to use only one sink node is that the customers should be able to find stable routes to the single broker node.
- The simulated network area is 2500mX2500m with 100 source nodes. Each transmitting source node has a bandwidth of 1 Mb/s and attempts to transmit one data packet of length 40 bits to the sink node.
- The simulation driver simulates a random interval repeatedly for each transmitting node. Each node stays awake for the input awake percentage of the simulated random interval and notifies the routing module about its next earliest time to sleep each time when it awakes.
- Each simulation was run for 5 trials, with a full range of awake percentages from 0% to 100% with an interval of 10%. Each trial was run for 500 seconds. Each transmitting source node attempts to send one data packet to the sink node.
- Nodes intermittently turn off their network cards for power-saving in absence of significant communication, which leads to link failures and low network connectivity. SA-DSR does not consider mobility as the cause of link breakage. The nodes used in the simulation are therefore stationary.

5.2 Results

We evaluate and compare the performance of SA-DSR to that of standard DSR in terms of five metrics as follows:

In Figure 2(a), we measure *the route discovery success rate* for both of the protocols. We define route discovery success rate as the ratio of number of successful source nodes getting RREP packets back to the number of source nodes sending a RREQ packet. This graph gives an impressive performance improvement of SA-DSR over standard DSR (up to 60%). It is expected because SA-DSR uses both the min-hop and stability metrics in route discovery and finds more routes than standard DSR, which never takes into account the link down times of nodes along the source route. At awake percentage of 100%, DSR should have the same route discovery success rate, the slight difference between the DSR and SA-DSR performance is due to the difference in network connectivity during various simulation trials.

[2] http://jist.ece.cornell.edu/

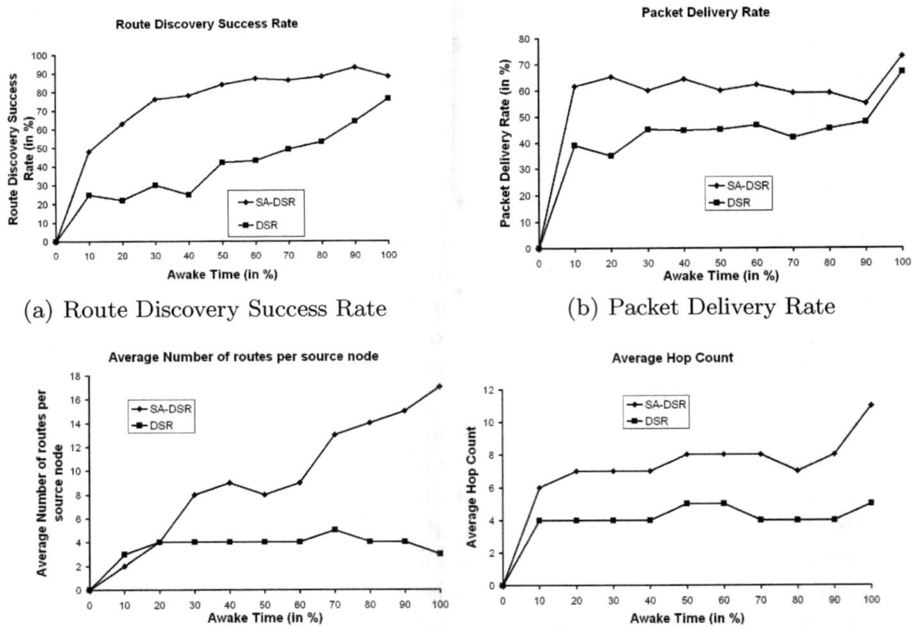

(a) Route Discovery Success Rate

(b) Packet Delivery Rate

(c) Average number of routes per source nodes

(d) Average hop count

Fig. 2. Simulation results for SA-DSR

In Figure 2(b), we measure *the packet delivery rate* of both protocols. We define packet delivery rate as the ratio of number of data packets received at the sink node to the number of data packets transmitted by source nodes having a route in their routing table after a successful route discovery. In all of the cases SA-DSR provides better packet delivery rate than DSR (up to 31%). Even with lower awake percentage such as 10% SA-DSR gives 20% better packet delivery ratio than standard DSR. Three noticeable patterns are revealed in this figure:

1. Figure 2(b) shows that for lower awake percentages standard DSR has nearly 40% of packet delivery rate, but Figure 2(a) shows that it has only 25% of route discovery success rate. Although these two measurements seem to contradict each other, this anomaly can be explained as follows. When an intermediate nodes on a particular source route forwards a RREP back in standard DSR, it updates its own routing table with the partial route from itself to the target node. Thus the intermediate nodes often get a route to the target node without having to initiate any route discovery. Since in our simulation scenario there is only one sink node, such occurances are naturally expected. But since such nodes do not initiate any route discovery themselves, they are not considered in the route discovery success rate. But they can successfully send data packets to the target node, thus leading to some cases where packet delivery rate is higher than route

discovery success rate. But SA-DSR still offers better performance than DSR instead of inflated packet delivery rate of the latter protocol.
2. Figure 2(b) shows that SA-DSR offers nearly 65% packet delivery rate, but Figure 2(a) shows that it offers nearly 85% of route discovery success rate. This drop occurs because the MRSP algorithm is not implemented at the RREP phase and stability predictions during the RREQ phase do not consider the length of the route, leading to some false route stability predictions. With MRSP implemented at the RREP phase, SA-DSR performance in Figure 2(b) should be like that in Figure 2(a).

In Figure 2(c), *the average number of routes per node* has been depicted. It shows how SA-DSR finds more routes per node (almost more than 3 times in some cases) with increasing awake percentage.

In Figure 2(d), the *average hop count* has been measured for both protocols. It shows quite expectedly that the hop count for SA-DSR is much higher than standard DSR since the former finds multiple possible alive routes instead of only the min-hop one as in the case of standard DSR.

6 Conclusion and Future Work

In this paper, we introduce Stability Aware DSR (SA-DSR), which introduces DPM awareness into the routing decisions and finds multiple stable routes to the target node. We compare it to standard DSR, which does not consider power-saving but optimizes routing for shortest delay. We show that SA-DSR provides a significant increase in successful packet transmissions with comparable route establishment and maintenance overheads.

Future work will consider the following cases:

- *Probabilistic Stability:* We understand that there might be some specific scenarios in which the RSP algorithm in an intermediate node finds all routes to the sink node to be unstable and hence discards all corresponding RREQs, so that the source node does not get any stable path to the sink node in response to its route request. But in case of non-time-critical applications even a less aggressive approach could be acceptable, i.e., an intermediate node measures a probabilistic stability of the routes and forwards all the RREQs which gives stability guarantees above the desired level. The desired level could be input by the users.
- *Multi Constraint Routing:* SA-DSR finds multiple stable routes for a particular target node and picks the min-hop one from the set of stable routes. While SA-DSR combines the two metrics stability and hop count, our future work will study how the combination of multiple metrics such as real-time deadline-aware or residual-energy aware routing with stability aware routing influences the performance of the protocol, i.e. we plan to make the protocol a multi-constraint one.

References

1. Monteiro, J., Goldman, A., Ferreira, A.: Performance Evaluation of Dynamic Networks using an Evolving Graph Combinatorial Model. In: IEEE International Conference on Wireless and Mobile Computing, Networking and Communications (2006)
2. Chin, K.-W., Judge, J., Williams, A., Kermode, R.: Implementation Experience with MANET Routing Protocols. In: ACM SIGCOMM (2002)
3. Toh, C.K.: Maximum Battery Life Routing to support Ubiquitous Mobile Computing in Wireless Ad Hoc Networks. IEEE Communication Magazine (2001)
4. Tarique, M., Tepe, K., Naserian, M.: Energy Saving Dynamic Source Routing for Ad Hoc Wireless Networks. In: Third International Symposium on Modeling and Optimization in Mobile, Ad Hoc, and Wireless Networks (2005)
5. Marbukh, V., Subbarao, M.: Framework For Maximum Survivability Routing for a MANET. MILCOM (2000)
6. Venugopal, V., Bartos, R., Carter, M., Mupparapu, S.: Improvement of Robustness for Ad Hoc Networks Through Energy Aware Routing. Parallel and Distributed Computing and Systems (2003)
7. Maleki, M., Dantu, K., Pedram, M.: Lifetime Prediction Routing in Mobile Ad Hoc Networks. In: Proceedings of IEEE Wireless Communications and Networking Conference (2003)
8. Chiasserini, C.F., Rao, R.R.: A Distributed Power Management Policy for Wireless Ad Hoc Networks. In: Proceedings of IEEE Wireless Communication and Networking Conference, IEEE Computer Society Press, Los Alamitos (2000)
9. Singh, S., Woo, M., Raghavendra, C.S.: Power-Aware Routing in Mobile Ad Hoc Networks. In: Proceedings of MobiCom 1998 Conference (1998)
10. Johnson, D.B., Maltz, D.A.: Maltz: Dynamic Source Routing in Ad Hoc Wireless Networks. In: Imielinski, T., Korth, H. (eds.) Mobile Computing. ch. 5, pp. 153–181. Kluwer Academic Publishers, Dordrecht (1996)
11. Perkins, C., Belding-Rower, E.M., Das, S.: Ad Hoc On Demand Vector (AODV) Routing. IETF Internet Draft, draft-itef-manet-aodv-09.txt (2001)
12. Perkins, C., Bhagwat, P.: Highly Dynamic Destination Sequenced Distance Vector Routing (DSDV) for Mobile Computers. In: Proceedings of ACM SIGCOMM, ACM Press, New York (1994)

Reliable Event Detection and Congestion Avoidance in Wireless Sensor Networks

Md. Mamun-Or-Rashid, Muhammad Mahbub Alam, Md. Abdur Razzaque, and Choong Seon Hong*

Networking Lab, Department of Computer Engineering, Kyung Hee University
Giheung, Yongin, Gyeonggi, 449-701 South Korea
{mamun,mahbub,razzaque}@networking.khu.ac.kr,
cshong@khu.ac.kr

Abstract. Due to dense deployment and innumerable amount of traffic flow in wireless sensor networks (WSNs), congestion becomes more common phenomenon from simple periodic traffic to unpredictable bursts of messages triggered by external events. Even for simple network topology and periodic traffic, congestion is a likely event due to time varying wireless channel condition and contention caused due to interference by concurrent transmissions. Congestion causes huge packet loss and thus hinders reliable event perception. In this paper, we present a congestion avoidance protocol that includes *source count* based hierarchical medium access control (HMAC) and weighted round robin forwarding (WRRF). Simulation results show that our proposed schemes avoid packet drop due to buffer overflow and achieves more than 90% delivery ratio even under bursty traffic condition, which is good enough for reliable event detection.

1 Introduction

Wireless Sensor Networks (WSNs) are densely deployed for a wide range of applications in the military, health, environment, agriculture and smart office domain. These networks deliver numerous types of traffic, from simple periodic reports to unpredictable bursts of messages triggered by sensed events. Therefore, congestion happens due to contention caused by concurrent transmissions, buffer overflows and dynamically time varying wireless channel condition [1][2][3]. As WSN is a multi-hop network, congestion taking place at a single node may diffuse to the whole network and degrade its performance drastically [4]. Congestion causes many folds of drawbacks: (i) increases energy dissipation rates of sensor nodes, (ii) causes a lot of packet loss, which in turn diminish the network throughput and (iii) hinders fair event detections and reliable data transmissions. Therefore, congestion control or congestion avoidance has become very crucial to achieve reliable event detection for the practical realization of WSN based envisioned applications.

* This research was supported by the MIC , Korea, under the ITRC support program supervised by the IITA, Grant no-(IITA-2006-(C1090-0602-0002)).

R. Perrott et al. (Eds.): HPCC 2007, LNCS 4782, pp. 521 – 532, 2007.

We find that one of the key reasons of congestion in WSN is allowing sensing nodes to transfer as many packets as they can. This is due to the use of opportunistic media access control. The high amount of data transferred by sensing nodes can overwhelm the capacity of downstream nodes, particularly the nodes near to sink. Hence, we propose *source count* (defined in section 3) based hierarchical medium access control (HMAC) that gives proportional access, i.e. a node carrying higher amount of traffic gets more access to the medium than others. Therefore, downstream nodes obtain higher access to the medium than the upstream nodes. This access pattern is controlled with local values and is made load adaptive to cope up with various application scenarios.

Congestion due to buffer overflow is not insignificant [9][10]. To avoid congestion, before transmitting a packet each upstream node must be aware whether there is sufficient free buffer space at the downstream node. To implement this notion, we restrict an upstream node from delivering packets when its downstream node has not sufficient amount of free buffer space. This is achieved by our proposed *source count* based weighted round robin forwarding (WRRF).

Even though the sensor network can tolerate a certain percentage of packet loss, data transmission in such network is termed as unreliable if the packet delivery ratio decreases to very low value so that the event can not be detected reliably. We thus seek an efficient way to avoid congestion within the sensor network to ensure good delivery ratio for reliable event detection. Integrated effort of our proposed *source count* based HMAC and WRRF reduces packet drop due to collision and avoids packet drop due to buffer overflow and finally achieves more than 90% delivery ratio which is good enough for reliable event perception.

The rest of the paper is organized as follows: section 2 briefly discusses the related works, section 3 articulates the network model and assumption, section 4 describes our proposed protocol, section 5 discusses the performance evaluation and finally we conclude in section 6.

2 Related Work

A good number of transport layer protocols have been proposed for wireless sensor network [1]-[10]. These works aimed to provide reliability guarantee either by congestion detection and control or by congestion avoidance [1][9][14]. Few of these techniques are described below:

ESRT [4] allocates transmission rate to sensors such that an application-defined number of sensor readings are received at a base station, while ensuring the network is not congested. On reception of packets with congestion notification bit high, sink node regulates the reporting rate by broadcasting a high energy control signal so that it could reach to all sources. This high powered congestion control signal may disrupt some other transmissions. Also the assumption of congestion notification by the sink node is very optimistic. CODA [1] uses a combination of the present and past channel loading conditions and the current buffer occupancy, to infer accurate detection of congestion at each receiver with low cost. As long as a node detects congestion, it sends backpressure messages to upstream nodes for controlling reporting rate hop-by-hop. It is also capable of asserting congestion control over multiple sources from a single sink in the event of persistent congestion. Even though it overcomes some of the limitations of ESRT [4], it doesn't consider the event fairness and packet reliability at all. PSFQ [11] is scalable and reliable transport protocol that deals with strict data delivery guarantees rather than desired event reliability as it is done in ESRT. However, this approach involves highly specialized parameter tuning and accurate timing configuration that makes it unsuitable for many applications. Also PSFQ has several disadvantages (i) it cannot detect single packet loss since they use only NACK, (ii) it uses statically and slowly pump that result in large delay and finally (iii) it requires more buffer as hop-by-hop mechanism is used. As defined in Many-to-One Routing [2], event fairness is achieved when equal number of packets are received from each node. In this proposal, individual nodes divide its effective available bandwidth equally amongst all upstream nodes. This, in turn ensures fairness. Several disadvantages are including: (i) It provides no reliability guarantee. (ii) The effective throughput may decrease due to implementation of ACK in transport layer. RMST [12] is a transport layer paradigm designed to complement directed diffusion [13] by adding a reliable data transport service on top of it. It's a NACK based protocol like PSFQ, which has primarily timer driven loss detection and repair mechanisms. It does not provide with any congestion control mechanism. TARA [14] discusses the network hotspot problem and presents a topology aware resource adaptation strategy to alleviate congestion in sensor network.

In our approach two key reasons of packet loss have been taken into account: loss due to collision and loss due to buffer overflow. Our proposed HMAC scheme takes care of hierarchical medium access and thereby reduces packet drops due to collision. WRRF controls the number of packets to be received from upstream nodes in each round (single-hop control). Round operation is controlled by estimating buffer status at each individual downstream node using exponential moving average (EWMA). A downstream node allows packet from its upstream nodes only if there is available buffer and thereby avoid drops due to buffer overflow.

3 Network Model and Assumptions

We consider a network of N sensing nodes, deployed with uniform random distribution over an area A . Node density is defined as $\rho = N / A$. Therefore, the approximate number of nodes within the sensing radius of a particular event is calculated as:

$$N_s = \pi \rho R_s^2 \qquad (1)$$

Where, R_s is the sensing range of each node. We consider a single sink in the network, placed at anywhere within the terrain. All sensors are static; we do not consider mobile sensors that form a dynamic ad-hoc network.

All sensors are static and the network is homogeneous i.e., all nodes have the same processing power and equal sensing and transmission range. Data generation rate of each sensing node is also assumed to be equal. Since receiving explicit ACK from the downstream node incurs huge overhead on energy constraint sensor nodes [6][7] [8], we have used snoop-based implicit acknowledgement. Modified CSMA/CA is used as MAC protocol. We do not need binary exponential backoff, as we exclude ACK packets in response of reception of data packets. Therefore, the backoff value is uniformly random within the range $0 \sim (W-1)$, where W is the size of contention window. The value of W is dynamically updated as traffic load varies (subsection 4.1). We consider that all data packets have the same size and the amount of buffer at each node is represented by the number of packets it can store. We also consider that congestion does not occur if there is no data transmission in the network.

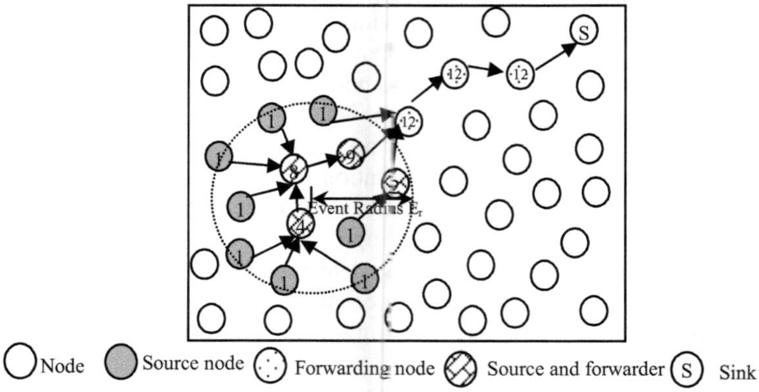

Fig. 1. Event to sink routing path and source count value of nodes

We have considered a tree based hierarchical static routing protocol. Hence, the route from each source to the sink is predetermined and unchanged during the data delivery of a certain event. The network is event driven; nodes within the event radius generate traffic and the sensed data eventually reach to the sink, forwarded by intermediary downstream nodes. Downstream node may also generate its own data by sensing the vicinity of its sensing range. As defined in subsection 1.1, *source count* value of any node *i*, denoted as SC_i, is the total number of source nodes for which it is forwarding data. If S_u represents the set of single hop upstream nodes of *i*, its *source count* value is calculated as follows

$$SC_i = \sum_{\forall k \in S_u} SC_k + I \tag{2}$$

Where, I is the source indicator function; $I=1$, if the node *i* is generating data, 0 otherwise. We do not use any type of control packets in designing different schemes of the proposed protocol. Since a downstream node requires knowing its *source count* value whenever it has some data packets to send, it is sufficient to propagate *SC* value along with the data packet. While transmitting data packets, each upstream node inserts its *source count* value in the packet header and the downstream node can easily calculate its *SC* value using Eq. 2. An upstream node learns the *source count* value of its downstream by snooping packets transmitted by the latter. Note that, a transient state exists between the event occurrence and the *SC* values of all downstream nodes are being stabilized. *SC* value of a downstream node is stabilized whenever it receives at least one packet from all of its upstream nodes and therefore the network enters into steady state when the sink node receives at least one packet from each source node. Since the duration of transient state is very short (less than a second in our simulation), the effectiveness of the proposed protocol is not hampered. It is notable that, *source count* values of each node along the routing path are updated without transferring any additional control packets. Fig. 1 shows the updated *source count* values for each node in the routing path. This *source count* parameter works as a driving entity for all schemes of our proposed protocol.

4 Proposed Protocol

The key idea of our proposed congestion avoidance protocol is as follows. We do not allow any node the opportunistic access to the medium; rather we grant proportionate access that does not overwhelm the capacity of downstream nodes. From each downstream node, we allow upstream nodes to transfer their weighted-share number of packets in a round robin fashion. An upstream node forwards packets if its downstream node has sufficient buffer space.

4.1 Hierarchical Medium Access Control (HMAC)

Due to many-to-one routing generalization [2] in sensor network (shown in Fig. 1), downstream nodes have to carry more traffic than upstream nodes. Therefore, as simple CSMA/CA gives equal opportunity to all contending nodes, it might cause huge loss of packets due to collision and increase media contention.

Sensing nodes must not transfer so high amount of data that can overwhelm the capacity of downstream nodes, particularly the nodes near to sink. Hence, we propose hierarchical medium access control (HMAC) that gives proportional access based on *source count* value, i.e. a node carrying higher amount of traffic gets more accesses than others. Each node then calculates its contention window using equation (3).

$$W(i) = CW_{\min} \times \frac{N_s}{SC_i} \tag{3}$$

We consider N_s, a global system parameter, in calculating W as because it has noteworthy impact in handling bursty traffic condition as well as aggregated load on downstream nodes. As we use implicit ACK, the transmitting nodes do not use binary exponential backoff procedure; instead they choose a uniformly random backoff value using equation (4).

$$backoff = rand(0 \sim (W-1)) \tag{4}$$

This proportional media access significantly reduces the media contention and congestion due to collision. The value of W calculated from equation (3) is the approximate value of N_s, which may not be equal to the approximate number of nodes in a practical sensor network all the times. This may lead to low medium utilization or overshoot the network capacity. To ensure the optimal contention window value, we incorporate the packet loss rate of each individual node for calculating W. So, load adaptive equation is expressed as follows:

$$W(i) = CW_{\min} \times \frac{N_s}{SC_i} \times \frac{1}{\alpha} \tag{5}$$

Where, α is a scaling factor that ranges from 0.5 to 1.5 based on channel contention. When a node has a packet to transmit, it gets the *backoff* value using Eq. 4 and transmits the packet when *backoff* value reduces to zero. Therefore, if the number of contending neighbors of a transmitting node is very low, lower value of α simply increases the medium access delay and reduces the network throughput. On the other hand, if the number of contending neighbors of a transmitting node is very high, a higher value of α increases the collision probability and thereby increases packet loss. The value of α is initialized to 1, which nullify its effect. Later on, to ensure efficient medium utilization, the value of α should be set carefully. A sharp increase or decrease of the value of α may also hinder the throughput of the network. Therefore, we have divided the range of α into 10 discrete values. Each node can easily identify the number of contending neighbors during the time between *backoff* assignment and packet transmission. In a round, if a node experiences collision with the current number of contending neighbors, it decreases the current value of α by 0.1 for the next round. Accordingly, if the node does not experience any collision, the α value is increased by 0.1.

4.2 Weighted Round Robin Forwarding (WRRF)

Even though the HMAC gives more accesses to nodes with higher *source count* values than others, it does not guarantee that upstream nodes will transfer their weighted-share amount of packets. Probability exists that a node may get multiple chances in succession and injects packets more than its share, depriving other nodes and worsening the fairness. Highly imbalance number of packets received at sink from different nodes engenders several potential problems: (i) event detection may be biased, (ii) nodes' battery will be drained quickly resulting early node failure and (iii) increased channel contention.

To address this issue, we propose *source count* based weighted round robin forwarding (WRRF) that implements hop-by-hop fair packet scheduling. In each round, a downstream node allows all of its upstream nodes to transmit their weighted-share amount of packets. If the downstream node allows R packets to be transmitted by all of its upstream nodes in a round and SC_d is its *source count* value, the weighted-share number of packets of any of its upstream node i, S_w, is calculated as follows:

$$S_w(u) = R \times \frac{SC_u}{SC_d} \tag{6}$$

Round is controlled by downstream node and a single bit field, *round control*, is appended with each packet forwarded from it. Rounds cycle with 0 and 1 values, a new round is started with 0 in *round control* bit and it remains unchanged until downstream node receives weighted-share number of packets from all of its upstream nodes. Upstream nodes get round value by snooping packets transmitted by its downstream node. An upstream node restricts itself from transmitting any further packet if

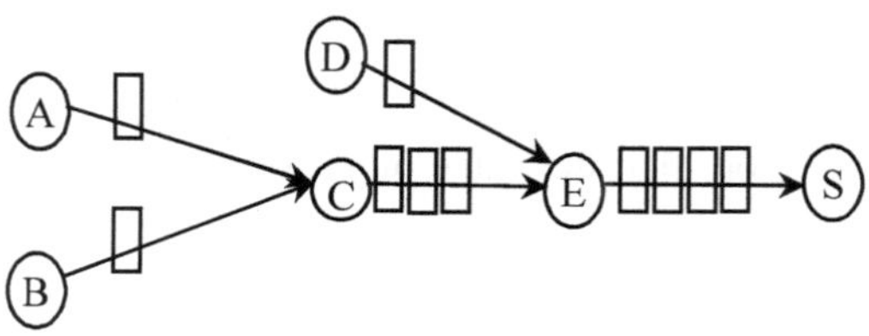

Fig. 2. Weighted round robin packet forwarding

it completes its share in that round. Thereafter, the downstream node switches *round control* bit to 1 and transmits further packets.

For instance, we consider R=6 in Fig. 2, the downstream node C allows 2 packet from A and 2 packet from B in each round. Similarly E allows 4 packets from C and 1 packet from D in a round. The value of R and the *round control* mechanism are much related with the amount of buffer space in a node and are explicitly discussed in section 4.3.

Thus controlling packet transmission in round robin fashion provides fair packet delivery in each routing path. It decreases channel contention and also takes care of nodes energy by allowing equal packets from all nodes.

4.3 Congestion Avoidance

The proposed mechanism avoids packet drops due to congestion by not allowing upstream nodes to transmit if there is not available buffer. This is achieved by controlling the transmission round explained in section 4.2. A downstream node changes the round based on its buffer status. In a round, the downstream node allows R packets to be transmitted by all the upstream nodes. Given the condition that the downstream node has proportionally higher probability to access the medium, even then the downstream node may not be able to forward all R packets. So, in the next round, the downstream node will have some packets from the previous round, which might cause congestion within few successive rounds. Therefore, two important things to be considered are:

- The value of R and its relations with buffer size
- After getting R packets from the upstream node whether the downstream node immediately change the round or not. More specifically, should the downstream node forward all R packets before changing the round?

Definitely, this will guarantee no packet drop due to congestion. However, such strict round robin forwarding will block the upstream nodes to transmit, even though there may be empty buffer in the downstream node. Therefore, it is important to know R_e the number of empty buffers in the downstream node, where $R_e \le R$ is required to change the current round. If the number of packets forwarded by the downstream node is F, when it receives the last packet of that round from one of the downstream nodes, then the downstream node should have at least $R_e = R - F$ empty buffers.

Similarly, in the next round, the downstream node requires R_e empty buffers to change the round and thus can avoid congestion. But R_e is not a constant and depends on the network load. We therefore find the value of R_e using the exponential weighted moving average (WEMA) and given by:

$$R_{e\_estimated} = (1-\beta)\times R_{e\_estimated} + R_{e\_current} \tag{9}$$

Finally, as long as there are not at least $R_{eestimited}$ number of empty buffers, the downstream node will not change the transmission round. This ensures near to zero packet drops and at the same time ensures efficient buffer utilization.

5 Performance Evaluation

To evaluate the performance of our proposed schemes we have performed extensive simulations using *ns-2*[15]. Proposed protocol implementation includes a tree based hierarchical static routing protocol, HMAC and WRRF. The tree based hierarchical static routing module creates parent (downstream) and child (upstream) hierarchy among the nodes in the network. The routing tree is constructed using Warshall's algorithm so that the sensed data could reach the sink with shortest number of hops. It may be mentioned here that, the choice of downstream nodes does not depend on any traditional parameters of sensor network routing *e.g.*, energy or delay. An event is generated at a random location and we have assigned source IDs randomly to the nodes within the event radius. We have modified the CSMA/CA MAC implementation of *ns-2* as follows. We have added two additional fields in the MAC frame header: *source count* and *round control*. At each downstream node, virtual queues are created using link list data structure for storing packets from individual upstream nodes. Dissemination and update operation of *source count* value is described in section 2. When a node has data to transmit, HMAC takes care of assigning its *backoff* value and WRRF calculates the weighted-share number of packets to transmit.

Following matrices are used to realize the performance of proposed schemes:

- Delivery Ratio: It indicates the ratio of number of packets sent by the sources to the number of packets successfully received at sink.
- Packet Drop: The ratio of dropped packets due to collision and buffer overflow to the number of sent packets
- Efficiency: Number of hops traveled by each successful reception of a packet at the sink divided by the total number of transmission required for the packet in entire path.
- Energy Dissipation: Amount of energy dissipated per node per unit time, measured in Joule

We have compared our protocol with the following four mechanisms:

- No Congestion Control (NoCC): Under this scheme packets are transmitted using a hierarchical routing without controlling transmission rates at the sources and forwarders.

- No Congestion Control with Implicit ACK (NoCC-IA): This is the same scheme as NoCC without RTS-CTS-DATA-ACK handshake. It uses snoop based implicit ACK.
- Backpressure: It is a hop-by-hop rate control based congestion control mechanism explained in CODA [1]. If a sensor gets congested, the mechanism advertises congestion using explicit congestion notification bit to reduce the transmission rate of its upstream sensors by a factor of 0.5. If an upstream neighbor is a data source, the neighbor reduces the rate at which it generates new data by the same percentage.
- Proposed Protocol: It includes load adaptive hierarchical medium access (HMAC) and weighted round robin forwarding (WRRF).

5.1 Simulation Setup Parameters

Table 1. Simulation parameters

Parameter	Value
Total Area	100m X 100m
Number of nodes	100
Initial Energy	5 Joule/Node
Transmission power	5.85e-5
Receive signal threshold	3.152e-20
Data rate	300 kbps
Transmission Range	30m
Packet size	64 bytes
Initial α value	1.0
Range of α	0.5~1.5
Buffer size	20
Data Sources	1~15
Offered load	4~6 pkts. per sec. (pps)
Sink location	[3.6148, 99.2246]
Simulation Time	50 Sec

5.2 Simulation Results

Fig. 3.a shows collision drop rate for various protocols. In case of NoCC, drop rate increases sharply with the increased data sources. Since nodes within the event radius generate bursts of data packets and opportunistic medium access (CSMA/CA) provides equal share to all nodes, huge collision occurs and it becomes severe with the increase of number of sources. Algorithms using backpressure also cannot reduce the collision drop rate by a significant amount, since they also use opportunistic medium access for all nodes. On the other hand, the proposed protocol exhibits very less collision drop due to the use of hierarchical and controlled medium access which is implemented by the integrated employment of HMAC and WRRF.

Buffer drop rates of different protocols are plotted in Fig. 3.b In fact, buffer drop is relatively much less than collision drops [5]. Uncontrolled rate of transmission mechanism in NoCC is the main reason of higher packet drops. Backpressure

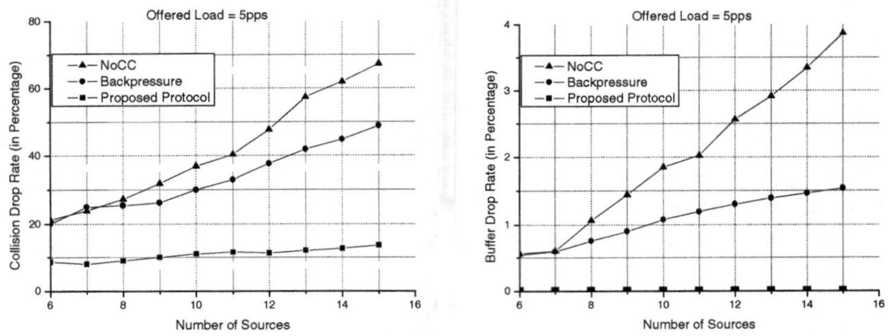

a. Drops due to collision **b.** Drops due to buffer overflow

Fig. 3. Packet drop rate due to collision and buffer overflow with increasing number of sources

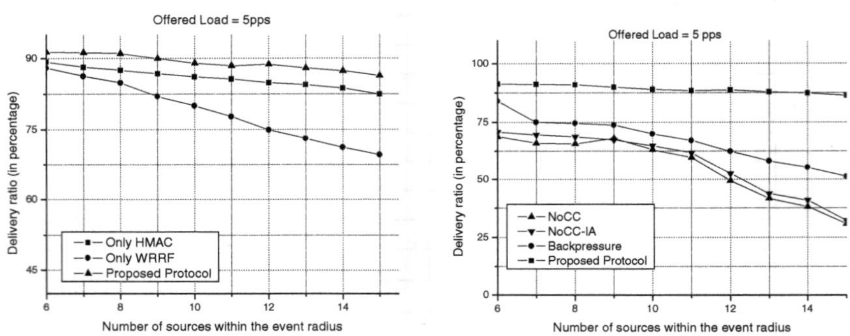

a. Comparison among individual schemes **b.** Comparison with other protocols

Fig. 4. Delivery ratio of individual schemes as well as their integrated effort and other protocols with a load of 5 pps

algorithms experience lower packet drops than NoCC, since they use rate control mechanism to control congestion. However, our algorithm completely avoids packet loss due to buffer overflow.

Fig. 4.a depicts the effectiveness of proposed HMAC and WRRF schemes individually and their combined effort in terms of delivery ratio. Only WRRF can achieve the least delivery ratio since in this case all nodes equally contend for the medium irrespective of their source count values and, thereby, increase the collision drop rate. HMAC exhibits lower delivery ratio than the combined effort of HMAC and WRRF as it does not care about the buffer drops. Our proposed protocol can achieve around 90% delivery ratio. According to Fig. 3.a and Fig. 3.b, since both NoCC and backpressure algorithms experience comparatively large amount of collision and buffer drops, their ultimate delivery ratio is very poor. While, the integrated employment of

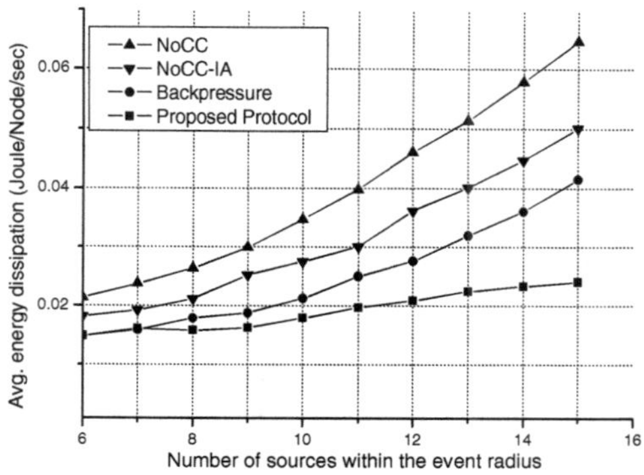

Fig. 5. Average energy dissipation with a load of 5 pps

HMAC and WRRF in our proposed protocol provides better delivery ratio than other protocols (NoCC, NoCC-IA and Backpressure) as depicted in Fig. 4.b.

Average energy dissipation of individual nodes for various protocols is depicted in Fig. 5. Proposed protocol achieves better energy efficiency than NoCC, NoCC-IA and backpressure algorithms, on an average, approximately by a factor 1.998, 1.586 and 1.808 respectively. The rationale behind this result can be explained as follows: *firstly*, loss of energy due to packet drops (collision and buffer drops) is greatly reduced in the proposed protocol as compared to the existing ones and *secondly*, number of retransmissions at each downstream node is also reduced which in turn saves energy. Finally, the use of snoop based acknowledgement further reduces energy consumption.

6 Conclusions

Reliable event perception is essential for collaborative actions in many envisioned applications of sensor networks. Congestion in WSNs causes huge packet loss and thereby hinders reliable event detection. We found that two major reasons of congestion are (i) congestion due to collision and (ii) congestion due to buffer overflow. To reduce packet loss and achieve a fair delivery ratio we propose a *source count* based hierarchical medium access control (HMAC) and weighted round robin forwarding (WRRF). HMAC ensures a hierarchical access of the medium according to the *source count* value. While WRRF assigns a weighted-share of packet delivery to the downstream node. These two schemes together greatly reduce media contention and thereby congestion due to collision. Congestion due to buffer overflow is completely avoided. We have utilized the source count value as a driving parameter for the schemes of our proposed protocol. Our proposed protocol greatly reduces packet loss due congestion, exhibits a higher delivery ratio, which is very necessary for reliable event detection.

References

1. Wan, C.Y., Eisenman, S.B., Campbell, A.T.: CODA: Congestion Detection and Avoidance in Sensor Networks. In: the proceedings of ACM SenSys, Los Angeles, pp. 266–279. ACM Press, New York (2003)
2. Ee, C.T., Bajcsy, R.: Congestion Control and Fairness for Many-to-One Routing in Sensor Networks. In: the proceedings of ACM SenSys, Baltimore, pp. 148–161. ACM Press, New York (2004)
3. Wang, C., Sohraby, K., Li, B., Daneshmand, M., Hu, Y.: A survey of transport protocols for wireless sensor networks. IEEE Network Magazine 20(3), 34–40 (2006)
4. Sankarasubramaniam, Y., Ozgur, A., Akyildiz, I.: ESRT Event-to-Sink Reliable Transport in wireless sensor networks. In: the proceedings of ACM Mobihoc, pp. 177–189. ACM Press, New York (2003)
5. Hull, B., Jamieson, K., Balakrishnan, H.: Mitigating Congestion in Wireless Sensor Networks. In: the proceedings of ACM SenSys, pp. 134–147. ACM Press, New York (2004)
6. Woo, A., Culler, D.: A Transmission Control Scheme for Media Access in Sensor Networks. In: the proceedings of ACM MobiCom, pp. 221–235. ACM Press, New York (2001)
7. Xie, P., Cui, J.H.: SDRT: A Reliable Data Transport Protocol for Underwater Sensor Networks. UCONN CSE Technical Report: UbiNet-TR06-03 (2006)
8. Park, S., Vedantham, R., Sivakumar, R., Akyildiz, I.: A Scalable Approach for Reliable Downstream Data Delivery in Wireless Sensor Networks. In: the proceedings of ACM MobiHoc, pp. 78–89. ACM Press, New York (2004)
9. Chen, S., Yang, N.: Congestion Avoidance Based on Lightweight Buffer Management in Sensor Networks. IEEE Transaction on Parallel and Distributed Systems 17(9), 934–946 (2006)
10. Zhang, H., Arora, A., Choi, Y., Gouda, M.G.: Reliable Bursty Convergecast in wireless Sensor Networks. In: the proceedings of ACM MobiHoc, pp. 266–276. ACM Press, New York (2005)
11. Wan, C.Y., Campbell, A.T., Krishnamurthy, L.: Pump-slowly, fetch-quickly (PSFQ): a reliable transport protocol for sensor networks. IEEE Journal of Selected Areas in Communication 23(4), 862–872 (2005)
12. Stann, R., Heidemann, J.: RMST Reliable data transport in sensor networks. In: the proceedings of IEEE SPNA, Anchorage, Alaska, pp. 102–112. IEEE Computer Society Press, Los Alamitos (2003)
13. Intanagonwiwat, C., Govindan, R., Estrin, D.: Directed diffusion: a scalable and robust communication paradigm for sensor networks. In: the proceedings of ACM MobiCom 2000, Boston, MA, pp. 56–67. ACM Press, New York (2000)
14. Kang, J., Zhang, Y., Nath, B.: TARA Topology Aware Resource Adaptation to Alleviate Congestion in Sensor Networks. IEEE Transaction on Parallel and Distributed Systems 18(7), 919–931 (2007)
15. The Network Simulator – ns-2, http://www.isi.edu/nsnam/ns/index.htm

Systolic Routing in an Optical Ring with Logarithmic Shortcuts

Risto T. Honkanen and Juha-Pekka Liimatainen

University of Kuopio, Department of Computer Science
P.O. Box 1627, FIN-70211 Kuopio, Finland
{rthonkan,jpliimat}@cs.uku.fi

Abstract. We present an all-optical ring network architecture with logarithmic shortcuts and a systolic routing protocol for it. An r-dimensional optical ring network with logarithmic shortcuts ($\mathcal{ORLS}$) consists of $n = 2^r$ nodes and $r2^r$ optical links. We study a systolic routing protocol that is based on cyclic changes of the states of routers and scheduled sendings of packets. The protocol ensures that no electro-optical conversions are needed in the intermediate routing nodes and all the packets injected into the routing machinery reach their targets without collisions. A work-optimal routing of an h-relation is achieved with a reasonable size of $h \in \Omega(n \log n)$.

1 Introduction

Most of the parallel computers on the market use an electronic communication network to transmit packets between processors. A possibility to increase the efficiency of communication is the use of optics. Optical communication offers several advantages in comparison with its electronic counterpart, for example, a possibility to use broader bandwidth and insensitivity to external interferences. For example, Saleh and Teich have presented opto-electronic components in detail [1].

Our work is motivated by another kind of communication problem, namely, the emulation of shared memory with distributed memory modules [2]. If a parallel algorithm has enough parallel *slackness*, the implementation of shared memory can be reduced to the efficient routing of an h-relation [3]. An h-relation is a routing problem where each processing node has at most h packets to send and it is the target of at most h packets [2]. An implementation of an h-relation is said to be *work-optimal* at *cost c*, if all the packets arrive their targets in time ch. A precondition for work-optimality is that $h \in \Omega(\phi)$, where ϕ is the diameter of a network (or the maximum of hop-distances), and that the network can move $\Omega(n\phi)$ packets in each step, where n is the number of processors. Otherwise slackness cannot be used to "hide" the latency influenced by the diameter.

In recent years, optical ring networks have been widely deployed as regional and enterprise networks [4]. Advantages of ring networks are, e.g., simplicity of the system hardware and of routing algorithms and easiness of restoration. The

R. Perrott et al. (Eds.): HPCC 2007, LNCS 4782, pp. 533–544, 2007.

average diameter of an n-node ring network is, however, rather high ($\Theta(n)$). In order to decrease the diameters of ring networks, a number of strategies have been studied. First, embedding of virtual topologies by using wavelength-division multiplexing can be applied [5,4]. Second, networks can be supplied with a number of shortcut links [6,7].

Small et al. have presented the use of Vortex networks using semiconductor optical amplifiers (SOA) as switching elements [8,9]. In their construction an optical packet consists of payloads of multiple WDM channels that are propagated through cascaded SOA-based switching nodes [8]. Barker et al. have considered the feasibility of optical circuit switching network network handling long-lived bulk data transfers [10]. We use, however, a predefined routing bit sequence and scheduled sendings of packets to make routing decisions for a large number of small packets. Liben-Nowell et al. have presented Chord, a scalable Peer-to-Peer lookup protocol for Internet applications that can find in logarithmic routing hops the node responsible for a given hash key [11,12,13]. In this work we use the idea of lookup protocol of Chord in a logarithmic all-optical network.

In this work we present an all-optical ring network architecture and a systolic routing protocol for it. An r-dimensional optical ring network with logarithmic shortcuts, $\mathcal{ORLS}$, consists of $n = 2^r$ nodes and $r2^r$ optical links of equal length. The routing time of the system is divided into time slots, whose length t_p equal to the bypass time of a packet between two consecutive nodes. The system operates synchronously under a common clock. Every node has r incoming and r outgoing links, therefore the routing machinery can move $r2^r$ packets in each time slot. For an r-dimensional $\mathcal{ORLS}$ having $n = 2^r$ processing nodes, the diameter $\phi = r$ fulfills the preconditions for work-optimality when $h \in \Omega(n \log n)$. Additionally, an advantage of our design is that the overall number of optical links is $\Theta(n \log n)$. Honkanen presented the systolic routing protocol for Sparse Optical Torus ($\mathcal{SOT}$) in his paper [14]. For $\mathcal{SOT}$, the number of optical links is $\Theta(n^2)$.

We present a packet routing protocol, called the systolic routing protocol. Packets are routed address-freely. Additionally, when a packet is injected into the routing machinery, neither electro-optic conversions are needed during its path from a source to the target processor, nor any collisions may happen between two distinct packets. Section 2 presents the structure of $\mathcal{ORLS}$. In Section 3 we introduce routing in the network proposed. Section 4 presents an analysis of the routing protocol for the $\mathcal{ORLS}$. Conclusions and future work are presented in Section 5.

2 Optical Ring with Logarithmic Shortcuts

We study the r-dimensional structure of $\mathcal{ORLS}$ having the diameter $\phi = r$ and having $n = 2^r$ nodes. Each node consists of a router and a processor. Section 2.1 presents the design of nodes. In Section 2.2 we introduce the construction of an $\mathcal{ORLS}$. Section 2.3 discusses the feasibility of our construction.

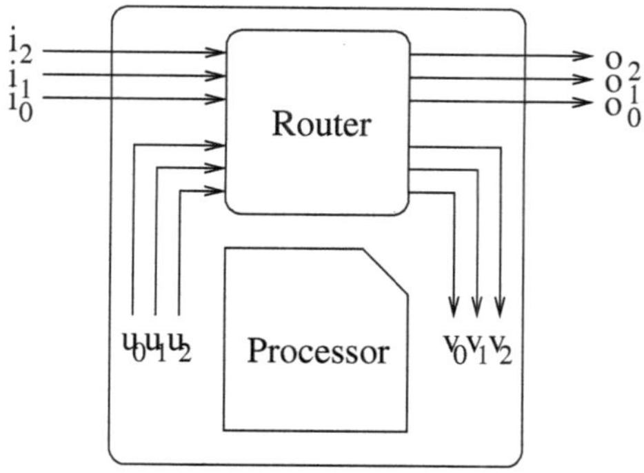

Fig. 1. A node of a 3-dimensional $\mathcal{ORLS}$

2.1 Construction of Nodes for $\mathcal{ORLS}$

An r-dimensional $\mathcal{ORLS}$ consists of $n = r2^r$ nodes $N_i = (R_i, P_i)$, $0 \leq i < n$, where R_i and P_i are the router and the processor of a node N_i. Each router R_i of an r-dimensional $\mathcal{ORLS}$ has r incoming links $i_0, i_1, \ldots, i_{r-1}$ from its predecessors, r outgoing links $o_0, o_1, \ldots, o_{r-1}$ to its successors, r inputs $u_0, u_1, \ldots, u_{r-1}$ from the transmitters of a processor P_i, and r outputs $v_0, v_1, \ldots, v_{r-1}$ to the receivers of the processor P_i. The basic component of the routers is an electrically controlled all-optical 2×2 switch. Switches can be implemented by $LiNbO_3$ technology [1]. We can construct a router of any degree with edge-disjoint paths, e.g., by using Beneš networks structure [15]. In Figure 1, a node of a 3-dimensional $\mathcal{ORLS}$ is presented.

Each router implements an injective function $\pi : \{i_0, i_1, \ldots, i_{r-1}\} \cup \{u_0, u_1, \ldots, u_{r-1}\} \rightarrow \{o_0, o_1 \ldots, o_{r-1}\} \cup \{v_0, v_1, \ldots, v_{r-1}\}$. The definition of the mapping π simply means that signals never collide at the outgoing links. The mapping can be different at different moments, corresponding to the state of the router. All the routers are set in the same state at a certain moment.

2.2 Design of $\mathcal{ORLS}$

An r-dimensional $\mathcal{ORLS}$ consists of $n = 2^r$ nodes and $r2^r$ edges. Two nodes N_i and $N_{i'}$ are connected with an unidirectional edge (optical link) if

(1) $i' = (i + 1) \bmod n$ or
(2) $i' = (i + 2^j) \bmod n, 1 \leq j < r$.

We call edges by condition (1) ring edges and edges by condition (2) shortcut edges. A consequence of the definition is that the routing machinery looks exactly the same from the viewpoint of each node. Owing to the definition of $\mathcal{ORLS}$, a larger network can be recursively built by combining two smaller ones. That is, an r-dimensional $\mathcal{ORLS}$ can be constructed by joining two $(r-1)$-dimensional $\mathcal{ORLS}$s together. In Figure 2 a 2-dimensional $\mathcal{ORLS}$ is built from

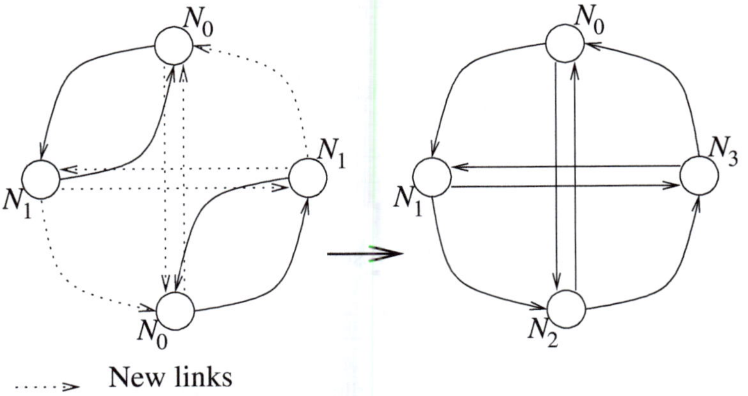

Fig. 2. Construction of a 2-dimensional $\mathcal{ORLS}$

two 1-dimensional $\mathcal{ORLS}$s by rearranging connections and relabeling the nodes. In Figure 2 a circle presents a node and the dotted arrows new links to be formed.

Lemma 1. *The average hop-distance in an r-dimensional* $\mathcal{ORLS}$ *is* $r/2$.

Proof. By induction of r. When $r = 1$, the result is true by an inspection. Let us assume that the result is true for $r-1$. The average hop-distance from the node N_0 to the other nodes in an $(r-1)$-dimensional $\mathcal{ORLS}$ having $n = 2^{r-1}$ nodes can be evaluated by $\overline{d}_{r-1} = (|N_0| + |N_1|, \ldots, |N_{n-1}|)/n$, where $|N_i|$ means the minimum of the hop-distances from the node N_0 to a node N_i. When two $(r-1)$-dimensional $\mathcal{ORLS}s$ $\mathcal{NW}$ and $\mathcal{NW}'$ are connected to form an r-dimensional $\mathcal{ORLS}$, each node N_i of $\mathcal{NW}$ is connected to the node $N_{(i+n) \bmod 2n}$ of $\mathcal{NW}'$ in the new network. A consequence is that the average hop-distance of an r-dimensional $\mathcal{ORLS}$ can be evaluated by $\overline{d}_r = (2|N_0| + 1 + 2|N_1| + 1+, \ldots, +2|N_{n-1}| + 1)/2n = (n + 2(|N_0| + |N_1|+, \ldots, +|N_{n-1}|))/2n = \overline{d}_{r-1} + 1/2 = (r-1)/2 + 1/2 = r/2$.

Lemma 2. *The maximum of hop-distances (diameter) between any two nodes* N_i *and* N_j $(i \neq j)$ *in an r-dimensional* $\mathcal{ORLS}$ *is* r.

Proof. Lemma 2 is a direct consequence of the definition of $\mathcal{ORLS}$.

2.3 Feasibility of $\mathcal{ORLS}$ with Systolic Routers

The switching time of $LiNbO_3$ switches lies in the range of 10–15 ps [1]. The length of a packet (l_p) can be evaluated by equation $l_p = \frac{N_p \times v_c}{B \times r}$, where N_p is the size of a packet in bits, $v_c = 0.3$ m/ns is the speed of light in vacuum, $r = 1.5$ is the refractive index of a fiber [1], and B is the link bandwidth. Assuming the bandwidth to be B=100 Gb/s, the length of a bit in the fiber is $\frac{v_c}{B \times r} = 2$ mm.

In order to estimate the feasibility of a 6-dimensional $\mathcal{ORLS}$ (having 64 processing nodes) let us assume the link bandwidth to be $B = 100$ Gb/s, and the size of packets to be $N_p = 256$ b. The corresponding length of a packet in the fiber is $l_p \simeq 512$ mm, and the length of a time slot is $t_p \simeq 2.6$ ns. Assuming the length of a clock cycle of the processors to be $t_{cc} = 1$ ns (corresponding to the frequency of 1 GHz), it will take 2.6 clock cycles for a packet to travel between two adjacent routing nodes. The overall amount of fibers is $L_f \simeq 200$ m, and the average routing time of a packet is $t_r \simeq 8$ clock cycles.

However, a couple of drawbacks arise, when the system is scaled up. First, the degree of the routing nodes in an $\mathcal{ORLS}$ increases logarithmically with respect to the size of the network. Second, putting M elements in the physical space requires at least volume of size $\Omega(\sqrt[3]{M})$ [16,17]. We consider the requested parameters to be reasonable and the design to be feasible if the number of processing nodes is restricted up to a few hundreds of nodes.

3 Routing in $\mathcal{ORLS}$

We develop a routing algorithm for $\mathcal{ORLS}$. Routing in $\mathcal{ORLS}$ can be divided into two phases, namely, a preprocessing phase and a routing phase. Section 3.1 presents the preprocessing phase. Section 3.2 presents the routing algorithm for our design.

3.1 Preprocessing Phase of Routing

The preprocessing phase consists of determining routing paths for packets, determining the states of routers, and determining routing tables for the nodes. Section 3.1 presents determining routing paths. Section 3.1 introduces determining routing tables and the states of routers.

Determining Routing Paths. Let us consider an "arithmetic distance" from the node N_0 to a node N_i in an r-dimensional $\mathcal{ORLS}$. It can be expressed as $i = x_{r-1}2^{r-1} + \ldots + x_0 2^0$, $x_l \in \{0,1\}$. For example, the distance from the node N_0 to the node N_{27} in an 5-dimensional $\mathcal{ORLS}$ can be expressed as $27 = 1 \times 2^4 + 1 \times 2^3 + 0 \times 2^2 + 1 \times 2^1 + 1 \times 2^0$. The series $2^4 + 2^3 + 2^1 + 2^0$ corresponds to the optical link path $N_0 \rightarrow N_{16} \rightarrow N_{24} \rightarrow N_{26} \rightarrow N_{27}$ in the 5-dimensional $\mathcal{ORLS}$ using the longest shortcut at each hop step. Let $\langle 4,3,1,0 \rangle$ denote the corresponding routing path of the length of four from N_0 to N_{27} using three times shortcut edges and once a ring edge.

Figure 3 presents the routing groups of a 3-dimensional $\mathcal{ORLS}$. Table 1 presents the corresponding routing paths between the nodes presented in Figure 3. Consider routing from the node N_0. Paths are divided in $r = 3$ groups. Group $\mathcal{G}_0$ consists of the routing path $\langle 0 \rangle$ leading to the node N_1, group $\mathcal{G}_1$ consists of the routing paths $\langle 1 \rangle$ and $\langle 1,0 \rangle$ leading to the nodes N_2 and N_3, and finally group $\mathcal{G}_2$ consists of the routing paths $\langle 2 \rangle$, $\langle 2,0 \rangle$, $\langle 2,1 \rangle$, and $\langle 2,1,0 \rangle$ leading to the nodes N_4, N_5, N_6 and N_7 correspondingly. Routing paths between nodes can be evaluated by Algorithm `Routing Paths`.

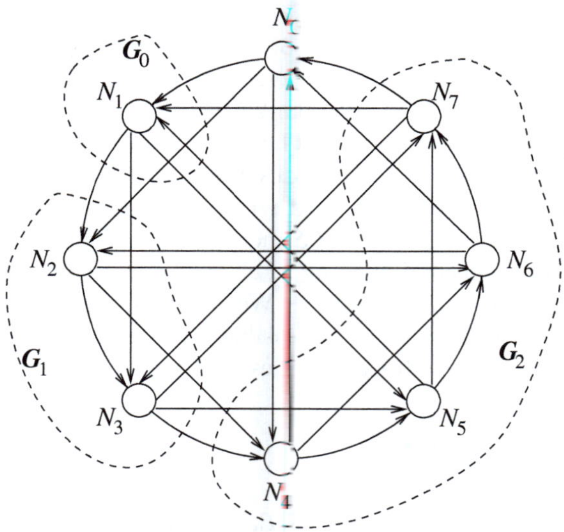

Fig. 3. Routing groups of a 3-dimensional $\mathcal{ORLS}$

Algorithm `Routing Paths`
 Input: `The set of nodes` $\mathcal{N}$ `of an` r`-dimensional` $\mathcal{ORLS}$
 Output: `Groups of routing paths` $\mathcal{G}_0 \ldots \mathcal{G}_{r-1}$
 for each $N_s \in \mathcal{N}$ **pardo** {`each node evaluates`}
  for $0 \leq d < 2^r$ **do**
   `Find the binary presentation` s_k `for which`
    $|s-d| = s_k = x_{r-1}2^{r-1} + \ldots + x_1 2^1 + x_0 2^0$;
   `Evaluate the corresponding routing path`
    $\langle \alpha \rangle = \langle z_1, z_2, \ldots, z_k \rangle$ `for which` z_i `is`
    `the exponent of the` i`th term in` s_k `having` $x_i = 1$
    `and add it into the routing path` $\mathcal{G}_{z_1}$

Obviously, Algorithm `Routing Paths` divides paths in r groups (if the path leading to the processor itself is omitted) so that the size of the ith group is 2^i. Let $\langle z_1, \alpha_k \rangle \in \mathcal{G}_i$ denote the kth routing path of group $\mathcal{G}_i$ and α_k denote the "tail" of it. For a half of routing paths of group $\langle z_1, \alpha_k \rangle \in \mathcal{G}_i$ $(1 < i < r)$ holds true $\langle \alpha_k \rangle \in \mathcal{G}_{i-1}$.

Determining Routing Tables and the States of Routers. The structure of $\mathcal{ORLS}$ is node and edge symmetric. That is why it is enough to consider routing from the node N_0 to the other nodes in the network. Routing in $\mathcal{ORLS}$ has a number of properties. First, at each time step t all the routers of the network are set in the same state corresponding to an injective mapping function π_i. Second, any two packets sent at the same time may never collide if those are chosen from different routing groups and their routing paths differ from each path positions. Finally, a packet can be sent into an output link o_i if there is no packet sent

Table 1. Routing paths in 3-dimensional $\mathcal{ORLS}$

	N_0	N_1	N_2	N_3	N_4	N_5	N_6	N_7
N_0	$\langle\rangle$	$\langle 0\rangle$	$\langle 1\rangle$	$\langle 1,0\rangle$	$\langle 2\rangle$	$\langle 2,0\rangle$	$\langle 2,1\rangle$	$\langle 2,1,0\rangle$
N_1	$\langle 2,1,0\rangle$	$\langle\rangle$	$\langle 0\rangle$	$\langle 1\rangle$	$\langle 1,0\rangle$	$\langle 2\rangle$	$\langle 2,0\rangle$	$\langle 2,1\rangle$
N_2	$\langle 2,1\rangle$	$\langle 2,1,0\rangle$	$\langle\rangle$	$\langle 0\rangle$	$\langle 1\rangle$	$\langle 1,0\rangle$	$\langle 2\rangle$	$\langle 2,0\rangle$
N_3	$\langle 2,0\rangle$	$\langle 2,1\rangle$	$\langle 2,1,0\rangle$	$\langle\rangle$	$\langle 0\rangle$	$\langle 1\rangle$	$\langle 1,0\rangle$	$\langle 2\rangle$
N_4	$\langle 2\rangle$	$\langle 2,0\rangle$	$\langle 2,1\rangle$	$\langle 2,1,0\rangle$	$\langle\rangle$	$\langle 0\rangle$	$\langle 1\rangle$	$\langle 1,0\rangle$
N_5	$\langle 1,0\rangle$	$\langle 2\rangle$	$\langle 2,0\rangle$	$\langle 2,1\rangle$	$\langle 2,1,0\rangle$	$\langle\rangle$	$\langle 0\rangle$	$\langle 1\rangle$
N_6	$\langle 1\rangle$	$\langle 1,0\rangle$	$\langle 2\rangle$	$\langle 2,0\rangle$	$\langle 2,1\rangle$	$\langle 2,1,0\rangle$	$\langle\rangle$	$\langle 0\rangle$
N_7	$\langle 0\rangle$	$\langle 1\rangle$	$\langle 1,0\rangle$	$\langle 2\rangle$	$\langle 2,0\rangle$	$\langle 2,1\rangle$	$\langle 2,1,0\rangle$	$\langle\rangle$

during the previous time steps preferring the same output and other packets bypassing or to be sent differ from target address by each path positions.

Consider routing in a 3-dimensional $\mathcal{ORLS}$. At the time step 0 the node N_0 may send packets using the routing paths $\langle 2,1,0\rangle$, $\langle 1,0\rangle$, and $\langle 0\rangle$. The corresponding mapping would be $\pi_0 : (i_0, i_1, i_2, u_0, u_1, u_2) \rightarrow (\star, \star, \star, o_0, o_1, o_2)$, where $\star$ denotes that we do not know the state yet. Tentative mappings π_1, π_2, and π_3 for the next three rounds would be $\pi_1 : (i_0, i_1, i_2, u_0, u_1, u_2) \rightarrow (v_0, o_0, o_1, \star, \star, \star)$, $\pi_2 : (i_0, i_1, i_2, u_0, u_1, u_2) \rightarrow (v_0, o_0, \star, \star, \star, \star)$, and $\pi_3 : (i_0, i_1, i_2, u_0, u_1, u_2) \rightarrow (v_0, \star, \star, \star, \star, \star)$. During the time step 1 the output port o_2 is free. In principle, we could choose any of packets using the routing paths $\langle 2\rangle$, $\langle 2,0\rangle$, or $\langle 2,1\rangle$ to be routed. During the time step 2, however, the output port o_0 is reserved to route a packet using the routing path $\langle 2,1,0\rangle$ and a packet using the routing path $\langle 2\rangle$ can be sent at any time. A packet using the routing path $\langle 2,1\rangle$ is chosen to be sent and the mappings π_1, π_2, and π_3 are updated to $\pi_1 : (i_0, i_1, i_2, u_0, u_1, u_2) \rightarrow (v_0, o_0, o_1, \star, \star, o_2)$, $\pi_2 : (i_0, i_1, i_2, u_0, u_1, u_2) \rightarrow (v_0, o_0, o_1, \star, \star, \star)$, $\pi_3 : (i_0, i_1, i_2, u_0, u_1, u_2) \rightarrow (v_0, v_1, \star, \star, \star, \star)$. The mappings are updated until there are no packets left to route. The last mapping having only one-hop paths to route can be merged with the mapping π_0 because in it all the connections to the receivers are free. The final four mappings needed are

$$\begin{aligned}
&\pi_0 : (i_0, i_1, i_2, u_0, u_1, u_2) \rightarrow (v_0, v_1, v_2, o_0, o_1, o_2),\\
&\pi_1 : (i_0, i_1, i_2, u_0, u_1, u_2) \rightarrow (v_0, o_0, o_1, \star, \star, o_2),\\
&\pi_2 : (i_0, i_1, i_2, u_0, u_1, u_2) \rightarrow (v_0, o_0, o_1, \star, \star, o_2), and\\
&\pi_3 : (i_0, i_1, i_2, u_0, u_1, u_2) \rightarrow (v_0, v_1, o_0, \star, o_1, o_2).
\end{aligned}$$

The corresponding states of the routers are presented in Figure 4 and the corresponding timing diagram of the 3-dimensional $\mathcal{ORLS}$ is presented in Figure 5.

Consider routing from N_0 to other nodes N_d in an r-dimensional $\mathcal{ORLS}$. It is worth noting that a half of the packets must be sent using the shortcut link leading to the node set $\{N_{n/2}, \ldots, N_{n-1}\}$, a quarter of the packets using the shortcut link leading to the node set $\{N_{n/4}, \ldots, N_{n/2-1}\}$, etc. On the other hand, the router of the node N_0 is an intermediate node for a number of packets from nodes in the upstream. The algorithm determining the mappings and routing table is `Routing Table` and it can be presented as follows:

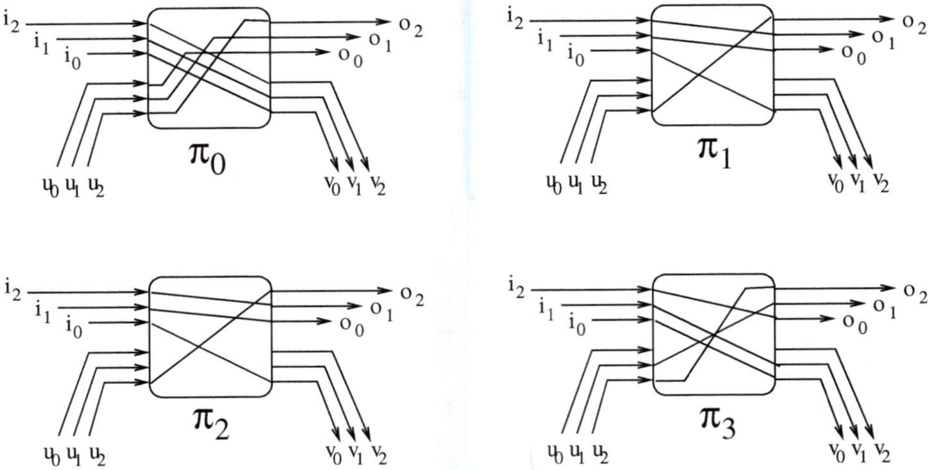

Fig. 4. Routing states of a 3-dimensional $\mathcal{ORLS}$

Algorithm `Routing Table`
  Input: `Groups of routing paths` $\mathcal{G}_0 \ldots \mathcal{G}_{r-1}$ `of node` N_s
  Output: `Routing table` R_s `and states of routers`
  for each $N_s \in \mathcal{N}$ **pardo** `{each node evaluates}`
    for `phase = 0` **to** $n/2$ **do**
      for `group = r-1` **downto** `0` **do**
        if $\mathcal{G}_{group} \neq \emptyset$ **then**
          `Pick up the longest path` $\langle\alpha\rangle$ `from` $\mathcal{G}_{group}$
          if $\langle\alpha\rangle$ `can be embedded into mappings` Π_{phase} **then**
            `Embed` $\langle\alpha\rangle$ `into mappings` Π_{phase}
            `Update routing table` R `so that packets using` $\langle\alpha\rangle$
              `are sent at phase` *phase*
          else `return` $\langle\alpha\rangle$ `into` $\mathcal{G}_{group}$
    `Update mappings` Π_0 `and` $\Pi_{n/2}$

Algorithm `Routing Table` needs $\Theta(n \log n)$ iteration rounds to execute the routing table. To see why, let us consider the sizes of routing groups and postfixes of routing paths. Obviously, the size of the largest group $\mathcal{G}_r$ $n/2 = 2^{r-1}$ is the minimum for iteration rounds. At the first round, the longest routing path is selected from each routing group. The next $n/4 - 1$ sendings from the largest group $\mathcal{G}_r$ prevent sendings from the group $\mathcal{G}_{r-1}$ because the second term in routing paths of group $\mathcal{G}_r$ equals to the first term of the routing paths of group $\mathcal{G}_{r-1}$. After $n/4 = 2^{r-2}$ rounds sendings from group $\mathcal{G}_{r-1}$ can be started overlapped with the rest of the group $\mathcal{G}_r$. Recursively, sendings from the groups $\mathcal{G}_{r-2} \ldots \mathcal{G}_1$ can be started after rounds $2^{r-2} + 2^{r-3} + \ldots + 2^0$. The overall number of iteration rounds is $2^{r-2} + 2^{r-3} + \ldots + 2^0 = 2^{r-1} - 1 = n/2$. The inner loop needs $\Theta(r) = \Theta(\log n)$ iteration rounds to be executed. The routing table of the node N_0 in a 3-dimensional $\mathcal{ORLS}$ is presented in Table 2.

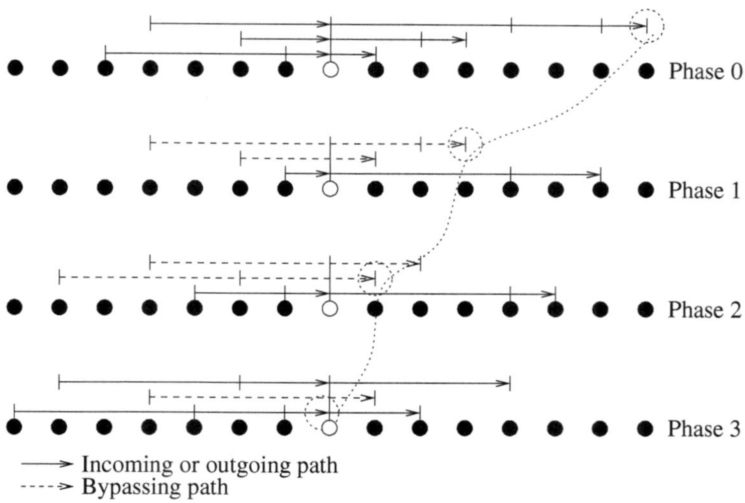

Fig. 5. Timing diagram of a 3-dimensional $\mathcal{ORLS}$

Table 2. Routing table of N_0 in a 3-dimensional $\mathcal{ORLS}$

N_i	Path	Phase	Output
N_1	$\langle 0 \rangle$	0	o_0
N_2	$\langle 1 \rangle$	3	o_1
N_3	$\langle 1, 0 \rangle$	0	o_1
N_4	$\langle 2 \rangle$	3	o_2
N_5	$\langle 2, 0 \rangle$	2	o_2
N_6	$\langle 2, 1 \rangle$	1	o_2
N_7	$\langle 2, 1, 0 \rangle$	0	o_2

3.2 Routing Algorithm for $\mathcal{ORLS}$

At the preprocessing phase each node N_s determines routing paths to the other nodes N_d of the system, routing table, and the states of the routers. This is done only once when the system is set up. At the beginning of each routing phase each node of the $\mathcal{ORLS}$ has a number of packets to send. Each node N_s inserts packets destined to node N_d into sending buffer $B_{phase,output}$ according to the phase to be sent and the output port for the packet.

At each time step t each node picks up packets from sending buffers $B_{t \bmod (n/2+1),o_i}$ according to the routing table and injects them into outgoing links. At the same time the routers are set in state $\Pi_{t \bmod (n/2+1)}$.

4 Analysis of Routing

In the preprocessing phase, each of h packets of processor P_i are inserted into the sending buffers according to their phase and output link to be sent. Clearly, all

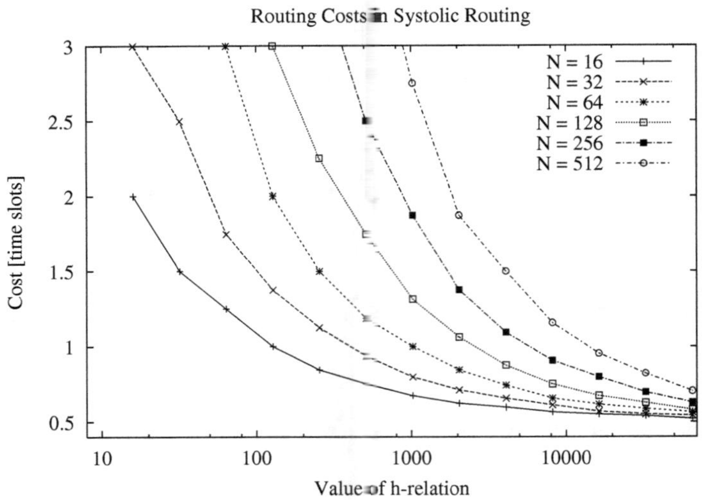

Fig. 6. Routing costs in $\mathcal{ORLS}$ when the size of h-relation varies

of the packets have been routed after time $O(S \cdot n/2)$, where S is the maximum size of the all buffers. According to Mitzenmacher et al. [18], supposing that we throw n balls into n bins each ball choosing a bin independently and uniformly at random, then the *maximum load* is approximately $\log n/\log\log n$ with high probability[1]. The maximum load means the largest number of balls in any bin. Correspondingly, if we have n packets to send and n sending buffers during a simulation step, then the maximum load of sending buffers is approximately $\log n/\log\log n$ *whp*. The overall routing time of those packets is $n \log n/2\log\log n + \Theta(1)$, which is not work-optimal according to the definition of work-optimality.

If the size of an h-relation is enlarged to $h \geq n\log n$, the maximum load is $\Theta(h/n)$ *whp* [19]. Assuming that $h = n\log n$, the maximum load is $\Theta(\log n)$ and the corresponding routing time is $\Theta(n\log n)$. A work-optimal result is achieved according to the definition of work-optimality.

Routing an h-relation in time $\Theta(h)$ implies work-optimality. We ran some experiments to get an idea about the cost. We ran 100 simulation rounds for each occurence using a C program. Packets were randomly put into output buffers and the average value of the routing time over all 100 simulations were evaluated. The average cost were evaluated using equation $c = t/h$, where t is the average routing time. Figure 6 gives support to the idea that h does not need to be extremely high to get a reasonable routing cost.

5 Conclusions and Future Work

We have presented an all-optical ring network design with logarithmic shortcuts and a systolic routing protocol for it. No electro-optical conversions are needed

[1] We use *whp*, *with high probability* to mean with probability at least 1-$O(1/n^{\alpha})$ for some constant α.

during the transfer and all the packets injected into the machinery are guaranteed to reach their destinations. The design and the systolic routing protocol offer a work-optimal routing of an h-relation if $h \in \Omega(n \log n)$. We believe that the simple structure presented and the systolic routing protocol are useful and realistic if the number of processing nodes is restricted up to a few hundreds of nodes. There still remain a number of open questions which we shall attempt to answer in the future, e.g. how to deal with node fails?

Acknowledgments

We express our gratitude to professor Pekka Kilpeläinen and professor Martti Penttonen for their valuable discussions on the topic.

References

1. Saleh, B., Teich, M.: Fundamentals of Photonics. John Wiley & Sons, Inc. Chichester (1991)
2. Adler, M., Byers, J., Karp, R.: Scheduling parallel communication: The h-relation problem. In: Proceedings of Mathematical Foundations of Computer Science, Prague, Czech Republic, pp. 1–20 (1995)
3. Valiant, L.: 18. Handbook of Theoretical Computer Science. Algorithms and Complexity, vol. A, pp. 943–971. Elsevier Science Publishers BV, Amsterdam (1990)
4. Ramaswami, R., Sivarajan, K.: Optical Networks: A Practical Perspective. Morgan Kaufmann, San Francisco (1998)
5. Aggrawal, A., Bar-Noy, A., Coppersmith, D., Ramaswami, R., Shieber, B., Sudan, M.: Efficient routing in optical networks. Journal of the ACM 42(6), 973–1001 (1996)
6. Narayanan, L., Opatrny, J., Sotteau, D.: All-to-all optical routing in chordal rings of degree four. In: Proceedings of the Symposium on Discrete Algorithms, pp. 695–703 (1999)
7. Takao, M., Hitoshi, O.: Effect of a shortcut link supplemented to ring networks. In: Proceedings of IEEE International Conference on Communications, pp. 1330–1334. IEEE Computer Society Press, Los Alamitos (2003)
8. Liboiron-Ladouceur, O., Small, B., Bergman, K.: Physical layer scalability of wdm optical interconnection network. Journal of Lightwave Technology 24(1), 262–270 (2006)
9. Small, B., Lee, B., Bergman, K.: Flexibility of optical packet format in a complete 12×12 data vortex network. IEEE Photonics Technology Letters 18(16), 1693–1695 (2006)
10. Barker, K.J., et al.: On the feasibility of optical circuit switching for high performance computing systems. In: SC 2005, pp. 16–38 (2005)
11. Liben-Nowell, D., Balakrishnan, H., Karger, D.: Analysis of the evolution of peer-to-peer systems. In: PODC, pp. 233–242 (2002)
12. Liben-Nowell, D., Balakrishnan, H., Karger, D.: Observations on the dynamic evolution of peer-to-peer networks. In: IPTPS, pp. 22–33 (2002)
13. Stoica, I., Morris, R., Liben-Nowell, D., Karger, D.R., Kaashoek, M.F., Dabek, F., Balakrishnan, H.: Chord: a scalable peer-to-peer lookup protocol for internet applications. IEEE/ACM Trans. Netw. 11(1), 17–32 (2003)

14. Honkanen, R.: Systolic routing in sparse optical torus. In: Kilpeläinen, P., Päivinen, N. (eds.) Proceedins of the Eight Symposium on Programming Languages and Software Tools, pp. 14–20 (2003)
15. Leighton, F.: Introduction to Parallel Algorithms and Architectures: Arrays, Trees, Hypercubes. Morgan Kaufmann, San Francisco (1992)
16. Vitányi, P.: Locality, communication, and interconnect length in multicomputers. SIAM Journal of Computing 17(4), 659–672 (1988)
17. Vitányi, P.: Multiprocessor architectures and physical law. In: Proceedings of 2nd Workshop on Physics and Computation, pp. 24–29 (1994)
18. Mitzenmacher, M., Richa, A., Sitaraman, R.: The power of two random choices: A survey of the techniques and results. In: Handbook of Randomized Computing, pp. 255–305. Kluwer Academic Publishers, Dordrecht (2001)
19. Raab, M., Steger, A.: "balls into bins"—a simple and tight analysis. In: Rolim, J.D.P., Serna, M.J., Luby, M. (eds.) RANDOM 1998. LNCS, vol. 1518, pp. 159–170. Springer, Heidelberg (1998)

On Pancyclicity Properties of OTIS Networks

Mohammad R. Hoseinyfarahabady and Hamid Sarbazi-Azad

Sharif University of Technology, Tehran, Iran, and
IPM School of Computer Science, Tehran, Iran
Hoseiny@sharif.edu, azad@ipm.ir

Abstract. The OTIS-Network (also referred to as two-level swapped network) is composed of n clones of an n-node original network constituting its clusters. It has received much attention due to its many favorable properties such as high degree of scalability, regularity, modularity, package-ability and high degree of algorithmic efficiency. In this paper, using the construction method, we show that the OTIS-Network is Pancyclic if its basic network is Hamiltonian-connected. The study of cycle embeddings with different sizes arises naturally in the implementation of a number of either computational or graph problems such as those used for finding storage schemes for logical data structures, layout of circuits in VLSI, etc. Our result is resolving an open question posed in [6] and generalizing a number of proofs in the literature for specific Hamiltonian properties of similar networks.

1 Introduction

The Optical Transpose Interconnection System (OTIS), which was proposed in [1, 2, 3], generates a wide class of high-performance scalable interconnection networks. It offers a new optoelectronic computer architecture that takes benefits from both optical and electronic technologies. In this architecture, processors are divided into groups where electronic interconnects are used to connect processors within each group, while optical interconnects are used for inter-group communication [3]. In such an OTIS system, an optical link connects processor p of group g to processor g of group p. It has been shown in [4] that when the number of processors in a group equals the number of groups, the bandwidth and the power consumption in OTIS-Networks are optimized and system area and volume are minimized. Thus, in an N^2-processor OTIS network, processors are partitioned into N groups of N processors. Besides the electronic connections between the processors in each group, processor i in group j is optically connected to processor j of group i. The OTIS-hypercube and OTIS-mesh are two of the most widely studied instances of the OTIS architecture [5-17]. A number of algorithms have been developed for these networks, such as routing, selection, and sorting [5, 8, and 11], data rearrangement [8], matrix multiplication [13], and broadcasting [6]. Many of topological properties of these systems, such as node degree, diameter, β-cut, and bisection width are addressed in previous studies [6, 8]. Also, in [9, 10], it was proved that if G is a Hamiltonian-connected graph, so is the OTIS-*G*. Moreover, the fault tolerance of OTIS-Networks has been addressed and the

R. Perrott et al. (Eds.): HPCC 2007, LNCS 4782, pp. 545–553, 2007.

fault diameter of OTIS-*G* is derived with respect to the fault diameter of G [9]. In [15, 17] the performance merits of the OTIS-*hypercube* and the effect of different structural and workload parameters on the overall performance are investigated. Their result reveals that the OTIS multi-computers are good candidates for interconnection networks of future generation parallel computers.

In this paper, we prove the conjecture proposed in [6]. According to this conjecture if G contains a cycle of length *L*, then there exists a cycle of length L^2 in OTIS-*G*. We also generalize this result and demonstrate that if there exists a Hamiltonian path between every two arbitrary nodes of graph *G* (or if *G* is Hamiltonian-connected), then all cycles with length 7, 8 … $|V(G)|^2$ can be constructed in OTIS-*G*.

The organization of the paper is as follows. In Section 2, OTIS-Network is formally defined and some basic properties of this network are reported. In Section 3, the Pancyclicity property of the OTIS-*G* with regard to the Hamiltonian property of the basic network G is proved. Finally, Section 4 concludes this paper.

2 The OTIS Network: Definition and Basic Properties

The reader is referred to [6] for an in-depth account of basic concepts and properties of the OTIS networks such as topology, routing algorithms, broadcasting, embedding of graphs, etc. In this section, more specific concepts are described.

Definition 1. Let $G = (V_G, E_G)$ be an undirected graph. The OTIS-$G = (V_{OT}, E_{OT})$ network is an undirected graph defined by:

$V_{OT} = \{\langle g, p\rangle \mid g, p \in V_G\}$ and

$E_{OT} = \{(\langle g, p_1\rangle, \langle g, p_2\rangle) \mid g \in V_G, (p_1, p_2) \in V_G\} \cup \{(\langle g, p\rangle, \langle p, g\rangle) \mid g, p \in V_G, g \neq p\}$.

The graph *G* is called the factor (or basic) network of OTIS-*G*. If *G* has N nodes, the OTIS-*G* is composed of *N* node-disjoint sub-networks $G_1, G_2, \ldots, G_N$, called groups. Each of these groups is isomorphic to the factor graph *G*. A node $\langle g, p\rangle$ in OTIS-*G* corresponds to the node of address p in group G_g. An intra-group edge of the form $(\langle g, p_1\rangle, \langle g, p_2\rangle)$ corresponds to an electronic link, while an inter-group edge of the form $(\langle g, p\rangle, \langle p, g\rangle)$ corresponds to an optical link and passing over such a link is referred to as an OTIS movement [6]. Figure 1 displays some examples of the OTIS-*G* where *G* is either a Q_3 (3-dimensional hypercube) or a K_5 (complete graph of 5 nodes) [15].

A generalization of the OTIS-Networks is an *l*-level hierarchical swapped network [9], denoted *SW(G, l)*, which is based on a nucleus (factor) graph *G*, too. To build an *l*-level swapped network, *SW(G, l)*, we use $N_{l-1} = |SW(G, l-1)|_V$ identical copies (clusters) of *SW(G, l-1)*. Each copy of *SW(G, l-1)* is viewed as a level-*l* cluster. Node *i* in cluster *j* is connected to node *j* in cluster *i* for all $i \neq j$, $0 \leq i, j \leq N_{l-1}$. It is clear that *SW(G, 2)* is topologically equivalent to the OTIS-*G* network. The *SW(G, 3)* is topologically isomorphic to the OTIS-OTIS-*G* network and this holds for higher level networks.

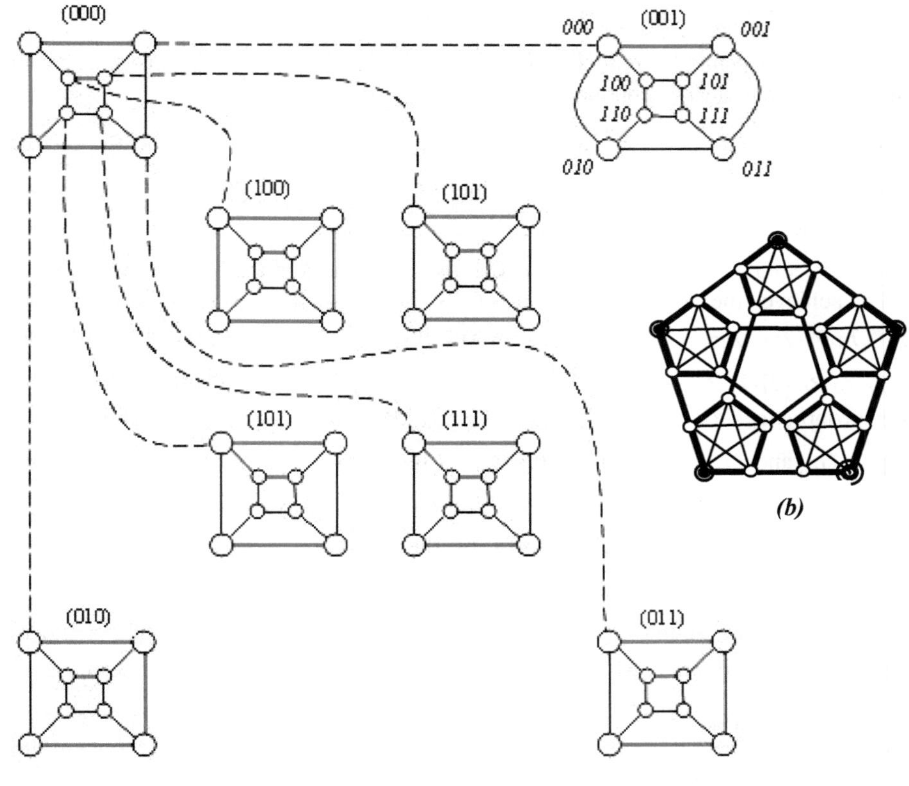

Fig. 1. (a) A 3-dimensional OTIS-hypercube with the optical connections exiting one of the sub-graphs (numbers inside parenthesis are the group numbers) (b) The topologies of OTIS-K_5

Theorem 1. [9] If *G* has node degree *d* and diameter *D*, the degree and diameter of *SW(G, l)* are $d_{sw} = d + l$ and $D_{sw} = 2^{l-1}(D + 1) - 1$, respectively.

3 Pancyclicity of OTIS-Networks

In this section, we propose a solution to the open question stated in [6]. Let *G* be a graph of $n \geq 3$ vertices. A *k*-cycle is a cycle of length *k*. A Hamiltonian cycle (path) of *G* is a cycle (path) containing every vertex of *G*. A Hamiltonian graph is a graph containing a Hamiltonian cycle. We prove that if there is an *L*-cycle in the factor graph *G*, this would give a cycle of length L^2 in OTIS-*G*. In a special case, it would imply that if *G* is Hamiltonian, so is OTIS-*G*. Studying Hamiltonicity and related properties of interconnection networks has attracted much attention [5, 9, 14, 18-20].

Theorem 2. [6] If there exists an *L*-cycle (a cycle of length *L*) in *G*, then there exists a cycle of length L^2-L in OTIS-*G*.

Theorem 3. If graph *G* contains an *L*-cycle, then there exists an L^2-cycle in OTIS-*G*.

Proof. We prove this by induction on *L*, the length of the cycle. By considering the parity of *L*, we regard 2 separate cases.

Case 1. *L is even*: Let *L* = 2*S* and $\gamma_0, \gamma_1, \gamma_2, ... \gamma_{2S-2}, \gamma_{2S-1}, \gamma_{2S} = \gamma_0$ be the circular sequence of node addresses corresponding to an *L*-cycle in *G*, $L \le |V(G)|$. We obtain such a cycle in each of 2*S* groups namely $G_{\gamma_0}, G_{\gamma_1}, G_{\gamma_2}, ..., G_{\gamma_{2S-2}}, G_{\gamma_{2S-1}}$ of OTIS-*G*. When *L* is equal to 4, one can easily check the correctness of the base of induction.

To achieve the L^2-cycle within OTIS-*G*, for *L*>*4*, we remove from each cycle $\gamma_0, \gamma_1, \gamma_2, ... \gamma_{2S-2}, \gamma_{2S-1}, \gamma_{2S} = \gamma_0$ of sub-graph $G_{\gamma_i}, 5 \le i \le 2S-1$, the two edges $(<\gamma_i, \gamma_0>, <\gamma_i, \gamma_1>)$ and $(<\gamma_i, \gamma_2>, <\gamma_i, \gamma_3>)$, and replace them by the optical inter-group edges $(<\gamma_i, \gamma_j>, <\gamma_j, \gamma_i>)$ for 5≤*i*≤2*S*-1, 0≤*j*≤3. As well, within the sub-graphs G_{γ_0} and G_{γ_1}, we remove all edges of the form $(<\gamma_0, \gamma_{2t}>, <\gamma_0, \gamma_{2t+1}>)$ and $(<\gamma_1, \gamma_{2t}>, <\gamma_1, \gamma_{2t+1}>)$ for 2≤*t*≤*S*-1. In a similar way, within the sub-graphs G_{γ_2} and G_{γ_3}, we remove all links of the form $(<\gamma_2, \gamma_{2t-1}>, <\gamma_2, \gamma_{2t}>)$ and $(<\gamma_3, \gamma_{2t-1}>, <\gamma_3, \gamma_{2t}>)$ for 3≤*t*≤*S*. In addition, let us remove edges $(<\gamma_0, \gamma_2>, <\gamma_0, \gamma_3>)$ and $(<\gamma_4, \gamma_0>, <\gamma_4, \gamma_1>)$ from the corresponding subgroup and replace them by the inter-group edges $(<\gamma_0, \gamma_2>, <\gamma_2, \gamma_0>)$, $(<\gamma_0, \gamma_3>, <\gamma_3, \gamma_0>)$, $(<\gamma_0, \gamma_4>, <\gamma_4, \gamma_0>)$, and $(<\gamma_1, \gamma_4>, <\gamma_4, \gamma_1>)$. Therefore, all of L^2 nodes which have already belonged to separate cycles within separate groups now create a unique L^2-node cycles within the OTIS-*G*.

Case 2. *L is odd:* Let *L* = 2*S*-1 and $\gamma_0, \gamma_1, \gamma_2, ... \gamma_{2S-2}, \gamma_{2S-2}, \gamma_{2S-1} = \gamma_0$ be the L-cycle in G, $L \le |V(G)|$. We have such a cycle in each of 2*S*-1 groups namely $G_{\gamma_0}, G_{\gamma_1}, G_{\gamma_2}, ..., G_{\gamma_{2S-2}}, G_{\gamma_{2S-2}}$ in the OTIS-*G*. The following construction method can build all cycles of length *L*≥3: Remove edge $(<\gamma_i, \gamma_0>, <\gamma_i, \gamma_1>)$ from each sub-graph G_{γ_i}, 2≤*i*≤2*S*-2, and replace it with an inter-group edges of the form $(<\gamma_0, \gamma_i>, <\gamma_i, \gamma_0>)$ and $(<\gamma_1, \gamma_i>, <\gamma_i, \gamma_1>)$ for 2≤*i*≤2*S*-2. Also, remove all edges of form $(<\gamma_0, \gamma_{2t-1}>, <\gamma_0, \gamma_{2t}>)$ and $(<\gamma_1, \gamma_{2t}>, <\gamma_1, \gamma_{2t+1}>)$, for 1≤*t*≤*S*-1, within sub-graphs G_{γ_0} and G_{γ_1}. Finally, add the edge $(<\gamma_0, \gamma_1>, <\gamma_1, \gamma_0>)$ to close the cycle. □

Theorem 4. If *G* is Hamiltonian, then OTIS-*G* is also Hamiltonian.

Proof. Theorem 3 implies that if *G* has a cycle of length $|V(G)|$, then OTIS-*G* has a cycle of length $|V(G)|^2 = |V(OTIS-G)|$. □

In the rest of this section, the relation of two important properties of Hamiltonian-connectedness of the factor graph *G* and pancyclicity of the OTIS-*G* is studied.

Precisely, it will be proved that the OTIS-G network possesses all cycles of length 7, 8, …, and $|V(G)|^2$, provided that the factor graph G is Hamiltonian-connected.

Definition 2. For a given network $G=(V, E)$, G is said to be Hamiltonian-connected if it contains a Hamiltonian path starting from any node $x\in V$ and ending at any node $y\in V-\{x\}$, i.e. a path that leads from x to y and traverses every node of G exactly once.

Definition 3. For a given network $G=(V, E)$ and a given set $\Sigma \subseteq \{3, 4, ..., |V(G)|\}$, G is called to be $\Sigma-pancyclic$ provided that G contains all cycles of length $\delta \in \Sigma$. Comparably, G is said to be $\overline{\Sigma}-pancyclic$ if G contains all cycles of length $\delta \in \{3,4 \ldots |V(G)|\} - \Sigma$.

Definition 4. A graph G is said to be *pancyclic*, if it is $\Sigma-pancyclic$ where $\Sigma=\{3,4 \ldots |V(G)|\}$. Graph G is said to be *weakly pancyclic* [20] if it contains cycles of all lengths between a minimum and a maximum cycle length that can be embedded within G.

Pancyclicity is an important property determining if the topology of a network is suitable for an application in which mapping rings of different lengths into the host network is required. Such cycle embedding in various networks has attracted much attention and been a challenging research issue in recent years [18- 20]. For OTIS-networks, the only result about its Hamiltonian property is the one in [9, 10]. There, authors showed that if the factor graph G is Hamiltonian, so is the OTIS-G. However, the following theorem presents a remarkable relation between Hamiltonian-connectedness of the factor graph G and pancyclicity of the OTIS-G.

Theorem 5. The OTIS-G network is $\overline{\Sigma}-pancyclic$, for $\Sigma=\{3,4,5,6\}$, provided that the factor graph G is Hamiltonian-connected (i.e. OTIS-G is weakly pancyclic).

Proof. To construct an L-cycle within the OTIS-G graph, for $7 \le L \le |V(G)|^2$, we consider two different cases. First, we deal with the problem of finding the cycles of length L, $7 \le L \le 2\times|V(G)|+1$, and then we consider other remaining cycle lengths.

Case 1. Since G is Hamiltonian-connected, we are able to find a Hamiltonian path between two arbitrary neighboring nodes u and v in G. Let $\gamma_u=\gamma_0,\gamma_1,\ldots\gamma_{|V(G)|-2},\gamma_{|V(G)|-1}=\gamma_v$ be the nodes address sequence corresponding to this $|V(G)|-$cycle in G. Two following methods construct all cycles of length $7 \le L \le 2\times|V(G)|+1$ within OTIS-G.

When L is odd: Let $L=2\xi+1$, for $3\le\xi\le L$. We build the L-cycle within OTIS-G graph as follows. (Note that these cycles only use three different groups in OTIS-G)

$C_{2\xi+1}$: $(<\gamma_0,\gamma_1>,<\gamma_1,\gamma_0>) \| (<\gamma_1,\gamma_0>,<\gamma_1,\gamma_1>) \| (<\gamma_1,\gamma_1>,<\gamma_1,\gamma_2>) \| \ldots \| (<\gamma_1,\gamma_{\xi-2}>,<\gamma_1,\gamma_{\xi-1}>) \| (<\gamma_{\xi-1},\gamma_1>,<\gamma_{\xi-1},\gamma_0>) \| (<\gamma_0,\gamma_{\xi-1}>,<\gamma_0,\gamma_{\xi-2}>) \| \ldots \| (<\gamma_0,\gamma_2>,<\gamma_0,\gamma_1>)$.

When L is even: Let $L = 2\xi + 4$, for $2 \le \xi \le L-2$. We build the cycle of length L within OTIS-G as follows (we assume $|V(G)| > 3$):

$$C_{2\xi+4}: (<\gamma_0,\gamma_2>,<\gamma_2,\gamma_0>) \,||\, (<\gamma_2,\gamma_0>,<\gamma_2,\gamma_1>) \,||\, (<\gamma_1,\gamma_2>,<\gamma_1,\gamma_3>) \,||\, \ldots \,||$$
$$(<\gamma_1,\gamma_\xi>,<\gamma_1,\gamma_{\xi+1}>) \,||\, (<\gamma_{\xi+1},\gamma_1>,<\gamma_{\xi+1},\gamma_0>) \,||\, (<\gamma_0,\gamma_{\xi+1}>,<\gamma_0,\gamma_\xi>) \,||$$
$$\ldots \,||\, (<\gamma_0,\gamma_3>,<\gamma_0,\gamma_2>).$$

The above-mentioned cycle consists of joining two paths of length ξ within the groups g_0 and g_1, and two path of length 2 within groups g_2 and g_ξ, respectively. For the case of $|V(G)| = 3$, we compose the cycle of length 8 as follows:

$$C_8: (<\gamma_0,\gamma_2>,<\gamma_2,\gamma_0>) \,||\, (<\gamma_2,\gamma_0>,<\gamma_2,\gamma_2>) \,||\, (<\gamma_2,\gamma_2>,<\gamma_2,\gamma_1>)$$
$$||\, (<\gamma_2,\gamma_1>,<\gamma_1,\gamma_2>) \,||\, (<\gamma_1,\gamma_2>,<\gamma_1,\gamma_1>) \,||\, (<\gamma_1,\gamma_1>,<\gamma_1,\gamma_0>) \,||$$
$$(<\gamma_1,\gamma_0>,<\gamma_0,\gamma_1>) \,||\, (<\gamma_0,\gamma_1>,<\gamma_0,\gamma_2>).$$

Case 2. To build a cycle of length L, for $2 \times |V(G)| + 1 \le L \le |V(G)|^2$, L could be written as $L = K|V(G)| + \xi$, where $2 < \xi < |V(G)|$ and $3 \le K < |V(G)|$ (case of $K = |V(G)|$ could be treated according to theorem 3). The following method generates a cycle of length $K|V(G)| + \xi$ within OTIS-G. The sequence $\gamma_u = \gamma_0, \gamma_1, \ldots \gamma_{|V(G)|-2}, \gamma_{|V(G)|-1} = \gamma_v$ forms a Hamiltonian path between two arbitrary nodes of u and v in G showed by HP(u v). Furthermore, we choose group G_m to construct a path of length ξ within it, where $m = max(K, \xi)$.

$$C_{K|V(G)|+\xi}: (<\gamma_m,\gamma_{m-1}>,<\gamma_m,\gamma_{m-2}>) \,||\, (<\gamma_m,\gamma_{m-2}>,<\gamma_m,\gamma_{m-3}>) \,||\, \ldots \,||$$
$$(<\gamma_m,\gamma_{m-\xi+1}>,<\gamma_m,\gamma_{m-\xi}>) \,||\, (<\gamma_m,\gamma_{m-\xi}>,<\gamma_{m-\xi},\gamma_m>)$$
$$||\, HP(<\gamma_{m-\xi},\gamma_m>,<\gamma_{m-\xi},\gamma_{m-\xi+1}>) \,||\, (<\gamma_{m-\xi},\gamma_{m-\xi+1}>,<\gamma_{m-\xi+1},\gamma_{m-\xi}>) \,||$$
$$HP(<\gamma_{m-\xi+1},\gamma_{m-\xi}>,<\gamma_{m-\xi+1},\gamma_{m-\xi+2}>) \,||\, (<\gamma_{m-\xi+1},\gamma_{m-\xi+2}>,<\gamma_{m-\xi+2},\gamma_{m-\xi+1}>)$$
$$||\, \ldots \,||\, HP(<\gamma_{m-\xi+\zeta+1},\gamma_{m-\xi+\zeta}>,<\gamma_{m-\xi+\zeta+1},\gamma_{m-\xi+\zeta+2}>) \,||$$
$$(<\gamma_{m-\xi+\zeta+1},\gamma_{m-\xi+\zeta+2}>,<\gamma_{m-\xi+\zeta+2},\gamma_{m-\xi+\zeta+1}>) \,||\, \ldots \,||$$
$$HP(<\gamma_{m-\xi+k-2},\gamma_{m-\xi+k-3}>,<\gamma_{m-\xi+k-2},\gamma_{m-1}>) \,||$$
$$(<\gamma_{m-\xi+k-2},\gamma_{m-1}>,<\gamma_{m-1},\gamma_{m-\xi+k-2}>) \,||\, HP(<\gamma_{m-1},\gamma_{m-\xi+k-2}>,<\gamma_{m-1},\gamma_m>) \,||$$
$$(<\gamma_{m-1},\gamma_m>,<\gamma_m,\gamma_{m-1}>).$$

In the above cycle, variable ζ changes between 0 and k-3.

For the special case of $L = K|V(G)|+1$, $3 \le K < |V(G)|$, the following construction method could be used. Let $BackP(<\gamma_i,\gamma_j>,<\gamma_i,\gamma_{j'}>)$, for $0 \le j < j' \le |V(G)|-1$ denote a path within group G_i which starts from node $<\gamma_i,\gamma_j>$, ends at node $<\gamma_i,\gamma_{j'}>$, and does not include any node of the form $<\gamma_i,\gamma_l>$ for all $j < l < j'$. This path can be easily constructed as follows:

$$BackP(<\gamma_i,\gamma_j>,<\gamma_i,\gamma_{j'}>): <\gamma_i,\gamma_j> \| <\gamma_i,\gamma_{j-1}> \| \dots \| <\gamma_i,\gamma_0> \| <\gamma_i,\gamma_{|V(G)-1|}> \| \dots \| <\gamma_i,\gamma_{j'+1}> \| <\gamma_i,\gamma_{j'}>.$$

Now, we can easily state our method to build a cycle of length $L = K|V(G)|+1$:

$$C_{K|V(G)|+1}: BackP(<\gamma_0,\gamma_{|V(G)|-1}>,<\gamma_0,\gamma_1>) \| (<\gamma_0,\gamma_1>,<\gamma_1,\gamma_0>) \| BackP(<\gamma_1,\gamma_0>,<\gamma_1,\gamma_2>) \| \dots \| BackP(<\gamma_i,\gamma_{i-1}>,<\gamma_i,\gamma_{i+1}>) \| (<\gamma_i,\gamma_{i+1}>,<\gamma_{i+1},\gamma_i>) \| \dots \| BackP(<\gamma_{K-2},\gamma_{K-3}>,<\gamma_{K-2},\gamma_{K-1}>) \| (<\gamma_{K-2},\gamma_{K-1}>,<\gamma_{K-1},\gamma_{K-2}>) \| BackP(<\gamma_{K-1},\gamma_{K-2}>,<\gamma_{K-1},\gamma_{|V(G)|-1}>) \| (<\gamma_{K-1},\gamma_{|V(G)|-1}>,<\gamma_{|V(G)|-1},\gamma_{K-1}>) \| HP(<\gamma_{|V(G)|-1},\gamma_{K-1}>,<\gamma_{|V(G)|-1},\gamma_0>) \| (<\gamma_{|V(G)|-1},\gamma_0>,<\gamma_0,\gamma_{|V(G)|-1}>).$$

The abovementioned cycle consists of concatenating K-1 paths of length $|V(G)|$-1, one path of length K, and one Hamiltonian path of length $|V(G)|$, contributing in a cycle of length $K|V(G)|+1$ in the OTIS-G. □

The above cases show that the Hamiltonian-connectedness of factor graph G implies the weak Pancyclicity of the OTIS-G for $\Sigma = \{7,8 \dots |V(G)|^2\}$.

4 Conclusions and Future Work

The OTIS structure is an attractive inter-node communication network for large multiprocessor systems as it offers benefits from both optical and electrical technologies. A number of suitable properties of OTIS-G networks including basic topological properties, embedding, routing, broadcasting, and fault tolerance have been studied in the past. As well, a number of parallel algorithms like parallel sorting and Lagrange interpolation algorithm have been studied [5-14]. In this paper, we addressed an open question stated in [6]; it says that the OTIS-G possesses an L^2-cycle provided that the factor graph G contains an L-cycle. Moreover, embedding of different size cycles into networks is an important issue for development of some algorithms of control/data flow for distributed computation, which is also known as Pancyclicity. We have expressed in mathematical terms a general definition for Pancyclicity of a graph which can cover all the different cases previously stated in the literature. In this connection, we have proved the Weak Pancyclicity of OTIS-*G* when the factor graph *G* is Hamiltonian-connected.

The future work, in this line, includes identifying the necessary and sufficient conditions of factor graph *G* for the Pancyclicity of the OTIS-*G*.

References

1. Marsden, G., Marchand, P., Harvey, P., Esener, S.: Optical transpose interconnection system architectures. Optics Lett. 18, 1083–1085 (1993)
2. Yeh, C.-H., Parhami, B.: Swapped networks: unifying the architectures and algorithms of a wide class of hierarchical parallel processors. In: Proc. Int'l Conf. Parallel and Distributed Systems, pp. 230–237 (1996)
3. Zane, F., Marchand, P., Paturi, R., Esener, S.: Scalable network architectures using the optical transpose interconnection system. J. Parallel Distribute Computing 60, 521–538 (2000)
4. Krishnamoorthy, P., Marchand, F., Kiamilev, F., Esener, S.: Grain–size considerations for optoelectronic multistage interconnection networks. Applied Optics 31(26), 5480–5507 (1992)
5. Wang, C., Sahni, S.: Basic operations on the OTIS-mesh optoelectronic computer. IEEE Transactions on Parallel and Distributed Systems 9(12), 1226–1236 (1998)
6. Day, K., Al-Ayyoub, A.: Topological properties of OTIS-networks. IEEE Trans. Parallel Distrib. Systems 13, 359–366 (2002)
7. HoseinyFarahabady, M., Sarbazi-Azad, H.: The grid-pyramid: A generalized pyramid network. Journal of Supercomputing 37(1), 23–45 (2006)
8. Rajasekaran, S., Sahni, S.: Randomized routing, selection, and sorting on the OTIS-mesh. IEEE Trans. on Parallel and Distributed Systems 9(9), 833–840 (1998)
9. Parhami, B.: Swapped interconnection networks: Topological, performance, and robustness attributes. J. Parallel Distrib. Comput. 65(11), 1443–1452 (2005)
10. Hoseiny Farahabady, M., Sarbazi-Azad, H.: Algorithmic Construction of Hamiltonian Cycles in OTIS-Networks. In: IASTED 2005 Networks and Communication Systems, pp. 464–110 (2005)
11. Osterloh, A.: Sorting on the OTIS-mesh. In: IPDPS 2000, pp. 269–274 (2000)
12. Prasanta, K.: Polynomial interpolation and polynomial root finding on OTIS-mesh. Parallel Computing 32, 301–312 (2006)
13. Wang, C.-F., Sahni, S.: Matrix multiplication on the OTIS-mesh optoelectronic computer. In: MPPOI 1999, pp. 131–138 (1999)
14. Hoseiny Farahabady, M., Imani, N., Sarbazi-Azad, H.: Some Topological and Combinatorial Properties of WK-Recursive Mesh and WK-Pyramid Interconnection Networks. Technical Report, IPM School of Computer Science, Tehran, Iran (2007)
15. Hashemi Najaf-abadi, H., Sarbazi-Azad, H.: Comparative Evaluation of Adaptive and Deterministic Routing in the OTIS-Hypercube. In: Yew, P.-C., Xue, J. (eds.) ACSAC 2004. LNCS, vol. 3189, pp. 349–362. Springer, Heidelberg (2004)
16. Hashemi Najaf-abadi, H., Sarbazi-Azad, H.: Multicast communication in OTIS-hypercube multi-computer systems. International Journal of High Performance Computing and Networking 4(3-4), 161–173 (2006)
17. Hashemi Najaf-abadi, H., Sarbazi-azad, H.: On routing capabilities and performance of cube-based OTIS-systems. In: SPECTS 2004, pp. 161–166 (2004)

18. Xu, J., Ma, M.: Cycles in folded hypercubes. Applied Mathematics Letters 19, 140–145 (2006)
19. Araki, T.: Edge-pancyclicity of recursive circulants. Information Processing Letters 88(6), 287–292 (2003)
20. Hsieh, S., Chang, N.: Hamiltonian Path Embedding and Pancyclicity on the Mobius Cube with Faulty Nodes and Faulty Edges. IEEE Transactions on Computers 55(7), 854–863 (2006)
21. Flandrin, E., Li, H., Marczyk, L., Schiermeyer, I., Wo´zniak, M.: Chvtal–Erdos condition and pancyclism. Disc. Math. Graph Theory 26, 335–342 (2006)
22. Li, T.: Cycle embedding in star graphs with edge faults. Applied Mathematics and Computation 167, 891–900 (2005)
23. Fan, J., Lin, X., Jia, X.: Node-Pancyclicity and edge-Pancyclicity of crossed cubes. Information Processing Letters 93, 133–138 (2005)

MC2DR: Multi-cycle Deadlock Detection and Recovery Algorithm for Distributed Systems

Md. Abdur Razzaque, Md. Mamun-Or-Rashid, and Choong Seon Hong

Department of Computer Engineering, Kyung Hee University
1, Seocheon, Giheung, Yongin, Gyeonggi, Korea, 449-701
m_a_razzaque@yahoo.com, mamun@networking.khu.ac.kr, cshong@khu.ac.kr

Abstract. Even though there have been many research works on distributed deadlock detection and recovery mechanisms, the multi-cycle deadlock problems are not extensively studied yet. This paper proposes a multi-cycle deadlock detection and recovery mechanism, named as MC2DR. Most existing algorithms use edge-chasing technique for deadlock detection where a special message called *probe* is propagated from an initiator process and echoes are sent back to it that carries on necessary information for deadlock detection. These algorithms either can't detect deadlocks in which the initiator is indirectly involved or a single process is involved in multiple deadlock cycles. Some of them often detect phantom deadlocks also. MC2DR defines new structures for *probe* and *victim* messages, allows any node to detect deadlock dynamically, which overcomes the aforementioned problems and increases the deadlock resolution efficiency. The simulation results show that our algorithm outperforms the existing probe based algorithms.

1 Introduction

A distributed deadlock can be defined as cyclic and inactive indefinite waiting of a set of processes for exclusive access to local and/or remote resources of the system. Such a deadlock state persists until a deadlock resolution action is taken. Persistence of a deadlock has two major deficiencies: *first*, all the resources held by deadlocked processes are not available to any other process and the *second*, the deadlock persistence time gets added to the response time of each process involved in the deadlock. Therefore, the problem of prompt and efficient detection and resolution of a deadlock is an important fundamental issue of distributed systems [1, 3, 10]. [1] The state of a distributed system that represents the state of process-process dependency is dynamic and is often modeled by a directed graph called *Wait-for-Graph* (WFG), where each node represents a process and an arc is originated from a process waiting for another process holding that resource [13, 14]. A cycle in the WFG represents a deadlock. From now on, processes in the distributed system will be termed as nodes in this paper.

[1] This research was supported by the MIC, Korea, under the ITRC support program supervised by the IITA, Grant no - (IITA-2006 - (C1090-0602-0002)).

R. Perrott et al. (Eds.): HPCC 2007, LNCS 4782, pp. 554–565, 2007.

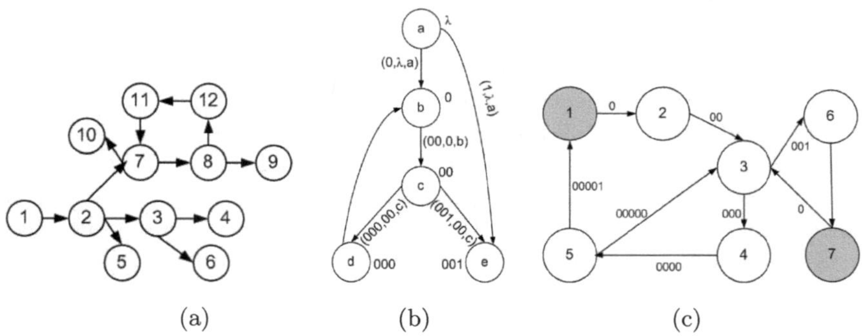

Fig. 1. Wait-for-Graphs, some example node-node dependency scenarios

As shown in Fig. 1(a), node '2' is called *successor* of *parent* node '1' and {7, 8, 11, 12} is called a *deadlocked* set of processes. Such state graph is distributed over many sites, may form multiple dependency cycles and thereby many nodes are blocked indefinitely [12, 13].

The most widely used distributed deadlock detection scheme is *edge-chasing* that uses a short message called *probe.* If a node suspects the presence of a deadlock, it independently initiates the detection algorithm, creates a *probe* message and propagates it outward to all of its *successor* nodes. Deadlock is declared when this *probe* message gets back to the initiator *i.e.*, forming a dependency cycle [1, 2, 3, 4, 5, 6]. The key limitation of these algorithms is that they are unable to detect deadlocks whenever the initiator is not belong to the deadlock cycle. In the worst case, this may result in transmission of almost N^2 messages to detect a deadlock, where N represents the number of blocked nodes in the WFG. Algorithms proposed in [7, 8, 9, 10, 11] overcome this problem but some of them detect phantom deadlocks and the rests can't detect deadlocks in the case that a single node is involved in multiple deadlock cycles.

Our proposed algorithm, MC2DR, introduces a modified *probe* message structure, a *victim* message structure and for each node a *probe* storage structure. The contributions of MC2DR includes: (i) it can detect all deadlocks reachable from the initiator of the algorithm in single execution, even though the initiator does not belong to any deadlock, (ii) it can detect multi-cycle deadlocks *i.e.*, deadlocks where a single process is involved in many deadlock cycles, (iii) it decreases the deadlock detection algorithm initiations, phantom deadlock detections, deadlock detection duration and the number of useless messages and (iv) it provides with an efficient deadlock resolution method.

The rest of the paper is organized as follows. A thorough study and critics of state-of-the-art probe based algorithms are presented in section 2. Section 3 introduces the network and computation models, section 4 and 5 describe the proposed algorithm and its correctness proof respectively. Simulation and performance comparisons are presented in section 6 and section 7 concludes the paper.

2 Related Works

The key concept of CMH algorithm [1, 2] is that the initiator propagates *probe* message in the WFG and declares a deadlock upon receiving its own *probe* gets back. Sinha and Natarajan [3] proposed the use of priorities to reduce the no. of *probe* messages. Choudhary et. al. [4] found some weaknesses of this algorithm and Kashemkalyani and Singhal [5] proposed further modifications to this and provided with a correctness proof. Kim Y.M. et. al. [6] proposed the idea of *barriers* to allow the deadlock to be resolved without waiting for the token to return, thereby reducing the average deadlock persistence time considerably.

None of the above algorithms can detect deadlocks in which the initiator is not directly involved. Suppose in Fig. 1, node '1' initiates algorithm execution, the deadlock cycle {7, 8, 12, 11, 7} can't be detected by any of the above algorithms as because in all those algorithms, deadlock is declared only if the initiator ID matches with the destination ID of the *probe* message.

S. Lee in [7] proposed a *probe* based algorithm that exploits reply messages to carry the information required for deadlock detection. As a result, the *probe* message does not need to travel a long way returning back to its initiator and thereby time and communication costs are reduced up to half of those of the existing algorithms. Even though this algorithm can detect deadlocks where the initiator node is not directly involved, but except the initiator no other nodes will be able to detect deadlocks. For instance in Fig. 1(a), if node 1, 7 and 12 initiate algorithm executions one after another in order with little time intervals, then the same deadlock cycle {7,8,12,11,7} will be detected by all of them, which is a system overhead.

S. Lee and J. L. Kim in [8] proposed an algorithm to resolve deadlock in single execution even though the initiator doesn't belong to any deadlock. This is achieved by building a tree through the propagation of the probes and having each tree node collects information on dependency relationship (route string) among its subtree nodes to find deadlocks based upon the information. But, we found several drawbacks and incapabilities of this algorithm. *First*, all deadlocks reachable from the initiator may not be resolved by a single execution of the algorithm since a deadlock may consist of only tree and cross edges in the constructed tree. *Second*, deadlock detection algorithm works correctly for single execution of the algorithm, but it would detect phantom deadlocks in case of multiple executions. To prove the above statement, let we consider in Fig. 1(b), node 'e' initiates algorithm at sometime later than node 'a' and in addition to dependency edges shown in the figure, there are two other edges, one from 'e' to 'c' and another from 'c' to 'b'. Node 'b' forwarded *probe* message initiated by 'a' and then as the steps of algorithm [8] phantom deadlock will be detected if it receives *probe* message initiated by 'e' before receiving one from 'd'. This is happened due to appending unique bits for each successor (0, 1, 2,...,m) to its own path string. Our algorithm resolves this problem by appending system wide unique ID of individual nodes.

The algorithm [9] proposed by the same authors, criticized [8] for not giving any deadlock resolution method and proposed priority based victim detection. This may lead to starvation for low priority nodes. N. Farajzadeh et. al. in [10, 11] considered simultaneous execution of many instances of the algorithm but what happens if a single process is involved in multiple deadlock cycles was not illustrated. Simulation has not also been carried out. Hence, their algorithm is so weak that even in simple example scenarios it can't detect deadlocks.

Suppose in Fig. 1(c), node '3' stores the initiator ID (1) and route string (00) of the *probe* message initiated by node '1', forwards the message with necessary modification to node '4' and '6', and then receives another *probe* message initiated by node '7', at this stage according to their algorithm node '3' replaces the stored route string with new one (0). Due to this incorrect replacement, node '3' will not be able to detect deadlock cycle {3, 4, 5, 3} although it does exist. It is not necessary to unfold that such incorrect replacement of existing route string might also cause the probe message infinitely moving around the cycle and increase the number of algorithm initiations as well as message passing.

3 Network and Computation Model

3.1 Network Model

We assume that each data object in our distributed system is given a unique lock that can't be shared by more than one node. A node can make request for locks residing either at local or remote sites. A request for a lock is processed by the lock manager to determine whether the lock can be granted. If the requested lock is free, it is granted immediately; otherwise, the lock manager sends a reject message to the requesting node and inserts the requesting node ID into the waiting list for the lock. A reject message carries the identifier of the node which is currently holding the resource. Upon receiving a reject message for any of the locks requested, the node remains *blocked* until the lock is granted, and inserts the lock holder's ID into the successor list.

It is assumed that there might be one or more nodes running on each site as we are concerned with both distributed and local deadlocks. There is no shared memory in the system, nodes communicate with each other by message passing. Each node is uniquely identified by its {site id:process id} pair. But for simplicity of explanation of the proposed algorithm, we have assigned unique integer numbers (0, 1, 2, 3,...,m) to all nodes as shown in Fig. 3.

Another important characteristic of our network model is that the underlying channel is FIFO, *i.e.*, messages are arrived at the destination nodes in the order in which they were sent from the source nodes without any loss or duplication. Message propagation delay is arbitrary but finite. MC2DR is proposed for multi-resource model where a single process may make multiple lock requests at a time. In this model, the condition for a *blocked* node to get *unblocked* is expressed as a predicate involving the requested resources. In the WFG, there will be no self-loop *i.e.*, no node make requests for resources held by itself.

InitID	VictimID	DepCnt	RouteString

(a) Probe Message

InitID	VictimID

(b) Victim Message

Fig. 2. Structure of Probe and Victim Messages

3.2 Computation Model

A node can be in any of the two states at any time instant: *active* and *blocked*. If the requested lock is not available *i.e.*, it is being used by some other node then the requesting node will enter the *blocked* state until the resource is obtained. Deadlock occurs when a set of nodes wait for each other for an indefinite period to obtain their intended resources.

The *probe* message used for deadlock detection in MC2DR consists of four fields as shown in Fig. 2(a). The first field *InitID* contains the identity of initiator of the algorithm. *VictimID* is the identity of the node to be victimized upon detection of the deadlock and *DepCnt* of a node represents the number of successors for which it is waiting for resources. The fourth field, *RouteString*, contains the node IDs visited by *probe* message in order.

At each node there will be a *probe* message storage structure, named *ProbeStorage*, same as that of probe message for temporary storage of probes. At most one probe message is stored in *ProbeStorage* at a particular time. MC2DR is history independent and upon detection of a deadlock, respective *probe* message is erased from storage. The node that detects the deadlock sends a *victim* message to the node found to be victimized for deadlock resolution. This message will also be used for deleting *probes* from respective storage entries. This short message contains just first two fields of *probe* message as shown in Fig. 2(b).

4 Proposed Algorithm

4.1 Informal Description of the Algorithm

We have found in our study that a correct and efficient deadlock detection algorithm needs to consider a number of parameters and corresponding strategies.

Strategies for algorithm initiation. A node initiates the deadlock detection algorithm execution if it waits for one or more resources for a predefined timeout period, T_0 and its *probe* storage is empty. But if T_0 is shorter, then many nodes may be aborted unnecessarily, and if it is longer, then deadlocks will persist for a long time. Choosing appropriate value of T_0 is the most critical issue as because it does depend on several system environment factors such as process mix, resource request and release patterns, resource holding time, and the average number of locks held (locked) by nodes. As the above dependency factors change dynamically, the value of T_0 is also set dynamically in our algorithm. If the value of T_0 is decreased (or increased) at a node for each increase

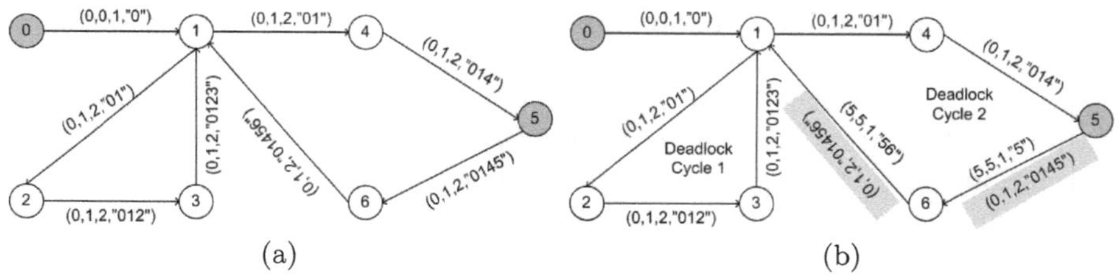

Fig. 3. Example WFGs in our network model with edges labeled by probe messages

(or decrease) of *DepCnt* (defined in section 3.2) our simulation results show that average deadlock detection duration and the average number of algorithm executions are decreased. Dynamic updating of T_0 value also increases the probability that the node involves in multiple deadlock cycles would initiate the algorithm execution.

Probe message forwarding policy. On reception of probe message, a node first checks the emptiness of its *ProbeStorage*. If it is found to be empty (*i.e.*, till now no *probe* message is forwarded by this node), then it compares its own *DepCnt* value with probe's *DepCnt* value. If this node's *DepCnt* is higher, then probe's *VictimID* and *DepCnt* values are updated with this node's ID and *DepCnt* values respectively; otherwise the values are kept intact. Before forwarding the *probe* message to all successors of this node, probe's *RouteString* field is updated by appending this node's ID at last of existing string (*i.e.*, concatenate operation). One copy of updated *probe* message is saved in *ProbeStorage* of this node. For example, in Fig. 3(a), node '0' has initiated execution and send *probe* message (0,0,1,"0") to its successor node 1. As node 1's *ProbeStorage* is empty and *DepCnt* value is 2, it has updated the *probe* message, stored the modified *probe* (0,1,2,"01") in *ProbeStorage* and forwarded to its successors 2 and 4. Nodes 2, 3, 4, 5 and 6 have updated only the *RouteString* field of the *probe* message and forwarded to their successors.

Deadlock detection. If the *ProbeStorage* is nonempty, the node first goes for checking whether the stored route string is a prefix of the received probe's route string. If it is, deadlock is detected and otherwise the *probe* message is discarded. *Probe* message is also discarded by a node that has just detected a deadlock. So, MC2DR can detect deadlock cycle at any node right at the moment the traveled path of *probe* message makes a dependency cycle. Node '1' in Fig. 3(a) has eventually got back its forwarded *probe* and detected one of the two deadlock cycles {1,2,3,1} and {1,4,5,6,1}. If the *probe* message for deadlock cycle {1,2,3,1} is received first then that from node '6' is discarded or vice-versa. Again, if node '4' would be the successor of node '6' then two deadlock cycles {1,2,3,1} and {4,5,6,4} would be detected (by a single probe) by node 1 and 4 respectively, even though none of them is the initiator.

Strategies for deadlock resolution. A deadlock is resolved by aborting at least one node involved in the deadlock and granting the released resources to other nodes. When a deadlock is detected, the speed of its resolution depends on how much information about it is available, which in turn depends on how much information is passed around during the deadlock detection phase. We opine that rather than victimizing initiator node or node with lowest priority, it is better to victimize a node which is more likely to be responsible for multiple deadlocks. To make the above notion a success, MC2DR selects the node with highest *DepCnt* value as victim and the deadlock detector node sends a *victim* message to all successors. If the detector node is not the initiator, it also sends the *victim* message to all *simply blocked* (node that is blocked but not a member of deadlock cycle) nodes. On reception of this message, the *victim* node first forwards it to all of its successors and then releases all locks held by it and kills itself, other nodes delete deadlock detection information from their ProbeStorage memories. Node '1' in Fig. 3(a) has killed itself as because it has the highest *DepCnt* value amongst the members in any of the cycles. Node '1' is not the initiator, so it has also sent the *victim* message to *simply blocked* node '0'. Node '3' and node '6' stop further propagation of *victim* message.

One exceptional condition. Even though the probability of appearing the following exceptional condition in our algorithm is very low, we have to block it for the sake of algorithm's correctness. One exception of the above mechanism in *probe* message discard policy is that only if any node Q is the initiator of another *probe* message then rather than discarding, Q will keep the message in a buffer space and waits for Q's probe to return back. If Q's probe gets back to it or Q receives a *victim* message within average deadlock detection period T_d (computed as in [9]), then the buffered *probe* is discarded, otherwise it is forwarded to Q's all successors. The last one is a worst case situation in MC2DR where node Q defers from detecting any deadlock. Let we consider in Fig. 3(b), immediately after node '0' has initiated execution, node '5' starts another one and node '1' receives second *probe* from node '6' after it has forwarded the first one, so the second *probe* is discarded. Node '5' has blocked first *probe* from further forwarding as it is the initiator of second *probe* and waited for T_d. In this scenario, the chance is very high that within T_d, node '5' receives *victim* message sent by node '1 (on detection of deadlock cycle 1) and discards first *probe* message, deletes second *probe* from it's *ProbeStorage* and forwards *victim* message to its successor node '6'; otherwise, node '5' forwards first *probe* as described in previous part and keeps itself away from detecting any deadlock.

4.2 Deadlock Detection and Resolution Algorithm

For a particular node *i*, pseudo code for deadlock detection and recovery algorithm is presented in Fig. 4.

```
Algorithm_Initiation() {
    int W; //waiting time for a particular resource
    probe p; allocate memory for p;
    if (W > To && ProbeStorage == NULL) {
        p = Create_Probe(i); Send_Probe(i, p);}
}

probe Create_Probe(node i) {
    p.InitID = i.ID;
    p.VictimID = i.ID; p.DepCnt = i.DepCnt;
    p.RouteString = i.ID; return (p);
}

Send_Probe(node i, probe p){
    int j = i.DeptCnt;
    while (j){ //sends probe to all successors, j
        send (j, p); j--; }
}

Receive_Probe(probe p){
    if (ProbeStorage == NULL){
        if( p.DepCnt < i.DepCnt) { p.VictimID = i.ID; p.DepCnt = i.DepCnt;}
        p.RouteSting = p.RouteString + i.ID;
        Send_Probe(i,p);  }
    else if( i.RouteString is prefix of p.RouteString){
        Deadlock is detected.
        //send victim message to all successors and simply blocked nodes
        Send_Victim(j, p.VictimID);  }
    else if (i is the initiator of another probe)
        Exception_Handling(p);
    else { Discard (p);} //probe message is discarded
}

Receive_Victim(int VictimID){
    //forward victim message to all successors
    Send_Victim(j, VictimID);
    if(VictimID == i.ID){  // this node is vicitimized
        Release (All locks held by this node);
        Kill (this node); }
    else {Erase Probe message from ProbeStorage;}
}

Exception_Handling(probe p){
    int Td; //avg. deadlock detection period
    put p in a buffer space;
    wait for Td and check for i's receiving probe
    if(i's probe is received){ Discard (p);}
    else { p.RouteSting = p.RouteString + i.ID;
            Store(p); //Store p in ProbeStorage
            Send_Probe(i, p); }
}
```

Fig. 4. Pseudo code of MC2DR

5 Correctness Proof

Theorem 1. *If a deadlock is detected, the corresponding nodes are really in deadlocked state. Phantom deadlocks are not detected.*

Proof. Let's prove it using proof by contradiction. A set of nodes could be detected as in a phantom deadlock, when the detection algorithm misinterprets the existence of a deadlock cycle. This type of misinterpretation can be taken place in MC2DR only and if only at least any two nodes have the same ID. In the case, a false deadlock is detected even though the traveling path of *probe* message does not make a cycle. But this contradicts with our network model, described in section 3.1.

Theorem 2. *A single deadlock cycle would never be detected by more than one node.*

Proof. Again, we use proof by contradiction. A single deadlock cycle could be detected by multiple nodes if and only if any detection algorithm allows multiple probes to be forwarded by a single node But, MC2DR defers it by storing other probes into node's *ProbeStorage*. Only in exceptional condition case (described in section 4.1), it is allowed but at the same time the forwarder node is kept away from detecting any deadlock. Hence, there is no chance of multiple detections of a deadlock. This is a distinctive contribution of MC2DR.

Theorem 3. *Multiple deadlocks can be detected by a single probe message.*

Proof. Let's prove it by using proof by contradiction. Detection of multiple deadlock cycles by a single *probe* is prohibited whenever a detection algorithm does not allow any node other than the initiator to detect deadlock cycles. It is further restricted by priority based probing, where the higher priority nodes discard the *probe* message initiated by low priority nodes. As described in section 4.1 under the heading "Deadlock Detection", MC2DR defers both the above methods and here the intermediary nodes forward the *probe* message towards multiple directions, which enables MC2DR to detect multiple real deadlocks exist in the system.

6 Simulation and Performance Comparison

We have run the simulation program using fixed sites (20) connected with underlying network speed of 100Mbps, but with varying multiprogramming level (MPL), ranges from 10 to 40. The interarrival times for lock requests and the service time for each lock are exponentially distributed for each node. Only write operations to data objects are considered. To increase the degree of lock conflicts, a relatively small size database in comparison with the transaction size has been chosen. Experiments have been carried out in both the light and heavily loaded environments. We have found that deadlocks increases linearly with the

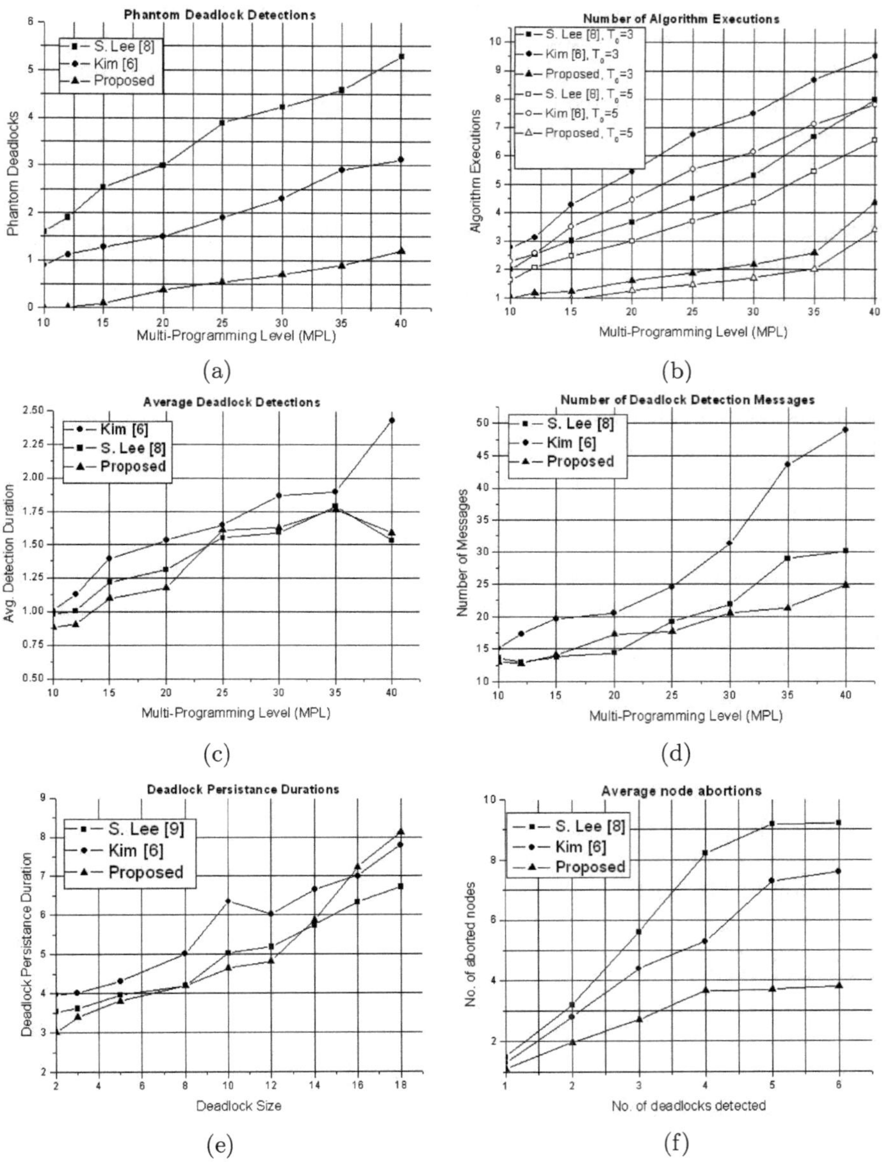

Fig. 5. Simulation results showing performance comparisons

degree of multiprogramming and exponentially with average lock holding time. We have compared the performance of MC2DR with that of Y.M. Kim et. al.'s algorithm [6] and S. Lee's algorithm [8].

As shown in Fig. 5(a), the number of phantom deadlocks detected by MC2DR is 5 times less than Lee's algorithm [8] and almost 3 times less than Kim's algorithm [6]. This is because phantom deadlocks can only be detected by MC2DR in the case of unusual excessive large size of deadlocks, not for any misinterpretation of nonexistence deadlock cycles. Fig. 5(b) shows the mean number of algorithm initiations with varying timeout periods and multiprogramming levels, MC2DR requires about 56% less number of initiations than [6] and about 45% less than that of [8]. This result shows congruence with the theoretical expectation as because MC2DR dynamically controls algorithm initiations (see section 4.1).

Average deadlock detection duration resulted from Kim's algorithm [6] and our algorithm is almost same for higher MPL values (>25) and is slightly less than that from Lee's algorithm [8], as shown in Fig. 5(c). For low to medium MPL values (<20) MC2DR takes 30% to 50% less time than Kim's algorithm [6]. It is observed that the deadlock detection duration increases with MPL until the number of nodes reaches to 30 for most graphs and then become almost flat. The reason behind this could be the increase of *simply blocked* nodes with MPL and the increased chance of algorithm initiations by nodes having higher *DepCnt* values. As shown in Fig. 5(d), Kim's algorithm passes 2 times more messages than MC2DR and almost 1.5 times than Lee's algorithm [8] for higher MPL values (>33). This is because in some cases Kim's algorithm requires multiple executions for detecting a single deadlock. For lower to medium MPL values (10-30), it is observed that proposed algorithm and [8] need almost same number of messages to detect deadlocks.

Fig. 5(e) indicates that the mean deadlock persistence duration is nonlinear for all the algorithms. In case of exceptional conditions arisen in MC2DR, the persistence time may be much longer. Graphs of Fig. 5(f) shows that the number of aborted nodes is very less in MC2DR as compared to [6] (almost 50% decreased) and [8] (almost 65% decreased). This is happened as because in MC2DR, node having highest *DepCnt* value is aborted and it is more likely that abortion of single node might untie more than one deadlock cycles. It's a key contribution of MC2DR.

7 Conclusion

Even though the deadlock persistence duration of MC2DR is increased highly in some rarely occurred exceptional conditions, it is more reliable and robust as because it does detect real deadlocks on single execution of the algorithm and ensures detection of all deadlocks reachable from the initiator including multi-cycle deadlocks. Algorithm initiation policy and deadlock resolution mechanism of MC2DR make it more efficient.

References

1. Chandy, K.M., Misra, J., Haas, L.M.: Distributed Deadlock Detection. ACM Transaction on Computer Systems. 144–156 (1983)
2. Chandy, K.M., Misra, J.: A Distributed Algorithm for Detecting Resource Deadlocks in Distributed Systems. In: Proceedings of the ACM Symposium on Principles of Distributed Computing, Ottawa, Canada, pp. 157–164. ACM Press, New York (1982)

3. Sinha, M.K., Natarajan, N.: A Priority Based Distributed Deadlock Detection Algorithm. IEEE Trans. Software Engg. 11(1), 67–80 (1985)
4. Choudhary, A.N., Kohler, W.H., Stankovic, J.A., Towsley, D.: A Modified Priority Based Probe Algorithm for Distributed Deadlock Detection and Resolution. IEEE Transactions on Software Engg. 15(1), 10–17 (1989)
5. Kashemkalyani, A.D., Singhal, M.: Invariant Based Verification of a Distributed Deadlock Detection Algorithm. IEEE Transactions on Software Engineering 17(8), 789–799 (1991)
6. Kim, Y.M., Lai, T.W., Soundarajan, N.: Efficient Distributed Deadlock Detection and Resolution Using Probes, Tokens, and Barriers. In: Proc. Int'l Conf. on Parallel and Distributed Systems, pp. 584–591 (1997)
7. Lee, S.: Fast Detection and Resolution of Generalized Distributed Deadlocks. In: EUROMICRO-PDP 2002 (2002)
8. Lee, S., Kim, J.L.: An Efficient Distributed Deadlock Detection Algorithm. In: Proc. 15th IEEE Int'l Conf. Distributed Computing Systems, pp. 169–178 (1995)

FROCM: A Fair and Low-Overhead Method in SMT Processor*

Shuming Chen and Pengyong Ma

School of Computer Science and Technology,
National University of Defense Technology, Changsha. 410073 China
mpy9608@yahoo.com.cn
mapy@sohu.com

Abstract. Simultaneous Multithreading (SMT)[1][2] and chip multiprocessors (CMP) processors [3] have emerged as the mainstream computing platform in major market segments, including PC, server, and embedded domains. However, prior work on fetch policies almost focuses on throughput optimization. The issue of fairness between threads in progress rates is studied rarely. But without fairness, serious problems, such as thread starvation and priority inversion can arise and render the OS scheduler ineffective. The fairness research methods always disturb the threads running Simultaneous, such as single thread sampling [4]. In this paper, we propose an approach FROCM (Fairness Recalculate Once Cache Miss) to enhance the fairness of running multithreads in SMT processor without disturbing their running states. Using FROCM, every thread's $IPC_{approximately}$ is re-calculated in SMT processor Once Cache Miss, $IPC_{approximately}$ is the approximately value of IPC when the thread runs alone. Using $IPC_{approximately}$, the instructions' issue priority may be changed in due course. We can hold the Fairness value (Fn) higher. Fn is fairness metric defined in this paper, when it is equal to 1, it means utterly fair. Results show that using FROCM, we can hold the most of Fn larger than 0.95, and the throughput hasn't larger change. It needs less hardware to realize the FROCM, including 4 counters, 1 shifter and 1 adder.

1 Introduction

During the last two decades, different architectures were introduced to support multiple threads on a single die (chip). Simultaneous Multi-Threading (SMT) is the active one, in which instructions from multiple threads are fetched, executed and retired on each cycle, sharing most of the resources in the core. SMT processors improve performance by running instructions from several threads at a single cycle. Co-scheduled threads share some resources, such as issue queues, functional units, and on chip cache. The way of allocating shared resources among the threads will affect throughput and fairness. Prior work on fetch policies almost focuses on throughput optimization, such as ICOUNT2.8[1], PDG[5], and so on. The issue of fairness between threads in progress rates is studied rarely. But fairness is a critical

* This work is supported by National Natural Science Foundation of China(No. 60473079) and the Research Fund for the Doctoral Program of Higher Education (No. 20059998026).

R. Perrott et al. (Eds.): HPCC 2007, LNCS 4782, pp. 566–576, 2007.

issue because the Operating System (OS) thread scheduler's effectiveness depends on the hardware to provide fairness to co-scheduled threads. Without fairness, serious problems, such as thread starvation and priority inversion, may arise and render the OS scheduler ineffective.

In multithread processor, we can achieve fairness by many methods. Currently, shared resources allocation is mainly decided by the instruction fetch policy, especially in SMT processor with VLIW architecture.

Ron Gabor and Shlomo Weiss research the Fairness and Throughput in Switch on Event Multithreading[6]. In coarse gain multithreads processors, there is only one thread running in processor simultaneously, every thread's IPC_{alone} (IPC of thread when executed alone) can be accounted easily. IPC_{SOE} (IPC of each thread when executed using Switch on Event with other threads) can get every clock cycle. Every thread's acceleration can be easily worked out. So they can control the processors' fairness according to the thread's acceleration. But in SMT processor, the IPC_{alone} can't be accounted by this method.

Caixia Sun, Hongwei Tang, and Minxuan Zhang proposed a Fetch Policy ICOUNT2.8-fairness with Better Fairness for SMT Processors [4][7]. They take fairness as the main optimization goal. In this policy, relative progress rates of all co-scheduled threads are recorded and detected each cycle. If the range of relative progress rates is lower than a threshold, fairness is met approximately and ICOUNT2.8 is used as the fetch policy. Relative progress rates are used to decide fetch priorities. The lower a thread's relative progress rate is, the higher its fetch priority is. Unfair situation is corrected by this way. But in this policy, in order to get IPC_{alone} dynamically, they employ two phases: sample phase and statistic phase. During the sample phase, the processor has to run in single-thread mode. It will disturb other running threads and the throughput will be reduced.

There are other methods to insure the fairness of system, for example static and dynamic resource partition[11][12], Handling long-latency load[13], and so on.

In this paper, we propose a method FROCM to correct unfairness without sample phase, all the threads run from beginning to end without interrupt. Every thread's $IPC_{approximately}$ is re-calculated in SMT processor Real-timely, $IPC_{approximately}$ is the approximately value of IPC when the thread runs alone. Using $IPC_{approximately}$, we can hold the Fn (fairness metric defined in this paper, when it is equal to 1, it means utterly fair) larger than a threshold. Results show that we can hold the most of Fn larger than 0.95.

The rest of the paper is organized as follows. Section 2 presents the methodology and gives the definition of system's fairness. In Section 3, we detail how to realize FROCM with less hardware in SMT processor with VLIW architecture. Section 4 illustrates the results. Finally, concluding remarks are given in Section 5.

2 Model Proposed and Fairness Definition

Fairness is very important in multi-user or multi-task system. No user would like to wait long respond time. In SMT processor, OS assigns more time slices to threads with higher priority, thus priority-based time slice assignment can work in a SMT processor system as effectively as in a time-shared single-thread processor system. So the hardware must support fairness, otherwise priority inversion may arise. In this paper, we detail how to support fairness by hardware.

In order to judge whether the performance of multithreads in a SMT processor is fair, at first we define several conceptions.

IC[i]: the instruction counters of thread i

IPC_{alone}: IPC of thread when executed alone.

$IPC_{alone}[i]$: the IPC_{alone} of thread i.

$IPC_{SMT}[i]$: IPC of thread i when executed in SMT processor with other threads.

$T_{alone}[i]$: the runtime of thread i when executed alone,

$$T_{alone}[i] = \frac{IC[i]}{IPC_{alone}[i]}. \tag{1}$$

$T_{SMT}[i]$: the runtime of thread i when executed in SMT processor with other threads,

$$T_{SMT}[i] = \frac{IC[i]}{IPC_{SMT}[i]}. \tag{2}$$

$IPC_{approximately}[i]$: when thread i executed in SMT processor with other threads, we calculate the approximately value of its IPC_{alone}, $IPC_{approximately}[i] \approx IPC_{alone}[i]$.

Deceleration[i]: the deceleration of thread [i]'s performance when executed in SMT processor. Because of sharing resource with other threads in SMT processor, it will cause resource confliction. The threads' runtime will enlarge, and then the performance will be decelerated.

$$Deceleration[i] = \frac{T_{alone}[i]}{T_{SMT}[i]}. \tag{3}$$

Apparently $Deceleration[i] \in [0, 1]$.

$Deceleration_{ratio}[i, j]$: the ration of performance deceleration when thread i and thread j run in SMT processor,

$$Deceleration_{ratio}[i, j] = Min(\frac{Deceleration[i]}{Deceleration[j]}, \frac{Deceleration[j]}{Deceleration[i]}), \tag{4}$$

Apparently $Deceleration_{ratio}[i, j] \in [0, 1]$.

Fn: The fairness value we defined in SMT processor,

$$Fn = \underset{\forall i, j}{Min}(Deceleration_{ratio}[i, j]). \tag{5}$$

It is two threads' deceleration ratio that they deceleration value difference is maximal, because of $\forall (i, j), Deceleration_{ratio}[i, j] \in [0, 1]$, so $Fn \in [0, 1]$.

If we want that the system is fair, we should try our best to let the drop of every thread's performance equally. In other words, if we ensure the Fn approximate to 1, the system is fair. We can know from Equation 5:

$$
\begin{aligned}
& Fn = 1 \\
& \Rightarrow Deceleration_{ratio}[i, j] = 1 \\
& \Rightarrow \frac{Deceleration_{ratio}[i]}{Deceleration_{ratio}[j]} = 1 \\
& \Rightarrow \frac{T_{alone}[i]}{T_{SMT}[i]} = \frac{T_{alone}[j]}{T_{SMT}[j]} \\
& \Rightarrow \frac{IPC_{SMT}[i]}{IPC_{alone}[i]} = \frac{IPC_{SMT}[j]}{IPC_{alone}[j]} \\
& \Rightarrow \frac{IPC_{SMT}[i]}{IPC_{SMT}[j]} = \frac{IPC_{alone}[i]}{IPC_{alone}[j]}
\end{aligned} \tag{6}
$$

In other words, if we insure that the ratio of random two threads' IPC in SMT processor is equal to the ratio of the IPC when they run alone, the system is fair. Because $IPC_{alone}[i]$ and $IPC_{alone}[j]$ are the character of thread i and j, they don't change when running in SMT processor. So if we get every thread's IPC_{alone}, we can control the system's fairness accurately.

Usually there are two ways to acquire the thread's IPC_{alone}, static method and dynamic sampling. Static method is that every thread run alone in processor at the beginning, then we can get the IPC_{alone} statistically. This way is unpractical in most of application. In dynamic sampling, the processor runs in single-thread mode. Each thread runs alone for a certain interval respectively. This method has disadvantage too, if the sample phase is too frequently or too long, it will degrade the throughput of the SMT processor. On the other contrary, if the sample phase is infrequently or too short, the statistical value will largely differ from the real IPC_{alone}. We can't get the accurate IPC_{alone}, and then we can't ensure the system's fairness

Now we analysis the threads' run model in SMT and uni-core processor. The IPC_{alone} is correlative with the issue cycle and the memory miss delay.

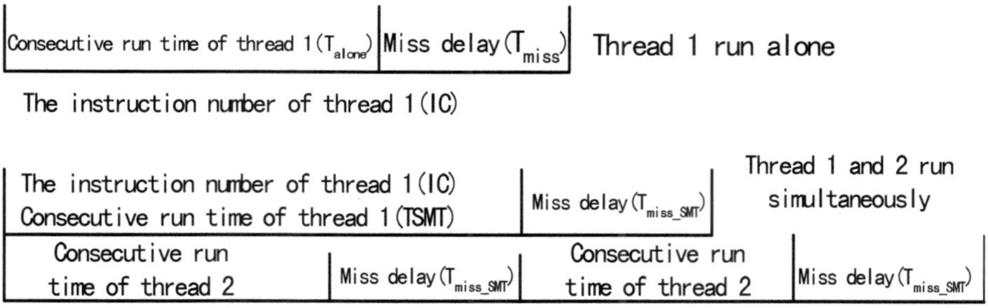

Fig. 1. The sequence of thread 1 run alone and with other thread simultaneously

As Fig. 1 shows, the time of thread 1 is divided into 2 parts per cache miss, the execute time(T_{alone}) and miss delay (T_{miss}). So we can get the IPC of thread 1.

$$IPC_{alone} = \frac{IC}{T_{alone} + T_{miss}}. \tag{7}$$

The T_{miss} is the processor's character, and it is a constant.

When thread 1 run in SMT processor with other threads, the run schedule as the bottom of Fig. 1 (suppose that the cache miss sequence as same as it when thread 1 run alone in uni-core processor.). When cache miss occurs, the instruction counter is same as which when thread run alone. T_{miss} is the character of the uni-core processor and we can acquire it at the beginning. So if we want get the IPC_{alone}, nothing but that we acquire the T_{alone}. As Fig. 1 show, T_{SMT} is larger than T_{alone} in evidence because of resource sharing, so it can't replace the T_{alone}. But in SMT processor, we can statistic the IC of thread 1, the IC is equal to the issue cycle for single issuing processor, that is to say IC=T_{alone}. But it doesn't fit for supper scalar or VLIW processor. In super scalar or VLIW processor, the IC or instruction issuing cycle is not equal to T_{alone}.

As Fig. 2 shows, in VLIW SMT processor, there are five instructions of thread 2 may be issued in time T0, because of share 8 execute units and thread 1 with high priority, only SUB and MPY were issued in T0. The remainder instructions ADD, ADD and MPY were issued in T1. If we counter the issue cycle, the issue cycle of these five instructions is 2. But when thread 2 run alone in uni-core processor, the execute time of these 5 instructions is only 1 cycle. So no matter how many clock cycle one execute packet issued, we only counter once. When thread run alone in uni-core processor with VLIW architecture, one cycle issue one execute packet (EP), so in SMT processor, the number of EP is equal to T_{alone}. EP = T_{alone}.

Substituting EP = T_{alone} into equation 7,

$$IPC_{approximately} = \frac{IC}{EP + T_{miss}} \tag{8}$$

In SMT processor, because of source shared by several threads, the Cache miss ratio may be increased. So in equation 8, we use $IPC_{approximately}$ instead of IPC_{alone}. If system ensures Cache partition is fair, then the increase ratio of threads Cache miss are close to each other. The decrease ratios of threads' IPC_{alone} are approximately, so the system is fair too.

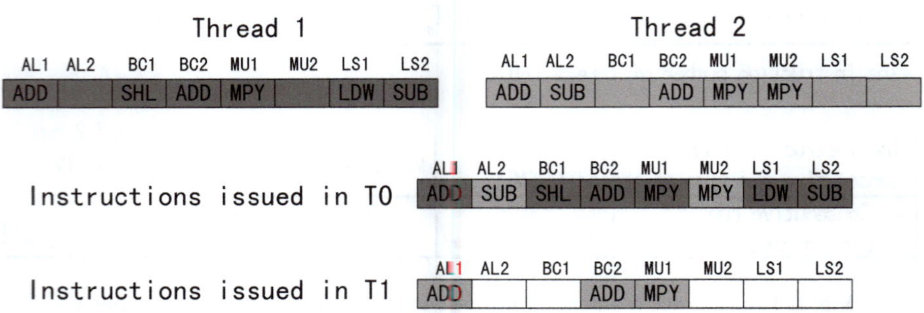

Fig. 2. Separate issue the instructions of thread 2

Substituting $IPC_{approximately}=IPC_{alone}$ into equation 6,

$$\frac{IPC_{SMT}[i]}{IPC_{SMT}[j]}=\frac{IPC_{approximately}[i]}{IPC_{approximately}[j]}. \tag{9}$$

If system ensure equation 9, It is a fair system.

3 Realization with Low-Hardware

Equation 8 shows that if we want acquire the approximate IPC_{alone} of thread: $IPC_{approximately}$, it needs a division unit. It needs lots hardware to realize a division unit, so we should avoid using it. YHFT DSP/900 is an instance of SMT processor. It is a two-ways SMT with 8 execute units shared. Because that it issues 8 instructions one cycle at most, so the $IPC_{approximately}$ value is a real number between 0 and 8, we use a integer register to store it, in order to enhance the precision, we enlarge both threads' $IPC_{approximately}$ to 8 times. It is a value between 0 and 64, so a 6-bits integer register is enough. We may use subtraction instead of division. When we subtract ($EP+T_{miss}$) from IC, if the remainder larger than 0, add 1 to $IPC_{approximately}$. Repetition until the remainder is less than 0. Fig. 3 shows the arithmetic flow of $IPC_{approximately}$.

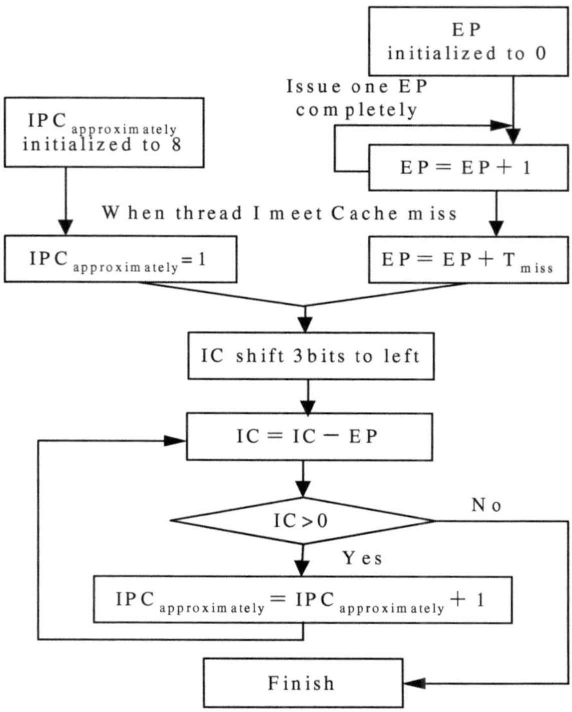

Fig. 3. FROCM arithmetic flow

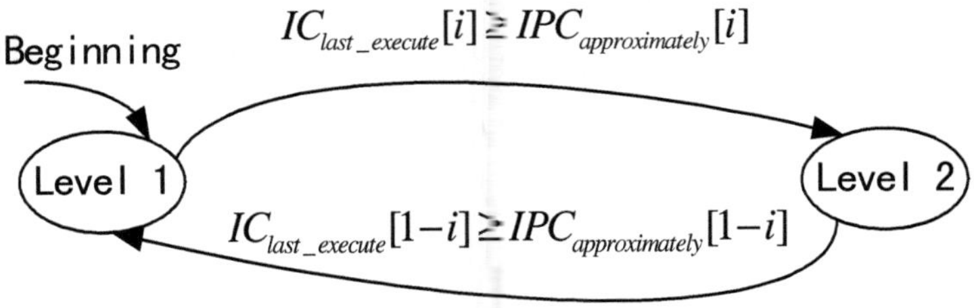

Fig. 4. Transition of thread[i] issue priority

For each thread, processor set a 6-bits register to record the number of issued instructions $IC_{last\_execute}$ after the last modifying issuing priority. Once $IC_{last\_execute}$ is larger than or equal to $IPC_{approximately}$, $IC_{last\_execute} = IC_{last\_execute} - IPC_{approximately}$, and then the thread reduces its instruction issued priority to level 2, enhance the other thread's instruction issued priority to level 1. Fig. 4 shows the transition of thread[i] issue priority.

For example, thread 0 and 1 run in SMT processor simultaneously. $IPC_{approximately}[0]$ is equal to 2.5, and $IPC_{approximately}[1]$ is equal to 3.7. After enlarge 8 times, the values are 20 and 30. After instructions issued in one clock cycle, the $IC_{last\_execute}[0]$ is 24, and $IC_{last\_execute}[1]$ is 29, this means that thread 0 run rapidly. $IC_{last\_execute}[0] = IC_{last\_execute}[0] - IPC_{approximately}[0] = 24 - 20 = 4$, and then thread 0 reduces its instruction issued priority to level 2, enhance the priority of thread 1 to level 1 simultaneously.

Now we analyze the cost of hardware.

Counter IC: in order to record the number of issued instructions per Cache miss. A 16-bits register is enough; it can record 2^{16}=65536 instructions.

Counter EP: in order to record the number of execute packet per Cache miss. It is less than IC, so a 16-bits register is enough.

Shifter: in order to enhance the precision, IC shift 3-bits to left.

Adder: use it to calculate $EP=EP+T_{miss}$ and IC=IC-EP.

Counter $IPC_{approximately}$: it is about 8 times of IPC_{alone}, because IPC_{alone} is a value between 0 and 8, so $IPC_{approximately}$ is between 0 and 64. A 6-bits register is enough.

Counter $IC_{last\_execute}$: the number of issued instructions after last modifying issuing priority. In FROCM, it is less than $IPC_{approximately}$, so a 6-bits register is enough.

4 Results

4.1 Environment Establishing

In order to test the performance of FROCM, we realize RR(round robin) and FROCM of issuing priority in YHFT DSP/900[8][9] simulator respectively. Several program combination run in the simulator with both RR and FROCM respectively.

Now we introduce the architecture of YHFT DSP/900 at first. It is a two-ways SMT processor with VLIW architecture. Each thread hold private fetch instructions unit, register files, and so on. They share 8 execute units, Cache on chip, peripheral function, and so forth. The pipeline is divided into 3 stages, fetch, dispatch and execute. Every clock, it can issue 8 instructions at most.

Fig. 5 is the diagram of YHFT DSP/900 core. The gray area in Fig. 5 is the shared resource. It has two levels Cache. Once level one data or Instruction Cache miss occurred, the delay T_{miss} is 5 clock cycles. The delay of level 2 Cache miss T_{miss} is 60 clock cycles.

Table 1 shows the character of YHFT DSP/900's memory system. It lists the architecture and the bandwidth of data bus. The delay of L2 and the delay of memory is the value in uni-core processor YHFT DSP/700. In other words, it is the delay when only one thread miss request.

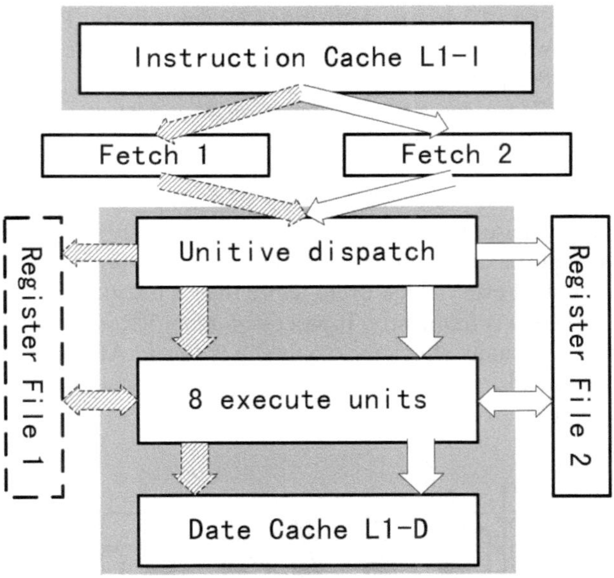

Fig. 5. The diagram of YHFT DSP/900 core

Table 1. The character of memory system

Fetch width	8×32bits instructions
Bandwidth between L1 I-Cache and L2	256bits
Bandwidth between L1 D-Cache and L2	128bits
L1 I-Cache	4KB, direct mapping, 512bits per line, single port, read miss allocation
L1 D-Cache	4KB, 2 ways, 256bits per line, two ports, read miss allocation
L2	64KB, 4 ways, 1024bits per line, single port, read/write miss allocation
latency of L2	5 cycles
latency of memory	60 cycles

4.2 Test Benchmarks

The program is divided into 2 types. One type is the program, which don't aim at any type processor. It includes FFT, the decode/encode of ADPCM in Mibench, the decode/encode of G721 in MediaBench. The other type is the benchmarks that aim at DSP processor, these benchmarks has high parallel and includes dotp_sqr, fir, matrix, and fftSPxSP. When two threads run in YHFT DSP/900 simultaneously, the runtimes aren't equal to each other. In order to test the performance of FROCM exactly, we ensure there are two threads running in processor from beginning to end. When the first thread finished, it will restart again until the second thread finish too. Then we take the time that two threads finished respectively as the threads' runtime in SMT T_{SMT}.

Fig. 6 shows the Fn of various instructions issued priority, RR(round robin) and FROCM. Fig. 7 shows the throughput of system when they adopt RR and FROCM. In Fig. 7, we take the IPC as the system's throughput.

From Fig. 6 and 7, we can find that the fairness of system is enhanced 5% after adopting FROCM policy. The most of Fn are larger than 0.95 without dropping the system's throughput. From Fig. 6, we find the Fn of four program-combinations enhanced greatly. There are FFT+fftSPxSP, dotp_sqr+fir, matrix+fftSPxSP, G721_D+matrix. Because that dotp_sqr, fir, matrix and fftSPxSP are the assemble code, they have high parallel. When they run with other programs, there are many instructions may issued every clock cycle, more than 8 instructions sometime. So the conflict of execute units is frequently. If processor adopt RR policy, it always result in overbalance of execute units allocation, so the Fn is lower. After use FROCM policy, processor control threads run speed by the IPC_{alone}, and allocate the units fairly. So the Fn is enhanced largely.

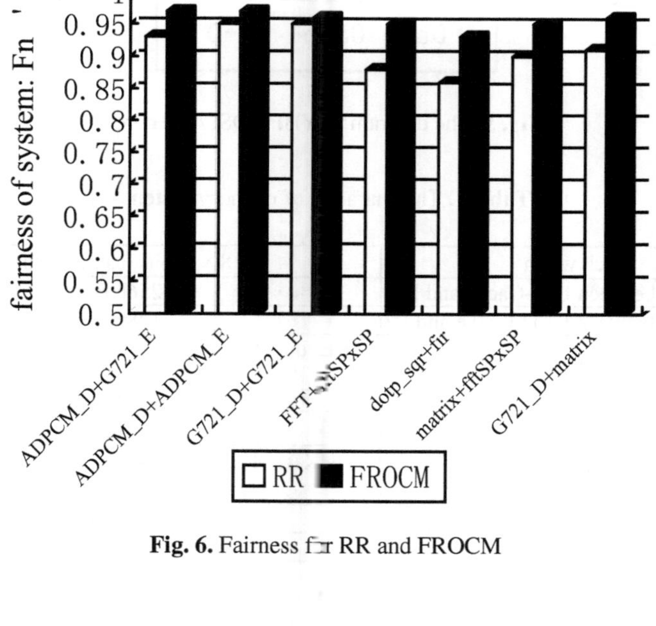

Fig. 6. Fairness for RR and FROCM

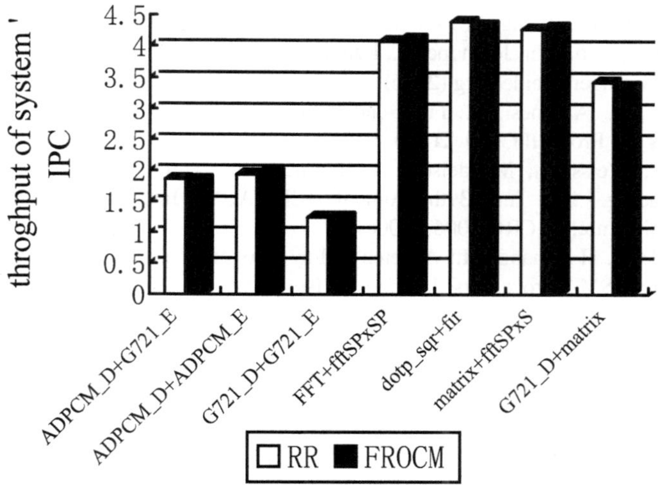

Fig. 7. Throughput for RR and FROCM

5 Conclusions

Because that we can't get the IPC of thread running alone in uni-core process when it runs with other threads in SMT processor, then the system's fairness is difficult to balance. This paper proposes a low and less-hardware fairness policy FROCM. In FROCM, the IPC_{alone} will be recalculated again once the thread meets a Cache miss, then the system will adjust the threads' priority depending on the IPC_{alone} and balance the processor's fairness. The FROCM policy has several advantages. The first is that it doesn't need sample phase; it doesn't interrupt the simultaneous running threads, so the throughput will not be dropped. The second is that it can recalculate the thread's IPC_{alone} at the real time, and then adjust the threads' priority. So it can control the fairness accurately. The last is the low-hardware. Such as a two-ways thread SMT processor with 8words VLIW architecture, it only needs two 16bits counter, two 6bits counter, one shifter and a 16bits adder. The simulations show that FROCM can ensure the most Fn of program-combination larger than 0.95, this indicates that it is an effective fairness policy.

References

1. Tullsen, D.M., Eggers, S., Emer, J., Levy, H., Lo, J., Stamm, R.: Exploiting choice: Instruction fetch and issue on an implementable simultaneous multithreading processor. In: In Proc. of ISCA-23, pp. 191–202 (1996)
2. Tullsen, D.M., Eggers, S., Levy, H.: Simultaneous multithreading: maximizing on-chip parallelism. In: ISCA 1998, pp. 533–544 (1998)
3. Olukotun, K., et al.: The Case for a Single-Chip Multiprocessor. In: Intl. Conf. on Architectural Support for Programming Languages and Operating Systems, pp. 2–11 (1996)

4. Sun, C., Tang, H., Zhang, M.: Enhancing ICOUNT2.8 Fetch Policy with Better Fairness for SMT Processors. In: Jesshope, C., Egan, C. (eds.) ACSAC 2006. LNCS, vol. 4186, pp. 459–465. Springer, Heidelberg (2006)
5. El-Moursy, A., Albonesi, D.: Front-end policies for improved issue efficiency in SMT processors. In: Proc. HPCA-9 (2003)
6. Gabor, R., Weiss, S., Mendelson, A.: Fairness and Throughput in Switch on Event Multithreading. In: The 39th Annual IEEE/ACM International Symposium on Microarchitecture, MICRO 2006. (2006)
7. Sun, C., Tang, H., Zhang, M.: A Fetch Policy Maximizing Throughput and Fairness for Two-Context SMT Processors. In: Cao, J., Nejdl, W., Xu, M. (eds.) APPT 2005. LNCS, vol. 3756, pp. 13–22. Springer, Heidelberg (2005)
8. Shuming, C., Zhentao, L.: Research and Development of High Performance YHFT DSP. Journal of Computer Research and Development, China 43 (2006)
9. Jiang-Hua, W., Shu-Ming, C.: MOSI: a SMT Microarchitecture Based On VLIW Processors. Chinese Journal of Computer 29 (2006)
10. Dan-Yu, Z., Peng-Yong, M., Shu-Ming, C.: The Mechanism of Miss Pipeline Cache Controller based on Two Class VLIW Architecture. Journal of Computer Research and Development, China (2005)
11. Raasch, S.E., Reinhardt, S.K.: The impact of resource partitioning on SMT processors. In: Proc. of PACT-12, p. 15 (2003)
12. Luo, K., Gummaraju, J., Franklin, M.: Balancing throughput and fairness in SMT processors. In: Proc. of the International Symposium on Performance Analysis of Systems and Software, pp. 164–171 (2001)
13. Tullsen, D.M., Brown, J.: Handling long latency loads in a simultaneous multithreading processor. In: Proc. Of MICRO-34, pp. 318–327 (2001)

A Highly Efficient Parallel Algorithm for H.264 Encoder Based on Macro-Block Region Partition

Shuwei Sun, Dong Wang, and Shuming Chen

Computer School, National University of Defense Technology, Changsha, China
shuwei97@sina.com, nudtjum@163.com, smchen@nudt.edu.cn

Abstract. This paper proposes a highly efficient MBRP parallel algorithm for H.264 encoder, which is based on the analysis of data dependencies in H.264 encoder. In the algorithm, the video frames are partitioned into several MB regions, each of which consists of several adjoining columns of macro-blocks (MB), which could be encoded by one processor of a multi-processor system. While starting up the encoding process, the wave-front technique is adopted, and the processors begin encoding process orderly. In the MBRP parallel algorithm, the quantity of data that needs to be exchanged between processors is small, and the loads in different processors are balanced. The algorithm could efficiently encode the video sequence without any influence on the compression ratio. Simulation results show that the proposed MBRP parallel algorithm can achieve higher speedups compared to previous approaches, and the encoding quality is the same as JM 10.2.

1 Introduction

Compared with previous standards, H.264 developed by the JVT(Joint Video Team formed by ISO MPEG and ITU-T VCEG) achieves up to 50% improvement in bit-rate efficiency and more than 4 times of the computational complexity [1], due to many new features including quarter-pixel motion estimation (ME) with variable block sizes and multiple reference frames (up to 16), intra-prediction, integer transformation based on discrete cosine transform (DCT), alternative entropy coding mode Context-based Adaptive Variable Length Coding (CAVLC) or Context-Based Adaptive Binary Arithmetic Coding (CABAC), in-loop de-blocking filter and so on [2]. Therefore, the parallel structure and parallel algorithm are an alternative ways for real-time H.264 video application.

Multi-processor and multi-threading encoding system and parallel algorithms have been discussed in many papers [3][4][5][6]. In [3], the H.264 encoding algorithm is mapped onto multi-processor DSP chips, C3400 MDSP of CRADLE, in which, computing subsystem of 2 RISC core and 4 DSP are used to implement the baseline encoder, forming three pipeline stages, implementing a CIF resolution H.264 codec. However, as a traditional parallel mechanism, task pipelining is not appropriate for H.264 codec due to two reasons: 1) large amount of data need transferring between processors, which demand a lot for the system bandwidth; 2) functions in the H.264 codec have different computing complexity, and it is hard to map the algorithms among processors uniformly, so the final performance is always restricted by the processor with the heaviest load.

R. Perrott et al. (Eds.): HPCC 2007, LNCS 4782, pp. 577–585, 2007.

Because there are a lot of data parallelisms in the video codec, many parallel algorithms are proposed exploiting data parallelism [4][5]. Y. K. Chen, et al. give the parallel algorithms of H.264 encoder based on Intel Hyper-Threading architecture [4], they split a frame into several slices, which are processed by multiple threads, resulting speedups ranging from 3.74 to 4.53 on a system of 4 Intel Xeon processors with Hyper-Threading technique. However, this brings additional overheads on bit-rates: splitting frames into slices increases the bits for information of slice header; macro-blocks (MB) between slices do not need ME and MC, which increases the bits for the transformed coefficients.

Z. Zhao and P. Liang propose the parallel algorithms with wave-front technique based on the analysis of the data dependencies in the H.264 baseline encoder [5], data are mapped onto different processors at the granularity of frames or MB rows, and the final speedups are over 3.0 on software simulator with 4 processors. This method of data partition with the wave-front technique avoids the damage on compression ratio by splitting frames into slices. However, in the motion estimation of H.264 encoder, the search center is the predicted motion vector (PMV), which leads to the data dependencies introduced by ME are not confined by the search range and have the property of uncertainty, however, the analysis in [5] is unsatisfactory, in fact, their method confines the search center at the position of (0,0).

In this paper, we present a method of data partition based on the analysis of the data dependencies in the H.264 encoder, in which, frames are split into several MB regions, each of which includes several adjoining columns of MBs and is mapped onto a different processor, while starting up the encoding process, the wave-front technique is adopted. At last, we give a new efficient parallel algorithm for H.264 encoder based on the MB region partition, i.e. MBRP parallel algorithm.

The rest of the paper is organized as follows: Section 2 provides an overview of the data dependencies in H.264 encoder and the homologous limitative factor to exploit data parallelism. Section 3 details our data partition method based on the MB region and our MBRP parallel algorithm. Section 4 provides the simulation results and Section 5 concludes our work and gives the future work.

2 Data Dependencies in H.264

In the H.264 encoder, a MB is composed of 16×16 luma pixels and 8×8×2 chroma pixels. The reference software JM 10.2 provided by JVT processes each MB in sequence [7], which results in several types of data dependencies that should be avoided in parallel algorithm.

2.1 Data Dependencies Introduced by Inter-prediction

In inter-prediction, the PMV defines the search center of ME, which comes from the motion vectors (MV) of the neighboring sub-blocks, MV_A, MV_B, MV_C, and the corresponding reference indexes [7], as shown in Fig.1 (a). Only the difference between the final optimal MV and the PMV will be encoded. Accordingly, the ME processes of the left, top, and top-right neighboring MBs should be finished before encoding the current MB. Besides, data of the reconstructed pixels in the search

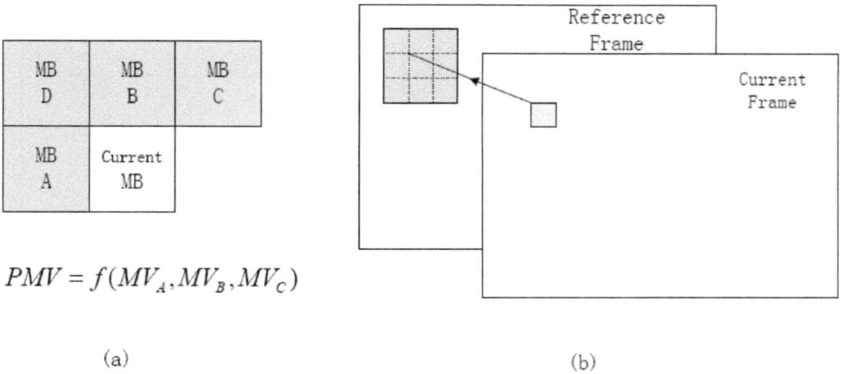

Fig. 1. Data dependencies introduced by inter-prediction

window should be available at that time, if the search range is 16, at least the reconstructed pixels of 9 MBs should be available, as shown in Fig.1 (b).

What should be noticed is, in H.264 encoder the search center of ME is the PMV, which results in that the data dependencies introduced by ME are not confined by the search range and have the property of uncertainty, especially when the RDO is used (there is no restriction for the search center), the final optimal MV may go far beyond the search range. We have collected the final MV of each MB using the JM10.2 software, and the evaluation is carried out on the standard video sequence "Foreman"(CIF, 300 frames), and the main encoder parameters are shown in Table 1. The results show that the biggest MV value is 61 pixels (in horizontal axis), which means that the final MV may be as four times long as the search range. The value may be bigger for sequence with high amount of movement. If RDO is not used, the search center is restricted to be smaller than search range; therefore the final MV could be as two times long as the search range at most.

Table 1. Encoder parameters used for evaluation

Search range	16	QP	28
ME algorithm	FS	Hadamard transform	Used
Number of reference frames	1	RDO mode decision	Used
Sequence type	IPPP	Entropy coding method	CAVLC

2.2 Data Dependencies Introduced by Intra-prediction and In-Loop Filter

In the intra-prediction, the reconstructed pixels of the neighboring MBs in the same frame should be available before encoding the current MB, as shown in Fig. 2 (a), in which the white field figures the current MB and the grey field figure the reconstructed pixels, which are distributed in the left, top-left, top and top-right neighboring MBs. What should be noticed is that these reconstructed data are not the same as those used in the inter-prediction; the latter are produced after the former are in-loop filtered.

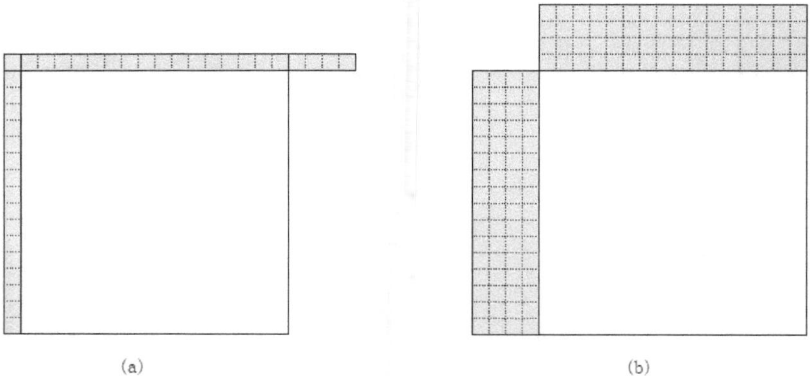

Fig. 2. Data dependencies introduced by intra-prediction and in-loop filter

In the process of in-loop filtering the current MB, the reconstructed and half-filtered data of the neighboring MB should be available, as shown in Fig. 2 (b), in which, the white field figures the current MB and the grey field figures the reconstructed and half-filtered pixels, which are distributed in the left and top neighboring MBs. And we must know that these data in the neighboring MBs are not full-filtered.

2.3 Data Dependencies Introduced by CAVLC

CAVLC is the entropy-coding tool in the baseline of H.264 encoder. In the process, the total number of non-zone transformed coefficients and the number of tailing ones are coded firstly, and there are 4 look-up tables to choose from according to the number of none-zero transformed coefficients of the left and top neighboring MB, which means the number of none-zero transformed coefficients of the left and top neighboring MB must be available before coding the current MB.

2.4 Data Dependencies Summary

As analyzed above, there are 3 types of data dependencies: 1) data dependencies between frames, which means MB could not be processed until the reconstructed and filtered pixels needed for inter-prediction are available; 2) data dependencies between MB rows in the same frame, which means MB could not be processed until the three neighboring MBs above are encoded and reconstructed; 3) data dependencies in the same MB row, which means MB could not be processed until the left neighboring MB is encoded and reconstructed. The three types of data dependencies must be avoided to exploit the data parallelisms in the H.264 encoder.

3 MBRP Parallel Algorithm

Firstly, we partition a frame into several MB regions, as shown in Fig.3, in which each little square stands for a MB, and each MB region comprises several adjoining

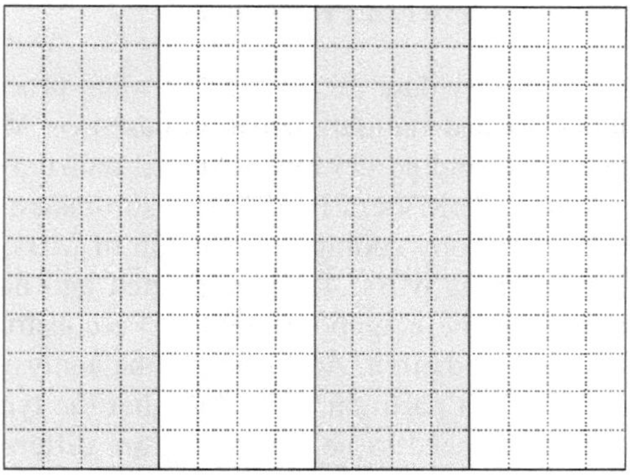

Fig. 3. Partition for MB region

columns of MBs; then, these MB regions are mapped onto different processors, and the data is exchanged appropriately according to the three types of data dependencies; at last, we propose our MBRP parallel algorithm with the wave-front technique.

3.1 The MB Region Partition

In order to achieve high performance, we must try to minimize the amount of data that needs to be exchanged between processors. In H.264 encoder, the data must be exchanged are those which are involved in the data dependencies but encoded by different processors.

Firstly, the data dependencies between frames are uncertainty because of the uncertainty of the search center in inter-prediction, and all the position in the reference frame may be the candidate to be checked in theory. Thereby, the reconstructed and filtered data of a MB region must be transferred to all the other processors as the reference data, no matter how the data is partitioned.

As for the data dependencies between MB rows in the same frame, because the three neighboring MB above are all involved, it seems that no matter how to partition the data, a processor always needs to exchange data with the processors which encode the neighboring MB regions. However, considering the data dependencies in the same MB row, if a MB region comprises only one column of MBs, the neighboring MB regions could not be processed simultaneously, which is opposite to our objective, thereby each MB region should comprise two columns of MBs at least.

On the other hand, when we implement the parallel H.264 encoder, the number of the processors in the multi-processor system is another important factor to determine the MB region partition. If the processors number is not too large (smaller than half of the number of MB columns of a frame), the number of MB columns each MB region comprises should be equal or close as much as possible, so that the load of each processor could be balanced; if the processor number is larger than half of the number of MB columns of a frame, each MB region should comprise two columns of MBs, and we should combine our MBRP algorithm with some other parallel algorithms, e. g., methods in [4] or [5], to improve the performance further.

3.2 The Data Exchanging Between Processors

Now, we analyze the necessary data exchanging between processors due to the three types of data dependencies and the data partition based on MB region. There are 5 types of data that need exchanging: 1) reconstructed and unfiltered data, which are used in the intra-prediction; 2) MVs of the MBs or sub-macro-block, which are used in the inter-prediction; 3) entropy coding information of MBs, which are used in the entropy coding for neighboring MBs; 4) reconstructed and half-filtered data, which are used in in-loop filter for the neighboring MB; 5) reconstructed and filtered data, which are used in the inter-prediction. According to the analysis of data dependencies and the description of the data partition, it is clear that the types of exchanging-data and the quantity of data that need to be exchanged are different, therefore, we must arrange the time and manner to exchange these data appropriately.

3.3 The Wave-Front Technique

After the data partition, we can map the MB regions to different processors. Considering the data dependencies in the same MB row, MBs in a MB row must be encoded in sequence, therefore we adopt the wave-front technique, starting up the encoding process of processors orderly to encode MBs in the same MB row but different regions. After adopting the wave-front technique, processors start to encode data after a short time one by one, and during the time a processor could encode a row of MBs in a MB region and transfer required reconstructed data to the next adjoining processor, thus avoiding the data dependencies in the same MB row, and processors could encode MBs in different MB rows of respective MB regions synchronously, as shown in Fig.4, squares with single diagonal denote the encoded MBs while squares with chiasm diagonal stand for the MBs being encoded by processors synchronously.

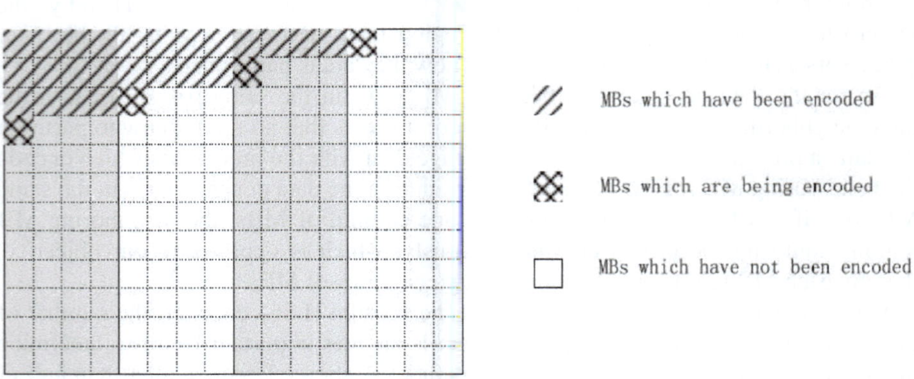

Fig. 4. Wave-front technique based on MB region partition

3.4 The Parallel Algorithm Based on the MB Region Partition

At last, we present our MBRP parallel algorithm for H.264 encoder: Firstly, we determine the MB region partition according to the encoder parameters and the processor number of the multi-processor system, insuring that each MB region comprises equal or close number of columns of MBs. Then, we map the MB regions onto different processors to be encoded. While starting up the encoding process, we adopt the wave-front technique, and the processors begin to encode data orderly after a short time to avoid the data dependencies in the same MB row. Each processor encodes a MB region in the order of MB rows, after encoding the boundary MB (the first and the last columns of MB in MB region), required reconstructed and unfiltered data, the MVs and the number of non-zero transformed coefficients of the boundary MB are transferred to the neighboring processors (which encode the neighboring MB regions). After the in-loop filtering of the MB region, the half-filtered data are transferred to the next neighboring processor (which encodes the right neighboring MB region), and the filtered data are broadcasted to all the other processors. When the required data are available, all the processors work synchronously, encoding different MB regions of video frames respectively.

4 Simulation Results

The simulator of our MBRP parallel algorithm for H.264 encoder is developed using C language and implemented on a PC with a P4 1.7GHz processor and a 512MB memory. The simulation results are compared with those from JM10.2, which is a sequential encoding structure. In our software simulation of H.264 encoder, four processors are simulated. The main encoder parameters are shown in Table 1. The simulator collects the maximal encoding time among every 4 concurrently processed MB regions and the corresponding time spent on data exchanging. Some of the simulation results are presented in Table.2 and Table.3. For Table.2, we used "Foreman" (CIF) as video source, and 300 frames were encoded. And for Table.3, we used "Foreman" (SDTV) as video source, and 300 frames were encoded. Experiments show that the speedups higher than 3.3 are achieved and the encoding quality is the same as JM10.2.

Table 2. Simulation results for “Foreman” (CIF)

	Encoder time /frame	PSNR (Y)	PSNR (U)	PSUN (V)	Bits/frame	Speedup
JM10.2	5.5656 s	36.12	40.48	42.02	15295.87 bits	1
MBRP Algorithm	1.67045 s	3612	40.48	42.02	15295.87 bits	3.33

Table 3. Simulation results for “Foreman” (SDTV)

	Encoder time /frame	PSNR (Y)	PSNR (U)	PSUN (V)	Bits/frame	Speedup
JM10.2	22.343 s	37.84	42.67	44.14	58437 bits	1
MBRP Algorithm	6.73535 s	37.84	42.67	44.14	58437 bits	3.32

In figure 5, we give the relationship between the speedup and the number of processors in multi-processor system, and we used the "Foreman" (CIF) and the "Foreman" (SDTV) as the video source. As seen in Fig.5, when the number of processors is small, the compressing performance improves linearly approximately as the number increases; when the numbers are of some values, the speedups keep unchanged, this is because the loads of different processors are imbalanced, and the most heavy load is the same for these number values; when the numbers reach certain values, 11 and 23 for "Foreman" (CIF) and "Foreman" (SDTV) respectively, the speedups get the highest value.

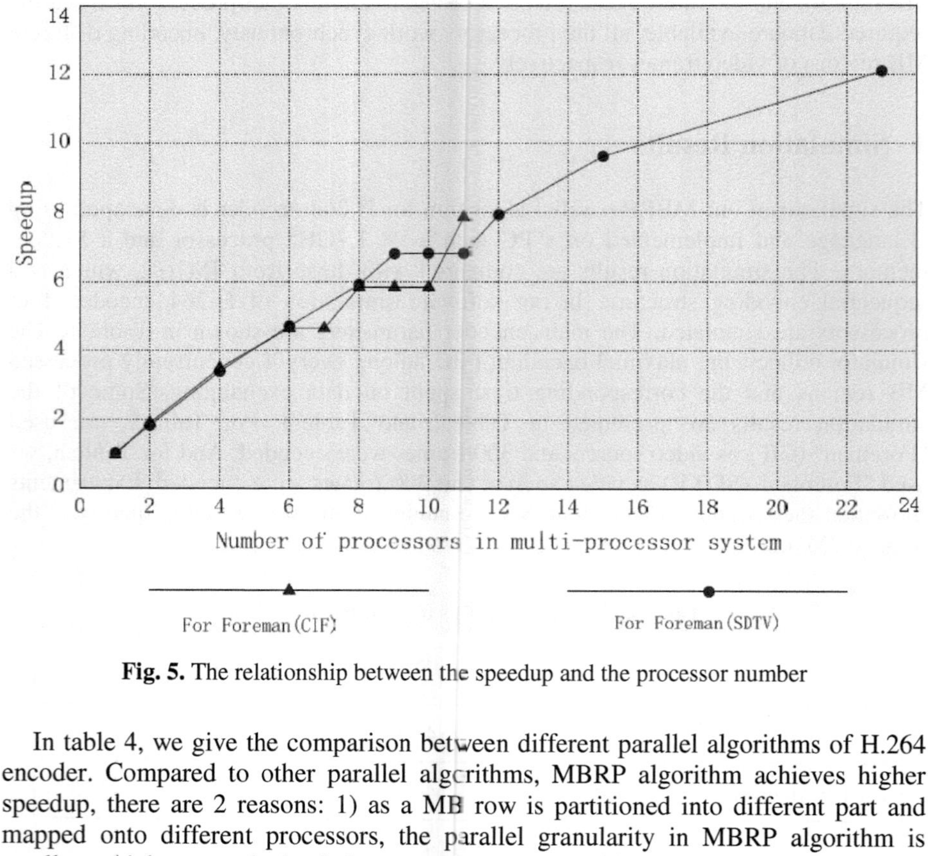

Fig. 5. The relationship between the speedup and the processor number

In table 4, we give the comparison between different parallel algorithms of H.264 encoder. Compared to other parallel algorithms, MBRP algorithm achieves higher speedup, there are 2 reasons: 1) as a MB row is partitioned into different part and mapped onto different processors, the parallel granularity in MBRP algorithm is smaller, which means the loads between processors could be balanced more easily; 2) each MB region comprises several adjoining columns of MBs, which means the amount of data that need to be exchanged between processors are smaller. On the other hand, since each MB region should comprise 2 columns of MBs at least, the highest speedup our MBRP parallel algorithm could achieve is restricted by the resolution of the video source. However, our algorithm could be combined with some other parallel algorithms easily to improve the performance further.

Table 4. Comparison between different parallel algorithms

Parallel Algorithm	Parallel Granularity	Number of Processors	Compression Ratio degradation	Speedup
Algorithm in [4]	Frame and Slice	4 (8 logic processors)	Yes	3.74~4.53
Algorithm in [5]	Frame and MB Row of frame	4	No	3.1
MBRP Algorithm	MB row of MB region	4	No	3.3

5 Conclusions and Future Work

We propose the MBRP parallel algorithm for H.264 encoder in this paper, in which frames are split into several MB regions; each MB region includes several adjoining columns of MBs and is mapped onto a different processor to be encoded; with the wave-front technique, data in different MB regions could be encoded synchronously. The simulation results show that the MBRP parallel algorithm is quite efficient, and the compressing performance improves linearly approximately as the number of the processors in the multi-processor system increases.

Future work includes the implementation of the algorithm on a multi-core SoC and the investigation on parallel algorithm of H.264 encoder with CABAC entropy coding mode.

References

1. Chen, T.C., Huang, Y.W., Chen, L.G.: Analysis and design of macro-block pipelining for H.264/avc VLSI architecture. In: Proceedings of the 2004 International Symposium on Circuits and Systems (2004)
2. JVT Draft recommendation and final draft international standard of joint video specification. ITU-T Rec. H.264 and ISO/IEC 14496-10 AVC (2003)
3. Gulati, A., Campbell, G.: efficient mapping of the H.264 encoding algorithm onto multi-processor DSPs. In: Proceedings of SPIE-IS&T Electronic Imaging (2005)
4. Chen, Y-K., Ge, S., T. X., G.M.: towards efficient multi-level threading of H.264 encoder on intel hyper-threading architectures. In: 18th International Parallel and Distributed Processing Symposium (2004)
5. Zhao, Z., Liang, P.: A Highly Efficient Parallel Algorithm for H.264 Video Encoder. In: 31st IEEE International Conference on Acoustics, Speech, and Signal Processing (2006)
6. Jacobs, T.R., Chouliaras, V.A., Nunez-Yanez, J.L.: A Thread and Data-Parallel MPEG-4 Video Encoder for a SoC Multiprocessor. In: Proceedings of 16th International Conference on Application-Specific Systems, Architecture and Processors (2005)
7. JM10.2, http://bs.hhi.de/suehring/tml/download/jm10.2.zip

Towards Scalable and High Performance I/O Virtualization – A Case Study

Jinpeng Wei[1], Jeffrey R. Jackson[2], and John A. Wiegert[2]

[1] Georgia Institute of Technology, 801 Atlantic Drive
30332 Atlanta, Georgia, USA
weijp@cc.gatech.edu
[2] Intel Corporation, 2111 NE 25th Ave
97124 Hillsboro, Oregon, USA
{jeff.jackson,john.a.wiegert}@intel.com

Abstract. I/O Virtualization provides a convenient way of device sharing among guest domains in a virtualized platform (e.g. Xen). However, with the ever-increasing number and variety of devices, the current model of a centralized driver domain is in question. For example, any optimization in the centralized driver domain for a particular kind of device may not satisfy the conflicting needs of other devices and their usage patterns. This paper has tried to use IO Virtual Machines (IOVMs) as a solution to this problem, specifically to deliver scalable network performance on a multi-core platform. Xen 3 has been extended to support IOVMs for networking and then optimized for a minimal driver domain. Performance comparisons show that by moving the network stack into a separate domain, and optimizing that domain, better efficiency is achieved. Further experiments on different configurations show the flexibility of scheduling across IOVMs and guests to achieve better performance. For example, multiple single-core IOVMs have shown promise as a scalable solution to network virtualization.

1 Introduction

I/O Virtualization provides a way of sharing I/O devices among multiple guest OSes in a virtualized environment (e.g., Xen [2[illegible]]). Take network virtualization for example (see Fig. 1), here the platform has one Gigabits/second network link, but a guest OS may only need 100Mbps network bandwidth, so it would be very cost-efficient to share this Gigabit link among several guest OSes. In order to support this kind of device sharing, Xen has employed a split-driver design, where the I/O device driver is split into a *backend* and a *frontend*. The physical device is managed by a *driver domain* which acts as a proxy between the guest OSes and the real device. The driver domain creates a device backend, and a guest OS which needs to access the device creates a frontend. The frontend talks to the backend in the driver domain and creates an illusion of a physical device for the guest OS. Multiple guest OSes can share a physical device in this way through the backend in the driver domain.

R. Perrott et al. (Eds.): HPCC 2007, LNCS 4782, pp. 586–598, 2007.

The most common way of deploying driver domains in Xen is to have the service OS (e.g. domain 0) as the single driver domain (called *centralized* I/O Virtualization in this paper), as shown in Fig. 1. However, this centralized architecture is not scalable when the platform has many devices, as in a server consolidation environment. Specifically, as the number and variety of shared devices increase, the centralized I/O Virtualization has the following problems. First, the Service OS can be easily overloaded by I/O virtualization. Second, any optimization of the Service OS for better I/O virtualization performance needs to consider all other tasks in the service OS (e.g., service daemons and management tasks), so it may not be easy to do such optimizations. Third, different devices may have different needs for better performance. For example, a graphics device may be more sensitive to latency while a network device may be more interested in throughput, so it may be difficult to find an optimization that satisfies both graphic devices and network devices at the same time.

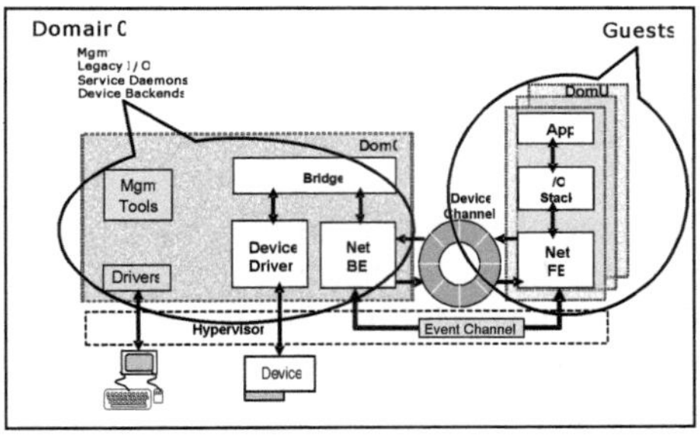

Fig. 1. Centralized I/O Virtualization Architecture

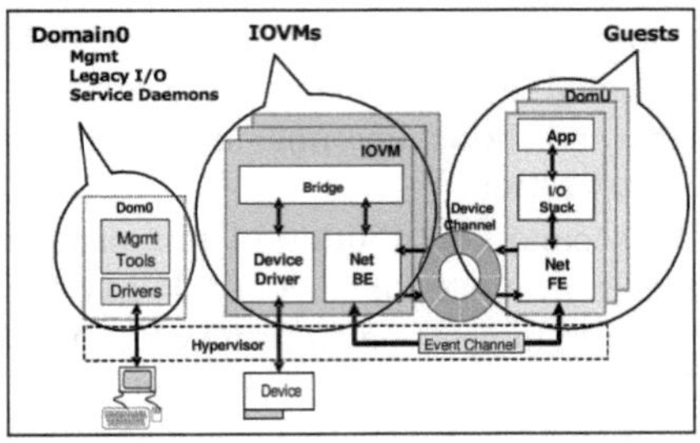

Fig. 2. IOVM Architecture

One possible solution to the scalability problems with *centralized* I/O Virtualization is *scale-up* (e.g., adding more resources to the Service OS). However, our experience with Xen (version 3.0.2) shows that allocating more computing resources (e.g., more CPUs) does not necessarily translate into better I/O virtualization performance. For example, we found that a Service OS domain (uniprocessor Linux) with only one CPU saturates before it can support 3 Gigabits/second NICs (Network Interface Cards) at full capacity. Adding one more CPU to this domain (SMP Linux) does not show much improvement - the receive throughput increases by 28%, but at the cost of reduced efficiency (by 27%); the transmit throughput even decreases by 6.6%, and the transmit efficiency drops by 88%. The definition of efficiency can be found in Section 4.1.

The difficulties of *scale-up* lead us to consider another way of deploying driver domains: *scale-out,* e.g., using dedicated guest domains other than the service OS as driver domains. This idea was suggested by Fraser [5] and Kieffer [7], and we have also proposed IOVMs [1]. The IOVM approach can solve the three problems with centralized I/O Virtualization. The first problem is solved by moving the I/O Virtualization out of the service OS, so it will not be overloaded by I/O Virtualization. Second, any possible optimization is performed in a different domain from the service OS so it does not affect the existing tasks. Third, when different devices have different requirements for high performance, we can create different IOVMs and optimize each one in a different way, according to the characteristics of the device and workload. Therefore the IOVM approach opens up opportunity for scalable I/O virtualization given that proper optimizations are done in the IOVMs and the addition of IOVMs to the platform does not consume excessive resources.

This paper presents our initial experience with IOVMs for *network virtualization* on a multi-core platform. The question we want to address is - Given a platform equipped with multi-core, how can its support for network virtualization *scale* as the number of network devices increases? By 'scale' we mean 'aggregated bandwidth increases linearly with the increase of computing resources (CPU cycles) on the platform'. We believe that the software architecture (including the network device driver and protocol stack) is a key to the solution. In this paper, we show how IOVMs enable the division of work load among cores in such a way that scalability can be achieved by incrementally starting IOVMs on new cores. Specifically, this paper makes the following contributions:

- It proposes a novel way of using IOVMs for scalable I/O Virtualization. Although dedicated driver domains were initially proposed for fault isolation [8][5], we observe and show experimentally in this paper that it is also beneficial to scalable I/O Virtualization.
- The second contribution of this paper is a comprehensive set of experimental results to evaluate the idea of IOVMs. It compares the performance of three different configurations of network IOVMs: Monolithic IOVM, Multiple Small IOVMs, and Hybrid IOVMs. And it concludes that the Hybrid IOVM configuration offers a promising balance between scalable throughput and efficient use of core resources.

The rest of this paper is organized as follows. Section 2 describes the IOVM architecture. Section 3 briefly outlines the optimizations that we have carried out in a Network IOVM. Section 4 presents a series of experimental results which evaluate the performance and scalability of the Network IOVM. Related work is discussed in Section 5 and we draw conclusions in Section 6.

2 IOVM Architecture

This section describes our IOVM architecture (See Fig. 2). The main idea is to move I/O Virtualization work out of the service OS (domain 0) and into dedicated driver domains.

In Xen an IOVM is a specialized guest operating system, so it has the basic features and structure of a modern operating system (e.g., memory management and process management). In addition, it has the following components. First, it contains the native device driver for the physical device(s) that it virtualizes. Second, it runs the backend drivers for guest domains interested in sharing the physical device. Finally, it runs any multiplexer/demultiplexer that glues the first two together (e.g., code for routing). A network IOVM essentially has the structure of a switch or router.

2.1 Different IOVM Configurations

Three different IOVM configurations can be used to virtualize devices: Monolithic, Multiple Small IOVMs, and Hybrid.

Monolithic IOVMs (Fig. 3)**:** All devices are assigned to a single IOVM. As a result, the platform only has one IOVM, but this IOVM can be very heavy-weight due to a large number of devices to be virtualized.

Multiple Small IOVMs (Fig. 4)**:** Each device is assigned to a dedicated IOVM. So the number of IOVMs is equal to the number of physical devices being shared. When there are many devices, this configuration can result in many IOVMs.

Hybrid IOVMs (Fig. 5)**:** This configuration is a compromise between the first two configurations. In this configuration, there are multiple IOVMs, and each IOVM is assigned a subset of the physical devices being shared. The IOVMs are medium-sized, so they are larger than those in the Multiple Small IOVMs configuration, but they are smaller than a Monolithic IOVM. A Hybrid IOVMs configuration results in a smaller number of IOVMs in the system compared with the Multiple Small IOVMs configuration, but a larger number of IOVMs compared with the Monolithic IOVM configuration.

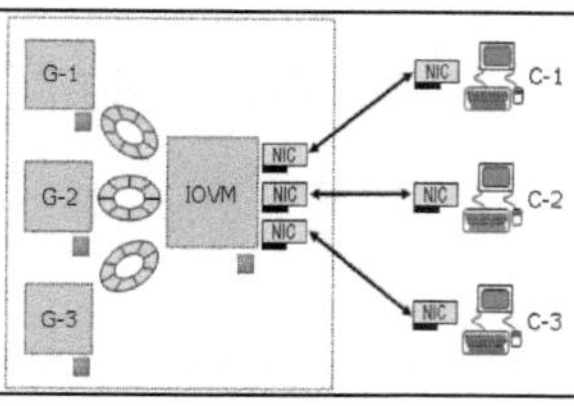

Fig. 3. Monolithic Configuration

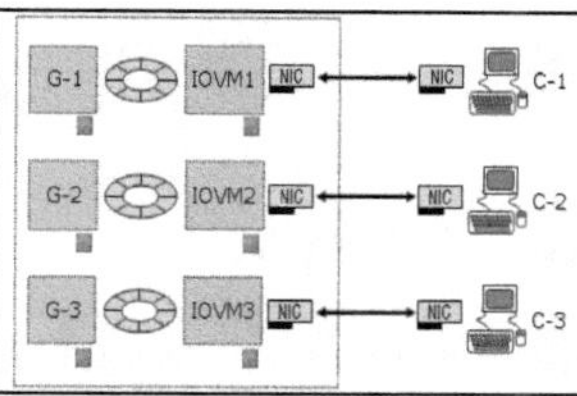

Fig. 4. Multiple Small IOVMs Configuration

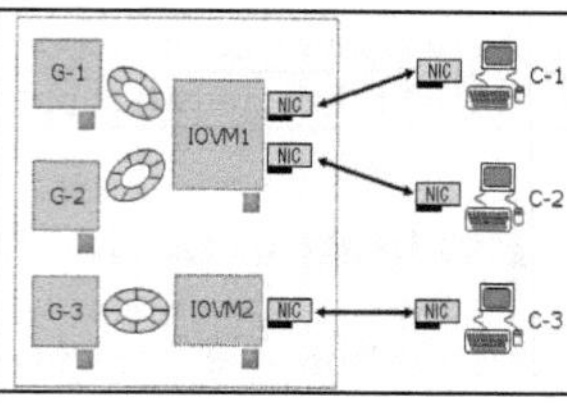

Fig. 5. Hybrid Configuration

3 An IOVM for Network Virtualization

Having a separate IOVM gives us much freedom in terms of constructing it. For example, we can start from a commodity OS (such as Linux) and specialize it; we can also build a custom OS from scratch. In this project we choose the first approach. Specifically, we customize the para-virtualized Linux for Xen for a network IOVM. The customizations (optimizations) that we perform fall into three categories: kernel, network protocol stack, and runtime.

Minimal kernel for a Network IOVM: We use the Linux configuration facility to remove irrelevant modules or functionalities from the kernel, e.g., most of the device drivers, IPv6, Cryptography, and Library Routines. The essential part of the kernel (e.g., memory and process management) is kept. Besides, we keep the list of functionalities shown in **Table 1**. The network interface card driver is configured to use polling for receive. Due to the inefficiency of SMP Linux on network virtualization, we turned off SMP support in the Network IOVM for the rest of the paper. By performing this step, we reduce the compiled size of the "stock" kernel by 44%.

Minimal network protocol stack: To make the network IOVM fast we use Ethernet Bridging in Linux kernel to forward packets between the guest OS and the NICs (Network Interface Cards). An Ethernet bridge processes packets at the data link layer (e.g. only looking at the MAC addresses to make forwarding decisions). We do not use layer 3 or above forwarding, and we do not use iptables. As another special note, we found that bridged IP/ARP packets filtering is very CPU intensive: For example, although it happens at the data link layer, it computes IP checksum, looks up routing table, and replicates packets. In other words it adds significant processing to the critical path of every packet, even if no filtering rules are defined. So it is disabled in the network IOVM for better performance.

Minimal IOVM runtime: In this part we shutdown most of the irrelevant services in the network IOVM (e.g., *sendmail*) to save CPU cycles. As a result only *network*, *syslog* and possibly *sshd* are needed to support a network IOVM. We also start the network IOVM in a simple run-level (multiuser without NFS).

Table 1. Kernel Functionality Required for the Network IOVM

Basic TCP/IP networking support	802.1d Ethernet bridging
Packet socket	Unix domain socket
Xen Network-device backend driver	Ext2 , ext3, /proc file system support
Initial RAM disk (initrd) support	Network Interface Card driver (PCI, e1000)

4 Evaluation of the Network IOVM

In this section, we present several experiments which lead to a method for scalable and high performance network virtualization.

4.1 Experiment Settings

We test the idea of network IOVMs on an Intel platform with two Core Duo processors (4 cores total). This platform has several gigabit NICs. Each NIC is directly connected to a client machine, and is exclusively used by a guest OS to communicate with the client machine (Figures 3-5). The maximum bandwidth through each NIC is measured by the iperf benchmark [6]. There are 4 TCP connections between each guest OS and the corresponding client machine, which start from the guest OS, traverse through the IOVM and the NIC, and reach at the client machine. The connections are all transmitting 1024 byte buffers. The combined bandwidth of the NICs is considered the aggregated bandwidth (throughput) supported by the platform. Both Xen 3.0.2 and Xen-unstable (a version before Xen 3.0.3) are used as the virtual machine manager, and a para-virtualized Linux (kernel 2.6.16) is used as the guest OS. The guest OSes and IOVMs have 256MB memory each.

We me asured two metrics: **throughput** and **efficiency**. Throughput is measured using the microbenchmark iperf. Efficiency is calculated by dividing the aggregate throughput transferred by the IOVM(s) by the number of processor cycles used by the IOVM(s) during the workload run. The processor cycles used by the IOVM(s) are measured using xentop, a resource measurement tool included in Xen. This results in a value measured in bits transmitted per processor cycle utilized or bits/Hz for short.

Network bandwidth through each NIC is measured in two directions: (1) from the client to the guest OS (denoted as Rx, for receive), (2) from the guest OS to the client (denoted as Tx, for transmit).

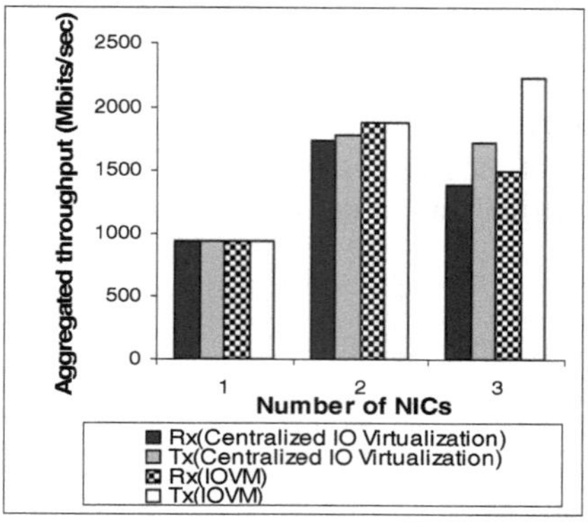

Fig. 6. Comparison of Throughput

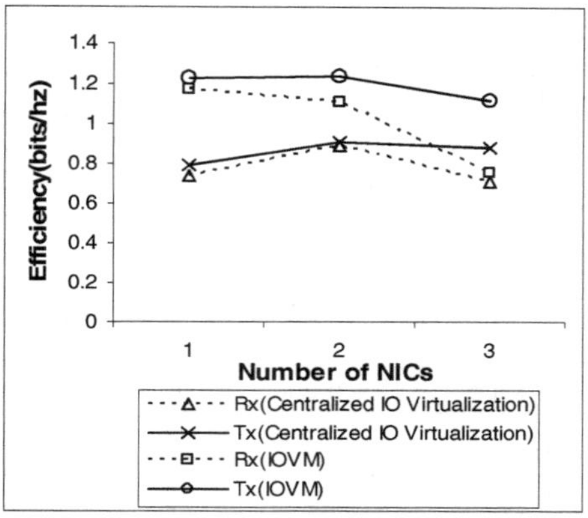

Fig. 7. Comparison of Efficiency

4.2 Making a Case for Network IOVMs

The first experiment compares the performance of centralized network virtualization versus a network IOVM. Here we use a Monolithic configuration (Fig. 3) with 3 NICs. Fig. 6 shows the throughput results for different number of NICs and different virtualization architectures (Rx means receive and Tx means transmit in this paper). From Fig. 6 we can see that the throughputs are the same for centralized network virtualization and IOVM network virtualization when only one NIC is used by iperf. It is so because at this load the CPU is not a bottleneck so both architectures can support nearly line rate (e.g. 940Mbits/sec). When we move on to 2 NICs, we can see that the IOVM network virtualization continues to deliver nearly line rate (1880 Mbits/sec), but the centralized network virtualization can not (1760Mbits/sec). This result shows that at 2 NICs, the Service OS in centralized network virtualization starts to be saturated. The IOVM is doing better because it is optimized for networking as described in Section 3. When we move on to 3 NICs, it becomes more apparent that IOVM network virtualization can support higher throughput. E.g., 2230Mbits/sec versus 1700Mbits/sec for transmit (denoted as Tx), and 1500Mbits/sec versus 1400Mbits/sec for receive (denoted as Rx). Comparatively the benefit of IOVM is not as big for receive as it is for transmit. The reason is that receive is more CPU intensive in current implementation (polling is used). As a result the optimization in the network IOVM is still not enough to meet the increase in CPU demand when there are 3 iperf loads. By carrying out more aggressive optimization in the network IOVM we may be able to support higher throughput. But the point of Fig. 6 is that using a network IOVM has advantage in terms of throughput.

Fig. 7 shows the result for efficiency (Section 4.1), which is more revealing about the benefit of a network IOVM. For example, when there is only one NIC used by iperf, the efficiency is about 1.2 bits/Hz for IOVM and about 0.8 bits/Hz for centralized network virtualization, meaning that network IOVM is 50% more efficient than centralized network virtualization. In other words, although the two architectures appear to support the same level of throughput (940Mbits/sec in Fig. 6) at one NIC level, the network IOVM is actually using much less CPU cycles to support that throughput. So a network IOVM is more efficient in terms of CPU usage. This claim is also true for the 2 NICs case in Fig. 7. When there are 3 NICs, the efficiency of IOVM is still much higher than that of centralized network virtualization in terms of transmit, but the efficiency difference in terms of receive becomes small. This is because in both IOVM and centralized network virtualization the CPU is saturated, but the throughput is nearly the same.

Table 2. The Static Core Assignment Schemes

co-locate	Each guest and its corresponding IOVM are on the same core.
separate	Each guest and its corresponding IOVM are on different cores, but a guest shares a core with a different IOVM.
IOVM-affinity	All IOVMs are on the same core, and the guests are each on one of the remaining cores.

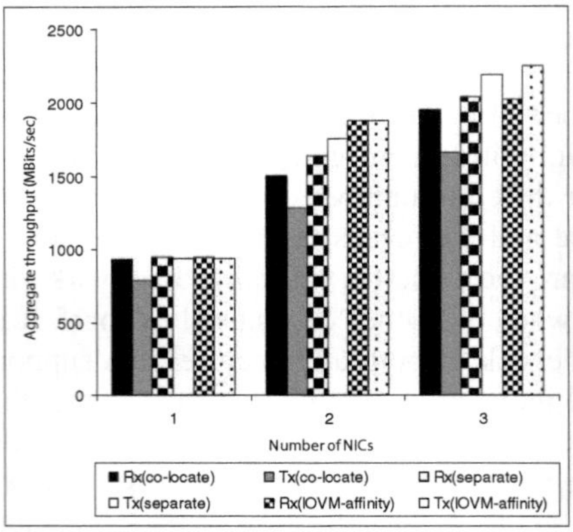

Fig. 8. Throughput Results for Different Static Core Assignment Schemes Using the Multiple Small IOVMs Configuration

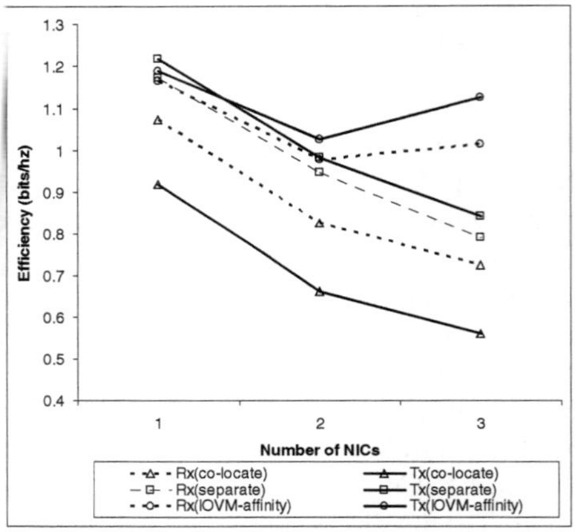

Fig. 9. Efficiency Results for Different Static Core Assignment Schemes Using the Multiple Small IOVMs Configuration

4.3 Multiple Small IOVMs and Static Core Assignment

This section evaluates the performance of using Multiple Small IOVMs (Fig. 4) for network virtualization, e.g., assigning each NIC to a dedicated IOVM. One concern about this configuration is the overhead (e.g., memory and scheduling overhead) of having a large number of simple IOVMs (domains). For example, the platform that we used only has 4 cores, but we have 7 domains in total (3 guests, 3 IOVMs and domain 0) when there are 3 test loads. Obviously some of the domains must share a physical CPU core. Xen 3.0.2 allows the assignment of a domain to a physical core, and this section shows that the way that the IOVMs and the guest domains get assigned is very important to the virtualization performance.

Specifically, we have tried 3 different assignment schemes as shown in **Table 2**. Fig. 8 and Fig. 9 show the evaluation results.

Fig.8 compares the 3 different assignment schemes in terms of throughput. We can see that given a certain number of NICs (test loads), no matter if it is transmit or receive, **co-locate** has the lowest throughput, **IOVM-affinity** has the highest throughput, and **separate** has a throughput in between the other two schemes.

That **co-locate** performs worse than **separate** is somewhat surprising: one would expect the other way around because if a guest and its corresponding IOVM are on the same physical core, there will be no need for InterProcessor Interrupts (IPIs) when packets are sent from the guest to the IOVM. However, the benefit of eliminating such IPIs is offset by a more important factor: the elimination of opportunities for parallelism between the guest and the corresponding IOVM. Specifically, there are pipelines of packets between the guest and the IOVM, so when they are assigned on different cores (as in the **separate** assignment scheme), they can run concurrently so that while the IOVM is processing the nth packet the guest can start sending the $n+1^{th}$ packet. However, when these two domains are assigned on the same core, they lose such opportunities. As a result, there is no pipeline for **co-locate** and the throughput is lower.

But why is the throughput of **IOVM-affinity** the best? We found that a guest needs more CPU cycles than the IOVM to drive the same test load because the guest needs to run the network stack as well as iperf, which does not exist in an IOVM. Thus the bottleneck for CPU occurs in the guest. By assigning the IOVMs on the same core we reserve the most cores possible for the guests (each guest is running on a dedicated core in this case). As a result, the guests are able to drive more network traffic and we get the highest throughput. But this assignment can not be pushed too far, because if too many IOVMs are assigned to a same core, eventually they will run out of CPU cycles so there will be no further increase in throughput. In such cases, the bottleneck will move to the IOVMs.

Fig. 9 compares the 3 different assignment schemes in terms of efficiency. The overall result is the same: **Co-locate** has the worst efficiency, **IOVM-affinity** has the best efficiency, and **separate** has efficiency in between the other two schemes. **Co-locate** is the least efficient because of too much context switching overhead: for example, whenever the guest sends out one packet, the IOVM is immediately woken up to receive it. There are 2 context switches for each packet sent or received. One may argue that **separate** should have the same context switching overhead, but this is not the case, because after the guest sends out one packet, the IOVM on a *different* core may be woken up, but the guest can continue to send another packet without being suspended. The Xen hypervisor is able to deliver packets in batches to the IOVM; running the guest and IOVM on different cores allows multiple packets to be received by the IOVM within one context switch.

Finally, the workload of an IOVM is different from that of a guest, and a guest is more CPU intensive, so mixing them together on the same core (as in **co-locate** and **separate**) may result in more complicated, thus negative, interferences (e.g., cache and TLB misses due to context switches) to an IOVM than putting the homogeneous workloads of the IOVMs on the same core. Therefore, IOVM-affinity scheme is the most efficient.

One more note about Fig. 9 is that the efficiency drops with the increase of the number of NICs for **co-locate** and **separate**, which indicates that they are not scalable schemes. The reason is that each IOVM uses almost the same amount of CPU cycles under these schemes, so that when the number of NICs increases from 1 to 2 for example, the CPU cycles used by the corresponding IOVMs are nearly doubled, but the throughput is much less than doubled (Fig. 8). As a result, the efficiency drops from 1 NIC to 2 NICs. Similarly, efficiency drops off when moving from 2 NICs to 3 NICs. On the other hand, the efficiency for **IOVM-affinity** remains fairly constant as the number of NICS increases and is therefore much more scalable.

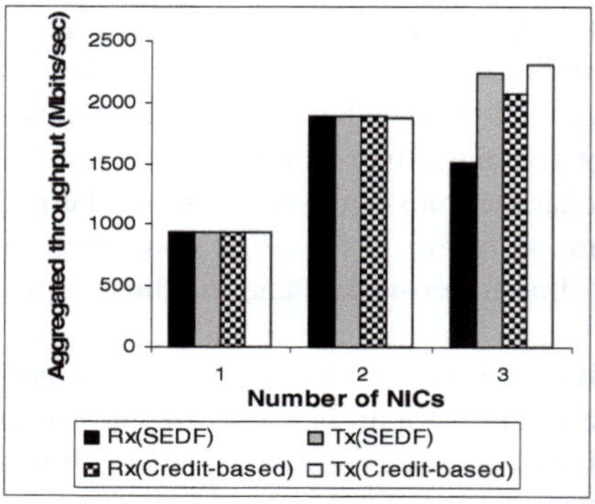

Fig. 10. Throughput Comparison between SEDF and the Credit-based Scheduler

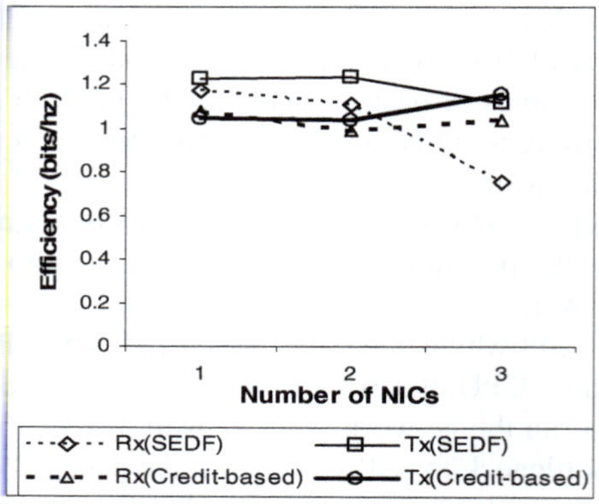

Fig. 11. Efficiency Comparison between SEDF and the Credit-based Scheduler

4.4 Credit-Based Scheduling and IOVM Configuration

Core Scheduling and IOVM Performance

In the evaluation so far we have been using static core schedulers - static in the sense that the CPU core that a domain runs on is fixed during its lifetime. The advantage of such schedulers is simplicity of implementation and reduced overhead of domain migration across cores. However, the drawback is that workload is not balanced across multiple cores, so that the platform resources are not efficiently utilized to achieve better overall performance (e.g. throughput). For example, while the core that an IOVM is running on is saturated, another core where a guest is running on may be idle for 30% of the time. Obviously the IOVM becomes the bottleneck in this case and thus the throughput can not improve, but there are free cycles on the platform that are not used. So if we could give the 30% free cycles to the IOVM, we can mitigate the bottleneck and have higher overall throughput as a result. This is the idea of the credit-based scheduler [13]. In a nutshell, a credit-based scheduler allows a domain to change the core where it runs on dynamically. The credit-based scheduler supports load balancing across the cores and is helpful for better overall performance.

Fig. 10 and Fig. 11 show the throughput and efficiency comparison of a static scheduler (SEDF, or Simple Earliest Deadline First) and the credit-based scheduler, on a Monolithic configuration (Fig. 3). Fig. 10 shows that using credit-based scheduler can achieve equal or higher throughput than using a static scheduler. Especially, when there are 3 test loads, the overall 'receive' throughput is 2070Mbits/sec for the credit-based scheduler, which is significantly higher than 1500Mbits/sec where the static scheduler is used. The credit-based scheduler does especially well for receive because receive is more CPU-intensive.

Fig. 11 shows that the credit-based scheduler is not as efficient as a static scheduler when there are one or two test loads. This is understandable because moving the domains around incurs overhead, e.g., a moving domain needs to transfer its context to the new core and warm up the cache at the new core. When the CPU is not the bottleneck, this overhead makes the network IOVM less efficient. However, when the network IOVM is saturated (3 NICs), the credit-based scheduler results in better efficiency than the static scheduler, especially for receive.

Credit-based Scheduling and IOVM Configuration

Credit-based scheduling provides a solution to scalable network virtualization. This is because it virtualizes the core resources in a transparent way. In this sub-section we try to find out how to make the best use of this scheduler. We use the 3 configurations mentioned in Section 2.1 as the controlled variable.

In this experiment, 3 test loads are used. We run the experiment using 3 different configurations: Monolithic, Multiple Small IOVMs, and a Hybrid configuration with 2 IOVMs, where the first IOVM is assigned 2 NICs, and the second IOVM is assigned 1 NIC. Fig. 12 and Fig. 13 show the throughput and efficiency for the 3 configurations, respectively.

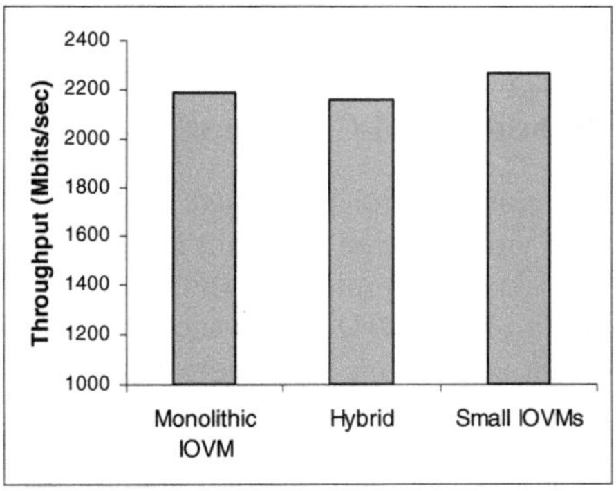

Fig. 12. Throughput of Different IOVM Configurations under the Credit-based Scheduling

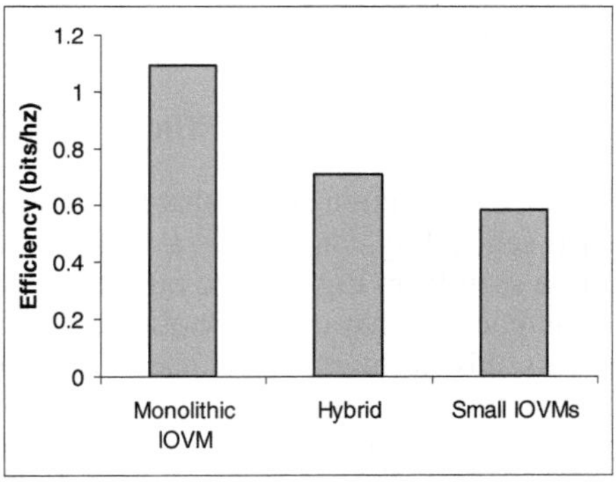

Fig. 13. Efficiency of Different IOVM Configurations under the Credit-based Scheduling

From Fig. 12 and Fig. 13 we can see that the Monolithic IOVM configuration is the most efficient, the Multiple Small IOVMs configuration has the highest throughput but is the least efficient, and the Hybrid configuration has good enough throughput and is more efficient than the Multiple Small IOVMs configuration.

The better efficiency of the Monolithic and Hybrid configurations is due to larger packet batch size in their IOVMs. It turns out that Xen has optimized the network implementation such that the network backend exchanges packets with the frontend in batches to reduce the number of hypervisor calls. The larger the batch size, the more savings in terms of hypervisor calls, and thus the more efficient. At runtime, the batch size is influenced by workload, or the number of NICs in our case. For example, as Fig. 14 shows, an IOVM with 2 NICs has larger batch size than an IOVM with only 1 NIC, so an IOVM with 2 NICs is more efficient. In our experiment, the Monolithic IOVM has the largest number of NICs (3), each of the Multiple Small IOVMs has the smallest number of NICs (1), and each of the IOVMs in the Hybrid configuration has 1.5 NICs on average, so we get the efficiency result as shown in Fig. 13.

Although the Monolithic IOVM is the most efficient, obviously it can not scale because it only has one core (it uses a uniprocessor kernel, see Section 3), so it can not support higher throughput beyond the core limit. On the other hand, the Multiple Small IOVMs configuration is a scalable solution because it can utilize more cores, but it is the least efficient. The Hybrid configuration is also scalable for reasons similar to the Multiple Small IOVMs configuration, but it is more efficient. So the Hybrid configuration is the best configuration in terms of scalability and efficiency.

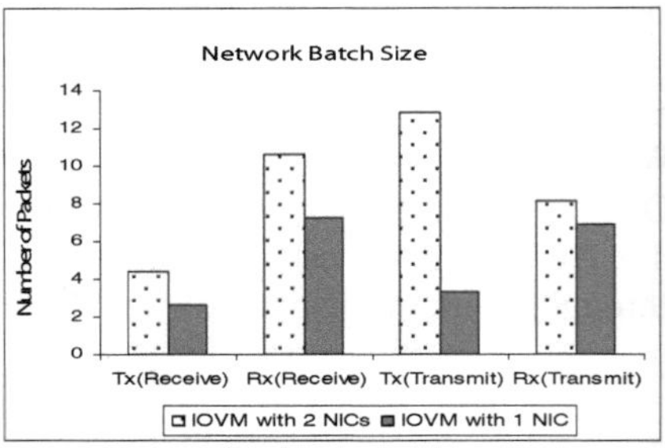

Fig. 14. Network Batch Size

Table 3. Combining Static Core Assignment and Credit-based Scheduler Yields Better Performance

	Throughput (Mbits/sec)	Efficiency (bits/Hz)
Default	2,106	0.72
Core Pinning	2,271	0.82

4.5 Further Improve the Efficiency of the Hybrid IOVM Configuration

The previous section shows that a Hybrid IOVM configuration combined with Credit-based scheduling can give us a scalable network virtualization solution. However, as can be seen from Fig. 13, the efficiency of a Hybrid configuration (0.7bits/Hz) is still not as good as that of a Monolithic configuration (1.1 bits/Hz). This section tries to address this problem. Specifically, we found that combining static core assignment and Credit-based scheduling can give a Hybrid configuration better efficiency.

In this experiment, we have 4 NICs and we use a Hybrid configuration where 2 network IOVMs are assigned 2 NICs each. We first run 4 test loads under the Credit-based scheduling and measure the throughput and efficiency. We call this test result "Default" in **Table 3**. Then we change the core scheduling a little bit: instead of letting the Credit-based scheduler schedule all the domains on the 4 cores, we manually pin the first IOVM to core 3 and the second IOVM to core 2, and let the Credit-based scheduler schedule the other domains on the remaining cores. We do the 4 guest experiment again and put the result in **Table 3** denoted as "Core Pinning". As we compare the two sets of results, we can see that using limited core pinning (or static assignment) results in improved efficiency as well as higher throughput (a pleasant side effect). This result suggests that it is not efficient to move the IOVMs across different cores. So it is more efficient to combine static core assignment (for the IOVMs) and Credit-based scheduling when using a Hybrid configuration for scalable network virtualization.

5 Related Work

I/O virtualization can be carried out in the VMM [3][4], or the host OS [12], in addition to dedicated guest domains [1][5][8][10]. VMM-based I/O virtualization requires nontrivial engineering effort to develop device drivers in the VMM, provides inadequate fault-isolation, and is not flexible for driver optimization. HostOS-based I/O virtualization takes a further step by reusing existing device drivers, but does not support fault isolation and is still inflexible in terms of driver optimization. The IOVM approach supports driver reuse, fault isolation, and flexible driver optimization at the same time.

There has been some related work in improving the performance of I/O virtualization. For example, VMM-bypass IO [9] achieves high performance I/O virtualization by removing the hypervisor and the driver domain from the normal I/O critical path. But this approach heavily relies on the intelligent support provided by the device hardware to ensure isolation and safety, and it does not address the scalability issue. The IOVM approach that we proposed does not rely on such hardware support, and we have put much emphasis on scalability. In another work, Menon [11] proposes three optimizations for high performance network virtualization in Xen: high-level network offload, data copying instead of page remapping, and advanced virtual memory features in the guest OS. These optimizations are orthogonal to what we are doing in this paper, since our main concern is scalability, and incorporating such optimizations into our implementation may further improve the efficiency. Wiegert [14] has explored the scale-up solution by increasing an IOVMs compute resources to improve scalability.

Utility Computing has been a hot research area in recent years. From a service provider point of view, one of the goals is to achieve optimal overall resource utilization. Our work addresses the problem of optimizing the utilization of core (CPU) resources. Concerns about other resources (such as memory and disks) have not been a problem for us, but they can be added into our future work.

6 Conclusions and Future Work

Scalable I/O virtualization is very important in a server consolidation environment. This paper proposes IOVMs as a software solution to this problem. An IOVM is a guest OS dedicated to and optimized for the virtualization of a certain device. IOVMs are good for scalability and flexible for high performance.

The first contribution of this paper is a novel way of using a hybrid configuration of IOVMs to achieve scalable and high-performance I/O virtualization. It makes the scalability and efficiency tradeoff: the scalability is achieved by spawning more IOVMs to utilize more core resources, and the efficiency is achieved by making full use of the core resource within each IOVM.

The second contribution of this paper is a comprehensive set of experiments to evaluate the performance of network IOVMs. They show that IOVMs result in higher throughput and better efficiency compared to the centralized IO virtualization architecture, that a combination of static assignment and credit-based scheduling offers better efficiency, and that a hybrid configuration of IOVMs is a choice for scalable network virtualization.

Future work: The physical constraint of the multi-core platform nowadays limits the scope of our experiments. For example, if we could have a platform with 32 cores, we can gather more data points to do a more comprehensive analysis. Second, we have studied network virtualization only as a starting point; applying IOVM architecture to other kinds of devices (e.g., storage devices) may further test the validity of IOVM approach. Finally, we plan to use other kinds of work load besides iperf to further evaluate the network IOVMs.

References

1. Abramson, D., Jackson, J., et al.: Intel Virtualization Technology for Directed I/O. Intel Technology Journal 10(03) (2006)
2. Barham, P., et al.: Xen and the Art of Virtualization. In: Proceedings of the 19th ACM Symposium on Operating Systems Principles, pp. 164–177. ACM Press, New York (2003)
3. Borden, T., Hennessy, J.P., Rymarczyk, J.W.: Multiple operating systems on one processor complex. IBM Systems Journal 28, 104–123 (1989)
4. Bugnion, E., et al.: Disco: Running Commodity Operating Systems on Scalable Multiprocessors. ACM Transactions on Computer Systems 15(4) (1997)
5. Fraser, K., et al.: Safe Hardware Access with the Xen Virtual Machine Monitor. In: OASIS. Proceedings of the 1st Workshop on Operating System and Architectural Support for the on demand IT InfraStructure, Boston, MA (2004)
6. Iperf, http://dast.nlanr.net/Projects/Iperf/
7. Kieffer, M.: Windows Virtualization Architecture, http://download.microsoft.com/download/9/8/f/98f3fe47-dfc3-4e74-92a3-088782200fe7/TWAR05013_WinHEC05.ppt
8. LeVasseur, J., et al.: Unmodified Device Driver Reuse and Improved System Dependability via Virtual Machines. In: OSDI (2004)
9. Liu, J., et al.: High Performance VMM-Bypass I/O in Virtual Machines. In: Proceedings of the USENIX Annual Technical Conference (2006)
10. McAuley, D., Neugebauer, R.: A case for Virtual Channel Processors. In: Proceedings of the ACM SIGCOMM, ACM Press, New York (2003)
11. Menon, A., et al.: Optimizing Network Virtualization in Xen. In: Proceedings of the USENIX'06 Annual Technical Conference (2006)
12. Sugerman, J., et al.: Virtualizing I/O Devices on VMware Workstation's Hosted Virtual Machine Monitor. In: Proceedings of the USENIX Annual Technical Conference (2001)
13. http://wiki.xensource.com/xenwiki/CreditScheduler
14. Wiegert, J., et al.: Challenges for Scalable Networking in a Virtualized Server. In: 16th International Conference on Computer Communications and Networks (2007)

A Proactive Method for Content Distribution in a Data Indexed DHT Overlay

Bassam A. Alqaralleh, Chen Wang, Bing Bing Zhou, and Albert Y. Zomaya

School of Information Technologies
University of Sydney, NSW 2006, Australia
{bassam,cwang,bbz,zomaya}@it.usyd.edu.au

Abstract. In a data-indexed DHT overlay network, published data annotations form distributed databases. Queries are distributed to these databases in a non-uniform way. Constructing content distribution networks for popular databases is effective for addressing the skew problem. However, various crucial replication decisions such as which data object should be replicated, how many replicas should be created and what replica placement policy should be used, may affect the performance of replication based content distribution and load balancing mechanisms in the presence of non-uniform data and access distribution. Particularly, the impact of the propagation speed of replicas on the performance of such type of overlay networks is not well studied. In this paper, a proactive method is given to tackle this problem. The method can adaptively adjust the number of replicas to create based on the changing demand. It works in a fully distributed manner. Our experiments show that the approach is able to adapt quickly to flash query crowds and improve the query service quality of the systems.

1 Introduction

Peer-to-Peer (P2P) technologies are effective for flexible information sharing among a variety of applications. This is mainly because autonomous data sources can flexibly annotate their data (called data indexing in this paper) and the data can be located through decentralized search. A structured P2P overlay network achieves this through mapping data, data index and computers (content nodes) that store the data into the same ID space [1]. The ID of a data item is normally calculated from the attributes [2] of the data object or the hash key of the data itself. However, this mapping can cause *data skew* problem as data objects are unevenly distributed among content nodes. It can also cause *access skew* problem where the workload of nodes that serve queries to these data objects are unevenly distributed due to different data popularity. For a file-sharing P2P system, the access skew problem is not as severe as that observed in Web traffic due to the "fetch-at-most-once" behavior of P2P file-sharing users [3], however, for P2P overlays mainly used for data indexing, the problem is hard to ignore. In a data indexed overlay, data annotations are grouped and stored in nodes selected by underlying DHT mapping algorithms. These stored data annotations (or indexes) form a number of distributed databases. We consider a group of annotations

R. Perrott et al. (Eds.): HPCC 2007, LNCS 4782, pp. 599–610, 2007.

mapped into the same ID as a data object. Data objects in these nodes are frequently changed; therefore "fetch-at-most-once" behavior is not a common phenomenon in this scenario. Due to the database search cost, the non-uniform access distribution is likely to overwhelm the nodes that are responsible for popular data IDs. Solving the skew problem is crucial to the scalability of systems built on P2P technologies.

Constructing content distribution networks (CDN) through replicating popularly accessed data objects is a common technique to address the *access skew* problem. However, fundamental decisions must be made on issues such as which objects should be replicated, how many replicas should be created whenever the new replicas creation conditions apply, what replica placement mechanism should be used, and how fast these replica should be created and distributed.

In this paper, we give a self-organized content distribution system as well as a proactive content distribution algorithm in order to address the access skew problem. Determining the number of replicas is important for a content distribution system to efficiently use resources and reduce the cost for maintaining consistency among replicas. However it is a non-trivial issue to decide the number of replicas in a dynamic environment. Existing replication strategies for structured P2P networks often leave this problem un-tackled. They either create one new replica at a time when the load exceeds the capacity of the a content node [4,5], or create a fixed number of replicas at a time for caching and fault-tolerant purpose[6], or create replicas along query coming path in order to distribute content as fast as possible and serve flash crowds [7]. As shown in [4], too many replicas can make a system perform poorly due to high overhead. On the other hand, creating replicas slowly will result in poor query serving quality and long query queuing delays. In this paper, we propose a proactive CDN expansion mechanism to make replica creation adapt to the incoming query demands. Meanwhile, we give mechanisms to optimize the CDN performance by reducing the queuing delay and improving the content node utilizations.

The rest of the paper is organized as follows: Section 2 is related work; Section 3 is our CDN construction mechanism and system model; in Section 4 we describe our proactive content distribution mechanism; Section 5 presents experiment results, and Section 6 concludes the paper.

2 Related Work

Replication methods have long been used in distributed systems to exploit the data locality and shorten the data retrieval time. Content distribution networks such as Akamai[8] demonstrate the use of replication methods in the Internet environment. However, Internet content distribution networks often require global load information and manual deployment.

Recently, there are many replication strategies in the context of both unstructured [9, 10] and structured overlay networks [4,6,7]. They can be classified into three different types based on the number of replicas created whenever new replica creation conditions apply.

The aggressive strategy does not control the replica number. Some unstructured overlays [7] replicate data on all peers along the query forwarding route between the peer requesting the data and the peer having the data. This method, called path replication is not efficient and may result in a large amount of replicas with low utilization. The overhead may also drag down the performance of the whole system. On the other hand, the cost of maintaining consistency among replicas is high and this strategy is not suitable for a data index overlay where the replicated data objects are dynamic database tables. Another replication method called *owner replication* [10,11] replicates a data items to the peer that has successfully received the service through a query. The effect of this method is limited by the bandwidth and system performance of the receiving node.

Another strategy creates replicas based on a fixed replication ratio at a time. For example, paper [12] proposed two replication-based methods derived from *path replication* and replicate data on selected nodes on query path. *Path adaptive replication* determines the probability of the replication in each peer based on a predetermined replication ratio and its resource status. *Path random replication* is a combination of path replication method coupled with a replication ratio. Based on the probability of the pre-determined replication ratio, each intermediate peer randomly determines whether or not a replica is to be created and placed there. Apparently, a fixed replication ratio can not adapt to the change in incoming queries.

Most replication strategies create one replica at a time when new replica creation conditions apply [4, 5]. A new replica is created only when exiting replicas can not catch up with the increase in incoming queries. However, creating a new replica takes time depends on the network condition and replica size. This strategy may cause large amount of queries loss due to that the increasing speed of the query serving capacity can not catch up with that of the query incoming rate.

The algorithm given in this paper differs from these algorithms mainly in that it can determine the number of replicas for proactive content distribution to adapt to the incoming query rate. Determining the number of replicas is normally discussed in the context of reducing search cost [1] and system recoverability. However, it is important for addressing access skew problem. Recently, the work in [13] proposes to use historical data to predicate whether a data key is hot or not and make multiple replicas for hot data, however, it lacks proper metric to determine how many replicas are sufficient.

3 System Description

In a structured overlay, an ID is associated with a data object or a group of data objects who share the same data index. A data object here can be a data file or a data entry annotating a data file. Replicas of data objects that have the same ID form the content distribution network of the data ID.

As shown in Fig. 1, each data ID may have its own content distribution network. Each content node in the overlay network runs a process for handling queries to the data objects it stores. The query processing is FCFS based. We assume a lossless queue with a *capacity* defined as the number of queries the node can process in a certain time frame. When the capacity is reached, the node is overloaded and the queries coming subsequently may suffer long delay. They may be forwarded to a node that might be able to solve the query. In its simplest form, the content distribution network of a data ID is fully connected and load information is exchanged

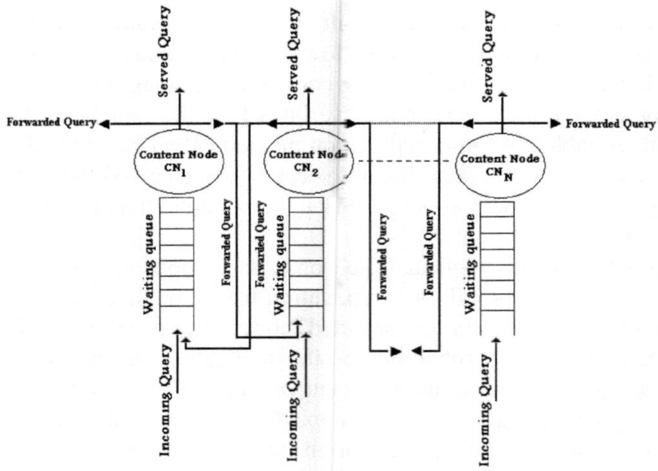

Fig. 1. Content distribution model

periodically along these links so that the forwarding destination can be easily obtained. When the data hosted by a node is popular, it is likely that the queue size is close to or over the capacity most of the time. When all the nodes in a content distribution network are overloaded, the network will be expanded by creating a new replica. The new replica node is selected from available nodes in the underlying network. Each node in the system can autonomously create replicas of its local data objects onto nodes selected by replica placement mechanism.

The system contains the following components for achieving the above:

- Query routes and access history collection: this component captures the temporal locality of incoming queries, and to measure the request arrival rate for every data ID over some measurement time interval.
- Content distribution network construction: this component uses a proactive content distribution algorithm to create new replicas.
- Load balancing: this mechanism dispatches queries in the content distribution network in order to make efficient use of the content nodes.

A. Query Routes and Access History Collection

The analyses of traces collected from Gnutella (http://www.gnutella.com) revealed that access locality exist in P2P systems. A popular file that has been accessed recently is also likely to be requested in the near future. We give a mechanism to collect useful information for exploiting query locality, in which each node manages its own data access history. This mechanism maintains the access history for every data ID hosted by a node as follows: Firstly, it obtains the search paths from recent queries in order to determine the pattern of paths recent queries come from. Each node maintains a *QueryStat* table to record the last-hop nodes queries travel through before they arrive in the destination node, as well as the count of queries coming from these last-hop nodes in the latest time frame. A sample *QueryStat* table is shown in Table 1. This table records top-3 frequently query routing nodes. Only top-K nodes are kept in each list to limit the size of *QueryStat* table. In this table, K0 has been queried 65 times recently, while 25 of them are routed via node N1, 20 of them are routed via N2, 10 are routed via node N4 and the rest 10 are routed via other nodes. Secondly, this mechanism measures request arrival rate for every data ID hosted by a node over some measurement time interval, in order to estimate the number of replicas needed for an ID. In the simplest form, a *KeyStat* table records the number of requests node *n* has received for data ID *k* in the latest time frame.

Table 1. QueryStat table of an overloaded Node

ID (req#)	**Node-Id** (#Requests)		
K0(65)	N1(25)	N2(20)	N4(10)
K1(30)	N2(18)	N3(7)	N6(5)
K2 (5)	N3(5)	-	-
K3(3)	N2(2)	N5(1)	-

B. CDN Construction

Our replica placement mechanism places replicas on nodes where queries frequently come from. *QueryStat* table is used to select nodes to host replicas. Node *n* which does not have a copy of the data ID *k* to replicate and passes most queries to ID *k* will be selected to replicate data objects of *k*. Replicas are therefore created along the query incoming paths. Our algorithm most likely places replicas on adjacent nodes which are one hop a way from the overloaded node. However, if there was no candidate nodes available in the *QueryStat* table to host new replicas; new content node will be selected randomly from the ID space of the overlay.

Fig. 2 shows the protocol used by an overloaded content node to recruit new node to join the CDN. An overloaded node can send CDN requests to multiple candidate content nodes simultaneously. However, the data distribution is done sequentially to each candidate nodes that accept the CDN request as different node may replicate different data IDs. We do not assume multicast support in the underlying networks.

C. Load Balancing in the CDN

Nodes storing the same data ID form the CDN of the ID. We denote a set of nodes that hold the data objects with ID *k* as cdn_k. As mentioned above, a content node's capacity of processing queries is described by the number of queries the node can process within a certain time frame. A content node is considered as overloaded if the number of queries in its queue reaches the capacity value. Before initiating CDN expansion, the overloaded node will forward the queries it can not process in time to lightly loaded nodes in cdn_k to ensure a certain query processing quality. A content node will update other nodes in the same CDN its load when the load change is above a pre-defined threshold. This cost can be high when the size of CDN becomes large. We gave a method in [5] to reduce the load information updating cost. cdn_k needs expansion when there is no lightly loaded node available to share the load.

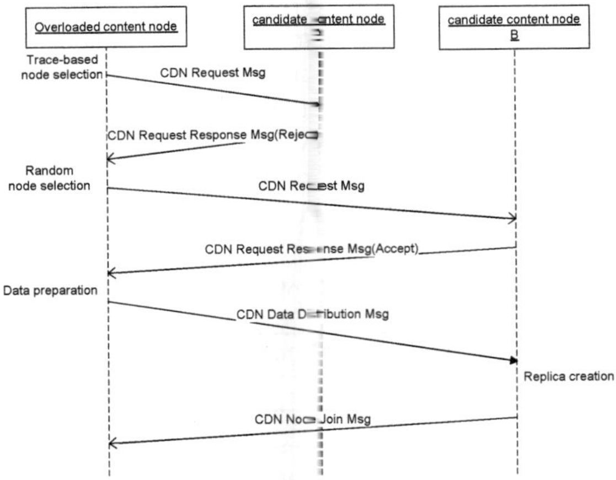

Fig. 2. CDN expansion example

A CDN node can withdraw from the CDN it belongs to if the CDN contains at least two nodes. This happens when the node keeps a low utilization on the stored content for a long time. A hand-off process is performed during the departure of a CDN node.

4 Proactive Content Distribution

In this section, we describe a proactive mechanism for CDN expansion in order to improve the query serving quality, especially for flash query crowds.

Normally, queries are unevenly distributed among data IDs. Similar phenomena are often modelled using Zipf distribution which indicates that the number of queries to the *i*th most popular data ID, denoted by *n*, is inversely proportional to *i*, i.e.,

$n \sim i^{-\alpha}$, in which α is the shape parameter.

In another words, assuming the expected number of queries to the *i*th popular data ID is $\bar{n}$, there will be *i* data IDs that have $\bar{n}$ or more queries. According to [14], the probability that a data ID has $\bar{n}$ or more queries is as below:

$$\Pr\{X > \bar{n}\} \sim \bar{n}^{-1/\alpha} \tag{1}$$

in which X is the number of queries to the data ID.

This is a Pareto distribution. We further have the following conditional probability when we observe there are *a* queries to the data ID:

$$\Pr\{X > b \mid queries = a\} \sim \left(\frac{a}{b}\right)^{1/\alpha} \tag{2}$$

in which a is the observed number of queries to the data ID and b is the total number of queries to this ID. Equation (2) indicates that the more queries to a data ID we observe, the more likely that the total number of queries is greater than b.

When $\alpha = 1.0$, we have the following:

$$\Pr\{X > N\} \sim \frac{1}{N} \tag{3}$$

$$\Pr\{X > b \mid n = a\} \sim \left(\frac{a}{b}\right) \tag{4}$$

Equation (2) and (4) gives a way to predicate the popularity of a data ID based on the number of observed queries to the ID. Early research on the number of visits to web sites further revealed that the difference in the number of visits to a site in two successive time frames is proportional to the total visits to the site [15]. This can be applied to our case: the difference in the number of queries to a data ID in two successive time frames is proportional to the total number of queries to the ID. With this, we can estimate the changing popularity of among data IDs as follows:

We denote the number of queries data ID k receives at time t in current time frame as n_k^t. Assume n_k^{ti} is the number of queries k received in previous time frame ending at ti. $\underset{ti}{\Delta n_k} = n_k^t - n_k^{ti}$ can therefore be used to estimate the number of queries to k by the end of the current time frame, which is

$$\Pr\{n_k^{t(i+1)} > b \mid \underset{ti}{\Delta n} = a\} \sim \left(\frac{a}{b}\right) \tag{5}$$

When the capacity of a content node reaches, the number of new content nodes need to create is calculated using algorithm 1 based on equation (5). In algorithm 1, each data ID k hosted in the node is associated with the number of queries to it in previous time frames, denoted as n_k^{tp} and the average number of queries to it in the current time frame n_k^{tc}. The two values are obtained through data ID access history collection mechanism. The number of replicas to create contains two parts. The first part is the number of replica needed by the end of current time frame, which is calculated using n_k^{tp} and n_k^{tc}. The second part is the number of replicas needed for reducing the current workload in the node, which is calculated using the current queries waiting in the queue.

Algorithm 1 assumes homogeneous nodes in the overlay. It is not difficult for us to extend the algorithm to heterogeneous environment.

Algorithm 1: Data ID Selection and Replica Number Calculation

Input: S={(ki, n_k^{tp}, n_k^{tc}) | ki is stored in the local node}

c – the query processing capacity of the local node

q – the current request queue length, normalized by the query processing capacity

thr – the probability threshold

Output: *(k, r)*: k is the data ID to replicate, r is the number of replicas to create.

for each *ki* in *S* do

$\Delta n_{ki} = n_k^{tc} - n_k^{tp}$

$r_{ki} = \lfloor \Delta n_{ki} / c \rfloor$; $m_{ki} = \Delta n_{ki} - r_{ki} \cdot c$

$P_{ki} = m_{ki} / c$

if $P_{ki} > thr$

$r_{ki} = r_{ki} + 1$

end if

$R = R \cup \{r_{ki}\}$

end for

sort R in descending order

get r_{ki}^0 from R

$r_{ki} = r_{ki}^0 + (q - c) / c$

return (k, r_{ki})

5 Evaluation

To study the performance of the proactive replica creation mechanism on structured overlays, we implement a simulator using FreePastry 1.4.1. The overlay has the following parameters unless stated otherwise. It consists of 3000 nodes with ID randomly generated in 128 bit ID space There are 200 data IDs published in the overlay and the total number of published data objects is 5000. Replicating a data ID incurs replica node negotiating cost, data dissemination cost and data processing cost. The average latency between two overlay nodes is set to 5 milliseconds. The average bandwidth between two overlay nodes is 10Mbps. The average data processing time including constructing a data table and populating its contents is 500 ms in average. Queries in an overlay network follow Poisson distribution. These queries can be sent from any node to published data IDs. Query sources are randomly selected. Queries select data IDs based on Zipf distribution with $\alpha = 1.0$. The query processing time is randomized between 10 to 100 ms. The time frame for counting queries is set to

100ms. The node capacity is set to 30 queries in the queue and workload change threshold is set to 30% of the capacity value. The probability threshold in algorithm 1 is set to 0.8.

A. Effect of the proactive CDN expansion mechanism

Our CDN expansion algorithm is adaptive to the changes of query arrival rate. In our experiment, queries arrival in three batches. The first batch contains 2000 queries with mean generating interval 2ms; the second batch contains 4000 queries with mean generating interval 1.4ms and the third batch contains 6000 queries with mean interval 0.8ms. These three batches of queries have an increasing demand to the CDN. Fig. 3. shows the overall query processing capacity change of the CDN that hosts the most popular data ID. The overall query processing capacity and the demands from incoming queries are normalized using the average query processing speed in a node. Fig. 3 shows that the CDN expansion mechanism is capable of identifying the popular data ID. The CDN expansion mechanism is also sensitive to the changes in query arrival patterns and it adapts to the demand of incoming queries well.

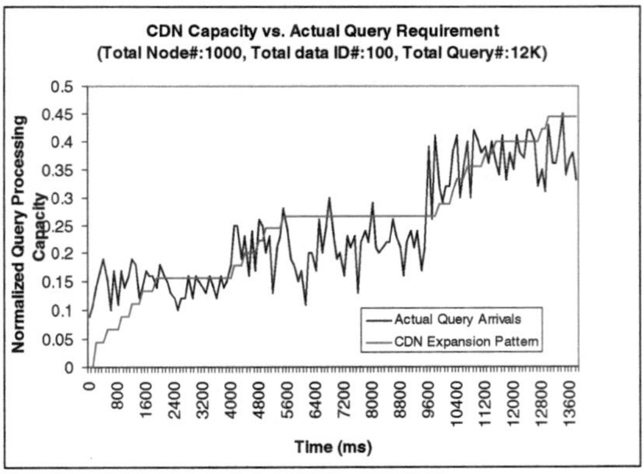

Fig. 3. Proactive CDN expansion vs. actual query arrivals: mean arrival interval- first 2000 queries: 2ms; next 4000 queries: 1.4ms; last 6000 queries: 0.8ms

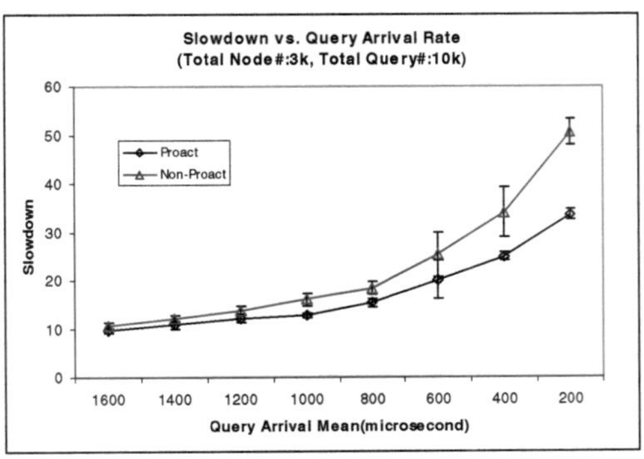

Fig. 4. Average query slowdown vs. query arrival rate

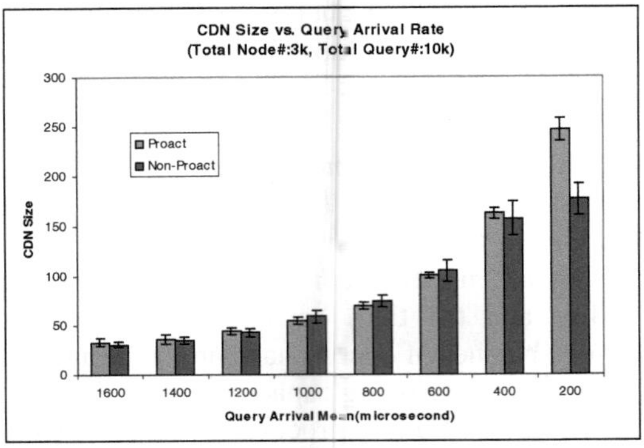

Fig. 5. CDN size vs. query arrivals

B. CDN performance comparison

To further quantify the effect of the data ID selection and replica number calculation mechanism, we compare the CDN performance under the proactive mechanism (*Proact*) to that under conservative CDN expansion mechanism (*Non-Proact*). A conservative mechanism adds one replica at a time when the CDN is overloaded. It is commonly used as an adaptive replication method. Fig. 4. compares the average query slowdown changes with the query arrival rates under the two mechanisms. The slowdown of a query is defined as below:

$$slowdown = \frac{query\_processing\_time + query\_queuing\_delay}{query\_proc\_time}$$

It is clear that *Proact* outperforms *Non-Proact* in terms that queries suffer shorter queuing delay. As the increase of query arrival rates, the slowdown gap also increases. *Proact* apparently creates replicas faster than *Non-Proact* does when the query arrival rate is high, as shown in Fig. 5. Note that under *Proact*, the average

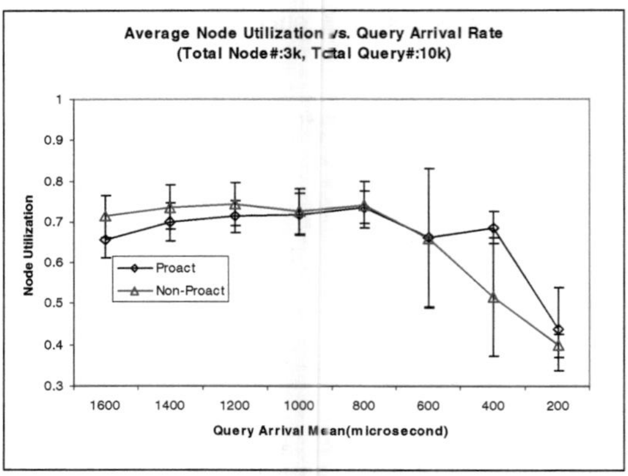

Fig. 6. Average node utilization vs. query arrival rate

query slowdown increases along with the query arrival rate. This is mainly due to that the cost of replica creation makes the CDN expansion speed can not keep up with the increasing query arrival rate. Even though *Proact* tends to create more replicas than *Non-Proact* does, their average CDN node utilizations are comparable as Fig. 6. shows, which means the proactive replica creation mechanism can efficiently control the CDN expansion speed.

C. Query forwarding hops

To study the load-balancing cost of our proactive mechanism, we compare the average query forwarding hops within a CDN under different replica creation mechanisms, namely *NonProact-Route*, *Proact-Route* and *NonProact-Random*. *NonProact-Route* uses conservative CDN expansion mechanism and new content node is selected based on *QueryStat* table discussed in section 3. *Proact-Route* uses proactive CDN expansion mechanism and new content node is selected based on *QueryStat* table. *NonProact-Random* mechanism uses conservative CDN expansion mechanism and new content node is randomly selected from the overlay ID space. Table 2 shows the average query forwarding hops in the CDN. It is clear that the proactive mechanism does not incur extra load-balancing cost.

Table 2. Query forwarding hops comparison

Mean Query Arrival	NonProact-Routes	Proact-Routes	NonProact-Random
1600 (microsecond)	0.183141	0.186367	0.2552
1200 (microsecond)	0.2392	0.194386	0.3304
800 (microsecond)	0.315944	0.276759	0.396439
400 (microsecond)	0.465948	0.416457	0.504721

6 Conclusion

We have presented a proactive content distribution method for data indexed DHT networks, which effectively addressed the access skew problem in such type of overlay networks. The method is capable of estimating the number of replicas needed for unevenly distributed demands to data IDs. Our simulation evaluation shows that the mechanism can significantly reduce the query slowdown when the query arrival rate is very high. Meanwhile, it uses resources efficiently by keeping the content node utilization reasonably high and load-balancing cost low.

References

1. Balakrishnan, H., Kaashoek, M.F., Karger, D.R., Morris, R., Stoica, I.: Looking up data in P2P systems. Communications of the ACM 46(2), 43–48 (2003)
2. Garces-Erice, L., Felber, P.A., Biersack, E.W., Urvoy-Keller, G., Ross, K.W.: Data indexing in Peer-to-Peer DHT networks. In: ICDCS 2004. Proc. of the 24th International Conference on Distributed Computing Systems, pp. 200–208 (2004)

3. Gummadi, K.P., Dunn, R.J., et al.: Measurement, modeling, and analysis of a Peer-to-Peer file-sharing workload. In: SOSP 2003. Proc. of the 19th ACM Symposium on Operating System Principles, pp. 314–329. ACM Press, New York (2003)
4. Gopalakrishnan, V., Silaghi, B., Bhattacharjee, B., Keleher, P.: Adaptive replication in Peer-to-Peer systems. In: ICDCS 2004. Proc. of the 24th International Conference on Distributed Computing Systems, pp. 360–369 (2004)
5. Wang, C., Alqaralleh, B., Zhou, B., Brites, F., Zomaya, A.Y.: Self-organizing content distribution in a data indexed DHT network. In: P2P 2006. Proc. of the 6th IEEE International Conference on Peer-to-Peer Computing, pp. 241–248. IEEE Computer Society Press, Los Alamitos (2006)
6. Rowstron, A., Druschel, P.: Storage management and caching in PAST, a large-scale, persistent peer-to-peer storage utility. In: SOSP 2001. Proc. of the 18th ACM Symposium on Operating System Principles, pp. 188–201. ACM Press, New York (2001)
7. Dabek, F., Kaashoek, M.F., Karger, D., Morris, R., Stoica, I.: Wide-area cooperative storage with CFS. In: SOSP 2001. Proc. of the 18th ACM Symposium on Operating System Principles, pp. 202–215. ACM Press, New York (2001)
8. Akamai (2006), http://www.akamai.com
9. Cohen, E., Shenker, S.: Replication strategies in unstructured peer-to-peer networks. In: Proc. of 2002 SIGCOMM conference, pp. 177–190 (2002)
10. Lv, Q., Cao, P., Cohen, E., Li, K., Shenker, S.: Search and replication in unstructured peer-to-peer networks. In: Proc. of the 16th ACM Conf. on Supercomputing Systems, pp. 84–95. ACM Press, New York (2002)
11. Iyer, S., Rowstron, A., Druschel, P.: Squirrel: A decentralized peer-to-peer web cache. In: PODC 2002. the 21st Annual ACM Symposium on Principles of Distributed Computing, pp. 213–222. ACM Press, New York (2002)
12. Yamamoto, H., Maruta, D., Oie, Y.: Replication methods for load balancing on distributed storages in P2P networks. In: SAINT 2005. Proc. 2005 Symposium on Applications and the Internet, pp. 264–271 (2005)
13. Xu, Z., Bhuyan, L.: Effective load balancing in P2P systems. In: CCGrid 2006. Proc. of 6th IEEE International Symposium on Cluster Computing and the Grid, pp. 81–88. IEEE Computer Society Press, Los Alamitos (2006)
14. Adamic L. A.: Zipf, Power-laws, and Pareto - a ranking tutorial. Available online, http://www.hpl.hp.com/research/idl/papers/ranking/ranking.html
15. Adamic, L.A., Huberman, B.A.: The nature of markets in the World Wide Web. Quarterly Journal of Electronic Commerce 1(1), 5–12 (2000)

CDACAN: A Scalable Structured P2P Network Based on Continuous Discrete Approach and CAN

Lingwei Li[1], Qunwei Xue[2], and Deke Guo[1]

[1] School of Information System and Management
National University of Defense Technology, Changsha, Hu Nan, 410073, P.R. China
[2] China Institute of Geo-Environment Monitoring, Beijing 100081, P.R. China
lilingwei_82104@126.com, qwxue@hotmail.com, guodeke@gmail.com

Abstract. CAN is a famous structured peer-to-peer network based on d-dimensional torus topology with constant degree and logarithmical diameter, but suffers from poor scalability when $N>>2^d$, N is the number of peers. To address this issue, we proposes a novel scalable structured peer-to-peer overlay network, CDACAN that embeds the one-dimensional discrete distance halving graph into each dimension of CAN. The out-degree and average routing path length of CDACAN are $O(d)$ and $O(log\ (N))$, respectively, and are better than that of CAN. On the other hand, we analyze the optimal value of dimensions and the smooth division method of d-dimensional Cartesian coordinate space when handling the dynamic operations of peers. The smooth division of multidimensional space can improve the routing performance, and also is helpful to keep load balance among peers. Those properties and protocols are carefully evaluated by formal proofs or simulations. Furthermore, we present a layered improving scheme to decrease the out-degree of each peer in the future work. The expected topology will keep 8 out-degree and $O(log_2(N)+d)$ routing path length.

1 Introduction

Structured peer-to-peer networks, abbreviated as P2P, have been proposed as a good candidate infrastructure for building large scale and robust network applications, in which participating peers share resources as equals. They impose a certain structure on the overlay network and control the placement of data, and thus exhibit several unique properties that unstructured P2P systems lack.

Several interconnection networks have been used as the desired topologies of P2P networks. Among existing P2P networks, CAN [1] is based on the d-dimensional torus topology; Chord[2] Pastry[3] and Kademlia [4] are based on the hypercube topology; Viceroy[5] is based on the butterfly topology; Koorde [6], Distance Halving network [7], D2B [8], ODRI [9] and Broose [10] are based on the de Bruijn topology; FissionE[11], MOORE[12] are based on the Kautz topology.

CAN is a famous structured peer-to-peer network based on d-dimensional torus topology with constant degree and logarithmical diameter. It supports multi-dimensional exact and range queries in a natural way, but suffers from poor scalability when $N>>2^d$, N is the number of peers. To address this issue, we introduce

R. Perrott et al. (Eds.): HPCC 2007, LNCS 4782, pp. 611–622, 2007.

CDACAN, an improved structured P2P network based on CAN and the CDA (Continuous-Discrete Approach). CDACAN is a constant degree DHT with $O(log(N))$ lookup path length and inherits the properties of Cartesian coordinate space. The main contributions of this paper are as follows:

- We extend one-dimensional discrete distance halving graph to multi-dimensional space, abbreviated as *DGd*, to serve as the topology of CDACAN. The *DGd* can keep the good properties of Cartesian coordinate space.
- We improve the routing algorithms proposed in literature [7] to support *DGd* . Thus, CDACAN achieves a less value of the average routing path length than that of the traditional CAN.
- We design algorithms to calculate the necessary precision of destination code to perform the search. We reveal that the smoothness and regularity of the spatial division can reduce the hazard in the estimation of precision and neighbor forwarding can amend the routing.
- We design the peer joining and departing protocols based on an associate sampling mechanism to improve the smoothness and regularity of all cells of a whole Cartesian coordinate space after division.
- We point out that the out-degree of each peer in CDACAN is still relative high, and propose a layered scheme to reduce the value of out-degree. It achives a better tradeoff between the routing table size and the routing delay just as the Cycloid does.

The rest of this paper is organized as follows. Section 2 introduces some useful conceptions in CDA. Section 3 proposes the topology construction mechanism, the routing algorithm, and the joining, departing protocols for CDACAN. Section 4 evaluates the characteristics of CDACAN by comprehensive simulations. Conclusion and future work are discussed in Section 5.

2 Basic Concepts of Continuous Discrete Approach

Continuous Distance Halving Graph ***Gc*****:** The vertex set of *Gc* is the interval $I= [0, 1)$. The edge set of *Gc* is defined by the following functions: $L(y)=y/2$, $R(y)=y/2+1/2$, $B(y)=2y \ mod \ 1$, where $y \in I$. L abbreviates 'left', R abbreviates 'right' and B abbreviates 'backwards'.

Discrete Distance Halving Graph ***Gd*****:** X denotes a set of n points in I. A point X_i may be the ID of a processor V_i or a hashing result of the value of ID. The points of X divide I into n segments. Let us define the segment of X_i to be $S(X_i) = [X_i, X_{i+1})$, $i=1,2,\ldots n-1$, and $S(X_n)=[X_n, 1) \cup [0, X_1)$. Each processor V_i is associated with the segment $S(X_i)$. If a point y is in $S(X_i)$, we say that V_i covers y. A pair of vertices (V_i, V_j) is an edge of *Gd* if there exists an edge (y, z) in the continuous graph, such that $y \in S(X_i)$ and $z \in S(X_j)$. The edges (V_i, V_{i+1}) and (V_n, V_1) are added such that *Gd* contains a ring. In figure 1, the upper diagram illustrates the edges of a point in *Gc* and the lower diagram shows the mapped-intervals of an interval that belongs to *Gd*.

The greedy routing and distance halving routing algorithms apply to the graph, and are abbreviated as GA and DHA, respectively. If the latter adopts the binary string representation of the destination X_d to perform the first step, we call it as direct distance halving algorithm, abbreviated as DDHA. We also use continuous discrete routing algorithm, abbreviated as CDRA, as a general name for both GA and DDHA.

3 CDACAN

In order to improve the scalability of CAN, we embed a d-dimensional discrete distance halving graph into CAN, and achieve a novel P2P network.

3.1 Topology Construction Mechanism

We extend the conceptions of *Gc* and *Gd* to multidimensional space.

d*-Dimensional Continuous Distance Halving Graph** *DGc:** The vertex set of *DGc* is the unit *d*-dimensional Cartesian coordinate space $S=\{(X_0,X_1,\ldots\ldots X_{d-1})|X_i \in [0,1), i=0,1,\ldots\ldots d-1\}$. Given $Y=(Y_0, Y_1,\ldots\ldots Y_{d-1})$, the edge set of *DGc* is defined by the following functions: $DL(Y)=\{(X_0,X_1,\ldots X_j,\ldots X_{d-1})|X_i=Y_i,\ i=0,1,\ldots j-1,\ j+1\ldots d-1, X_j=Y_j/2,\ j=0,1,\ldots\ldots d-1\}$; $DR(Y)=\{(X_0,X_1,\ldots X_j,\ldots X_{d-1})|X_i=Y_i,\ i=0,1,\ldots j\text{-}1,\ j+1\ldots d-1, X_j=Y_j/2+0.5,\ j=0,1,\ldots\ldots d-1\}$; $DB(Y)=\{(X_0,X_1,\ldots X_j,\ldots X_{d-1})|X_i=Y_i,\ i=0,1,\ldots j-1, j+1\ldots d-1, X_j=2Y_j \bmod 1, j=0,1,\ldots\ldots d-1\}$;

d*-Dimensional Discrete Distance Halving Graph** *DGd:** *P* denotes a set of *N* processors, divide *S* into *N* zones and each zone Z_j is charged by P_j in *P*, $j=1,2,3,\ldots N$. A pair of vertices (P_i, P_k) is an edge of *DGd* if there exists an edge (y, z) in *DGc*, such that $y \in Z_i$ and $z \in Z_k$. The CAN-like edges are included in *DGd* too.

Let *Z* denote the zone set of *DGd*, Z_i is a zone that $Z_i \in Z$ and has a serial number of i, $i=1,2,3\ldots N$. $Z_i.j$ denotes the interval of the *j*th dimension of Z_i, $j=0,1,2,\ldots d-1$.

CLAIM1: let N_i is a node in *DGd*, whose zone is Z_i with volume *V*. Z_i is mapped to *d* *L*-mapped zones, *d* *R*-mapped zones and *d* *B*-mapped zones, and their volumes are *V*/2, *V*/2 and 2*V* respective.

The definitions, corollaries and lemmas mentioned below will be used in later.

Definition 1. The **length** of a binary decimal *a* is ***L*(*a*)**, where *L*(*a*) is the number of bits behind the radix point of *a*. If $L(a)=i$, *a* is a ***i*-regular-point**.

Definition 2. Let an interval $I=[a, b)$, where *a* is a binary decimal and $L(a) \le i$, $b=a+2^{-i}$. We call it ***i*-regular-interval**. Both *a* and *b* are the **boundaries** of *I*.

We can infer a corollary and two lemmas on the base of *definition* 2. The related proofs are omitted due to the page limitation.

Corollary 1. In *Gd*, the mapped intervals of *i*-regular-intervals induced by *R*(*y*) or *L*(*y*) are (*i*+1)-regular-intervals; and that induced by *B*(*y*) are (*i*−1)-regular-intervals.

Lemma 1. Any *i*-regular-interval $I_a=[a, b)$ possesses a unique binary prefix *Sp* whose length is *i*. Every point in [0,1) whose binary form with prefix *Sp* is included in I_a; and the *i*-prefix of every point *p* in (*a*, *b*) is *Sp* and *p* satisfies $L(p)>i$.

Lemma 2. There must be only one boundary point *c* of *i*-regular-interval *I* satisfies: $L(c) = i$, the length of other boundary point *e* satisfies $L(g) < i$.

Definition 3. The **smoothness of *i*th dimension** and **smoothness of volume** are denoted as $SM_i(Z)$ and $SM_V(Z)$, respectively. They are defined to be the existing number p that satisfies $MAX_{j,k}\ |Z_j.i|/|Z_k.i|<p$, $i = 0,1,2,\ldots\ldots d-1$, and $MAX_{j,k}\ |Z_j|/|Z_k|<p$ for any network size N, respectively.

Definition 4. The **regularity of *Z*** means that all the $Z_i.j$ are regular-intervals, i=1,2,3…N, j=0,1,2,…d−1.

The smoothness problem has been discussed in [7]. Its effect on routing algorithms, however, is still an open problem. We show that CDA can generate a series of regular intervals, and its essential objective is to gain the unique prefix of the interval (zone).

Theoretic Analysis of Degree distribution: supposing the spatial division is **smooth and regular**. For *d*-CDACAN, the degree of a node N_i is about 6*d*, the edges including: about 2*d* CAN-like edges to the neighbor of Cartesian coordinate space; about *d*, *d* and 2*d* edges induced by *DR*, *DL* and *DB* functions in *DGd* respectively.

Figure 2 shows a 2-*DGd*, the overstriking area is an actual zone Z_i whose owner is N_i; red areas are *R*-mapped zones and *L*-mapped zones of Z_i; green areas are *B*-mapped zones of Z_i. The number of degree of N_i is 16, including CAN-like edges.

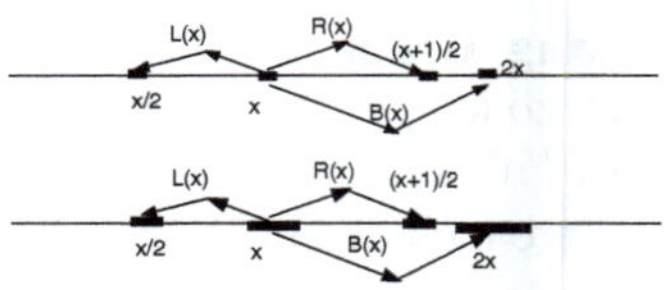

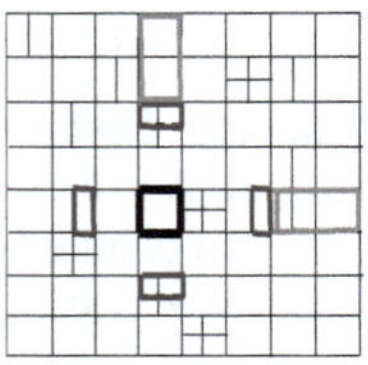

Fig. 1. node degree in *Gc* and *Gd* **Fig. 2.** A possible 2-DGd

It is hard to get the upper bound (mentioned in *Theorem1* in [7]) of the number of edges of any node in *DGd*. And it is clear that the spatial division of high dimensional *DGd* without smooth volumes and regular shapes may lead to a high value of degree.

3.2 Routing Protocol

3.2.1 Routing Algorithm

Let X_d denote a lookup destination, $I_d = [I_x, I_y)$ is the interval that contains X_d and V_d is the processor that owns I_d. I_s is the interval of the lookup initiator V_i. I_r is the longest regular-interval in I_d and $L(I_r)$=*h*, and I_L is the longest regular-interval in I_s and $L(I_s)$=*v*. Our discussion in 3.2.1 and 3.2.2 is general, and does **not base on the hypothesis of smoothness and regularity**. We will reveal three problems that have not been pointed out and clarified in [7].

- In *DGd*, how does CDRA calculate I_L and I_r?
- They did not reveal why the CDRA is prone to fail in distributed scenario, and how the smoothness and regularity of spatial division reduce the hazard of routing, and how neighbor forwarding amends the routing.
- They did not reveal the key factors that affect the routing path length in CDRA.

The three problems are interconnected and will be settled later.

The routing algorithm of CDACAN is **routing on every dimension obeyed to the improved CDRA respectively**. The improved CDRA include three steps: calculating appropriate I_r (for both GA and DDHA) and I_L (for GA only); routing by CDRA; neighbor forwarding (if necessary). We will introduce our contributions, which are the first and third steps.

Because the improved CDRA is performed on every dimension of CDACAN, we adopt *Gd* as the background of improved CDRA for convenience and the adjective of "improved" may be omitted in the latter.

3.2.2 Calculation of I_r (I_L) Under the Global Division Information

We assume that the interval division states are known by all the peers in this section and devise an algorithm to calculate I_r (I_L) under the hypothesis. We will introduce a lemma first.

Lemma 3. Let x and y denote two binary decimals and $1>y>x\bullet 0$, there must exist one binary decimal m in I = $[x, y)$ whose length $L(m)$ is the smallest, We call m the **Shortest Normative Point**.

Then we will propose an algorithm to calculate m. The functions of binary string operations in the pseudo code are quite straightforward.

```
Algorithm 1. FindShortestBinaryDecimal(x, y)
Require: The inputs must satisfy that x< y
t • y - x
z • t.PreserveLargest1Bit()
L • z.GetLength()
if y = y.PreserveBits(L) then
     m • y - 2^-L
else
     m • y.PreserveBits(L)
p • m - 2^-L
if p • x and m.GetLength()>p.GetLength() then
      m • p
return m
```

Corollary 2. The Shortest Normative Point of a regular interval I=[a, b) is a.

Theorem 1. I_r can be calculated by the algorithms mentioned as follows.

```
Algorithm 2. DetermineRoutingLengthLowerBound (I_d)
Require: I_d=[I_x, I_y) what has been defined in 3.2.1.
t • I_y - I_x
z • t.PreserveLargest1Bit()
h • z.GetLength()
StartPoint • FindShortestBinaryDecimal(I_x, I_y)
I_r1 = SearchLeft(StartPoint, h, I_x)
I_r2 = SearchRight(StartPoint, h, I_y)
return I_r = max(I_r1, I_r2)
```

Algorithm 3. SearchLeft(*StartPoint, h,* I_x)

Require: *StartPoint* is a point in I_d, and *h* is a positive integer.

if (*StartPoint* – 2^{-h}) < I_x) **then**

    I_{r1}= *SearchLeft*(*StartPoint, h* + 1, I_x)

else

    I_{r1}=[*StartPoint* – 2^{-h}, *StartPoint*)

return I_{r1}

Algorithm 4. SearchRight(*StartPoint, h,* I_y)

Require: *StartPoint* is a point in *I*, and *h* is a positive integer.

if (*StartPoint* + 2^{-h}) • I_y) **then**

    I_{r2}= *SearchRight* (*StartPoint, h+1,* I_y)

else

    I_{r2}= [*StartPoint, StartPoint* + 2^{-h})

return I_{r2}

Proof. let the Shortest Normative Point of I_d =[I_x, I_y)is *m* and $I_1 = m - I_x$, $I_2 = I_y - m$. We can decompose I_1 and I_2 into the formulas whose form is $x_1 2^{-1} + x_2 2^{-2} + \ldots x_i 2^{-i} + \ldots$, x_i = 0 or 1, *i*=1,2,……. If *L*(*m*)=*L* there must be

$$I_1 = x_1 2^{-L} + x_2 2^{-L-1} + \ldots x_{i+1-L}\, 2^{-i} + \ldots,\ x_i = 0 \text{ or } 1,\ i=L,\ L+1\ldots\ldots \quad (1)$$

$$I_2 = y_1 2^{-L} + y_2 2^{-L-1} + \ldots y_{i+1-L} 2^{-i} + \ldots,\ y_i = 0 \text{ or } 1,\ i=L,\ L+1\ldots\ldots \quad (2)$$

The relationship between the decomposition of I_1 and *Algorithm3* is that *Algorithm3* can find the largest regular-interval I_{r1} at the left of *m*, corresponding to the first number appeared in (1).

And *SearchRight* can find a regular-interval I_{r2} at the right of *m* too. □

It is obvious that DDHA will adopt the unique prefix of I_r to perform the routing when the interval division state is known by all peers. GA can get I_r and I_L by the same method too. The detailed explanation of DDHA and GA can be found in [7].

Corollary 3. if I_d and I_s are regular-intervals, then $I_r = I_d$, $I_L = I_s$, respectively.

3.2.3 Neighbor Forwarding and the Effects of Smoothness and Regularity

The Algorithm in 3.2.2 bases on a hypothesis that the interval division is known by all peers. The hypothesis can not be satisfied because of the distributed property of P2P network. If V_i wants to query a resource whose ID is X_d. Firstly V_i have to estimate I_d what contains X_d to perform *Algorithm2*, the estimation including the length and the beginning point of I_d, and the estimation of I_d is denoted by I_d'; Secondly V_i adopt *algorithm2* to calculate I_r' and I_L by I_d' and I_s respectively; Thirdly V_i perform DDHA by I_r' or GA by I_r' and I_L respectively; Fourthly, the searching massage is sent to a processor denotes by V_d'; At last, V_d' is not always V_d because of the uncertainty of the estimation of I_d, and the method for amending is **neighbor forwarding**.

Let suppose L_{ave} is the average interval length in the network. The effect of the smoothness is making the calculation of L_{ave} accurately, thus $L(I_d')$ too; and the effect of regularity is making the position of the entire intervals regular. In CDACAN, the regularity is achieved by the CAN-like divisional scheme, and the smoothness is achieved by the sampling operation in the joining and departing stages.

However, smoothness problem can not be addressed by a determinate method. So V_d' may be not V_d and we design a **neighbor forwarding** scheme for V_d', i.e. **forwarding a message to Cartesian-coordinate-neighboring nodes**, to amend the routing failure of CDRA what is resulted from a low-estimated I_r'.

3.2.4 Routing Path Length of Continuous Discrete Routing Algorithm

We can infer three corollaries when consider the problem under a **distributed scenario**.

Corollary 4. The key factor that affects the routing path length of GA is v.

Proof. In GA, V_i can calculate logical beginning point P_s as follows: gets the prefix of P_s by the unique prefix of I_L, and then gets the suffix of P_s by the unique prefix of I_r'. Because $P_s \in I_S$, the routing path length of original GA which starts at P_s is v. And I_r' affect the size of the routed message. □

Corollary 5. The key factor that affects the routing path length of DDHA is $h'=L(I_r')$.

Corollary 6. If the spatial division is smooth and regular, then $| h'-v |<log_2(p)$.

Proof. According to *Corollary3,* if the division is regular, there must be $I_r' = I_d'$ and $I_L = I_s$, and $|I_d'|=L_{ave}=2^{-h'}$, $|I_s|=2^{-v}$. According to the definition of smooth we know that existing a number p>1 what satisfies $1/p<|I_d'|/|I_s|<p$. So $|h'-v| <log_2(p)$. □

If the spatial division is smooth and regular, *Corollary6* implies that the routing path lengths of DDHA and GA are close. Else, it is difficult for V_i to determine appropriate I_d' to perform CDRA because V_i doesn't know I_d in distributed circumstance. GA can adopt a little larger I_d' to increase the probability of successful lookup and avoid neighbor forwarding because it does not waste too much bandwidth according to *Corollary4*; DDHA can not adopt too large I_d' because it increases the entire routing path length unnecessarily according to *Corollary5*. So the risk what comes from the uncertainty of the calculation of I_d' can do more harm to DDHA than GA.

Theoretic Analysis of Routing path length distribution: In smooth and regular division, the average routing path length and the diameter of DDHA and GA are both about L_{ave}, and there may be some lengths larger than L_{ave} because of neighbor forwarding caused by the impossibility of even spatial division. The hop number of neighbor forwarding is determined by smoothness.

3.3 Node Joins

As for most P2P networks, we assume there are some existing nodes as entry points of CDACAN, which can receive and process the node joining message. The joining procedure includes three stages: selects a zone to divide, redistributes resources, and updates routing tables. We will introduce the first step because it is unacquainted

The process is inspired by the join algorithm of *Gd* in [7]:

1. Estimate log_2N. The method to estimate log_2N is simple and is omitted here.
2. Sample $tlog_2N$ random points in d-dimensional unit space S, t is a constant and can be known by *Theorem11* [7], we also can select an appropriate t by simulation.

3. Check all zones that contain those points. Let N_l denote the node that contains the largest zone Z_l. Then divides Z_l into two parts in a similar way used by CAN.

Borrowing the *Theorem11* from *Gd* in [7], we can deduce the theorem as follow under the background of *DGd*.

Theorem 2. For any initial state, the largest zone would be of volume at most $1/2n$ after inserting n points.

The proof of theorem2 is omitted because it can be transplanted from [7] easily. This theorem provides the basis to keep smooth in *DGd* by the original sample scheme. We have improved the sample scheme by using the **local search** rule. It can reduce the number of messages used to carry out the sampling process and keep smooth.

Samples with Local Search: The sampled nodes check all the neighbors and select the neighbor whose zone is the largest. It can improve the efficiency and reduce the number of sampled points remarkably at the same time.

3.4 Node Departs

Because of the dynamic nature of P2P network, CDACAN need to consider not only node joining operation but also node departure operation. Updating the routing table and the reassigning zones are omitted because they are familiar. The departing protocol includes four steps as follows:

1. If N_D, whose zone is Z_D, wants to depart immediately, it selects the smallest neighbor node N_S to take over Z_D. So N_S manages two zones temporarily.
2. N_S samples number of $tlog_2N$ points in d-dimensional unit space S and send $tlog_2N$ messages to their owners, where the value of t is assigned in a same method used by the peer joining algorithm.
3. Every sampled node P_s searches "partition tree" to find a "leaf nodes" N_L by a similar algorithm used by the topology maintenance mechanism in CAN. Then P_s sends N_L back to N_S.
4. N_S selects the smallest leaf node N_{Le}, and then N_{Le} hand its zone Z_{Le} to its "geminate node" to merge to a valid zone and take over Z_D.

The first step ensures that the departure operation can be finished successfully. The last two steps can keep the regularity of the division for the same reason in CAN.

Our major contribution is the step two. The sample-based selection can improve the smoothness and does not bring too much sampling messages. By comparing with CDACAN, CAN adopt a simple scheme to find N_S and it can not deal with mass adversarial departure; the method in CDA needs to keep additional structure (bucket or Cyclic Scheme).

4 Evaluation

4.1 Query Path Length

In this section, we will verify the theoretical analysis results by comprehensive simulations. The query path length is measured in terms of overlay network hops. We

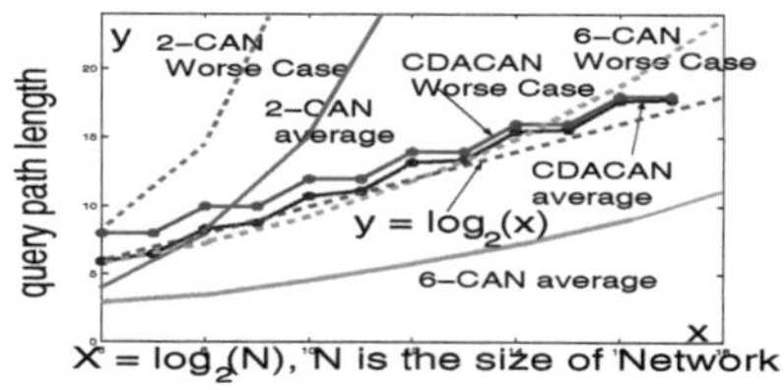

Fig. 3. Routing path length of CDACAN

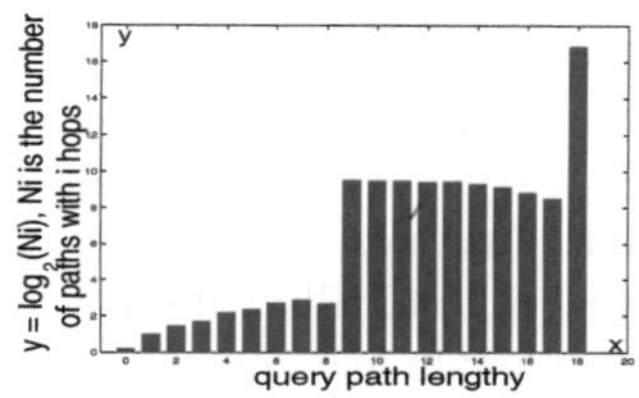

Fig. 4. Query path length distribution of CDACAN with 128k nodes

assume a simple model for the offered query load: queries are generated randomly and uniformly at each node, the desired query results are uniformly distributed in the key-space.

We plot the routing path length of 2-CDACAN in figure 3, and adopt 2-CAN and 6-CAN as the benchmark for our experiments since 2-CDACAN is based on 2-CAN and has the same degree with 6-CAN. We have simulated the worst-case and average-case path lengths as a function of the number of nodes N (x-axis, in the log scale). Figure 3 demonstrates that the routing path length of 2-CDACAN is close to the $y=log_2N$ function. The "step-like-increase" on both CDACAN routing path curves is due to the increasing of the estimated system size when N becomes larger. As expected, curves for the worse-case and average-case of 2-CAN and 6-CAN are coincide with that of functions $n^{1/2}$, $(1/2)n^{1/2}$, $3n^{1/6}$ and $(3/2)n^{1/6}$ respective. We also discover that 2-CDACAN is better than 2-CAN and inferior to 6-CAN. The simulation results also expose that the degree of each peer in CDACAN is relative high, and should decrease as less as possible. Figure 4 shows the distribution of path lengths in CDACAN with 128k nodes. We can observe that most paths with the length of $1+log_2(N)$. It is because our simulations adopt higher precision of I_r than the normal one for one bit to avoid neighbor search.

4.2 Sample in the Join Operation

The smoothness and degree problems of CDACAN have impact on the sample algorithm. We will analyze it by simulations.

We choose average degree to reveal the effects of local search as shown in figure 5. The upper line and lower line represent the average degree of peers in CDACAN that

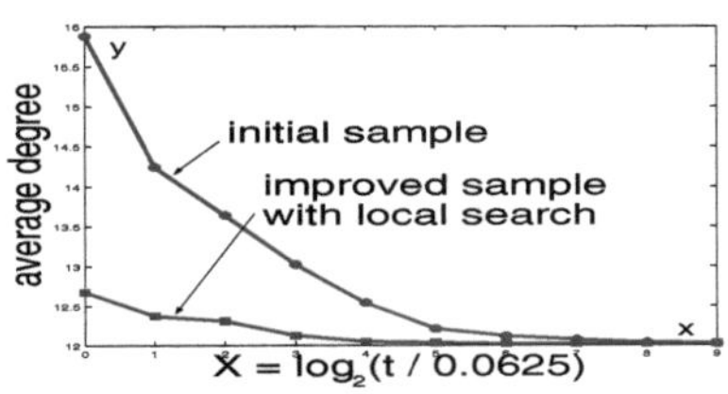

Fig. 5. Average degree of CDACAN with 1024 nodes

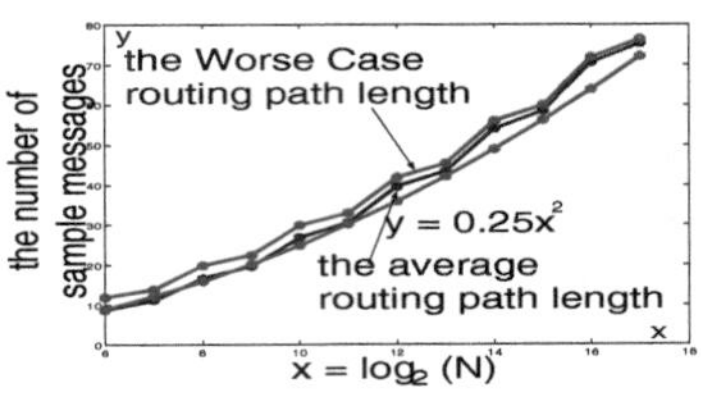

Fig. 6. The number of sample messages in CDACAN with $t = 0.25$

is generated by the original sampling and improved sampling with local search, respectively. We discover that the local search can improve the smoothness (Figure 9 and Figure 10) and reduce the degree (Figure 5) obviously.

Figure 6 illustrates that the average number of sample messages in a join operation increases logarithmically with respect to the size *N*. The number is the product of the routing path length and amount of samples. This experiment adopts t=0.25, which is a relative large value. The green line assumes the worse case of routing path and the blue line adopts the average. They are both close to $y=0.25x^2$ because the routing path length is close to $log_2(N)$ as shown in figure 3. The average number of overlay messages what are generated by the sample operation is less than 80 in CDACAN with 128k nodes. It illuminates that sample-based scheme in CDACAN is practical.

4.3 Volume and Smoothness

Volume distribution is an important network trait in CDACAN-like protocols what based on spatial division and spatial mapping. Figure 7 and Figure 8 demonstrate the effect of regular division. Figure 7 shows that irregular division leads to trivial distribution of volume and the ratio of the biggest zone and the smallest zone is not less than 20; Figure 8 shows that the regular scheme generates smooth division and the ratio of the biggest volume and the smallest volume is 4.

Figure 9 and Figure 10 demonstrate that the volume distribution of CDACAN becomes smooth when t increases. And they illustrate the volume distribution what is generated by the sample with local search and without local search respectively. Comparing figure 9 with figure 10, we discover that local search can improve the smoothness remarkably. And they uncover the causation of figure 5 too.

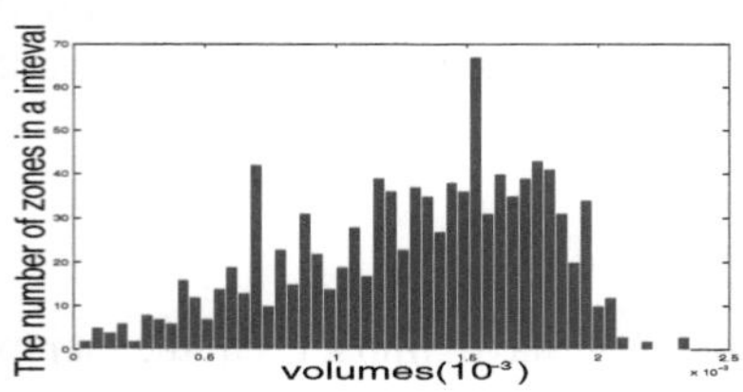

Fig. 7. Volume Distribution of CDACAN with irregular division

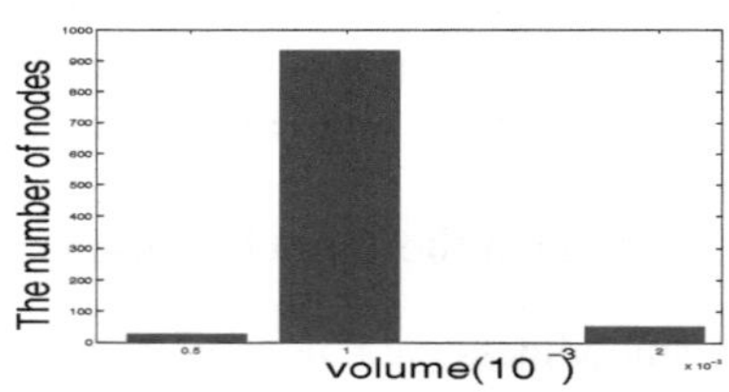

Fig. 8. Volume Distribution of CDACAN with regular division

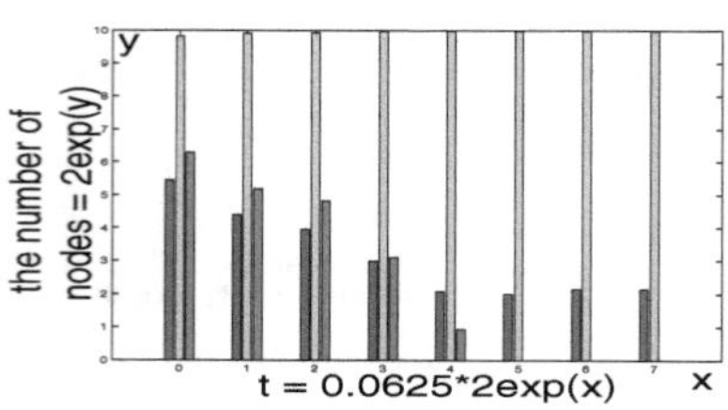

Fig. 9. The volume distribution in CDACAN without local search

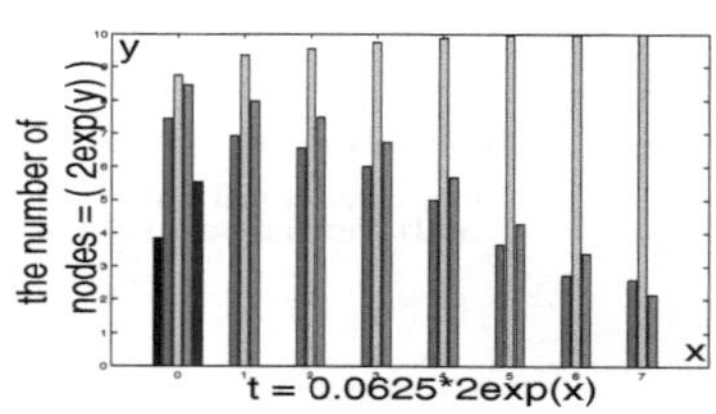

Fig. 10. The volume distribution in CDACAN with local search

4.4 Distribution of Degree

The distribution of degree in CDACAN is an important trait that represents the routing table overhead. The experiment which corresponds to Figure 11 is called E1, and the experiments which correspond to the green, blue and red bars in figure 12 are called E2, E3 and E4 respectively.

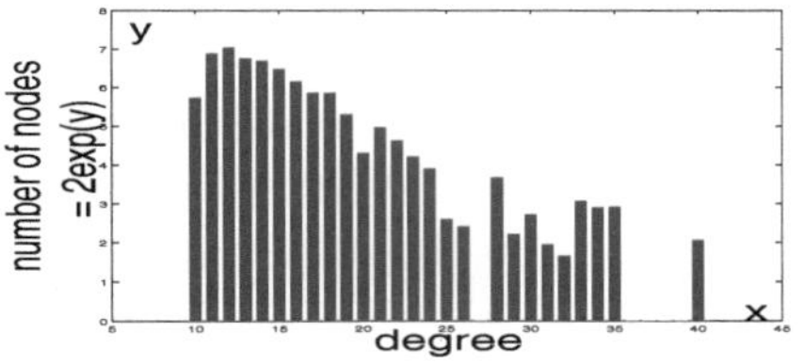

Fig. 11. Degree distribution in 1024 CDACAN by t=0.0625 without local search

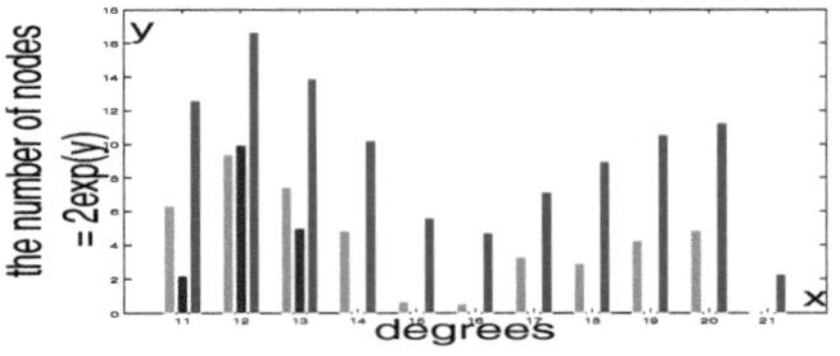

Fig. 12. Comparison of degree distribution

Figure 11 and the green and blue bars in figure 12 illustrate the degree distribution of experiments what have 1024 nodes. E1 adopts t=0.0625 without local search, and its volume distribution is shown with the first bar cluster in figure 9. E2 adopt t=0.0625 with local search; E3 adopt t=2 with local search. E2 and E3 correspond to the first and sixth bar cluster in figure 10 respectively. We can discover that **the distribution of degree is determined by the distribution of volume**. And the degree distribution is discrete because the volume distribution is discrete too.

Figure 5 shows that the average degree decreases when t increases. It is because the volume distribution becomes smooth as shown in figure 9 and figure 10. And the average degree of 2-CDACAN is about 12 what accords with the theoretical analysis. E4 adopt t=0.0625 with local search and 128k nodes. The setting of E4 and E2 is the same except for the network size. And we discover that their degree distributions are nearly the same. By reverse reasoning, we discover that their volume distributions are similar and CDACAN keep smoothness well under large network size.

5 Conclusions and Future Work

In this paper we have presented CDACAN, a scalable structured P2P network based on Continuous Discrete Approach and CAN. We firstly introduce the related conceptions in CDA. Then we introduce the design of our proposal detailedly: presenting the topology construction and the routing algorithm; In the routing algorithm, we propose algorithm to calculate the necessary precision of the destination, devise neighbor forwarding scheme to deal with the uncertainty of the routing in distributed circumstance, analyze the key factors that affect the routing path length of CDRA; introducing the join and departure protocols of CDACAN. At last we verify our protocols by simulation.

The main problem of CDACAN is its high degree. In CDACAN, the average routing path length and degree are about $log_2(N)$ and $6d$ respectively. We discover that the high degree can not reduce the path length. Our anticipant solution is layered improvement, named LCDACAN, whose **scheme is dividing *d*-CDACAN into *d* layers and the nodes in *i*-layer keep the edges on the *i*th dimension, *i*=0,1,...*d*−1**. The topology has $O(log_2(N)+d)$ routing path length and about 8 degrees on every node. LCDACAN keeps Cartesian coordinate space property and its tradeoff between routing path length and size of routing table is close to Cycloid [13].

References

1. Ratnasamy, S., Francis, P., Handley, M., Karp, R.M., Shenker, S.: A scalable content-addressable network. In: Proc. ACM SIGCOMM, San Diego, CA, pp. 161–172. ACM Press, New York (2001)
2. Stoica, I., Morris, R., Karger, D., Kaashoek, M., Balakrishnan, H.: Chord: A scalable peer-to-peer lookup service for internet applications. IEEE/ACM Trans. Networking 11(1), 17–32 (2003)
3. Rowstron, A., Druschel, P.: Pastry: Scalable, decentralized object location, and routing for large-scale peer-to-peer systems. In: Guerraoui, R. (ed.) Middleware 2001. LNCS, vol. 2218, pp. 329–350. Springer, Heidelberg (2001)
4. Maymounkov, P., Mazieres, D.: Kademlia: A peer-to-peer information system based on the xor metric. In: Proc. International Peer-to-Peer Symposium, Cambridge, MA, USA, pp. 53–65 (2002)
5. Malkhi, D., Naor, M., Ratajczak, D.: Viceroy: A scalable and dynamic emulation of the butterfly. In: Proc. the 21st ACM PODC, Monterey,CA, pp. 183–192. ACM Press, New York (2002)
6. Kaashoek, F., Karger, D.: Koorde: A simple degreeoptimal distributed hash table. In: Proc. International Peer-to-Peer Symposium, Berkeley,CA,USA, pp. 98–107 (2003)
7. Naor, M., Wieder, U.: Novel architecture for p2p applications: the continuous-discrete approach. In: Proc. ACM Symposiumon Parallel Algorithms and Architectures, San Diego, California, USA, pp. 50–59. ACM Press, New York (2003)
8. Fraigniaud, P., Gauron, P.: D2B: a de bruijn based content-addressable network. Theor. Comput. Sci. 355(1), 65–79 (2006)
9. Loguinov, D., Casas, J., Wang, X.: Graph-theoretic analysis of structured peer-to-peer systems: Routing distances and fault resilience. IEEE/ACM Trans. Networking 13(5), 1107–1120 (2005)
10. Gai, A., viennot, L.: Broose: a practical distributed hashtable based on the de bruijn topology. In: Proc. the International Conference on Peer-to-Peer Computing, Switzerland, pp. 167–174 (2004)
11. Li, D., Lu, X., Wu, J.: Fissione: A scalable constant degree and low congestion dht scheme based on kautz graphs. In: Proc. IEEE INFOCOM, Miami,Florida,USA, pp. 1677–1688. IEEE Computer Society Press, Los Alamitos (2005)
12. Guo, D., Wu, J., Chen, H., Luo, X.: Moore: An extendable peer-to-peer network based on incomplete kautz digraph with constant degree. In: Proc. 26th IEEE INFOCOM, IEEE Computer Society Press, Los Alamitos (2007)
13. Shen, H., Xu, Ch., Chen, G.: Cycloid: A constant-degree and lookup-efficient P2P overlay network. J. Performance Evaluation 63, 195–216 (2006)

Multi-domain Topology-Aware Grouping for Application-Layer Multicast*

Jianqun Cui[1], Yanxiang He[2], Libing Wu[2], Naixue Xiong[3],
Hui Jin[2], and Laurence T. Yang[4]

[1] Department of Computer Science, Huazhong Normal University, Wuhan, 430079, China
[2] School of Computer, Wuhan University, Wuhan, 430072, China
[3] Information Science School, Japan Advanced Institute of Science and Technology, Japan
[4] Department of Computer Science, St. Francis Xavier University, Canada
jqcui@126.com, {yxhe,wu}@whu.edu.cn, naixue@jaist.ac.jp,
hjin6@iit.edu, ltyang@stfx.ca

Abstract. Application-layer multicast (ALM) can solve most of the problems of IP-based multicast. Topology-aware approach of ALM is more attractive because it exploits underlying network topology data to construct multicast overlay networks. In this paper, a novel mechanism of overlay construction called Multi-domain Topology-Aware Grouping (MTAG) is introduced. MTAG manages nodes in the same domain by a special node named domain manager. It can save the time used to discover topology information and execute the path matching algorithm if there are some multicast members in the same domain. The mechanism can lower the depth of the multicast tree too. Simulation results show that nodes can acquire multicast service more quickly when the group size is large or the percentage of subnet nodes is high.

1 Introduction

IP Multicasting [1] implemented at the network layer can provide an efficient delivery service for multiparty communications. It is the most efficient way to perform group data distribution, as it eliminates traffic redundancy and improves bandwidth utilization on the wide-area network. However, although IP Multicasting has been available for many years, today's Internet service providers are still reluctant to provide a wide-area multicast routing service due to some reasons such as forwarding state scalability, full router dependence and so on [2, 15]. In order to achieve global multicast, application-layer multicast (ALM) has recently been proposed, where the group members form an overlay network and data packets are relayed from one member to another via unicast. ALM shifts multicast support from core routers to end systems.

Application-layer multicast can solve most of the problems of IP-based multicast. It can easily be deployed because it does not require any modification to the current Internet infrastructure. However, application-layer overlay multicast technology is not

* Supported by the Important Research Plan of National Natural Science Foundation of China (No. 90104005).

R. Perrott et al. (Eds.): HPCC 2007, LNCS 4782, pp. 623–633, 2007.

as efficient as IP multicasting. It will incur some delay and bandwidth penalties and less stability of multicast tree. Till now, many ALM protocols [3-9] have been proposed to utilize overlay technique to provide scalable and high quality multicast service to applications. Among these protocols, topology-aware approach is more attractive because it exploits underlying network topology data to construct multicast overlay networks. The constructed tree has low relative delay penalty and a limited number of identical copies of a packet on the same link. TAG (Topology-Aware Grouping) [10, 11] is a typical ALM protocol of this approach. Experiments show that TAG is efficient in reducing delays and duplicating packets with reasonable time and space complexities.

But TAG considers the Internet as a non-domain network. Each node in the multicast tree executes the same algorithms to join or leave the tree. In fact, the Internet is a multi-domain network. Here domain can be considered as an autonomous system or a subnet. Millions of self-similar autonomous systems or subnets form the Internet. TAG will construct a high-depth multicast tree when the multi-domain feature of the Internet is ignored. For example, if there are 10 multicast members in the same domain, the last join node who wants to acquire the multicast data has to wait for the other 9 nodes' forwarding. It will cost a lot and the node will spend more time waiting for the multicast service. Therefore a new mechanism of the overlay construction based on TAG, called Multi-domain Topology-Aware Grouping (MTAG), is introduced in this paper. The mechanism takes the multi-domain feature into account to provide more quickly service for multicast members.

The paper is structured as follows. Section 2 provides an overview of related work and section 3 introduces the proposed MTAG mechanism. Performance of the mechanism is verified by simulation in section 4. Finally, we conclude the paper with a summary of conclusion and future work.

2 Related Work

Many application-layer multicast solutions have been put forward with their respective strengths and weaknesses [12]. Several work propose overlay construction schemes that take the topology of the physical network into account. In [13], a distributed binning mechanism is proposed where overlay nodes partition themselves into bins such that nodes that fall within a given bin are relatively close to one another in terms of network latency. In TAG [10], the information about overlap in routes to the sender among group members is used to guide the construction of overlay tree. TAG selects the node that shares the longest common *spath* prefixes, subject to bandwidth, degree, and possibly delay constraints as parent for a new node. The *spath* of a node refers to a sequence of routers comprising the shortest path from source node to this node according to the underlying routing protocol. The resulting tree of TAG has low relative delay penalty, and introduces a limited number of identical packets on the same link.

TAG considers that all nodes in the multicast tree have no difference. However, the Internet is connected with many domains. Most multicast applications have a characteristic that there are more than one multicast members in the same domain because the spread of a new multicast application is often related to geographic location. Users who live in the same building or work at the same company have more chances to enjoy the same multicast service. In TAG, supposing there are 10 nodes in the same domain, we can find that the next node out of this domain will have the depth more than 10 even if it is very close to the source node. The multicast tree can not match the real topology very well. For this purpose, our MTAG mechanism modifies the overlay construction of TAG and provides a domain manager to manage the nodes in the same domain. The mechanism can lower the average depth of the whole multicast tree and provide more quickly service for multicast members.

3 Multi-domain Topology-Aware Grouping

3.1 TAG Review

Let us review the overlay construction process of TAG before our mechanism is introduced. To simplify the operation, we adopt complete the path matching algorithm to construct the multicast tree.

Fig. 1 shows a multicast tree produced by member join algorithm of TAG. We use a more complex example than [10] so that you can find the disadvantages of TAG.

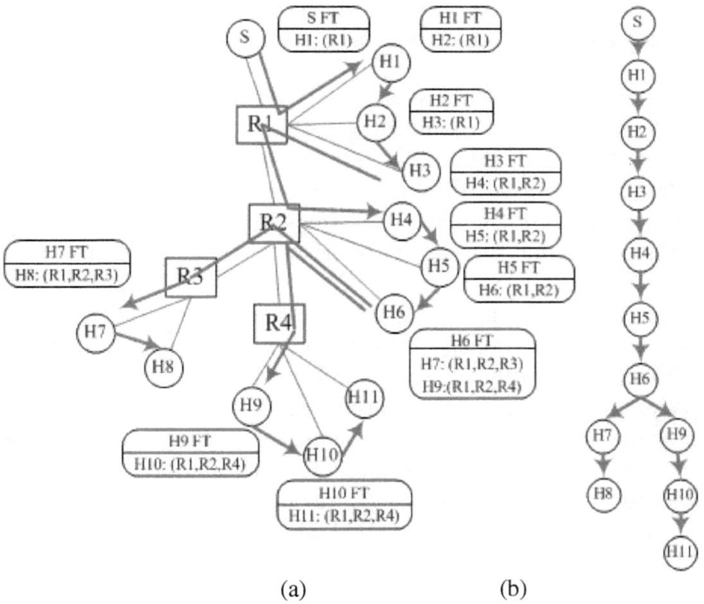

Fig. 1. Multicast overlay constructed by TAG

Source S is the root of the multicast tree, R1 through R4 are routers, and H1 through H11 are hosts in Fig. 1. Thick solid lines denote the current data delivery tree. Each node of the multicast tree maintains a family table (FT) defining parent-child relationship for this node (showing only the children in Fig. 1).

Fig. 1(a) shows the real topology and FT of each non-leaf node. The new node who wants to join the multicast tree will send a JOIN message to S first. When S receives the message, it computes the *spath* to the new node by network path-finding tools such as *traceroute*. After S gets the *spath* of the new node, it starts the path matching algorithm execution. The path matching algorithm will be executed by next nodes until the best parent is found. The details of the algorithm can be found in [10]. In Fig. 1(b), we ignore the topology and other information so that we can get a clear look on the multicast tree. There are just 3 routers from the source node to the node H11, while the depth of the node H11 has already been 9. The data forwarding latency will increase with the depth's increase. Another shortcoming of TAG is that it must compute the same *spath* for many times if there are some nodes (such as H9, H10 and H11) in the same domain. The topology discovery procedure will cost a lot and lead to poor performance.

3.2 Mechanism of MTAG

To overcome the weaknesses of TAG mentioned above, we propose another mechanism MTAG to construct and maintain the multicast overlay. We introduce a new role named domain manager to manage nodes in the same domain. The source node will maintain another table called domain table (DT) to save the information of domain managers and the DT is empty at the very beginning. To implement the new mechanism, we add the node's path information into its FT (the old FT just saves the information of its parent and its children).

Once a new node sends a JOIN message to the source node, the source node will get the domain information of the node first. How can we know which domain the node belongs to? We can identify the node's domain by the network number of its IP address which can be obtained from the JOIN message if a subnet is regarded as a domain. Otherwise, if there are some other rules to distinguish domains, we need to add the domain information into the JOIN message. Here, for simplicity, we think a domain is a subnet.

Then the source node checks the DT to make sure whether the domain manager of the new node exists. If the domain manager is found, the source node will directly send a CHILD message with the node's information to the domain manager. The domain manager takes the new node as its child and sends a PARENT message to the node. The domain manager provides the multicast service for the node at the same time. What the new node should do is to add the parent information into its FT table when it receives the PARENT message. In another way, if the node's domain manager does not exist, the source node will add the node's information into the DT and send a MANAGER message to the node. Then the source node starts the topology discovery procedure and the path matching algorithm similar to TAG to find the best parent of the node. The node who receives a MANAGER message will identify itself as the domain manager.

A normal member who wants to leave the multicast tree will execute the same algorithm as TAG. The node just needs to send a LEAVE message with its FT to the parent. The parent will remove the node from its FT and add FT entries for the children of the leaving node. If the domain manager wants to leave the tree, it should send a LEAVE message to the source node in addition to the parent. The LEAVE message includes its FT (if its children are all in other domains) or the new domain manager candidate chosen from its children (if it has any children in the domain). If the parent receives the LEAVE message with the FT, it will execute the same algorithm with TAG, otherwise it will replace the node information with the candidate in its FT. The source node will update or delete the node information from the DT based on the information of the LEAVE message. The last step is the old domain manager sends an UPDATE message to inform its children who is the new domain manager. The new domain manager takes the other nodes as its children and provides the multicast service right now when it receives the UPDATE message. Other children just need to update their parents of the FT. The member leaving process of the domain manager seems more complex than TAG. However, one domain just has one domain manager, so the probability is small.

The mechanism of MTAG is illustrated in Fig. 2.

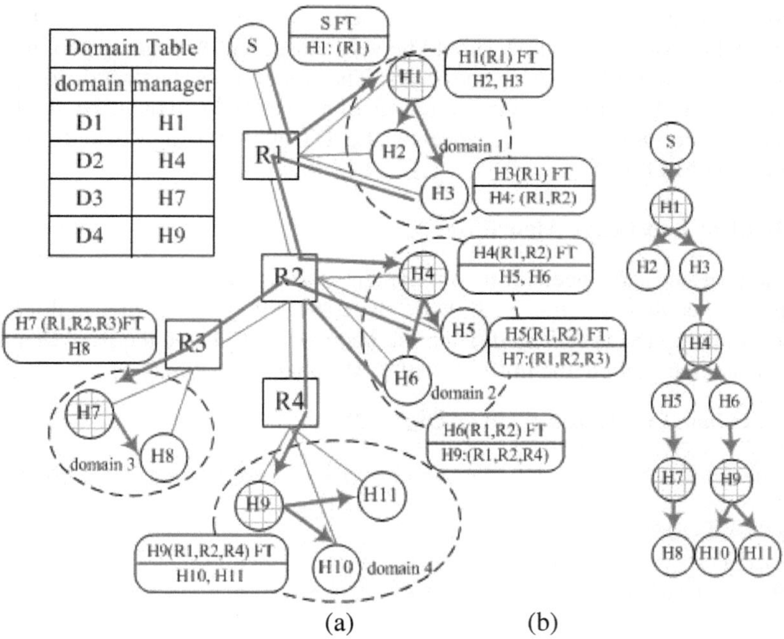

Fig. 2. Multicast overlay constructed by MTAG

We can get a number of advantages from MTAG:

1. The topology discovery and path matching algorithm are not necessary if the domain manager exist: Just when the first node in the domain joins the multicast tree, the two steps (cost a lot) need to be executed. This mechanism will shorten the waiting time of the new node if it is not the first member in the domain.

2. MTAG can lower the average depth of the multicast tree: From Fig. 2(b), we can find the depth of the multicast tree is 6 instead of 9 (Fig. 1(b)) under the same topology. The reason is that nodes in the same domain all take the domain manager as their parent. The domain manager can afford so many children because the bandwidth is often high enough in a LAN. The lower the depth of the tree is, the sooner the node acquires the multicast data.

3. MTAG can balance the traffic of the domain: In TAG, the node chosen to be the parent of the next node (out of the domain) must be the last join node. The node H6 in Fig. 1 is such a node. However, in MTAG, the domain manager has more than one child, so that it can choose randomly its children to be the parent of the next node. We can find the node H5 and H6 in Fig. 2 act as the different nodes' parents respectively. We should not let one node be the parent because the connection is inter-domain.

4. The domain manager does not need to save the *spath* of its children: Nodes in the same domain have the same *spath* so that the domain manager just needs to record its own *spath* instead of each child's *spath*. The domain manager can use less memory to store the FT.

Compared with TAG, MTAG has the only obvious drawback: The source node need to maintain the DT. In MTAG the source node just needs to forward the multicast data to one child which should be very close to it. Another task of the source node is to process JOIN/LEAVE messages. The source node just needs to update its DT in most cases because the domain manager has shared some management tasks for the source node. So the maintenance of the DT will not significantly affect the performance of MTAG.

3.3 Member Join/Leave Algorithm

Three algorithms are needed to implement the MTAG mechanism. Member_Joining(S) executed on the new node is used for joining the multicast tree. Member_Leaving (S) is executed when the node wants to leave the tree. Members of the multicast tree execute the third algorithm Member_executing() as a daemon.

```
Algorithm 1: Member_Joining (S)
send(S, JOIN_msg);
msg = receive_msg() //receive_msg() will be blocked
till a message is received or timeout
If (isPARENT_msg(msg))
    add_FT_parent(msg.address);
Else If (isMANAGER_msg(msg))
     isManager = true;
Endif

Algorithm 2: Member_Leaving (S)
if (isManager)
  newManager = getDomainChild(FT);
  If (newManager = = NULL)
      LEAVE_msg = new_LEAVE_Msg(FT);
  Else
      LEAVE_msg = new_LEAVE_Msg(newManager);
  Endif
  send(S, LEAVE_msg);
Endif
send(FT.parent,LEAVE_msg);
```

```
Algorithm 3: Member_executing( )
Do while true
   msg = receive_msg();
   If (isJOIN_msg(msg)) //just the source node can
receive the JOIN message
      manager = checkInDT(msg.domain);
      If (manager = = NULL) //there are not any
nodes in the domain
          updateDT(msg.domain,msg.sourceAddr);
          send(msg.sourceAddr, MANAGER_msg);
          spath = topo_discovery(msg.sourceAddr);
          PathMatch(myself, spath); //execute path
matching algorithm of TAG
      Else
          send(manager, CHILD_msg);
      Endif
  Else
      If (isCHILD_msg(msg))
          add_FT_child(msg. sourceAddr);
          send(msg.sourceAddr, PARENT_msg);
          send(msg.sourceAddr, multicast_data);
      Else
          If (isLEAVE_msg(msg))
             If (msg.newManager= =NULL)
                If (isSourceNode(myself)
                  deleteFromDT(msg.domain);
               Else //other nodes
                  update_FT_child(msg.FT.children);
               Endif
            Else //has new domain manager
                If (isSourceNode(myself)
                  updateFromDT(msg.domain,msg.newManager);
               Else //other nodes
                  update_FT_child(msg.sourceAddr,
msg.newManager);
               Endif
            Endif
        Else
            If (isUPDATE_msg(msg))
               If (msg.newManager= =myself)
                   add_FT_child(msg.FT.children);
                   send(msg.FT.children,multicast_data);
               Else
                  update_FT_parent(msg.newManager);
               Endif
            Endif
        Endif
    Endif
   Endif
Enddo
```

4 Evaluation

In this section, we present the simulation results to evaluate the proposed MTAG mechanism and compare it to the original TAG protocol.

4.1 Simulation Environment

We generate a Transit-Stub network topology with 500 nodes using GT-ITM [14]. On top of the physical network, the multicast group members are attached to the stub nodes. Nodes belong to the same stub are thought to be in the same subnet. We can control the percentage of nodes in the same subnet. The multicast group size ranges from 50 to 500 members. The transit-to-transit, transit-to-stub, and stub-to-stub link bandwidth are assigned between 10 Mbps and 100 Mbps. The links from edge routers to end systems have the bandwidth between 1 Mbps and 10 Mbps.

4.2 Performance Metrics

A simulation is carried out to contrast the performance of our proposed MTAG and TAG. Because the main idea of the overlay construction has not been changed by MTAG, the mean RDP (Relative Delay Penalty) and Link Stress are same as TAG. So we will not evaluate these two performance metrics which can be found in [10]. The performance metrics used for our experiments are Member Join Latency (MJL) and Data Acquisition Latency (DAL).

Member Join Latency denotes how long it takes for the join operation. If T_{JOIN} denotes the time when the new node sends the first JOIN message and T_{data} denotes the time when it receives the first packet of multicast data from the parent, the value of the MJL can be measured by T_{data}- T_{JOIN}. We report the mean of the latency of all n members:

$$\text{Mean Member Join Latency}=\frac{1}{n}\sum_{i=1}^{n}(T_{data\,i}-T_{JOIN\,i}) \quad (1)$$

Data Acquisition Latency of a node is an expression of how much time it takes for a packet of multicast data to get from the source node to this node. To get the DAL of each node, we add the additional mechanism into all nodes. It is not necessary for ALM. The source node sends a PROBE message (very small) to its children in a certain interval. The node who receives the message will send a HELLO message to its parent first and forward the PROBE message to its children at the same time. So the PROBE message can be received by all nodes in the multicast tree. Because the path of the round trip is same, the source node can calculate approximately the DAL by *RTT*/2. The source node will probe each node for m times to get the average value. We also use the mean of the DAL to evaluate the performance:

$$\text{Mean Data Acquisition Latency}=\frac{1}{n}\sum_{i=1}^{n}\left(\frac{1}{m}\sum_{j=1}^{m}(RTT_{i\,j}/2)\right) \quad (2)$$

4.3 Results and Analysis

We will evaluate the two metrics in two different conditions. We attach the multicast group members to the stub nodes randomly and adjust the multicast group size from 50 to 500 in the first experiment. The relationship between the performance metrics and the group size is shown in Fig. 3.

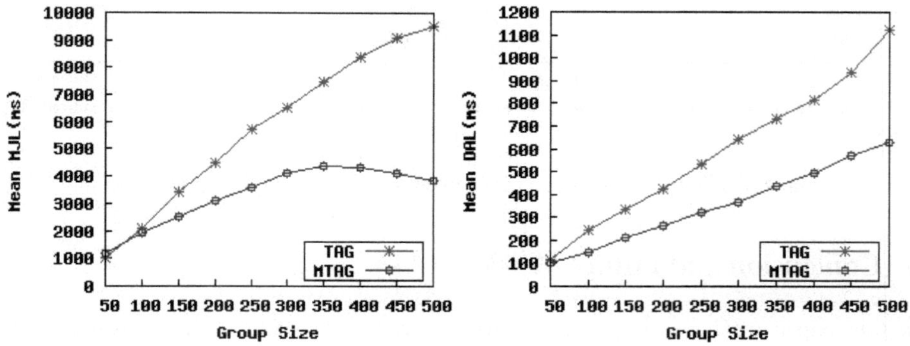

(a) (b)

Fig. 3. Performance metrics versus group size

We find the mean MJL of TAG become higher when the group size increases from the experimental data showed in Fig. 3 (a). However, MTAG can get lower MJL when the group size reaches a certain number. The reason is that we have more nodes in one subnet when the group size is large. MTAG will save more join time if there are more nodes in a subnet.

Fig. 3 (b) shows that the mean DAL will increase with the group size in both mechanisms. We also find the mean DAL of MTAG is much less than that of TAG. MTAG can build a multicast tree with lower depth if there are some nodes in the same domain. The low depth multicast tree will shorten the latency of data forwarding.

In the second experiment, we give a more obvious percentage of the nodes in the same subnet and fix the group size of 300. For example, 20% nodes in the same subnet means there are 20% nodes in such subnet where there are two or more member nodes. These 20% nodes needn't be in one subnet. Considering the size of stub nodes and member nodes, the percentage ranges from 20% to 100%. The result is shown in Fig. 4.

Fig. 4 shows that the percentage of the nodes in the same subnet has little effect on TAG's performance while the mean MJL and DAL of MTAG decrease sharply when the percentage increases.

It can be concluded from the simulation results that MTAG can decrease the mean MJL and DAL than TAG when the group size is large or the percentage of the subnet nodes is high.

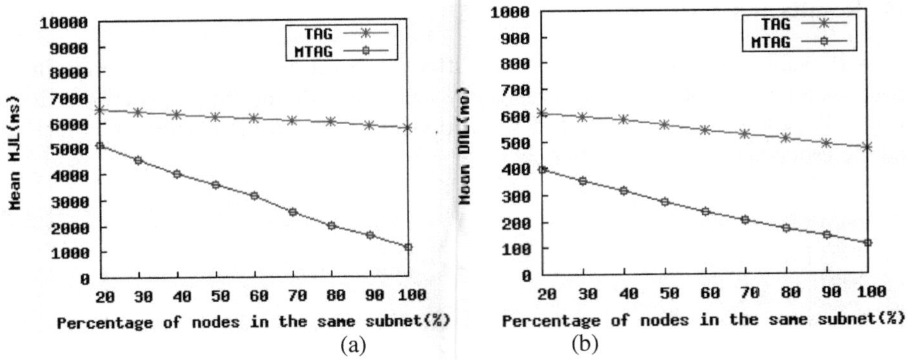

Fig. 4. Performance metrics versus percentage of nodes in the same subnet

5 Conclusion and Future Work

In this paper we have proposed a novel overlay construction mechanism MTAG based on the topology-aware application-layer multicast. We introduce the domain manager to manage nodes in the same domain. The mechanism can shorten the join time of a new node and lower the depth of the multicast tree. The results show that the new nodes can acquire the multicast service more quickly when the group size is large or percentage of the subnet nodes is high.

For simplicity, we do not take the bandwidth into account when we use the path matching algorithm of TAG. The topology of simulation environment is not complex and sweeping enough. Future work will be done on testing and optimizing MTAG in a more complex environment.

References

1. Kosiur, D.: IP Multicasting: The Complete Guide to Interactive Corporate Networks. John Wiley & Sons, Inc. Chichester (1998)
2. Chu, Y., et al.: A Case for End System Multicast. IEEE Journal on Selected Areas in Communication (JSAC), Special Issue on Networking Support for Multicast (2002)
3. Chawathe, Y.: Scattercast: An Architecture for Internet Broadcast Distribution as an Infrastructure Service. Ph.D. Thesis, University of California, Berkeley (2000)
4. Chu, Y.-H., Rao, S.G., Zhang, H.: A Case for End System Multicast. In: Proceedings of ACM SIGMETRICS, ACM Press, New York (2000)
5. Francis, P.: Yoid: Extending the Multicast Internet Architecture,White paper (1999), http://www.aciri.org/yoid/
6. Jannotti, J., Gifford, D., Johnson, K., Kaashoek, M., O'Toole, J.: Overcast: Reliable Multicasting with an Overlay Network. In: Proceedings of the 4th Symposium on Operating Systems Design and Implementation (2000)
7. Pendarakis, D., Shi, S., Verma, D., Waldvogel, M.: ALMI: An Application Level Multicast Infrastructure. In: Proceedings of 3rd Usenix Symposium on Internet Technologies & Systems (2001)

8. Ratnasamy, S., Handley, M., Karp, R., Shenker, S.: Application-level multicast using content-addressable networks. In: Proceedings of 3rd International Workshop on Networked Group Communication (2001)
9. Zhuang, S.Q., Zhao, B.Y., Joseph, A.D., Katz, R., Kubiatowicz, J.: Bayeux: An architecture for scalable and fault-tolerant wide-area data dissemination. In: NOSSDAV 2001 (2001)
10. Kwon, M., Fahmy, S.: Path-aware overlay multicast, Computer Networks. The International Journal of Computer and Telecommunications Networking 47(1), 23–45 (2005)
11. Kwon, M., Fahmy, S.: Topology-Aware Overlay Networks for Group Communication. In: Proc. of ACM NOSSDAV, pp. 127–136. ACM Press, New York (2002)
12. Yeo, C.K., Lee, B.S., Er, M.H.A.: Survey of application level multicast techniques. Computer Communications , 1547–1568 (2004)
13. Ratnasmy, S., et al.: Topology-Aware Overlay Construction and Server Selection. In: IEEE INFOCOM 2002 (2002)
14. Doar, M.: A better model for generating test networks. In: GLOBECOM 1996, London, UK, pp. 83–96. IEEE Computer Society Press, Los Alamitos (1996)
15. Xiong, N., Défago, X., Jia, X., Yang, Y., He, Y.: Design and Analysis of a Self-tuning Proportional and Integral Controller for Active Queue Management Routers to Support TCP Flows. In: IEEE Infocomm 2006, Barcelona, Spain (April 23-29, 2006)

A Generic Minimum Dominating Forward Node Set Based Service Discovery Protocol for MANETs

Zhenguo Gao[1], Sheng Liu[1], Mei Yang[2], Jinhua Zhao[3], and Jiguang Song[1]

[1] College of Automation, Harbin Engineering University, Harbin, 150001, China
{gag,liusheng,songjg}@hrbeu.edu.cn
[2] Department of Electrical and Computer Engineering University of Nevada, Las Vegas, NV 89154, USA meiyang@egr.unlv.edu
[3] School of Astronautics, Harbin Institute of Technology, Harbin, 150001, China
zhaojinhua@hit.edu.cn

Abstract. Service discovery is a crucial feature for the usability of mobile ad-hoc networks (MANETs). In this paper, Minimum Dominating Forward Node Set based Service Discovery Protocol (MDFNSSDP) is proposed. MDFNSSDP has the following characteristics: 1)It minimizes the number of coverage demanding nodes in the current node's 2-hop neighbor set. 2) It minimizes the size of dominating forward node set used to cover all coverage demanding nodes. Forward nodes are selected based on local topology information and history information piggybacked in service request packets. Only these forward nodes are responsible for forwarding service request packets. 3) The coverage of service request packets is guaranteed. 4) Multiple service requests can be fulfilled in just one service discovery session. Simulations show that MDFNSSDP is an effective, efficient, and prompt service discovery protocol for MANETs.[1]

1 Introduction

Mobile Ad-Hoc Networks (MANETs)[1] are temporary infrastructure-less multi-hop wireless networks that consist of many autonomous wireless mobile nodes. Flexibility and minimum user intervention are essential for such future communication networks[2]. Service discovery which allows devices to advertise their own services to the rest of the network and to automatically locate network services with requested attributes, is a major component of MANETs.

In the context of service discovery, service is any hardware or software features that can be utilized or benefited by any node; Service description of a service is the information that describes the service's characteristics, such as types and

[1] This work is supported by National Postdoctoral Research Program (No.NPD060234), Postdoctoral Research Program of HeiLongJiang (No.HLJPD060234), Fundamental Research Foundation of Harbin Engineering University (No.HEUFT06009), Fostering Fundamental Research Foundation of Harbin Engineering University (No.HEUFP07001).

R. Perrott et al. (Eds.): HPCC 2007, LNCS 4782, pp. 634–646, 2007.

attributes, access methods, etc; A server is a node that provides some services; A client is a node that requests services provided by other nodes. When a node needs services from others, it generates a service request packet. When receiving request packets, each node that provides matched services responds with a service reply packet. Nodes without matched services forward the packet further. All these packet transmissions, including request packets and reply packets, form a service discovery session. Coverage preservability of a service discovery protocol is the property that the coverage of service discovery requests when the protocol is used is the same to that when flood policy is used.

The objective of service discovery protocol is to reduce service request packets redundancy while retaining service discoverability. Usually this is approached by searching for more efficient policy of service request packet forwarding.

Existing service discovery protocols targeted at MANETs are not very applicable for their variety of problems. Flood-based protocols (Konark[3], Proximity SDP[4]) face the risk of broadcast storm problem[5]. Some other flood-like protocols (FFPSDP[6], RICFFP[7]) can not guarantee the coverage of service request packets, which leads to lower service discovery ratio. Two group-based protocols (GSD[8], Allia[9]) cause serious redundancy of unicast request packets. 2-layer protocols[10][11] construct upper logic layer by selecting out some nodes with variety of criterions. Upper logic layer requires costly maintenance. DSDP[12] is superior to other 2-layer protocols for its lower topology maintenance overhead. In multi layer protocols (ServiceRing[13], MultiLayerSDP[14]), their hierarchy architectures are hard to maintain, which has been verified through simulations[15]. Additionally, no existing protocols consider the requirement of fulfilling multiple service discovery requests in one service discovery session. For a more detailed survey please ref to ref[16].

Noticing above problems and the importance of coverage preservability property of service discovery protocols, Minimum Dominating Forward Node Set based Service Discovery Protocol (MDFNSSDP) is proposed in this paper. The new scheme minimizes the number of coverage demanding nodes in the current node's 2-hop neighbor set, and finds a minimum dominating forward node set to cover all these coverage demanding nodes. In this way, service request packet overhead is greatly reduced. MDFNSSDP preserves the coverage of service discovery sessions, and it can fulfill multiple service discovery requests in just one service discovery session. Besides, it offers some user-definable parameters, which can be used to adjust its operation.

Dominating set scheme has been applied to many aspects in MANET, including routing, broadcasting, and applications, but no similar work is found used in service discovery protocol. Besides, most works are either omitted the necessity to minimize the set to be covered or lack of coverage preservability proof. Some work using dominating set can be found in ref[17].

The rest of the paper is organized as follows. In Section 2, data structures of MDFNSSDP and some definitions are given. A problem called dominating forward node set (DFNS) problem is proposed and its NP-completeness is proved. An heuristic to solve the DFNS problem is proposed, and its approximation ratio is analyzed. In Section 4, operations of MDFNSSDP are described and demonstrated. In Section 5, some properties of MDFNSSDP, such as the existence of dominating forward node set and coverage preservability, are proved. In Section 6, the performance of MDFNSSDP is analyzed through extensive simulations. Finally, in Section 7, a conclusion is made.

2 Preliminaries of MDFNSSDP

2.1 Data Structures

In MDFNSSDP, each node broadcasts hello packets periodically. The structure of hello packets is shown in Fig. 1(a). The *packet-type* field indicates the type of the packet. The *sender-id* field indicates the sender of the packet. The *service-list* field stores the list of local services' descriptions. The *life-time* field indicates the time that the packet can be cached by others. The *neighbor-list* field stores the list of the current node's 1-hop neighbors.

Hello packets received will be cached in NLSIC for a period. The structure of NLSIC item is shown in Fig. 1(b).

packet type	sender id	service list	neighbor list	life time

(a) Hello packet

sender id	service list	neighbor list	life time

(b) NLSIC entry

packet type	packet id	source id	visited list	sender id	receiver list	remain hop	request list

(c) Service request packet

predecessor id	packet id	source id

(d) RRT entry

packet type	source id	packet id	receiver id	replier id	matched service list

(e) Service reply packet

Fig. 1. Data structures in MDFNSSDP protocol

The structure of service request packets is shown in Fig. 1(c). The *packet-id* filed is a number that increases monotonically with each service request packet generated by a node. This field identifies different service request packets from the same node. The *source-id* filed indicates the client that generates the service request packet. The *visited-list* filed stores the list of nodes that the packet has passed. Depends on the value of protocol parameter, the *visited-list* field may also store the list of 1-hop or 2-hop neighbors. The *sender-id* filed indicates the direct sender of the packet. The *receiver-list* filed stores the list of so called forward nodes that are responsible for forwarding the request packet further. The *remain-hop* filed indicates the number of hops that the packet can still travel. The *request-list* field describes the list of the descriptions of the requested services as well as whether any matched services has been found for each requested service.

Each node maintains a RRT which is used in two tasks: 1) check for duplicated request packets, and 2) forward service reply packets reversely to the corresponding client node. Fig. 1(d) shows the structure of an RRT entry. The predecessor-id field indicates the node from which the service request packet is received. It is just the next hop node that a corresponding reply packet should be forwarded to. The *packet-id* and *source-id* field are the same to that of the service request packet

The structure of service reply packets is shown in Fig. 1(e). The *packet-type* field indicates the type of the packet. The *source-id* field stores the destination of the reply packet, which is just the node that generates the corresponding request packet. The *packet-id* field stores the packet-id of the corresponding request packet. The *receiver-id* field indicates the next-hop node of the service reply packet. The *replier-id* field indicates the node that generates the reply packet. The *matched-service-list* field stores the description of the matched services.

2.2 Notations and Definitions

The following notations are used in the rest of the paper.

u: the current node.
s: the client that generates the service request packet.
v: the direct sender of the current service request packet.
$N_x(u)$: the set of nodes that are at most x-hop away from node u, i.e., node u's x-hop neighbor set, including u itself.
$H_x(u)$: the set of nodes that are just x-hop away from node u.
$V(v)$ the set of nodes in the *visited-list* field of the request packet sent by v.
$F(v,u)$: $F(v,u) = H_1(u) - N_1(v)$.
$R(u)$: The set of nodes in the *receiver-list* field of request packet sent by u.

We make the following definitions.

Def. 1 (Forward Node): Each node in $R(u)$ is a forward node of node u.

Def. 2 (Candidate Forward Node): Each node in $F(v,u)$ is a candidate forward node of node u. Obviously, $R(u) \subseteq F(v,u)$ since that forward nodes are all selected from $F(v,u)$.

Def. 3 (Coverage of a service discovery session): All nodes that could receive or have received service request packets of a service discovery session are called as in the coverage of the service discovery session. It can also be said that these nodes are covered by the service discovery session.

Def. 4 (Extendedly Covered): For a node w, if $\exists x \in R(u)$ meets $w \in H_1(x)$, then we say that node w is extendedly covered by node u.

Def. 5 (Coverage Demanding Node): To obtain coverage preservability, when forwarding service request packets, all node in $H_2(u)$ should be guaranteed to be covered by the current service discovery session. Some nodes in $H_2(u)$ have already been covered or to be covered through other ways. Hence, only a subset of $H_2(u)$ should be guaranteed to be extendedly covered by the current node. These nodes are called as coverage demanding nodes. The set of coverage demanding nodes is denoted as $H_{CDN}(v, u)$. Obviously, $H_{CDN}(v, u) \subseteq H2(u)$. In MDFNSSDP, $H_{CDN}(v, u) = H_2(u) - N_2(V(v))$.

Def. 6 (Dominating Forward Node Set): Given a set of nodes $R(u)$, if for $\forall w \in H_{CDN}(v, u)$, $\exists x \in R(u)$ that meets $w \in H_1(x)$, or in other words , $H_{CDN}(v, u) \subseteq H_1(R(u))$, then $R(u)$ is a dominating forward node set, denoted as $F_{DFNS}(v, u)$.

3 Find a Minimum Dominating Forward Node Set

When receiving a service request packet, each forward node will have to forward the packet further, unless all requested services are found or the packet's hop-limit is reached. Hence, in order to reduce request packet overhead, the size of DFNS should be minimized. The task of finding a DFNS with minimum size is called as DFNS problem.

Using H to represent the set of coverage-needed hidden servers $H_{CDN}(v, u)$, C to represent the family of sets $\{H_1(w)|w \in F(v, u)\}$, and F to represent $F(v, u)$, the DFNS problem can be defined formally as follows:

DFNS problem: Given H, C, and F, find a dominating forward node set $F_{DFNS}(v, u) \subseteq F$ with minimum size.

Since that $\bigcup_{w \in F(v,u)} H_1(w) \supseteq H_2u - N_2V(v) = H_{CDN}(v, u)$(refer to Lemma 3 in the following text), and there is a 1-to-1 correspondence between C and F, the decision version DFNS Problem can be defined formally as follows:

DFNS problem (decision version): Given H=$\{h_1, \cdots, h_n\}$, C=$\{C_1, \cdots C_m\}$, $\bigcup_{C_i \in C} C_i \supseteq H$, and a positive integer k, decides whether there is subset $B \subseteq C$ with size k such that $\bigcup_{C_i \in B} C_i \supseteq H$.

Lemma 1. *The decision version of the DFNS problem is NP-complete.*

Proof. The proof is omitted for the limit of space. □

Since that DFNS problem is NP-complete, the following greedy heuristics is proposed in MDFNSSDP to select a dominating forward node set $F_{DFNS}(v, u)$ from H, C, and F.

Heuristic: greedy Minimum DFNS heuristic.

1. Let $F_{DFNS}(v, u) = \Phi$(empty set), $H_{RC}(v, u) = \Phi(H_{RC}(v, u)$ is a temporal set used to store node set already covered by $F_{DFNS}(v, u))$.
2. Find node $w \in F(v, u)$ with maximum $|H_1(w) \bigcap H_{CDN}(v, u) - H_{RC}(v, u)|$. In case of a new tie, select w with smallest ID.
3. $F_{DFNS}(v, u) = F_{DFNS}(v, u) + w, F(v, u) = F(v, u) - w$.
4. $H_{RC}(v, u) = H_{RC}(v, u) \bigcup H_1(w)$. If $H_{RC}(v, u) \supseteq H_{CDN}(v, u)$, exit. Otherwise, go to step 2.

Lemma 2. *The approximation ratio of the greedy minimum DFNS heuristic is* $ln(max_{z \in F(v,u)} |H_1(z) \bigcap H_{CDN}(v,u)|)$.

Proof. The proof is omitted for the limit of space. □

4 The Operations of MDFNSSDP Protocol

MDFNSSDP protocol has three basic operations: Hello packet exchanging, service request packet forwarding, and service reply packet routing.

4.1 Hello Packet Exchanging

In MDFNSSDP, each node in a MANET should broadcast hello packets periodically. Node can cache the information in the hello packets into their NLSIC. This information will be kept valid for a user-defined period. Hello packets can only travel 1 hop, i.e., nodes should not forward a reply packet further. New hello packets are constructed basing the information in NLSIC.

The content of the *service-list* field of a hello packet is determined by protocol parameter S_{TYPE} as follows.

- If S_{TYPE}=NONE, this field contains nothing.
- If S_{TYPE}=SELF, this field contains the descriptions of all services provided by the node.
- If S_{TYPE}=HOP1, this field contains not only the descriptions of the services provided by the node, but also those of the services provided by the current node's neighbors.
- If S_{TYPE}=HOP2, the field contains the descriptions of the services of three sources: 1)the current node, 2)1-hop neighbors of the current node, and 3) 2-hop neighbors of the current node.

4.2 Service Request Packet Forwarding

When needing services, a node firstly checks to see if there are any matched services for each requested service, either provided by the node itself or found from its NLSIC. If yes, the service discovery request succeeds. If no, the node constructs a service request packet and sent it out. The *request-list* field of a request packet can contain multiple service requests. Hence, in MDFNSSDP, multiple service discovery requests can be fulfilled in one service discovery session. The maximum number of service discovery requests in *request-list* field is determined by parameter CSize.

Request Packet Forward Decision. When receiving a service request packet from other nodes, or the current node needs services, the node should determine whether to forward the packet or not. If the following 4 conditions are all met, the packet should be forwarded. The 4th condition is important to the property of fulfilling multiple service discovery requests in just one service discovery session, called as multi task mode.

- This is a new service request packet for the node. Whether a service request packet is new or not for the receiver can be determined basing on the current node's RRT.
- The *remain-hop* field of the service request packet is bigger than 0.
- The current node is in the list of forward nodes in the *receiver-list* field of the service request packet.
- There are some unmatched service requests yet.

Request Packet Forwarding Procedure. Following 5 steps are used to forward a service request packet.

1.	Determine the set of coverage demanding nodes, $H_{CDN}(v,u) = H_2(u) - N_2(V(v))$
2.	Find a minimum dominating forward node set. Using the previous greedy-based DFNS heuristic to find a dominating set $F_{DFNS}(v,u)$ from $H_{CDN}(v,u)$, $F(v,u)$ and $H_1(w)\|w \in H_1(u)$.
3.	Enclose nodes in $F_{DFNS}(v,u)$ into the the packet's *receiver-list* field.
4.	Update the contents of other fields of the request packet (section 3.2.3).
5.	Send the service request packet out in broadcast mode.

Request Packet Content Update. Before forwarding a service request packet, some fields of the service request packet should be updated as follows. The content of the *visited-list* field depends on the parameter V_{TYPE}, whereas the size of the items in *visited-list* field is determined by V_{SIZE}.

- If V_{TYPE} =NONE, *visited-list*field contains nothing. If V_{TYPE}=SELF, *visited-list* field contains the identity of the current node.If V_{TYPE}=HOP1, *visited-list* field contains $N_1(u) = u + H_1(u)$.If V_{TYPE}=HOP2, *visited-list* field contains $N_2(u) = u + H_1(u) + H_2(u)$.
- If the number of items in the *visited-list* field is bigger than V_{SIZE}¿0, the earliest item is removed from the *visited-list* field to make room for the latest item. If V_{SIZE}=-1, the number of items is not limited.
- In *request-list* field, all matched service requests are designated as "matched".
- The *remain-hop* field is decreased by 1.

4.3 Service Reply Packet Routing

When receiving a service request packet, each node that finds matched services should construct a service reply packet and send it out, no matter what the source of the matched services are, the node itself or its NLSIC.

When receiving a service reply packet, a node firstly checks if this node is just the destination of the packet. If yes, the node caches the service information in reply packet, and this service discovery session finishes. If no, the node then checks if there are any matched services not seen in previous reply packets. If yes, the service reply packet will be forwarded further. Otherwise it will be discarded.

To forward a service reply packet, the node searches for the RRT item that corresponds to the service reply packet. Then the service reply packet is forwarded to the node indicated by the predecessor-id field of the founded item. In this way, a service reply packet will be relayed to the source of the corresponding service request packet along the reverse path.

5 Property Analysis of MDFNSSDP

5.1 Existence of Dominating Forward Node Set

In MDFNSSDP, $H_{CDN}(v, u)$ is determined by removing $N_2(V(v))$ from $H_2(u)$, and then a set $F_{DFNS}(v, u)$ is selected from $F(v, u)$ to cover $H_{CDN}(v, u)$.

This section focuses on two aspects: 1) the possibility of finding a qualified $F_{DFNS}(v, u)$, 2) the property of coverage preservability of MDFNSSDP.

Lemma 3. *A qualified $F_{DNFS}(v, u)$ can be found in $F(v, u)$ to cover $H_{CDN}(v, u)$.*

Proof. The proof is omitted for limited space. □

5.2 Coverage Preservability of MDFNSSDP

In MDFNSSDP, the set of coverage demanding nodes, $H_{CDN}(v, u)$, is determined as $H_2(u) - N_2(V(v))$. This configuration can still guarantee the coverage of service discovery sessions. This is to say, the coverage of service discovery requests when MDFNSSDP is used is the same to that when flood policy is used. About the correctness of this property, there are the following lemmas. All these lemmas are based on the following assumptions:

Assumption 1: The underlining MAC protocol is ideal. That is to say, each packet sent by a node will be received by its neighbors correctly and timely, and there are no collisions.

Assumption 2: The value of the remain-hop field of a request packet is large enough to cover the network, i.e., the restriction of remain-hop is not considered.

Assumption 3: The effect of cancellation of service request packet forwarding by a node where all requested services have been matched is not considered.

Lemma 4. *In MDFNSSDP, if a node has received a service request packet of a service discovery session, then all neighbors of the node must be able to receive a service request packet of the service discovery session.*

Proof. The proof is omitted for limited space. □

Lemma 5. *In MDFNSSDP, if a MANET is connected, then all nodes in the MANET must be able to receive a service request packet of a service discovery session.*

Proof. In a connected MANET, for each node pair, there are paths of 1 or more hops. Suppose the client of a service discovery session is node s, then for each node w in the MANET, there must be a path. Suppose $s \to x_1 \to x_2 \to \cdots \to x_n \to w$ is a path from node s to w, now we prove that node w must be able to receive a service request packet of the service discovery session.

According to Assumption 1, $x_1 \in H_1(s)$ must be able to receive the service request packet sent by node s. Then using Lemma 4 repeatedly along the path $s \to x_1 \to x_2 \to \cdots \to x_n \to w$, nodes $x_1, x_2, \cdots, x_n, w$ must also be able to receive request packets of the service discovery session. Hence, the lemma follows. □

6 Simulation Analysis

6.1 Performance Metrics

Four performance metrics are considered in our simulations.

- **Request-Packet-Number:** It measures the number of service request packets sent in one simulation. It reflects the efficiency the policy of forwarding service request packets.
- **Succeeded-SDP-Number:** It is the number of service discovery sessions in which the client has received at least one successful reply packet. It reflects the effectiveness (service discoverability) of service discovery protocols.
- **First-Response-Time:** It is the interval between the arrival of the first reply packet and the generation of the corresponding request packet. This metric is averaged over all succeeded service discovery sessions. It measures the promptness of service discovery protocols. It also reflects the average distance between clients and the corresponding first repliers.
- **Ratio of Succeeded-SDP-Number to Total-SDP-packet-number (Suc2Total)**: This metric is the ratio of Succeeded-SDP-number to the sum of service request packets and service reply packets. It reflects the efficiency of service discovery protocols. The number of periodical packets, such as hello packets in MDFNSSDP, is sensitive to protocol parameters, and almost all service discovery protocols generates such packets. Hence, to make a more discriminative inspection on protocol performance, periodical packets are not calculated in this metric.

6.2 Simulation Models

Simulation studies are performed using Glomosim[19]. The distributed coordination function (DCF) of IEEE 802.11 is used as the underlying MAC protocol. Random Waypoint Model (RWM) is used as the mobility model.

In RWM mobility model, nodes move towards their destinations with a randomly selected speed $V \in [V_{min}, V_{max}]$. When reaching its destination, a node keeps static for a random period $T_P \in [T_{min}, T_{max}]$. When the period expires, the node randomly selects a new destination and a new speed, then moves to the new destination with the new speed. The process repeats permanently. In our simulations, $T_{min} = T_{max} = 0, V_{min} = V_{max} = V$.

6.3 Select Comparative Solutions

To make a comparative study, we implement MDFNSSDP and other three typical service discovery protocols for MANETs: BASIC, GSD[8], SSDP[2].

BASIC represents the most straightforward service discovery protocol where each node should forward each service request packet it received, unless the packet reaches its hop limit. BASIC method is widely accepted as a benchmark in evaluating service discovery protocols.

In GSD, services are classified into groups. Each server generates service advertisement packets periodically. Through advertisement packets, each node knows the services provided by its adjacent nodes as well as the group information of some services that these adjacent nodes have seen in received service advertisement packets. When forwarding a service request packet, a node can intelligently forward the packet towards some selected nodes in unicast mode. These selected nodes have seen some services of the same group as the requested service.

In DSDP, some nodes are selected out according to node degree and link stability to construct an upper layer logic backbone. DSDP works in three stages: hello packet collection, backbone node selection, backbone maintenance. They are all based on periodical hello packets. After the construction of backbone, all service discovery packets will spread along the backbone, and thus service request packets will be reduced.

6.4 Simulation Results

Seven simulaiton sets were performed with radio range set to 100m, 125m, 150m, 175m, 200m, 225m, 250m, respectively. Each simulation set contains 4 subsets, which adopts the four protocols, especially. Each subset includes 100 similar simulations with different random seeds. Simulation scenarios are created with 100 nodes initially distributed according to the steady state distribution of random waypoint mobility model. At the beginning of each simulation, 100 nodes are randomly distributed in the scenario area, and predetermined number of nodes are randomly selected as servers. These selected servers provide randomly selected services. During each simulation, 100 service discovery sessions are started at

Table 1. Basic parameters in simulation study

Parameters	Value	Parameters	Value
Scenario	1000m×1000m	Number of service groups	2
Number of nodes	100	Number of services in each group	5
Simulation time	1000s	Hop of broadcast packets(GSD)	1
bandwidth	1Mbps	Hello(broadcast) packet interval	20s
Session number	100	Valid time of cache item	30s
SType(MDFNSSDP)	SELF	Node speed(m/s)	0m/s
VType(MDFNSSDP)	SELF	Hop of request packets	3
VSize(MDFNSSDP)	-1	Number of servers	100
CSize	1		

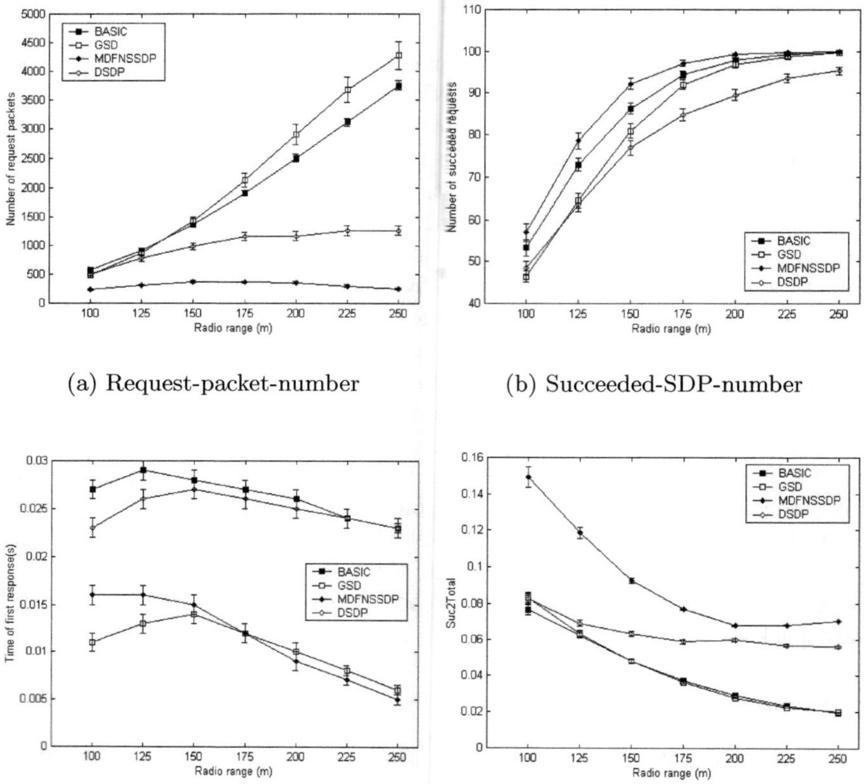

(a) Request-packet-number (b) Succeeded-SDP-number

(c) First-response-time (d) Suc2Total

Fig. 2. Protocol performance under different radio range

randomly selected time by randomly selected nodes.Some basic simulation parameters are listed in Table 1, and simulation results are shown in Fig. 5 with error bars report 95% confidence.

Fig.2 shows that under different radio range, MDFNSSDP has the least request packet overhead (Fig.2(a)), the highest service discoverability (Fig. 5(b)), the quickest response (Fig.2(c)), and the highest efficiency (Fig.2(d)).

7 Conclusions

In this paper, Minimum Dominating Forward Node Set based Service Discovery Protocol (MDFNSSDP) for MANETs is proposed. MDFNSSDP protocol has the following properties:

- MDFNSSDP can fulfill multiple service discovery requests in just one service discovery session. The request-list field can store multiple service requests, and protocol operations are tailored elaborately for multiple-request tasks.

- MDFNSSDP preserves the coverage of service discovery sessions.
- MDFNSSDP minimizes the size of coverage demanding node set $H_{CDN}(v, u)$.
- MDFNSSDP minimizes the number of selected forward nodes by finding a minimum dominating forward node set to cover all coverage demanding nodes using a greedy-based DFNS heuristic.

Simulation results show that MDFNSSDP is an efficient, effective, and prompt service discovery protocol for MANETs.

References

1. IETF, Mobile ad-hoc network (MANET) working group. [Online]. Available: http://www.ietf.org/html.charters/manet-charter.html
2. Kozat, U.C., Tassiulas, L.: Service discovery in mobile ad hoc networks: an overall perspective on architecture choices and network layer support issues. Ad Hoc Networks, 23–44 (2004)
3. Helal, S., Desai, N., Verma, V., Lee, C.: Konark - a service discovery and delivery protocol for ad-hoc networks. In: WCNC 2003. New Orleans, USA pp. 2107–2133 (2003)
4. Azondekon, V., Barbeau, M., Liscano, R.: Service selection in networks based on proximity confirmation using infrared. In: ICT 2002. Beijing, China, pp. 116–120 (2002)
5. Tseng, Y.C., Ni, S.Y., Chen, Y.S., Sheu, J.P.: The broadcast storm problem in a mobile ad hoc network. ACM Wireless Networks, 153–167 (2002)
6. Gao, Z.G., Yang, X.Z., Cai, S.: FFPSDP: flexible forward probability based service discovery protocol. Journal of Harbin Institute of Technology 1265–1270 (2005)
7. Gao, Z.G., Yang, X.Z., Ma, T.Y., Cai, S.B.: RICFFP an efficient service discovery protocol for MANETs. In: Yang, L.T., Guo, M., Gao, G.R., Jha, N.K. (eds.) EUC 2004. LNCS, vol. 3207, pp. 786–795. Springer, Heidelberg (2004)
8. Chakraborty, D., Joshi, A., Yesha, Y., Finin, T.: GSD: a Novel Group-based Service Discovery Protocol for MANETs. In: MWCN 2002. Stockholm, Sweden, pp. 140–144 (2002)
9. Ratsimor, O., Chakarborty, D., Joshi, A., Finin, T.: Allia: alliance-based service discovery for ad-hoc environments. In: Păun, G., Rozenberg, G., Salomaa, A., Zandron, C. (eds.) Membrane Computing. LNCS, vol. 2597, pp. 1–9. Springer, Heidelberg (2003)
10. Nordbotten, N.A., Skeie, T., Aakvaag, N.D.: Methods for service discovery in bluetooth scatternets. Computer Communications, 1087–1096 (2004)
11. Liu, J.C., Zhang, Q., Zhu, W.W., Li, B.: Service locating for large-scale mobile ad hoc network. International Journal of Wireless Information Networks, 33–40 (2003)
12. Yoon, H.J., Lee, E.J., Jeong, H., Kim, J.S.: Proximity-based overlay routing for service discovery in mobile ad hoc networks. In: Aykanat, C., Dayar, T., Körpeoğlu, İ. (eds.) ISCIS 2004. LNCS, vol. 3280, pp. 176–186. Springer, Heidelberg (2004)
13. Klein, M., Ries, B.K., Obreiter, P.: Service rings - a semantic overlay for service discovery in ad hoc networks. In: Mařík, V., Štěpánková, O., Retschitzegger, W. (eds.) DEXA 2003. LNCS, vol. 2736, pp. 180–185. Springer, Heidelberg (2003)
14. Klein, M., Ries, B.K.: Multi-layer clusters in ad-hoc networks - an approach to service discovery. In: IWP2PC'02. Pisa, Italy, pp. 187–201 (2002)

15. Klein, M., Hoffman, M., Matheis, D. et al.: Comparison of overlay mechanims for service trading in ad hoc networks. TR 2004-2, University of Karlsruhe (2004)
16. Gao, Z.G., Yang, Y.T., Zhao, J., Cui, J.W., Li, X.: Service discovery protocols for MANETs: a survey. In: Cao, J., Stojmenovic, I., Jia, X., Das, S.K. (eds.) MSN 2006. LNCS, vol. 4325, pp. 232–243. Springer, Heidelberg (2006)
17. Brad, W., Tracy, C.: Comparison of broadcasting techniques for mobile ad hoc networks. In: MobiHoc 2002, Lausanne, Switzerland, pp. 194–205 (2002)
18. Chvatal, V.: A greedy heuristic for the set-covering problem. Mathematics of Operations Research, 233–235 (1979)
19. Glomosim: a scalable simulation environment for wireless and wired network system. Available: http://pcl.cs.ucla.edu/projects/domains/glomosim.html

Parallel Performance Prediction for Multigrid Codes on Distributed Memory Architectures

Giuseppe Romanazzi and Peter K. Jimack

School of Computing, University of Leeds, Leeds LS2 9JT, UK
{roman,pkj}@comp.leeds.ac.uk

Abstract. We propose a model for describing the parallel performance of multigrid software on distributed memory architectures. The goal of the model is to allow reliable predictions to be made as to the execution time of a given code on a large number of processors, of a given parallel system, by only benchmarking the code on small numbers of processors. This has potential applications for the scheduling of jobs in a Grid computing environment where reliable predictions as to execution times on different systems will be valuable. The model is tested for two different multigrid codes running on two different parallel architectures and the results obtained are discussed.

1 Introduction

Multigrid is one of the most power numerical techniques to have been developed over the last 30 years [1,2,13]. As such, state-of-the-art parallel numerical software is now increasingly incorporating multigrid implementation in a variety of application domains [6,9,11]. In this work we seek to model the performance of two typical parallel multigrid codes on distributed memory architectures. The goal is to be able to make accurate predictions of the performance of such codes on large numbers of processors, without actually executing them on all of these processors. This is of significant potential importance in an environment, such as that provided by Grid computing, where a user may have access to a range of shared resources, each with different costs and different levels of availability, [4,7,10].

A vast literature on performance models exists, varying from analytical models designed for a single application through to general frameworks that can be applied to many applications on a large range of HPC systems. This latter approach is typically based upon a convolution of an application trace with some benchmarks of the HPC system used. Both approaches have been demonstrated to be able to provide accurate and robust predictions, although each has its potential drawbacks too. In the former, for example, significant expertise is needed in deriving the analytic model, which is extremely code specific, whereas in the latter approach, a large amount of computer time is typically required for tracing the application.

Considering these limitations, the choice between these two approaches will depend primarily on the goal of the predictions. For example, when it is most

R. Perrott et al. (Eds.): HPCC 2007, LNCS 4782, pp. 647–658, 2007.

important to predict the run-time of large-scale applications on a given system, as opposed to just comparing their relative performance, it is preferable to build and apply a detailed analytic model for the available set of HPC systems, as in [8] for example. On the other hand, when it is more important to compare the performance of some real applications on different machines, the latter approach is preferable: in which case different benchmarks metrics can be used and convoluted with the application trace file, as in [3] or [5].

Our approach lies between these two extremes. We use relatively crude analytic models that are applicable to a general class of algorithms (multigrid) and through simulations of the application on a limited number of CPUs we attempt to evaluate the parameters of these models. In comparison with the first approach the sophistication of the analytic model is much less (but also much less dependent on the specific code or implementation). In comparison with the second approach, there is no need for tracing the application nor running large numbers of HPC benchmarks on the HPC facility: our benchmarking is simply based upon execution of the code on small numbers of processors of the HPC system.

The layout of the remainder of the paper is as follows. In the next section we provide a very brief introduction to parallel multigrid algorithms for the solution of elliptic or parabolic partial differential equations (PDEs) in two space dimensions. This is followed by an analysis of the performance of two such codes on an abstract distributed memory architecture. The analysis is then used to build a predictive model for this class of codes, that is designed to allow estimates of run times to be obtained for large numbers of processors, based upon observed performance on very small numbers of processors. The paper concludes with a description of some numerical tests to assess the accuracy and robustness of these predictions and a discussion of the outcomes obtained. Further extensions of the work are also suggested.

2 Multigrid and Parallel Implementation

The general principal upon which multigrid is based is that when using many iterative solvers for the systems of algebraic equations that result from the discretization of PDEs, the component of the error that is damped most quickly is the high frequency part [2,13]. This observation leads to the development of an algorithm which takes a very small number of iterations on the finite difference or finite element grid upon which the solution is sought, and then restricts the residual and equations to a coarse grid to solve for an estimate of the error on this grid. This error is then interpolated back onto the original grid before a small number of further iterations are taken and the process repeated. When the error equation is itself solved in the same manner, using a still coarser grid, and these corrections are repeated recursively down to a very coarse base grid, the resulting process is known as multigrid.

Any parallel implementation of such an algorithm requires a number of components to be implemented in parallel:

- application of the iterative solver at each grid level
- restriction of the residual to a coarse level
- exact solution at the coarsest level
- interpolation of the error to a fine level.
- a convergence test

There is also a variant of the algorithm (primarily designed with nonlinear problems in mind), known as FAS (full approximation scheme) [13], which requires the solution as well as the residual to be restricted to the coarser grid at each level.

In this work we consider the parallel implementation of two multigrid codes: one is standard and the other uses the FAS approach. In both cases they partition a two-dimensional finite difference grid across a set of parallel processors by assigning blocks of rows to different processors. Note that if the coarsest mesh is partitioned in this manner, then if all finer meshes are uniform refinements of this they are automatically partitioned too: see Fig. 1 for an illustration.

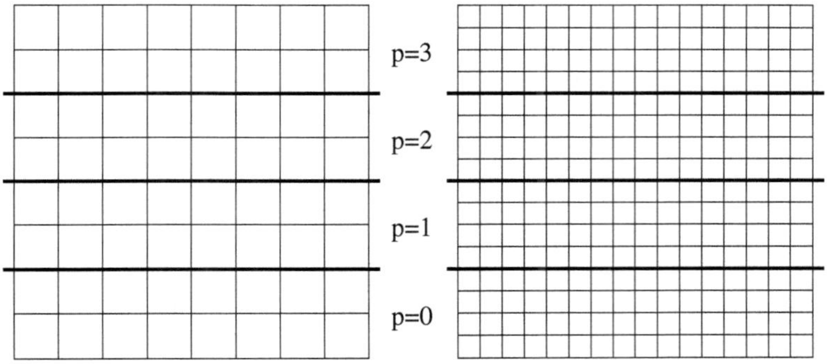

Fig. 1. Partitioning of a coarse and a fine mesh across four processors by assigning a block of rows to each processor

It is clear from inspection of the meshes in Fig. 1 that each stage of the parallel multigrid process requires communication between neighbouring processes (iteration, restriction, coarse grid solution, interpolation, converge test). The precise way in which these are implemented will vary from code to code, however the basic structure of the algorithm will remain the same. In this work we consider two different implementations, referred to as m1 and m2.

2.1 The Algorithm m1

This algorithm solves the steady-state equation

$$\begin{aligned} -\nabla^2 u &= f \quad \text{in } \Omega, \\ \Omega &= [0,1] \times [0,1], \\ u_{|\partial\Omega} &= 0. \end{aligned}$$

The discretization is based upon a centred finite difference scheme on each grid. The iterative solver employed is the well-known Red-Black Gauss-Seidel (RBGS) method [9], which is ideally suited to parallel implementation. The partitioning of the grids is based upon Fig. 1 with both processors p and $p+1$ owning the row of unknowns on the top of block p. Each processor also stores an extra, dummy, row of unknowns above and below its own top and bottom row respectively (see Figure 2): this is used to duplicate the contents of the corresponding row on each neighbour. After each red and black sweep of RBGS there is an inter-processor communication in which these dummy rows are updated. This is implemented with a series of non-blocking sends and receives in MPI. Similar inter-processor communication is required at the restriction step but the interpolation step does not require any message passing. A global reduction operation is required to test for convergence.

2.2 The Algorithm m2

This algorithm, described in more detail in [9], uses an unconditionally-stable implicit time-stepping scheme to solve the transient problem

$$\begin{aligned} &\frac{\partial u}{\partial t} = \nabla^2 u + f \text{ in } (0,\mathrm{T}] \times \Omega, \\ &\Omega = [0,1] \times [0,1], \\ &u_{|\partial\Omega} = 0, \\ &u_{|t=0} = u_0. \end{aligned}$$

As for m1, the discretization of the Laplacian is based upon the standard five point finite difference stencil. Hence, at each time step it is necessary to solve an algebraic system of equations for which multigrid is used. Again RBGS is selected as the iterative scheme and so communications are required between neighbour processors after each red and black sweep. Different from m1, here only processor p owns the row of unknowns at the top of block p, see Fig. 2. This means that the total memory requirement is slightly less than with algorithm m1 but that, unlike m1, communications are also required at the interpolation phase, as well as the restriction and convergence test phases. Also different from m1, the inter-processor sends and receives are based upon a mixture of MPI blocking and non-blocking functions. Finally, as mentioned above, m2 is implemented using the FAS algorithm [13] and so the current solution must also be interpolated from the fine to the coarser grid at each level.

3 The Predictive Model

The goal of most parallel numerical implementations is to be able to solve larger problems than would be otherwise possible. For the numerical solution of PDEs this means solving problems on finer grids, so as to be able to achieve higher accuracy. Ideally, when increasing the size of a problem by a factor of np and solving it using np processors (instead of a single processor), the solution time

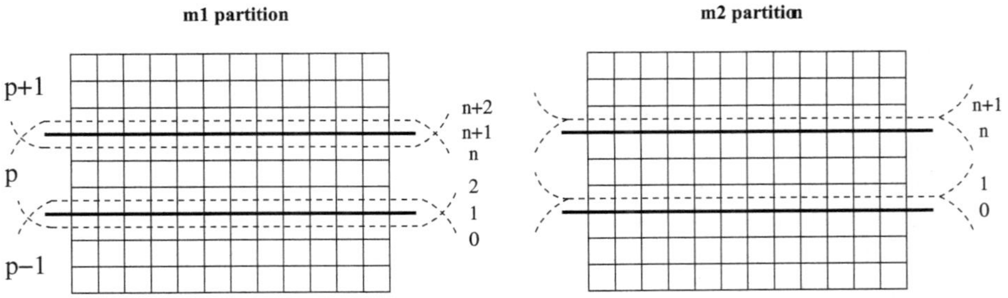

Fig. 2. Domain partitions for algorithms m1 and m2. The dummy rows of processor p have indexes 0, $n+2$ and 0, $n+1$ for m1 and m2, respectively

should be unchanged. This would represent a perfect efficiency and is rarely achieved due to the parallel overheads such as inter-processor communications and any computations that are repeated on more than one processor. In this research our aim is to be able to predict these overheads in the situation where the size of the problem (the number of discrete unknowns, N^2, on the finest mesh) scales in proportion to the number of processors. Consequently, the basic assumption that we make is that the parallel solution time (on np processors) may be represented as

$$T_{parallel} = T_{comp} + T_{comm}. \tag{1}$$

In (1), T_{comp} represents the computational time for a problem of size N^2/np ($= N(1)^2$, say) on a single processor, and T_{comm} represents all of the parallel overheads (primarily due to inter-processor communications).

The calculation of T_{comp} is straightforward since this simply requires execution of a problem of size $N(1)^2$ on a single processor. One of the major attractions of the multigrid approach is that this time should be proportional to the size of the problem on the single processor. As demonstrated in [12] this is indeed the case for both of the implementations, m1 and m2, considered in this paper. The key consequence of this property is that in situations where we scale the problem size with the number of processors, np, then T_{comp} remains independent of np.

The more challenging task that we have, therefore, is to model T_{comm} in a manner that will allow predictions to be made for large values of np. Note that any additional work that is undertaken when computing in parallel is associated with the dummy rows that are stored on each processor, as is the communication overhead. When solving on a fine mesh of size N^2 the value of T_{comm} may be estimated, to leading order, as

$$T_{comm} \propto T_{start} + kN. \tag{2}$$

Here T_{start} relates to the start-up cost for the communications whilst kN represents the number of tasks of size N at each multigrid cycle (which will depend upon the number of communications and the number of additional iterative solves on duplicated rows). Furthermore, it is important to note that k itself may not necessarily be independent of N. In fact, slightly more than np times the work is undertaken by the parallel multigrid algorithm on an N^2 fine mesh than by the sequential multigrid solver on an N^2/np fine mesh. This is because in the former case either there are more mesh levels, or else the coarsest grid must be finer. In all of the examples discussed in this paper we consider each problem using the same coarse level N_0^2 (with N_0 a power of 2) and the same work per processor at the finest level (N^2 on np processors, where $N = \sqrt{np}N(1)$). Therefore, the number of grid levels used, $nlevels$, depends on the number of processors np, that is

$$nlevels = log_2(N/N_0) + 1. \quad (3)$$

We consider two different models for T_{comm}, based upon (2). In the first of these we do assume that k is approximately constant to obtain

$$T_{comm} = a + bN. \quad (4)$$

In the second model we add a quadratic term $\gamma N(1)^2$ that allows a nonlinear growth of the overhead time

$$T_{comm} = \alpha + \beta N + \gamma N(1)^2. \quad (5)$$

The quadratic term introduced in (5) is designed to allow a degree of nonlinearity to the overhead model, reflecting the fact that k in (2) may not necessarily be a constant. As will be demonstrated below, the importance (or otherwise) of this nonlinear effect depends upon the specific characteristics of the processor and communication hardware that are present on each particular parallel architecture. In particular, the effects of caching and of a multicore architecture appear to require such a nonlinear model.

In order to obtain values for (a, b) and (α, β, γ) in (4) and (5) respectively, the parallel performance of a given code must be assessed on each target architecture. Note from Fig. 1 that the communication pattern is identical regardless of np: requiring only neighbour to neighbour communications. Hence our next assumption is that (a, b) or (α, β, γ) may be determined using just a small number of processors. In this work we choose $np = 4$ in order to approximate (a, b) or (α, β, γ) by fitting (4) or (5) respectively to a plot of $(T_{parallel} - T_{comp})$ against N. Note that, from (1), $T_{comm} = T_{parallel} - T_{comp}$ and T_{comp} is known from the data collected on a single processor.

A summary of the overall predictive methodology is provided by the following steps. In the following notation np is the target number of processors and N^2 is the largest problem that can be solved on these processors without swapping effects causing performance to be diminished. Similarly, $N(1)^2 = N^2/np$ is the largest such problem that can fit onto a single processor.

1. For $\ell = 1$ to m
   Run the code on a single processor with a fine grid of dimension $(2^{1-\ell}N(1))^2$.
   In each case collect T_{comp} based upon average timings over at least 5 runs.
2. For $\ell = 1$ to m

Run the code on 4 [or 8] processors, with a fine grid of dimension $(2^{2-\ell}N(1))^2$ [or $(2^{2-\ell}N(1) \times 2^{3-\ell}N(1))$].

In each case collect $T_{parallel}$ based upon average timings over at least 5 runs.

3. Fit a straight line of the form (4) through the data collected in steps 1 and 2 to estimate a and b, or fit a quadratic curve through the data collected to estimate α, β and γ in (5).
4. Use either model (4) or model (5) to estimate the value of T_{comm} for larger choices of np and combine this with T_{comp} (determined in step 1) to estimate $T_{parallel}$ as in (1).

4 Numerical Results

The approach derived in the previous section is now used to predict the performance of the two multigrid codes, m1 and m2, on the two parallel architectures WRG2 and WRG3. These computers form part of the University of Leeds' contribution to the White Rose Grid [4]

- WRG2 (White Rose Grid Node 2) is a cluster of 128 dual processor nodes, each based around 2.2 or 2.4GHz Intel Xeon processors with 2GBytes of memory and 512 KB of L2 cache. Myrinet switching is used to connect the nodes and Sun Grid Engine Enterprise Edition provides resource management.
- WRG3 (White Rose Grid Node 3) is a cluster of 87 Sun microsystem dual processor AMD nodes, each formed by two dual core 2.0GHz processors. Each of the $87 \times 4 = 348$ batched processors has L2 cache memory of size 512KB and 2GBytes of physical memory. Again, both Myrinet switching and Sun Grid Engine are used for communication and resource management respectively.

Table 1 shows values of (a, b) and (α, β, γ) obtained by following steps 1 to 4 described above, using $N(1) = 2048$ and $m = 4$. Each of the codes is run on each of the selected parallel architectures. Clearly the precise values obtained for (a, b) and (α, β, γ) depend upon a number of factors.

- Because users of WRG2 and WRG3 do not get exclusive access to the machines, or the Myrinet switches, there is always some variation in the solution times obtained in steps 1 and 2.
- On WRG2 there are (75) 2.4GHz and (53) 2.2GHz processor nodes, hence the parameters will depend on which processors are used to collect execution times in steps 1 and 2.
- On WRG3 the situation is made more complex by the fact that each node consists of two processors and 4 cores. Users may therefore have access to a core on a node where other users are running jobs or else they may have an entire node to themselves. This creates significant variations in the timings for both sequential and small parallel runs.

As outlined in steps 1 and 2 above, a simple way to reduce the effects of these variations is simply to take average timings over five runs (say). Such a crude approach, whilst accounting for the the relatively minor effects of sharing resources such as access to the communications technology, are not generally sufficient on their own however. For example, for a heterogeneous architecture such as WRG2, which has processors of two different speeds, it should also be recognised that a parallel job using a large number of processors will typically make at least some use of the slower 2.2GHz nodes. It is therefore essential to ensure that in step 1 a slower processor is used to compute the sequential timings, and in step 2 at least one slower processor should be used in the parallel runs. If only the faster processors are used in steps 1 and 2 above then the resulting model will inevitably under-predict solution times on large numbers of processors when any of these processors are 2.2GHz rather than 2.4GHz. This use of the slower processors has been imposed for the two WRG2 columns of Table 1.

Table 1. Parameters (a,b) and (α,β,γ) of the T_{comm} models (4) and (5) respectively, obtained for $np = 4$ and for $N(4) = 512, \ldots, 4096$ with coarse grid of size 32

		m2-WRG2	m2-WRG3	m1-WRG2	m1-WRG3
1 node	$a =$	0.1540	-0.1658	$-7.957e-02$	-0.2250
	$b =$	$6.616e-05$	$5.139e-04$	$1.261e-04$	$2.943e-04$
2 nodes	$a =$		$-8.518e-03$		-0.2316
	$b =$		$4.083e-04$		$3.191e-04$
1 node	$\alpha =$	-0.3825	-0.2079	$-9.245e-02$	$6.012e-02$
	$\beta =$	$7.863e-04$	$5.704e-04$	$1.434e-04$	$-8.833e-05$
	$\gamma =$	$-6.075e-07$	$-4.765e-08$	$-1.458e-08$	$3.228e-07$
2 nodes	$\alpha =$		-0.4211		$-1.698e-02$
	$\beta =$		$9.621e-04$		$3.106e-05$
	$\gamma =$		$-4.671e-07$		$2.430e-07$

Furthermore, in order to better control the effects of multiple cores on WRG3, we have chosen to undertake all of the sequential runs using four copies of the same code: all running on the same node. Again, this decision is made bearing in mind the situation that will exist for a large parallel run in which all of the available cores will be used. In addition to this, for WRG3, two sets of predictive timings are produced. The first of these is obtained by running the 4 process parallel job on a single (four core) node, whilst the second is obtained by running an 8 process parallel job across two (four core) nodes. The latter is designed to allow both intra-communication (communication between processes on the same node) and inter-communication (communication between processors on different nodes) overheads to be captured by our model (the former will only capture intra-communication costs). These two situations are denoted by "1 node" and "2 nodes" respectively in Table 1.

There are a number of interesting observations to make about the parameters shown in Table 1. As indicated in step 3 above, these are obtained by fitting curves (either linear or quadratic) through the averaged data collected in steps 1 and 2 (using (4) or (5) respectively, with $T_{comm} = T_{parallel} - T_{comp}$). For the linear model it is important to state that the fit to the data is not particularly good which provides a clear indication that the model may be deficient. For the quadratic cases the fits to the data are, perhaps not surprisingly, a lot better. For WRG3, the effect of undertaking the small parallel runs (step 2) using cores on two nodes rather than one is not particularly significant for the linear model but is much more noticeable for the quadratic model. This might suggest that other inaccuracies are dominant in the former case.

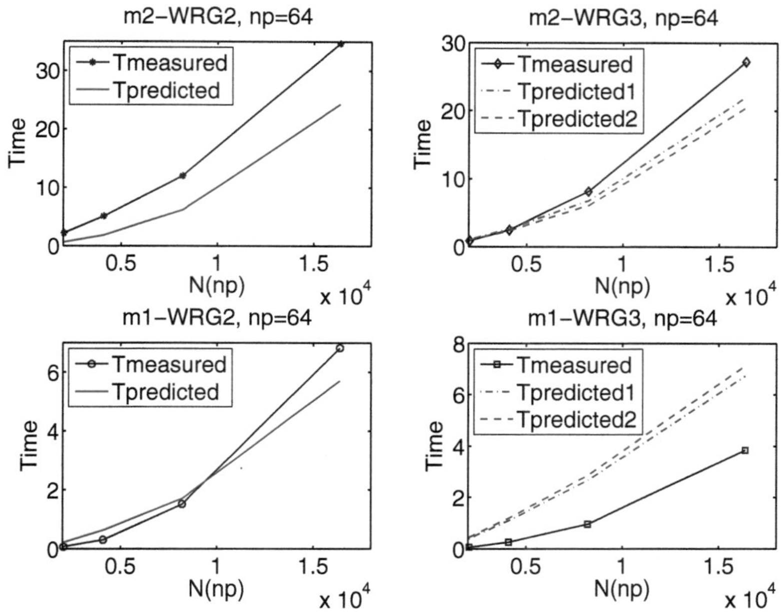

Fig. 3. Linear model $T = T_{comp} + a + bN$ for Time predicted, and Time measured (both in seconds) on $np = 64$ processors, $N(64) = 2048, \ldots, 16384$

Given the values shown in Table 1, it is now possible to make predictions for the performance of the multigrid codes on greater numbers of processors. In this paper, we consider executing the codes on $np = 64$ processors (the maximum queue size available to us), with the grids scaled in size in proportion to np. Figure 3 shows results using the linear model, (4), whilst Figure 4 shows similar results based upon the quadratic model (5). In each case all four combinations of algorithms (m1,m2) and computer systems (WRG2,WRG3) are presented. Note that for the runs on WRG3, following Table 1, two predictions are presented (Tpredict1 and Tpredict2): these are based upon the best-fit parameters (α, β and γ) obtained when 1 node (4 cores) or 2 nodes (8 cores) are used in step 2 respectively.

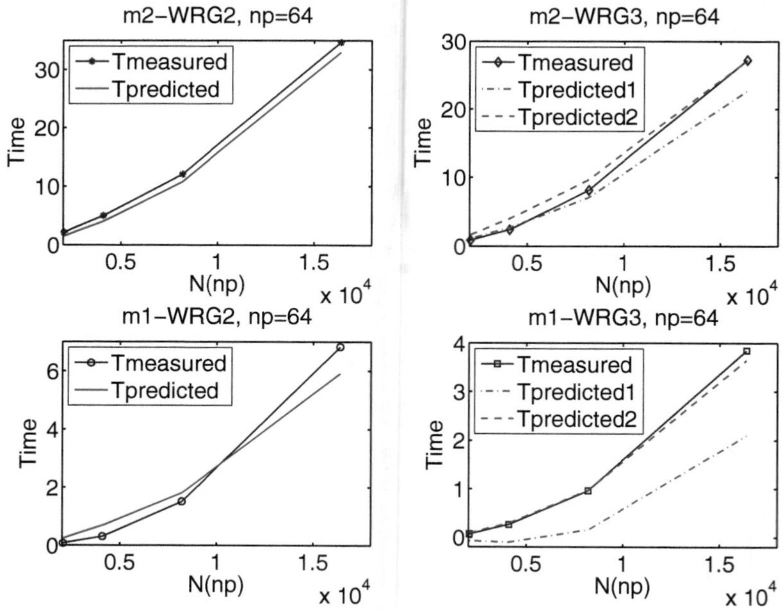

Fig. 4. Quadratic model $T = T_{comp} + \alpha + \beta N + \gamma N(1)^2$ for Time predicted, and Time measured (both in seconds) on $np = 64$ processors, $N(64) = 2048, \ldots, 16384$

It is clear from Fig. 3 that the linear model provides disappointing predictions in almost all cases. It is noticeable that this model under-predicts the solution times for algorithm m2 (which includes blocking as well as non-blocking communication) whilst it tends to over-predict the solution times for algorithm m1. It is hard to discern any obvious pattern from these results other than the fact that the qualitative behaviour seems to have been captured in each case, even though the quantitative predictions are unreliable. This suggests that nonlinear effects are important in the parallelization, either due to communication patterns (e.g. switch performance and/or the effects of non-blocking communication) or nonlinear cache effects (with multi-core processors for example).

From Fig. 4 it is apparent that using the quadratic model to capture the predicted nonlinear effects can be highly effective. This model provides significantly better predictions for both multigrid implementations and on both parallel architectures (provided that Tpredict2 is used on WRG3). In the case of the results obtained on WRG2 it is important to emphasize again that steps 1 and 2 of the algorithm were undertaken using at least one slower processor in each run. Without this restriction the predictions were of a much poorer quality: providing significant under-estimations of the parallel run times on 64 processors (see [12] for further details). In the case of WRG3 recall that Tpredict2 is based upon the use of 8 cores in step 2 of the methodology described in the previous section (as opposed to 4 cores for Tpredict1). This clearly demonstrates that, since the large parallel jobs typically use all of the available cores on each node, both intra- and inter-communication costs must be captured by the predictive model. Perhaps not surprisingly, when this model fails to capture all of the communication patterns present in the full parallel code the resulting predictions (Tpredicted1) become unreliable.

5 Conclusion and Future Work

In this paper we have proposed a simple methodology for predicting the performance of parallel multigrid codes based upon their characteristics when executed on small numbers of processors. The initial results presented are very encouraging, demonstrating that remarkably accurate and reliable predictions are possible provided that sufficient care is taken with the construction of the model and the evaluation of its parameters. In particular, it has been possible to demonstrate that the effects of both heterogeneous and multicore architectures can be captured, and that the models proposed can be applied to two quite different multigrid codes.

It is clear from the results presented in the previous section that, despite the communication costs being $O(N)$ and the $O(N)$ complexity of the multigrid approach (as illustrated in [12]), the simple linear model for the growth in the parallel overhead is not sufficient to capture the practical details of scalability to large numbers of processors. There are numerous possible causes of this (e.g. nonlinear communication patterns, caching effects, etc.) however it is demonstrated that the addition of a quadratic term to the model, to capture the nonlinear effects to leading order, improves its predictive properties substantially. Unsurprisingly, care must be taken to deal with non-homogeneous architectures or multicore architectures in an appropriate manner. If these effects are ignored then even the quadratic model fails to yield realistic predictions.

It has been observed in this work that the quality of the best fit that is made in order to determine the values of the parameters in Table 1 appears to provide a useful indication as to the reliability of the resulting model. For example, the linear fits in the top half of the table are relatively poor, as are the predictions in Fig. 3. In future work it would be interesting to investigate this phenomenon further in order to attempt to produce a reliability metric for the predictions that are made. Recall that one of the primary motivations for this work is to provide information on expected run-times on different numbers of processors on different architectures in order to allow optimal (or improved) scheduling of jobs in a Grid-type environment where a variety of potential resources may be available. Clearly, providing additional information, such as error bounds for these expected run times, will further assist this process. An additional factor that should also be included in this modelling process is the ability to consider different domain decomposition strategies (e.g. partitioning the data into blocks rather than strips, [6]) and predict their relative performance for a given problem on a given architecture.

Finally, we observe that, in addition to deciding which single computational resource to use in order to complete a given computational task, there will be occasions when multiple Grid resources are available and might be used together. Consequently, in future work we also intend to extend our models to provide predictions that will allow decisions to be made on how best to split the work across more than one resource and to determine the likely efficiency (and cost-effectiveness) of so doing.

References

1. Brandt, A.: Multi-level adaptive solutions to boundary value problems. Mathematics of Computation 31, 333–390 (1977)
2. Briggs, W.L., Henson, V.E., McCormick, S.F.: A Multigrid Tutorial. SIAM (2000)
3. Carrington, L., Laurenzano, M., Snavely, A., Campbell, R., Davis, L.P.: How well can simple metrics represent the performance of HPC applications? In: Proceedings of SC 2005 (2005)
4. Dew, P.M., Schmidt, J.G., Thompson, M., Morris, P.: The White Rose Grid: practice and experience. In: Cox, S.J. (ed.) Proceedings of the 2nd UK All Hands e-Science Meeting. EPSRC (2003)
5. DIMEMAS, `http://www.cepba.upc.es/dimemas/`
6. Goodyer, C.E., Berzins, M.: Parallelization and scalability issues of a multilevel ellastohydrodynamic lubrication solver. Concurrency and Computation: Practice and Experience 19, 369–396 (2007)
7. Huedo, E., Montero, R.S., Llorente, I.M.A.: module architecture for interfacing Pre-WS and WS Grid resource management services. Future Generation Computing Systems 23, 252–261 (2007)
8. Kerbyson, D.J., Alme, H.J., Hoisie, A., Petrini, F., Wasserman, H.J., Gittings, M.: Predictive Performance and Scalability Modeling of a Large-Scale Application. In: Reich, S., Tzagarakis, M.M., De Bra, P.M.E. (eds.) Hypermedia: Openness, Structural Awareness, and Adaptivity. LNCS, vol. 2266, Springer, Heidelberg (2002)
9. Koh, Y.Y.: Efficient Numerical Solution of Droplet Spreading Flows. Ph.D. Thesis, University of Leeds (2007)
10. Krauter, K., Buyya, R., Maheswaran, M.A.: taxonomy and survey of Grid resource management systems. Int. J. of Software: Practice and Experience 32, 135–164 (2002)
11. Lang, S., Wittum, G.: Large-scale density-driven flow simulations using parallel unstructured grid adaptation and local multigrid methods. Concurrency and Computation: Practice and Experience 17, 1415–1440 (2005)
12. Romanazzi, G., Jimack, P.K.: Performance prediction for parallel numerical software on the White Rose Grid. In: Proceedings of UK e-Science All Hands Meeting (2007)
13. Trottenberg, U., Oosterlee, C.W., Schüller, A.: Multigrid. Academic Press, London (2003)

Netgauge: A Network Performance Measurement Framework

Torsten Hoefler[1,2], Torsten Mehlan[2], Andrew Lumsdaine[1], and Wolfgang Rehm[2]

[1] Open Systems Lab, Indiana University, Bloomington IN 47405, USA
{htor,lums}@cs.indiana.edu
[2] Chemnitz University of Technology, 09107 Chemnitz, Germany
{htor,tome,rehm}@cs.tu-chemnitz.de

Abstract. This paper introduces **Netgauge,** an extensible open-source framework for implementing network benchmarks. The structure of Netgauge abstracts and explicitly separates communication patterns from communication modules. As a result of this separation of concerns, new benchmark types and new network protocols can be added independently to Netgauge. We describe the rich set of pre-defined communication patterns and communication modules that are available in the current distribution. Benchmark results demonstrate the applicability of the current Netgauge distribution to to different networks. An assortment of use-cases is used to investigate the implementation quality of selected protocols and protocol layers.

1 Introduction

Network performance measurement and monitoring plays an important role in High Performance Computing (HPC). For many different network protocols and interconnects, numerous tools are available to perform functionality and correctness tests and also benchmark two fundamental network parameters (i.e., latency and bandwidth) . While these features are generally useful for system administrators and end users, for high-end performance tuning, the HPC community usually demands more detailed insight into the network.

To satisfy the demands of application and middleware developers, it is often necessary to investigate specific aspects of the network or to analyze different communication patterns. The simplest measurement method is a "ping-pong" benchmark where the sender sends a message to a server and the server simply reflects this message back. The time from the send to the receive on the sender is called Round Trip Time (RTT) and plays an important role in HPC. Another benchmark pattern, called ping-ping, can be used to analyze pipelining effects in the network (cf. Pallas Microbenchmarks [1]). However, more complicated communication patterns are necessary to simulate different communication situations, such as one-to-many or many-to-one patterns to analyze congestion situations.

The most advanced network benchmarks are parametrizing network models, such as LogP [2], LogGP [3] or pLogP [4]. These require special measurement

R. Perrott et al. (Eds.): HPCC 2007, LNCS 4782, pp. 659–671, 2007.

methods such as described in [4,5,6,7]. Those models can be used to predict the running time of parallel algorithms. The LogP model family can also be used to predict the potential to overlap communication and computation for different network protocols (cf. Bell et al. [7]).

1.1 Related Work

Existing portable tools, such as netperf or iperf, can readily measure network latency and bandwidth. However, it is often necessary to measure other network parameters (e.g., CPU overhead to assess the potential for computation and communication overlap or performance in threaded environments). Such parameters can be measured, but the tools to do so are generally specific either to the parameter or hardware being measured. HUNT[8], White's implementation[9] and COMB[10] are able to assess the overlap potential of different network protocols. Other tools like Netpipe [11,12] simply benchmark latency and bandwidth for many different interconnection benchmarks. Another interesting tool named coNCePTuaL [13,14] can be used to express arbitrary communication patterns, but the support for different low-level networks is limited and the addition of new networks is rather complicated. The MIBA [15] tool reports a number of different detailed timings, but is specific to the InfiniBand network.

Our work combines the advantages of previous work and enables users to easily implement their own communication patterns and use low-level communication modules to benchmark different network stacks in a comparable way. The interface of the communication modules is kept as simple as possible to ensure an easy addition of new networks or protocols.

1.2 Challenges and Contributions

After presenting a motivation and an overview about related projects, we will discuss the challenges we faced in the design and implementation phase and then highlight the main contributions of our project to the scientific community.

Challenges

The biggest challenge we faced was to design a single interface between the communication layer and the benchmark layer that supports many different communication networks. A main distinction has to be made between one-sided protocols where the sender writes directly into the passive receiver's memory and two-sided communication where the receiver fully takes part in the communication. Other issues come up when networks with special requirements, (i.e., the communication-memory must be registered before any communication can take place) need to be supported. We describe our methodology to deal with the unification of those different interfaces and networks into a single simple-to-use interface that supports easy addition of new benchmark patters. Therefore, we avoid all semantics that are not required for our benchmarking purposes (such as message tagging), even though they may be helpful or even required for real parallel applications. Furthermore, in order to reflect real-world applications as accurately as possible, the benchmark implementor must be able to "simulate" application behavior in his implementation by using the abstract communication interface provided. Portability is achieved by limiting the number of external dependencies. Thus, Netgauge is written in plain C and needs only MPI (a portable MPI implementation is available with Open MPI [16]).

Contributions

Our main contribution to the scientific community is the design and implementation of an abstract interface to separate the communication part from the benchmark part in a network benchmark. We show that our framework is able to support many different kinds of network benchmarks (including those that simulate applications). The interface is kept as simple as possible to allow an easy addition of benchmarks (for researchers interested in performance of applications or communication algorithms) or communication modules (for researchers interested in the performance of a particular network or hardware, e.g., different parameters).

Merging those two groups makes Netgauge a useful tool for research. For example, if a new network has to be tested, the researcher needs only implement a communication module and will then immediately have all benchmarks available, including network parameter assessment of well-know network models, flow control tests and, of course, also the simple latency and bandwidth tests. Another scenario supports more theoretical research in network modeling. If a new network model is designed, the researcher simply implements the parameter assessment routine as a benchmark and thus measures the parameters for all networks supported by Netgauge.

The current version of Netgauge ships support for many communication networks and benchmark algorithms that can be used as templates for the addition of new modules. Netgauge also offers an abstract timer interface which is able to support system-dependent high-performance timers (e.g., RDTSC [17]) to enable very accurate time measurements.

2 The Performance Measurement Framework

Netgauge uses a component architecture [18] similar to LAM/MPI or Open MPI which consists of different "frameworks". A framework defines a particular interface to the user or other frameworks and a "module" is a specific instance of a framework.

Fig. 1 shows the overall structure of Netgauge. Different types of benchmarks are implemented in the "pattern" framework. The low-level communication layer is represented by the "communication" framework. The pattern implementor is free to use the functionality offered by the communication framework to implement any benchmark pattern. The communication framework abstracts the different network types. Netgauge currently supports a simplified two-sided interface similar to MPI. This interface assures an easy implementation of new communication modules.

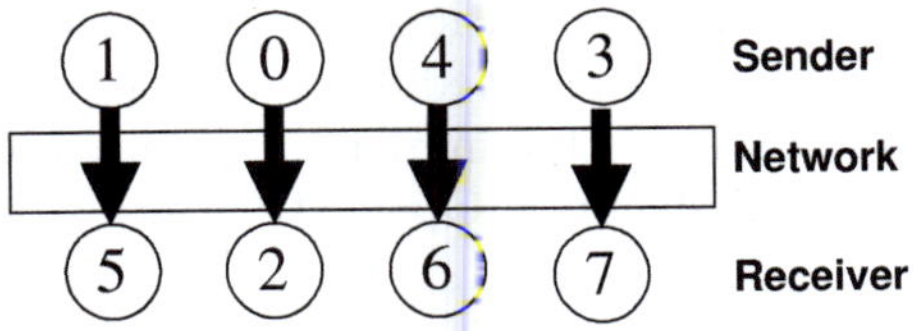

Fig. 1. Netgauge Framework Structure

2.1 The Pattern Framework

The pattern framework is the the core of every benchmark with Netgauge. A pattern interface is very simple and consists of a name for the pattern, a short description, the requirement flags and a benchmark function pointer. The user selects a specific pattern via the command line and Netgauge checks that the communication module supports all requirements indicated by the pattern's flags (e.g., if the pattern requires non-blocking communication). If so, Netgauge calls the specified module's benchmark function and passes a reference to the user-selected communication module.

2.2 The Communication Framework

The communication framework is used by the user-selected pattern module to benchmark the communication operations. The framework interface contains a name, a short description, flags and different communication functions. A module can also indicate a maximum message-size (e.g., for UDP) and how many bytes it adds as additional header (e.g., for RAW Ethernet). Different module flags indicate if the module offers reliable transport, channel semantics and/or requires memory registration. Optional `init()`, `shutdown()` and `getopt()` functions can be offered to initialize, shut the module down or read additional module-specific command line options. All optional function-pointers may be set to `NULL` which means that the module does not offer this functionality.

Every module must at least offer blocking send and receive calls. The `sendto()` function accepts destination, a buffer and a message-size as parameters. There is no tag and so the ordering of messages is significant for the message matching. The `recvfrom()` function gets a source, a buffer and a size as parameters and blocks until some data is received. The `recvfrom()` function returns the number of received bytes. Macros to send or receive a complete message (`send_all()` and `recv_all()`) are also available.

An optional non-blocking interface can be used to benchmark asynchronous communication or analyze overlap capabilities of the communication protocols. This interface is the same as the blocking interface except it contains an additional request handle. Some of the patterns that require non-blocking communication (e.g., `1:n`) will not work with communication modules that do not offer this functionality.

Memory Registration Issues
Some types of networks need registered memory areas for communication. Typically the interface to those networks exposes functions to register memory. The communication functions have to ensure data transmission only from registered memory to registered memory. Thus the communication module has a member function to allocate memory for data transmission. Patterns have to use this function instead of malloc(). The communication module has to perform all the tasks to setup memory for communication. In many cases this includes the use of some hash table to store additional information (i.e. LKEY and RKEY for InfiniBand).

2.3 Control Flow

Netgauge's main routine is the only part that interacts directly with the user. All modules for all frameworks are registered at the beginning of the program. The selected module is initialized and its parameters are passed in the module's `getopt()` function (this enable communication module implementors who add network-specific command line arguments). General internal variables like e.g., maximum message size and test repetitions are set and the user-selected pattern module is called. The pattern module performs all benchmarks and may either print the results directly or use helper functions provided by the statistics and output module to process and output data. All modules may use MPI calls (e.g., to exchange address information or parameters in order to establish initial connections). However, the implementor should try to avoid excessive MPI usage so that the modules can also be used in environments without MPI (which is supported by a basic set of modules).

2.4 Other Available Communication Patterns

Simple Microbenchmark Patterns
This section describes simple patterns that are available in Netgauge.

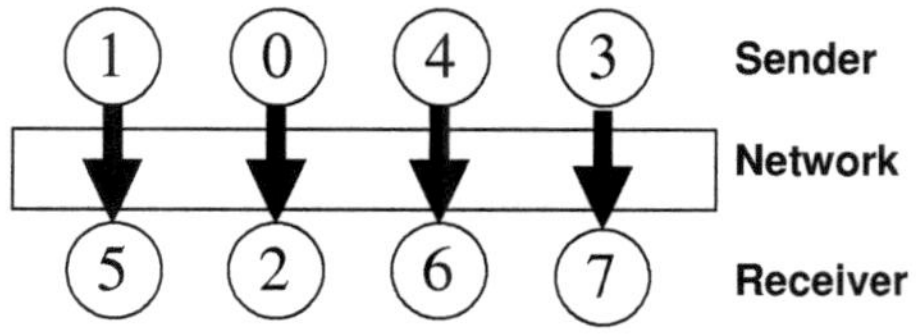

Fig. 2. The one–to–one communication pattern of Netgauge

Pattern 1:1. The main purpose of the one–to–one communication pattern, shown in Fig. 2, is to test a bisectional bandwidth[1] of a given network. This pattern communicates data between an even number of Netgauge processes. The setup stage splits all processes randomly into a group of servers and a group of clients of equal sizes. One member of each group gets associated with exactly

[1] The bisection is determined randomly, i.e., client/server pairs are chosen randomly.

one partner of the other group. After these steps the benchmark performs a barrier synchronization and synchronously starts sending data between all pairs of server and client processes. Thus, one may assume that all pairs communicate approximately at the same time. The special case of two communicating processes represents the well–known ping–pong test.

Pattern 1:n and n:1. The communication patterns one–to–many and many–to–one, shown in Fig. 3 perform an asymmetric network benchmark. In contrast to the most conventional approaches many processes send data to a single process or a single process sends data to many processes. This pattern is most suitable to determine the behavior of the network's flow control or congestion control. Netgauge randomly selects one of the processes to be the server and the remaining processes (n) work as clients. Before the actual test starts, a barrier synchronization is done. All client processes start sending data to the server process which receives the entire data volume of the clients and sends a reply as soon as it received all data. The current implementation uses non-blocking communication to ensure a realistic benchmark. The clients measure the time of this operation. In each subsequent round, more data is sent, up to the user specified limit. Thus, this time measurement allows the user to rate the efficiency of flow control and congestion control.

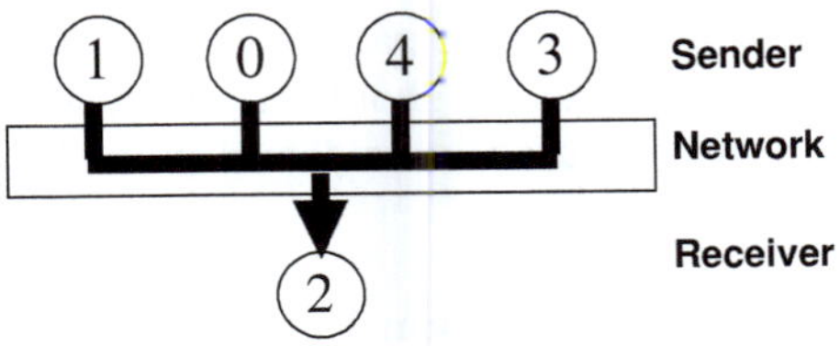

Fig. 3. The many–to–one communication pattern of Netgauge

Network Model Parametrization Patterns

Model parametrization patterns are much more complex than microbenchmark patterns. They use elaborate methods to assess parameters of different network models. They are not described in detail due to space restrictions but references to the original publications which describe the methodology in detail are provided.

Pattern LogGP. The LogGP pattern implements the LogGP parameter assessment method described in [19] for blocking communication. This method uses a mixture of 1:1 and 1:n benchmarks with inter-message delay to determine the parameters g, G and o as exactly as possible. The parameter L can not be measured exactly (cf. [19]).

Pattern pLogP. The pLogP pattern implements the measurement of o_s and o_r in terms of Kielmann's pLogP model. Details about the measurement method can be found in [4].

2.5 Available Communication Modules

Two-sided Communication Modules

Two-sided communication modules implement two-sided communication protocols where both, the sender and the receiver are involved in the communication. The protocol is relatively simple and maps easily to Netgauge's internal communication interface. The receiver has to post a receive and the sender a matching send. It is important to mention that the simplified module interface does not offer message tagging, i.e., the message ordering determines the matching at the receiver.

Module MPI. Netgauge is able to use the well–known Message Passing Interface to perform the actual data transmission. Since MPI is frequently used in parallel programs this feature is useful for many users. The MPI module of Netgauge makes use of the conventional send and receive calls, known as point–to–point communication in the MPI specification. Blocking semantics are supported by calls to `MPI_Send()` and `MPI_Recv()`. The functions `MPI_Isend()` and `MPI_Irecv()` are used to implement the non–blocking interface. Because of its simplicity and complete implementation the MPI module serves as baseline for implementing other network modules. The semantics of the corresponding module functions of Netgauge are very close to those of MPI.

Socket-Based Modules. Several modules supporting standard Unix System V socket-based communications are available. The "TCP" module creates network sockets prepared for the streaming protocol [20]. The "UDP" module implements communication with UDP [21] sockets. It is able to send data up to the maximum datagram size (64 kiB). The "IP" module of Netgauge sends and receives data using raw IP sockets [22]. The raw Ethernet ("ETH") module opens a raw socket and sends crafted Ethernet packets to the wire. The opening of raw sockets requires administrator access. Standard Posix `sendto()` and `recvfrom()` calls are used to transmit data..

Support for two special socket-based low-overhead cluster protocols is also available. The Ethernet Datagram Protocol ("EDP") and the Ethernet Streaming Protocol ("ESP") are described in [23,24] in detail. They aim at the reduction of communication overhead in high-performance computing cluster environments.

This support for many different protocol layers inside the operation system enables the user to determine the quality of the higher-level protocol implementations (e.g., TCP, UDP) by comparing with raw device performance.

Module Myrinet/GM. The Myrinet/GM ("GM") module implements an interface to the Glenn's Messages API of Myrinet [25]. It supports RDMA and send/receive over GM. Different memory registration strategies, such as registering the buffer during the send, copying the data into a pre-registered buffer and sending out of a registered buffer without copying are supported.

Module InfiniBand. The communication module for the native InfiniBand [26] interface ("IB") invokes the API from the OpenFabrics project to transmit data over InfiniBand network adapters. Currently the module supports the four transport types of InfiniBand: Reliable Connection, Unreliable Connection, Reliable Datagram and Unreliable Datagram. The unreliable transport types do not protect against packet loss since additional protocol overhead would affect the performance measurements. Blocking send posts a work queue element according to the InfiniBand specification and polls for completion of this request and returns. The blocking receive function works in a similar way.

One-sided Communication Modules

In one–sided communication the sender directly writes to the memory of the passive receiver. To test for new data, the receiver may poll for a flag, although this action does not belong to the task of pure data transmission. Usually one–sided communication requires synchronization between sender and receiver. The sender has to get the information about the memory area of the receiver. Moreover, some flag may be required to notify the receiver about the completion of the data transfer. The protocol of Netgauge maintains sequence numbers for both the sender and the receiver. The sender increments the sequence count of a public counter variable at the receiver after data transmission. Accordingly, the receiver checks the public sequence count against a local sequence count to test for completion of the data transmission. The protocol is illustrated in Fig. 4.

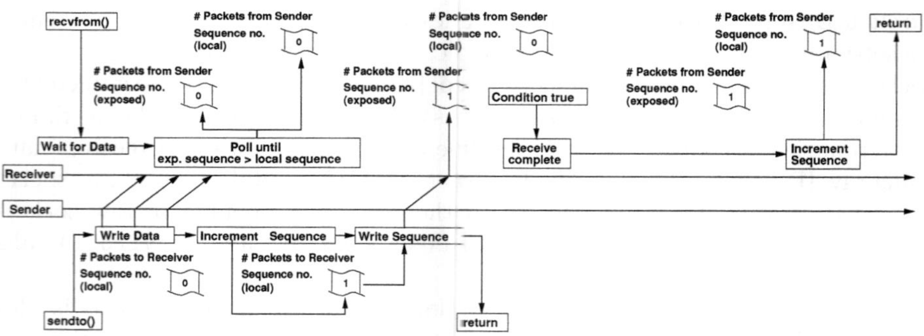

Fig. 4. The protocol for one–sided communication modules

Module ARMCI. The "ARMCI" module uses the ARMCI API [27] to communicate. The blocking send interface utilizes ARMCI_Put() to copy data to the remote side and the receive polls the completion flag.

Module MPI-2 One Sided. MPI 2.0 features the one–sided communication functions to access remote memory without participation of the target process [28,29]. The corresponding Netgauge module performs data transmission using these MPI functions. Any send operation corresponds to an access epoch according to the MPI 2.0 specification. In the same way any receive operation is associated with one exposure epoch.

Module SCI. The Scalable Coherent Interface ("SCI") [30] is a network type providing shared memory access over a cluster of workstations. Data is transmitted by writing and reading to and from remote memory segments mapped into the address space of a process. The SCI module of Netgauge sends data to a receiver by writing to the remote memory.

3 Benchmark Examples and Results

To demonstrate the effectiveness of Netgauge, we present a small selection of benchmark graphs that we recorded with different transport protocols and patterns.

Two test clusters were used to run the benchmarks. On system **A**, a dual Intel Woodcrest 2 GHz cluster with 1 InfiniBand HCA Mellanox MT25204 InfiniHost III and 2 Gigabit Ethernet ports driven by Intel 82563, we run all tests regarding Ethernet, InfiniBand and ARMCI. System **B**, a dual AMD Athlon MP 1.4 GHz with 1 Myrinet 2000 network card and 1 Gigabit Ethernet port SysKonnect SK-98, was used to benchmark Myrinet.

3.1 Benchmarks of the 1:1 Communication Pattern

The one–to–one communication pattern, described in Section 2.4 provides pairwise communication between an arbitrary number of processes.

Fig. 5 shows the latency and bandwidth measurements of an InfiniBand network. The results are given for MPI over InfiniBand and TCP using IPoIB. This benchmark allows to assess the performance and implementation quality of the different protocol layers. The slope of IPoIB transmission falls behind the performance of MPI over InfiniBand using Open MPI 1.1.3.

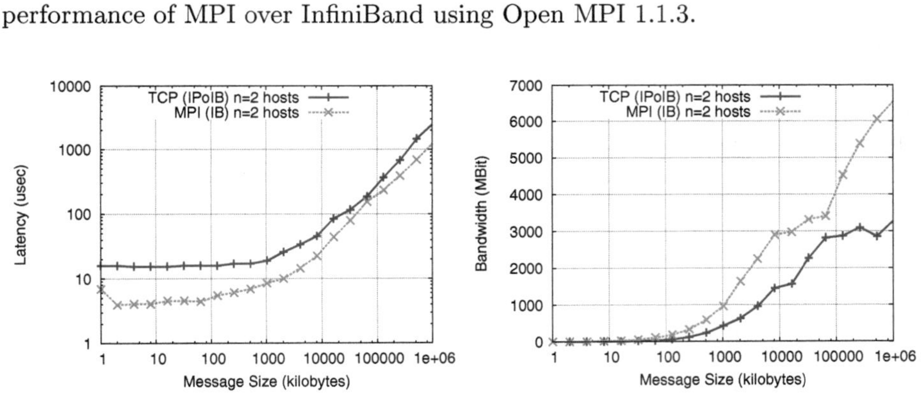

Fig. 5. Comparison of the Latency (left) and Bandwidth (right) of a one–to–one communication via MPI IB and IPoIB

3.2 Benchmarks of the 1:n Communication Pattern

The 1:n communication pattern, described in Section 2.4, was used to compare the flow-control implementation of the specialized ESP [24] protocol with the traditional TCP/IP implementation. The results are shown in Fig. 6.This shows that the ESP implementation achieves a reasonably higher bandwidth and lower latency than TCP/IP for this particular communication pattern.

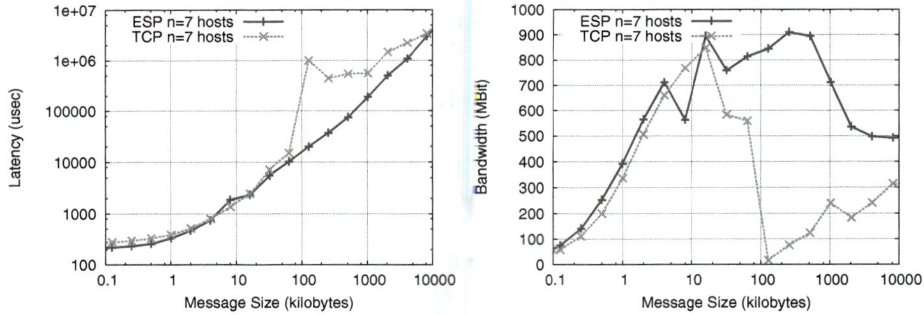

Fig. 6. Comparison of the Latency (left) and Bandwidth (right) of a congested 1:8 communication for TCP and ESP

3.3 Benchmarks of the LogGP Communication Pattern

The LogGP communication pattern implements the LogGP parameter measurement method presented in [19]. The left diagram in the following figures shows the message size dependent overhead of the communication protocol and the right part shows the network transmission graph T. This graph can be used to derive the two parameters g (value at size=1 byte) and G (slope of the graph) and is discussed in detail in [19]. It results from different $PRTT$ measurements, and is calculated as $T(s) = (PRTT(n, 0, s) - PRTT(1, 0, s))/(n - 1)$ where we chose $n = 16$ and s represents the message size.

Fig. 7 compares the InfiniBand IP over IB implementation with Open MPI 1.1/OFED. We see that the overhead is as high as the network transmission time. This is due to our use of blocking communication to benchmark this results. Two different protocol regions for the MPI implementation and the pattern can be seen ($g = 1.82, G = 0.0012$ for $s < 12kiB$ and $g = 19.75, G = 0.0016$ for $s > 12kiB$). The IPoIB implementation results in $g = 7.79$, $G = 0.0061$.

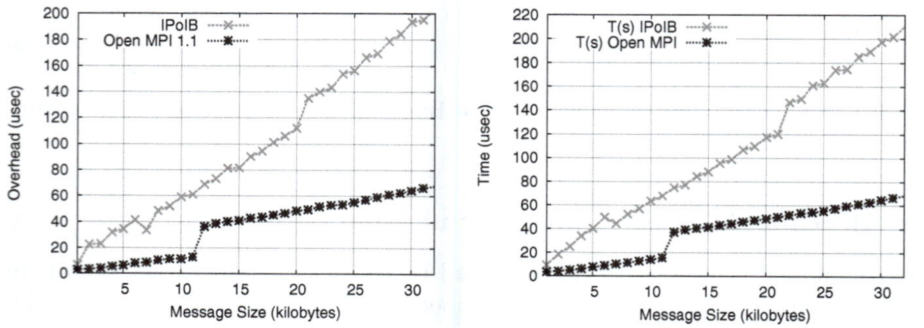

Fig. 7. LogGP graphics for different InfiniBand based transport protocols

4 Conclusions and Future Work

We introduced the Netgauge network performance analysis tool with support for many different network communication protocols. The modular structure and clear interface definition allows for an easy addition of new network modules to support additional protocols and benchmarks. Netgauge offers an extensible set of different simple and complex network communication patterns. The large number of protocol implementations enables a comparison between different transport protocols and different transport layers. Different benchmark examples and other tests demonstrated the broad usability of the Netgauge tool.

We are planning to add new communication modules for different networks and to design more complex communication patterns.

Acknowledgments

Pro Siobhan. The authors want to thank Matthias Treydte, Sebastian Rinke, René Oertel, Stefan Wild, Robert Kullmann, Marek Mosch and Sebastian Kratzert for their contributions to the Netgauge project and Laura Hopkins for editorial comments. This work was partially supported by a grant from the Saxon Ministry of Science and the Fine Arts, a grant from the Lilly Endowment and a gift the Silicon Valley Community Foundation on behalf of the Cisco Collaborative Research Initiative.

References

1. Pallas GmbH: Pallas MPI Benchmarks - PMB, Part MPI-1. Technical report (2000)
2. Culler, D., Karp, R., Patterson, D., Sahay, A., Schauser, K.E., Santos, E., Subramonian, R., von Eicken, T.: LogP: towards a realistic model of parallel computation. In: Principles Practice of Parallel Programming, pp. 1–12 (1993)
3. Alexandrov, A., Ionescu, M.F., Schauser, K.E., Scheiman, C.: LogGP: Incorporating Long Messages into the LogP Model. Journal of Parallel and Distributed Computing 44(1), 71–79 (1995)
4. Kielmann, T., Bal, H.E., Verstoep, K.: Fast Measurement of LogP Parameters for Message Passing Platforms. In: IPDPS 2000. Proceedings of the 15 IPDPS 2000 Workshops on Parallel and Distributed Processing, London, UK, pp. 1176–1183. Springer, Heidelberg (2000)
5. Culler, D., Liu, L.T., Martin, R.P., Yoshikawa, C.: LogP Performance Assessment of Fast Network Interfaces. IEEE Micro (February 1996)
6. Iannello, G., Lauria, M., Mercolino, S.: Logp performance characterization of fast messages atop myrinet (1998)
7. Bell, C., Bonachea, D., Cote, Y., Duell, J., Hargrove, P., Husbands, P., Iancu, C., Welcome, M., Yelick, K.: An Evaluation of Current High-Performance Networks. In: IPDPS 2003. Proceedings of the 17th International Symposium on Parallel and Distributed Processing, p. 28.1. IEEE Computer Society, Washington, DC, USA (2003)

8. Iancu, C., Husbands, P., Hargrove, P.: Hunting the overlap. In: PaCT 2005. Proceedings of the 14th International Conference on Parallel Architectures and Compilation Techniques (PACT 2005), pp. 279–290. IEEE Computer Society, Washington, DC, USA (2005)
9. III, J.W., Bova, S.: Where's the Overlap? - An Analysis of Popular MPI Implementations (1999)
10. Lawry, W., Wilson, C., Maccabe, A.B., Brightwell, R.: Comb: A portable benchmark suite for assessing mpi overlap. In: CLUSTER 2002. 2002 IEEE International Conference on Cluster Computing, Chicago, IL, USA, September 23-26, 2002, pp. 472–475. IEEE Computer Society Press, Los Alamitos (2002)
11. Turner, D., Chen, X.: Protocol-dependent message-passing performance on linux clusters. In: CLUSTER 2002. Proceedings of the IEEE International Conference on Cluster Computing, p. 187. IEEE Computer Society, Washington, DC, USA (2002)
12. Turner, D., Oline, A., Chen, X., Benjegerdes, T.: Integrating new capabilities into netpipe. In: Dongarra, J.J., Laforenza, D., Orlando, S. (eds.) Recent Advances in Parallel Virtual Machine and Message Passing Interface. LNCS, vol. 2840, pp. 37–44. Springer, Heidelberg (2003)
13. Pakin, S.: Reproducible network benchmarks with conceptual. In: Danelutto, M., Vanneschi, M., Laforenza, D. (eds.) Euro-Par 2004. LNCS, vol. 3149, pp. 64–71. Springer, Heidelberg (2004)
14. Pakin, S.: Conceptual: a network correctness and performance testing language. In: Proceedings 18th International Parallel and Distributed Processing Symposium 2004, pp. 79–89. IEEE, Los Alamitos (2004)
15. Chandrasekaran, B., Wyckoff, P., Panda, D.K.: Miba: A micro-benchmark suite for evaluating infiniband architecture implementations. Computer Performance Evaluation / TOOLS, pp. 29–46 (2003)
16. Gabriel, E., Fagg, G.E., Bosilca, G., Angskun, T., Dongarra, J.J., Squyres, J.M., Sahay, V., Kambadur, P., Barrett, B., Lumsdaine, A., Castain, R.H., Daniel, D.J., Graham, R.L., Woodall, T.S.: Open MPI: Goals, Concept, and Design of a Next Generation MPI Implementation. In: Proceedings, 11th European PVM/MPI Users' Group Meeting, Budapest, Hungary (September 2004)
17. Intel Corporation: Intel Application Notes - Using the RDTSC Instruction for Performance Monitoring. Technical report, Intel (1997)
18. Squyres, J.M., Lumsdaine, A.: A Component Architecture for LAM/MPI. In: Dongarra, J.J., Laforenza, D., Orlando, S. (eds.) Recent Advances in Parallel Virtual Machine and Message Passing Interface. LNCS, vol. 2840, pp. 379–387. Springer, Heidelberg (2003)
19. Hoefler, T., Lichei, A., Rehm, W.: Low-Overhead LogGP Parameter Assessment for Modern Interconnection Networks (March 2007)
20. Postel, J.: Transmission Control Protocol. RFC 793 (Standard), Updated by RFC 3168 (September 1981)
21. Postel, J.: User Datagram Protocol. RFC 768 (Standard) (August (1980)
22. Postel, J.: Internet Protocol. RFC 791 (Standard) Updated by RFC (1349) (September 1981)
23. Reinhardt, M.: Optimizing Point-to-Point Ethernet Cluster Communication. Master's thesis, TU-Chemnitz (2006)
24. Hoefler, T., Reinhardt, M., Mietke, F., Mehlan, T., Rehm, W.: Low Overhead Ethernet Communication for Open MPI on Linux Clusters. CSR-06(06) (July 2006)
25. Myricom, Inc.: The GM Message Passing System. Myricom, Inc. (2000)

26. The InfiniBand Trade Association: Infiniband Architecture Specification Volume 1, Release 1.2. InfiniBand Trade Association (2004)
27. Nieplocha, J., Carpenter, B.: ARMCI: A portable remote memory copy library for distributed array libraries and compiler run-time systems. In: Rolim, J.D.P. (ed.) Parallel and Distributed Processing. LNCS, vol. 1586, p. 533. Springer, Heidelberg (1999)
28. Gropp, W., Lusk, E., Thakur, R.: Using MPI–2 Advanced Features of the Message–Passing Interface. MIT Press, Cambridge (1999)
29. Message Passing Interface Forum: MPI-2: Extensions to the Message-Passing Interface. Technical Report, University of Tennessee, Knoxville (1997)
30. IEEE standard: IEEE standard for Scalable Coherent Interface (1993)

Towards a Complexity Model for Design and Analysis of PGAS-Based Algorithms

Mohamed Bakhouya[1], Jaafar Gaber[2], and Tarek El-Ghazawi[1]

[1] Department of Electrical and Computer Engineering
High Performance Computing Laboratory
The George Washington University
{bakhouya,tarek}@gwu.edu
[2] Universite de Technologies de Belfort-Montbeliard
gaber@utbm.fr

Abstract. Many new Partitioned Global Address Space (PGAS) programming languages have recently emerged and are becoming ubiquitously available on nearly all modern parallel architectures. PGAS programming languages provide ease-of-use through a global shared address space while emphasizing performance by providing locality awareness and a partition of the address space. Examples of PGAS languages include the Unified Parallel C (UPC), Co-array Fortran, and Titanium languages. Therefore, the interest in complexity design and analysis of PGAS algorithms is growing and a complexity model to capture implicit communication and fine-grain programming style is required. In this paper, a complexity model is developed to characterize the performance of algorithms based on the PGAS programming model. The experimental results shed further light on the impact of data distributions on locality and performance and confirm the accuracy of the complexity model as a useful tool for the design and analysis of PGAS-based algorithms.

1 Introduction

PGAS implicit communication and fine-grain programming style make application performance modelling a challenging task [2], [5]. As stated in [2] and [5], most of the efforts have gone into PGAS languages design, implementation and optimization, but little work has been done for the design and analysis of PGAS-based algorithms. Therefore, a complexity model for PGAS algorithms is an interesting area for further research and design, and this field, to the best of our knowledge, had not yet been adequately defined and addressed in the literature. The challenge is to design, analyze, and evaluate PGAS parallel algorithms before compiling and running them on the target platform.

Recall that in sequential programming, assuming that sufficient memory is available, execution time of a given algorithm is proportional to the work performed (i.e., number of operations). This relationship between time and work makes the performance analysis and comparison very simple. However, in parallel performance models proposed in literature, other platform-dependent parameters, such as communication overhead, are used in developing parallel programs

R. Perrott et al. (Eds.): HPCC 2007, LNCS 4782, pp. 672–682, 2007.

[8]. By changing these parameters, developers can predict the performance of the programs on different parallel machines. However, these parameters are not program-dependent and considering them in the design phase can complicate the analysis process. In other words, having platform-dependent parameters in the model makes it quite difficult to obtain a concise analysis of algorithms. To meet this requirement, complexity models are useful for programmers to design, analyze, and optimize their algorithms in order to get better performance. They also provide a general guideline for programmers to choose the better algorithm for a given application.

This paper addresses a complexity model for the analysis of the intrinsic efficiency of the PGAS algorithms with disregard to machine-dependent factors. The aim of this model is to help PGAS parallel programmers understand the behavior of their algorithms and programs, making informed choices among them, and discovering undesirable behaviors leading to poor performance.

The rest of the paper is organized as follows. In section 2, we present related work. Section 3 presents an overview of PGAS programming model by focusing on one language implementing it, UPC. In section 4, the complexity model for PGAS-based algorithms is presented. Experimental results are given in section 5. Conclusion and future work are presented in section 6.

2 Related Work

There has been a great deal of interest in the development of performance models, also called abstract parallel machine models, for parallel computations [12]. The most popular is the Parallel Random Access Machine (PRAM) model, which is used for both shared memory and network-based systems [10]. In shared memory systems, processors execute in concurrently and communicate by reading and writing locations in shared memory spaces. In network-based systems, processors coordinate and communicate by sending messages to their neighbors in the underlying network (e.g., array, hypercube, etc) [6], [10]. While the PRAM model does not consider the communication cost [10], it is considered by many studies to be high level and unable to accurately model parallel machines. New alternatives such as Bulk Synchronous Parallel (BSP) [9], LogP and its variants [4],[8], and Queuing Shared Memory (QSM) [10] have been proposed to capture the communication parameters. These models can be grouped into two classes: shared memory-based models and message passing-based models. LogP and its variants, and BSP are message passing-based models that directly abstract distributed memory and account for bandwidth limitations. Queuing Shared Memory (QSM) is a shared memory-based model [10], [11]. Despite its simplicity it uses only machine-dependent parameters, and in addition the shared memory is global and not partitioned, i.e. it provides no locality awareness.

Recently, approaches to predict the performance of UPC programs and compiler have been proposed in [2] and [5]. In [5], authors include machine-dependent parameters in their model to predict UPC program performance. The model proposed in [2], offers a more convenient abstraction for algorithms design and development. More precisely, the main objective of this modelling approach is to help PGAS designers to select the most suitable algorithm regardless of the implementation or the target machine. The model presented in this paper extends the modelling approach proposed in [2] by deriving analytical expressions to measure the complexity of PGAS algorithms. As stated in [2], considering explicitly platform parameters, such as the latency the bandwidth, and the overhead, during the design process could lead to platform-dependent algorithms. Our primary concern in this paper is a complexity model for the design and analysis of PGAS algorithms.

3 PGAS Programming Model: An Overview

PGAS programming model uses fine-grain programming style that makes application performance modelling and analysis a challenging task. Recall that unlike coarse-grain programming model where large amounts of computational work are done between communication events, fine-grain programming style relatively small amounts of computational work can be done between communication events. Therefore, if the granularity is too fine it is possible that the overhead required for communications takes longer than the computation. PGAS programming model provides two major characteristics namely data partition and the locality that should be used to reduce the communication overhead and get better performance. UPC is one of partitioned global address space programming language based on C and extended with global pointers and data distribution declarations for shared data [1].

A PGAS-based algorithm depends on the number of threads and how they are accessing the space [1], [2]. A number of threads can work independently in a Simple Program Multiple Data (SPMD) model. Threads communicate through the shared memory and can access shared data while a private object may be accessed only by its own thread. The PGAS memory model, used in UPC for example, supports three different kinds of pointers: private pointers pointing to the shared address space, pointers living in shared space that also point to shared data, and private pointers pointing to data in the thread's own private space. The speed of local shared memory accesses will be slower than that of private accesses due to the extra overhead of shared-to-local address translation [5]. Also, remote accesses in turn are significantly slower because of the network overhead and address translation process.

According to these PGAS programming features, the fine-grained programming model is simple and easy to use. Programmers need to only specify the data to be distributed across threads and reference them through special global pointers. However, a PGAS program/algorithm can be considered efficient (compared to another algorithm) if it achieves the following objectives [2]. The first objective is to minimize the inter threads communication overhead incurred by remote memory accesses to shared space. The second objective is to maximize the efficient use of parallel threads and data placement or layout together with data locality (i.e., the tradeoff between the number of allocated threads and the communication overhead). The third objective is to minimize the overhead caused by the synchronization points.

Recently, all efforts are focused on compiler optimization, i.e. without the control of the programmer, in order to increase the performance of the PGAS parallel programming language such as UPC. In [3] and [7], many optimization suggestions have been proposed for incorporation into UPC compilers, such as message aggregation, privatizing local shared accesses, and overlapping computation and communication to hide the latencies associated with remote shared accesses.

4 A Complexity Model for PGAS Algorithms

A PGAS algorithm consists of a number of threads that use a SPMD model to solve a given problem. Each thread has its own subset of shared data and can coordinate and communicate with the rest of threads by writing and reading remote shared objects or data to perform its computation. In addition, each thread performs local computations and memory requests to its private space. The algorithm complexity can be expressed using algorithm-dependent parameters to measure the number of basic operations and the number of read/write requests. More precisely, the algorithm complexity can be expressed using T_{comp} and T_{comm}, where T_{comp} denotes the computational complexity, and T_{comm} the number of read and write requests. The computational complexity depends on the problem size, denoted by N, and the number of threads, T. For example, in the case where all threads execute the same workload (i.e., the workload is uniform), the number of computation operations for all threads is similar. However, in irregular applications, certain threads could perform more or less computational operations. To includes the idle time induced by waiting, e.g. for other threads, we consider that T_{comp} is the maximum of T^i_{comp} for $1 \leq i \leq T$, where T is the number of threads. It is worth notice that each thread should be executed by one processor; the number of threads T is equal to the number of processors P. In order to calculate T^i_{comp}, we consider that a PGAS algorithm can be composed of a sequence of phases eventually separated by synchronization points. At a given phase, a thread can compute only or compute and communicate. Therefore, alike sequential programming, the computational complexity of a thread i in all elementary phases j, $j \in \phi$, is $T^i_{comp} = max_{j=1:\phi}(T^i_{comp}(j))$, where ϕ is the number of all elementary phases. The computational complexity T_{comp} of the algorithm is the highest computational complexity of all threads and is determined as follows: $T_{comp} = max_{i=1:T}(T^i_{comp})$. The complexity is dominated by the thread that has the maximum amount of work. Unlike the computational complexity, the communication complexity of a thread i is the sum (over all phases) of $T^i_{comm}(j)$ as follows: $T_{comm} = \sum_{j=1}^{\phi} T^i_{comm}(j)$. The communication complexity of the algorithm is the highest communication complexity overall threads and is determined as follows: $T_{comm} = max_{i=1:T}(T^i_{comm})$. T_{comm} is an upper bound on the number of memory requests made by a single thread.

The communication complexity $T^i_{comm}(j)$ at each phase j and for each thread i, depends on the number of requests to private and shared memory spaces, called the pointer complexity [2]. The pointer complexity is defined to be the total number of pointer-based manipulations (or references) used to access data. There are three sorts of pointers in a PGAS algorithm represented by $N_p^i(j)$, the number of references to the private memory space, $N_\ell^i(j)$ the number of references to the local shared memory space, and $N_r^i(j)$ the number of references to remote shared memory spaces. Since accesses to remote shared memory spaces are more expensive than accesses to private and local shared memory spaces [1], [2], [5], $N_p^i(j)$ and $N_\ell^i(j)$ can be neglected due to their insignificance relative to the cost of remote references accesses to remote shared spaces $N_r^i(j)$. Therefore, the communication complexity depends mainly on the number of references to remote shared spaces, and hence we have: $T^i_{comm}(j) = O(N_r^i(j))$. This number of remote references (i.e., read and write requests) can be easily and directly computed from the number and layout of data structures declared and used in the algorithm. More precisely, it depends on the following parameters: the number of shared Data elements D declared and used in the algorithm, the number of Local shared data elements L to a thread ($L \leq D$), the number of threads T, and the size N of the shared Data structures . In what follows, we consider the worst case when any computational operation could equally involve global or local shared requests to shared memory spaces. Each thread also needs to write or read from other threads' partitions. Using these parameters, the number of remote references represents the communication complexity (i.e., number of read and write requests), at each phase j, as follows:

$$T^i_{comm}(j) = O(N_r^i(j)) = O(\frac{(D^i_j - L^i_j)(T-1)}{T} T^i_{comp}(j)) \tag{1}$$

$T^i_{comp}(j)$ captures the number of computational operations, at the phase j, and depends only on the average number of elements in shared data structures N and the number of threads T. D^i_j is the number of all shared pointers to remote spaces (i.e., local spaces of other threads) used in the algorithm, for the thread i in the phase j. L^i_j captures the number of shared references (i.e., pointers) to a local space of the thread i in the phase j.

Let us consider that $\alpha^i_j = \frac{D^i_j - L^i_j}{D-L}$. Equation 1, which determines the communication complexity of thread i at each phase j, can be rewritten as follows:

$$T^i_{comm}(j) = O(\frac{\alpha^i_j (D-L)(T-1)}{T} T_{comp}(j)) \tag{2}$$

To estimate the communication complexity of the algorithm, let us now consider, that a certain thread i has the highest computational complexity $T_{comp} = T^i_{comp}(k)$ overall threads and overall phases k. According to equation 2, the communication complexity of the algorithm is:

$$T^i_{comm} = \max_{i=1:T}(\sum_{j=1}^{\phi} T^i_{comm}(j)) = O(\frac{\alpha(D-L)(T-1)}{T} T_{comp}) \tag{3}$$

where $\alpha = \max_{i=1:T}(\sum_{j=1}^{\phi} \alpha^i_j)$. It should be noted here that the computation complexity depends all time on the size of the problem N and the number of threads T, $T_{comp} = f(N, T)$. In addition, the communication complexity depends on the number of shared data elements D declared and used in the algorithm, the number of local shared data elements L to a thread, the size of the shared data structures N, the number of threads T (T is equal to the number of processors P), and the parameter α, $f(D, L, N, T, \alpha)$, where α is a function of T. For example, let us consider the following $N \times N$ matrix multiplication algorithm, $C = A \times B$:

Matrix multiplication algorithm

Input: a $N \times N$ matrix A and a $N \times N$ matrix B
Output: the $N \times N$ matrix $C = A \times B$
Data Distribution: A , B, and C are distributed in round-robin using four cases
//Loop through rows of A with affinity (e.g., i)
For $i \leftarrow 1$ and $affinity = i$ to N parallel Do
 For $j \leftarrow 1$ to N Do *// Loop through columns of B*
  $C(i, j) \leftarrow 0$
  For $k \leftarrow 1$ to N Do *// perform the product of a row of A and a column of B*
   $C(i, j) \leftarrow C(i, j) + A(i, k) \times B(k, j)$
  End for k
 End for j
End for i

This algorithm takes two matrix A and B as an input and produces a matrix C as an output. These matrices are $N \times N$ two-dimensional arrays with N rows and N columns. The data distribution allows us to define how these matrices will be partitioned between threads. Four cases of data distribution are considered in this example and will be described in section 5. The first loop allows the distribution of independent work across the threads, each computing a number of iterations represented by the field $affinity$. This field determines which thread executes a given iteration. In this example, the affinity i indicates that the iteration i will be performed by thread (i mod T). Using this field, iterations will be distributed across threads in round-robin manner and each thread will process only the elements that have affinity to it avoiding costly remote requests to shared spaces. This parallel loop can be translated into the parallel construct of the considered PGAS language, UPC for example [1].

In this algorithm, the computational complexity T_{comp} requires $O(\frac{N^3}{T})$ multiplication and addition operations, and the communication complexity T_{comm} requires $O(\frac{\alpha(D-L)(T-1)}{T^2} N^3)$ requests to remote shared spaces, where D=3, and $L \in \{0, 1, 2, 3\}$. The value of L depends on the case of data distribution considered (see section 5). According to this analysis, the communication complexity of a given PGAS algorithm achieves a lower bound when $L = D$, which means that threads compute on their own data without performing any communication to access remote ones. In other words, a communication complexity of a PGAS algorithm achieves its lower bound when $L = 0$; all data are accessed with remote shared pointers.

The speedup of a PGAS algorithm, denoted by S_{PGAS}, is defined as the ratio between the time taken by the most efficient sequential algorithm, denoted by

T_{seq}, to perform a task using one thread and the time, denoted by T_{par}, to perform the same task using T threads (i.e., $T = P$). This speedup can be estimated as follows:

$$S_{PGAS} = \frac{T_{seq}}{T_{par}} = O(\frac{T_{seq}}{1 - \alpha(D-L)\frac{(T-1)}{T}T_{comp}}) \quad (4)$$

Since the most optimal sequential algorithm yields to $T_{comp} \leq \frac{T_{seq}}{T}$, the speedup, given in the equation 4, can be rewritten as follows:

$$S_{PGAS} = O(\frac{T}{1 + a(T-1)}) \quad (5)$$

where $a = \frac{\alpha(D-L)}{T}$ is a key parameter that emphasizes the influence of the locality degree of the algorithm in function of the number of threads, i.e., number of processors. According to this equation, the PGAS speedup is optimal, i.e., linear, when S_{PGAS} is close to T (i.e., $S_{PGAS} = O(T)$). This linear speedup can be obtained when $D = L$; i.e., there are no remote requests to shared spaces. Therefore the designer should simultaneously maximize the locality degree L, and minimize the usage of remote shared pointers represented by α. When T approaches infinity, $a(T-1)$ is bounded by $a(D-L)$, where α is not constant, but depends on the number of threads. Hence, even if we increase the number of threads, in parallel the number of processors, to run the algorithm, the performance is limited by the impact of the number of requests to remote shared memory spaces a. Therefore, the designer of PGAS algorithms should attempt to reduce this parameter to the smallest possible value. It should be noted also that since the model considers the asymptotic complexity of the computation and the communication, the overlapping mechanism that allows the computation to proceed without waiting for the entire data transfer to be completed is also captured by the model.

5 Experimental Results

In this section, to validate the performance model presented in this paper, we have implemented a UPC program to perform matrix multiplication algorithm with different data distribution schemes. The matrix multiplication is selected because it is simple to show the effectiveness of the model. In addition, matrix multiplication represents one of the most important linear algebra operations in many scientific applications..

The experiments were done using the GCC UPC toolset running on Origin 2000. The SGI Origin 2000 platform consists of 16 processors and 16GB of SDRAM interconnected with a high speed cache coherent Non Uniform Memory Access (ccNUMA) link in hypercube architecture. The GCC UPC toolset provides a compilation and execution environment for programs written in UPC language.

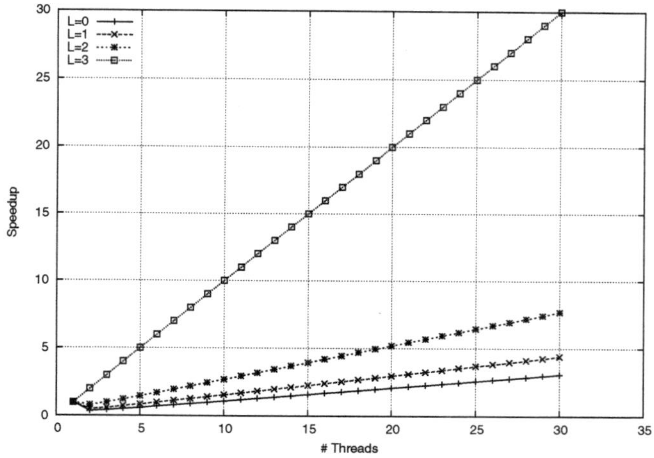

Fig. 1. The theoretical speedup vs. L

In this experimentation, three data structures A, B, and C ($D = 3$) are considered. We consider the case where all matrices are of size ($N = 128$) and the number α is equal to 3. Four different data distribution cases ($L = 0$, $L = 1$, $L = 2$, and $L = 3$) are also considered in these experiments. In the first case, $L = 0$, matrices A, B, C are partitioned among T threads (i.e., processors) by assigning, in round robin, each thread $\frac{N}{T}$ columns of A, B, and C respectively. More precisely, elements of these matrices will be distributed acroses the threads element-by-element in round-robin fashion. For example, the first element of the matrix A is created in the shared space that has affinity to thread 0, the second element in the shared space that has affinity to thread 1, and so on. In the second case, $L = 1$, the compiler partitions computation among T threads by assigning each thread $\frac{N}{T}$ rows of A and $\frac{N}{T}$ columns of B and C. More precisely, the elements of matrices A and B are distributed element-by-element in round-robin fashion, i.e., each thread gets one column in round robin fashion. At the end, each thread up with $\frac{N}{T}$ columns of B and C, where ($N \bmod T = 0$). The elements of the matrix A will be distributed across the threads N-elements by N-elements in round-robin fashion. In the case where $L = 2$, the compiler partitions computation among T threads by assigning each threads $\frac{N}{T}$ rows of A, and C, and $\frac{N}{T}$ columns of B. This case is similar to the second case except that the elements of the matrix C will be distributed across the threads N-elements by N-elements in round-robin fashion. The last case, $L = 3$, is similar to the third case with the exception that each thread has the entire B.

Recall that from the model, the designer should maximize the locality degree L, and minimize α, i.e. minimize the usage of remote shared pointers. Figure 1 presents the theoretical speedup, calculated from the model (eq. 5), in function of the locality degree, $L = 0$, $L = 1$, $L = 2$, and $L = 3$ ($\alpha = 3$). This figure shows that as we increase L, the speedup increases and tends to become linear when $D = L$, i.e., good data locality. More precisely, the algorithm (matrix multiplication) without remote communication ($L = 3$) performs better than the algorithm with

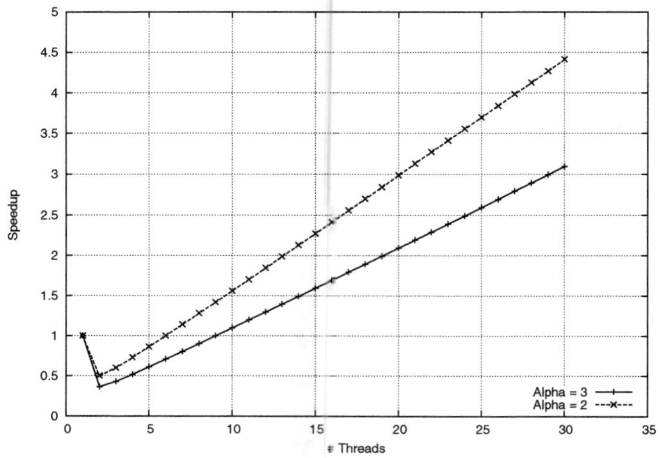

Fig. 2. The theoretical speedup vs. α

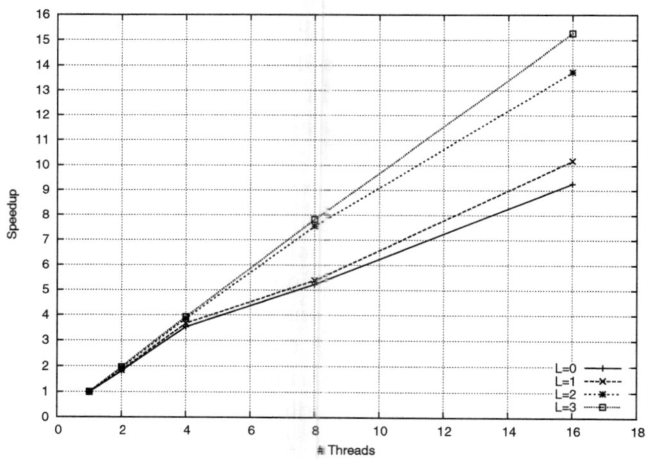

Fig. 3. The experimental speedup vs. L

a minor remote communication ($L = 2$) that performs better than the algorithm with an intermediate remote communication ($L = 1$). This last case performs better than the algorithm with a larger remote communication ($L = 0$).

According to the model, the designer should also attempt to reduce the parameter α to the smallest possible value. To illustrate this point, let us consider the matrix multiplication algorithm with three data structures ($D = 3$), and L is equal to 0. Figure 2 presents the theoretical speedup in function of the number α. This figure shows that as we decrease the the usage of remote shared pointers (i.e., from $\alpha = 3$ to $\alpha = 2$), the speedup increases.

The objective of the experimentation is to prove this statement, which is given from the complexity model. Each thread is executed in one processor. The

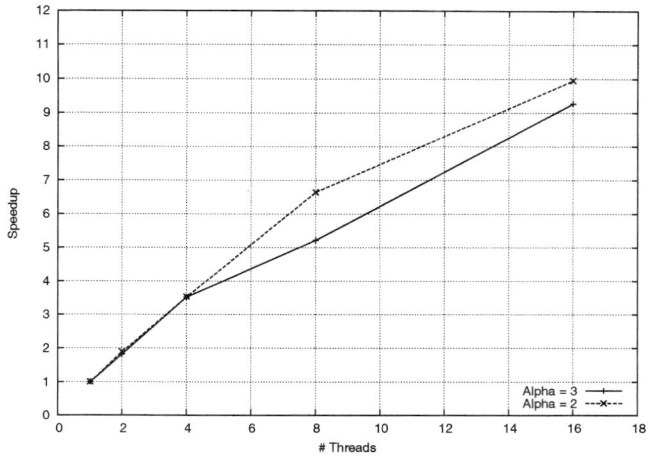

Fig. 4. The experimental speedup vs. α

results depicted in figure 3 illustrate that as we increase L, the speedup increase to become linear when $L = D$. We can see also in figure 4 that decreasing the usage of remote shared pointers (i.e., from $\alpha = 3$ to $\alpha = 2$), the speedup increases. It should be noted that the objective of these experiments is not to compare the performance of this program according to the literature but to show the following behavior: as we increase L and decrease α there is an increase in the speedup.

These primary experimental results corroborate the result obtained from the complexity model described above and show that to improve the algorithms' performance, by decreasing the communication cost, the programmer must increase as much as possible the value of the locality parameter L and minimize the usage of remote shared pointers represented by α.

6 Conclusion and Future Work

In this paper, we present a performance model based on the complexity analysis that allows programmers to design and analyze their algorithms independent of the target platform architecture. Given an algorithm, we have shown that we can extract the program-dependent parameters to calculate the computation and the communication complexity. According to this model, to obtain algorithms that perform well, the remote communication overhead has to be minimized. The choice of a good data distribution is of course the first step to reduce the number of requests to remote shared spaces. Consequently, the programmer is able to ensure good data locality and obtain better performance using this methodology as a tool to design better algorithms. The utility of this model was demonstrated through experimentation using matrix multiplication program under different data distribution schemes. According to the complexity model and the experimental results, the most important parameters are the number D and the size N

of data structures, the locality degree L, the number α used in the algorithm, and the number of threads. Therefore, the algorithm designer should simultaneously minimize D and α, and maximize L to get better performance.

Future work addresses additional experiments with larger applications and other platforms. The objective is to provide a tool for the comparison of alternate algorithms to the same application without necessarily resorting to an actual implementation. In addition, we will extend this complexity model to predict the computational and the communication times by including machine-dependent parameters such as the latency and the overhead.

References

1. El-Ghazawi, T., Carlson, W., Sterling, T., Yelick, K.: UPC: Distributed Shared Memory Programming. Book. John Wiley and Sons Inc., New York (2005)
2. Gaber, J.: Complexity Measure Approach for Partitioned Shared Memory Model, Application to UPC. Research report RR-10-04. Universite de Technologie de Belfort-Montbeliard (2004)
3. Cantonnet, F., El Ghazawi, T., Lorenz, P., Gaber, J.: Fast Address Translation Techniques for Distributed Shared Memory Compilers. In: International Parallel and Distributed Processing Symposium IPDPS 2006 (2006)
4. Cameron, K.W., Sun, X.-H.: Quantifying Locality Effect in Data Access Delay: Memory logP. In: IPDPS 2003. Proc. of the 17th International Symposium on Parallel and Distributed Processing, p. 48.2 (2003)
5. Zhang, S., Seidel, R.Z.: A performance model for fine-grain accesses in UPC. In: 20th International Parallel and Distributed Processing Symposium, p. 10 (2006) ISBN: 1-4244-0054-6
6. Juurlink Ben, H.H., Wijshoff Harry, A.G.: A quantitative comparison of parallel computation models. In: Proc. 8th ACM Symp. on Parallel Algorithms and Architectures, pp. 13–24. ACM Press, New York (1996)
7. Chen, W-Y., Bonachea, D., Duell, J., Husbands, P., Iancu, C., Yelick, K.: A Performance Analysis of the Berkley UPC Compiler. In: Annual International Conference on Supercomputing (ICS) (2003)
8. Culler, D., Karp, R., Patterson, D., Sahay, A., Schauser, K.E., Santos, E., Subramonian, R., von Eicken, T.: LogP: Towards a Realistic Model of Parallel Computation. In: PPOPP 1993: ACM SIGPLAN, pp. 1–12. ACM Press, New York (1993)
9. Gerbessiotis, A., Valiant, L.: Direct Bulk-Synchronous Parallel Algorithms. J. of Parallel and Distributed Computing 22, 251–267 (1994)
10. Gibbons, P.B., Mattias, Y., Ramachandran, V.: Can a Shared-Memory Model Serve as a Bridging Model for Parallel Computation? In: SPAA 1997. 9th Annual ACM Symposium on Parallel Algorithms and Architectures, pp. 72–83. ACM Press, New York (1997)
11. Valiant, L.G.: A Bridging Model for Parallel Computation. Comm. of the ACM 33(8), 103–111 (1990)
12. Maggs, B.M, Matheson, L.R., Tarjan R.E.: Models of Parallel Computation: A Survey and Synthesis. In: Proceeding of the Twenty-Eight Hawaii Conference on System Sciences, vol. 2, pp. 61–70 (1995)

An Exploration of Performance Attributes for Symbolic Modeling of Emerging Processing Devices

Sadaf R. Alam, Nikhil Bhatia, and Jeffrey S. Vetter

Oak Ridge National Laboratory,
Oak Ridge, TN 37831, USA
{alamsr,bhatia,vetter}@ornl.gov

Abstract. Vector, emerging (homogenous and heterogeneous) multi-core and a number of accelerator processing devices potentially offer an order of magnitude speedup for scientific applications that are capable of exploiting their SIMD execution units over microprocessor execution times. Nevertheless, identifying, mapping and achieving high performance for a diverse set of scientific algorithms is a challenging task, let alone the performance predictions and projections on these devices. The conventional performance modeling strategies are unable to capture the performance characteristics of complex processing systems and, therefore, fail to predict achievable runtime performance. Moreover, most efforts involved in developing a performance modeling strategy and subsequently a framework for unique and emerging processing devices is prohibitively expensive. In this study, we explore a minimum set of attributes that are necessary to capture the performance characteristics of scientific calculations on the Cray X1E multi-streaming, vector processor. We include a set of specialized performance attributes of the X1E system including the degrees of multi-streaming and vectorization within our symbolic modeling framework called Modeling Assertions (MA). Using our scheme, the performance prediction error rates for a scientific calculation are reduced from over 200% to less than 25%.

1 Introduction

Computing devices like vector processors [3], homogeneous and heterogeneous multi-core processors [1, 2], Field Programmable Gate Arrays (FPGAs) [16], multi-threaded architecture processors [18] offer the potential of dramatic speedups over traditional microprocessor run times for scientific applications. Performance acceleration on these devices is achieved by exploiting the data, instruction and thread level parallelism of a program. However, an application needs to be mapped to the appropriate computing devices in order to exploit the unique performance enhancing features of a target system. Performance modeling studies have been employed to investigate and to understand this mapping and the achievable performance of an application on a target system. Efforts involved in developing a performance modeling strategy for unconventional and emerging systems are prohibitively expensive. Nevertheless, the high cost of high-performance computing (HPC) systems and the strategic importance of HPC applications make it imperative to predict the performance of these applications on these machines before their deployment.

R. Perrott et al. (Eds.): HPCC 2007, LNCS 4782, pp. 683–694, 2007.

In this study, we present our performance modeling approach for one such unconventional HPC architecture: the Cray X1E [3]. A Cray X1E processing unit is composed of vector processors and a complex memory subsystem. Using a technique called *multi-streaming*, groups of four X1E vector processors can cooperate to execute outer loop iterations; each group of four processors is called a Multi-Streaming Processor (MSP). We evaluated our approach within the Modeling Assertions (MA) framework [9] that allows a user to express and subsequently develop symbolic performance models of workload requirements in terms of an application's input parameters. We define and quantitatively validate a minimal set of performance attributes that are essential in expressing performance models symbolically and predicting runtime performance on the X1E system. As a result, we sacrifice neither the generality of the symbolic models that are developed for a microprocessor based system nor the accuracy of runtime performance predictions. We evaluated our technique using a set of programs from the NAS parallel benchmarks [4]. Using our approach, the runtime error rates are reduced from over 200% to just below 25% for the programs we considered.

The layout of the paper is as follows: Section 2 briefly outlines the related work in the area of performance modeling and prediction for scientific HPC applications. Background of the Cray X1E vector system and our modeling scheme is presented in Section 3. Implementation details are provided in Section 4. Section 5 presents experiments and results. Conclusions and future work are discussed in Section 6.

2 Related Work

Several techniques have been proposed and investigated for predicting the performance of applications on conventional architectures [11, 12, 14, 20]. Here we briefly survey the most recent performance modeling efforts. Snavely *et. al.* [14] predict applications' runtime on conventional processing architectures using an application's memory bandwidth requirements, and processing speed and bandwidth capabilities of the target architecture. The technique relies on obtaining application memory access patterns by collecting instruction traces for memory reference instructions, usually on a traditional microprocessor-based platform. Microprocessor-based modeling techniques have limited applicability for performance modeling of unconventional HPC systems due to the unique features of these architectures and the overheads involved in collecting and analyzing huge amount of trace data. Vendors of unconventional architectures may not provide any memory tracing toolkit for ISA level tracing; no such support is available on the X1E. Yang *et. al.* [20] describe a technique based on partial execution of an application on existing systems and then extrapolation of the results for unconventional architectures. Typically the extrapolation does not take into consideration the unique architectural features that enhance performance on these unconventional architectures, thereby inducing very high runtime error rates. A similar but exhaustive performance modeling approach is presented by Kerbyson *et. al.* [12], which involves manually developing an expert human knowledge base of the applications as well as the target systems. Our scheme combines a code developer's symbolic representation of an application and runtime hardware counter information and systematically feeds back execution-time information to improve model accuracy within the MA framework.

3 Background

3.1 Cray X1/X1E

The Cray X1E is distributed shared memory system with globally addressable memory. The primary functional building block of a X1E is a compute module. A compute module contains four multichip modules (MCMs), local memory, and System Port Channel I/O ports. Each MCM contains two multi-streaming processors (MSPs). Each MSP is comprised of 4 single-streaming processors (SSPs) as shown in Fig. 1. Each SSP contains two deeply-pipelined vector units running at 1.13 GHz and a single scalar unit running at 0.565 GHz. All SSPs within a MSP share a 2MB E-cache and each SSP has a 16KB Data cache and a 16KB instruction cache.

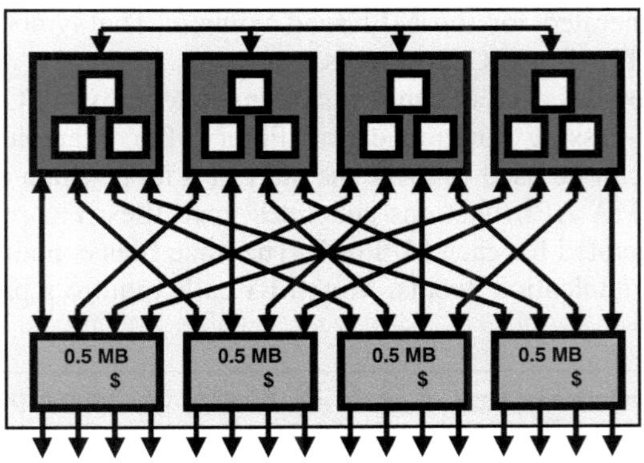

Fig. 1. Inside Cray X1 Multi-streaming processor

The Cray X1E compilers can exploit the data level parallelisms by vectorizing inner loops so they execute in the vector units of an SSP. The compiler can also parallelize outer loops such that the loop's iterations can be executed concurrently on each of the four SSPs within an MSP. Together, these two features have a theoretical peak performance of 18 GFLOPS/MSP. From the memory subsystem point of view, the memory hierarchy is different for scalar and vector memory references. Vector memory references are cached in the E-cache but not in the D-cache. The vector register space acts as a level-1 cache for vector memory references. On the other hand, the scalar memory references are cached in the E-cache as well as the D-cache. The E-cache acts a level-2 cache for scalar memory references.

3.2 The Modeling Assertions (MA) Framework

Because of the limited applicability of conventional modeling techniques for dealing with unconventional architectures, we have devised a modeling scheme that incorporates "application aware" as well as "architecture aware" attributes in model representation. We implement our approach using our Modeling Assertions (MA) framework. MA allows a user to develop hierarchical, symbolic models of applications using code annotation; the MA models can project performance requirements and allow us to conduct sensitivity analysis of workload requirements for future and larger problem instance of an application [10]. The MA models can be incrementally refined based on the empirical data that are obtained from application runs on a target system.

The MA framework has two main components: an API and a post-processing toolset. Fig. 2 shows the components of the MA framework. The MA API is used to annotate the source code. As the application executes, the runtime system captures important information in trace files, primarily to compare runtime values for annotated symbolic expressions to actual runtime data in order to validate symbolic models. These trace files are then post-processed to analyze, and construct models with the desired accuracy and resolution. The post-processing toolset is a collection of tools or Java classes for model validation, control-flow model creation and symbolic model generation. The modeling API is available on Linux clusters with MPICH, IBM pSeries systems and Cray X1E vector systems [3]. The symbolic model shown in the Fig. 2 is generated for the MPI send volume. The symbolic models can be evaluated with MATLAB and Octave tools.

The MA API provides a set of functions to annotate a given FORTRAN or C code with MPI message-passing communication library. For example, ma_loop_start, a MA API function, can be used to mark the start of a loop. Upon execution, the code instrumented with MA API functions generates trace files. For parallel applications, one trace file is generated for each MPI task. The trace files contain traces for ma_xxx calls and MPI communication events. Most MA calls require a pair of ma_xxx_start

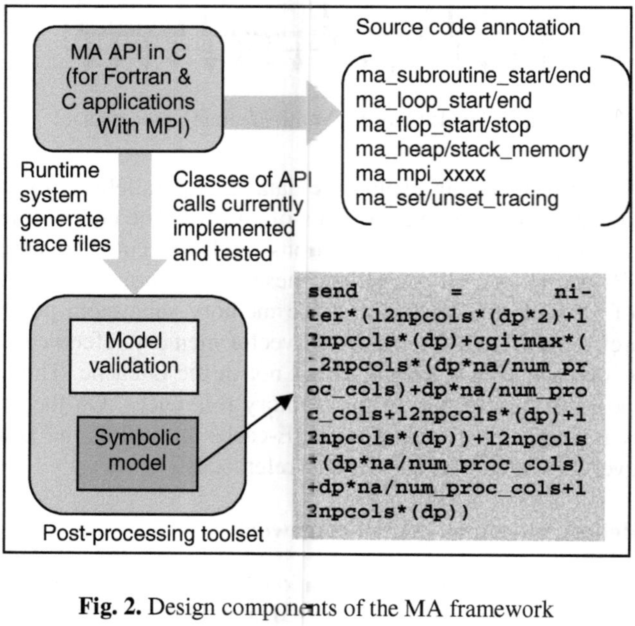

Fig. 2. Design components of the MA framework

and a ma_xxx_end calls. The ma_xxx_end traces are primarily used to validate modeling assertions against the runtime values. The assertions for hardware counter values, ma_flop_start/stop, invoke the PAPI hardware counter API [5]. The ma_mpi_xxx assertions on the other hand are validated by implementing MPI wrapper functions (PMPI) and by comparing ma_mpi_xxx traces to PMPI_xxx traces. Additional functions are provided in the MA API to control the tracing volume, for example, the size of the trace files, by enabling and disabling the tracing at compile time and also at runtime.

At runtime, the MA runtime system (MARS) tracks and captures the actual instantiated values as they execute in the application. MARS creates an internal control flow representation of the calls to the MA library as they are executed. It also captures both the symbolic values and the actual values of the expressions. The validation of an MA performance model is a two-stage process. When a model is initially being created, validation plays an important role in guiding the resolution of the model at various phases in the application. Later, the same model and validation technique can be used to validate against historical data and across the parameter space. Moreover, since models can be constructed and validated at multiple resolution levels, a user can keep track of error rate propagation from low level calculations to higher, functional levels. This process allows a user to refine the model resolution level and to identify possible causes of large error rates in model projections.

4 Implementation

We extended our MA framework not only to quantify X1E architectural optimizations but also to include a new set of performance metrics or attributes. Since metrics like memory bandwidth requirements cannot be formulated for a complex architecture without empirical information, we capture this information using standard memory benchmarks [7, 17]. To validate and to estimate runtime for our application from the data collected by the MA runtime system, we devise a set of mathematical expressions to formulate the degree with which these new metrics influence the runtime. The modifications to the MA framework for the X1E system include: a loopmark listing analyzer, X1E processor performance attributes, X1E memory bandwidth attributes analysis, and an infrastructure for runtime performance model validation and prediction.

4.1 Loopmark Listings Analyzer

The Cray X1E Fortran and C/C++ compilers generate text reports called loopmark listings that contain information about optimizations performed when compiling a program, such as whether a given loop was vectorized and multi-streamed. Loopmark listing generated for the most time-consuming loop from the NAS Conjugate Gradient (CG) serial benchmark:

```
583.  1 M----------<         do j=1, lastrow-firstrow+1
584.  1 M                       sum = 0.d0
585.  1 M V-------<            do k=rowstr (j), rowstr (j+1)-1
586.  1 M V                        sum = sum + a (k)*p (colidx(k))
587.  1 M V------->            enddo
```

```
588.  1 M                    q(j) = sum
589.  1 M---------->       enddo
```

To support our modeling approach, we have created a loopmark analyzer that can generate an analysis file. The tuples (shown below) indicate whether a loop has been multi-streamed (mflag=1) and/or vectorized (vflag=1). We introduce an abstract definition that includes both flags, which we call the "MV" score of a loop.

```
id: 17 func: conj_grad beginl: 583 endl: 589 mflag: 1 vflag: 0
id: 18 func: conj_grad beginl: 585 endl: 587 mflag: 1 vflag: 1
```

4.2 X1E Performance Attributes

Based on the information gathered from our loopmark listing analyzer, we introduce "architecture aware" metrics within the MA framework for predicting application run times on the X1E. Our first metric is the average vector length (AVL). Each Cray X1E SSP has two vector units; both contain vector registers that can hold 64 double-precision floating point elements, so the AVL is at most 64.

The peak memory bandwidth can be obtained if all 64 registers are utilized. In other words, if AVL for a loop is less than 64, it will be unable to utilize the peak memory bandwidth. Our performance models incorporate this penalty if AVL is less than AVL_{max}. The AVL of a loop can be computed using the loop bounds of a fully vectorized loop. The loop bounds of critical loops can be symbolically expressed as functions of input parameters of an application. Therefore, we can express AVL in the form of MA annotations. For example, the AVL of the CG loop shown above is:

$$AVL = \sum_{j=1}^{lastrow-firstrow} \left(\frac{rowstr(j+1) - rowstr(j)}{ceil(div((rowstr(j+1) - rowstr(j)),64)} \right)$$

In addition to AVL, there are certain performance attributes like memory bandwidth requirements that depend on an application's intrinsic properties. For instance, memory access patterns of critical loops determine spatial and temporal locality of a loop block and its demand for memory bandwidth. The achievable bandwidth depends on architectural complexities such as sizes and bandwidths of caches and the main memory. On the X1E, these architectural complexities in turn depend on compiler generated optimizations as specified by a loop's MV score. The peak memory bandwidth for multi-streamed vector memory references on Cray X1E is ~34 GB/s and for multi-streamed scalar memory references is ~4.5 GB/s.

Due to the lack of sufficient memory tracing tools for the Cray X1E system and a unique memory subsystem hierarchy for scalar and vector memory operations, we quantify an application's memory bandwidth through empirical methods. We use profile information of an application obtained from various performance tools like TAU [8], KOJAK [14] and CrayPAT (Performance Analysis Toolkit) [5] to make calculated assumptions about a loop's memory bandwidth.

On a Cray X1E, the maximum bandwidth per SSP is 8.5 GB/s. This bandwidth is obtained by running a unit-strided memory benchmark called "Stream" which exploits the spatial locality to obtain maximum available bandwidth [7]. The minimum bandwidth per SSP is 2.5 GB/s, which is obtained by running a random access memory benchmark called "Random Access" which spans through the memory by accessing random locations. The average or mixed bandwidth, a combination of stream and random memory access patterns, we considered as 5.5 GB/s.

5 Model Validation Results

After collecting the empirical performance data and characterizing performance attributes of the X1E processor, we formulate performance validation and runtime performance prediction techniques. MA performance attributes are validated with X1E hardware counter values. The X1E provides native counters to measure AVL, Loads and Stores (both vector and scalar), and floating point operations (vector and scalar). We use the PAPI library to capture these counter values.

We explain the validation process using the NAS SP, a simulated Computational Fluid Dynamics (CFD) application. The workload configurations can be changed by using different problem classes. The MA model for SP is represented in terms of an input parameter (`problem_size`), number of iterations, the number of MPI tasks that determines some derived parameters like the square-root of number of processors in the SP benchmark to simplify model representations. The SP benchmarks follow a Single Program Multiple Data (SPMD) programming paradigm. Hence, the workload and memory mapping and distribution per processor not only depend on the key input parameters but also on the number of MPI tasks.

5.1 Model Parameters

From the input parameters, `problem_size`, `niter` and `nprocs`, we compute two critical derived parameters: `ncells` and `size` for individual MPI tasks. A code listing example for SP with extended MA annotation is shown below:

```
              call maf_vec_loop_start(1,"tzetar","size^2*(size-1)*26", (size**2)*(size-1)*26," (size^2)*(size-1)*16", (size**2)*(size-1)*16,"(size-1)/(size/64+1)", (size-1)/(size/64+1),3)
M-------<        do   k = start(3,c), cell_size(3,c)-end(3,c)-1
M 2-----<          do   j = start(2,c), cell_size(2,c)-end(2,c)-1
 M 2 Vs--<            do     i = start(1,c), cell_size(1,c)-end(1,c)-1
M 2 Vs
M 2 Vs                 xvel = us(i,j,k,c)
M 2 Vs                 yvel = vs(i,j,k,c)
```

In this example, `tzetar` is the loop identifier, which is followed by symbolic expressions for vector load-store operations. Note that the two identical expressions is not an error but represent symbolic expression written in the trace file and another one is evaluated by the runtime system. The next two expressions represent vector floating-point operations. We then include AVL calculations and finally the MV score. Note that these workload parameters are represented in terms of a handful input parameters of the SP calculations.

5.2 Hardware Counter Measurements

The MA runtime system generates trace files that contain output for `ma_vec_loop_start` and `ma_vec_loop_stop` annotations when the instrumented application executes on the target X1E platform. This output file contents are:

```
  MA:2:ma_vec_loop_start:tzetar:(size^2)*(size-1)*26:1179360:
(size^2)*(size-1)*16: 725760:(size-1)/(size/64+1):35.000000:3
  MA:2:ma_vec_loop_stop:tzetar:1:244916:151434:1123632:708816:
8:177196:33.351922:0:2421:445
```

The runtime system empirically measure and report the hardware counter values in the following order in the `ma_vec_loop_stop` output: MA trace identifier, context depth of the call, MA API identifier, loop name, loop id, number of cycles, stalled cycles, vector flops, vector load-store operations, vector load stride, vector load allocated, vector length, scalar floating-point operations, scalar load-store operations, and data cache miss. Note that the values generated in `ma_vec_loop_stop` confirm results of our symbolic models. For instance, the number of FP operations, predicted from the model for Class W problem configurations with a single MPI tasks is expected to be 725,760 operations. The actual measurement shows that it is 708,816 operations. Similarly, the AVL computed using symbolic models (35) for the loop count and MV score confirms the measured value (33.35). Fig. 3 shows the error rates for hardware counter values for four different input configurations of the SP calculations with one and 4 MPI tasks. Note that the error rates are less than 10% for most cases. The error rates are higher for smaller workload configuration, class W, since these calculations are unable to saturate the CrayX1E MSP units. The percentage error rates for predicted and measured values are computed as:

$$\%error = \frac{(Value_{predicted} - Value_{measured})}{Value_{measured}}$$

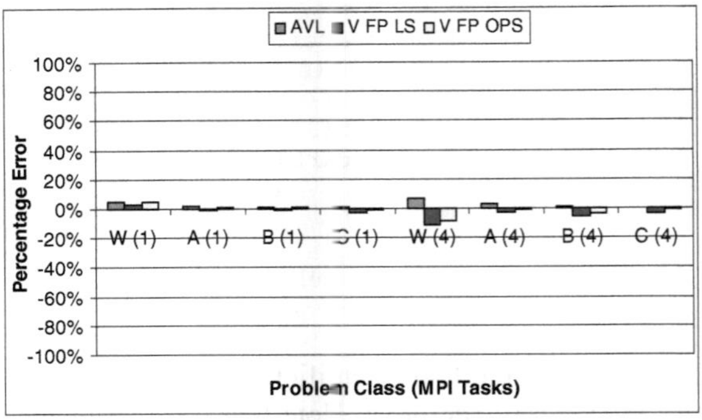

Fig. 3. Error rate calculations for MA model predictions as compared to runtime data

5.4 Runtime Calculations

We compute the execution times according to mathematical formulae that consider two components of execution time of a loop: first, the time taken to do the floating point computation; and second, the time taken to access the memory subsystem. On the Cray X1E system, the two runtime components (shown in Fig 4.) have two representations: one for vector and another for scalar execution: T_v and T_s. The memory access times are T_{vm} and T_{sm}, and the compute times are T_{vc} and T_{sc} respectively. These formulae are based on the MV score of a loop. If a loop is vectorized, the runtime is predicted using the formula for T_v. If the loop is also multi-streamed, the clock speed is 18GHz; otherwise it is 4.5GHz. If a loop is not vectorized, the runtime is predicted using the formula for T_s. If the loop is also multi-streamed the clock speed is 2.26 GHz; otherwise it is 0.565 GHz. In the Fig. 8, *VFLOPS* refers to the vector floating point operations, *VLOADS* refer to the vector load operations, *VSTORES* refers to the vector store operations, *SFLOPS* refers to the scalar floating point operations, *SLOADS* refers to the scalar load and *SSTORES* refers to the scalar store operations.

$$T_v = T_{vm} + T_{vc}$$
$$T_{vc} = \left(\frac{VFLOPS}{4.5GHz/18GHz}\right)$$
$$T_{vm} = \left(\frac{(VLOADS + VSTORES)*8*64}{BW*10e9*AVL}\right)$$

$$T_s = T_{sm} + T_{sc}$$
$$T_{sc} = \left(\frac{SFLOPS}{0.565GHz/2.26GHz}\right)$$
$$T_{sm} = \left(\frac{(SLOADS + SSTORES)*8}{BW*10e9}\right)$$

Fig. 4. T_v is the time for vector processing and T_s is the time for scalar processing

Using the quantitative information generated by the MA symbolic models, the expressions to calculate timings, and empirical assumptions about the bandwidth of a particular code region, we can predict runtime of that code region on the target system. Fig. 5 shows the error rates for predicted runtimes using the MA models with Cray X1E attributes and three memory bandwidth values: minimum (2.5 GB/s), average/mixed (5.5 GB/s) and maximum (8.5 GB/s) per SSP. We show an error bar of 20% in measured runtimes since the execution times for a given calculation, `tzetar` for this example, vary from one time step iteration to other and during multiple runs. We observe that the error rates are substantially higher with the minimum bandwidth values across multiple runs. We therefore conclude that the memory access pattern for this calculation is not random. The lowest error rates are obtained with highest bandwidth ranges, except for the smallest input configurations that are not capable of saturating the X1E execution and memory bandwidth resources.

After validating the performance model and running performance prediction experiments, we investigate the impact of Cray X1E performance enhancing attributes that are introduced in the MA framework. For instance, we take into account the MV score and AVL of a given loop in our performance prediction framework and runtime prediction. We ran experiment by removing the multi-stream optimization for the compiler application by using "`-O stream0`" compiler flag. Since the model does not contain MV information, the runtime prediction is performed without knowledge of target resource utilization. Fig. 6 shows error rates for this experiment. As shown in the figure, we over-predict runtime because our model and subsequently the runtime prediction method are unaware that only one of the 4 SSP units within a Cray X1E MSP is utilized for this calculation. Hence, we conclude that there are several advantages of introducing a minimal set of performance enhancing features in the modeling

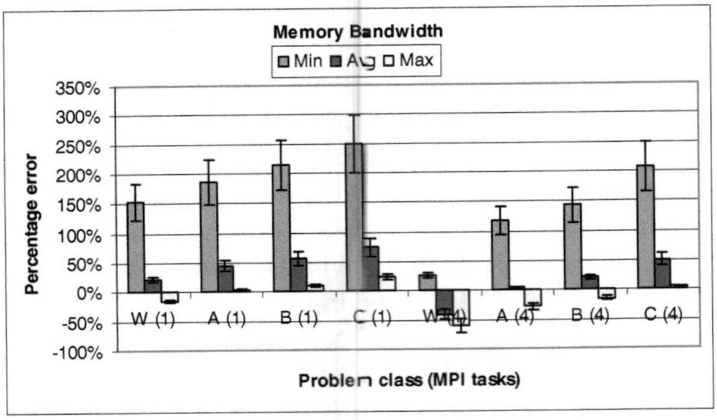

Fig. 5. Error rates for runtime prediction with different memory bandwidth values

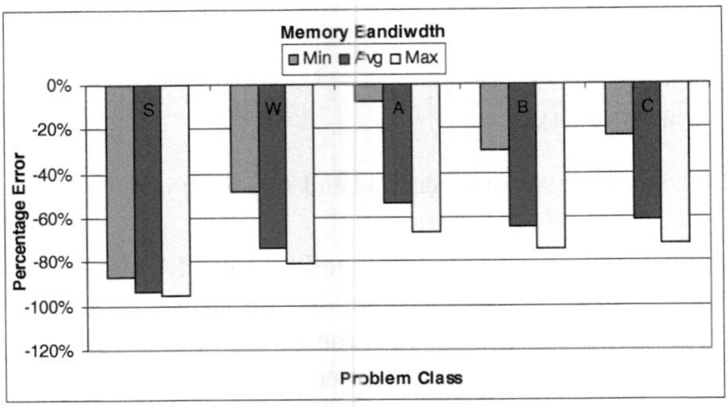

Fig. 6. Impact on performance prediction for not utilizing system specific attributes

scheme. These include an improved runtime prediction capability as well as identification of potential compile and runtime problems. In this particular example, there are no data and control dependencies in a given code, therefore, a code developer can assert that this loop is expected to be vectorized and multi-streamed by the X1E compiler. Using the MA scheme, we can identify this problem even before we the runtime experiments are performed. Since the MA annotations contain user-specified information, the code developer can locate the problem quickly and inspect the loopmark listing for potential compiler issues.

6 Conclusions and Future Directions

We have developed a new performance modeling approach that augments Modeling Assertions (MA) with information about the performance-enhancing features of unconventional architectures. We demonstrated that our performance modeling approach enables us to reduce the performance prediction error rates significantly by incorporating a minimal set of "architecture aware" attributes in the MA framework. Furthermore, we identified that an insight into the understanding of the Cray X1E memory hierarchy is critical to performance modeling strategies for this platform. An inappropriate selection of the achievable memory bandwidth values resulted in error rates as high as 250%, but using our modeling approach we observed error rates of less than 25% for a representative scientific algorithm. Based on our success in augmenting MA annotations for X1E architectural features, we plan to extend the MA framework for performance modeling of emerging systems with multi-core processors and accelerator devices by identifying and incorporating the target architecture-specific performance attributes into our modeling framework.

Acknowledgements

The submitted manuscript has been authored by a contractor of the U.S. Government under Contract No. DE-AC05-00OR22725. Accordingly, the U.S. Government retains a non-exclusive, royalty-free license to publish or reproduce the published form of this contribution, or allow others to do so, for U.S. Government purposes. This research used resources of the Center for Computational Sciences at Oak Ridge National Laboratory, which is supported by the Office of Science of the U.S. Department of Energy under Contract No. DE-AC05-00OR22725.

References

1. AMD Dual-Core and Multi-Core Processors, http://www.amdboard.com/dual_core.html
2. The Cell Project at IBM Research, http://www.research.ibm.com/cell/
3. Cray X1E supercomputer, http://www.cray.com/products/x1e/
4. NAS Parallel Benchmarks, http://www.nas.nasa.gov/Software/NPB/
5. Optimizing Applications on Cray X1 Series Systems, available at http://docs.cray.com
6. Performance Application programming Interface (PAPI), http://icl.cs.utk.edu/papi/
7. Stream Memory Benchmark, http://www.streambench.org
8. Tuning and Analysis Utilities (TAU), http://www.cs.uoregon.edu/research/tau/
9. Alam, S.R., Vetter, J.S.: A Framework to Develop Symbolic Performance Models of Parallel Applications. In: PMEO-PDS 2006. 5th International Workshop on Performance Modeling, Evaluation, and Optimization of Parallel and Distributed Systems (2006) (to be held in conjunction with IPDPS)
10. Alam, S.R., Vetter, J.S.: Hierarchical Model Validation of Symbolic Performance Models of Scientific Applications. In: European Conference on Parallel Computing (Euro-Par) (2006)
11. Bailey, D., Snavely, A.: Performance Modeling: Understanding the Present and Prediction the Future. In: European Conference on Parallel Computing (Euro-Par) (2005)
12. Kerbyson, D.J., et al.: A Comparison Between the Earth Simulator and AlphaServer Systems using Predictive Application Performance Models. In: International Parallel and Distributed Processing Symposium (IPDPS) (2003)
13. Luk, C., et al.: Pin: Building Customized Program Analysis Tools with Dynamic Instrumentation. In: ACM SIGPLAN Conference on Programming Language Design and Implementation, ACM Press, New York (2005)
14. Mohr, B., Wolf, F.: KOJAK - A Tool Set for Automatic Performance Analysis of Parallel Applications. In: European Conference on Parallel Computing (EuroPar) (2003)
15. Snavely, A., et al.: A Framework for Application Performance Modeling and Prediction. In: ACM/IEEE Supercomputing Conference, ACM Press, New York (2002)

16. Storaasli, O.F., et al.: Cray XD1 Experiences and Comparisons with other FPGA-based Supercomputer Systems. In: Cray User Group (CUG) Conference (2006)
17. Strohmaier, E., Shan, H.: Apex-MAP: A Global Data Access Benchmark to Analyze HPC Systems and Parallel Programming Paradigms. In: ACM/IEEE Supercomputing Conference (2005)
18. Vetter, J.S., et al.: Characterizing Applications on the Cray MTA-2 Multi-threaded Architecture. In: Cray User Group Conference (2006)
19. Weinberg, J., et al.: Quantifying Locality in the Memory Access Patterns of the HPC Applications. In: ACM/IEEE Supercomputing Conference (2005)
20. Yang, T., et al.: Predicting Parallel Applications' Performance Across Platforms using Partial Execution. In: ACM/IEEE Supercomputing Conference (2005)

Towards Scalable Event Tracing for High End Systems

Kathryn Mohror and Karen L. Karavanic

Department of Computer Science
Portland State University
P.O. Box 751
Portland, OR 97207-0751
{kathryn,karavan}@cs.pdx.edu

Abstract. Although event tracing of parallel applications offers highly detailed performance information, tracing on current leading edge systems may lead to unacceptable perturbation of the target program and unmanageably large trace files. High end systems of the near future promise even greater scalability challenges. Development of more scalable approaches requires a detailed understanding of the interactions between current approaches and high end runtime environments. In this paper we present the results of studies that examine several sources of overhead related to tracing: instrumentation, differing trace buffer sizes, periodic buffer flushes to disk, system changes, and increasing numbers of processors in the target application. As expected, the overhead of instrumentation correlates strongly with the number of events; however, our results indicate that the contribution of writing the trace buffer increases with increasing numbers of processors. We include evidence that the total overhead of tracing is sensitive to the underlying file system.

1 Introduction

Event tracing is a widely used approach to performance tuning that offers a highly detailed view of parallel application behavior. It is used to diagnose an important range of performance problems, including inefficient communication patterns, capacity bottlenecks on communication links, and imbalanced message sends and receives. However, the extreme scale of existing and near-future high-end systems, which include tens or hundreds of thousands of processors, introduces several difficult challenges that must be solved for event-based tracing to continue to be useful. First, pushing tracing technology to higher and higher scales drives up the overall perturbation to the application, yielding less accurate results. Second, tracing across the processors of a large-scale run results in unmanageably large trace files, in many cases breaking the measurement tools completely. Third, many trace analysis and visualization tools fail when trying to process trace files from more than a few hundred processors, or from a run duration greater than a few minutes.

Application developers currently work around these tracing limitations by ad hoc techniques for data reduction, or by restricting measurement to profiling instead of tracing. Common methods for reducing the amount of data collected in a trace include: starting and stopping tracing around a particular section of code of interest; collecting trace data for only certain operations, e.g. message passing calls; and tracing a

R. Perrott et al. (Eds.): HPCC 2007, LNCS 4782, pp. 695–706, 2007.

scaled down version of the application with a greatly reduced number of processes. These approaches all introduce the risk that the performance problem may be missed or misdiagnosed. The performance problem might be missed, for example if the problem only shows up with runs using greater than 2000 processes [10], or the problem occurred in time-steps that were not measured. In addition, they increase the burden on the tool user to identify the approximate location of the problem, and to make code changes to control which events are captured.

Profiling is more scalable than tracing; however, event-based traces are still required for successful diagnosis of an important class of performance problems, because the needed information cannot be deduced from an execution profile [5]. A profile gives an overall picture of the performance of the code, but can't give information such as relative timings of events across processes. For example, a profile may show that a message-passing program is spending too much time in a receive call; however, trace data can reveal whether this is due to a "late sender" situation (the receiving process executes a blocking receive call in advance of the sending process executing a matching send call, causing the receiving process to waste time in the receive call), or if the send operation occurred on time, but network congestion caused the message to be received late.

The current situation drives the need for new, more scalable approaches to event tracing. The necessary first step in such an effort is to gain a detailed understanding of the interactions between current approaches and typical runtime environments. In this paper we present the results of a series of experiments designed to identify and to quantify the overheads incurred when tracing the execution of an application on a representative current architecture. Our particular focus is the runtime portion of the tracing process; scaling issues with merging processor-level traces, or post-mortem analysis and visualization tools, are outside of the scope of this study. Our goal was to characterize the scaling behavior by measuring the separate overheads of two distinct components: the overhead of just the trace instrumentation, and the overhead of flushing the trace buffer contents to files. Breaking down the total trace overhead allows us to demonstrate that there is an important difference in the scaling characteristics between the instrumentation overhead and the writing overhead: the instrumentation overhead scales with the number of events, but the write overhead scales with the number of processes, even if the number of events remains fairly stable. Our results also demonstrate that behavior of the parallel file systems typically installed with high-end systems today is a significant factor in determining the performance of a tracing tool.

2 Experiment Design

Our experiments are designed to focus on separating runtime tracing overhead into two distinct components: the overhead of just the trace instrumentation, and the overhead of flushing the trace buffer contents to files. We performed runs with and without trace instrumentation (*instr* and *noInstr*); and with and without buffer flush to file enabled (*write* or *noWrite*), then calculated the overheads using the following metrics:

- *Wall clock time*: MPI_Wtime, measured after MPI_Init and before MPI_Finalize. The following are not included in this measurement:

instrumentation overhead for tool setup, finalization, and function calls before/after the timer is started/stopped; and writing overhead for trace file creations, final trace buffer flushes before file closure, trace file closure, and, in the case of MPE, trace file merging.

- *Write overhead*: Average total wall clock time of the *write* runs minus average total wall clock time of the *noWrite* runs
- *Instrumentation overhead*: Average total wall clock time of the runs that did not write the trace buffer minus average total wall clock time of the *noBuff_noInstr_noWrite* runs

Given our goal of pushing to the current scaling limits of tracing, we wanted to measure an application with a very high rate of communication, so that trace records for a high number of MPI communication events would be generated. We picked SMG2000 (SMG) [1] from the ASC Purple Benchmark suite. SMG is characterized by an extremely high rate of messages: in our four process runs, SMG executed 434,272 send and receive calls in executions that took approximately 15 seconds. For comparison, we also included another ASC Purple Benchmark, SPhot (SP) [18]. SP is an embarrassingly parallel application; in a four-process, single-threaded execution of 512 runs with a total execution time of 350 seconds, the worker processes pass 642 messages, and the master process passes 1926 messages. We configured both applications with one thread per MPI process.

To vary the number of processes, we used weak scaling for the SMG runs. As we increased the number of processors, we altered the processor topology to $P * 1 * 1$, where P is the number of processors in the run, and kept the problem size per processor, $nx * ny * nz$, the same, thereby increasing the total problem size. We used both weak and strong scaling for the SP runs, referred to as SPW and SPS respectively. We configured these approaches by changing the Nruns parameter in the input file input.dat, which controls the total amount of work done in a single execution. For strong scaling, we kept Nruns constant at 512 for all processor counts; for weak scaling, we set Nruns equal to the number of MPI ranks.

We used SMG's built-in metrics to measure Wall Clock Time, summing the values reported for the three phases of the execution: Struct Interface, SMG Setup, and SMG Solve. We used the native SPhot wall clock time values for Wall Clock Time. Figure 1 shows the scaling behavior of the uninstrumented applications. As expected, the execution time of SPS decreases with increasing numbers of processors, since we are keeping the total problem size constant.

In some sense the choice of a particular tracing tool was irrelevant to our goals: we wanted to investigate a "typical" tracing tool. However, we wanted to avoid results that were in some way an artifact of one tool's particular optimizations. Therefore, we used two different robust and commonly used tracing tools for our experiments: TAU and MPE.

We built several versions of TAU version 2.15.1 [19]. For the *noWrite* versions we commented out the one line in the trace buffer flush routine of the TAU source that actually calls the `write` system call. We altered the number of records stored in the trace buffer between flushes, by changing the #define for TAU_MAX_RECORDS

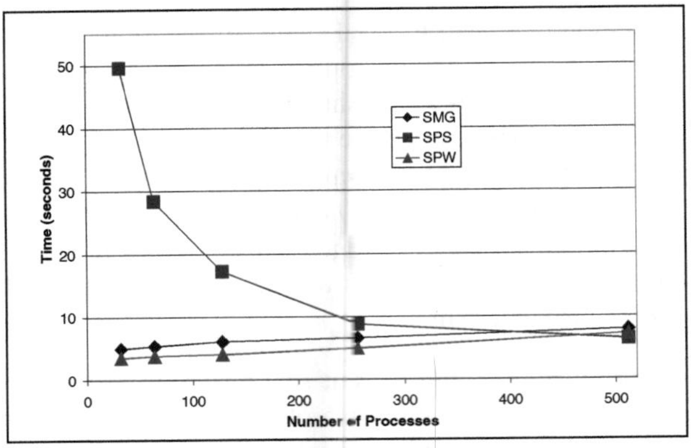

Fig. 1. Performance of Uninstrumented Executions

in the TAU source for each size and rebuilding, to test two different buffer sizes: 0.75 MB (32,768 TAU events); 1.5 MB (default size for TAU; 65,536 TAU events); 3.0 MB (131,02 TAU events); and 8.0 MB (349,526 TAU events). We used the default level of instrumentation for TAU, which instruments all function entries and exits.

MPE (the MultiProcessing Environment (MPE2) version 1.0.3p1 [26]) uses the MPI profiling interface to capture the entry and exit time of MPI functions as well as details about the messages that are passed between processes, such as the communicator used. To produce an MPE library that did not write the trace buffer to disk, we commented out three calls to `write` in the MPE logging source code. We also had to comment out one call to `CLOG_Converge_sort` because it caused a segmentation fault when there was no data in the trace files. This function is called in the MPE wrapper for `MPI_Finalize`, so it did not contribute to the timings reported in the SMG metrics. We altered the buffer sizes by changing the value of the environment variable CLOG_BUFFERED_BLOCKS. We also set the environment variable MPE_LOG_OVERHEAD to "no" so that MPE did not log events corresponding to the writing of the trace buffer. In MPE, each MPI process writes its own temporary trace file. During `MPI_Finalize`, these temporary trace files are merged into one trace file, and the temporary trace files are deleted. The temporary and merged trace files were written in CLOG2 format. We used two different buffer sizes: 1.5 MB (24 CLOG buffered blocks), and 8.0 MB (default size for MPE; 128 CLOG buffered blocks). For SPW only, we altered the SPhot source to call MPE logging library routines to log events for all function calls, to correspond to the default TAU behavior more directly. We refer to this as "MPc" for MPE with customized logging. For the SPW MPc experiments, we disabled the trace file merge step in `MPI_Finalize`, because it became quite time consuming with larger trace files.

We collected all of our results on MCR, a 1152-node Linux cluster at LLNL running the CHAOS operating system [8] (See Figure 2). Each node comprises two 2.4 GHz Pentium Xeon processors and 4 GB of memory. All executions ran on the batch partition of MCR. The trace files, including any temporary files, were stored using the Lustre file system [3]. This platform is representative of many high end Linux clusters in current use.

Each of our experiment sets consisted of thirty identical executions.

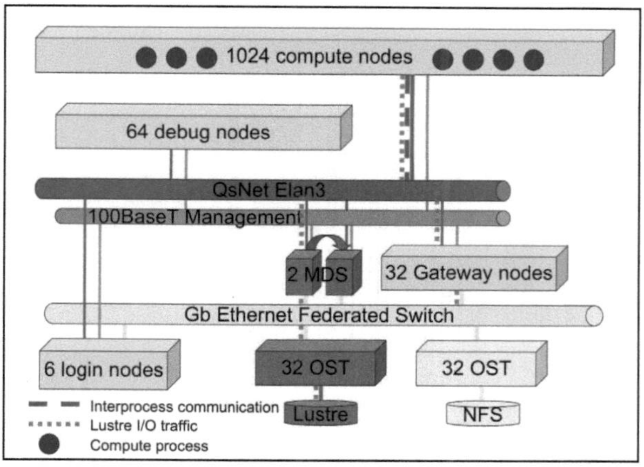

Fig. 2. Experiment Environment. The MPI processes in our experiments, represented by purple circles in the diagram, ran on a subset of the 1024 compute nodes of MCR. MPI communication between the processes traveled over the Quadrics QsNet Elan3 interconnect, shown by the purple dashed line. The I/O traffic for the Lustre file system, represented by the blue dotted line, also traveled over the Quadrics interconnect. Metadata requests went to one of two metadata servers (MDS), a fail-over pair. File data requests first went through the gateway nodes to an object storage target (OST), which handled completing the request on the actual parallel file system hardware.

3 Results

In this section, we present results from two studies: The first is an investigation of the relationship between system software and tracing efficiency; and the second is a study of tracing overheads as we scale up the number of application processes. These results are part of a larger investigation; full details are available as a technical report [14].

3.1 Impact of Runtime System on Tracing Overheads

We ran sets of executions of SMG with 4, 8, 16, 32, and 64 processes, measured with TAU, using three different buffer sizes: 0.75 MB (*smBuff*), 1.5MB (*defBuff – the default setting for TAU*), and 3.0 MB (*bigBuff*). Surprisingly, we found that there was a noticeable difference between *bigBuff* runs before and after a system software upgrade. As shown in Figure 3, running under CHAOS 3.0 and Lustre 1.4.3.3 resulted in much higher overhead for the *bigBuff* runs than running under CHAOS 3.1 and Lustre 1.4.5.8, regardless of whether or not the buffer contents were flushed to file. There were no statistically significant differences between the runs for the Struct Interface metrics, or for the *smBuff_write* case. However, the *defBuff_write*, *defBuff_noWrite*, *bigBuff_write*, and *bigBuff_noWrite* sets of runs all showed statistically significant differences due to system version. We performed a two-factor ANOVA test for each buffer size, with the two factors being the system differences and whether or not the trace buffer was written to disk. The results showed a striking difference for the two

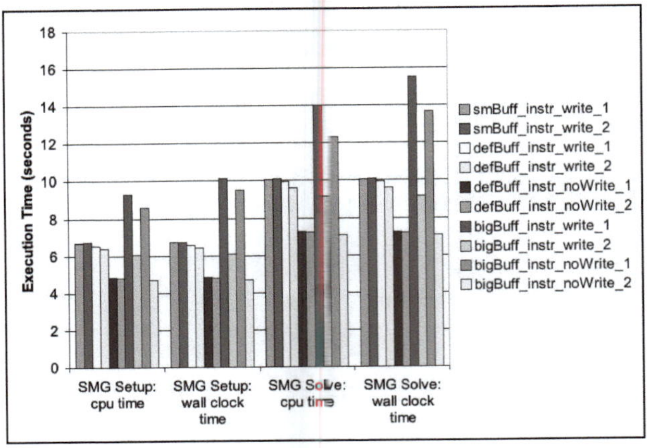

Fig. 3. Comparison of Executions with Different Buffer Sizes Showing System Differences. The chart shows the average values for the SMG Setup and Solve metrics. The y-axis is time in seconds. The legend shows the system configuration for each bar. The names that end in "_1" were run under CHAOS 3.0/Lustre 1.4.3.3. The names that end in "_2" were run under and CHAOS 3.1/Lustre 1.4.5.8.

buffer sizes: the major contributor (> 90%) to variation between the *defBuff* runs was whether or not the trace buffer was written to disk, whereas the major contributor (> 87%) for the *bigBuff* runs was the system differences.

Operating system performance regression tests run before and after the upgrade showed no significant effect on OS benchmark performance: network and memory bandwidth, message passing performance and I/O performance (Mike Haskell, personal communication, June 2006). We conclude that the performance differences we saw were due to changes to the Lustre file system.

3.2 Scaling Behavior of Tracing Overhead

In this study, we examined how the overheads of tracing change as the application scales. We ran sets of experiments with 32, 64, 128, 256, and 512 processes, traced with TAU and MPE, using buffer sizes of 1.5 and 8.0 MB.

3.2.1 Event Counts and Trace File Sizes

Here we describe the event counts generated while tracing the applications. Complete details can be found in the technical report [14]. For SMG, the counts for TAU and MPE exhibit similar trends, but are different by roughly an order of magnitude. As the numbers of processors double, the per-process event counts and trace data written by each process increase slightly (in part due to increased communication), while the total number of events and resulting trace file sizes double. For SPS, there are markedly different results between TAU and MPE; the event counts differ by six orders of magnitude. This is because with TAU we are measuring all function entries and exits, whereas with MPE we measure only MPI activity. For both TAU and MPE, doubling the number of processors results in the per-process event counts decreasing by half.

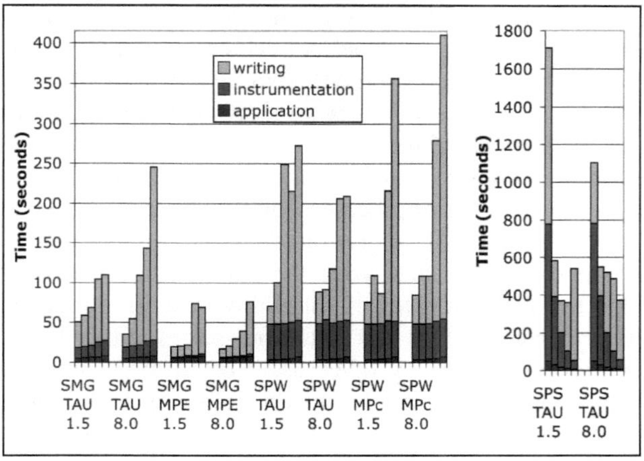

Fig. 4. Performance of Instrumented Executions. Here we show the total execution time for SMG measured with TAU and MPE, and SPhot measured with TAU. The colors in the bars indicate the time spent in application code, time in trace instrumentation, and time writing the trace buffer. Each bar in a set represents the average behavior of executions with 32, 64, 128, 256, and 512 processes, respectively. The set labels include (top to bottom): the benchmark name, the measurement tool, and the buffer size.

For TAU only, the total event count and resulting trace file sizes remain constant, whereas for MPE, the maximum per-process event count, the total event count, and resulting trace file sizes increase slightly. For SPW, the counts for TAU and MPc are nearly identical, while the counts for MPE differ. Again, this is because of differences in what was measured by the tools. The total event count and trace file sizes for MPE are roughly six orders of magnitude less than those of TAU and MPc.

We use this information to derive an expectation for tracing overheads for the different applications and tools. For the weakly-scaled SMG and SPW, we expect that the overheads of tracing would remain relatively constant with increasing numbers of processors because the amount of data being collected and written per-process remains relatively constant. However, for SPW with MPE, we expect to see very little overheads due to the small amount of data collected. For SPS and TAU, we expect the overheads of tracing to decrease with increasing numbers of processors, because the amount of data being collected and written per-process decreases with increasing processes. For SPS with MPE, we expect to see very little overhead because of the small amount of data collected.

3.2.2 Execution Time

Figure 4 shows the average wall clock times for our experiments broken down into time spent in application code, trace instrumentation, and writing the trace buffer. The graph on the left shows the measurements for SMG with TAU and MPE, and SPW with TAU and MPc. In each run set, we see the same trend; as the number of processes increases, the total execution time increases, largely due to the time spent writing the trace buffer. The time spent in the application code and in trace instrumentation remains relatively constant. The graph on the right shows the execution times of SPS with TAU. Here, as the numbers of processes increase, the total execution time decreases. However, even though the time spent in writing the trace buffer decreases with increasing processors, it does not decrease as rapidly as the time spent in instrumentation or application code. For SPS and SPW with MPE, the differences between the *write* and *noWrite* executions were indistinguishable due to the very small amounts of data collected and written.

We computed the percentage contribution to variation using three-factor ANOVA, with the buffer size, the number of processes, and whether or not the trace buffer was written to disk as the factors [14]. In general, there was quite a bit of variation in the running times of the executions that wrote the trace buffer, which explains the high contribution of the residuals. Sources of variability in writing times for the different executions include: contention for file system resources, either by competing processes in the same execution, or by other users of Lustre; contention for network resources, either by other I/O operations to Lustre, or by MPI communication; and operating system or daemon interference during the write. Any user of this system gathering trace data would be subject to these sources of variation in their measurements. For SMG measured with TAU and MPE, the largest contributing factor was whether or not the buffer was written, at 33% and 26%, respectively. The largest contributing factor for SPS with TAU was the number of processes in the run (19%), followed closely by whether or not the trace buffer was written (14%). SPS with MPE had the number of processes as the dominating factor at 51%. SPW with TAU and MPc both had writing the trace buffer as the largest contributor, at 34% and 24%, while SPW with MPE had the number of processes as the largest, at 81%. The differences in the dominating factors for the SP runs with MPE are attributed to the comparatively very small amount of data collected.

3.2.3 Execution Time vs Event Counts

Table 1 shows the correlation of the average total wall clock time with the maximum event count over all ranks. SPS with MPE had a relatively weak negative correlation with the maximum event count, because as the process count increases, the number of messages that the master process receives increases, and the execution time decreases, giving a negative correlation. In general, executions that did not write the trace buffer to disk had a higher correlation with the event count than did the executions that did write the trace buffer to disk.

Figure 5 shows the overheads of writing and instrumentation as the maximum number of events in a single rank increases. For SMG with TAU and MPE, we see a clear pattern. The instrumentation overhead appears to vary linearly with the number

Table 1. Correlation of Total Wall Clock Time with Maximum Event Count in a Rank

		SMG		SPS		SPW		
Buffer Sz	**Write?**	**TAU**	**MPE**	**TAU**	**MPE**	**TAU**	**MPE-C**	**MPE**
1.5	yes	0.96	0.85	0.91	-0.78	0.69	0.80	0.98
8.0	yes	0.97	0.90	0.95	-0.81	0.61	0.76	0.98
1.5	no	0.98	0.98	0.99	-0.70	0.81	0.55	0.96
8.0	no	0.98	0.98	0.99	-0.79	0.74	0.77	0.95

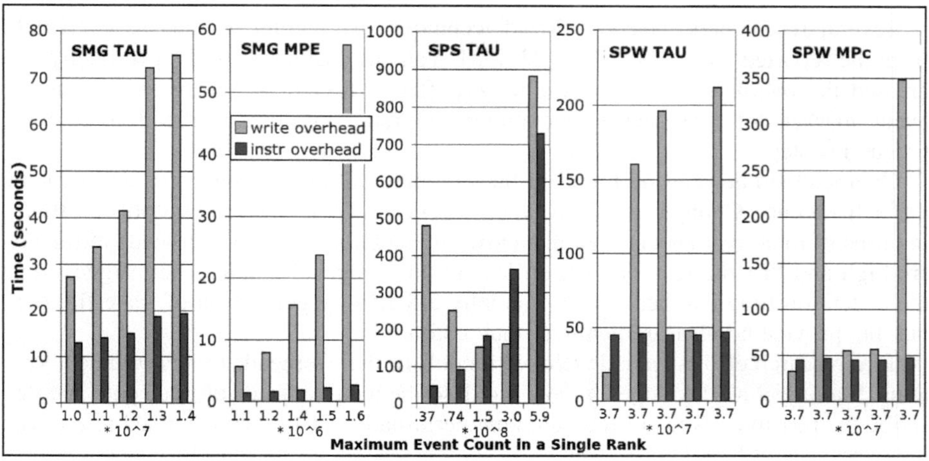

Fig. 5. Tracing Overhead Behavior with Maximum Event Count in a Single Rank. The groups of bars from left to right in the charts represent different processor counts: for SMG they represent 32, 64, 128, 256, and 512 processes; for SPS they represent 512, 256, 128, 64, and 32 processes; for SPW, they represent 32, 256, 128, 64, and 512 processes.

of events, while the overhead of writing the trace increases much more rapidly, and does not appear to have a linear relationship with the event count. The behavior of SPS is different, because in this application, as the number of events increases, the number of processes decreases; however, the instrumentation overhead still appears to have a linear relationship with the event count. The write overhead is high at higher event counts, but also at the low event counts, when the number of writing processes is higher. For SPW, the instrumentation overhead is relatively constant, as expected since the number of events does not change much between the run sets. However, the writing overhead fluctuates widely. The reason for this is that the maximum event count in a rank does not monotonically increase or decrease with increasing processors as it does for SMG or SPS.

4 Related Work

Because perturbation is intrinsic to measurement [6], research focuses on techniques to lower or limit the overheads, remove the overheads in the resulting data, and to measure and model the overheads.

Researchers have investigated methods to lower the overheads of tracing [11],[15],[16],[19],[24]. The Event Monitoring Utility (EMU) was designed to allow the user to adjust how much data was collected in each trace record, thereby altering the amount of overhead in each measurement [11]. The authors found the writing overhead to be the largest monitoring overhead. TAU [19] addresses instrumentation overhead by allowing users to disable instrumentation in routines that are called very frequently and have short duration. TAU also includes a tool that uses profile data to discover which functions should not be instrumented, and feeds this information to the automatic source instrumentor.

Several researchers have developed techniques to attempt to remove overheads from the reported data [4],[7],[23],[22],[25]. Yan and Listgarten [25] specifically addressed the overhead of writing the trace buffer to disk in AIMS by generating an event marker for these write operations and removing the overhead in a post-processing step.

Overhead studies can be found in the literature, although their focus and content differ from ours. Chung et al [2] evaluate several profiling and tracing tools on BG/L in terms of total overhead and write bandwidth, and note that the overheads of tracing are high and that the resulting trace files are unmanageably large. They suggest that the execution time overhead is substantially affected by generation of trace file output, but provide no measurements for their claim.

Two research efforts have developed models of the overheads in measurement systems. Malony et al. developed a model to describe the overheads of trace data and describe the possible results of measurement perturbation [13], then extended it to cover the overheads of SPMD programs [17]. They noted, as we do, that the execution time of traced programs was influenced by other factors than just the events in each processor independently. However, they did not explore this further. Waheed et al. [21] explored the overheads of trace buffer flushing and modeled two different flushing policies [20]. They found that the differences between the policies decreased with increased buffer sizes.

A primary difference between our results and prior work is that we identify a previously unexplored scalability problem with tracing. To the best of our knowledge, while others have noted that the largest overhead of tracing is writing the data, none have shown how this overhead changes while increasing the scale of application runs.

5 Conclusions and Future Work

In our scaling experiments, the execution times of the *noWrite* runs tended to scale with the maximum number of events. However, the execution times of the *write* runs did not scale as strongly with the number of events, and tended to scale with increasing numbers of processors, possibly due to contention caused by sharing the file system resource. Our results suggest that the trace writes will dominate the overheads more and more with increasing numbers of processes. They indicate that the trace overheads are sensitive to the underlying file system. They also imply that scaling by increasing the number of threads, which does not increase the total number of trace buffers, will primarily scale up the overheads associated with instrumentation, as opposed to writing, while increasing the number of trace buffers will quickly drive up the cost of writing.

Our research proceeds in two directions: modeling and data reduction.

Previously developed models of the overheads of tracing make some simplifying assumptions and do not account for the overheads we saw in our experiments. The early work done by Malony et al. assumes that in the case of programs that do not communicate, the perturbation effect for each processor is only due to the events that occur on that processor [17]. Our results dispute this assumption. If the application we used did not contain communication, the times for writing the trace file to disk could have been extended due to resource contention on the network by the other processes trying to write the trace file to disk, and could have increased with increasing numbers of processes sharing the resource. The model presented by Waheed et al. also does not account for this interaction when modeling the buffer flushing policies [20] – instead, they assume a constant latency for all writes of the trace buffer. We are developing a new model that accounts the characteristics of high end clusters and parallel file systems.

Realization of a scalable approach to tracing will require an overall reduction in the total amount of data. Data reduction is needed not only to reduce runtime overhead, but also to address the difficulties of storing and analyzing the resulting files. We are currently incorporating the results of our measurement studies into the design of an innovative approach to scalable event tracing.

Acknowledgements

This work was funded in part by a PSU Maseeh Graduate Fellowship and an Oregon Sports Lottery Graduate Fellowship. The authors acknowledge Mike Haskell and Richard Hedges for their help with CHAOS/Lustre information; Bronis de Supinski for feedback on an early draft; and Lisa Zurk, Michael Rawdon, and our CAT team for emergency office space and tech support during construction. We would also like to thank our anonymous reviewers whose helpful suggestions improved this work.

References

1. Brown, P., Falgout, R., Jones, J.: Semicoarsening multigrid on distributed memory machines. SIAM Journal on Scientific Computing 21(5), 1823–1834 (2000) (also available as Lawrence Livermore National Laboratory technical report UCRL-JC-130720)
2. Chung, I., Walkup, R., Wen, H., Yu, H.: MPI Performance Analysis Tools on Blue Gene/L. In: Proc. of SC2006, Tampa, Florida (November 11-17, 2006)
3. Cluster File Systems, Inc.: Lustre: A Scalable, High-Performance File System. Cluster File Systems, Inc. whitepaper (2002), available at (June 2006) http://www.lustre.org/docs/whitepaper.pdf
4. Fagot, A., de Kergommeaux, J.: Systematic Assessment of the Overhead of Tracing Parallel Programs. In: Proc. of 4th Euromicro Workshop on Parallel and Distributed Processing, pp. 179–185 (1996)
5. Fahringer, T., Gernt, M., Mohr, B., Wolf, F., Riley, G., Traff, J.: Knowledge Specification for Automatic Performance Analysis. APART Technical Report Revised Edition (2001), available at (October 5, 2006), http://www.fz-juelich.de/apart-1/reports/wp2-asl.ps.gz
6. Gait, J.: A Probe Effect in Concurrent Programs. Software - Practice and Experience 16(3), 225–233 (1986)
7. Gannon, J., Williams, K., Andersland, M., Lumpp, J., Casavant, T.: Using Perturbation Tracking to Compensate for Intrusiuon Propagation in Message Passing Systems. In: Proc. of the14th International Conference on Distributed Computing Systems, Poznan, Poland, pp. 141–412 (1994)
8. Garlick, J., Dunlap, C.: Building CHAOS: an Operating Environment for Livermore Linux Clusters. Lawrence Livermore National Laboratory, UCRL-ID-151968 (2002)
9. Hollingsworth, J., Miller, B.: An Adaptive Cost Model for Parallel Program Instrumentation. In: Fraigniaud, P., Mignotte, A., Bougé, L., Robert, Y. (eds.) Euro-Par 1996. LNCS, vol. 1123, pp. 88–97. Springer, Heidelberg (1996)

10. Kale, L., Kumar, S., Zheng, G., Lee, C.: Scaling Molecular Dynamics to 3000 Processors with Projections: A Performance Analysis Case Study. In: Sloot, P.M.A., Abramson, D., Bogdanov, A.V., Gorbachev, Y.E., Dongarra, J.J., Zomaya, A.Y. (eds.) ICCS 2003. LNCS, vol. 2660, pp. 23–32. Springer, Heidelberg (2003)
11. Kranzlmüller, D., Grabner, S., Volkert, J.: Monitoring Strategies for Hypercube Systems. In: Proc. of the Fourth Euromicro Workshop on Parallel and Distributed Processing, pp. 486–492 (1996)
12. Lindlan, K., Cuny, J., Malony, A., Shende, S., Mohr, B., Rivenburgh, R., Rasmussen, C.: A Tool Framework for Static and Dynamic Analysis of Object-Oriented Software with Templates. In: Proc. of SC2000, Dallas (2000)
13. Malony, A., Reed, D., Wijshoff, H.: Performance Measurement Intrusion and Perturbation Analysis. IEEE Transactions on Parallel and Distributed Systems 3(4), 433–450 (1992)
14. Mohror, K., Karavanic, K.L.: A Study of Tracing Overhead on a High-Performance Linux Cluster. Portland State University CS Technical Report TR-06-06 (2006)
15. Ogle, D., Schwan, K., Snodgrass, R.: Application-Dependent Dynamic Monitoring of Distributed and Parallel Systems. In: IEEE Transactions on Parallel and Distributed Systems, pp. 762–778. IEEE Computer Society Press, Los Alamitos (1993)
16. Reed, D., Roth, P., Aydt, R., Shields, K., Tavera, L., Noe, R., Schwartz, B.: Scalable Performance Analysis: the Pablo Performance Analysis Environment. In: Proc. of the Scalable Parallel Libraries Conference, Mississippi State, MS, pp. 104–113 (1993)
17. Sarukkai, S., Malony, A.: Perturbation Analysis of High Level Instrumentation for SPMD Programs. In: Proc. of the 4th ACM SIGPLAN Symposium on Principles and Practice of Parallel Programming, San Diego, CA, pp. 44–53. ACM Press, New York (1993)
18. SPHOT Benchmark (2006), available at (Decemver 8, 2006), http://www.llnl.gov/asci/purple/benchmarks/limited/sphot/
19. Shende, S., Malony, A.: The TAU Parallel Performance System. the International Journal of High Performance Computing Applications 20(2), 287–331 (2006)
20. Waheed, A., Melfi, V., Rover, D.: A Model for Instrumentation System Management in Concurrent Computer Systems. In: Proc. of the 28th Hawaii International Conference on System Sciences, pp. 432–441 (1995)
21. Waheed, A., Rover, D., Hollingsworth, J.: Modeling and Evaluating Design Alternatives for an On-line Instrumentation System: A Case Study. IEEE Transactions on Software Engineering 24(6), 451–470 (1998)
22. Williams, K., Andersland, M., Gannon, J., Lummp, J., Casavant, T.: Perturbation Tracking. In: Proc. of the 32nd IEEE Conference on Decision and Control, San Antonio, TX, pp. 674–679. IEEE Computer Society Press, Los Alamitos (1993)
23. Wolf, F., Malony, A., Shende, S., Morris, A.: Trace-Based Parallel Performance Overhead Compensation. In: Yang, L.T., Rana, O.F., Di Martino, B., Dongarra, J.J. (eds.) HPCC 2005. LNCS, vol. 3726, Springer, Heidelberg (2005)
24. Yaghmour, K., Dagenais, D.: Measuring and Characterizing System Behavior Using Kernel-Level Event Logging. In: Proc. of the USENIX Annual 2000 Technical Conference, San Diego, CA, pp. 13–26 (2000)
25. Yan, J., Listgarten, S.: Intrusion Compensation for Performance Evaluation of Parallel Programs on a Multicomputer. In: Proc. of the 6th International Conference on Parallel and Distributed Systems, Louisville, KY (1993)
26. Zaki, O., Lusk, E., Gropp, W., Swider, D.: Toward Scalable Performance Visualization with Jumpshot. High-Performance Computing Applications 13(2), 277–288 (1999)

Checkpointing Aided Parallel Execution Model and Analysis

Laura Mereuta and Éric Renault

GET / INT — Samovar UMR INT-CNRS 5157
9 rue Charles Fourier, 91011 Évry, France
Tel.: +33 1 60 76 45 56, Fax: +33 1 60 76 47 80
{laura.mereuta,eric.renault}@int-evry.fr

Abstract. Checkpointing techniques are usually used to secure the execution of sequential and parallel programs. However, they can also be used in order to generate automatically a parallel code from a sequential program, these techniques permitted to any program being executed on any kind of ditributed parallel system. This article presents an analysis and a modelisation of an innovative technique — CAPE — which stands for Checkpointing Aided Parallel Execution. The presented model provides some hints to determine the optimal number of processors to run such a parallel code on a distributed system.

1 Introduction

Radical changes in the way of taking up parallel computing has operated during the past years, with the introduction of cluster computing [1], grid computing [2], peer-to-peer computing [3]... However, if platforms have evolved, development tools remain the same. As an example, HPF [4], PVM [5], MPI [6] and more recently OpenMP [7] have been the main tools to specify parallelism in programs (especially when supercomputers were the main issue for parallel computing), and they are still used in programs for cluster and grid architectures. Somehow, this shows that these tools are generic enough to follow the evolution of parallel computing. However, developers are still required to specify almost every information on when and how parallel primitives (for example sending and receiving messages) shall be executed.

Many works [8,9] have been done in order to automatically extract parallel opportunities from sequential programs in order to avoid developers from having to deal with a specific parallel library, but most methods have difficulties to identify these parallel opportunities outside nested loops. Recent research in this field [10,11], based on pattern-maching techniques, allows to substitute part of a sequential program by an equivalent parallel subprogram. However, this promising technique must be associated an as-large-as-possible database of sequential algorithm models and the parallel implementation for any target architectures for each of them.

At the same time, the number of problems that can be solved using parallel machines is getting larger everyday, and applications which require weeks (or

R. Perrott et al. (Eds.): HPCC 2007, LNCS 4782, pp. 707–717, 2007.

months, or even more...) calculation time are more and more common. Thus, checkpointing techniques [12,13,14] have been developed to generate snapshots of applications in order to be able to resume the execution from these snapshots in case of problem instead of restarting the execution from the beginning. Solutions have been developed to resume the execution from a checkpoint on the same machine or on a remote machine, or to migrate a program in execution from one machine to another, this program being composed of a single process or a set of processes executing in parallel.

Through our researches, we try to address a different problem. Instead of securing a parallel application using checkpointing techniques, checkpointing techniques are used to introduce parallel computing inside sequential programs, i.e. to allow the parallel execution of parts of a program for which it is known these parts can be executed concurrently. This technique is called CAPE which stands for Checkpointing Aided Parallel Execution. It is important to note that CAPE does not detect if parts of a program can be executed in parallel. We consider it is the job of the developer (or another piece of software) to indicate what can be executed in parallel. CAPE consists in transforming an original sequential program into a parallel program to be executed on a distributed parallel system. As OpenMP already provides a set of compilation directives to specify parallel opportunities, we decided to use the same in order to avoid users from learning yet another API. As a result, our method provides a distributed implementation of OpenMP in a very simple manner. As a result, CAPE is a good alternative to provide a distributed implementation of OpenMP in a very simple manner.

Moreover, CAPE provides three main advantages. First, there is no need to learn yet another parallel programming environment or methodology as the specification of parallel opportunities in sequential programs is performed using OpenMP directives. Second, CAPE inherently introduces safety in the execution of programs as tools for checkpointing are used to run concurrent parts of programs in parallel. Third, more than one node is used only when necessary, i.e. when a part of the program requires only one node to execute (for example if this part is intrinsincly sequential), only one node is used for execution.

Other works have presented solutions to provide a distributed implementation of OpenMP [15]. Considering that OpenMP has been designed for shared-memory parallel architectures, the first solution was consisting in executing OpenMP programs on top of a distributed shared memory based machine [16]. More recently, other solutions have emerged, all aiming at transforming OpenMP code to other parallel libraries, like Global Arrays [17] or MPI [18].

This article aims at presenting the model we developed to analyze the behaviour of CAPE implementations. Using this model, we determine the optimal number of nodes one should use to minimize the execution time of a paralle program based on CAPE. The article is organized as follows. First, we present CAPE, our method to make a parallel program from a sequential program and demonstrate that if Bernstein's conditions are satisfied then the execution of a program using CAPE is equivalent to the sequential execution of the same program. Then, we introduce a model to describe the different steps involved in the execution of a program with CAPE; from the temporal equation derived from the model, we determine the optimal number of processors one should use to minimize the execution time of a given program.

2 CAPE

CAPE, which stands for Checkpointing Aided Parallel Execution, consists in modifying a sequential program (for which parts are recognized as being executable in parallel) so that instead of executing each part the one after the other one on a single machine, parts are automatically spread over a set of machines to be executed in parallel. In the following, we will just present the principle of CAPE to make the rest of the article readable. A deeper presentation is available in [19,20], which includes a proof of the concept.

As lots of works are on the way for distributing a set of processes over a range of machines, in the following we consider that another application (like Globus [21], Condor [22] or XtremWeb [23]) is available to start processes on remote nodes and get their results from there. In this section, this application is called the "dispatcher". CAPE is based on a set of six primitives:

- create (filename) stores in file "filename" the image of the current process. There are two ways to return from this function: the first one is after the creation of file "filename" with the image of the calling process; the second one is after resuming the execution from the image stored in file "filename". Somehow, this function is similar to the fork ( ) system call. The calling process is similar to the parent process with fork ( ) and the process resuming the execution from the image is similar to the child process with fork ( ). Note that it is possible to resume the execution of the image more than once; there is no such equivalence with the fork ( ) system call. The value returned by this function has a similar meaning as those of the fork ( ) system call. In case of error, the function returns -1. In case of success, the returned value is 0 if the current execution is the result of resuming its execution from the image and a strictly positive value in the other case (unlike the fork ( ) system call, this value is not the PID of the process resuming the execution from the image stored in the file).
- diff (first, second, delta) stores in file "delta" the list of modifications to perform on file "first" in order to obtain file "second".
- merge (base, delta) applies on file "base" the list of modifications from file "delta".
- restart (filename) resumes the execution of the process which image was previously stored in "filename". Note that, in case of success, this function never returns.
- copy (source, target) copies the content of file "source" to file "target".
- wait_for (filename) waits for any merges required to update file "filename" to complete.

The following presents two cases where CAPE can be applied automatically with OpenMP: the first case is the "parallel sections" construct and the second case is the "parallel for" construct.

In order to make the piece of code in Fig. 1 and Fig. 2 readable, operations for both the modified user program and the dispatcher have been included, and operations executed by the dispatcher are *emphasized.* Operations performed by the dispatcher are initiated by the modified user program while sending a message to the dispatcher. Operations are then performed as soons as possible but asynchronously regarding the modified user program.

Error cases (especially when saving the image of the current process) are not represented in Fig. 1(b). In some cases, a test is performed just after the creation of a new image in order to make sure it is not possible to resume the execution from the image direcly. This kind of image is used to evaluate variables updated by a specific piece of code and its execution is not intended to be effectively resumed.

2.1 The "Parallel for" Construct

In the case of for loops (see Fig. 1(a) and Fig. 1(b)), the first step consists in creating an "original" image for both having a reference for the "final" image

```
# pragma omp parallel for
for ( A ; B ; C )
    D
```

(a) OpenMP

→

```
parent = create ( original )
if ( ! parent )
    exit
copy ( original, target )
for ( A ; B ; C )
    parent = create ( before_i )
    if ( parent )
        ssh host_x restart ( before_i )
    else
        D
        parent = create ( after_i )
        if ( ! parent )
            exit
        diff ( before_i, after_i, delta_i )
        merge ( target, delta_i )
        exit
parent = create ( final )
if ( parent )
    diff ( original, final, delta )
    wait_for ( target )
    merge ( target, delta )
    restart ( target )
```

(b) CAPE

Fig. 1. Templates for "Parallel for"

and having a basis for the "target" image which is used at the end to resume the execution of the program sequentially after having executed the whole loop in parallel. Thus, once created, the "original" image is copied to a "target" image which will be updated step by step using deltas after the execution of each loop iteration.

The second step consists in executing loop iterations. After executing instruction A and checking condition B is true, a "before" image is created. When returning from function create (before), two cases are possible: either the function returns after creating the image, or the function returns after resuming from the image. In the first case, the dispatcher is asked to restart the image on a remote part and then the execution is resumed on the next loop iteration, i.e. executing C, evaluating B... In the second case, the loop iteration is executed and another image is saved in order to determine what are the modifications involved by the execution of this loop iteration (these modifications are then merged asynchronously to the "target" image by the dispatcher). Once all loop iterations have been executed, the third step consists in creating the "final" image which difference from the "original" image is used in order to set in image "target" modifications involved when executing C and evaluating B for the last time. Moreover, it ensures that the current frame in the execution stack is correct. This is the reason why it is necessary to wait for all other merges to be performed before including the one from "final". When restarting image "target", the execution resumes after create (final). As the process resumes its execution from an image, the last four lines in Fig. 1(b) are not executed the second time.

```
# pragma omp parallel sections
{
    # pragma omp section
    P1
    # pragma omp section
    P2
}
```

(a) OpenMP

→

```
parent = create ( original )
if ( parent )
    copy ( original, target )
    ssh host_x restart ( original )
    P1
    parent = create ( after1 )
    if ( parent )
        diff ( original, after1, delta1 )
        merge ( target, delta1 )
        wait_for ( target )
        restart ( target )
else
    P2
    parent = create ( after2 )
    if ( parent )
        diff ( original, after2, delta2 )
        merge ( target, delta2 )
```

(b) CAPE

Fig. 2. Templates for "Parallel sections"

2.2 The "Parallel Sections" Construct

Let P_1 and P_2 be two parts of a sequential program that can be executed concurrently. Fig. 2(a) presents the typical code one should write when using OpenMP and Fig. 2(b) presents the code to substitute to run P_1 and P_2 in parallel with CAPE.

The first step consists in creating an "original" image used to resume the execution on a distant node, calculate the delta for each part executed in parallel and build the "target" image to resume the sequential execution at the end.

The second step consists in executing parts and generating deltas. Thus, the local node asks the dispatcher to resume the execution of the "original" image on a distant node. Parts are executed, two "after" images are generated to produce two "delta" files; then, these "delta" files are merged to the "target" image; all these operations are executed concurrently. The main difficulty here is to make sure that both the current frame in the execution stack and the set of processor registers are consistent. However, this can be easily achieved using a good checkpointer.

The last step consists in making sure all "delta" files have been included in the "target" image and then restarting the "target" image in the original process.

3 Parallel Model

This section presents an exact analysis of the model of our distributed implementation of OpenMP using CAPE. We are interested in the optimal number of processors for which the processing time is minimal. The model presented below has been done on modelling the temporal equations for each state of the model under various assumptions regarding the master-slave paradigm.

In our model, there are two kinds of processors. The master is in charge of distributing and collecting the data and the slaves are devoted to process the data. By convention, the master node is allocated process #0 and the slaves are allocated process number 1 to P. As a result, the total number of processes is $P + 1$.

Let $\Delta^{'}$ (resp. $\Delta^{''}$) be the amount of memory sent to (resp. received from) the slaves. In order to simplify the model, it is considered that these amounts of data are the same for all slaves. This is effectively the case for $\Delta^{'}$ as the image of the original process do not change a lot from one loop iteration to another; we also consider it is the case for $\Delta^{''}$ considering data and processes are equitably shared among slaves processes.

In the model, let C be the time to create a checkpoint, S be the time to send a message, N be the network latency (N is inversely proportional to the throughput) and I be the time to receive a message to update the local image of the process. Finally, let t_1 be the sequential execution time and t_P be the execution time for each of the P processors. Considering that the amount of work is equitably shared among processors, $t_P = \frac{t_1}{P}$.

Fig. 3 presents the different steps involve in the processing of the work through several processes and the relationship between the master and the slaves. Let

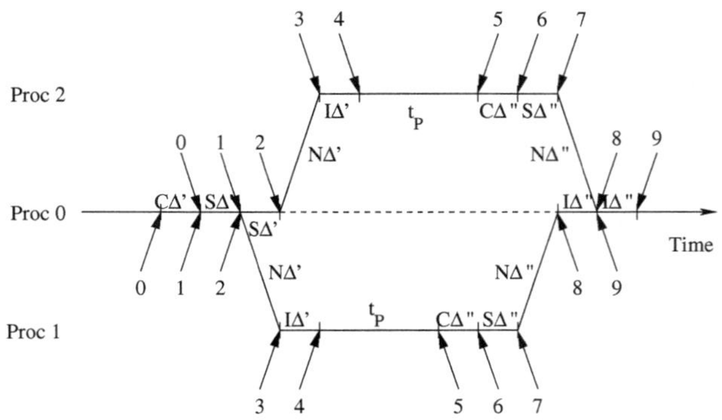

Fig. 3. The parallel model

$\Pi_{p,e}$ be the time when processor p ($p \in [0, P]$) reaches state e. As a result, with our model, we want to find the optimal value for P that makes $\Pi_{P,9}$ minimum.

From the analysis of the behaviour of our implementation of OpenMP for distributed memory architecture with CAPE and in accordance to Fig. 3, we can define $\forall p \in [0, P]$ the temporal equations (1) to (10).

$$\Pi_{p,0} = 0 \tag{1}$$
$$\Pi_{p,1} = \Pi_{p,0} + C\Delta^{'} \tag{2}$$
$$\Pi_{p,2} = \Pi_{p-1,2} + S\Delta^{'} \tag{3}$$
$$\Pi_{p,3} = \Pi_{p,2} + N\Delta^{'} \tag{4}$$
$$\Pi_{p,4} = \Pi_{p,3} + I\Delta^{'} \tag{5}$$
$$\Pi_{p,5} = \Pi_{p,4} + t_P \tag{6}$$
$$\Pi_{p,6} = \Pi_{p,5} + C\Delta^{''} \tag{7}$$
$$\Pi_{p,7} = \Pi_{p,6} + S\Delta^{''} \tag{8}$$
$$\Pi_{p,8} = \max(\Pi_{p,7} + N\Delta^{''}, \Pi_{p-1,9}) \tag{9}$$
$$\Pi_{p,9} = \Pi_{p,8} + I\Delta^{''} \tag{10}$$

Equations (3) and (9) can be normalized with: $\Pi_{0,2} = \Pi_{1,1}$ and $\Pi_{0,9} = \Pi_{P,2}$. In fact, both relations are due to the fact that all messages from the master must have been sent to the slaves before the master can receive the returning data from the slaves.

The set of equations from (1) to (10) can be reduced to the following ones:

$$\Pi_{p,2} = (C + pS + N)\Delta^{'}$$
$$\Pi_{p,7} = \Pi_{p,2} + (N + I)\Delta^{'} + t_P + (C + S)\Delta^{''}$$
$$\Pi_{p,9} = \max(\Pi_{p,7} + N\Delta^{''}, \Pi_{p-1,9}) + I\Delta^{''}.$$

4 Program Execution Time

The time needed to perform the execution of a program with CAPE is given by $\Pi_{P,9}$, ie. it is required the last processor has finished its work and returned it to the master node. In the following, we provide an expression for $\Pi_{P,9}$ with no recursion and that do not depend upon the value of p.

First, we need to determine the value for $\Pi_{P,8}$. From (9) and (10), we can see that the value for $\Pi_{P,8}$ can be expressed using the following recursion:

$$\Pi_{P,8} = \max \begin{cases} \Pi_{P,7} + N\Delta'' \\ \Pi_{P-1,8} + I\Delta'' \end{cases} \tag{11}$$

After replacing the value for $\Pi_{P-1,8}$ by the corresponding expression provided for $\Pi_{P,8}$ in (11), we obtain (12) for $\Pi_{P,8}$.

$$\Pi_{P,8} = \max_{i=0}^{P-2} \begin{cases} \Pi_{P-i,7} + (N+iI)\Delta'' \\ \Pi_{1,8} + (P-1)I\Delta'' \end{cases} \tag{12}$$

Then, $\Pi_{1,8}$ can be substituated its value using (9). After reduction, we have the following expression for $\Pi_{1,8}$:

$$\Pi_{P,8} = \max_{i=0}^{P-1} \begin{cases} \Pi_{P-i,7} + (N+iI)\Delta'' \\ \Pi_{P,2} + (P-1)I\Delta'' \end{cases}$$

After performing a variable change ($P-i \rightarrow i$), $\Pi_{P-i,7}$ and $\Pi_{P,2}$ are substituated by their respective values. From the new expression for $\Pi_{1,8}$, $\Pi_{1,9}$ can be provided $\forall p \in [0, P]$ using (10):

$$\Pi_{P,9} = \max \begin{cases} (C+pS+N+I)\Delta' + \frac{t_1}{P} + (C+S+N+(P-p+1)I)\Delta'' \\ (C+PS)\Delta' + PI\Delta'' \end{cases} \tag{13}$$

The set of expressions that depends upon the value of p in (13) can be reduced to two equations only. In fact, while taking p as a variable, this set of equations can be written:

$$(S\Delta' - I\Delta'')p + (C+N+I)\Delta' + \frac{t_1}{P} + (C+S+N+(P+1)I)\Delta'' \tag{14}$$

This is the equation of an affine function which maximum depends on the coefficient of p. As a matter of fact, we have two different interpretations for expression (14) according to the sign of $S\Delta' - I\Delta''$. If $S\Delta' < I\Delta''$, expression (14) is maximized when $p = 0$; if $S\Delta' > I\Delta''$, expression (14) is maximized when $p = P$.

As result, (15) for $\Pi_{P,9}$ does not depend upon the value of p and is not recursive.

$$\Pi_{P,9} = \max \begin{cases} \max(S\Delta', I\Delta'')P + \frac{t_1}{P} + (C+N+I)(\Delta' + \Delta'') + S\Delta'' \\ (S\Delta' + I\Delta'')P + C\Delta' \end{cases} \tag{15}$$

5 Optimal Number of Nodes

In the previous sections, we have defined a model to evaluate the performance of the execution of a parallel program based on the framework developed with CAPE and we have determine the time needed to perform the execution of a program. In this section, we use the model defined above to determine the optimal number of processors that one should use in the parallel machine in order to minimize the execution time of the parallel program.

Let P^* be the optimal number of processors that minimizes the execution time, ie. the number of processors that satisfies the following expression:

$$\Pi_{P^*,9} \leq \Pi_{P,9} \qquad \forall P$$

Let $f_1(P)$ and $f_2(P)$ be the first and the second expressions in (15) respectively, ie:

$$f_1(P) = \max(S\Delta^{'}, I\Delta^{''})P + \frac{t_1}{P} + (C+N+I)(\Delta^{'} + \Delta^{''}) + S\Delta^{''} \tag{16}$$
$$f_2(P) = (S\Delta^{'} + I\Delta^{''})P + C\Delta^{'}$$

The derivation in P of f_1 leads to the following expression:

$$f_1'(P) = \max(S\Delta^{'}, I\Delta^{''}) - \frac{t_1}{P^2} \tag{17}$$

One can demonstrate that there are only two zeros for this expression. Moreover, only one of both resides in $\mathcal{R}^+$. This means that there is just one inflection point in $\mathcal{R}^+$ for f_1. As f_2 is an affine function and as the coefficient for P in this expression is bigger than the one in f_1, there is one and only one intersection point between f_1 and f_2. As a result, the maximum for (15) is provided by f_1 before the intersection of both expressions and is provided by f_2 in the other case. In a more formal way:

$$P^* = \min \begin{cases} f_1 = f_2 \\ f_1' = 0 \end{cases}$$

In order to determine the intersection of both expressions, we compute $f_1(P) - f_2(P) = 0$ which leads to expression (18).

$$\min(S\Delta^{'}, I\Delta^{''})P^2 - \left[(C+S)\Delta^{''} + (N+I)(\Delta^{'} + \Delta^{''})\right] P - t_1 = 0 \tag{18}$$

Note that the maximum in (16) has been transformed into a minimum in (18) using the following relation:

$$x + y = \min(x,y) + \max(x,y) \qquad \forall x, y$$

Equation (18) is a second degree equation in P. Considering that C, S, N, I, $\Delta^{'}$, $\Delta^{''}$ and t_1 are all positive constants, the associated discrimant is necessarily strictly positive and there is one and only one solution in $\mathcal{R}^+$.

In order to determine the value of P^* in the second case, we have to determine the zero of $f_1'(P) = 0$. From (17), $f_1'(P) = 0$ can be transformed into a second degree equation of the form $ax^2 + b = 0$ which provides a single obvious solution in $\mathcal{R}^+$.

6 Conclusion

This article is based on an original way of transforming a sequential program into a parallel program, CAPE for Checkpointing Aided Parallel Execution, that uses checkpointing techniques rather than shared memory or any message-passing library. Then, we introduced a model to evaluate the performane of the distributed execution. In order to do so, we defined the parameters of the model and provided the temporal equations. Finally, we determined the optimal number of processors that should be used to minimize the execution time.

In the current model, we considered that the time spent by each slave processor is the same on all slaves. In the future, we will consider that these time may vary a little bit from one process to another.

References

1. Buyya, R.: High Performance Cluster Computing: Architectures and Systems, vol. 1. Prentice-Hall, Englewood Cliffs (1999)
2. Foster, I., Kesselman, C., Tuecke, S.: The Anatomy of the Grid: Enabling Scalable Virtual Organizations. The International Journal of High Performance Computing Applications 15(3), 200–222 (2001)
3. Leuf, B.: Peer to Peer. In: Collaboration and Sharing over the Internet, Addison-Wesley, Reading (2002)
4. Loveman, D.B.: High Performance Fortran. IEEE Parallel & Distributed Technology: Systems & Applications 1(1), 25–42 (1993)
5. Geist, A., Beguelin, A., Dongarra, J., Jiang, W., Manchek, R., Sunderam, V.S.: PVM: Parallel Virtual Machine: A Users' Guide and Tutorial for Network Parallel Computing. Scientific and Engineering Computation Series. MIT Press, Cambridge (1994)
6. Snir, M., Otto, S., Huss-Lederman, S., Walker, D., Dongarra, J.: MPI: The Complete Reference, 2nd edn. Scientific and Engineering Computation Series. MIT Press, Cambridge (1998)
7. OpenMP Architecture Review Board: OpenMP Application Program Interface, Version 2.5 Public Draft (2004)
8. Allen, J.R., Callahan, D., Kennedy, K.: Automatic Decomposition of Scientific Programs for Parallel Execution. In: Proceedings of the 14th ACM SIGACT-SIGPLAN symposium on Principles of programming languages, Munich, West Germany, pp. 63–76. ACM Press, New York (1987)
9. Feautrier, P.: Automatic parallelization in the polytope model. In: Perrin, G.-R., Darte, A. (eds.) The Data Parallel Programming Model. LNCS, vol. 1132, pp. 79–103. Springer, Heidelberg (1996)
10. Barthou, D., Feautrier, P., Redon, X.: On the Equivalence of Two Systems of Affine Recurrence Equations. In: Monien, B., Feldmann, R.L. (eds.) Euro-Par 2002. LNCS, vol. 2400, pp. 309–313. Springer, Heidelberg (2002)
11. Alias, C., Barthou, D.: On the Recognition of Algorithm Templates. In: Knoop, J., Zimmermann, W. (eds.) Proceedings of the 2nd International Workshop on Compiler Optimization meets Compiler Verification, Warsaw, Poland, pp. 51–65 (2003)
12. Web page: Ckpt (2005), `http://www.cs.wisc.edu/~zandy/ckpt/`

13. Osman, S., Subhraveti, D., Su, G., Nieh, J.: The Design and Implementation of Zap: A System for Migrating Computing Environments. In: Proceedings of the 5th USENIX Symposium on Operating Systems Design and Implementation, Boston, MA, pp. 361–376 (2002)
14. Plank, J.S.: An Overview of Checkpointing in Uniprocessor and Distributed Systems, Focusing on Implementation and Performance. Technical Report UT-CS-97-372, Department of Computer Science, University of Tennessee (1997)
15. Merlin, J.: Distributed OpenMP: extensions to OpenMP for SMP clusters. In: EWOMP 2000. 2nd European Workshop on OpenMP, Edinburgh, UK (2000)
16. Karlsson, S., Lee, S.W., Brorsson, M., Sartaj, S., Prasanna, V.K., Uday, S.: A fully compliant OpenMP implementation on software distributed shared memory. In: Sahni, S.K., Prasanna, V.K., Shukla, U. (eds.) HiPC 2002. LNCS, vol. 2552, pp. 195–206. Springer, Heidelberg (2002)
17. Huang, L., Chapman, B., Liu, Z.: Towards a more efficient implementation of OpenMP for clusters via translation to global arrays. Parallel Computing 31(10–12), 1114–1139 (2005)
18. Basumallik, A., Eigenmann, R.: Towards automatic translation of OpenMP to MPI. In: Proceedings of the 19th annual international conference on Supercomputing, pp. 189–198. ACM Press, Cambridge, MA (2005)
19. Renault, E.: Parallel Execution of For Loops using Checkpointing Techniques. In: Skeie, T., Yang, C.S. (eds.) Proceedings of the 2005 International Conference on Parallel Processing Workshops, Oslo, Norway, pp. 313–319. IEEE Computer Society Press, Los Alamitos (2005)
20. Renault, E.: Automatic Parallelization made easy using Checkpointing Techniques. In: Oudshoorn, M., Rajasekaran, S. (eds.) Proceedings of the 18th International Conference on Parallel and Distributed Computing Systems, Las Vegas, NV (2005)
21. Foster, I., Kesselman, C.: Globus: A Metacomputing Infrastructure Toolkit. The International Journal of Supercomputer Applications and High Performance Computing 11(2), 115–128 (1997)
22. Litzkow, M., Livny, M., Mutka, M.: Condor - A Hunter of Idle Workstations. In: The 8th International Conference on Distributed Computing Systems, San Jose, CA, pp. 104–111. IEEE Computer Society Press, Los Alamitos (1988)
23. Fedak, G., Germain, C., Néri, V., Cappello, F.: XtremWeb: A Generic Global Computing System. In: Buyya, R., Mohay, G., Roe, P. (eds.) Proceedings First IEEE/ACM International Symposium on Cluster Computing and the Grid, Brisbane, Australia, pp. 582–587. IEEE Computer Society Press, Los Alamitos (2001)

Throttling I/O Streams to Accelerate File-IO Performance

Seetharami Seelam[1,*], Andre Kerstens[2,*], and Patricia J. Teller[3]

[1] IBM Research, Yorktown Heights, NY 10598, USA
sseelam@us.ibm.com
[2] SGI, Sunnyvale, CA 94085, USA
kerstens@sgi.com
[3] The University of Texas at El Paso (UTEP), El Paso, TX 79968, USA
pteller@utep.edu

Abstract. To increase the scale and performance of high-performance computing (HPC) applications, it is common to distribute computation across multiple processors. Often without realizing it, file I/O is parallelized with the computation. An implication of this is that multiple compute tasks are likely to concurrently access the I/O nodes of an HPC system. When a large number of I/O streams concurrently access an I/O node, I/O performance tends to degrade, impacting application execution time. This paper presents experimental results that show that *controlling* the number of file-I/O streams that concurrently access an I/O node can enhance application performance. We call this mechanism file-I/O stream throttling. The paper (1) describes this mechanism and demonstrates how it can be implemented either at the application or system software layers, and (2) presents results of experiments driven by the cosmology application benchmark MADbench, executed on a variety of computing systems, that demonstrate the effectiveness of file-I/O stream throttling. The I/O pattern of MADbench resembles that of a large class of HPC applications.

1 Introduction

Many parallel high-performance computing (HPC) applications process very large and complex data sets. Often the file I/O needed to store and retrieve this data is parallelized along with the computation. This generates multiple I/O streams (e.g., one for each MPI task) that concurrently access the system's I/O nodes. Because the number of I/O nodes usually is much smaller than the number of compute processors, even when only one such application is executing on a system, often an I/O node is concurrently accessed by multiple streams. Due to the advent of multicore processors, as the number of processors in next-generation systems grows to solve larger problem sizes and as the ratio of compute processors to I/O nodes increases, the number of I/O streams concurrently accessing an I/O node is destined to increase. Currently, this ratio is greater than 40 for some high-end systems [1].

* This work was done while the author was affiliated with UTEP.

R. Perrott et al. (Eds.): HPCC 2007, LNCS 4782, pp. 718–731, 2007.

In some cases, e.g., when the I/O nodes are RAIDs, I/O parallelization can result in improved I/O performance. But, as indicated in [2] and as confirmed by our experimental results, I/O performance drops rather precipitously in many cases. When the number of I/O streams exceeds a certain threshold, application execution times are impacted negatively – this is due to events that prevent full I/O utilization, e.g., seek and rotational latencies, failure to take advantage of prefetching, and/or insufficient I/O request frequency.

For HPC applications similar to MADCAP CMB [3], traditional parallelization of file I/O can result in this behavior, which will be exacerbated in next-generation systems. The challenge is, given a workload, schedule I/O so that as much I/O-node bandwidth as possible is realized. Although this paper does not address this challenge optimally, it describes and gives evidence of a potential solution, i.e., dynamic selection of the number of concurrently-active, synchronous, file-I/O streams. We call this control mechanism *file-I/O stream throttling.*

The potential effectiveness of file-I/O stream throttling is demonstrated in Figure 2(a). The figure compares MADbench[1] execution times when one of 16, four of 16, eight of 16, and 16 of 16 MPI tasks generate I/O streams that concurrently access an I/O node. The depicted experimental results show that careful selection of the number of concurrently-active MADbench file-I/O streams can improve application execution time: in this case, the optimal number of I/O streams is four of 16. In comparison, with one of 16 or 16 of 16 (default parallelization), execution time increases by 18% and 40%, respectively.

The paper describes this potential I/O performance problem in detail, shows its impact on MADbench, which represents the I/O behavior of a large application class, presents the file-I/O stream throttling mechanism, and demonstrates the effect of this mechanism on I/O performance and application execution time. I/O-stream throttling can be performed at the application or system software layers. The former is achieved by having the application code specify the number of streams concurrently accessing an I/O node; this is the method that is incorporated in the MADbench code. Throttling via system software could be implemented by stream-aware I/O schedulers or file-system policies. The strengths and weaknesses of each method are described in the paper.

The remainder of the paper is organized as follows. Section 2 describes related work, Section 3 describes two different stream throttling methods, and Section 4 describes MADbench. Section 5 presents the systems used for experimentation, while the results and their implications are presented in Section 6. Section 7 concludes the paper.

2 Related Work

There are several studies that concentrate on optimizing the parallel file-I/O performance of HPC applications by providing libraries and parallel file systems, and

[1] MADbench is a benchmark that represents the I/O behavior of the MADCAP CMB application.

exploiting advancements in storage technologies [4,5,6]. Other optimization techniques include exploiting access patterns to assist file-system prefetching, data sieving, and caching [7], overlapping computation with I/O [8], and employing asynchronous prefetching [9]. The latter technique is particularly suitable for HPC applications, which generally have very regular data access patterns that facilitate the identification of the data needed for subsequent computations. Given sufficient disk system bandwidth, prefetching may minimize the effect of I/O-node performance on application performance. However, prefetching too aggressively increases memory traffic and leads to other performance problems [10]. In addition, as mentioned above, when bandwidth is limited, the problem of multiple streams concurrently accessing an I/O node is likely; this can result in a loss of the benefits of prefetching.

Our contribution differs from these methods but is complementary to the data sieving, data prefetching, and caching approaches. The focus of our work is on runtime control of the number of file-I/O streams that concurrently access an I/O node. Such control can minimize expensive disk-head positioning delays and, thus, increase both I/O and application performance.

3 File-I/O Stream Throttling

File-I/O stream throttling controls the number of synchronous I/O streams concurrently accessing the I/O system. In general, it must consider various *application characteristics*, e.g., request characteristics (sequential or random) and *system characteristics*, e.g., the numbers of I/O and compute nodes, numbers of processors and tasks on a compute node, storage system configuration, compute-node memory size, and application data size. The large number of characteristics and the complexities associated with them make dynamic I/O-stream throttling a challenging task, one that has not yet been accomplished.

However, many HPC applications (see, e.g., [3,11,12]) have sequential data layouts, where each requested data stream is sequentially accessed from the storage system, and tend to read in an iterative fashion. For such data layouts and read behaviors, the number of streams that should concurrently access an

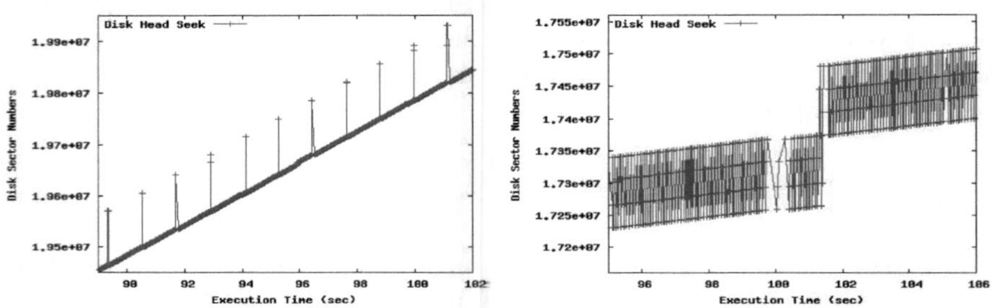

(a) Sequential stream (b) Concurrent streams

Fig. 1. Seek behavior of the disk head when using one or multiple I/O streams

I/O node is dependent only on the I/O node configuration; the complexities associated with application characteristics need not be considered. In this case, given a single-disk I/O node, I/O performance can be enhanced by minimizing expensive disk-head positioning delays; our experiments demonstrate this. For example, consider an application with a computational loop that is iterated multiple times by four MPI tasks that access unique files stored in an interleaved fashion on a shared disk. When all four streams concurrently access the disk with no coordination the disk head continuously seeks between files; this behavior is illustrated in Figure 1(b). In contrast, as shown in Figure 1(a), when the number of streams that concurrently access the disk is one (the four streams access the disk one after another) the disk head hardly seeks. In this particular example, one stream results in the best performance because the storage system is one simple disk, the bandwidth of which is well utilized by the one I/O stream. In contrast, an advanced storage system with multiple disks needs multiple streams to utilize the available bandwidth. Thus, one should not conclude from this example that using one stream is the best for all storage systems.

3.1 Application-Layer Stream Throttling

One way to change (throttle) the number of streams concurrently accessing an I/O node is to wrap the parallelized I/O routines with code that implements a token-passing mechanism [3] – this approach is used by some HPC applications, including MADbench. If one token is used by all MPI tasks then access to storage is sequentialized. At the other extreme, if the number of tokens is equal to the number of MPI tasks, then all I/O streams can concurrently access storage.

The token-passing approach has two problems. First, it involves changes to the application code, which often is targeted to run on various compute- and I/O-node configurations. As shown in this paper, the selection of the number of concurrently-active I/O streams depends on application I/O behavior and system characteristics. Thus, selecting the number requires extensive knowledge of both, as well as the ability to develop configuration-specific code that is tailored to the various systems. As demonstrated in [3], it is not impossible to throttle the number of I/O streams at runtime within an application. However, the use of higher-level I/O libraries such as MPI-I/O can make this very challenging.

Second, I/O throttling is applicable only to synchronous I/O streams – not to asynchronous I/O streams. A synchronous request blocks the execution of the requesting task until it is satisfied by the underlying I/O system, while an asynchronous request does not block task execution. In fact, asynchronous requests often are queued and delayed in the system's underlying memory buffer cache and scheduled to the relatively slower I/O system at times that provide opportunities for improved I/O performance. It is well known that larger asynchronous request queues provide better opportunities for optimizing disk head movement and improving I/O system performance. Thus, throttling the number of asynchronous request streams is dangerous – it may constrain the number of streams that generate data to the memory buffer cache and, thus, the number of outstanding asynchronous requests in the buffer cache queue. Due to these two problems, application-layer I/O-stream throttling is not always the easiest and most scalable option. Other alternatives, such as throttling in the OS, file system, and/or middleware, should be explored to improve I/O-node performance. In the next section, we discuss how I/O-stream throttling can be realized in the OS layer of an I/O node.

To properly understand the effect of stream throttling at the application layer, we must ensure that the number of synchronous I/O streams are not throttled at other layers between the application and storage system. For example, I/O scheduling algorithms in the OS often tend to minimize the positional delays on the disk by servicing requests in a pseudo shortest-seek-first manner, rather than a fair first-come-first-served manner – this is true of our test OS, Linux. Fortunately, the latest versions of Linux allow users to select among four I/O scheduling algorithms, one of which is a completely fair queuing algorithm (CFQ); we use CFQ to demonstrate our application throttling approach.

3.2 System-Layer Stream Throttling

In Linux, the Anticipatory Scheduler (AS), unlike CFQ and other I/O schedulers, has the ability to detect sequentiality in an I/O stream and allow a stream that exhibits spatial locality of reference to continue to access the storage system for a certain period of time. This throttling reduces disk-head positioning delays even when multiple streams compete for the disk. In addition, the AS detects asynchronous and synchronous streams and throttles only the synchronous streams. We use AS to demonstrate the effect of system-software throttling.

The main advantage of the system-layer approach, over the application-layer approach, is that it is transparent to the application and application programmer. There are many ways to implement I/O-stream throttling at the system software layer, including the scheduler approach that we use in our experiments. Disk controllers on modern I/O systems have intelligence built into them. This means that I/O-stream throttling implemented at the controller could be more beneficial than if implemented at the I/O-node OS software layer. Also, due to adaptive prefetching policies, some disk controllers can handle multiple streams very efficiently. For instance, [2] shows that storage devices can handle as many as three streams efficiently. However, when the number of disk cache segments is less than the number of streams, performance degrades significantly. Unfortunately, the number of concurrently-active I/O streams that the device can handle effectively is not always communicated to the OS software layer. Thus, in comparison to throttling at the device, throttling at the OS layer may lose some benefits. Also, since the AS is designed only to throttle down the number of streams to one and is not able to judiciously change the number of concurrently-active I/O streams, this scheduler will not be optimal for all system configurations. Currently, a scheduler capable of judiciously selecting the number of concurrently-active I/O streams per I/O node is under investigation.

An experimental evaluation of our system-layer throttling approach is described in Section 6.2. These experiments use the AS and are driven by the unthrottled, concurrently-active file-I/O streams of MADbench – every MPI task of MADbench concurrently generates I/O.

4 MADbench Cosmology Application Benchmark

MADbench [3] is a light-weight version of the Microwave Anisotropy Dataset Computational Analysis Package (MADCAP) Cosmic Microwave Background (CMB) power spectrum estimation code [13]. It was developed using MPI and is one of a set of benchmarks used by NERSC at LBNL to make procurement decisions regarding large-scale DoE computing platforms. MADbench represents a large class of applications that deal with Cosmic Microwave Background Analysis. A significant amount of HPC resources are used by cosmologists, e.g., about 17% at NERSC. The I/O characteristics of MADbench are similar to those of many other applications and benchmarks, such as SMC (electron molecule collision), SeisPerf/Seis (seismic processing), CAM, FLASH [12], and codes based on the Hartree Fock algorithm like NWChem and MESSKIT [14,11]. This indicates that this study, which is based on MADbench, has relevance to a large class of applications and can be used to improve the I/O performance of applications with I/O behavior that is amenable to file-I/O stream throttling.

MADbench has three distinct I/O phases: (1) `dSdC` (write only): Each MPI task calculates a set of dense, symmetric, positive semi-definite, signal correlation derivative matrices and writes these to unique files. (2) `invD` (read only): Each MPI task reads its signal correlation matrices from the corresponding unique files and for each matrix it calculates the pixel-pixel data correlation (dense, symmetric, positive definite) matrix D and inverts it. (3) `W` (read only): Each MPI task reads its signal correlation matrices and for each matrix performs dense matrix-matrix multiplications with the results of `InvD`. Each of these three I/O phases are comprised of the following five steps: (1) computation, (2) synchronization (3) I/O, (4) synchronization, and (5) computation (except for `dSdC`). The time that each task spends in the first (second) sync step can be used to identify computational (I/O) imbalance among the tasks.

MADbench includes the following parameters that can be used to generate configurations that differ with respect to the stress they apply on the I/O system:

- `no_pix`: size(in pixels) of the pixel matrix. The size, S, in KBytes is $\frac{1}{8}(no\_pix)^2$. With P tasks, each task accesses a file of $\frac{S}{P}$ KBytes.
- `no_bin`: number of chunks of data in a file, where the chunk size, C, is $(S/P)/no\_bin$. The choice of `no_bin` impacts the layout and the interleaving of the files on the disk. The number of chunks of data in a file equals the number of iterations of the three phases.
- `no_gang`: number of processors for gang scheduling. In our case there is no gang scheduling, i.e., `no_gang` $= 1$.
- `blocksize`: ScaLAPACK block size.

In our experiments we use `no_bin` $= 16$ and `blocksize` $= 32$KB. These values are comparable to the values used in the original MADbench study [3]. The value for `no_pix` is based on the system's memory size and must be set large enough to stress the system's storage subsystem. If `no_pix` is too small, all the data will fit in memory and the I/O subsystem will not be stressed.

Because of the MADbench lock-step execution model, all tasks concurrently read or write their data from or to storage, creating multiple concurrently-active I/O streams. Two additional input parameters, `RMOD` and `WMOD`, can be used to control the number of concurrently-active reader and concurrently-active writer streams, respectively. At most, one out of `RMOD` (`WMOD`) tasks are allowed to concurrently access the I/O system. With N tasks, the total number of concurrent readers is $\lceil \frac{N}{RMOD} \rceil$. MADbench can be executed in two modes: computation and I/O, or I/O only [3]; we use both in our experiments.

5 Experimental Systems

Table 1 describes the three systems used in our experiments (Intel Xeon, IBM p690, and Scyld Beowulf cluster) in terms of the number of processors, memory capacity per node, I/O system configuration, and I/O system throughput. The Intel Xeon system, which runs Linux 2.6, contains dual processors (2.80 GHz Pentium 4 Xeons with Hyper Threading) and is attached to a 7,200 RPM 10.2GB EIDE disk. The IBM p690 is a 16-way POWER4 SMP, which also runs Linux 2.6. Experiments on the p690 are of two types w.r.t. memory size and I/O system configuration: memory size is either 16GB or 2GB, while the I/O system is either a 7,200 RPM 140GB SCSI disk or a 350GB DS4300 RAID-5 comprised of six 15,000 RPM SCSI disks. Each node of the Scyld Beowulf cluster contains two 2.0 GHz AMD Opteron processors and 4GB memory; it runs a custom Linux 2.4 kernel from Scyld. Experiments on this system also are of two types, but w.r.t. the I/O system configuration: either a 1.4TB NFS-mounted RAID-5 disk system comprised of six 7,200 RPM 250GB SATA disks or a local 7,200 RPM 120GB SATA scratch disk per node. On all the systems, to eliminate the interference of I/O requests from OS activities, MADbench is configured to access a storage system different than that hosting the OS. In addition, on all systems, to remove buffer cache effects, before an experiment is started, the MADbench storage system is unmounted and remounted.

Table 1. I/O- and compute-node configurations of systems

System	Nodes	CPUs/ Node	Memory/ Node(GB)	I/O System(s)	Throughput (MB/s)
Intel Xeon	1	4	1	EIDE disk	23
IBM p690	1	16	16/2	SCSI disk/RAID-5	35/125
Beowulf Cluster	64	2	4	SATA/RAID-5	45

6 Results of Experiments

To understand the importance of I/O performance, we conducted an experiment with MADbench running in computation-and-I/O mode on the Xeon system. In this case, I/O time, which depends on the number of concurrently-active readers and writers, ranges from 14% to 37% of total execution time. Borrill, et al. [3] report that on larger systems with >= 256 processors I/O consumes 80% of execution time. This indicates that for this class of applications I/O contributes significantly to application execution time.

To understand the true impact of I/O performance, one must ensure that the data footprint does not fit in the main memory buffer cache. Otherwise, requests are satisfied by the buffer cache rather than the I/O system. With the explosive growth of the data needs of HPC applications, it is rare that the data ever fits in main memory. Thus, the remainder of our experiments ensure that the data footprint is larger than the combined size of the main memory of the participating compute nodes.

We conducted two sets (base and base+) of experiments driven by MADbench in I/O-only mode. Each experiment in the *base set* has one or all readers and writers. The *base+* set is the base set plus the following experiments: (1) two readers/one writer, (2) four readers/one writer, and (3) eight readers/one writer.

6.1 Application-Layer Stream Throttling

The experiments discussed here use application throttling to control the number of concurrently-active MADbench readers and writers. To isolate the effect of application throttling, these experiments use the CFQ I/O scheduler, which provides relatively fair disk access to all concurrently-active MADbench I/O streams and does not inherently perform any I/O-stream throttling, as does AS.

The base set of experiments was conducted on four system configurations, while the base+ set was conducted on two. Rows E1 through E3 of Table 2 show the input parameters for the base set of experiments and rows E4 and E5 show this for the base+ set. Table 2 also shows the best and second-best number of readers/writers for all experiments. Results of each experiment are presented in terms of total execution time and I/O times (LBSTIO and TIO). Since each MPI task waits at a barrier after each iteration of each MADbench phase, LBSTIO represents the average load balancing time, including barrier synchronizations. TIO is the average task I/O time. Note that LBSTIO depends on the number of concurrently-active I/O streams and the underlying I/O scheduler. With a "fair" I/O scheduler, if all tasks are allowed to access the I/O system concurrently, all finish at approximately the same time, hence, the amount of time each waits at a barrier for other tasks to finish will be minimal. In contrast, as the number of concurrently-active I/O streams decreases, tasks scheduled earlier finish earlier and experience longer delays at a barrier waiting for other tasks that are scheduled later. Because of this, we expect an increase in LBSTIO, indicating that some tasks finish their I/O earlier than others.

Table 2. Parameter values for experiments

Exp. #	System/ Storage	# of tasks	no_pix	Data Size (GB)	Memory (GB)	Two Best Readers/Writers
E1	p690/single disk	16	5,000	2.98	16	16/16 and 1/16
E2	Xeon/single disk	4	3,000	1.1	1	1/1 and 1/4
E3	p690/single disk	16	5,000	2.98	2	1/1 and 1/16
E4	Cluster/RAID-5/NFS	16	25 000	74.5	64	8/1 and 4/1
E5	p690/RAID-5	16	10 000	11.9	2	2/1 and 4/1

Intel Xeon System: The results of the base set of experiments run on this system are shown in Table 3. Since the system contains dual processors, the number of logical processors and, thus, the number of MPI tasks, is four. As shown in Table 3, the default practice of all readers/all writers (4/4), results in the longest application execution time. In contrast, employing one reader, reduces execution time by as much as 22%. Note, however, that this decrease in execution time is accompanied by a nine-fold increase in the LBSTIO time in the W (InvD) phase, resulting in load balancing issues.

Given a fixed number of readers, the number of writers has a negligible impact on total execution time – less than or equal to 5%. In contrast, given a fixed number of writers, one reader, versus "all readers", in this case four, improves execution time by as much as 19.5%.

Table 3. Application-layer stream throttling: Total execution and I/O times for base set of experiments run on Intel Xeon system

Number of Writers	Number of Readers	Time (s)						
		Total	dSdC		InvD		W	
			TIO	LBSTIO	TIO	LBSTIO	TIO	LBSTIO
4	4	243	12	9	83	3	77	1
1	4	235	12	5	83	2	77	1
4	1	199	12	12	53	13	45	10
1	1	189	18	9	48	10	42	10

IBM p690 – Single Disk/2GB Memory: For this system the number of processors (number of MPI tasks) is 16. Similar to the results associated with the Xeon system, given a fixed number of writers, one reader, versus "all readers", improves execution time. In this case, execution time increases by as much as 14% with 16 readers. Again, with this execution time improvement comes an increase in the I/O load-balancing time, LBSTIO; it increases from less than ten seconds to as much as several tens of seconds. Similar to the results of the experiments run on the Xeon system, given a fixed number of readers, the number of writers has a negligible impact (less than 3%) on total execution time.

Table 4. Application-layer stream throttling: Total execution and I/O times for base+ set of experiments run on Beowulf cluster with RAID-5

Number of Writers	Number of Readers	Time (s)						
		Total	dSdC		InvD		W	
			TIO	LBSTIO	TIO	LBSTIO	TIO	LBSTIO
16	16	6353	1330	66	2150	72	2210	68
16	1	8690	1320	67	1815	1596	1826	1606
1	1	6973	1559	1071	1148	793	1146	794
1	2	6406	1543	1000	1746	554	1128	568
1	4	5741	1583	1004	1073	273	1077	265
1	8	5627	1590	1035	1255	17	1249	17
1	16	6131	1781	1155	1374	12	1319	28

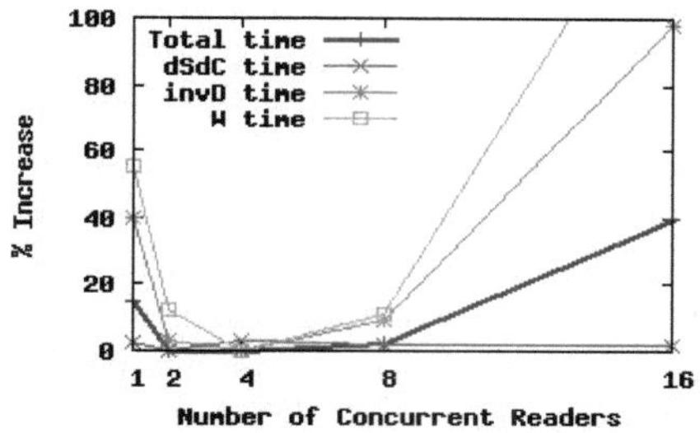

(a) IBM p690 with RAID-5

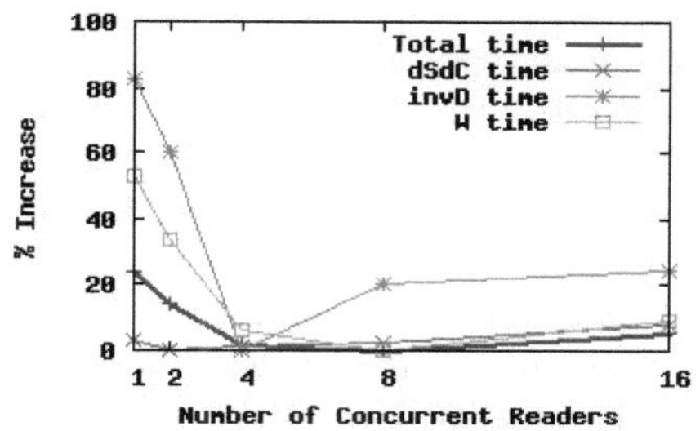

(b) Beowulf with RAID-5

Fig. 2. MADbench total and phase execution times for base+ set of experiments run with different numbers of concurrently-active readers on p690 with RAID-5 and Beowulf cluster with RAID-5

Beowulf Cluster with RAID-5: The results of the base+ set of experiments run on the Scyld Beowulf cluster with a 1.4TB NFS-mounted RAID-5 disk system comprised of six 7,200 RPM 250GB SATA disks are shown in Table 4. Only 16 processors, i.e., one processor on each of 16 compute nodes (16 MPI tasks) were employed in the experiments. As shown in row E4 of Table 2, each of the 16 nodes has 4GB memory (64GB total), thus, a larger problem size ($no\_pix = 25,000$) is used in order to force disk accesses.

Since each node runs one MPI task and the RAID-5 supports multiple concurrent streams, we expected that given a fixed number of readers, 16 (all) writers, rather than one, would result in better execution time for the write phase (dSdC). One writer is not likely to fully utilize the storage system bandwidth and "all writers" results in larger I/O queues of asynchronous requests, which is known to improve I/O performance [15]. As shown in Table 4, "all writers", as compared to one, does significantly reduce the `dSdC` write-phase I/O time (TIO) by over 25% and results in better use of storage system bandwidth. A total of 74.5 GB is written in 1,900 seconds, which translates to 40MB/s, which is 90% of peak bandwidth (45 MB/s). In contrast, one writer takes about 3,000 seconds, which translates to 25.5 MB/s, which is less than 60% of peak bandwidth. In addition, as shown in Table 4, "all writers" results in relatively smaller load imbalance times (LBSTIO) in the dSdC phase. However, the reduction in the dSdC phase execution time due to the use of "all writers" does not translate into a smaller application execution time because "all writers" has a negative impact on the I/O times of the subsequent read phases: given a fixed number of readers, "all writers" results in 60%, 69%, and 20% increases in the `InvD` and `W` TIO times, and total execution time, respectively. We speculate that this is due to data placement on the disk system. When multiple streams compete for storage, the file system tends to allocate chunks to the streams in pseudo round-robin fashion; this results in files that, although contiguous in memory, reside on non-contiguous disk blocks. Thus, with multiple writers block allocation seems to be random, which would significantly impact the subsequent read performance. To circumvent the random placement, we use only one writer for our other experiments. We anticipate that a single writer will remove randomness in block allocation and result in consistent execution times.

Given one writer, we next identify the optimal number of concurrently-active readers (1, 2, 4, 8, or 16) for this system. For each number, Figure 2(b) gives the percentage increase in MADbench's total and phase execution times; Table 4 contains the data. The sweet spot is eight. In comparison, with 16 readers, the total execution time is about 8% higher; with two, 13% higher; with four, less than 2% higher; and with one, more than 20% higher. With one reader the disk system is underutilized (less than 85% peak bandwidth) because there are not enough I/O requests; with 16, there are enough I/O requests but the disk system spends a significant amount of time seeking to different locations (positional delays) to service readers. In addition, as pointed out in [2], it is likely that 16 reader streams are more than the number of prefetch streams that the storage system can handle; modern storage arrays can prefetch four to eight sequential streams [2]. With eight readers, however, the disk system is well utilized (close to 100% of peak bandwidth), which only means that the positional delays are minimized and the storage system prefetching policies are effective. In summary, as compared to the all writers/all readers (one writer/one reader) case, with one writer and eight readers, application execution time improves by as much as 11% (by about 20%) – see Table 4 and Figure 2.

IBM p690: RAID-5/2GB Memory: We repeated the `base+` set of experiments described in the last section on the IBM p690 with a high-end 350GB RAID-5 I/O system. This is done to see if improved execution time results from throttling down the number of readers, i.e., not using "all readers", in other types of storage systems. Note that the manufacturer of the p690's storage system is different than that of the cluster's. In addition, the p690's storage system is comprised of more expensive hardware, has different disks, and has a faster disk spindle speed (15,000 RPM as compared to 7,200 RPM). The p690's storage system gets a peak bandwidth of 125MB/s, while that of the cluster gets 45 MB/s. However, both systems have the same number of disks (six) and use the same organization (RAID-5). Thus, we expected and got similar results. The results, shown in Figure 2(a), show that the sweet spot is four concurrently-active readers. In comparison, 16 readers increases the execution time by about 40%; one increases it by over 13%; and for four and two readers, the difference is less than 2%. Reasoning and analysis similar to that presented in Beowulf cluster section can be used to explain these performance results.

Two interesting observations can be made by comparing Figures 2(a) and 2(b). First, examining the slope of the execution curves between 8 and 16 readers (although, we do not have the data for the intermediate points, it is not required for this observation), one could easily conclude that for MADbench the cluster's storage system handles high concurrency better than the p690's storage system. In comparison, the performance of the p690's storage system drops rather dramatically. Second, examining the slope of the execution time curves between two and four readers, at a concurrency of two readers, the cluster's storage system seems more sensitive. We must acknowledge that since the file system plays a major role in performance, and since these systems have different file systems, this may not be attributable to the storage system alone.

6.2 System-Layer Stream Throttling

As discussed in Section 3.2, our experiments with system software I/O-stream throttling are implemented at the OS layer using the Linux 2.6 Anticipatory Scheduler (AS). This I/O scheduler does not throttle the number of concurrently-active writers; it throttles only the number of concurrently-active readers. In addition, it is designed to reduce the number of reader streams to one. We are working on a scheduler that allows throttling to an arbitrary number of streams.

Using the AS, which is available only under Linux, our system-layer throttling experiments reduce to experiments with "all writers" and one reader on the Intel Xeon and IBM p690 systems. For comparison purposes, we also present the no-throttling case, which is implemented using the CFQ I/O scheduler, and the best case from the application-layer throttling experiments. The results of our experiments indicate that the AS throttling behaves in a way that is consistent with the results of our application-layer throttling experiments. With few exceptions, the number of writers (fixed in this case and variable in application throttling) has no significant impact on application execution time and one reader is a good choice.

On the Intel Xeon system, although using the AS (System) decreases execution time by 12%, as compared to the unthrottled case (UT), application throttling performs best; its execution time is about 22% less than that of UT and 11% less than that of System. Thus, it appears that system throttling may provide a good alternative between no throttling and application throttling. Similarly, experiments conducted on the p690, for which the number of processors and MADbench MPI tasks is 16, shows that, although using the AS (System) decreases execution time by 5%, as compared to the UT, application throttling performs best; its execution time is about 12% less than that of UT and 8% (Xeon: 11%) less than that of System. However, in both experiments, system throttling results in much smaller LBSTIO times. This is because, as compared to application throttling, the AS provides fine-grained throttling of streams and, thus, results in much lower I/O load imbalances in MADbench's read phases.

This gives further credence to the indication that application-transparent, system throttling may provide a good alternative between no throttling and application throttling. It is worth remembering that a big advantage of system throttling is that it is transparent to the application programmer.

7 Summary and Conclusions

Our results indicate that synchronous I/O-stream throttling can improve execution time significantly for a large class of HPC applications. This improvement is at least 8% and at most 40% for different systems executing MADbench. The best number of concurrent readers is one for the single-disk case, four for the 16-processor SMP with a state-of-the-art RAID, and eight for a Beowulf cluster with a RAID. This shows that stream throttling depends on application I/O behavior as well as the characteristics of the underlying I/O system. On the other hand, we see that the effects of asynchronous I/O-stream throttling are indirect and are not very well understood. On systems with a single disk the number of writers has a negligible effect on total application execution time and on the write-phase time. However, on systems with a RAID, although the number of writers does not impact the write-phase time, it impacts the subsequent read performance and, hence, the total execution time.

We explored I/O-stream throttling at the application layer and, in a limited form, at the system software layer (i.e., via the Linux Anticipatory Scheduler, which can dynamically reduce the number of streams to one). The former requires changes to the application code, while the latter is transparent to the application and can be implemented in the OS, file system, and/or middleware.

In summary, our work indicates that too few or too many concurrently-active I/O streams can cause underutilization of I/O-node bandwidth and can negatively impact application execution time. A general mechanism within system software is needed to address this problem. We are developing a dynamically-adaptive stream throttling system that will reside in the OS or file system of I/O nodes. Because of the load balancing issues discussed in the paper, we plan to provide controls to enable and disable this adaptation.

Acknowledgements

This work is supported by DoE Grant No. DE-FG02-04ER25622, an IBM SUR grant, and UTEP. We thank D. Skinner, J. Borrill, and L. Oliker of LBNL for giving us access to MADbench and D. Skinner and the UTEP DAiSES team, particularly, Y. Kwok, S. Araunagiri, R. Portillo, and M. Ruiz, for their valuable feedback.

References

1. Oldfield, R., Widener, P., Maccabe, A., Ward, L., Kordenbrock, T.: Efficient data movement for lightweight I/O. In: HiPerI/O. Proc. of the 2006 Workshop on High-Performance I/O Techniques and Deployment of Very-Large Scale I/O Systems (2006)
2. Varki, E., Merchant, A., Xu, J., Qiu, X.: Issues and challenges in the performance analysis of real disk arrays. IEEE Trans. on Parallel and Distributed Systems 15(6), 559–574 (2004)
3. Borrill, J., Carter, J., Oliker, L., Skinner, D.: Integrated performance monitoring of a cosmology application on leading HEC platforms. In: ICPP 2005. Proc. of the 2005 Int. Conf. on Parallel Processing, pp. 119–128 (2005)
4. Chen, Y., Winslett, M., Cho, Y., Kuo, S.: Automatic parallel I/O performance optimization in Panda. In: Proc. of the 7th IEEE Int. Symp. on High Performance Distributed Computing, IEEE Computer Society Press, Los Alamitos (1998)
5. Thakur, R., Gropp, W., Lusk, E.: Data sieving and collective I/O in ROMIO. In: Proc. of the 7th Symp. on the Frontiers of Massively Parallel Computation, pp. 182–189 (February 1999)
6. Chen, P., Lee, E., Gibson, G., Katz, R., Patterson, D.: RAID: High performance, reliable secondary storage. ACM Computing Surveys 26(2), 145–185 (1994)
7. Madhyastha, T., Reed, D.: Exploiting global input/output access pattern classification. In: Proc. of SC 1997, pp. 1–18 (November 1997)
8. Ma, X., Jiao, X., Campbell, M., Winslett, M.: Flexible and efficient parallel I/O for large-scale multi-component simulations. In: Proc. of the Int. Parallel and Distributed Processing Symp. (April 2003)
9. Sanders, P.: Asynchronous scheduling of redundant disk arrays. IEEE Trans. on Computers 52(9), 1170–1184 (2003)
10. Aggarwal, A.: Software caching vs. prefetching. In: ISMM '02. Proc. of the 3rd Int. Symp. on Memory Management, pp. 157–162 (2002)
11. Kendall, R., et al.: High performance computational chemistry: an overview of NWChem a distributed parallel application. Computer Phys. Comm. 128, 260–283 (2000)
12. Weirs, G., Dwarkadas, V., Plewa, T., Tomkins, C., Marr-Lyon, M.: Validating the Flash code: vortex-dominated flows. APSS 298, 341–346 (2005)
13. Carter, J., Borrill, J., Oliker, L.: Performance characteristics of a cosmology package on leading HPC architectures. In: Bougé, L., Prasanna, V.K. (eds.) HiPC 2004. LNCS, vol. 3296, pp. 176–188. Springer, Heidelberg (2004)
14. Crandall, P., Aydt, R., Chien, A., Reed, D.: Input/output characteristics of scalable parallel applications. In: Proc. of SC 1995 (1995)
15. Seelam, S., Babu, J., Teller, P.: Rate-controlled scheduling of expired writes for volatile caches. In: QEST 2006. Proc. of the Quantitative Evaluation of Sys-Tems (September 2006)

A Fast Disaster Recovery Mechanism for Volume Replication Systems

Yanlong Wang, Zhanhuai Li, and Wei Lin

School of Computer Science, Northwestern Polytechnical University,
No.127 West Youyi Road, Xi'an, Shaanxi, China 710072
{wangyl,linwei}@mail.nwpu.edu.cn, lizhh@nwpu.edu.cn

Abstract. Disaster recovery solutions have gained popularity in the past few years because of their ability to tolerate disasters and to achieve the reliability and availability. Data replication is one of the most key disaster recovery solutions. While there are a number of mechanisms to restore data after disasters, the efficiency of the recovery process is not ideal yet. Providing the efficiency guarantee in replication systems is important and complex because the services must not be interrupted and the availability and continuity of businesses must be kept after disasters. To recover the data efficiently, we (1) present a fast disaster recovery mechanism, (2) implement it in a volume replication system, and (3) report an evaluation for the recovery efficiency of the volume replication system. It's proved that our disaster recovery mechanism can recover the data at the primary system as fast as possible and achieve the ideal recovery efficiency. Fast disaster recovery mechanism can also be applicable to other kinds of replication systems to recover the data in the event of disasters.

1 Introduction

With the widespread use of computers, data is becoming more and more important in human life. But all kinds of accidents and disasters occur frequently. Data corruption and data loss by various disasters have become more dominant, accounting for over 60% [1] of data loss. Recent high-profile data loss has raised awareness of the need to plan for recovery or continuity. Many data disaster tolerance technologies have been employed to increase the availability of data and to reduce the data damage caused by disasters [2].

Replication [3] is a key technology for disaster tolerance and is quite different from the traditional periodical data backup. It replicates business data on a primary system to some remote backup systems in primary-backup architecture dynamically and on-line. It can not only retain the replicas in remote sites, but also make one of the backup systems take over the primary system in the event of a disaster. Therefore, it is widely deployed in disaster tolerance systems defend against both data loss and inaccessibility.

Disaster recovery mechanism (abbreviated to DRM) is one of the focuses in replication research fields. It helps us restore data after a disaster. Designing a right disaster recovery mechanism is an important and complex task. Although some

R. Perrott et al. (Eds.): HPCC 2007, LNCS 4782, pp. 732–743, 2007.

mechanisms have been presented, e.g. full or difference disaster recovery mechanism, we still need to improve the recovery efficiency and to afford the trade-off between data loss and the efficiency. It is ideal to implement a fast disaster recovery mechanism with the minimal data loss.

A replication system for disaster tolerance can be designed to operate at different levels, i.e. application-, file- or block-level. Of the three levels, block-level replication systems operate just above the physical storage or logical volume management layer, and they can take the advantage of supporting many different applications with the same general underlying approach. So we design a fast disaster recovery mechanism for volume replication systems at the block-level in this paper, although the mechanism could readily be implemented at file- and application-level.

The remainder of the paper is outlined as follows. In Section 2, we put our work into perspective by considering related work. In Section 3, we describe a general model of volume replication system and then describe two key disaster recovery mechanisms used in it. In Section 4, we introduce a fast disaster recovery mechanism, and discuss its principle and process in detail. In Section 5, we implement fast disaster recovery mechanism to recover data of the volume replication system on Linux and give a simple evaluation for it. Finally, we conclude the paper by highlighting the advantages of this work in Section 6.

2 Related Works

While more and more disaster tolerance systems are established, the study of disaster recovery mechanisms is motivated for practical importance and theoretical interest.

Many practitioners' guides (e.g., [4][5]) offer rules of thumb and high-level guidance for designing dependable storage systems, and recovering these systems after disasters. These books focus mainly on the logistical, organizational and human aspects of disaster recovery. They do not treat detailed disaster recovery issues for any certain technology.

Several recent studies explore how to improve the ability of a computer system or a storage system to recover from disasters and to meet user-defined goals (e.g., [6][7][8][9]). Practical guidelines are outlined that close the gap between business-oriented analyses and technical design of disaster recovery facilities in [6]. Some methods for automatically selecting data protection techniques for storage systems to minimize overall cost of the system are given in [7]. An adaptive scheduling algorithm is proposed for disaster recovery server processes in [8] and several methods for finding optimal or near-optimal solutions to deal with the data recovery scheduling problem are presented in [9]. All these studies focus on evaluating the risk of computer systems or the dependability of data storage systems and then making a best choice among various data protection techniques or making a best scheduling. Unfortunately, they do not give a detailed disaster recovery solution for a replication-based disaster tolerance system.

In the replication research literatures, we can find that replication protocols are the focus. Authors have proposed several protocols, including IBM's Peer-to-Peer Remote Copy (PPRC) [10], EMC's Symmetrix Remote Data Facility (SRDF) [11], Hitachi's Remote Copy [12] and VERITAS's Volume Replicator (VVR) [13]. Some optimized replication protocols are presented, such as Seneca [14], SnapMirror [15] and RWAR [16]. Disaster recovery mechanisms are scarcely mentioned in these books and papers, and some of them simply introduce a planned Failback operation to restore each site to its pre-failure role.

Failback has already been researched in cluster computing [17] [18]. In order to build up a fault-tolerant cluster, the features of Failover and Failback services are demanded. Failback process is exercised when a system in the cluster is switched back from the normally secondary to the normally primary host. But Failback is mainly used to switch the role rather than to recover the data. So it is different between a cluster and a replication system.

To restore the data as soon as possible, we present a fast disaster recovery mechanism for volume replication systems based on referring to the technology of Failback. Our fast disaster recovery mechanism can not only switch the role of a replication system, but also restore its data fast.

3 Current Disaster Recovery Mechanisms for a Volume Replication System

3.1 The Volume Replication System

Volume management systems provide a higher-level view of the disk storage on a computer system than the traditional view of disks and partitions. This gives us much more flexibility in allocating storage to applications and users. A typical volume replication system is a remote and online replication system based on volume management systems. It is mainly composed of *Client*, *Primary* and *Backup* (or Secondary), and its architecture is shown in Fig. 1.

When *Primary* receives a request committed by *Client*, it writes a copy of the data of the request into the log volume and the storage volumes. At the same time, it sends the data to *Backup* and then *Backup* records the data.

Both *Primary* and *Backup* are a pair of peers to implement the process of replication. *Replicator* is their key component. It builds up the replication link

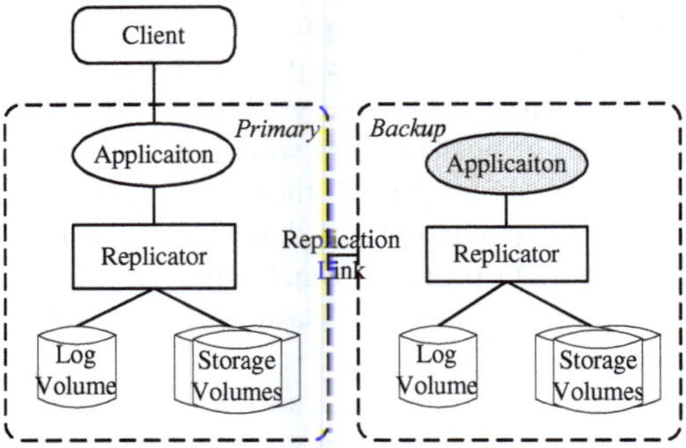

Fig. 1. Architecture of a typical volume system

between *Primary* and *Backup*, and runs the synchronous or asynchronous replication protocol. The log volume and the storage volumes are used to store the log and the data respectively. The volume replication system establishes one or more backup copies by replicating the data online and remotely. Once *Primary* suffers any disaster, *Backup* can take over the role of *Primary* and supply services to *Client*, which is referred to as Failover.

Obviously, the ability of the primary system (Primary) to tolerate disasters is increased by replicating. But if there is only a Failover mechanism, the primary system cannot restore its data according to the backup system (Backup) and be back to its primary role.

3.2 Current Disaster Recovery Mechanisms

Apart from the technology of Failback, there are some technologies to recover the data of the primary system, and they are unified as disaster recovery mechanisms. Full and difference disaster recovery mechanisms are the most key ones to recover the data of the primary system:

- Full disaster recovery mechanism: is a process to synchronize the former primary system with the current and new primary system (i.e. the former backup system) fully. That is, all the blocks at the former backup system are read and then sent to the former primary system. After receiving these blocks, the former primary system uses them to overwrite the local blocks.
- Difference disaster recovery mechanism: is a process to synchronize the former primary system with the current and new primary system by computing the differences between each pair of blocks at both the systems. That is, we compare a block at the former primary system with a corresponding one at the new primary system. If the pair of blocks are identical, the block at the new primary system does not need to be transferred to the former primary system; otherwise, the block must be sent to the former primary and be used to overwrite the local block. For all the blocks of the storage volumes at both the primary and backup systems have to be read, the differences are mostly gained by computing and comparing MD5 checksums for each pair of blocks to reduce the workload.

In order to describe the two mechanisms clearly, we define two rules as follow:

Definition 1: When restoring its data, a system must refer to the current and new primary system. It is defined as the newest referring rule (abbreviated to NRR).

Definition 2: After completing the process of restoring, a system must have the same image as the new primary system. It is defined as the data consistency rule (abbreviated to DCR).

At the same time, we define a set of symbols to describe the data sets at the different positions as follow:

Definition 3: D_{p-l} deontes the set of the log data at the primary system, D_{p-s} deontes the set of the storage data at the primary system, D_{b-l} deontes the set of the log data at the backup system, and D_{b-s} deontes the set of the storage data at the backup system. D_{p-s} and D_{b-s} also denote the image of the primary system and the one of the backup system respectively.

If the storage volume is considered as a set composed of N data blocks, we can describe the process of full and difference disaster recovery mechanisms in Table 1.

Table 1. The processes of current disaster recovery mechanisms

Full disaster recovery mechanism	Difference disaster recovery mechanism

BEGIN
i:=1;
REPEAT
B:=BackupReadBlock(i);
BackupSendBlock(B);
PrimaryRecieveBlock(B);
PrimaryWriteBlock(B,i);
i++;
UNTIL i==N+1;
END
	BEGIN
	i:=1;
	REPEAT
	B_p:=PrimaryReadBlock(i);
	M_p:=ComputeMD5Checksum(B_p);
	PrimarySendMD5Checksum(M_p);
	BackupReceiveMD5Checksum(M_p);
	B_b:=BackupReadBlock(i);
	M_b:=ComputeMD5Checksum(B_b);
	Tag:=Difference(B_b,B_p);
	IF Tag==0 THEN
	BEGIN
	BackupSendBlock(B_b);
	PrimaryRecieveBlock(B_b);
	PrimaryWriteBlock(B_b,i);
	END
	i++;
	UNTIL i==N+1;
	END

In Table 1, *Primary* denotes the former primary system and *Backup* denotes the former backup system (i.e. the new primary system). *BackupReadBlock(), BackupSendBlock(), PrimaryRecieveBlock(), PrimaryWriteBlock(), PrimaryRead Block(), PrimarySendMD5Checksum(), BackupReceiveMD5Checksum(), ComputeMD5 Checksum()* and *Difference()* are API functions to implement the process of disaster recovery mechanisms. It is easy for us to understand the former seven functions which involve four basic operations, i.e. *Read*, *Write*, *Send* and *Receive*. *ComputeMD5 Checksum()* and *Difference()* are two important functions for difference disaster recovery mechanism. *ComputeMD5Checksum()* is used to compute the MD5 checksum for a block. *Difference()* is used to compare two MD5 checksums. If the two MD5 checksums are identical, it returns *1*; otherwise, it returns *0*.

As shown in Table 1, full disaster recovery mechanism needs to transfer all the blocks from the backup system to the primary system, so it is very time-consuming. Moreover, although difference disaster recovery mechanism reduces the number of transferred blocks, it needs to compare the MD5 checksums for each pair of blocks and then increases the number of the operations of *Read*, *Send* and *Receive*. In addition, when running full and difference disaster recovery mechanisms, users have to stop replicating the data of all applications. Therefore, it is possible for us to optimize both the disaster recovery mechanisms.

4 Fast Disaster Recovery Mechanism (FDRM)

4.1 The Principle of FDRM

To optimize the disaster recovery mechanisms, we present a fast disaster recovery mechanism. It tracks the incremental changes as they happen. Only the changed blocks at the new primary system are read and sent back to the former primary system. Hence the number of transferred blocks is smaller than full disaster recovery mechanism, and the number of read operations required is smaller than difference disaster recovery mechanism. In addition, it does not need to stop the replication operation of the replication system.

Fast disaster recovery mechanism synchronizes the former primary system with the new primary system by finding out the changed blocks at the new primary system after the former primary system broke down. That means that we need record the changes at the new primary system after it took over the former primary system. But some changes may be omitted. For example, in a volume replication system, some blocks have been wrote on the data volumes at the primary system before disasters occur to the primary system, but they may fail to be replicated to the backup system, be written on the data storage at the primary system, or even be acknowledged to the primary system after being written at he primary system. Thus the corresponding blocks at the backup system are stored as an older version than those at the primary system. According to NRR in Definition 1, when recovering the primary system, these blocks at the new primary system (i.e. the former backup system) become a part of all the changes.

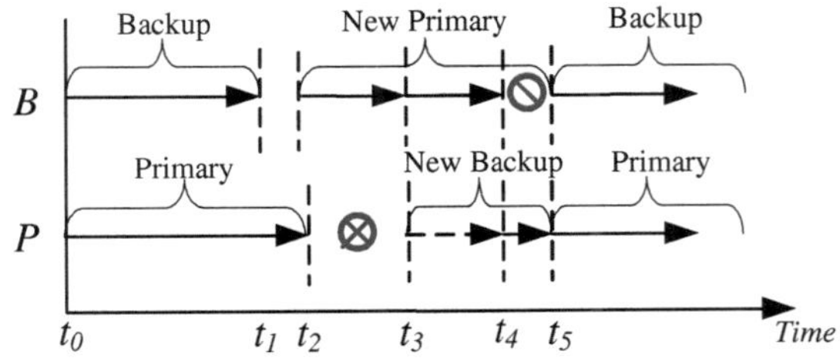

Fig. 2. Principle of fast disaster recovery mechanism

A simple description of fast disaster recovery mechanism is shown in Fig. 2. It assumes that *P* and *B* are a pair of systems for disaster tolerance. At first, *P* is the primary system and *B* is the backup system. At the time t_2, *P* suffers some disasters and *B* takes over *P*. At that time, the image of *P* is $D_{p-s}(t_2)$ and the one of *B* is $D_{b-s}(t_2) = D_{b-s}(t_1)$. At the time t_3, *P* comes back and becomes a new backup system for *B*. The data set of incremental changes at *B* from t_2 to t_3, $(D_{b-s}(t_3) - D_{b-s}(t_2)) \cup (D_{b-s}(t_2) - D_{b-s}(t_1))$, is transferred back to *P*. This process can be implemented without stopping applications and last out from t_3 to t_4, where

$D_{b-s}(t_3)-D_{b-s}(t_2)$ is the data set of real incremental changes at B from t_2 to t_4 and $D_{b-s}(t_2)-D_{b-s}(t_1)$ is the original data set of changes at P from t_1 to t_2, $D_{p-s}(t_2)-D_{p-s}(t_1)$. Then, the data set of incremental changes at B from t_3 to t_4, $D_{b-s}(t_4)-D_{b-s}(t_3)$, is transferred back to P. At the time t_5, according to DCR in Definition 2, the whole recovery process is finished and P has the same image as B. P also can rerun as the primary system and supply services again.

4.2 The Process of FDRM

According the above description, the key issue for fast disaster recovery mechanism is to compute and gain $D_{b-s}(t_3)-D_{b-s}(t_2)$, $D_{b-s}(t_2)-D_{b-s}(t_1)$ and $D_{b-s}(t_4)-D_{b-s}(t_3)$. In order to simplify the problem, we build a change bitmap table for each storage volume and use it together with the log volume to track writes in the volume replication system. When a block in the storage volume at the primary system is changed, its data are also stored temporarily and a flag bit in the change bitmap table is set. Based on the bitmap tables at P and B, we can gain $D_{b-s}(t_3)-D_{b-s}(t_2)$, $D_{b-s}(t_2)-D_{b-s}(t_1)$ and $D_{b-s}(t_4)-D_{b-s}(t_3)$ easily.

The process of fast disaster recovery mechanism is shown in detail in Table 2.

Table 2. The process of fast disaster recovery mechanism

Fast disaster recovery mechanism

```
BEGIN
  T_p:=PrimaryBuildBitmapTable(Log_p);//Build the bitmap table based on the
                         log volume Log_p at the former primary system,
                         and T_p denotes the bitmap table corresponding
                         to the data set of $D_{p-s}(t_2)-D_{p-s}(t_1)$
  PrimarySendBitmapTable(T_p);
  BackupReceiveBitmapTable(T_p);
  T_all:= BackupComputeOR(T_p, T_b); // T_b denotes the bitmap table for the
                         data set of $D_{b-s}(t_3)-D_{b-s}(t_2)$ and is obtained by
                         tracking the changes at the new primary system
                           //Compute the value of OR operation for T_p
                         and T_b, and T_all denotes the bitmap table for
                      the data set of $(D_{b-s}(t_3)-D_{b-s}(t_2))\cup(D_{b-s}(t_2)-D_{b-s}(t_1))$
  i:=1;                    //Scan the bitmap table T_all, transfer the
relevant
                 data set to the former primary system and recover it
  REPEAT
    Flag:=CheckBitmapTable(T_all);
    IF Flag==1 THEN
    BEGIN
      B_b:=BackupReadBlock(i);
      BackupSendBlock(B_b);
      PrimaryRecieveBlock(B_b);
      PrimaryWriteBlock(B_b,i);
   END
  i++;
  UNTIL i==N+1;
```

Table 2. *(continued)*

```
18  Tnew:=BackupBuildBitmapTable(Logb);//Build the bitmap table based on the
                                        log volume Logb at the new primary system,
                                        and Tnew denotes the bitmap table for the
                                        data set of D_{b-s}(t4) - D_{b-s}(t3)
19  j:=1;                              // Scan the bitmap table Tnew, transfer the
   relevant
                              data set to the former primary system and recover it
20  REPEAT
21    Flag:=CheckBitmapTable(Tnew);
22    IF Flag==1 THEN
23    BEGIN
24      Bb:=BackupReadBlock(j);
25      BackupSendBlock(Bb);
26      PrimaryRecieveBlock(Bb);
27      PrimaryWriteBlock(Bb,j);
28     END
29   j++;
30   UNTIL j==N+1;
31    END
```

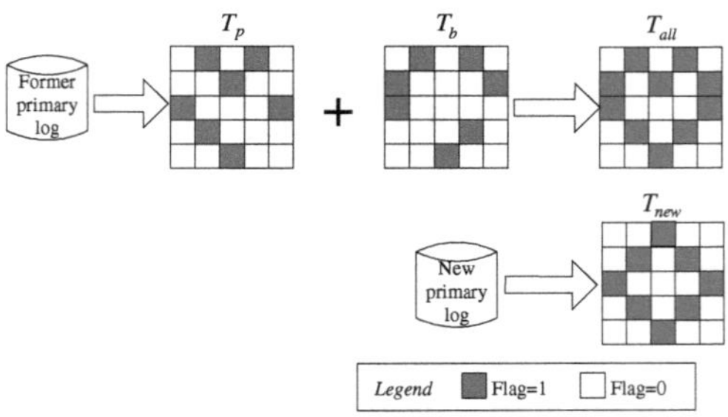

Fig. 3. An example of building change bitmap tables

In the process of fast disaster recovery mechanism, the change bitmap table plays an important role. In Table 2, Step 2, Step 5 and Step 18 are the key steps to build T_p, T_{all} and T_{new} (T_b is build by tracking the changes from t_2 to t_3 at the primary system). An example is illustrated in Fig. 3. After T_p, T_b, T_{all} and T_{new} are computed, the data sets transferred to the former primary system are easy to obtain and then the former primary system can be recovered as fast as possible.

5 Implementation and Evaluation

Logical volume replicator (LVR) is a remote and online volume replication system prototype developed by us based on logical volume manager (LVM) [19] on Linux. We can create one replicated volume group at the primary system, which includes a log volume and several storage volumes. We also can create the other replicated volume group at the backup system. Both the replicated volume groups are a pair of peers to build up the replication link and implement remote replication and disaster recovery.

In order to evaluate the fast disaster recovery mechanism in detail, we compare it with full and difference disaster recovery mechanisms by analyzing theoretically and doing experiments.

5.1 Analysis

Fast disaster recovery mechanism can be evaluated with many parameters. Of all the parameters, the number of data blocks recovered, time consumption and space consumption are the three main ones. The number of data blocks recovered means

Table 3. Comparison of FDRM and conventional mechanisms

Mechanism	The number of data blocks recovered	Time consumption	Space consumption
Full disaster recovery mechanism	N	$N(t_r + t_n + t_w)$	$N \times s_d$
Difference disaster recovery mechanism	M	$N(2t_r + 2t_{md5} + \frac{s_m}{s_d} t_n + t_{diff}) + M(t_n + t_w)$	$N \times (2s_d + 2s_m)$
Fast disaster recovery mechanism	$I + J - K$	$I \times t_{bitmap} + \frac{s_b}{s_d} \times t_n + t_{or} + (I + J - K)(t_r + t_n + t_w)$	$I \times s_d + 2s_b + (I + J - K)s_d$

how many data blocks are transferred to the former primary system to recover the data. Time consumption means how long it takes to implement the whole process of disaster recovery mechanism. And space consumption means the size of the memory space used at the primary and backup systems. The comparison of fast disaster recovery mechanism and the other two mechanisms are shown in Table 3.

In Table 3, we adopt some symbols to denote the element values of the three main parameters. They are explained as follow:

N is the number of the storage volume at the primary system or at the backup system, and M is the number of the different blocks at both the systems. I is the number of the blocks on the log volume at the primary system. These blocks were written to the storage volume but not replicated to the backup system, i.e. $I = |D_{p-s}(t_2) - D_{p-s}(t_1)|$. J is the number of the changed blocks at the backup system from the backup system taking over the primary system to now, i.e. $J = |D_{b-s}(t_3) - D_{b-s}(t_2)|$. K is the number of the same blocks between I blocks and J blocks, i.e. $K = |(D_{b-s}(t_3) - D_{b-s}(t_2)) \cap (D_{b-s}(t_2) - D_{b-s}(t_1))|$.

t_r, t_w and t_n are the time of *Read*, *Write* and *Transfer* respectively to deal with a block. *Transfer* includes two steps of *Send* and *Receive*. t_{md5} is the time to compute the MD5 checksum of a block. t_{diff} is the time to compare the MD5 checksum of one block at the primary system with that of the other corresponding block at the backup system. t_{bitmap} is the time to establish a bitmap table according to the log at the primary system. t_{or} is the time to compute the OR value of the primary bitmap table and the backup bitmap table.

In addition, s_d is the size of a data block. s_m is the size of a MD5 checksum. s_b is the size of a bitmap table.

From Table 3, we can find that the number of the recovered blocks in full disaster recovery mechanism is the largest one, and the number of the recovered blocks in different disaster recovery mechanism is equal or less than the one of fast disaster recovery mechanism. That is, $M \le I + J - K \le N$. We also can find that the time consumption and the space consumption in difference disaster recovery mechanism may be the largest of the three mechanisms because of computing the MD5 checksum for each block and comparing each pair of MD5 checksums. In addition, we can find that although the number of the recovered blocks in fast disaster recovery mechanism may be larger than different disaster recovery mechanism, the other two parameters in fast disaster recovery mechanism may be the smallest ones in all the three mechanisms.

When using MTBF (Mean Time Between Failures) and MTTR (Mean Time To Recover) to measure the reliability and the maintainability of the replication system, the failure rate λ is a constant for each disaster recovery mechanism, i.e. $MTBF = \frac{1}{\lambda}$ is a constant. At the same time, the maintenance rate μ in fast disaster recovery mechanism is the largest. Therefore, $Availability = \frac{MTBF}{MTBF + MTTR} \times 100\% = \frac{\mu}{\lambda + \mu} \times 100\%$ in fast disaster recovery mechanism is the largest and then it is the most ideal for a replication system.

5.2 Experiments

Obviously, it is not enough to compare the three disaster recovery mechanisms and to evaluate fast disaster recovery mechanism by analyzing. In order to understand the advantages of fast disaster recovery mechanism, we do some experiments based on LVR.

LVR is run on a pair of the primary and backup systems. Either of the two systems is an Intel Celeron 1.7 GHz computer with 512 MB of RAM. At the either system, the log volume is 1 GB and the storage volume is 2 GB. The versions of Linux kernel and LVM are 2.4.20-8 and 2.00.07 respectively. The block size used by LVM is set to 4KB and the Gigabit-Ethernet 1000baseT is used as connection between the primary system and the backup system. The replication protocol is the standard asynchronous one and the *random write* of IOzone-3.263 benchmark [20] is used to simulate the applications at the primary system.

With the cost of the memory becoming lower and lower, the space consumption is playing a less important role. So in the following two experiments, we do not consider the space consumption for the three mechanisms again.

In the first experiment, a manual disaster is made at a fixed time (i.e. t_2) after replication was started at the primary system. Disaster recovery occurs at a fixed time (i.e. t_3) after the backup system took over the primary system. Here t_2 is assumed to a moment when replication has been running for 100s and t_3 is assumed to a moment

Table 4. The number of recovered blocks and the time to recover for the three mechanisms

Mechanism	The number of data blocks recovered	The time to recover (s)
Full disaster recovery mechanism	524288	673
Difference disaster recovery mechanism	140243	464
Fast disaster recovery mechanism	173975	257

when the backup system has served as a new primary system for 30s. The results of comparing are shown in Table 4.

In the second experiment, we evaluate fast disaster recovery mechanism by testing the impact of t_3. In practice, t_1 and t_2 are the objective moments that can not be changed artificially. t_4 and t_5 are the moments that are affected by t_3. Here t_2 is still assumed to a moment when replication has been running for 100s and t_3 is the moment to start recovering the primary system where $t_3 - t_2$ is set with a series of values. The results are shown as Table 5.

Table 5. Impact of the moment to implement fast disaster recovery mechanism

$t_3 - t_2$ (s)	The number of data blocks recovered	The time to recover (s)
20	113821	127
40	320420	410
60	433975	543
80	500288	664
100	514337	729

From Table 4, we can find that the time to recover in fast disaster recovery mechanism is the smallest. From Table 5, we can find that it is ideal for fast disaster recovery mechanism to run as soon as possible after disasters occur.

In a word, it is proved by the above experiments that fast disaster recovery mechanism is worth deploying in the current replication systems.

6 Conclusion

With data becoming more and more important, all kinds of information systems require 24x7 availability. Replication is a rising data disaster-tolerance technique, and disaster recovery mechanisms play an important role in replication systems. How to recover the data at the primary system as soon as possible after the primary system suffers disasters is important to design a efficient replication system.

Fast disaster recovery mechanism is presented to recover the data at the primary system quickly and to improve the recovery efficiency of volume replication systems. Its principle and process are described in detail, and then it is implemented in logical volume replicator on Linux. It's proved by analyzing theoretically and doing experiments that fast disaster recovery mechanism makes the recovery process of volume replication systems fast and also guarantees the availability of the system. Therefore, it is helpful for volume replication systems to tolerate various disasters.

Acknowledgments. This work is supported by the National Natural Science Foundation of China (60573096).

References

1. Smith, D.M.: The cost of lost data. Journal of Contemporary Business Practice 6(3) (2003)
2. Chervenak, A., Vellanki, V., Kurmas, Z.: Protecting file systems: A survey of backup techniques. In: Proc. of Joint NASA and IEEE Mass Storage Conference, IEEE Computer Society Press, Los Alamitos (1998)
3. Yang, Z., Gong, Y., Sang, W., et al.: A Primary-Backup Lazy Replication System for Disaster Tolerance (in Chinese). Journal Of Computer Research And Development, 1104–1109 (2003)
4. Cougias, D., Heiberger, E., Koop, K.: The backupbook: disaster recovery from desktop to data center. Schaser-Vartan Books, Lecanto, FL (2003)
5. Marcus, E., Stern, H.: Blueprints for high availability. Wiley Publishing, Indianapolis, IN (2003)
6. Cegiela, R.: Selecting Technology for Disaster Recovery. In: DepCos-RELCOMEX 2006. Proc. of the International Conference on Dependability of Computer Systems (2006)
7. Keeton, K., Santos, C., Beyer, D., et al.: Designing for Disasters. In: FAST 2004. Proc. of the 3rd USENIX Conf on File and Storage Technologies, pp. 59–72 (2004)
8. Nayak, T.K., Sharma, U.: Automated Management of Disaster Recovery Systems Using Adaptive Scheduling. In: Proc. of the 10th IFIP/IEEE International Symposium on Integrated Network Management, pp. 542–545 (2007)
9. Keeton, K., Beyer, D., Brau, E., et al.: On the road to recovery: restoring data after disasters. In: EuroSys. Proc. of the 1st ACM European Systems Conference, pp. 235–248. ACM Press, New York (2006)
10. Azagury, A.C., Factor, M.E., Micka, W.F.: Advanced Functions for Storage Subsystems: Supporting Continuous Availability. IBM SYSTEM Journal 42 (2003)
11. Using EMC SnapView and MirrorView for Remote Backup. Engineering White Paper, EMC Corporation (2002)
12. Software Solutions Guide for Enterprise Storage. White Paper, Hitachi Data Systems Corporation (2000)
13. VERITAS Volume Replicator 3.5: Administrator's Guide (Solaris). White Paper, Veritas Software Corp. (2002)
14. Ji, M., Veitch, A., Wilkes, J.: Seneca: remote mirroring done write. In: USENIX 2003. Proc. of the 2003 USENIX Technical Conference, pp. 253–268 (2003)
15. Patterson, H., Manley, S., Federwisch, M., Hitz, D., Kleiman, S., Owara, S.: SnapMirror: File-System-Based Asynchronous Mirroring for Disaster Recovery. In: Proc. of the First USENIX conference on File and Storage Technologies (2002)
16. Wang, Y., Li, Z., Lin, W.: RWAR: A Resilient Window-consistent Asynchronous Replication Protocol. In: ARES 2007. Proc. of the 2nd International Conference on Availability, Reliability and Security, pp. 499–505 (2007)
17. Hwang, K., Chow, E., Wang, C.-L., et al.: Fault-Tolerant Clusters of Workstations with Single System Image. In: Cluster computing (1998)
18. McKinty, S.: Combining Clusters for Business Continuity. In: Cluster 2006. Proc. of the 2006 IEEE International Conference on Cluster Computing (2006)
19. Redhat (2005), http://sources.redhat.com/lvm2/
20. IOzone (2006), http://www.iozone.org/

Dynamic Preemptive Multi-class Routing Scheme Under Dynamic Traffic in Survivable WDM Mesh Networks

Xuetao Wei, Lemin Li, Hongfang Yu, and Du Xu

Key Lab of Broadband Optical Fiber Transmission and Communication Networks,
University of Electronic Science and Technology of China, Chengdu, 610054, China
xuetao.wei@gmail.com, lml@uestc.edu.cn, yuhf@uestc.edu.cn,
xudu@uestc.edu.cn

Abstract. A large-scale optical network will carry various traffic classes with different fault-tolerance requirements. Several previous papers suggest a preemptive multi-class routing scheme. They have proposed Integer Linear Programming (ILP) formulations to implement this routing scheme and simulate it under the assumption that the traffic holding on the network will not depart. However, in the real world, most of the traffic is the dynamic traffic: the traffic holding on the network will leave at the end of connection holding time. In this paper, we propose a detailed algorithm for Dynamic Preemptive Multi-class Routing (DPMR) scheme under dynamic traffic pattern. Furthermore, we compare DPMR scheme with DPP (Dedicated Path Protection) and SPP (Shared Path Protection), and evaluate their performances from resource utilization ratio and blocking probability. In addition, we propose a modified link cost function for Dynamic Preemptive Multi-class Routing scheme to find the routes.

1 Introduction

The growth in internet traffic has led to a tremendous demand for bandwidth. Wavelength-division -multiplexing (WDM) technology can meet this demand and is deployed in the next generation network infrastructures [1]. Due to per fiber bandwidths of the order of several hundred gigabits-per-second (Gb/s), enormous data loss may be experienced in the event of network failures such as node, link, or channel faults. So it is critical to consider networks survivability and design survivable WDM networks [2].

Survivability mechanisms proposed for wavelength routed WDM networks are typically classified as preplanned protection mechanisms and dynamic restoration mechanisms. In [3] and [5], the authors have introduced conventional protection and restoration schemes, and discussed the benefits of different survivable approaches. For example, path protection scheme (dedicated path protection and shared path protection[5]) which will assign a backup path to its primary path can provide quick and guaranteed recovery, but does not use resources efficiently. Correspondingly, the restoration scheme which will not assign backup resources

R. Perrott et al. (Eds.): HPCC 2007, LNCS 4782, pp. 744–754, 2007.

to its primary path can use resources efficiently, but does not guarantee 100% recovery when resources are inadequate for recovery path allocation.

In these studies [3,5], the researchers have focused their attention on a single class of traffic. In practical network services, different connections will have different service level requirements [4]. For example, critical applications, such as emergency applications and banking institution transactions, will coexist with non-critical applications like Internet radio, file transfer, web browsing and e-mail Therefore, it is greatly necessary to differentiate different services and meet different customers' demands in finite network resources. Papers [6,7,8] have suggested differentiated reliability in mesh networks with shared path protection. They assign different backup resources to primary paths according to the primary path's different reliability requirements. It is an efficient approach to meet customer's different requirements using finite network resources. However, they do not consider that certain traffic's primary paths can share resources with high priority traffic's backup path which will save more resources. Papers [9,10] just differentiate priority of traffic and suggest that high priority traffic's backup path can share resources with low priority traffic's working path. But they just propose Integer Linear Programming (ILP) formulations to implement the routing scheme and simulate it under the assumption that the traffic holding on the network will not depart at all.

However, the fact is that more and more connection requests from customers will change from time to time, so the simulation assumption in [9,10] may result in low resource utilization and less flexibility [11,12]. Because of the mentioned reasons before, the investigation of dynamic network behavior for WDM optical networks becomes much more essential and important to us. In addition, general link cost function used to compute the shortest paths is designed to increase the degree of wavelength sharing. Due to the definition of preemptive multi-class routing scheme-low priority traffic primary paths can share resources with high priority traffic's backup paths-link cost function for it will be modified.

This paper is organized as follows: Section 2 presents three schemes: DPP, SPP and DPMR. Section 3 introduces link cost function and network model. Section 4 details algorithms for DPP, SPP and DPMR under dynamic traffic. Section 5 analyses the simulation and compares the performances. Section 6 concludes.

2 Scheme Description

In this section, we will describe three routing schemes: Dedicated Path Protection, Shared Path Protection and Dynamic Preemptive Multi-class Routing schemes. Detailed description will be shown in the following.

- Dedicated Path Protection
  - only single class traffic is considered.
  - every traffic will be assigned two link-disjoint paths: primary path and backup path.
  - backup paths can not share resources with each other.

- Shared Path Protection
  - only single class traffic is considered.
  - every traffic will be assigned two link-disjoint paths: primary path and backup path.
  - backup paths can share resources with each other.
- Dynamic Preemptive Multi-Class Routing
  - multi-class traffic is considered.
  - not every traffic will be assigned two link-disjoint paths: primary path and backup path; only high priority traffic will have backup paths.
  - high priority traffic's backup paths can share resources each other.
  - high priority traffic's backup paths can share resources with low priority traffic's working paths.

3 Link Cost Function and Network Model

3.1 Link Cost Function

In this section, we will discuss two link cost functions: general link cost function and modified link cost function. General link cost function is used by Dedicated Path Protection scheme and Shared Path Protection scheme. Due to the difference of Dynamic Preemptive Multi-class Routing (DPMR) with DPP and SPP, we will modify the general link cost function to implement it into DPMR scheme.

General Link Cost Function. A large degree of wavelength resource sharing is the advantage of shared protection. In order to achieve this goal, we will define the link cost function in the following which will be used for computing the primary and backup paths to implement the wavelengths sharing.

The capacity of link i can be divided into three types:

1. Free capacity (f_i): the free capacities that can be used by the following primary or backup paths.
2. Reserved capacity (RC_i): the reserved capacities of some backup paths.
3. Working capacity (W_i): the working capacities of some primary paths and can not be used for any other purpose until the corresponding primary path is released. The link cost function CP for finding a primary path with requested bandwidth (RB) is calculated as

$$CP_i = \begin{cases} -\infty & f_i < RB \\ c_i & f_i \geq RB \end{cases} \quad (1)$$

where c_i is the basic cost of link i that is decided by physical length of the fiber link, the expense of fiber link installation, and so on.

In order to find a backup path for the primary path, we should define the corresponding link cost function firstly. Concerning the requested bandwidth (RB) and the found primary path, the reserved capacity (RC_i) on link i can be further divided into two types:

1. Sharable capacity (sc_i): the capacities reserved by some protected path(s) and is shared by the found primary paths, where the "some protected path(s)" should be link-disjoint with the found primary paths.
2. Non-Sharable capacity (none-sc_i): the capacities reserved by some protected path(s) and is not shared by the found primary paths, where the "some protected path(s)" should not be link-disjoint with the found primary paths.

Obviously, $RC_i = sc_i +$ none-sc_i. Fig.1 shows the capacity along link i.

<table>
<tr><td colspan="2">working capacity</td></tr>
<tr><td colspan="2">free capacity</td></tr>
<tr><td rowspan="2">reserved capacity</td><td>sharable</td></tr>
<tr><td>non-sharable</td></tr>
</table>

Fig. 1. Capacity along link i

The link cost function CB for finding a backup path for protected path with requested bandwidth (RB) is calculated as

$$CB_i = \begin{cases} \varepsilon & RB \leq sc_i \\ \varepsilon + \alpha \cdot \frac{RB - sc_i}{f_i} & sc_i < RB \leq sc_i + f_i \\ +\infty & sc_i + f_i < RB \end{cases} \quad (2)$$

where ε is a adequately small positive constant, for instance, 0.001 or 0.0001; α is a parameter that is a positive constant (In the simulation of this paper, set $\varepsilon = 0.001$ and $\alpha = 1$). In our algorithm, if there is enough sharable capacity available along the link, the backup path will take the sharable capacity on a link firstly. The value of link cost will have two cases: 1) If there is enough sharable capacities to meet the requested bandwidth (RB), then the sharable capacities will be reserved for the backup path of the new request without needing to allocate new wavelength resource, thus we make the cost of link i be a adequately small positive constant ε; 2) If there isn't enough sharable capacity along the link, then certain free capacities will be occupied and the link cost will be determined by how many free capacities will be taken. If the summation of the sharable and free capacities is less than the RB, then the link is unavailable for the backup path, so we make the link cost be infinite. Therefore, in (2), we can observe that those links with enough reserved capacities will have smaller link cost. If the backup paths traverse these links, then we do not need to reserve new backup wavelengths. Thus, the resource utilization ratio will be improved.

Modified Link Cost Function. Because in Dynamic Preemptive Multi-class Routing scheme, there has multi-class traffic: high priority traffic and low priority traffic. Furthermore, high priority traffic's backup path can share resources with low priority traffic. So we will modify the general link cost function to improve wavelength resource sharing.

1. Link cost function for low priority traffic's path(s)

$$CP_i^* = \begin{cases} +\infty & f_i < RB \\ c_i & f_i = RB \\ c_i \cdot (1 - \frac{f_i - 1}{|W|}) & f_i > RB \end{cases} \tag{3}$$

   where c_i,f_i,W and RB are the same meaning with the notations in the last subsection. We can observe that the link cost reduces as the free wavelength links increase. The shortest path algorithm will favor fiber links that have much more free wavelength links. Therefore, the low priority traffic's working capacities will be distributed to all the links, and the load will be more balanced. The load balancing may lead to more low priority traffic's paths link disjoint and result in more spare capacities to be shared by high priority traffic's backup path(s)[13].
2. Link cost function for high priority traffic's primary path
   We will still adopt (1) in general link cost function (see the last subsection) to compute the high priority traffic's primary path.
3. Link cost function for high priority traffic's backup path
   Because the high priority traffic's backup path can share resources with low priority traffic, so we must consider the factor into the link cost function for high priority traffic's backup path. With the requested bandwidth (RB) and the found high priority traffic's primary path(FHPP), the reserved capacity(RC_i^*) along link i can be further divided into three types:
   (a) Sharable capacity (sc_i^*): the capacities reserved by high priority traffic's protected path(s) and is shared by FHPPs, where the "high priority traffic's protected path(s)" should be link-disjoint with FHPPs.
   (b) Non-sharable capacity (none-sc_i^*): the capacities reserved by some protected path(s) and is not shared by FHPPs, where the "high priority traffic's protected path(s)" should not be link-disjoint with FHPPs.
   (c) Low priority traffic's capacity (lc_i): the capacity reserved by low priority traffic's working path(s) and is shared by high priority traffic's backup path(s).

   Obviously, $RC_i^* = sc_i +$ none-$sc_i^* + lc_i$. Fig.2 shows the capacity along link i.

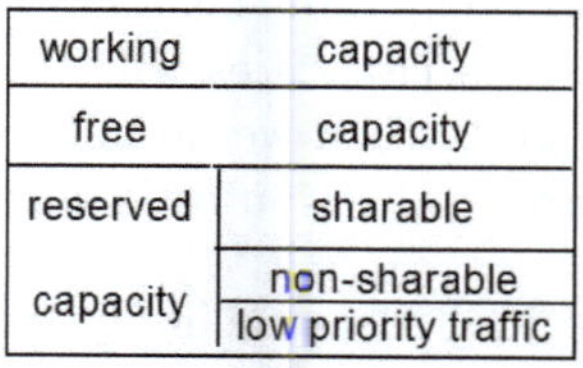

Fig. 2. Capacity along link i

The link cost function CB_i^* for finding a backup path for high priority traffic's primary path with requested bandwidth (RB) is calculated as

$$CB_i^* = \begin{cases} \varepsilon^* & RB \leq sc_i^* + lc_i \\ \varepsilon^* + \alpha^* \cdot \frac{RB - sc_i^* - lc_i}{f_i} & sc_i^* + lc_i < RB \leq sc_i^* + lc_i + f_i \\ +\infty & sc_i^* + lc_i + f_i < RB \end{cases} \quad (4)$$

where ε^* is also a sufficient small positive constant comparing with ε in the last subsection, such as 0.001 or 0.0001; α^* is a parameter that is a positive constant (in the simulation of this paper, set $\varepsilon^* = 0.001$ and $\alpha^* = 1$). The similar explanation process of (4) which has explained in the last subsection will present in the following. In our algorithm, the high priority traffic's backup path will take the sharable capacity and low priority traffic's capacity along a link firstly if there are enough sharable capacity and low priority traffic's capacity available on the link. If there are enough sharable capacities and low priority traffic's capacity to meet the requested bandwidth (RB), then the sharable capacities and low priority traffic's capacity would be reserved for the backup path of the new high priority traffic's request without allocating new wavelength, so we let the cost of link i to be a sufficient small positive constant ε^*. If there isn't enough sharable capacity and low priority traffic's capacity along the link, then certain free capacities will be taken and the link cost will be determined by how many free capacities will be taken. If the summation of the sharable, free and low priority traffic's capacities is less than the RB, the link is unavailable for the backup path, so we make the link cost to be infinite. From (4), we can observe that these links with enough reserved capacities will have smaller link cost. If the backup paths traverse these links, then we do not need to reserve new backup wavelengths. Thus, the resource utilization ratio will be improved.

3.2 Network Model

The network topology is denoted as $G(N, L, W)$ for a given survivable meshed WDM optical network, where N is the set of nodes, L is the set of bi-directional links, and the W is the set of available wavelengths per fiber link. $|N|$,$|L|$, and $|W|$ denote the node number, the link number and the wavelength number, respectively. We assume each required bandwidth is a wavelength channel and each node has the O/E/O wavelength conversion capacity. In some simple method, such as in [14], a standard shortest path algorithm (e.g., Dijkstra algorithm) can be used to find the connection's primary path and link-disjoint backup path. After finding the shortest primary path, the links traversed by the primary path will be removed. In the residual graph, the shortest backup path is selected by a standard shortest path algorithm.

4 Algorithms Description for DPP, SPP and DPMR Under Dynamic Traffic

4.1 Algorithm for DPP

- **Input:** the network graph $G(N, L, W)$ and the connection request will be routed, whose source node is s and destination node is d. The required bandwidth is a wavelength channel.
- **Output:** one primary path and one dedicated protected backup path for the connection request
- **Step1:** wait for a request, if one request comes, then go to step 2 , else update the network state.
- **Step2:** compute the primary path for this connection according (1) firstly, if the primary path is found, then record routes information and wavelength assignment. Update the network state and link cost function, go to step 3. If either the primary path or available wavelengths is not found, go to step 4.
- **Step3:** compute the backup path for this connection according to (1), if the backup path is found, then record routes information and wavelength assignment. Update the network state and link cost function, go to step 1. If either the backup path or available wavelengths is not found, go to step 4.
- **Step4:** block this request and go to step1.

4.2 Algorithm for SPP

- **Input:** the network graph $G(N, L, W)$ and the connection request will be routed, whose source node is s and destination node is d. The required bandwidth is a wavelength channel.
- **Output:** one primary path and one shared protected backup path for the connection request.
- **Step1:** wait for a request, if one request comes, then go to step 2 , else update the network state
- **Step2:** compute the primary path for this connection according to (1), if the primary path is found, then record routes information and wavelength assignment. Update the network state and link cost function, go to step 3. If either the primary path or available wavelengths is not found, go to step 4.
- **Step3:** compute the backup path for this connection according to (2), if the backup path is found, then record routes information and wavelength assignment. Update the network state and link cost function, go to step 1. If either the backup path or available wavelengths is not found, go to step 4.
- **Step4:** block this request and go to step 1.

4.3 Algorithm for DPMR

- **Input:** the network graph $G(N, L, W)$ and the connection request will be routed, whose source node is s and destination node is d. The required bandwidth is a wavelength channel.

- **Output:** one primary path and one shared protected backup path for the connection request.
- **Step1:** wait for a request, if the request is high priority traffic, then go to step 2; if the request is low priority traffic, then go to step4; otherwise, update the network state.
- **Step2:** compute the primary path for this high priority connection according to (1), if the primary path is found, then record routes information and wavelength assignment. Update the network state and link cost function, go to step 3. If either the primary path or available wavelengths is not found, go to step 5.
- **Step3:** compute the backup path for this high priority connection according to (4), if the backup path is found, then record routes information and wavelength assignment. Update the network state and link cost function, go to step 1. If either the backup path or available wavelengths is not found, go to step 5.
- **Step4:** compute the working path for this low priority connection according to (3), if the working path is found, then record routes information and wavelength assignment. Update the network state and link cost function, go to step 1. If either the working path or available wavelengths is not found, go to step 5.
- **Step5:** block this request and go to step 1.

5 Simulation Results and Analysis

All simulations have been performed on the 24 nodes USANET topology as shown in Fig.3, Each node pair is interconnected by a bi-directional fiber link and each fiber link has 8 wavelengths. We assume all nodes have full O/E/O wavelength convertible capacity. All connection requests are arriving by Poisson process and uniformly distributed among all resource-destination pairs. The holding time of each connection is normalized to unity and follows a negative exponential distribution. We simulate arrivals and departures of 100000 connections. Two main parameters: blocking probability and resource utilization will be performed.

5.1 Performance Parameters

Two main parameters: blocking probability and resource utilization.

1. Blocking probability
$$BP = B/A$$
   Blocking probability is the ratio of the total number of blocked connections (B) and the total number of arrivals (A). Obviously, the small value of BP means the small blocking probability.
2. Resource utilization
$$RU = H/CW$$

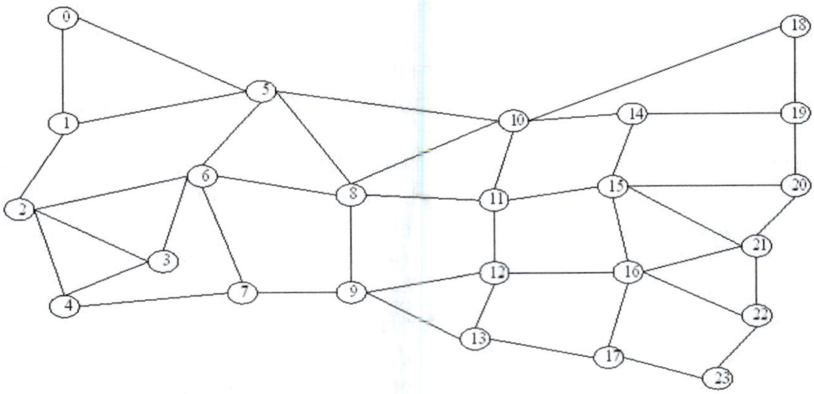

Fig. 3. USANET

Resource utilization is the ratio of the connections that are holding on the network (H) and the consumed wavelengths (CW). When the value of RU is big, the resource utilization ratio is high.

5.2 Analysis

We compare the performances of DPMR to SPP (Shared Path Protection) and DPP (Dedicated Path Protection). In Fig.4, we find that DPMR has larger values than DPP and SPP, and it means that resource utilization of DPMR is higher than that of DPP and SPP. The reason for this is that high priority traffic's backup paths can share resources of low priority traffic, thus DPMR can save more resources. As the network load increases, the value of DPMR increases which means the much higher resource utilization. It is because the chances of share resource between high priority traffic's backup paths and low priority traffic increase when more connections arrive in the network under much higher traffic intensity.

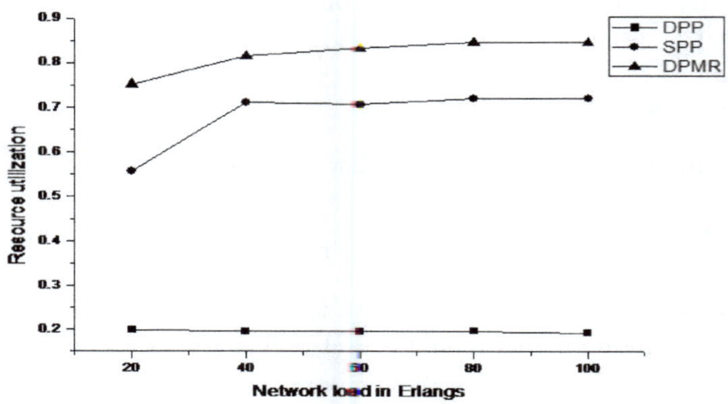

Fig. 4. Resource utilization vs. network load in Erlangs in USANET

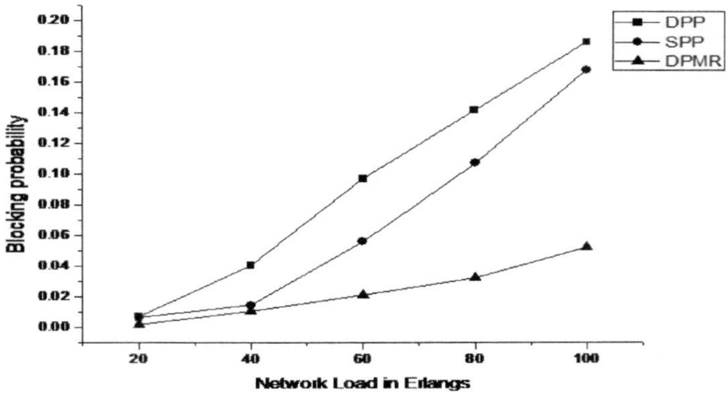

Fig. 5. Blocking probability vs. network load in Erlangs in USANET

In Fig.5, we also find that blocking probability of DPMR is much lower than that of DPP and SPP. The reason is that DPMR can save more wavelength resources as mentioned before. Then the network will have more resources to support the incoming connections and establish them successfully. When the network load increases, the advantage of DPMR is more significant.

6 Conclusions

In this paper, we have proposed an algorithm DPMR under dynamic traffic. We also have modified the general link cost function according to Dynamic Preemptive Multi-Class Routing scheme and used it to find the routes. Then, we have evaluated the performances from resource utilization and blocking probability. Simulations show that DPMR is better than DPP and SPP in resource utilization and blocking probability

Acknowledgments. This research is supported by National Key Basic Research Program of China (973 Program 2007CB307104 of 2007CB307100).

References

1. Mukherjee, B.: Optical Communication Networks. Mc-Graw-Hill, New York (1997)
2. Gerstel, O., Ramaswami, R.: Optical Layer Survivability-an Implementation Perspective. IEEE J. Sel. Areas. Commun. 18, 1885–1899 (2004)
3. Zang, H., Ou, C., Mukherjee, B.: Path-Protection Routing and Wavelength Assignment (RWA) in WDM Mesh Networks under Duct-Layer Constraints. IEEE/ACM Trans. Netw. 11, 248–258 (2003)
4. Gerstel, O., Ramaswami, R.: Optical Layer Survivability: a Services Perspective. IEEE Commun. Mag. 40, 104–113 (2002)
5. Ramamurthy, S., Sahasrabuddhe, L., Mukherjee, B.: Survivable WDM Mesh Networks. J. Lightwave Technol. 21, 870–883 (2003)

6. Andrea, F., Marco, T., Ferenc, U.: Shared Path Protection with Differentiated Reliability. In: Proc. of IEEE ICC 2002, pp. 2157–2161. IEEE Computer Society Press, Los Alamitos (2002)
7. Saradhi, C., Murthy, C.: Routing Differentiated Reliable Connections in WDM Optical Networks. Opt. Netw. Mag., 50–67 (2002)
8. Tacca, M., Fumagalli, A., Paradisi, A., et al.: Differentiated Reliability in Optical Networks: Theoretical and Practical Results. J. Lightwave Technol. 21 (2003)
9. Jaekel, A.: Priority-based Dynamic Lightpath Allocation for Survivable WDM Networks. Photonic Network Communications 11, 201–209 (2006)
10. Sen, A., Shen, B., Hao, B., et al.: On a Preemptive Multi-Class Routing Scheme with Protection Paths for WDM Networks. In: Proc. of IEEE ICC 2003, pp. 1417–1422. IEEE Computer Society Press, Los Alamitos (2003)
11. Zang, H., Jue, J., Mukherjee, B.: A Review of Routing and Wavelength Assignment Approaches for Wavelength-Routed Optical WDM Networks. SPIE Optical Networks Magazine 1, 47–60 (2000)
12. Spath, J.: Dynamic Routing and Resource Allocation in WDM Transport Networks. Elsevier Computer Networks 32, 519–538 (2000)
13. He, R., Wen, H., Li, L., et al.: Shared Sub-Path Protection Algorithm in Traffic-Grooming WDM Mesh Networks. Photon. Netw. Commun. 8, 239–249 (2004)
14. Jozsa, B., Orincsay, D., Kern, A.: Surviving Multiple Network Failures Using Shared Backup Path Protection. In: Proc. of ISCC 2003, pp. 1333–1340 (2003)

Quantification of Cut Sequence Set for Fault Tree Analysis

Dong Liu[1], Chunyuan Zhang[1], Weiyan Xing[2], Rui Li[1], and Haiyan Li[1]

[1] School of Computer, National University of Defense Technology,
Changsha, Hunan 410073, China
{LD5M,zcy89969,lirui7777,lh_lee}.163.com
[2] China Huayin Ordnance Test Center, Huayin 714200, China
weiyan.xing@tom.com

Abstract. A new evaluation method is presented that employs cut sequence set (CSS) to analyze fault trees. A cut sequence is a set of basic events that fail in a specific order that can induce top event. CSS is the aggregate of all the cut sequences in a fault tree. The paper continues its former researches on CSS and uses CSS, composed of sequential failure expressions (SFE), to represent the occurrence of top event. According to the time relationships among the events in each SFE, SFE can be evaluated by different multi-integration formulas, and then the occurrence probability of top event can be obtained by summing up all the evaluation results of SFE. Approximate approaches are also put forward to simplify computation. At last, an example is used to illustrate the applications of CSS quantification. CSS and its quantification provide a new and compact approach to evaluate fault trees.

1 Introduction

Since the first use in 1960s, fault tree analysis has been widely accepted by reliability engineers and researchers for its advantages of compact structure and integrated analyzing methods. There have been many methods developed for the evaluation of fault trees (FT) [1]. With the development of computer technology, former static fault tree (SFT) analysis is not applicable to some new situations, because SFT can not handle the systems that are characterized by dynamic behaviors, Hence, Dugan et al [2] introduced some new dynamic gates, such as FDEP, CSP, PAND and SEQ gates, and put forward dynamic fault trees (DFT) to analyze dynamic systems. In DFT, top event relies on not only the combination of basic events, but also on their occurrence order. The dynamic behaviors bring the difficulty to evaluate DFT.

Fussell et al [3] firstly analyzed the sequential logic of PAND gate, where they provided a quantitive method for PAND gate with no repair mechanism.

Tang et al [4] introduced sequential failures into the traditional minimal cut set for DFT analysis, and provided the concept of minimal cut sequence. Minimal cut sequence is the minimal failure sequence that causes the occurrence of top event, and it is the extension of minimal cut set. However, [4] did not indicate how to extend minimal cut set to minimal cut sequence and did not make quantitive analysis.

Long et al [5] compared Fussell's method [3] with Markov model, and got the same result for PAND gate with three inputs. Then, in [6], Long et al presented the

R. Perrott et al. (Eds.): HPCC 2007, LNCS 4782, pp. 755–765, 2007.

concept of sequential failure logic (SFL) and its probability model (SFLM), by which the reliability of repairable dynamic systems are evaluated by multi-integration. SFLM reveals the thought that dynamic systems can be translated into SFL, but it does not provide an integrated method to solve general DFT.

Amari et al [7] introduced an efficient method to evaluate DFT considering sequence failures. However, in the processes to evaluate spare gates, the method has to employ Markov model which will confront the problem of state explosion.

Our earlier work [8] indicated that cut sequence is intuitive to express the sequential failure behavior, and that it is an applicable way to make reliability analysis using cut sequence. So, in [8], we provided an integrated algorithm, named CSSA, to generate the cut sequence set (CSS), the aggregate of all cut sequences in DFT. This paper will continue our research and focuses on the quantification of CSS based on the analysis of SFLM. The purpose of the paper is to provide a new and compact quantification approach to calculate the reliability of dynamic systems.

Section 2 of this paper provides some concepts. Section 3 introduces the generation of CSS briefly. The quantification of CSS is detailed in section 4. In section 5, an example is used to illustrate our approach. And the conclusion is made in section 6.

2 Background and Concepts

2.1 Notation

x_i	an input to SFL.
λ_{x_i}	the failure rate of x_i.
μ_{x_i}	the repair rate of x_i.
τ_i	the duration of x_i in operation state.
τ_i'	the duration of x_i in warm standby state.
$F(t)$	the probability that top event occurs at time t.
α	the dormancy factor of warm standby component, and $0<\alpha<1$.
$f_{x_i}(t)$	the failure probability density of x_i in operation state.
$\overline{f_{x_i}}(t)$	the failure probability density of x_i in warm standby state.

2.2 Dynamic Fault Trees

Using dynamic gates, DFT obtain the ability to describe dynamic systems in a more feasible and flexible way. Fig. 1 shows the symbols of dynamic gates.

PAND gate is an extension of AND gate. Its input events must occur in a specific order. For example, if a PAND gate has two inputs, A and B, the output of the gate is true if:

- Both A and B have occurred, and
- A occurred before B.

FDEP gate has tree types of events:

- A trigger event: it is either a basic event or the output of another gate in the tree.

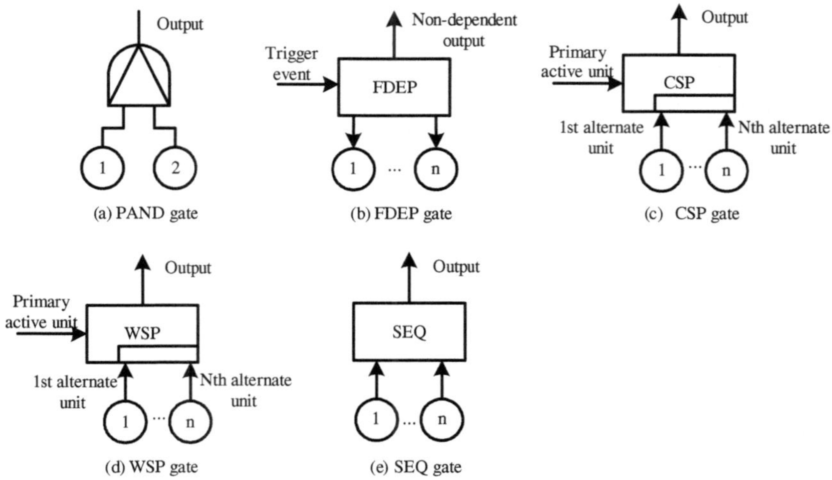

Fig. 1. Dynamic gates used in DFT

- A non-dependent output: it reflects the status of the trigger event.
- One or more dependent events: they functionally rely on the trigger event, which means that they will become inaccessible or unusable if the trigger event occurs.

CSP gate has a primary input and one or more alternate inputs. All inputs are basic events. The primary input is the one that is initially powered on, and the alternate input(s) specify the components that are used as the cold spare unit of the primary unit. The output of CSP gate becomes true if all inputs occur.

WSP gate is similar to CSP gate except that the alternate units in a WSP gate can fail independently before primary active unit.

SEQ gate forces events to occur in a particular order. The output of SEQ gate does not occur until all its inputs has occurred in the left-to-right order in which they appear under the gate. SEQ gate can be contrasted with PAND gate in that the inputs in PAND gate can occur in any order, whereas SEQ gate allows the events to occur only in a specified order. All the inputs, except the first one, of SEQ gate must be basic events, so, if the first input is a basic event, SEQ gate is the same to CSP gate.

2.3 Sequential Failure Logic Model

In SFLM [6], SFL is used to describe the failure relation among events. SFL is a qualitative expression of PAND gate. For example, if the inputs of SFL are x_1, x_2, ..., x_n, respectively, the output will occur when x_1, x_2, ..., x_n fail in a fixed sequence. If x_1 fails at time τ_1, and x_2 fails at time τ_2, ..., x_n fails at time τ_n, where $\tau_1<\tau_2<\ldots<\tau_n$, the output of SFL will occur at τ_n. If x_i fails after x_j, where $i < j$, output will not occur.

Long et al used the following formula to provide the probability that the output of SFL occurs at time t:

$$F(t) = \int_0^t \int_{\tau_1}^t \cdots \int_{\tau_{n-1}}^t f_{x_1}(\tau_1) f_{x_2}(\tau_2) \cdots f_{x_n}(\tau_n) d\tau_n d\tau_{n-1} \cdots d\tau_1 \tag{1}$$

In formula (1), the multi-integration implies the probability that inputs remain in failed states after their failing in a particular sequence, and that the sequences by which the occurrence of the output is not generated are excluded.

In this paper, we suppose that components and system can not be repaired, and that τ_i' is independent to τ_i. Besides, we suppose that components are exponentially distributed, then, the failure probability density function of x_i in operation state is $f_{x_i}(t) = \lambda_{x_i} e^{-\lambda_{x_i} t}$, and the failure probability density function of x_i in warm standby state is $\overline{f_{x_i}}(t) = \lambda_{x_i} e^{-\alpha_{x_i} \lambda_{x_i} t}$.

3 Generation of CSS

Using dynamic gates, DFT gain the ability to describe dynamic systems in a more feasible and flexible way (refer to [2] for more information about DFT). The top event of DFT depends not only on the combination of basic events, but also on the failure sequence of basic events. Tang et al expended the concept of minimal cut set for static fault trees, and provided the concept of minimal cut sequence for DFT [4]. Compared to cut set, which does not consider the sequences among basic events, cut sequence is a basic event sequence that can result in the occurrence of top event in DFT. For example, cut sequence {*A*->*B*} means that top event will occur if event *A* fails before event *B*.

Therefore, the top event of DFT can be described in a format composed of basic events and their sequential relations. In [8], we defined a new symbol, sequential failure symbol (SFS) "→", to express the failure sequence of events.

Connecting two events, SFS indicates that the left event fails before the right event. SFS and its two events constitute sequential failure expression, SFE, such as $A \rightarrow B$, where *A* and *B* can be basic events or their combination. Several events can be chained by SFS. For example, $A \rightarrow B \rightarrow C$ indicates that *A*, *B* and *C* fail according to their positions in the expression, namely, *A* fails first, *B* fails next, and *C* fails last.

Using SFS, the top event of DFT can be expressed by SFE, which are actually cut sequences. And all the cut sequences constitute the CSS. If there are more than one cut sequences, CSS can be expressed by the logic OR of some SFE.

For the AND gate that has *n* inputs, *n*! SFE can be obtained. For example, if the structure function of a fault tree is $\Phi = ABC$, it has 3! = 6 SFE, namely

$$\begin{aligned} CSS = \{ & (A \rightarrow B \rightarrow C) \cup (A \rightarrow C \rightarrow B) \cup (B \rightarrow A \rightarrow C) \cup \\ & (B \rightarrow C \rightarrow A) \cup (C \rightarrow A \rightarrow B) \cup (C \rightarrow B \rightarrow A)\} \end{aligned} \tag{2}$$

It is relatively simpler to get the CSS of a PAND gate, since we can use SFS to connect its inputs according to their sequence. For example, if a fault tree has the structure function $\Phi = A$ PAND B PAND C, its CSS is $\{A \rightarrow B \rightarrow C\}$.

The CSS generation for FDEP gates needs to be considered carefully, since it depends on two conditions, i.e., the condition that trigger event occurs and the condition that trigger event does not occur. Let E_1 denote the trigger event of a FDEP gate. If E_1 occurs, this means that all its dependent events are forced to occur. In this condition, we can analyze the fault tree and get one or more result SFE, which are denoted by E_2. If E_1 does not occur, we can also get one or more SFE in the new condition, which is denoted by E_3. Then, the CSS of the fault tree is $(E_1 \cap E_2) \cup E_3$.

In order to get the CSS of CSP gate, a new expression is needed, i.e., ${}_A^0B$. ${}_A^0B$ means that B is a cold spare unit of A, and that B does not start to work until A fails. The failure rate of B is zero before A fails. We can use $A \rightarrow {}_A^0B$ to express the failure relation between A and B, where B is a cold spare unit of A.

WSP gate is similar to CSP gate except that the alternate units in a WSP gate can fail independently before primary active unit. In order to show this kind of relation, additional two concepts are used: ${}_A^\alpha B$ and ${}^\alpha B$.

- ${}_A^\alpha B$ means that B is a warm spare unit of A, and that B do not turn to operate until A fails. The dormancy factor of B is α, which means that the failure rate of B is α times to its normal failure rate before A fails.
- ${}^\alpha B$ means that B fails independently and its dormancy factor is α before B fails.

Using the above two symbols, we can write down the SFE that express the dynamic behavior of WSP gate.

The detailed procedures about the CSS generation can be found in [8], where an integrated study of generating CSS for DFT is discussed.

4 Quantification of CSS

In this section, we will apply the analysis method about SFL introduced in [6] to SFE, and obtain a quantitive model for CSS.

Now that CSS is expressed by SFE, we can compute the occurrence probability of top event using the following formula:

$$\begin{aligned}\Pr(CSS) &= \Pr(SFE_1 \cup SFE_2 \cdots SFE_n) \\ &= \sum_{i=1}^{n} \Pr(SFE_i) - \sum_{i<j=2}^{n} \Pr(SFE_i \cap SFE_j) + \cdots + (-1)^{n-1} \Pr(SFE_1 \cap SFE_2 \cdots SFE_n)\end{aligned} \tag{3}$$

Each term of formula (3) can be decomposed recursively into one or several SFE. And eventually, the formula comes down to compute the probability of SFE set.

4.1 Quantification of Normal SFE

For AND, SEQ and PAND gates, their SFE can be expressed using SFE= $x_1 \rightarrow x_2 \rightarrow \ldots \rightarrow x_n$, the occurrence probability of which is

$$F(t) = \int_0^t \int_{\tau_1}^t \cdots \int_{\tau_{n-1}}^t \prod_{k=1}^{n} f_{x_k}(\tau_k) d\tau_n \cdots d\tau_1 = \int_0^t \int_{\tau_1}^t \cdots \int_{\tau_{n-1}}^t Q(\tau_1, \tau_2, \cdots, \tau_n) d\tau_n \cdots d\tau_1 \tag{4}$$

where

$$Q(\tau_1, \tau_2, \cdots, \tau_n) = \prod_{k=1}^{n} f_{x_k}(\tau_k) \tag{5}$$

Proof: SFE=$x_1 \rightarrow x_2 \rightarrow \ldots \rightarrow x_n$ means that x_j fails after x_i but before x_k, where $1 \le i < j < k \le n$. If x_i fails at time τ_i ($0 < \tau_i \le t$), we have $\tau_i < \tau_{i+1} \le t$. Therefore, the occurrence probability of SFE is

$$F(t) = \Pr\{0 < \tau_1 \le t, \tau_1 < \tau_2 \le t, \cdots, \tau_{n-1} < \tau_n \le t\} \tag{6}$$

From the conditional probability function, we have

$$\begin{aligned} F(t) = {} & \Pr\{\tau_{n-1} < \tau_n \le t \mid \tau_{n-2} < \tau_{n-1} \le t, \cdots, 0 < \tau_1 \le t\} \cdot \\ & \Pr\{\tau_{n-2} < \tau_{n-1} \le t \mid \tau_{n-3} < \tau_{n-2} \le t, \cdots, 0 < \tau_1 \le t\} \cdot \ldots \cdot \\ & \Pr\{\tau_1 < \tau_2 \le t \mid 0 < \tau_1 \le t\} \cdot \Pr\{0 < \tau_1 \le t\} \end{aligned} \tag{7}$$

Since

$$\begin{aligned} & \Pr\{0 < \tau_1 \le t\} = \int_0^t f_{x_1}(\tau_1) d\tau_1 \\ & \Pr\{\tau_{i-1} < \tau_i < t \mid \tau_{i-2} < \tau_{i-1} < t, \ldots, 0 < \tau_1 < t\} = \Pr\{\tau_{i-1} < \tau_i < t \mid \tau_{i-2} < \tau_{i-1} < t\} \\ & \qquad = \int_{\tau_{i-1}}^t f_{x_i}(\tau_i) d\tau_i \end{aligned} \tag{8}$$

we have the final expression:

$$\begin{aligned} F(t) = {} & \Pr\{\tau_{n-1} < \tau_n \le t \mid \tau_{n-2} < \tau_{n-1} \le t\} \cdot \Pr\{\tau_{n-2} < \tau_{n-1} \le t \mid \tau_{n-3} < \tau_{n-2} \le t\} \cdot \\ & \ldots \cdot \Pr\{\tau_1 < \tau_2 \le t \mid 0 < \tau_1 \le t\} \cdot \Pr\{0 < \tau_1 \le t\} \\ = {} & \int_0^t \int_{\tau_1}^t \cdots \int_{\tau_{n-1}}^t \prod_{k=1}^n f_{x_k}(\tau_k) d\tau_n d\tau_{n-1} \cdots d\tau_1 \end{aligned} \tag{9}$$

If the event number is large, the evaluation of formula (4) is difficult to perform. Then, we can turn to an approximate way that divides time t into M intervals and use $h=t/M$ as the integration step (just like Amari's approach [7]). The result is

$$F(t) \approx \sum_{i_1=1}^{M} \left(\sum_{i_2=i_1}^{M} \left(\sum_{i_3=i_2}^{M} \left(\cdots \sum_{i_n=i_{n-1}}^{M} Q(i_1 h, i_2 h, \cdots, i_n h) h^n \right) \right) \right) \tag{10}$$

4.2 Quantification of SFE for CSP Gates

For CSP gates, the SFE is $x_1 \rightarrow_{x_1}^{0} x_2 \rightarrow \cdots \rightarrow_{x_{n-1}}^{0} x_n$, the occurrence probability of which is

$$\begin{aligned} F(t) &= \int_0^t \int_0^{t-\tau_1} \int_0^{t-\tau_1-\tau_2} \cdots \int_0^{t-\tau_1-\tau_2-\cdots-\tau_{n-1}} \prod_{k=1}^n f_{x_k}(\tau_k) d\tau_n d\tau_{n-1} \cdots d\tau_1 \\ &= \int_0^t \int_0^{t-\tau_1} \int_0^{t-\tau_1-\tau_2} \cdots \int_0^{t-\tau_1-\tau_2-\cdots-\tau_{n-1}} Q(\tau_1, \tau_2, \cdots, \tau_n) d\tau_n d\tau_{n-1} \cdots d\tau_1 \end{aligned} \tag{11}$$

Proof: $x_1 \rightarrow_{x_1}^{0} x_2 \rightarrow \cdots \rightarrow_{x_{n-1}}^{0} x_n$ means that x_{i+1} does not start to work until x_i fails ($1 \le i \le n$). Therefore, the occurrence probability of SFE can be represented by

$$F(t) = \Pr\{\tau_1 + \tau_2 + \cdots \tau_n \le t\} \tag{12}$$

The probability can be computed by convolution function, which is

$$F(t) = \underset{\sum_{k=1}^{n} \tau_i \le t}{\int \ldots \int} \prod_{k=1}^{n} f_{x_k}(\tau_k) \tag{13}$$

The integral region is below the n-dimension hyperplane of $\sum_{k=1}^{n}\tau_i = t$. The iterated integral form is

$$F(t)=\int_0^t\int_0^{t-\tau_1}\int_0^{t-\tau_1-\tau_2}\cdots\int_0^{t-\tau_1-\tau_2-\cdots-\tau_{n-1}}\prod_{k=1}^{n}f_{x_k}(\tau_k)d\tau_n d\tau_{n-1}\cdots d\tau_1 \tag{14}$$

Thereby, we prove the formula (11). The approximation of formula (11) is

$$\begin{aligned}F(t)&\approx\sum_{i_1=1}^{M}(\sum_{i_2=1}^{M-i_1}(\sum_{i_3=1}^{M-i_1-i_2}\cdots\sum_{i_n=1}^{M-i_1-i_2-\cdots-i_{n-1}}(\prod_{k=1}^{n}f_{x_k}(i_k h)h^n)))\\&=\sum_{i_1=1}^{M}(\sum_{i_2=1}^{M-i_1}(\sum_{i_3=1}^{M-i_1-i_2}\cdots\sum_{i_n=1}^{M-i_1-i_2-\cdots-i_{n-1}}Q(i_1h,i_2h,\cdots,i_nh)h^n)\end{aligned} \tag{15}$$

4.3 Quantification of SFE for WSP Gates

For a WSP gate, its corresponding SFE has two forms, namely,

$$SFE=E\to_{E}^{\alpha_{x_1}}x_1\to_{x_1}^{\alpha_{x2}}x_2\cdots\to_{x_{n-1}}^{\alpha_{x2}}x_n \tag{16}$$

$$SFE=E\to^{\alpha_{y_1}}y_1\to_{E}^{\alpha_{x_1}}x_1\to\cdots\to^{\alpha_{y_{m-1}}}y_{m-1}\to^{\alpha_{y_m}}y_m\to_{x_{n-1}}^{\alpha_{x_n}}x_n \tag{17}$$

E is the trigger event of WSP gate, x_i is the component that fails after going to operation, and y_i is the component that fails independently in its warm standby state.

The occurrence probability of SFE (16) is

$$\begin{aligned}F(t)=\int_0^t\int_0^{t-\tau_E}\int_0^{t-\tau_E-\tau_1}\cdots\int_0^{t-\tau_E-\tau_1-\cdots-\tau_{n-1}}G(\tau_1',\tau_2',\cdots\tau_n')\\\times Q(\tau_E,\tau_1,\cdots,\tau_n)d\tau_n d\tau_{n-1}\cdots d\tau_1 d\tau_E\end{aligned} \tag{18}$$

where

$$\begin{aligned}G(\tau_1',\tau_2',\cdots,\tau_n')&=\int_{\tau_E}^{\infty}\overline{f_{x_1}}(\tau_1')d\tau_1'\int_{\tau_E+\tau_1}^{\infty}\overline{f_{x_2}}(\tau_2')d\tau_2'\cdots\int_{\tau_E+\tau_1+\cdots+\tau_{n-1}}^{\infty}\overline{f_{x_n}}(\tau_n')d\tau_n'\\&=(1-\int_0^{\tau_E}\overline{f_{x_1}}(\tau_1')d\tau_1')(1-\int_0^{\tau_E+\tau_1}\overline{f_{x_2}}(\tau_2')d\tau_2')\cdots(1-\int_0^{\tau_E+\tau_1+\cdots+\tau_{n-1}}\overline{f_{x_n}}(\tau_n')d\tau_n')\end{aligned} \tag{19}$$

Proof: The time relationships of SFE (16) can be expressed by Fig. 2. Therefore, its occurrence probability can be expressed by

$$\begin{aligned}F(t)=\Pr\{&\tau_E+\tau_1+\cdots+\tau_n\le t,\\&\tau_1'>\tau_E,\tau_2'>\tau_E+\tau_1,\cdots,\tau_n'>\tau_E+\tau_1+\cdots+\tau_{n-1}\}\end{aligned} \tag{20}$$

It easy to get the following probability

$$\Pr\{\tau_1'>\tau_E,\tau_2'>\tau_E+\tau_1,\cdots,\tau_n'>\tau_E+\tau_1+\cdots+\tau_{n-1}\}=G(\tau_1',\tau_2',\cdots,\tau_n') \tag{21}$$

Since the duration in warm standby state for a component is independent to that in operation state, utilizing formula (11), we have the final equation:

$$\begin{aligned}F(t)=\int_0^t\int_0^{t-\tau_E}\int_0^{t-\tau_E-\tau_1}\cdots\int_0^{t-\tau_E-\tau_1-\cdots-\tau_{n-1}}G(\tau_1',\tau_2',\cdots\tau_n')\\\times Q(\tau_E,\tau_1,\cdots,\tau_n)d\tau_n d\tau_{n-1}\cdots d\tau_1 d\tau_E\end{aligned} \tag{22}$$

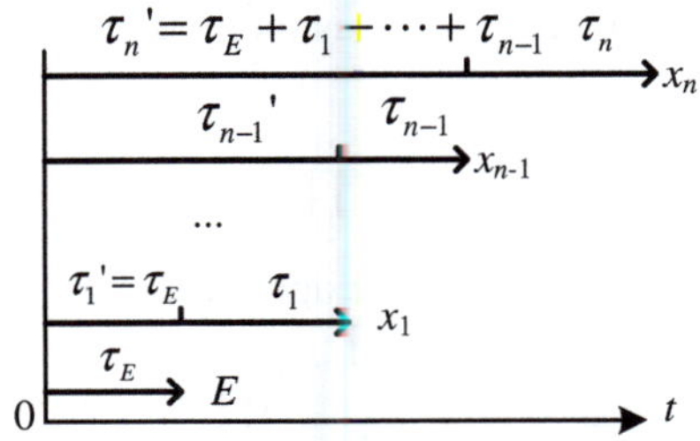

Fig. 2. Time relationships in $SFE = E \xrightarrow[E]{\alpha_{x_1}} x_1 \xrightarrow[x_1]{\alpha_{x2}} x_2 \cdots \xrightarrow[x_{n-1}]{\alpha_{x2}} x_n$

The approximation of formula (18) is

$$F(t) \approx \sum_{i_E=1}^{M} \Big(\sum_{i_1=1}^{M-i_E} \Big(\sum_{i_2=1}^{M-i_E-i_1} \cdots \sum_{i_n=1}^{M-i_E-i_1-\cdots-i_{n-1}} G \cdot Q(i_E h, i_1 h, \cdots, i_n h) h^{n+1} \Big)\Big) \tag{23}$$

where

$$G = \Big(1 - \sum_{i_1'=1}^{i_E} \overline{f_{x_1}}(i_1' h) h\Big)\Big(1 - \sum_{i_2'=1}^{i_E+i_1} \overline{f_{x_2}}(i_2' h) h\Big) \cdots \Big(1 - \sum_{i_n'=1}^{i_E+i_1+\cdots+i_{n-1}} \overline{f_{x_n}}(i_n' h) h\Big) \tag{24}$$

SFE (17) also derives from WSP gates, where $y_1, y_2, \ldots, y_n$ fail independently and $x_1, x_2, \ldots, x_m$ are warm spare units that fail in a fixed sequence. Let σ_i' denote the duration of y_i in warm standby state, then the occurrence probability of SFE (17) is

$$\begin{aligned} F(t) = &\int_0^t \int_0^{t-\tau_E} \int_0^{t-\tau_E-\tau_1} \cdots \int_0^{t-\tau_E-\tau_1-\cdots-\tau_{n-1}} G(\tau_1', \tau_2', \cdots \tau_n') \\ &\times Q(\tau_E, \tau_1, \cdots, \tau_n) K(\sigma_1', \sigma_2', \cdots, \sigma_m') d\tau_E d\tau_1 d\tau_2 \cdots d\tau_n \end{aligned} \tag{25}$$

where

$$K(\sigma_1', \sigma_2', \cdots, \sigma_m') = \int_{\tau_E}^{\tau_1+\tau_E} \overline{f_{y_1}}(\sigma_1') d\sigma_1' \cdots \int_{\sigma_{m-1}'}^{\tau_n+\tau_{n-1}+\cdots+\tau_1+\tau_E} \overline{f_{y_m}}(\sigma_m') d\sigma_m' \tag{26}$$

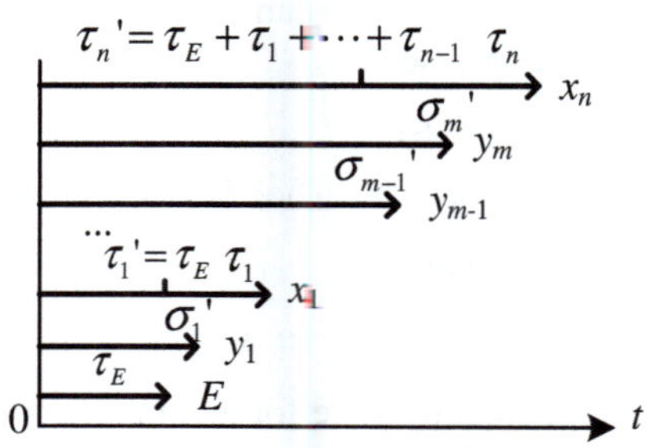

Fig. 3. Time relationships in SFE= $E \xrightarrow{\alpha_{y_1}} y_1 \xrightarrow[E]{\alpha_{x_1}} x_1 \rightarrow \cdots \xrightarrow{\alpha_{y_{m-1}}} y_{m-1} \xrightarrow{\alpha_{y_m}} y_m \xrightarrow[x_{n-1}]{\alpha_{x_n}} x_n$

Proof: The time relationship among the events in the SFE is shown in Fig. 3. Therefore, the probability of the SFE can be expressed by

$$F(t) = \Pr\{\tau_E + \tau_1 + \cdots + \tau_n \le t, \tau_1' > \tau_E, \cdots, \tau_n' > \tau_E + \tau_1 + \cdots + \tau_{n-1}, \\ \tau_E < \sigma_1' \le \tau_1 + \tau_E, \cdots, \sigma_{m-1}' < \sigma_m' \le \tau_n + \tau_{n-1} + \cdots + \tau_1 + \tau_E\} \tag{27}$$

It easy to get the following probability

$$\Pr\{\tau_E < \sigma_1' \le \tau_1 + \tau_E, \cdots, \sigma_{m-1}' < \sigma_m' \le \tau_n + \tau_{n-1} + \cdots + \tau_1 + \tau_E\} \\ = K(\sigma_1', \sigma_2', \cdots, \sigma_m') \tag{28}$$

Therefore, utilizing formula (18), we have the final equation (25).

The approximation of formula (25) is

$$F(t) \approx \sum_{i_E=1}^{M} (\sum_{i_1=1}^{M-i_E} (\sum_{i_2=1}^{M-i_E-i_1} \cdots \sum_{i_n=1}^{M-i_E-i_1-\cdots-i_{n-1}} G \cdot Q(i_E h, i_1 h, \cdots, i_n h) \cdot K \cdot h^{n+1})) \tag{29}$$

where

$$K = \sum_{k_1=i_E}^{i_E+i_1} \cdots \sum_{k_m=k_{m-1}}^{i_E+i_1+\cdots+i_n} \overline{f_{y_1}}(k_1 h) \cdots \overline{f_{y_m}}(k_m h) h^m \tag{30}$$

For any SFE that describes event sequential failures, it has a general formation like

$$z_1 \to z_2 \xrightarrow{\alpha_{y1}} y_1 \to E \xrightarrow{\alpha_{y2}} y_2 \xrightarrow[E]{\alpha_{x1}} x_1 \to \cdots z_p \xrightarrow{\alpha_{ym}} y_m \xrightarrow[x_{n-1}]{\alpha_{xn}} x_n \tag{31}$$

We can work out the probability of expression (31) using formula (25), and get its approximation using formula (29).

5 Case Study

In this section, we will present the quantification processes of CSS using a case system. The system is named Hypothetical Dynamic System, HDS.

HDS has four components, namely *A*, *B*, *C* and *S*. We would like to suppose that *A*, *B* and *C* are power supports of an equipment, and *S* is a switch that controls *C*. *C* is a cold spare unit that will take over *A* or *B* depending on which one fails first. If *S* fails before *A* or *B*, it will affect *C* that *C* can not switch into the system and thus can be thought failed. However, if *S* fails after it switches *C* into the system, it will no longer influence *C*. HDS requires at least two power supplies for operation. The fault tree of HDS is shown in Fig. 4, where FDEP gate indicates that its trigger event, which is the output of a PAND gate, will suspend *C*. The output of PAND gate becomes true if *S* fails before *A* or *B*.

The CSS of HDS is: [8]

$$CSS = \{(A \to B) \cup (A \xrightarrow[A]{0} C) \cup (B \xrightarrow[B]{0} C) \cup (B \to A) \cup (S \to A) \cup (S \to B)\} \\ \overset{\Delta}{=} \{SFE_1 \cup SFE_2 \cup SFE_3 \cup SFE_4 \cup SFE_5 \cup SFE_6\} \tag{32}$$

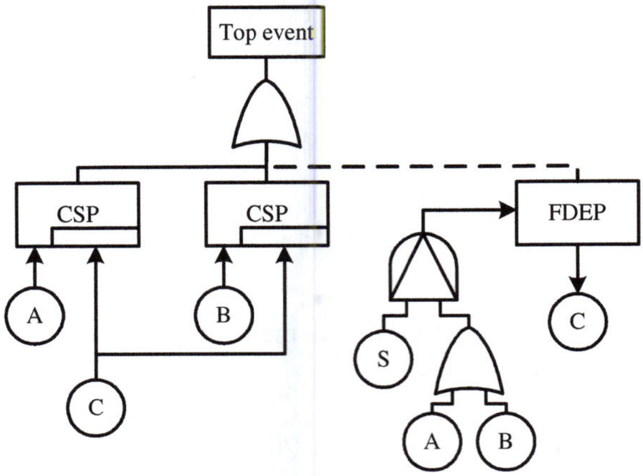

Fig. 4. The fault tree of HDS

The failure probability of HDS can be computed by the following formula:

$$\begin{aligned} F(t) &= \Pr(CSS) = \Pr(SFE_1 \cup SFE_2 \cdots SFE_6) \\ &= \sum_{i=1}^{6}\Pr(SFE_i) - \sum_{i<j=2}^{6}\Pr(SFE_i \cap SFE_j) + \sum_{i<j<k=3}^{6}\Pr(SFE_i \cap SFE_j \cap SFE_k) + \\ &\quad \cdots - \Pr(SFE_1 \cap SFE_2 \cdots SFE_6) \end{aligned} \tag{33}$$

SFE_1, SFE_4, SFE_5 and SFE_6 can be evaluated using formula (10) and SFE_2 and SFE_3 can be evaluated using formula (15).

All intersection results of $SFE_i \cap SFE_j$ are listed in Table 1. The probability of the SFE can be computed using formula (10), (15) or (29). From the conclusion in [6], we

Table 1. The results of $SFE_i \cap SFE_j$

$SFE_1 \cap SFE_2$	$(A \to B \to {}_A^0C) \cup$ $(A \to {}_A^0C \to B)$	$SFE_2 \cap SFE_6$	$(S \to B \to A \to {}_A^0C) \cup (A \to S \to B \to {}_A^0C) \cup$ $(A \to {}_A^0C \to S \to B) \cup (S \to A \to B \to {}_A^0C) \cup$ $(S \to A \to {}_A^0C \to B) \cup (A \to S \to {}_A^0C \to B)$
$SFE_1 \cap SFE_3$	$(A \to B \to {}_B^0C)$	$SFE_3 \cap SFE_4$	$(B \to {}_B^0C \to A) \cup (B \to A \to {}_B^0C)$
$SFE_1 \cap SFE_4$	Does not exist	$SFE_3 \cap SFE_5$	$(S \to A \to B \to {}_B^0C) \cup (B \to S \to A \to {}_B^0C) \cup$ $(B \to {}_B^0C \to S \to A) \cup (S \to B \to A \to {}_B^0C) \cup$ $(S \to B \to {}_B^0C \to A) \cup (B \to S \to {}_B^0C \to A)$
$SFE_1 \cap SFE_5$	$(S \to A \to B)$	$SFE_3 \cap SFE_6$	$(S \to B \to {}_B^0C)$
$SFE_1 \cap SFE_6$	$(A \to S \to B) \cup$ $(S \to A \to B)$	$SFE_4 \cap SFE_5$	$(B \to S \to A) \cup$ $(S \to B \to A)$
$SFE_2 \cap SFE_3$	*	$SFE_4 \cap SFE_6$	$(S \to B \to A)$
$SFE_2 \cap SFE_4$	$(B \to A \to {}_A^0C)$	$SFE_5 \cap SFE_6$	$(S \to A \to B) \cup$ $(S \to B \to A)$
$SFE_2 \cap SFE_5$	$(S \to A \to {}_A^0C)$		

Table 2. The unreliability of HDS

Time t[h]	h=10	h=1	Relex
300	0.0592	0.0496	0.0432
600	0.1548	0.1431	0.1395
1000	0.3152	0.2997	0.2947

can ignore the results of the third term and the latter terms in formula (33), because their values are very small.

Suppose that the components in HDS are exponential distributed and the failure rates are $\lambda_A=\lambda_B=0.0004$, $\lambda_C=0.0005$, $\lambda_S=0.0006$. We computed formula (33) using Matlab software given that h=1 and h=10 respectively. As a comparison, we also evaluated HDS using Relex software [9] that uses Markov model to solve dynamic gates. The unreliability of HDS for different time t is listed in Table 2. From the results, we can see that, with the decreasing of h, the evaluation result based on CSS becomes closer to that based on Markov model.

References

1. Amari, S.V., Akers, J.B.: Reliability analysis of large fault trees using the Vesely failure rate. In: Proceedings of Annual Symposium on Reliability and Maintainability, pp. 391–396 (2004)
2. Dugan, J.B., Bavuso, S., Boyd, M.: Dynamic fault tree models for fault tolerant computer systems. IEEE Transactions on Reliability 41, 363–377 (1992)
3. Fussell, J.B., Aber, E.F., Rahl, R.G.: On the quantitative analysis of priority-AND failure logic. IEEE Transactions on Reliability, 324–326 (1976)
4. Tang, Z., Dugan, J.B.: Minimal Cut Set/Sequence Generation for Dynamic Fault Trees. In: Proceedings of Annual Symposium on Reliability and Maintainability, LA (2004)
5. Long, W., Sato, Y.: A comparison between probabilistic models for quantification of priority-AND gates. In: Proc. PSAM-4, vol. 2, pp. 1215–1220 (1998)
6. Long, W., Sato, Y., Horigone, M.: Quantification of sequential failure logic for fault tree analysis. Reliability Engineering & System Safety 67, 269–274 (2000)
7. Amari, S., Dill, G., Howald, E.: A New Approach to Solve Dynamic Fault Trees. In: Proceedings of Annual Symposium on Reliability and Maintainability, pp. 374–379 (2003)
8. Liu, D., Xing, W., Zhang, C., et al.: Cut Sequence Set Generation for Fault Tree Analysis. In: Lee, Y.-H., et al. (eds.) ICESS 2007. LNCS, vol. 4523, pp. 592–603. Springer, Heidelberg (2007)
9. http://www.relex.com/

Improving a Fault-Tolerant Routing Algorithm Using Detailed Traffic Analysis

Abbas Nayebi, Arash Shamaei, and Hamid Sarbazi-Azad

Sharif University of Technology, Tehran, Iran, and
IPM School of Computer Science, Tehran, Iran
{nayebi,a_shamaei}@ce.sharif.edu, azad@{sharif.edu,ipm.ir}

Abstract. Currently, some coarse measures like global network latency are used to compare routing protocols. These measures do not provide enough insight of traffic distribution among network nodes in presence of different fault regions. This paper presents a detailed traffic analysis of f-cube routing algorithm achieved by a especially developed tool. Per-node traffic analysis illustrates the traffic hotspots caused by fault regions and provides a great assistance in developing fault tolerant routing algorithms. Based on such detailed information, a simple yet effective improvement of f-cube is suggested. Moreover, the effect of a traffic hotspot on the traffic of neighboring nodes and global performance degradation is investigated. To analyze the per-node traffic, some per-node traffic metrics are introduced and one of them is selected for the rest of work. In an effort to gain deep understanding of the issue of traffic analysis of faulty networks, this paper is the first attempt to investigate per-node traffic around fault regions.

1 Introduction

There exist several computational-intensive applications that require continued research and technology development to deliver computers with steadily increasing computing power [1]. The required levels of computing power can only be achieved with massively parallel computers, such as the Earth Simulator [2] and the Blue-Gene/L [3]. The long execution time of these applications requires keeping such systems running even in the presence of failures. However, the vast number of processors and associated devices significantly increases the probability of failure in the system. In particular, failures in the underlying interconnection network in a mulicomputer may isolate a large fraction of machine, wasting many healthy processors. It is also noteworthy that increasing clock frequencies leads to a higher power dissipation, which could lead to failures [4].

It has been reported that network-on-chips (NoCs) applications are prone to faults and have power dissipation restrictions [5-8]. The mesh network topology, due to its ideal simplicity and planarity, is widely used in NoCs [4,6]. Power dissipation is also an important issue in developing NoCs. Due to high level of complexity of VLSI circuit, uniform distribution of power dissipation and avoiding hotspots is so critical, too [9].

R. Perrott et al. (Eds.): HPCC 2007, LNCS 4782, pp. 766–775, 2007.

It is desirable for any fault-tolerant algorithm to be associated with a detailed traffic analysis. One drawback is that achieving per-node traffic is a time consuming process and requires advanced tools. Therefore, many of published works did not perform a detail analysis of traffic around fault regions [11-16,20]. There is only a brief analysis in [21] reported in the literature.

In this paper, we try to analyze traffic distribution around faulty regions in a most classical fault-tolerant routing algorithm, called f-cube [10], which is the basis of many other routing algorithms too. Here, some metrics for per-node traffic are provided and one of them is selected for the rest of the work. Based on the gained insight of traffic pattern around fault region, a simple but efficient improvement is proposed which has a noticeable effect on the performance measures of the f-cube routing algorithm.

The rest of paper is organized as follows. Section 2 provides an overview of related work. Section 3 presents f-cube routing algorithm briefly. Section 4 introduces some per-node traffic metrics which will be later used in our analysis. Section 5 gives the simulation methods used and the results on the distribution of traffic around fault regions. Our proposed improvement is explained in Section 6, and the last section, section 7, concludes the paper.

2 Related Work

Numerous deterministic fault-tolerant routing algorithms in meshes have been proposed in recent years [10-16, 20, 22] most of which augment the dimension-order routing algorithm to tolerate certain faults. Boppana and Chalasani proposed a fault-tolerant routing algorithm for mesh networks [10], and for mesh and torus networks [11]. The key idea of their algorithms is that, for each fault region, a fault ring or fault chain consisting of fault-free nodes and channels can be formed around it; if a message comes in contact with the fault region, the fault ring or chain is used to route the message around the fault region. This point implies that traffic around fault regions must be uneven and intensive. Deadlocks can be prevented by using four virtual channels per physical channel for deterministic (dimension-order) fault-tolerant routing. Their fault model is rectangle or special convex.

In this work, two measures have been used to compare the overall performance of the proposed routing algorithm: "latency" and newly introduced metric "AVC" (average number of used virtual channels in each node). There is nothing in the presented work giving insight about traffic intensity around fault region. Sui and Wang [15], Zhou and Lau[19], Tsai [17], and Wu [22] proposed some fault-tolerant wormhole routing algorithms using various number of virtual channels. Performance degradation reported in the above papers is mainly due to some bottlenecks in small areas of network especially at the corners of fault rings. We investigate this issue in details for the f-cube algorithm and show how a bottleneck in a corner of fault region could propagate traffic to neighboring nodes and increase the total network latency. This phenomenon is related to the nature of wormhole routing which has a great potential to propagate effects of regional bottlenecks into the entire network.

In [18], an appreciable analysis of traffic is performed and fairness of traffic is investigated but it is all about fault free networks.

3 The f-Cube Routing Algorithm

The f-cube routing algorithm was introduced in [10]. In this algorithm messages are assigned types based on the relative positions of the source and destination nodes and dimension-order routing. In a 2D mesh network, messages are typed as east-west (EW), west-east (WE), north-south (NS), or south-north (SN) based on the relative values of offsets in the first dimension. Routing is in dimension order until a message encounters a fault region. Depending on the type, the message is routed around the fault region (Figure 1).

The direction around the fault region is selected based on the relative position of the destination node. The WE and EW messages use c0 channels, and NS and SN messages use c1 channels as shown in Figure 1. Since the topology is mesh and there are no wraparound channels, there are no cyclic dependencies between c0 channels (and also between c1 channels). A brief description of the routing algorithm is presented in Figure 2.

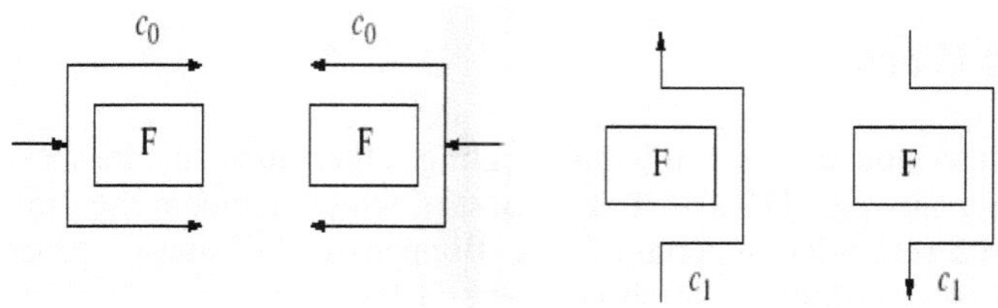

Fig. 1. Usage of virtual channels in f-rings

Algorithm f-cube2 (Fault-Tolerant Routing around Block Faults in 2D meshes)

1. Set and determine the message type (EW, WE, NS, or SN) based on the relative address of the destination.
2. At an intermediate node, a message is routed as follows:
   - If the message has reached the destination, deliver the message.
   - If the message has reached the destination column, set the message type to NS or SN.
   - The message is forwarded along the dimension-order path if fault-free.
   - The message has encountered a fault region. If this is the first hop in the fault ring, the message picks a direction to follow along the fault ring according to figure1and the relative location of the destination.
   - If a dimension-order path is not free, the message continues in the same direction along a fault ring.

Fig. 2. The f-cube2 algorithm

4 Per-Node Traffic Metric

In order to deal with local traffic in a node, a metric must be declared which depicts how a single node may block the incoming messages and lead to a bottleneck. Some of the metrics that can be considered are:

- AUC: Average Utilization of incoming and outgoing physical Channels of a node.
- ANB: Average Number of Blocked messages in routing element.
- AWB: Average Waiting time in Blocking queue.
- AVC: Average number of used Virtual Channels in the node switch.

In this paper, we use the AVC metric which shows the utilization level of node switch. Figure 3 shows the effect of global traffic on different metrics in a 16x16 mesh network with a Poisson node traffic generation distribution and uniform destination traffic pattern. Channel delay is assumed to be one cycle and switching/routing delay is set to zero. Switching method is wormhole with 8 virtual channels per physical channel. The above mentioned measures are captured for the

center node (8,8) in the network. These values are normalized to be comparable. As depicted in this figure, AUC metric is almost linear before saturation region. So, AUC indicates that the traffic is linear and is not sensitive when getting close to saturation point. However, AUC depicts the traffic intensity.

ANB metric has the sharpest curve. It's almost zero before saturation region and rise up rapidly near saturation point. This metric provides no insight on traffic propagation of hotspots because of ignoring the traffic levels lower than a threshold. However, ANB diagnose hotspots more clearly.

AWB is much similar to AUC. Note that AVC has the benefit of highlighting the hotspots with its exponential-like curve and also notices the low level values of traffic (unlike ANB).

5 Simulation Results

Simulation experiments are performed using XMulator [23]. XMulator (which is developed by the authors) is a complete, flit-level, event-based, and extensively detailed package for simulation of interconnection networks which can simulate different interconnection networks with arbitrary topology, switching methods, routing algorithms, and even in the presence of faults.

For the sake of this study, delays for switching and routing are ignored and only delay of physical channels is considered. Two virtual channels are used for each physical channel. Injection channels and ejection channels have 2 virtual channels, too. Message consumption and generation bandwidth is unlimited. Message generation rate for each node is 0.0003 which keeps the network in a normal functional situation in all the scenarios and does not make the network saturated. Physical channel delay is 1 cycle and message length is chosen to be 32 flits.

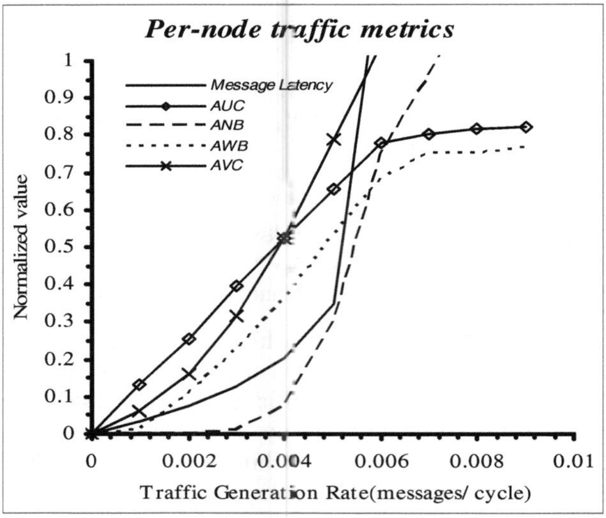

Fig. 3. Effect of global traffic on different per-node traffic measures in a 16x16 mesh

A 32 × 32 mesh is simulated, since we are interested in large parallel systems. Uniform traffic pattern is generated by each node and message generation interval has exponential distribution (a Poisson distribution of number of generated messages per a specified interval). As mentioned before, AVC (average number of used virtual channels in each node) is considered as the traffic metric. This is the time-averaged value of number of connected virtual channels in a switch element.

Four fault regions are simulated: A 1 × 8 region, an 8 × 1 region, and two 8 × 8 regions. Three of these fault regions are placed in the centre of the network and an 8 × 8 region is placed in the east side of the network to investigate the effect of fault region position on the traffic distribution The last fault region which placed near network edges is not stuck to network edge and there is a line of nodes to the edges. This space is considered for better comparison of results.

Results of simulation are depicted in Figure 4.a, c, e and g. Figure 4.a depicts the presence of 4 × 4 faulty region at the centre of network. As illustrated in the figure, the semi-flat surface of Figure 4.a is substantially changed here and two peaks at south-east and north-east of faulty area are appeared. Figure 4.e depicts the effect of a 1 × 8 horizontal fault region. Figure 4.g depicts the results for a centered 8 × 8 fault region while Figure 4.c shows similar results for an 8 × 8 region cited at the east side of the network.

There are two horizontal lines which rise just above and below the fault region. This is because of the nature of dimension-order routing. When an east to west or west to east message arrives at a fault region, it must go up or down and turn around the fault region but can not go back to its previous row unless it has been reached to destination column. As depicted in the figures, centered fault regions make stronger traffic hotspots than fault regions located in the corners. Moreover, comparing

Figure 4.e with Figure 4.a, c, and g reveals that f-cube algorithm performs poorly in case of horizontal fault regions. Comparing the overall latency values confirms this observation, too.

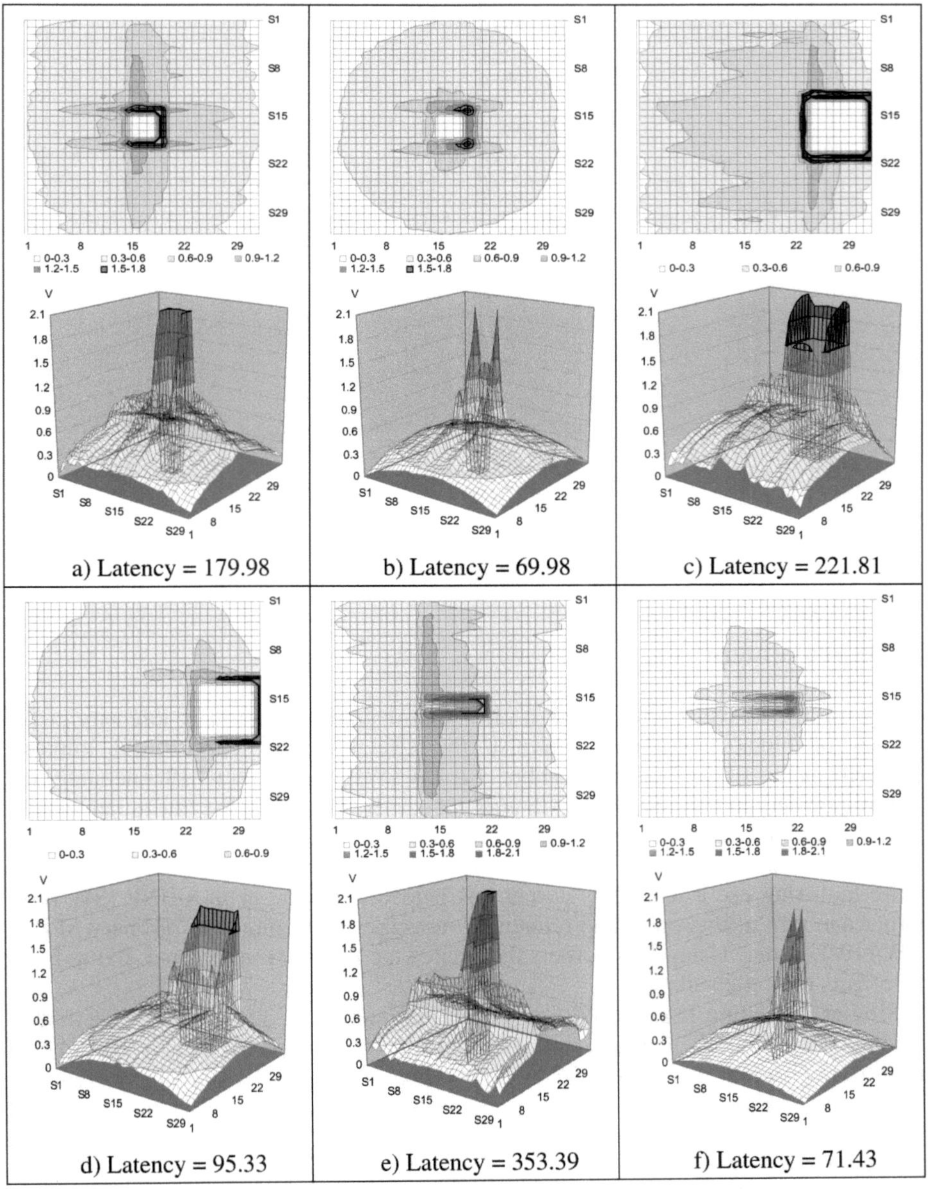

Fig. 4. Traffic level (AVC) for various fault regions. (a), (c), (e) and (g) are 4 × 4, 8 × 8(East), 1 × 8 and 8 × 8 regions respectively by f-cube2 algorithm and (b), (d), (f) and (g) are the same fault region applied by modified f-cube2 algorithm

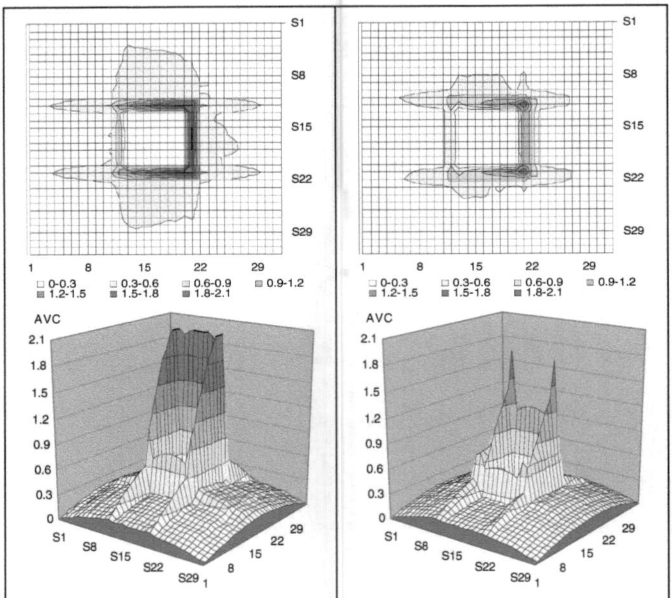

g) Latency = 83.45 h) Latency = 66.67

Fig. 4. (*continued*)

6 Improvement

Analyzing the above results makes it clear that degrading the traffic intensity at the corners of block faults may diminish the effects of bottlenecks and improve the performance of the f-cube algorithm. All EW (WE) messages which encounter the fault region should traverse SE/NE (SW/NW) corner. To reduce the traffic intensity at these corners, it is a good idea to divide the messages into two categories which turns at two different corners. Half of the messages with destinations absolutely outside the entire fault ring use a secondary corner just below/above the main SE/NE (SW/NW) corner and turn at this point and remaining messages go through the ordinary SE/NE (SW/NW) corner. Using two corners distributes the traffic at one corner and reduces the effects of bottleneck.

Similar procedure is proposed for NS and SN messages. The proposed method is deadlock free, since f-cube2 is deadlock free [10]. An example of the acyclic directed networks of virtual channels used by four types of messages is given in Figure 5. Comparing this figure with the corresponding figure in [10], it can be easily inferred that there is no cyclic dependency on the virtual channels acquired by the messages blocked in the network.

Comparing the two methods, for a range of message generation rate in Figure 6, shows that the proposed method has reduced the average message latency and improved the saturation point of the network.

Figure 4 also shows the effect of applying the proposed method for various fault regions. As depicted in the figure, the traffic intensity and bottleneck drawbacks are reduced significantly.

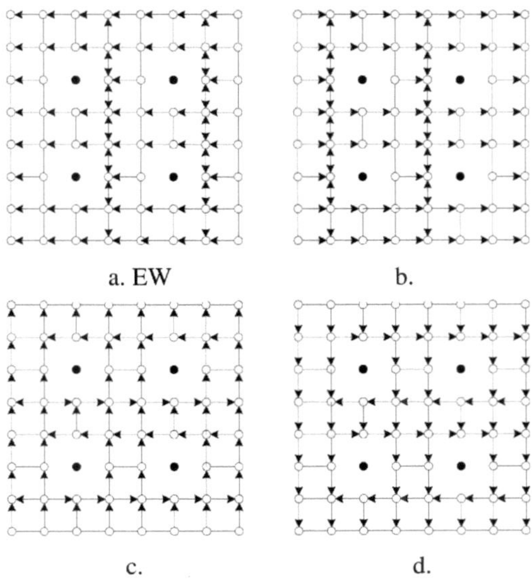

Fig. 5. Example of acyclic directed graph used by modified f-cube2 in an 8x8 mesh with four faulty regions. The channel graph in part (a) is used by East to West messages, part (b) by West to East messages, part (c) by South to North messages and part (d) by North to south messages.

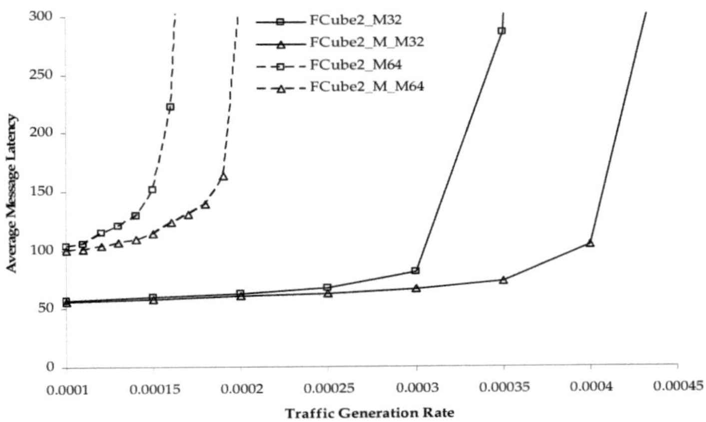

Fig. 6. Simulation result for a 32×32 mesh

7 Conclusions

In this paper, we analyzed the traffic distribution around faulty regions in mesh interconnection network. The f-cube algorithm is chosen as a classical fault-tolerant routing algorithm.

Four metrics of per-node traffic are proposed and the AVC (average number of used virtual channels in switch) is selected in the rest of the paper for the sake of this study. Simulations are performed on a typical 32 × 32 mesh network with uniform destination traffic pattern and with a moderate message generation rate at each node (that does not cause traffic saturation). Results show that the f-cube algorithm may lead to a traffic hotspot at north-east and south-east of the fault ring. In a simulated 8 × 8 fault region located in the centre of the network, traffic at the north-east corner was increased by a factor of 7.7 in comparison to a fault-free network. This result is so important and useful in designing power-aware NoCs. Moreover, it is demonstrated that f-cube algorithm makes two horizontal lines with intensive traffic on the top and bottom of fault region.

Our approach is the first attempt to provide insight on the traffic distribution around fault regions. Using the detailed information about the traffic hotspots, a simple but efficient improvement was made to f-cube algorithm which was shown through simulation experiments.

References

1. Ho, C.T., Stockmeyer, L.: A new approach to fault-tolerant wormhole routing for mesh-connected parallel computers. IEEE Trans. on Computers 53(4), 427–438 (2004)
2. Earth Simulator Center, http://www.es.jamstec.go. jp/esc/eng/index.html
3. IBM BG/L Team, An Overview of BlueGene/L Supercomputer, ACM Supercomputing Conference (2002)
4. Nilsson, E., Oberg, J.: Reducing Power and Latency in 2-D Mesh NoCs using Globally Pseudochronous Locally Synchronous Clocking. In: ISSS 2004, Stockholm, Sweden, pp. 176–181 (2004)
5. Dumitras, T., Kerner, S., Marculescu, R.: Towards On-Chip Fault-Tolerant Communication. In: Asia & South Pacific Design Automation Conf. ASP-DAC (2003)
6. Pirretti, M., Link, G.M., Brooks, R.R., Vijaykrishnan, N., Kandemir, I.M.: Fault Tolerant Algorithms for Network-On-Chip Interconnect. In: IEEE Computer Society Annual Symposium on VLSI Emerging Trends in VLSI Systems Design, pp. 46–52. IEEE Computer Society Press, Los Alamitos (2004)
7. Simunic, T., Mihic, K., De Micheli, G.: Reliability and Power Management of Integrated Systems. In: Eu-romicro Symposium on Digital System Design (DSD'04), pp. 5–11 (2004)
8. Wiklund, D.: Development and Performance Evaluation of Networks on Chip. Ph.D. thesis, Linköping University, Sweden (2005)
9. Bertozzi, D., Benini, L., Micheli, G.D.: Error Control Schemes for On-Chip Communication Links: The Energy–Reliability Tradeoff. IEEE Trans. On Computer-Aided Design of Integrated Circuits & Systems 24(6), 818–831 (2005)
10. Boppana, R.V., Chalasani, S.: Fault-tolerant wormhole routing algorithms for mesh networks. IEEE Trans. Computers 44(7), 848–864 (1995)
11. Chalasani, S., Boppana, R.V.: Communication in multicomputers with non-convex faults. IEEE Trans. Computers 46(5), 616–622 (1997)
12. Gomez, M.E., Duato, J., Flich, J., Lopez, P., Robles, A., Nordbotten, N.A., Skeie, T., Lysne, O.: A New Adaptive Fault-Tolerant Routing Methodology for Direct Networks. In: Bougé, L., Prasanna, V.K. (eds.) HiPC 2004. LNCS, vol. 3296, pp. 462–473. Springer, Heidelberg (2004)

13. Wu, J.: A simple fault-tolerant adaptive and minimal routing approach in 3-d meshes. J. of Computer Science and Technology 18(1), 1–13 (2003)
14. Wu, J.: Fault-tolerant adaptive and minimal routing in mesh-connected multi computers using extended safety levels. IEEE Trans. on Parallel and Distributed Systems 11(2), 149–159 (2000)
15. Wu, J., Wang, D.: Fault-tolerant and deadlock-free routing in 2-D meshes using rectilinear-monotone polygonal fault blocks. The International Journal of Parallel, Emergent and Distributed Systems, 1–14 (2005)
16. Sui, P.H., Wang, S.D.: An improved algorithm for fault-tolerant wormhole routing in meshes. IEEE Trans. Computers 46(9), 1040–1042 (1997)
17. Tsai, M.J.: Fault-tolerant routing in wormhole meshes. Journal of Interconnection Networks 4(4), 463–495 (2003)
18. Izu, C.: Throughput fairness in k-ary n-cube networks. In: Proc. of ACSC 2006 (2006)
19. Zhou, J.P., Lau, F.C.M.: Fault-tolerant Wormhole Routing in 2D Meshes. In: Proc. International Symposium on Parallel Architectures, Algorithms and Networks, Dallas/Richardson, Texas, USA, pp. 457–471 (2000)
20. Zhou, J.P., Lau, F.C.M.: Fault-tolerant Wormhole Routing Algorithm in 2D meshes without Virtual Channels. In: Cao, J., Yang, L.T., Guo, M., Lau, F. (eds.) ISPA 2004. LNCS, vol. 3358, pp. 688–697. Springer, Heidelberg (2004)
21. Varavithya, V., Upadhyay, J., Mohapatra, P.: An Efficient Fault-Tolerant Routing Scheme for Two-Dimensional Meshes. In: International Conference on High Performance Computing, pp. 773–778 (1995)
22. Wu, J.: A Fault-Tolerant and Deadlock-Free Routing Protocol in 2D Meshes Based on Odd-Even Turn Model. IEEE Trans. On computers 52(9), 1154–1169 (2003)
23. Nayebi, A., Meraji, S., Shamaei, A., Sarbazi-Azad, H.: XMulator: An Object Oriented XML-Based Simulator. In: Asia International Conference on Modeling & Simulation (AMS'07), pp. 128–132 (2007)
24. Duato, J., Yalamanchili, S., Ni, V.: Interconnection Networks, an Engineering Approach. Morgan Kauf-mann Publishers, USA (2003)

An Ontology for Semantic Web Services

Qizhi Qiu and Qianxing Xiong

Wuhan University of Technology, Wuhan, P.R. China
{qqz,xqx}@whut.edu.cn

Abstract. An ontology for Semantic Web Services is proposed in this paper, whose intention is to enrich Web Services description. Distinguish from the existing ontologies, the proposed ontology is based on both functionalities and performances, and it is organized as a layered construction. The discovery related with the proposed ontology is also discussed. Based on the Service Oriented Architecture, the proposed ontology is helpful for requesters to find their suitable services according to their own preference. Besides, as an example, an ontology for the learning resource is organized in the paper.

1 Introduction

As a novel Internet component, the significance of Web Services means enhanced productivity and quality by reuse and composition of Web Services. Developers use the services as fundamental elements for applications. As an emergent paradigm, Services Oriented Computing (SOC) is based on the service provider and service requester, where the provider builds a set of invocable applications that provide certain functionalities, while the requester invokes the services or composes suitable application.

With the development of network, there are more and more Web Services that are on different software and vendor platforms, and how to discover and compose them is one of the most challenges. All kinds of network environments require effective discovery and composition to satisfy the different kinds of requirements of requesters.

1.1 Existing Architecture—Service Oriented Architecture (SOA)

There are three main roles in SOA, which are services provider, services registry and services requester. In order to implement the basic operations (publication, discovery and binding), SOA should describe the services explicitly. The service descriptions are organized as a Service Directory. For example, UDDI (University Description, Discovery and Integration) Registry is a well-known tool.

However, UDDI organizes the services according to the defined categories (including the built-in NAICS, UN/SPSC and the user defined ones) without any semantic analysis, such as content or context. UDDI's search is based on syntax and relies on XML, which also enables syntactic. Syntax-based matching limits the reuse of Web services provided by the different kinds of providers. As we know, content and context play the important roles during the dynamic discovery and composition process, and semantic-based matching allows the requesters query through the content and context.

R. Perrott et al. (Eds.): HPCC 2007, LNCS 4782, pp. 776–784, 2007.

Furthermore, UDDI doesn't provide the quality description so that it can't enable the requesters to select the suitable ones. One of the paper's intentions is to enhance the semantic capabilities of UDDI by using tModel and so on.

1.2 Semantic Web Services

Fillman states that the Semantic Web intents to generate Web pages or services with formal declarative descriptions that programs can use to find the appropriate service and use it correctly [1]. To enrich Web Service description, the set of methods and tools in Semantic Web are employed. Ontology is an important part in the semantic web architecture proposed by Tim Burners-Lee in 2000 [2]. OWL-S is a well-known technology built inside the Web Services, and it is an ontology for web services that supplies Web Service providers with a core set of markup language constructs for describing the properties and capabilities of their Web Services in unambiguous, computer-interpretable form [3].

However, OWL-S also depends on WSDL (Web Services Description Language), an XML format for describing network services [4]. As one of main areas of Web service protocol stack, WSDL is used for describing the public interface to a specific web service. When OWL-S is introduced to describe services, the transformation from OWL-S to WSDL occurs. And there are semantic losses in the mapping stage. That means WSDL doesn't keep the rich semantics of OWL-S.

Although there are different ways to enhance the discovery of Web Services, there are several limitations in its description: (i) lack of semantics, such as pure WSDL, (ii) lack of non-functional attributes. All the above also bring forward new challenges to traditional SOC. The intention of this paper is to propose an ontology for Semantic Web Services in SOA so that UDDI can discover the more suitable services according to the preference of the requesters.

2 Ontology for Semantic Web Services

Ontology is a modeling tool for conceptual model, which describes the information system in terms of semantics and knowledge. The following is famous Guarino's definition of ontology: [5]

> *An ontology is a logical theory accounting for the intended meaning of a formal vocabulary, i.e. its ontological commitment to a particular conceptualization of the world.*

The definition shows that the intention of ontology is to capture, describe knowledge in explicit way, where the domain knowledge is represented as concepts and relationships. As a result, knowledge sharing is attained.

2.1 Ontology for General Purpose

Definition 1. Ontology O is defined as:

$$O = <c, p, a>$$

Where:

- *c* is a set of name concepts,
- *p* is a set of property concepts,
- *a* is a set of axioms.

In Definition 1, the axioms are the constraints and rules that are defined on the concepts and properties. There are four types of relationships between the concepts, which are part-of, kind-of, instance-of and attribute-of.

The triple in Definition 1 is corresponding to OWL (not OWL-S). The set *c* is equal to *Class* in OWL, whose elements are URI or expressions. The set *p* is equal to *Attributes* so that all the data types in XML Schema can be used in OWL.

In a word, *Ontology* is a static conceptual model for domain knowledge, which uses terminologies and their relationships agreed upon by wide communities to describe the domain knowledge and its structure.

2.2 Ontology for Web Services

Definition 2 (Web Services description model). A Web Service is defined as:

$$Service\ (d,\ f,\ p)$$

Where:

- *Service* is a Web Service that supports semantics;
- *d* is the basic information about *Service*, including name, ID, provider ID, taxonomy, version, etc.
- *f* is the functional information about *Service*, including related properties, the set of input/ output data and data-type, etc.
- *p* is the performance information about *Service*, including concerned non-functional properties.

At present, almost the web services descriptive languages are compliance with the model spontaneously, but they are all one-sided. For instance, WSDL addresses to describe the online communications in a structured way by defining an XML document for network services. It cannot deal with the semantic heterogeneity and not support the description for constraints that are important to composition.

As semantic markup for Web Services, OWL-S defines an upper ontology for services. Each instance of service will present a *ServiceProfile* description, be described by a *ServiceModel* description, and support a *ServiceGrounding* description. These descriptions answer respectively "what does the service provide for prospective clients?", "how is it used?", "how does one interact with it?". Besides the semantic loss that is mentioned above, OWL-S is regardless of quality of service [6]. For these matters, when semantic web service is proposed, there are several problems must be solved: (i) ontology-based representation, (ii) quality model, (iii) support of discovery and composition. The remainder of this paper will discuss the related issue.

2.3 A Layered Construction

According to the ontology methodology, certain service ontology can be organized into a layered construction (Shown in Fig. 1):

- Top level: there is the general knowledge or common sense, independent of any domain. In a word, they are shared by wide communities. For example, NAICS (The North American Industry Classification System) is a new industry classification.
- Domain knowledge level: there are domain ontology and upper ontology. The upper ontology is defined by *Definition* 2, which is a model to describe the attributes and operations of web service. Domain ontology aims to serve for the providers who are in certain domain. For example, a travel service. Both domain ontology and upper ontology is not invocable, just like an abstract class. What they provide is a framework for a concrete, invocable service. This level can also be called as a semantic level.
- Concrete level: there are invocable services that are described by OWL and organized according to the framework defined in the upper level. Besides, there are bindings to the specific implementations.

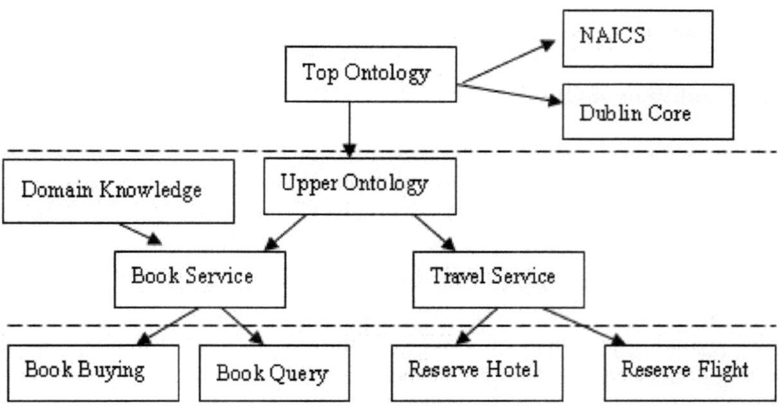

Fig. 1. A Layered Construction of Ontology

2.4 Quality Model

Definition 3 (Quality Model of Web Services). $QoS = \{t, c, a, s \ldots.\}$
Where:

- t: the response time of the certain Web service.
- c: the cost of the invoking of the certain Web service.
- a: the probability of availability of the certain Web service in a given interval, it is a statistic.
- s: the probability of success of the certain Web service in a given interval, it is also a statistic.

Vectors *t*, *a* and *s* are mature metric for quality of software, so their values are computed in the same way as that in quality of software [7].

(i) Vector *t* is numeric parameter, whose value is assigned when the ontology for a concrete service is built. For example, a provider assigns vector *t* according to the performance of his service when publishes his service, and a requester also assigns vector *t* according to his requirements when he summits his request.
(ii) Vector *a*, *s* can be computed by a counter in an *Agent* which is developed by the third party. Vector *a* is computed by the following formula, and Vector *s* can be computed by a similar way:

$$a = \frac{num(available\_invocation)}{num(total\_invocation)} \tag{1}$$

(iii) Vector *c* plays a critical role in discovery of SOC. Just like in the real world, the requesters pay more attentions on it than other vectors.

Definition 3 shows a general quality model. Considering the various domains, the model is defined as an open structure, where the users can define their own vectors about quality of service, and values of vectors vary with the domains. *tModel* in UDDI is exploited to support the model [8].

2.5 Implementation of Ontology

In order to implement the defined ontology, we select Protégé as the editor. Protégé is based on Java and provides a PnP environment. Protégé ontologies can be exported into a variety of formats including RDF (S), OWL, and XML Schema. Furthermore, RacerPro is a selection of inference engines as the back-end inference system (reasoner) for use with Protégé. The RacerPro system implements the description logic and can reason about OWL knowledge bases, together with some algebraic reasoning beyond the scope of OWL. With the help of Protégé, the defined framework of Service Ontology can be designed and the previous version of knowledge based can be exploited together.

3 An Example—Learning Resource Ontology

In this section, we will consider an example about service ontology. Learning Resources (for short: LR) have played an important role in modern education. While in terms of Web Services, all the items in the Internet, such software, hardware, information and so on, can be regarded as available *Services*. So is learning resources.

With reference to the main standard for learning object, we can define an ontology for LR compliance with *Definition* 2. *Table* 1 shows attributes of LR and related standards. Ontology LR is organized as shown in *Table* 1.

Table 1. Attributes of LR and Related Standard

Attribute	Sub-Element	Related Standard
d	identifier	IEEE LOM
	title	
	language	
	keywords	
	description	
	version	
	contribute	
f	format	
	size	
	requirement	
	duration	
	intended end	
	user role	
p	cost	
	scientificity	CELTS-24
	accessibility	
	update-interval	
	response time	
	probability of success	

The following give more details about LR:

- LR lies in the concrete level of *Fig*. 1, just like *book buying.*
- IEEE LOM (Learning Object Metadata) [9] is a standard developed by IEEE, which lies in the domain level of *Fig*. 1.
- CELTS (Chinese E-learning Technology Standardization) [10] is a standard developed by Chinese government, and CELTS-24 is its sub-standard for Service Quality of e-Learning.
- Except attribute *response time* and *succeed-probability*, other attributes are assigned values compliance with the standards.

4 Discovery Process

The intention of Service Ontology defined in Section 2 is to add semantics to WSDL and UDDI. In this section, we show how to use Service Ontology to find the suitable service for requesters or compound service developers.

4.1 Discovery Process

(i) Transform the requests of customers in forms of Service Template compliance with Service Ontology. All the vectors *f*, *d*, *p* in *Definition* 2 as well as their sub-elements, will be used in the following match. It is necessary for compound service developers to decompose the request into atomic process.

(ii) Search engine in UDDI matches the requests against Service Advertisement. IOPE (Input, Output, Precondition, Effect) is the major factors in this stage. As a result, a candidate set of services is gained.

(iii) Analyze the performance constraints and order the selected services in the candidate set so that customers can select the most suitable services according to their own preference. Matching algorithm will be exploited, which is introduced in following section.

4.2 Similarity Matching

During the discovery process, several Agents are needed. Those Agents aim to implement matching algorithms. With the increase of Web Services specification in UDDI, it is difficult to find out a suitable service. The traditional retrieval technologies, such as keyword-based match, are not applicable. So is the classical set theory, whose logic is based on either 0 or 1. In this paper, *Fuzzy Theory* and *Similarity Function* [11,12] are selected to match *Service Template* against *Service Advertisement*. The result of match computed by *Formula* 2 is on closed interval *[0,1]*. This is a quantified analysis, whose result is more accurate than that of some other algorithms.

$$s(a,b)=\Sigma\ (\mu_i s(a,b)) \qquad (\mu_i > 0, \sum \mu_i = 1) \tag{2}$$

Where:

- $s\ (a,\ b)$ is similarity between a and b, where a is the description given by a provider, b is that of a requester.
- μ_i is the weigh coefficient, they reflect the weightiness of attributes of service.

Apparently, it is easy to compute the value of similarity when IOPE is matched in step (ii) during discovery process. As for the match in step (iii), semantic relationships among services compliance with *Definition* 2 can be exploited so as to support the customers in discovery services that fit their requirements. A simplified form of Tversky function is applicable. We state more details about it in [13]. Furthermore, μ_i will be assigned by domain experts.

4.3 Implementation

The models and methods introduced are also used within SOA. This section states how to enable service discovery as described above.

4.3.1 Enhancing UDDI

Undoubtedly, all of the concrete services (mentioned in Section 2.3) are stored in UDDI Registry, while the existing UDDI Registry supports neither the description in OWL nor the performance vector in *Definition* 2. Therefore, one of our works is to extend the UDDI Registry, maintaining a full compatibility with it, so that the developers or the customers can either exploit the existing UDDI APIs or invoke the APIs provided.

Fortunately, one goal of UDDI is to make it easy to describe Web Services meaningfully, and its *tModel* structure provides the ability to describe compliance with certain specifications, concepts or other shared design. Besides *tModel*, UDDI v2 defines several elements for category: (i) *keyedReferenceGroup*, which is a simple list of *keyedReference* structures that logically belong together. It must contain a *tModelKey* attribute that specifies the structure and meaning of the *keyedReferences* contained in the *keyedReferenceGroup*. (ii) *catagoryBag*, which contains a list of business categories that each describes a specific business aspect of the *businessEntity*, such as industry category or a simple list of *keyedreferneceGroup*. It also can be used to categorize *tModel*. [14]

Making use of the extensive mechanism, UDDI can contain service description compliance with *Definition* 2. In the proposed method, we define a set of *tModel* to represent vector p in Definition 2, and some *tModel* for IOPE. All the sets are related with Service Ontology, while each *keyedReferernce* has its own *keyValue* that represents different ontology.

4.3.2 Annotated WSDL

Although Service Ontology is edited in OWL rather than OWL-S, WSDL is a de-facto standard for service functionality description. In the proposed method, we select Semantic Annotations for WSDL (for short: SAWSDL).

SAWSDL defines a set of extension attributes for WSDL that allows description of additional semantic of WSDL components. SAWSDL doesn't specify a language for representing the semantic models. Instead it provides mechanisms by which concepts from the semantic model can be referenced from within WSDL and XML Schema components using annotations. It also can be mapped into an RDF form compatible with previous versions. [15]

5 Conclusion and Further Work

An ontology for Semantic Web Services with both functional and non-functional attributes has been built in Protégé. By making use of the ontology in the model proposed in [13], much improvement will be made in Service discovery.

Service discovery is an ongoing research direction in Web Services community. Service Ontology is introduced to enrich the semantics of Web Service, including performance description. The proposed construction of Service Ontology makes it easy to identify and maintain the relationships among services. Exploiting such an ontology, as well as extension of UDDI and SAWSDL, service discovery techniques are used to match Service Template against Service Advertisement efficiently.

The further work includes:

- Do more competitive trials to prove the effect of the proposed method by recalling the services described in different way, such as WSDL, OWL-S and the proposed method.
- Do more study on mechanism in UDDI in order to enhance its semantics.
- Improve the match algorithm with reference to optimization technology in nonlinear systems.
- Optimize this method by take full advantage of traditional retrieval methods, such as Levenshtein Distance for string matching.

References

1. Fillman, F.R.: Semantic services. IEEE Internet Compute. 7 (2003)
2. Berners-Lee, T.: Semantic Web–XML2000 (2000), http://www.w3.org/2000/talks/ 1206-xml2k -th1/ silde10-0. html
3. Martin, D., et al.: OWL-S: Semantic Markup for Web Services (2004), http://www.w3.org/Submission/OWL-S/
4. Christensen, E., Curbera, F.: Web Services Description Language (WSDL)1.1. (2001), http://www.w3.org/TR/wsdl
5. Guarino, N.: Formal Ontology and Information Systems. In: Proceedings of FOIS 1998 (1998)
6. Ankolenkar, A., et al.: OWL-S: Semantic Markup for Web Services, OWL-S 1.0 Draft Release (2003)
7. Pressman, R.S.: Software Engineering: a Practitioner's Approach, 5th edn. China Machine Press (2005)
8. van Moorsel, A.: Metrics for the Internet Age: Quality of Experience and Quality of Business. In: Proceedings of the 5th Performability Workshop (2001)
9. Anonymity, Standard for Learning Object Metadata (2006), http://www.imsglobal.org/ metadata/mdv1p3/ imsmd_bestv1p3.html
10. Chinese E-Learning Technology Standardlization Committee, Specification for Service Quality Management System of e-Learning (2003), http://www.celtsc. edu.cn/680751c665875e93
11. Guoen, Y., Rong, A.: A Review of Theories and Models of Similarity in Categorization. Psychological Exploration (2005)
12. Wang, Y., et al.: Semantic structure matching for assessing web-Service Similarity. In: Orlowska, M.E., Weerawarana, S., Papazoglou, M.M.P., Yang, J. (eds.) ICSOC 2003. LNCS, vol. 2910, Springer, Heidelberg (2003)
13. Qizhi, Q., et al.: Study on Ontology-based Web Services Discovery. In: Proceeding of the 11th International Conference on Computer Supported Cooperative Work in Design (2007)
14. Bellwood, T., et al.: UDDI: Version 3.0.2 (October 2004), http://uddi.org/pubs/uddi_v3.htm# _Toc85908023
15. Farrel, J., et al.: Semantic Annotations for WSDL and XML Schema (2007), http://www.w3.org/TR/sawsdl

DISH - Dynamic Information-Based Scalable Hashing on a Cluster of Web Cache Servers*

Andrew Sohn[1], Hukeun Kwak[2], and Kyusik Chung[2]

[1] Computer Science Department, New Jersey Institute of Technology, Newark, NJ 07102, U.S.A.
[2] School of Electronic Engineering, Soongsil University, 511 Sangdo-dong, Dongjak-gu, Seoul 156-743, Korea
sohn@cs.njit.edu, {gobarian,kchung}@q.ssu.ac.kr

Abstract. Caching web pages is an important part of web infrastructure. The effects of caching services are even more pronounced for wireless infrastructures due to their limited bandwidth. Medium to large-scale infrastructures deploy a cluster of servers to solve the scalability problem and hot spot problem inherent in caching. In this report, we present Dynamic Information-based Scalable Hashing (DISH) that evenly hashes client requests to a cluster of cache servers. Three types of runtime information are used to determine when and how to cache pages, including cache utilization, CPU usage, and number of connections. Pages cached are stored and retrieved mutually exclusively to/from all the servers. We have implemented our approach and performed various experiments using publicly available traces. Experimental results on a cluster of 16 cache servers demonstrate that the proposed hashing method gives 45% to 114% performance improvement over other widely used methods, while addressing the hot spot problem.

1 Introduction

Caching web pages is an important part of web infrastructures [1, 2]. Those pages that are frequently requested can be saved in a way that they can be quickly handed out later without having to involve web, file and database servers, since reading and preparing files by these servers can take much time from the CPU's perspective. By saving these frequently requested files, the user can experience less response time and, at the same time, the servers can save the time for preparation, resulting in overall performance improvement. Efficient caching can help reduce the impact of the temporary hot spot problem and server down time when a deluge of a particular set of pages (files) is requested. Caching practice is particularly important in wireless infrastructure, where, unlike the land-line based infrastructure, the bandwidth is limited. Cache servers can be placed in various places in Web infrastructure, depending on the requirements of the enterprise. Three methods are typically used to place cache servers, including forward, transparent, and reverse caching [3]. Forward

* Hukeun Kwak and Kyusik Chung are supported in part by the Basic Research Program of the Korea Science & Engineering Foundation grant No.R01-2006-000-11167-0.

R. Perrott et al. (Eds.): HPCC 2007, LNCS 4782, pp. 785–796, 2007.

cache servers are placed on the edge of the enterprise, close to clients, while reverse cache servers are placed close to the web server(s). Transparent cache servers are placed between the clients and the web server, hence maintaining the transparency since the location is not explicitly known to the clients.

For small enterprises, one cache server would suffice, since the traffic would not warrant high performance. Or, conversely the web server itself can handle caching as well. However, for medium to large-scale enterprises, the roles and importance of cache servers can be critical, since loss of client requests means loss of business. For this reason, large-scale enterprises employ a cluster of cache servers to keep the response time tolerable. Deploying a cluster of cache servers presents technical challenges, mainly regarding performance scalability. Since multiple servers are deployed for caching, it is expected that the servers be highly available with tolerable response time, regardless of the traffic condition. These servers need to work together to achieve this common goal of performance scalability, hence *cooperative caching* [4, 5]. Central to cooperative caching is the storing and retrieval of cached pages. Managing cached pages can be classified into two types: *distributed* and *copied.* In the copy approach, each and every server can have exactly the same cached pages so that any server can serve any pages. Obviously this approach is against the use of cluster since the number of pages that can be cached is limited by the capability of an individual server. Furthermore maintaining the coherence of cached pages can be expensive, since any change must be reflected in all servers. It is, therefore, essential that the cached pages be distributed mutually exclusively across all servers.

Distributing cached pages presents its own challenges in identifying page location and resolving hot spots. Since pages are distributed, it is critical to quickly identify where the requested pages are cached. Secondly, since a particular page is cached in one server, a deluge of requests for that "hot" page(s) can overwhelm the server's capability. This is a typical situation encountered in the event of breaking news, as, if the server responsible for delivering this news becomes inaccessible, it defeats the very purpose of scalability. Therefore, cooperative caching needs to address these two challenges to be a scalable alternative to the copied approach. Numerous cooperative caching methods have been put into practice, which can be classified into two categories: protocol-based [6, 7] and hashing-based [8-14]. Protocol-based methods rely on the neighboring servers for cached pages. If a server does not have the requested page, it asks the neighboring servers. If none of the neighboring servers has the page, it would turn to the web server to produce the page. Hashing-based methods use the hashed value of the URL to locate a cache server. Since it is *fixed*, the hashing function always produces the same cache server for the URL. If the hashed server does not have the page, the server asks the web server for the page.

While hash-based methods are widely accepted, the main challenge of hashing-based caching is the distribution of URLs to cache servers. There is no guarantee that the hashing function will reasonably and evenly distribute URLs to servers. Some servers may have many while some have few, resulting in a load imbalance across the servers. As a result, performance scalability suffers. Hot pages will only amplify this problem. It is precisely the purpose of this paper to solve these two challenges of hot spot and performance scalability of cache servers. We use runtime server information to dynamically change how the pages are hashed to avoid the fixed nature of hashing.

The paper is organized as follows. Section 2 presents background materials on web page hashing. In Section 3, we explain our approach in detail. In section 4, we provide our experimental environment and results using publicly known Web traces. In Section 5, we analyze our results and compare them with the results of widely used methods. The last section concludes our work.

2 Background Work

Message Digest 5, introduced in 1992 [8], is a frequently used hashing method for messages, in particular mapping URLs to a cache server. Each message (Web page address or URL) of the size of up to 2^{64} bits is converted to a unique 128-bit hash value. The resulting 128-bit hash value is mapped to a cache server by taking a modular operation with the number of cache servers. MD5 generates a fixed-length hash value for every client request. Since it computes hash value at runtime, MD5 requires no memory to store this URL mapping to a cache server. A distinct advantage of MD5 is that cache servers can be added to and removed from the pool of servers without affecting the mapping. However, since hash value is computed each time, a large number of client requests can degrade the performance.

Consistent Hashing, introduced in 1997 [9], expands MD5 by organizing hash values between 0 and 1 in a circular fashion. The front-end load balancer keeps a table with entries to map a point in the circle. Each entry consists of hash value and its corresponding cache server. The client request is first converted to a fixed-sized hash value using, for example, MD5. If there is an entry in the table for this value, the corresponding server is used to serve the request. Otherwise, the nearest server in the clockwise direction is selected. The main advantages of consistent hashing include simple addition and removal of cache servers. However, it does not address the even distribution and hot spot problems.

Cache Array Routing Protocol (CARP), introduced by Microsoft in 1997 [10], combines both the URL and cache server name to determine the cache server. The front-end load balancer keeps an array consisting of rows with URL hash values and columns with cache server names. The main advantage of this method is that it allows flexible addition and removal of cache servers. And, at the same time, it addresses scalability to a limited extent, since it goes by hard-wired cache server names, provided the requests are reasonably distributed across different URL domains and cache servers. On the other hand, it suffers from the hot spot problem even more than other methods, since the server with hot pages is fixed.

Modified Consistent Hashing [11] is designed to fix the load imbalance problem of consistent hashing described above. Consistent hashing finds the nearest cache server in the clockwise direction. Modified consistent hashing attempts to find the nearest cache server in both clockwise and counter clockwise directions. The load imbalance can be spread over two directions (two servers) instead of one. While this method attempts to alleviate the load imbalance problem, it does not eliminate the problems of consistent hashing. (I.e., hot spot and load imbalance).

Adaptive Load Balancing [12] attempts to address mainly the hot spot problem by round-robin request distribution. The front-end load balancer keeps a list of popular pages using the Least Frequently Used policy. If it is on the list, the client request will be sent to a cache server in a round-robin fashion. Otherwise, it will go through MD5-like hashing to find a cache server. While this approach solves the hot spot problem, other problems, such as load imbalance, addition/removal of cache servers and performance scalability, are not addressed.

Extended Consistent Hashing [13] literally extends consistent hashing by providing a window of servers. Consistent and Modified Consistent caching methods find the nearest next server either clockwise or counter clockwise. Extended caching provides a window of servers in which the best suitable server is selected. URLs and pages that are frequently requested are assigned to a larger window of servers while others are assigned to a smaller number of servers. While it improves the distribution of requests to neighboring cache servers, extended caching does not address the inherent issues of consistent caching, namely scalability and overall distribution of requests.

Cache Clouds Dynamic Hashing [14] introduces a hierarchical organization of cache servers to utilize runtime server information. At the top of the hierarchy are multiple sets of servers, called Beacon Rings, each of which consists of three beacon points (servers). Client requests are first assigned to a particular Beacon Ring through MD5 hashing, followed by the second level hashing to eventually find a suitable cache server. The second level hashing uses CPU capacity and the current load information to find the best one within the ring. This method solves the hot spot problem by having a ring of multiple servers to serve hot pages. However, the method does not address the scalability in terms of even distribution of requests and server utilization across the rings since it inherits the characteristics of MD5 to begin with.

3 DISH - Dynamic Information-Based Scalable Hashing

3.1 The Proposed Method

The hashing methods described above translate hash values directly to cache servers. We propose in this report a new method that takes into account dynamic server information before translating to a cache server. Three types of dynamic server information are considered in this study: *cache utilization*, *CPU usage* and *number of connections*. Cache utilization refers to the frequency access to a particular cached page (or data). To control importance and granularity, each resource has weight associated with it, ranging from 0 to 10 with the increment of 1. The three usage types are combined together with individual weights to find the overall resource utilization U_{overall} for each cache server:

$$U = w_{\text{cache}} * u_{\text{cache}} + w_{\text{cpu}} * u_{\text{cpu}} + w_{\text{conn}} * u_{\text{conn}},$$

where w_{cache} = weight for cache utilization, w_{cpu} = weight for CPU usage, w_{conn} = weight for the number of connections between the front-end load balancer LVS and a cache server, u_{cache} = cache utilization / average cache utilization, u_{cpu} = CPU usage / average CPU usage, and u_{conn} = number of connections / average number of connections. Assigning client requests to cache servers is done by finding the server with the lowest overall utilization U at the time of request. This weight usage combined with the three dynamic resource types ensures that the hashed requests are spread equally across all cache servers.

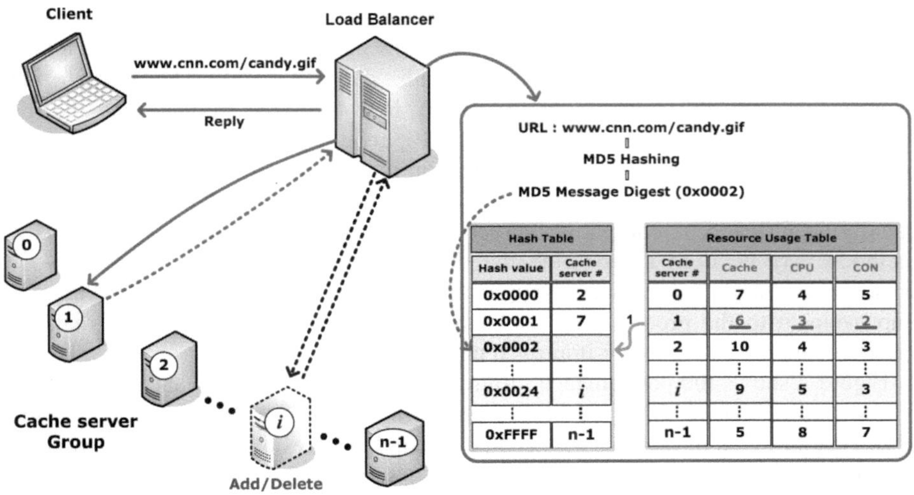

Fig. 1. DISH – Dynamic Server Information-based Distribution

Figure 1 illustrates the proposed approach DISH. The front-end load balancer maintains two tables: hash and resource usage with weights. The hash table consists of a list of pairs, with each designating a hash value and server. The resource usage table keeps a list of three weights for each cache server.

The example in Figure 1 illustrates how the proposed approach works. The LVS load balancer just received the URL www.cnn.com/candy.gif, and then generated the hash value 0x0002 using MD5. Suppose the entry 0x0002 is empty, indicating no cache server has been assigned to this hash value. The LVS looks up the resource usage table and computes the lowest overall usage as described above. Suppose further that the LVS finds Cache server 1 with the lowest weighted usage. This server is now assigned the hash value, as the curved arrow indicates. The mapping just established between the hash value and server is used to route future incoming requests that result in the same hash value. The hash table will eventually be filled, in which case the routing decisions will remain unchanged.

Hot spots are detected and handled in a round-robin fashion. When a cache server is asked to handle more than 400 requests a second, it is deemed that the server is overloaded with a hot spot. Should this situation persist for over a second, the hot page(s) will be copied to all servers, each of which will handle requests for the hot pages in a round-robin fashion. If and when the deluge of requests for the URL recedes and the hot spot situation is mitigated, client requests for the URL will be served by the one to which the URL was originally assigned. The fixed threshold of 400 requests is derived based on our experimental results for the given machine specifications. The threshold is determined by factors such as machine memory size and CPU speed, and is therefore subject to adjustment.

3.2 Finding Weights

Finding the best weight for each parameter is critical in determining cache servers, as inadequate weights may adversely affect the overall performance. To find the best weights, we have performed extensive experiments. The three parameters of cache, CPU, and connection each take on 0 to 10 with the increment of 1. There are a total of 11x11x11=1331 possible combinations with *duplicates*. We have performed experiments with all combinations to identify how the three weights play out. For example, the combination of w_{cache} =0, w_{cpu} =10, and w_{conn} =0 indicates that only CPU usage is used to compute the overall utilization of each server. On the other hand the combination of w_{cache} =5, w_{cpu} =5, and w_{conn} =5 indicates that all three parameters are equally significant in computing server utilization. While complex equations or methods may be used to combine the three weights more precisely, they would present new and unnecessary complications while, in turn, providing minimal improvement to the process. We only wish to identify the general trends that help combine the three weights for improved performance.

Figure 2 shows the performance difference between our method and MD5. The *x*-axis shows various combinations of weights while the *y*-axis shows the performance difference in terms of requests serviced per second. The results in the figure are based on the experimental setup which will be described shortly. Positive difference indicates that DISH serviced more requests per second than did MD5.

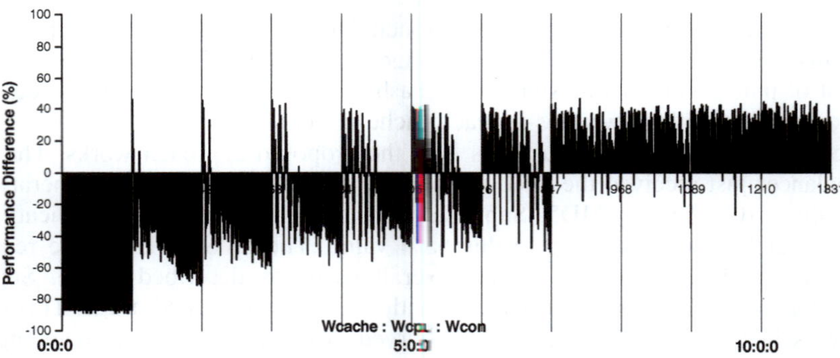

Fig. 2. Performance difference between DISH and MD5

When the weight for cache utilization is over 5, the performance of DISH improves. The best performance of DISH occurs when the weight for cache utilization is maximum, i.e., 10 out of 10. In other words, no information other than cache utilization is used. The main reason for this clear distinction is how current the "current" information is. Cache utilization is the most current information, while CPU usage and connection are less current than cache utilization. CPU usage and connections change much more often than cache utilization, since CPU usage is affected by many other runtime conditions, while cache utilization is fixed relative to CPU usage. For example, a clock tick of Linux kernel is 10 milliseconds while the LVS collects resource usage every second. The difference between the kernel tick of 10 milliseconds and the application time interval of 1 second is vast, indicating that the "current" information is not so current. Hashing decisions based on this less current information can adversely affect the performance. It is clear from the figure that CPU usage and connection information are not very effective in determining servers for processing client requests. For our main experiments, we use only cache utilization.

4 Experimental Results

4.1 Experiment Environment

To test the proposed approach, we have set up an experimental environment as shown in Figure 3. There are four types of machines: clients, LVS-based front-end server, a cluster of cache servers, and a web server. The machine specifications are listed in Table 1. The client machines generate requests using the publicly available traces of Webstone and Surge. The requests consist of HTML files and images. The front end LVS (Layer-4 switching) [18] distributes requests round-robin to the cache servers, while the Kernel TCP Virtual Server (KTCPVS for Layer-7 switching) [19] in each server decides where and how to send requests based on the hashing method. The six hashing methods discussed earlier have been implemented, and their results are presented.

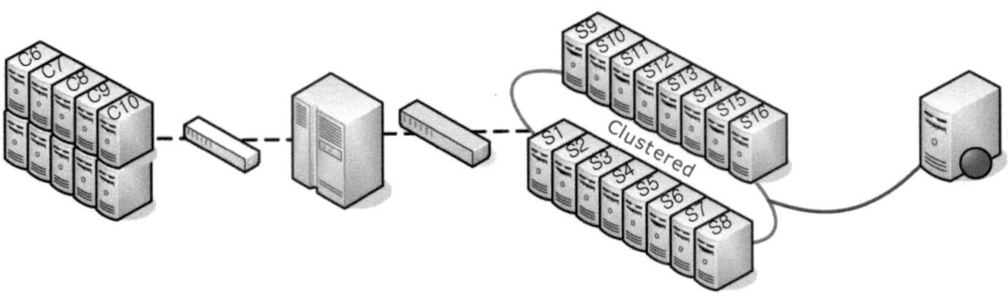

Fig. 3. Experimental setup

Table 1. Specifications of the machines

		Hardware		Software	#
		CPU (MHz)	RAM (MB)		
Client/Master		P-4 2260	256	Webstone [15] Surge [16]	10 / 1
LVS		P-4 2400	512	Direct Routing	1
Server	Cache	P-2 400	256	Squid [17]	16
	Distiller			JPEG-6b	
Web Server		P-4 2260	256	Apache	1

If the requested pages were cached in one of the 16 servers, the selected server would send them directly to the client without going through the LVS machine, hence *direct* routing. Therefore, the LVS will only handle one-way, incoming traffic. If the requested pages were not cached, the Web server would be called to provide the contents. The fresh contents received from the Web server are cached (stored) in the selected server that will subsequently send directly to the client. Squid is used to cache contents received from the Web server. Round-Robin is used mainly for comparison purposes since it gives optimal performance. When caching fresh pages received from the Web server, Round-Robin copies them to *every* server in the cluster while other methods do not. The other methods keep the data in only one server among the cluster. There is no copy involved, hence eliminating the cache coherence problem and enabling storage scalability. As all servers have exactly the same files in RR, no particular distribution strategy is necessary, and there are no hotspots in RR.

4.2 Distribution of Connections

Figure 4 shows the number of connections for *each* cache server, where connections refer to client requests. The x-axis is the cache server number while the y-axis is the number of connections. For all of the traces, the proposed method demonstrates that the number of connections is relatively constant in each of the 16 servers. On the other hand, other methods show that the numbers fluctuate due to inefficient hashing. These fluctuations will directly affect the cluster performance. Note that the y-axis is plotted to log scale for readability.

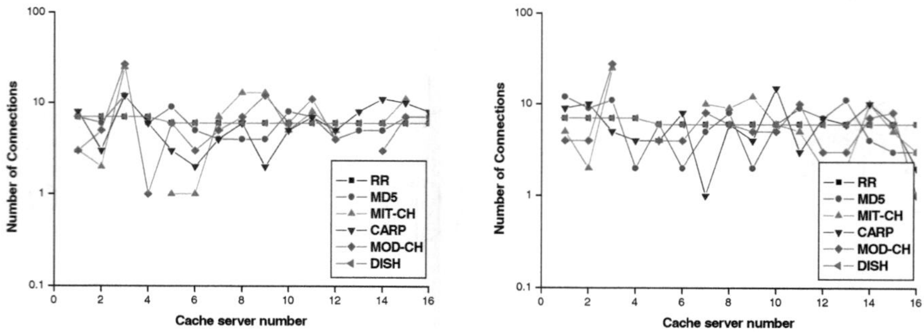

(a) Surge-100 (b) UNC-2001

Fig. 4. Distribution of connections

4.3 Distribution of URLs

Figure 5 shows the total number of URLs hashed to each cache server. The x-axis represents the *cache server number* while the y-axis represents the number of connections. A total of 100 different URLs drawn from the traces [20] are used in the experiment. It is clear from the figure that the proposed method shows a relatively constant number of URLs hashed to each and every cache server. On the other hand, other methods show a fluctuation in the numbers. Round-Robin results show exactly the same number of URLs, since every server has the same copy of files.

4.4 Standard Deviation

Standard deviation shows how far a sample is from the mean of a collection of samples. We wish to find exactly how efficient each hashing method is in comparison

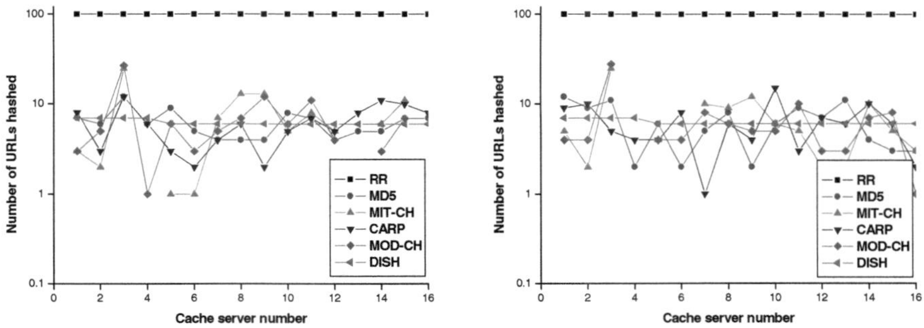

(a) Surge-100 (b) UNC-2001

Fig. 5. Distribution of URLs

to one another based on connections, URLs hashed and CPU usage. For each hashing method, the standard deviation is computed using 16 values, each taken from a server. As a result, standard deviation indicates how far each server is from the mean number of hashed URLs or connections. The smaller the standard deviation is, the better the hashing method is, because it indicates an even distribution of hashing to servers. Figure 6 depicts the standard deviations for all of the hashing methods.

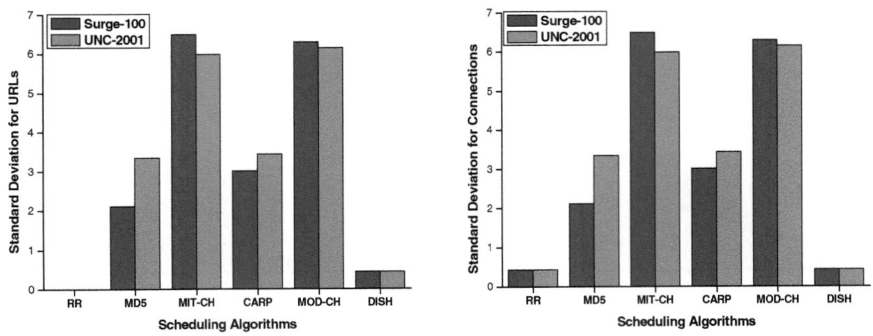

(a) Standard deviation for URLs (b) Standard deviation for connections

Fig. 6. Standard Deviation

Figure 6(a) shows the standard deviation for the URLs hashed to each server. As we can see from the figure, the proposed method (DISH) is the smallest among the six methods. Note that Round-Robin shows no bar charts, because the standard deviation for RR is zero. Round-Robin has exactly the same number of URLs in each server, resulting in zero deviation. The standard deviation for connections shown in Figure 6(b) is slightly different than the URL's. Here RR shows a slight standard deviation, mainly because the 100 total URLs cannot evenly be divided across 16 servers. Some servers will receive one more request than the others. The figure indicates the proposed method DISH outperforms all other methods except RR, which is included for comparison purposes only.

4.5 Scalability

Scalability is measured in terms of the number of requests serviced per second. Figure 7 shows the scalability of the methods using Surge-100 and UNC-2001 trace. The x-axis shows the number of cache server while the y-axis shows the requests served per second. Note that RR shows the highest scalability, as expected. Again, the main reason is that any server can handle any request since every server has exactly the same copy of files. It is used solely as a reference point.

For Surge-100 in Figure 7(a), the proposed method DISH yields higher performance scalability than other methods. When the number of servers is increased to two from one, the number of requests serviced per second also increases to 350 from 250, resulting in 40% improvement. On the other hand, the other methods show little improvement in terms of the number of requests serviced as the number of servers is increased. For UNC-2001, DISH still shows better performance than all other methods. In summary, DISH gives up to 6 fold improvement on 16 machines while the other methods show slightly over 2 fold improvement. The results demonstrate that DISH is scalable compared to the others. This performance scalability reaffirms that dynamic server information such as cache utilization is essential when hashing URLs to a cluster of cache servers.

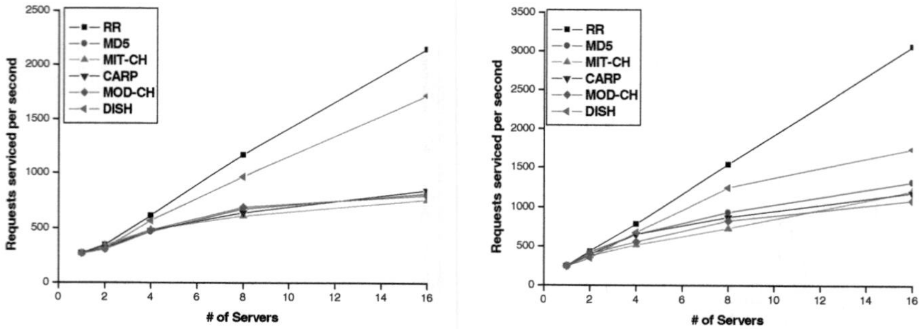

(a) Surge-100 (b) UNC-2001

Fig. 7. Number of requests serviced per second. Round-Robin (RR) shows an ideal performance since all servers have the same copy while other methods distributed files mutually exclusively across the servers. RR is included for reference.

4.6 Overall Comparison

As we have demonstrated through experimental results on 16 cache servers, the proposed method DISH provides performance scalability through balanced hashing and distribution of client requests. Table 2 summarizes the features of the methods discussed in this study.

Table 2. Summary of features and characteristics for the six hashing methods

Methods	Complexity	Server add/del	Scalability	Even distribution	Hot-Spot
MIT-CH [9]	2 steps	O	X	X	X
CARP [10]	3 steps	O	X	X	X
Modified CH [11]	2 steps	O	X	X	X
Adaptive LB [12]	3 steps	O	X	X	O
Extended CH [13]	3 steps	O	X	X	O
Cache Clouds [14]	3 steps	O	X	X	O
DISH	3 steps	O	O	O	O

5 Conclusions

Caching a large number of web pages requires multiple servers typically configured in a cluster. While multiple cache servers are designed to cache and quickly hand out frequently accessed pages, mapping these requests to cache servers presents several challenges in hot spot problems, request distribution and, ultimately, performance scalability. We have presented in this report a scalable hashing method that uses dynamic server information, including cache utilization, CPU usage and connections. Decisions as to where the requested pages are cached are made based on this dynamic data. All pages are mutually exclusively stored across all servers. Each request is directed to the server that has cached the pages. When a hot spot occurs, this particular page is distributed to all cache servers in the cluster in a way that all the servers can simultaneously hand out the page in a round-robin fashion.

To test out this approach, we set up an experimental environment consisting of a cluster of 16 cache servers, an LVS load-balancer, 10 client machines, and a Web server. Client machines generated requests based on the publicly available Web traces, including Surge-100 [16] and UNC-2001 [20], as well as various images and html pages. Our experiments are in two steps: First, determine the impact of each type of runtime information on performance. We have found that cache utilization affects the overall performance most among the three (cache, CPU, and connections). Second, the experiments are performed using the best performing runtime information. We ran two web traces on the cluster using cache utilization as the determining factor.

Experimental results have indicated that our method distributed the requests almost evenly across all of the cache servers while other methods did not, as evidenced by the standard deviation of requests. The standard deviation for URL distribution is approximately 0.5 for the proposed method DISH as opposed to 2 to 6 for others. This indicates that the proposed method is 4 to 12 times better in terms of how well the requests can be distributed. In fact, this even distribution of requests is the main reason why DISH has demonstrated the highest scalability among the six. Experimental results indicate that DISH is 114% more scalable than other methods using Surge-100 and 45% more scalable using UNC-2001 while, simultaneously, eliminating the hot-spot problem.

References

1. Zeng, D., Wang, F., Liu, M.: Efficient web content delivery using proxy caching techniques. IEEE Transactions on Systems, Man and Cybernetics 34(3), 270–280 (2004)
2. Challenger, J., Dantzig, P., Ivengar, A., Squillante, M., Zhang, L.: Efficient serving dynamic data at highly accessed web sites. IEEE/ACM Transactions on Networking 12(2), 233–246 (2004)
3. Trianfillou, P., Aekaterinidis, I.: ProxyTeller: a proxy placement tool for content delivery under performance constraints. In: Proceedings of the 4th International Web Information Systems Engineering, pp. 62–71 (2003)
4. Yin, L., Cao, G.: Supporting cooperative caching in ad hoc networks. IEEE Transactions on Mobile Computing 5(1), 77–89 (2006)
5. Ni, J., Tsang, D.: Large-scale cooperative caching and application-level multicast in multimedia content delivery networks. IEEE Communications Magazine 43(5), 98–105 (2005)
6. Fu, X., Yang, L.: Improvement to HOME based Internet caching protocol. In: IEEE 18th Annual Workshop on Computer Communications, pp. 159–165. IEEE Computer Society Press, Los Alamitos (2003)
7. Mei, H., Lu, C., Lai, C.: An automatic cache cooperative environment using ICP. In: International Conference on Information Technology: Coding and Computing, pp. 144–149 (2002)
8. Rivest, D.: The MD5 Message Digest Algorithm, RFC 1321 (1992)
9. Karger, D.: Web Caching with consistent hashing. In: WWW8 conference (1999)
10. Micorsoft Corp.: Cache Array routing protocol and microsoft proxy server 2.0, White Paper (1999)
11. Baboescu, F.: Proxy Caching with Hash Functions, Technical Report CS2001-0674 (2001)
12. Toyofumi, T., Satosi, K., Hidetosi, O.: Adaptive load balancing content address hashing routing for reverse proxy servers. In: IEEE International Conference on Communications, vol. 27(1), pp. 1522–1526. IEEE Computer Society Press, Los Alamitos (2004)
13. Lei, S., Grama, A.: Extended consistent hashing: an efficient framework for object location. In: Proceeding of 24th International Conference on Distributed Computing Systems, pp. 254–262 (2004)
14. Ramaswamy, L., Liu, L., Iyengar, A.: Cache Clouds: Cooperative Caching of Dynamic Documents in Edge Networks. In: Proceedings of 25th IEEE International Conference on Distributed Computing Systems, pp. 229–238. IEEE Computer Society Press, Los Alamitos (2005)
15. Mindcraft, Inc.: WebStone: The Benchmark for Web Server, http://www.mindcraft.com/web- stone
16. Barford, P., Crovella, M.: Generating Representative Web Workloads for Network and Server Performance Evaluation. In: Proc. ACM SIGMETRICS Conf., Madison, WI, ACM Press, New York (1998)
17. Squid Web Proxy Cache, http://www.squid-cache.org
18. LVS (Linux Virtual Server), http://www.linuxvirtualserver.org
19. KTCPVS (Kernel TCP Virtual Server), http://www.linuxvirtualserver..org/software/ktcpvs/ktcpvs.html
20. Felix, H., Jeffay, K., Smith, F.: Tracking the Evolution of Web Traffic. In: MASCOTS. Proceedings of the 11th IEEE/ACM International Symposium on Modeling, Analysis, and Simulation of Computer and Telecommunication Systems, pp. 16–25 (2003)

FTSCP: An Efficient Distributed Fault-Tolerant Service Composition Protocol for MANETs

Zhenguo Gao[1], Sheng Liu[1], Ming Ji[1], Jinhua Zhao[2], and Lihua Liang[1]

[1] College of Automation, Harbin Engineering University, Harbin, 150001, China
{gag,liusheng,jim,lianglh}@hrbeu.edu.cn
[2] School of Astronautics, Harbin Institute of Technology, Harbin, 150001, China
zhaojinhua@hit.edu.cn

Abstract. Service composition, which enables users to construct complex services from atomic services, is an essential feature for the usability of Mobile Ad hoc Networks (MANETs). Service composition in MANETs should be fault-tolerant and distributed. Efforts in this area are rare and sounds not meet the requirements very well. In this paper, we present a distributed Fault-Tolerant Service Composition Protocol (FTSCP) for MANETs. FTSCP has two main features. Firstly, it is efficient for that all atomic services consisted in a composite service are discovered in just one service discovery session. Secondly, it is fault-tolerant since that service composition process is entirely under the supervision of Execution Coordinator (EC) and inaccessible services can be rediscovered transparently. Both mathematical analysis and simulation results show that FTSCP outperforms another broker-based service composition protocol for MANETs in both terms of packet overhead and promptness.[1]

1 Introduction

Service composition refers to the process of combining simple services to form a larger, more complex service[1]. This offers users a great degree of transparency in discovery and selection of required services, instead of having to be cognizant of all the details about the simpler services that constitute the complex ones. A service can be any software or hardware entity that can be utilized by others.

In the context of service composition, an atomic service is a service provided by one single node that does not rely on any other services to fulfill user requests; a composite service is compound service constructed from other composite services or atomic services; the number of necessary atomic services consisted in the composite service is called as composite service size.

Service composition has been extensively studied in the context of wired networks where service composition always adopts a centralized approach[2,3,4]. In

[1] This work is supported by National Postdoctoral Research Program (No.NPD060234), Postdoctoral Research Program of HeiLongJiang (No.HLJPD060234), Fundamental Research Foundation of Harbin Engineering University (No.HEUFT06009), Fostering Fundamental Research Foundation of Harbin Engineering University (No.HEUFP07001).

R. Perrott et al. (Eds.): HPCC 2007, LNCS 4782, pp. 797–808, 2007.

centralized approaches, a particular node acting as a service composition manager supervises the whole process of service integration and execution.

With the popularity of portable devices, MANETs are abstracting more research efforts. Most efforts are focused on some other aspects, such as MAC protocol, routing, broadcasting, etc. Only very limited efforts has been performed on service composition in MANETs, such as [1,5]. Furthermore, these approaches are not satisfiable for variable reasons, such as more packet overhead, longer response time, etc.

Because of the highly dynamic topology of MANETs, service composition architectures should be distributed, fault-tolerant, and context aware.

In this paper, we present a Fault-Tolerant Service Composition Protocol (FTSCP) that meets the first two requirements. Context aware characteristics will be introduced in future work. Additionally, all atomic services can be discovered in just one service discovery session in FTSCP. Mathematical analysis and extensive simulations confirms the efficiency of FTSCP.

The organization of the rest of the paper is as follows: Section 2 gives an overview of related works. Section 3 shows the architecture of FTSCP and its operation procedure. Section 4 compares the performance of FTSCP and the approach proposed in[1] through mathematical analysis and extensive simulations. Section 5 concludes the paper.

2 Related Works

Research efforts in service composition follow two directions. One direction of research tries to define semantic-rich machine-readable description languages that can be used to describe services and workflows appropriately. The other direction aims in developing architectures that facilitate service composition. We focus on service composition architectures since we believe that it is critical in MANETs and requires a different approach from service composition architectures in wired context.

Current service composition architectures have been designed with the inherent assumption of fixed network infrastructure and high bandwidth communication channels, such as eFlow[2], CMI[3], Ninja[4], etc. These architectures are based on a centralized manager or broker that manages the various tasks of service composition, such as service discovery, combination of different services, and management of the information flow between services.

Because of the highly dynamic nature of MANETs, service composition architecture in MANETs calls for an alternate design approach. Notifying this, some research efforts have been performed in these areas[1][5]. Chakraborty et. al[1]proposed a distributed broker-based protocol for service composition in pervasive/ad hoc environments. In their approach, service composition is delegated to a broker selected during broker arbitration phase and then the broker searches for all necessary component services one by one. Obviously, broker arbitration and iterated service discovery scheme causes too much packet overhead, which is a great drawback since bandwidth is a critical resource in MANETs. Basu et al.[5] described a hierarchical task-graph based service composition protocol for

MANETs. In their work, a composite service is represented as a task-graph with leaf nodes representing atomic services. Sub trees of the graph are constructed in distributed mode, and service composition process is coordinated by the client. However, the coordinator uses flood scheme to search for all atomic services, which leads to larger packet overhead.

3 FTSCP

Our FTSCP service composition architecture consists of 7 components, as shown in Fig. 1. UnRecoverable Exception Manager (UREM) deals with all exceptions happened during the overall service composition period. It usually sends a notify message to the user application. Composite Service Request Parser (CSRP) extracts atomic services from composite service request received from user application. A notification will be sent to UREM in case of exception. Service Discovery Manager (SDM) finds all atomic services necessary to compose the requested composite service in just one service discovery session. Execution Coordinator (EC) schedules and supervises the executions of all atomic services. Under the supervision of EC, all atomic services cooperate in harmony to fulfill the requested composite service. Result Constructer (RC) combines the results of all atomic services and gets the final integrated results. Service Relationship Info Manager (SRIM) stores the information obtained by the CSRP when parsing composite service request. Context Manager (CM) collects and stores context information, which facilitates context-aware semantic service matching during service discovery sessions.

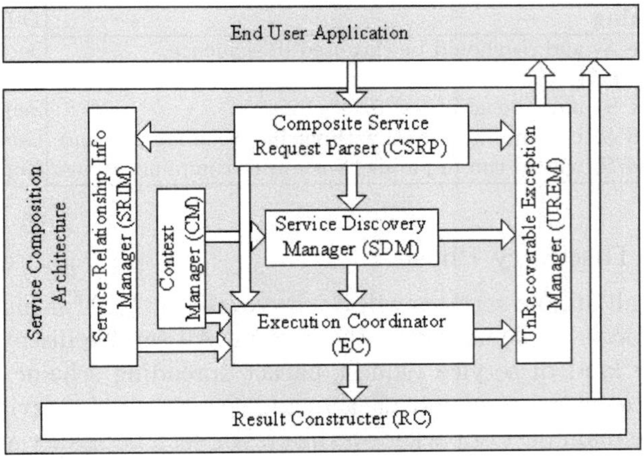

Fig. 1. Service Composition Architecture

FTSCP essentially consists of flowing four phases: 1) Composite service request parse phase, 2) Service discovery phase; 3) Service execution coordination phase; 4) Result construction phase.

3.1 Composite Service Request Parse Phase

In this phase, CSRP receives service composite request from end user application. CSRP extracts all atomic services necessary to fulfill the composite service request as well as the dependencies between these services. This information is stored in SRIM. This information is necessary for service scheduling. Composite service requests describes a task that requires coordination of atomic services as well as other composite services. We designed a new Document Type Definition (DDT) file for describing service composite request as follows.

```
<!ELEMENT Request (Service|AtomicService)*)>
<!ELEMENT Service (
        (Service|AtomicService, Element2Operator, Service|AtomicService)
        |(Factor, Element1Operator, Service|AtomicService)
        |AtomicService)*)>
<!ATTLIST Element2Operator name
        (SeqNormal|SeqArbitrary|SeqInitial|ConcurrentNoCom|ConcurrentCom)
        #REQUIRED>
<!ATTLIST Element1Operator name (Repeat) #REQUIRED>
<!ATTLIST Factor Value CDATA #REQUIRED "1">
<!ATTLIST CompositeService Name CDATA #IMPLIED >
<!ATTLIST AtomicService Name CDATA #REQUIRED>
```

Coordination relationships between atomic services are listed in Table 1.

Table 1. Notations used in the paper

Relation	Meaning	DTD Keyword
$S_1 \odot S_2$:	service S_1 and S_2 should be executed in sequence.	SeqNormal
$S_1 \diamond S_2$:	either service S_1 or S_2 can be executed first.	SeqArbitrary
$S_1 \vdash S_2$:	service S1 must be activated before S2.	SeqInitial
$S_1 \parallel S_2$:	S_1 and S_2 can run in parallel without inter-communications.	ConcurrentNoCom
$S_1 \Updownarrow S_2$:	S_1 and S_2 should run in parallel with inter-communications.	ConcurrentCom

3.2 Service Discovery Phase

In this phase, all atomic services will be discovered by SDM using some service discovery protocols implemented in the system. All service discovery protocols relies on some kind of service request packet spreading scheme to search for matched services. Service request packets and corresponding service reply packets makeup the main part of packet overhead. For each necessary atomic service, if one service discovery session is performed, more request packets will be generated. To reduce packet overhead, we implements a multi-task service discovery protocol (referred as MDFNSSDP)[6] which can find all atomic services in just one service discovery session. For each atomic service, there may be multiple matched services. The service with maximum priority is selected. All founded services are cached in SDM for possible latter use. During service discovery, context information can be get from CM to facilitate service matching operation.

3.3 Service Execution Coordination Phase

In this phase, all discovered atomic services will be performed under the supervision of EC. During this period, all atomic services cooperate in harmony to fulfill the requested composite service. The operation cycle in EC is as follows.

- **Step1.** From the information get from SRIM, EC knows the dependent relationships among atomic services. From the service descriptions of discovered atomic services from SDM, EC knows how to invoke services. Thus, EC can then schedule the execution of these services.
- **Step2.** EC invocates services according to the its schedule and supervise the operation of these services. During this process, some runtime information and temporary results are cached into Run Time DataBase (RTDB). The information in RTDB is used to facilitate the work of EC. If all services are finished successfully, results are passed to the RC component and this stage completes. Otherwise, in case of any exception during the execution, EC first checks if the exception is recoverable. If it is unrecoverable, an exception is thrown to UREM and the stage finished. If the exception is recoverable, such as that a service becomes unreachable, then goes to step 3.
- **Step3.** EC starts a run time service discovery session to find servers that provides the losted service. Recall that in service discovery phase, all founded service information are cached. Thus, a matched service that can replace the lost service may be found directly from SDM. Thus, new service discovery session is not necessary. Information in RTDB is used to in semantic service matching. Then it goes to step1.

Thus, the system is resistant to topology changes, and so it is suitable for MANETs where topology variation is frequent.

3.4 Result Construction Phase

Result Construction Phase When service execution coordination phase is finished successfully, it changes to result construction phase. In this phase, results of all atomic services are combined and the final integrated results are given to the user application. Till now, user application's service composition request is completed.

4 Performance Analysis and Simulations

4.1 Select Comparative Solutions

We implemented our service composition protocol in Glomosim[9]. To make comparative study, the Broker-based Distributed Service Composition Protocol (BDSCP) proposed in [1]is also implemented.

BDSCP consists of four phases. 1) In Broker Arbitration Phase, the client broadcasts a Broker Request Packet, and then in a following fixed period TBAP, the node waits for broker reply packets. The broker request packet floods in the network for limited hops. All nodes receiving the packet reply with a Broker Reply Packet enclosing its runtime priority value. The client then selects the neighbor with biggest priority value as its broker. 2) In Service Discovery Phase, the selected Broker uses the underlying GSD service discovery protocol to discover all required atomic services one by one. 3) In Service Integration Phase, atomic services are scheduled. 4) In Service Execution Phase, atomic services are executed according to the schedule.

4.2 Performance Metrics

The superiority of FTSCP over BDSCP mainly results from two aspects: 1) Broker arbitration phase is eliminated since broker arbitration is helpless in MANETs, and 2) One service discovery session searching for all atomic services is used instead of one service discovery session for each atomic service. Following performance metrics are used in our analysis and simulation study.

- **Number of Composition Preparation Packets:** In BDSCP, composition preparation packets include 1) broker request packets and broker reply packets, and 2) service request packets and service reply packets. Whereas in FTSCP, only the second part is included. This metric is averaged over all composite service requests successfully instantiated.
- **Delay of Composite Service Discovery:** It is the interval between the generation of a service composition request and the time when all necessary atomic services are founded. This metric is averaged over all composite service requests that are successfully instantiated. It measures the promptness of service composition protocols.
- **Composition Efficiency:** Composition efficiency refers to the fraction of service composite requests that all atomic services are discovered successfully.

4.3 Performance Analysis

In this section, mathematical analysis will be performed to estimate the performance of FTSCP and BDSCP in terms of performance metrics listed in previous section. Although MDFNSSDP outperforms GSD very much both in packet overhead and promptness[6], the analysis in the section is based on the following assumptions for simplicity: 1)M_{SDP}(averaged number of service request packets and service reply packets sent in one service discovery session)in FTSCP is equal to that in BDSCP, 2)T_{SDP}(averaged interval between the time that a service discovery session is started and the time the first reply packet with matched services is received) in FTSCP is equal to that in BDSCP, 3) Failure service discovery session is not retried.

Analysis on Number of Composition Preparation Packets. In FTSCP, composition preparation packets include service request packets and reply packets. Additionally, only one service discovery session is performed. Hence M_{FTSCP} (number of composition preparation packets in FTSCP) equals M_{SDP}.

In BDSCP, besides service request packets and service reply packets, composition preparation packets also include broker request packets and broker reply packets. Furthermore, for each atomic service, one service discovery session is performed. Hence, M_{BDSDP} (number of composition preparation packets in BDSCP) can be calculated as equation . Here M_{SDP} is the averaged number of service request packets and service reply packets sent in one service discovery session, M_{BAP} is averaged number of broker request packets and broker reply packets, M_{BQP} is averaged number of broker request packets, M_{BPP} is averaged number of broker reply packets).

$$M_{BDSCP} = M_{BAP} + M_{SDP} \times k = (M_{BQP} + M_{BPP}) + M_{SDP} \times k \quad (1)$$

In broker arbitration phase in BDSCP, broker request packets flood through the client's b-hop neighbor sets. Noticing that last-hop nodes do not rebroadcast packets, M_{BQP}(averaged number of broker request packets sent in broker arbitration phase in BDSCP) can be calculated as follows. Here ρ is averaged number of nodes in unit area, r is Radio range, b is the hop limit of broker request packets in BDSCP, and $n_1 = \rho\pi r^2$

$$M_{BQP} = \sum_{h=0}^{b-1} n(h) \approx \rho\pi(r(b-1))^2 = n_1(b-1)^2 \quad (2)$$

Each node receiving broker request packets will send back a broker reply packet in unicast mode. In case of $b > 1$, a broker reply packet will travel several hops to reach the source node that requires brokers. These relayed packets should also be calculated in this metric. Hence,

$$M_{BPP} = \sum_{h=1}^{b} hn(h) \approx \sum_{h=1}^{b} h(\rho\pi(rh)^2 - \rho\pi(r(h-1))^2) = n_1 \frac{4b^3 + 3b^2 - b}{6} \quad (3)$$

Combining Eqn. 1, 2, and 3, we have:

$$M_{BDSCP} = (M_{BQP} + M_{BPP}) + kM_{SDP} = n_1 \frac{4b^3 + 9b^2 - 13b + 6}{6} + kM_{SDP} \quad (4)$$

Compared with BDSCP, the number of saved composition preparation packets of FTSCP is:

$$M_{BDSCP} - M_{FTSCP} = n_1 \frac{4b^3 + 9b^2 - 13b + 6}{6} + (k-1)M_{SDP} > 0 \quad (5)$$

Analysis on Delay of Composite Service Discovery. T_{FTSCP} (delay of composite service discovery in FTSCP) and T_{BDSCP} (delay of composite service discovery in BDSCP) can be calculated as eqn. 6 and 7, respectively. Here T_{BAP} stands for the period that a client waits for broker reply packets in BDSCP.

T_{SDP} is the averaged interval between the time that a service discovery session is started and the time the first reply packet with matched services is received.

$$T_{FTSCP} = T_{SDP} \tag{6}$$

$$T_{BDSCP} = T_{BAP} + k \cdot T_{SDP} \tag{7}$$

Combining Eqn. 6, and 7, we have:

$$T_{BDSCP} - T_{FTSCP} = T_{BAP} + (k-1) \cdot T_{SDP} > 0 \tag{8}$$

Analysis on Composition Efficiency. A succeeded composite service request requires that all atomic services are discovered successfully. Hence, $P_{ComSuc}(k)$ (the probability of a composite service with size k being succeeded) can be calculated as eqn 9. Here k is composite service size, p is the probability of founding a requested atomic service in one service discovery session. There is no difference in this term between FTSCP and BDSCP.

$$P_{ComSuc}(k) = p^k \tag{9}$$

4.4 Simulation Models

In our simulation studies, the Distributed Coordination Function (DCF) of IEEE 802.11 is used as the underlying MAC protocol. Random Waypoint Mode (RWM) is used as the mobility model. In this model, a node moves towards its destination with a constant speed $v \in [V_{MIN}, V_{MAX}]$, which means a variable uniform distributed between V_{MIN} and V_{MAX}. When reached its destination, a node will keep static for a random period $t_P \in [T_{MIN}, T_{MAX}]$. Then the node will randomly select a new destination in scenario and move to the new destination with new speed. The process will repeat permanently. In our simulations, t_P is fixed to 0s.

The advantages of FTSCP over BDSCP that can be easily demonstrated through simulations rely in the phases before service schedule and execution. Service schedule and execution cause little impacts on evaluating the relative qualities of FTSCP and BDSCP. Hence, service schedule and execution are omitted in our simulation for simple.

In all simulations, some basic parameters are set as shown in Table. 2. Simulation scenarios are created with 100 nodes initially distributed according to the steady state distribution of random waypoint mobility model. At the beginning of each simulation, predefined number of nodes are randomly selected as servers. These selected servers provide randomly selected services. During each simulation, 10 service composition requests are started at randomly selected time by randomly selected nodes. Atomic services consisting of each service composition request are also selected randomly.

Basic parameters of the underlying service discovery protocols are set as shown in Table 3.

Table 2. Basic parameters

Parameters	Value	Parameters	Value
scenario area	1000m × 1000m	service composition request number	10
node number	100	composite service size	1∼7
radio range	250m	service discovery protocol in FTSCP	MDFNSSDP
simulation time	1000s	service discovery protocol in BDSCP	GSD
bandwidth	2Mbps	hop limit of broker packets in BDSCP	1
mobility	RWP	initial node placement	steady state
T_{BAP}	0.1s		

Table 3. Basic parameters in GSD and MDFNSSDP

Parameters	Value	Parameters	Value
service number	80	hop limit of advertisement packets	1
service group number	1	service number in each group	10
service number in each group	10	service advertisement packet valid time	30s
service advertisement interval	20s	hop limit of service request packets	3

4.5 Simulation Results

In this section, we inspect the impacts of composite service size on performance metrics through extensive simulations. In all the following figures, error bars report 95% confidence. In case of results of composition efficiency, the lower confidence limits of the single side confidence intervals are shown.

To inspect the impacts of composite service size, 2 other simulation sets are run. In all these simulations, node speed is fixed to 0m/s. Each simulation set

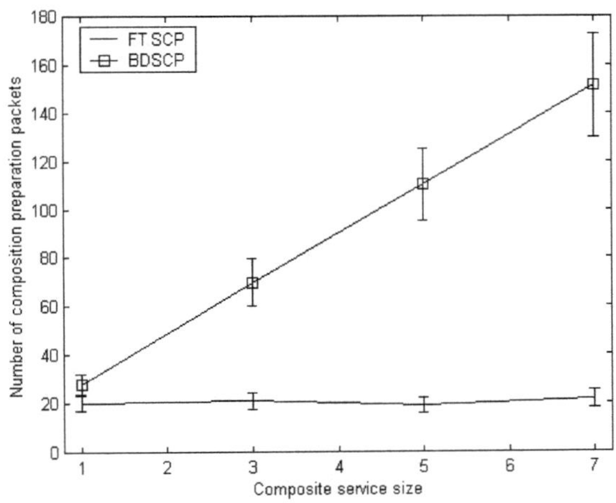

Fig. 2. number of composition preparation packets vs. composite service size

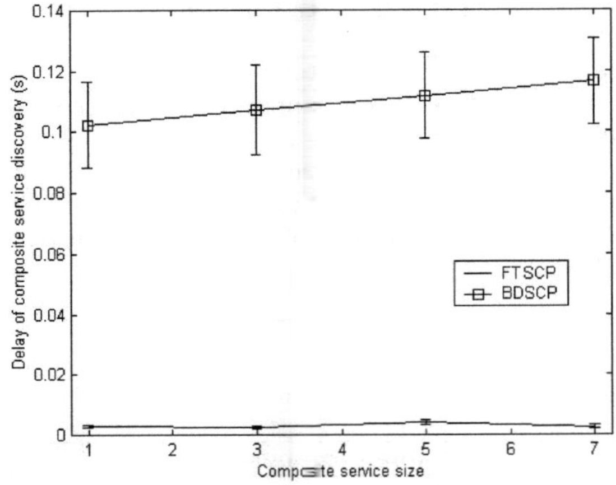

Fig. 3. delay of composite service discovery vs. composite service size

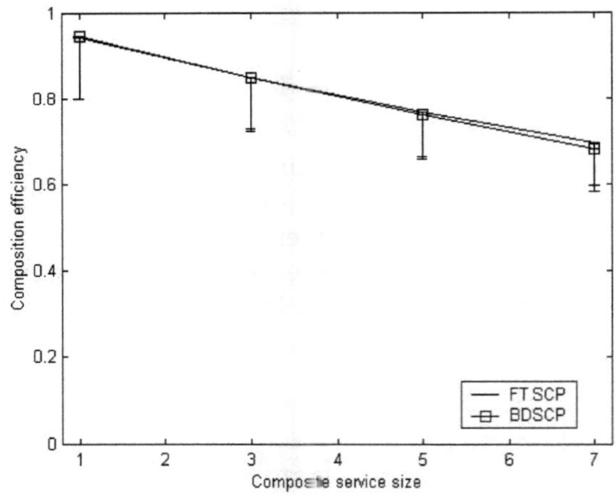

Fig. 4. composition efficiency vs. composite service size

includes 4 subsets where composite service size is set to 1, 3, 5, and 7, respectively. Each subset consists of 100 similar simulations.

Fig.2 shows the impact of composite service size on number of composition preparation packets. In BDSCP, for each atomic service, one service discovery session has to be performed. Thus, as composite service size increases, the number of composition preparation packets will increase. In FTSCP, however, no matter how many atomic services are consisted in the requested composite service, only one service discovery session will be performed. Hence, the number of these packets keeps almost constant.

Fig.3 shows that FTSCP is much more prompt than BDSCP. Delay of composition service discovery in BDSCP increases gradually as composite service size increases. This results from more service discovery sessions. While in FTSCP, this metric keeps almost constant. These results verify Eqn. 8.

Fig.4 shows the impact of composite service size on composition efficiency. In both FTSCP and BDSCP, as composition size increases, composition efficiency decreases quickly. This verifies the analysis result of Eqn.9.

5 Conclusions

In this paper, we have presented a distributed Fault-Tolerant Service Composition Protocol (FTSCP) for MANETS. In FTSCP, service composition process is entirely under the supervision of EC and inaccessible services can be rediscovered transparently. Hence, FTSCP is fault-tolerant. In service discovery phase, context information is used to perform semantic-rich service matching and select the best service. Hence, FTSCP is context-aware. In FTSCP, all atomic services consisted in a composite service are discovered in just one service discovery session. Therefore, it is efficiency. Mathematical analysis and simulation results confirm the superiority of FTSCP over another broker-based service composition protocol for MANETs in terms of packet overhead and promptness.

References

1. Chakraborty, D., Joshi, A., Yesha, Y., Finin, T.: Service composition for mobile environments. Journal on Mobile Networking and Applications (MONET), 435–451 (2005) (special issue on Mobile Services)
2. Casati, F., Ilnicki, S., Jin, L., Krishnamoorthy, V., Shan, M.: Adaptive and dynamic service composition in eflow. Technical Report, HPL-200039, Software Technology Laboratory, Palo Alto, USA (2000)
3. Schuster, H., Georgakopoulos, D., Cichocki, A., Baker, D.: Modeling and composing service-based and reference process-based multienterprise processes. In: Wangler, B., Bergman, L.D. (eds.) CAiSE 2000. LNCS, vol. 1789, pp. 247–263. Springer, Heidelberg (2000)
4. Katz, R.H., Brewer, E.A., Mao, Z.M.: Fault-tolerant, scalable, wide-area internet service composition. Technical Report. UCB/CSD-1-1129. CS Division. EECS Department. UC. Berkeley (2001)
5. Basu, P., Ke, W., Little, T.D.C.: A novel approach for execution of distributed tasks on mobile ad hoc networks. In: WCNC 2002. Proceedings of IEEE Wireless Communications and Networking Conference, Orlando, USA, pp. 579–585. IEEE Computer Society Press, Los Alamitos (2002)
6. Gao, Z.G., Liu, S., Liang, L.H., Zhao, J.H., Song, J.G.: A Generic Minimum Dominating Forward Node Set based Service Discovery Protocol for MANETs. In: HPCC 2007. 2007 International Conference on High Performance Computing and Communications (accepted 2007)

7. Chakraborty, D., Joshi, A., Yesha, Y., Finin, T.: GSD: a novel group-based service discovery protocol for MANETs. In: MWCN 2002. Proceedings of the 4th IEEE Conference on Mobile and Wireless Communications Networks, pp. 140–144. IEEE Computer Society Press, Los Alamitos (2002)
8. Dey, K.A., Abowd, D.G., Salber, D.: A context-based infrastructure for smart environment. In: MANSE 1999. Proceedings of the 1st International Workshop on Managing Interactions in Smart Environments, Dublin, Ireland, pp. 114–128 (1999)
9. Glomosim, Wireless Adaptive Mobility Lab. DEPT of Comp. SCI, UCLA, Glomosim: a scalable simulation environment for wireless and wired network system, Available: `http://pcl.cs.ucla.edu/projects/domains/glomosim.html`

CIVIC: A Hypervisor Based Virtual Computing Environment

Jinpeng Huai, Qin Li, and Chunming Hu

Trustworthy Internet Computing Lab, Beihang University, Beijing, China
huaijp@buaa.edu.cn, liqin@act.buaa.edu.cn, hucm@act.buaa.edu.cn

Abstract. The purpose of virtual computing environment is to improve resource utilization by providing a unified integrated operating platform for users and applications based on aggregation of heterogeneous and autonomous resources. With the rapid development in recent years, hypervisor technologies have become mature and comprehensive with four features, including transparency, isolation, encapsulation and manageability. In this paper, a hypervisor based virtual computing infrastructure, named CIVIC, is proposed. Compared with existing approaches, CIVIC may benefit in several ways. It offers separated and isolated computing environment for end users, and realizes hardware and software consolidation and centralized management. Beside this, CIVIC provides a transparent view to upper layer applications, by hiding the dynamicity, distribution and heterogeneity of underlying resources. Performance of the infrastructure is evaluated by an initial deployment and experiment. The result shows that CIVIC can facilitate installation, configuration and deployment of network-oriented applications.

1 Introduction

Along with the rapid development of IT infrastructures and Internet, the network has brought together large amount of resources, including computers, data, software and services. But the utilization of these resources is far from their full potential. Beside this, the internet-oriented applications demand for large-scale computing, massive data processing and global information services. At this point, how to organize and manage distributed, heterogeneous and autonomic resources, to share and coordinate resources across the autonomous domains, and finally to improve utilization of resource becomes a key issue of designing the software infrastructures for network-based computing systems.

One of the possible way to solve this problem is to build a virtual computing environment, on the open infrastructure of the Internet, providing harmonic, transparent and integrated services for end-users and applications[1]. Many interesting computing paradigm has been proposed from both academic and industrial communities, such as the Grid, Peer-to-Peer (P2P) systems, ubiquitous computing, and desktop computing[2-4]. Grid Computing is mainly focusing on the dynamic and across-domain network resource share and coordination; P2P is commonly used for end-to-end resource and information services; Ubiquitous computing is aiming at

R. Perrott et al. (Eds.): HPCC 2007, LNCS 4782, pp. 809–820, 2007.

accessing resources at anytime and anywhere; and the desktop computing is trying to integrate large numbers of idle resources to provide computing intensive applications with high-performance computing power. All of these approaches are trying to provide a unified resource virtualization environment to get better utilization of resource capacities.

In recent years, the virtual machine technology has got wider attention. Hypervisor or virtual machine[5] is a virtual layer between the hardware and software. It can provide applications with independent runtime environment by shielding the dynamic, distributed and heterogenic hardware resources. It can assign an independent and isolated computing environment to each individual user. And it can help system administrators to manage hardware and software resources in a virtual, centralized approach. According to these advantages, the virtual computing environment based on virtual machine has now become a hot spot, many virtual machine based mechanisms and systems has been proposed, such as Virtual Workspaces[6], VIOLIN[7], and Virtuoso[8]. However, the research of virtual computing environment based on virtual machine is still in its infancy. Existing approaches are usually focusing on single virtual machine, and lacked of the crossover of a variety of related technologies. In addition, management and monitoring features are not well supported in these systems, the status and the characteristics of the virtual computing environment.

In this paper, a hypervisor based computing infrastructure, named CIVIC, is proposed. Compared with existing approaches, CIVIC may benefit in several ways. It offers separated and isolated computing environment for end users, realizes hardware and software consolidation and centralized management. Beside this, CIVIC provides a transparent view to upper layer applications, by hiding the dynamicity, distribution and heterogeneity of underlying resources. Performance of the infrastructure is evaluated by an initial deployment and experiment. The result shows that CIVIC can facilitate installation, configuration and deployment of network-oriented applications.

This paper is organized as follows. In Section 2, a brief introduction on the hypervisor technology is given. In Section 3, several hypervisor based virtual computing systems are introduced. A hypervisor based software infrastructure for virtual computing, named CIVIC, is proposed in Section 4 with layered architecture and modules. Finally, the performance of the infrastructure is evaluated by an initial deployment and experiment in Section 5.

2 Virtual Machine and Hypervisor

The concept of virtual machine (VM) is still not clearly defined. When people are talking about *virtual machine*, sometimes, they refer to a type of software which can emulate a physical machine, like *Virtual PC* or *VMWare*. In this case, it is more accurate to use the word *hypervisor*, or *Virtual Machine Monitor* (VMM). Beside this, virtual machine can also refer to an instance of emulated (or virtual) machine with software installed inside. In this case, the word *virtual machine instance* is more appropriate. In this paper, we will follow this rule to avoid ambiguity.

2.1 Background

Hypervisor has a long history. In 1966, VM/360 (Virtual Machine) system[9] came into being in IBM Cambridge research center VM/360, which is used to share hardware resources for different user. In 1997, *Connectix* provided *Virtual PC*[10] able to run PC applications in a Mac machine, therefore transparent portability for legacy software could be archived. In 1998, *VMWare* Company released *VMware* hypervisor which is the first hypervisor software able to emulate PC on PC platform[11]. VMWare is now highly valued and widely deployed in enterprises.

In order to eliminate the impact of hardware failure and maintenance to enhance the stability and reliability of software, *VMWare* and *Xen* both proposed on-line migration technology in their own hypervisors, named *VMotion* [12] and Live Migration[13] respectfully. It can be used as a dynamic load balancing mechanism for hardware resources provision.

The performance issue is always the main obstacle to prevent the wider use of hypervisor. Hypervisor is preferred to be transparent to hardware and software, which also made its implementation complex and inefficient. Technologies like Xen[14] and Intel VT[15] greatly improved the performance of hypervisor while eliminate part of its transparency to software and hardware.

2.2 Features of Hypervisor

According to previous analysis of the hypervisor, we conclude there are four main features of hypervisor:

(1) Transparency. Transparency means software can execute in virtual machine environment directly, without being revised. Other features of hypervisor, such as hardware resource sharing, isolated environment, live migration, portability, etc, can be easily applied to any software running in virtual machine without modification.

(2) Isolation. Hypervisor enables multiple instances of virtual machine to be hosted in a physical machine. Not only software can share hardware resources via virtual machine, but also be protected and isolated by virtual machine. Each virtual machine can install its own version of software, without having to consider compatibility with the software installed in other virtual machines. Each virtual machine also maintains a separated runtime environment, while preventing software failure caused by software running in other virtual machines.

(3) Encapsulation. Whole system of a virtual machine is enclosed in a virtual hard disk, which is an ordinary file of the host machine. Through this form of encapsulation, virtual machine installation, backing up and restoring are as easy as copying and pasting a file in operating system, which can effectively reduce the difficulty of the system configuration and deployment, and increase the flexibility of software.

(4) Manageability. For virtual machine operations, such as boot, shutdown, sleep, and even add, modify, or remove virtual hardware, there all have programming interface. Via programming interface, virtual machine can be controlled by program completely. Therefore administrators have greater flexibility in controlling virtual machines, compared to manually controlling physical machines. Based on virtual machine, remote and centralized management of hardware resources can be achieved. The aforementioned live migration is another example of the manageability of virtual machine, software running in virtual machine can be controlled to change runtime environment without being interrupted.

3 Related Work

Virtual Workspaces[6] aims to provide a customizable and controllable remote job execution environment for Gird. Traditional grid mainly focuses on job exciting, but neglects the deployment and maintenance of execution environment. Virtual Workspaces implements web service interface that conform to WSRF and GSI standard for virtual machine remote management. Client can create and deploy virtual machine on-demand via the interface.

Virtual Workspaces supports unmanned installation of legacy applications, which can effectively reduce the deployment time, significantly minimize the artificial participation in it. By allocating and enforcing resources required by jobs with different priorities, virtual machine can realize fine-grained resource provision, including CPU, memory, network bandwidth, etc. However, in some cases, applications need to run in a network environment with multiple machines, Virtual Workspaces do not suit for the situation because it cannot provide a network environment for these applications.

VIOLIN[7] is another hypervisor-based project proposed by Purdue University, which can provides the isolated network environment for applications. In VIOLIN, users can design and create virtual network topology on physical machines and physical networks. Every virtual network is isolated and independent from each other. Each virtual machine and virtual network equipment can be customized. For example, VIOLIN can build a virtual mobile IP network using Mobile IP protocol stack to test the stability and performance without actual moving of the terminals.

Similar projects like In-VIGO [16] of University of Florida and Virtuoso[8] of Northwestern University, can facilitate configuration and management of virtual machine network for network applications, and serve as the basis for emulation and evaluation infrastructure of network software. However, these projects mostly focus on runtime support, but lack of designing support which can facilitate user's creating and customizing the virtual machine network.

Internet Suspend and Resume (ISR)[17] is another interesting hypervisor systems. It can provide a personal mobile computing environment for users, without having to carry any portable device. The personal computing environment is encapsulated in the virtual machine which is stored in a distributed file system. Whenever user closes or suspends the virtual machine, ISR will copy its state back to the distributed file system so that the virtual machine can be restored from anywhere later.

The demerit of ISR is that virtual machine must be suspended before user moves, therefore virtual machine will stop running during the movement. With the use of virtual machine live migration, this restriction can be removed somehow. However, live migration requires that the source and destination machine reside in the same link, which greatly limits the mobility.

SoftGrid[18] product provided by Softricity uses light-weight hypervisor technology to solve software management problem on the Windows platform. The light-weight hypervisor only need to intercept a small part of system call from application to emulate virtual file system and virtual registry. A similar technology under Linux platform is called jail or chroot. Since each application is isolated in a separate virtual file system and has separate registry namespace, software compatibility problem caused by file version conflicts and registry key conflicts can

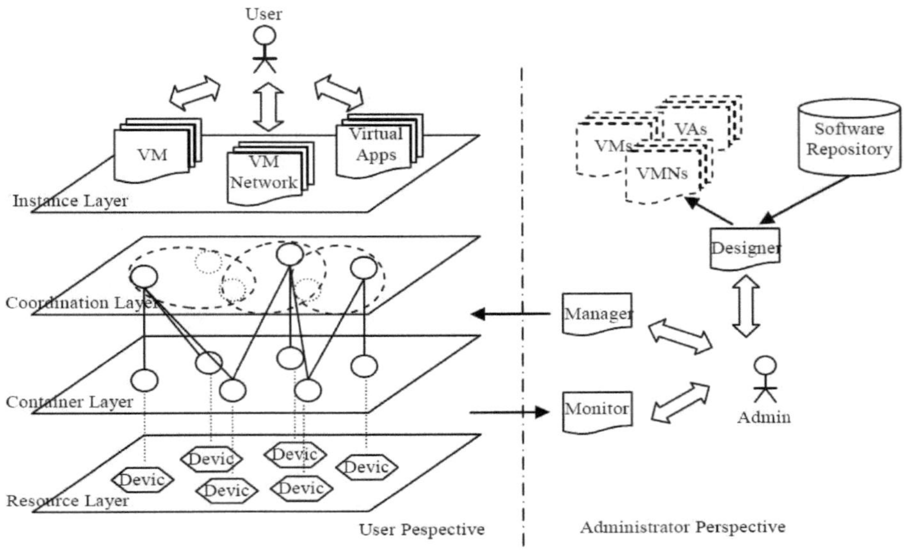

Fig. 1. CIVIC Perspectives

be reduced to minimum. Using SoftGrid, the image file only need to contain the application and its library dependency, which will be relatively smaller than normal virtual machine image which include full operating system. But this light-weight hypervisor has very limited isolation other than file system and registry, such that two applications running in SoftGrid hypervisor can interfere with each other.

4 CIVIC Architecture

In this section, we propose the CROWN-based Infrastructure for VIrtual Computing (CIVIC), a hypervisor-based computing environment. CROWN (China R&D Environment Over Wide-area Network) is a service-oriented Grid middleware system which enables resource sharing and problem solving in dynamic multi-institutional virtual organizations in for China e-Science community[19]. CIVIC is a core component of CROWN Grid Middleware version 3.0. CIVIC provides better support for grid service in CROWN with isolated runtime environment.

The benefit of CIVIC is listed as follows: Firstly, it can offer separated and isolated computing environment for users. Secondly, it can also realize hardware and software consolidation and centralized management for computer administrators. Thirdly, it can be transparent to upper layer applications, hiding the dynamicity, distribution and heterogeneity of underlying resources from applications.

As shown in figure 1, CIVIC can be viewed in two perspectives. From normal user perspective, CIVIC establishes a middleware infrastructure on top of raw hardware resources to provide hosting environment for virtual machine instance and virtual network instance. User can interact with the instance just like with a physical machine or physical network without having to know which specific physical machine hosts the instance. Using technologies like virtual machine live migration, CIVIC can keep virtual machine instance running from the infection of underlying hardware failure or maintenance.

From administrator perspective, CIVIC provides three modules. Firstly, CIVIC Monitor module can collect and present overall runtime status to administrator. Secondly, CIVIC Manage module can provide administrator a centralized management console for resources registered in CIVIC environment. Thirdly, CIVIC Designer module provides a visual editor which can facilitate the installation and configuration of virtual machine instance and virtual network.

As shown in figure 2, CIVIC can be divided into five layers: resource layer, container layer, coordination layer, instance layer, and interaction layer.

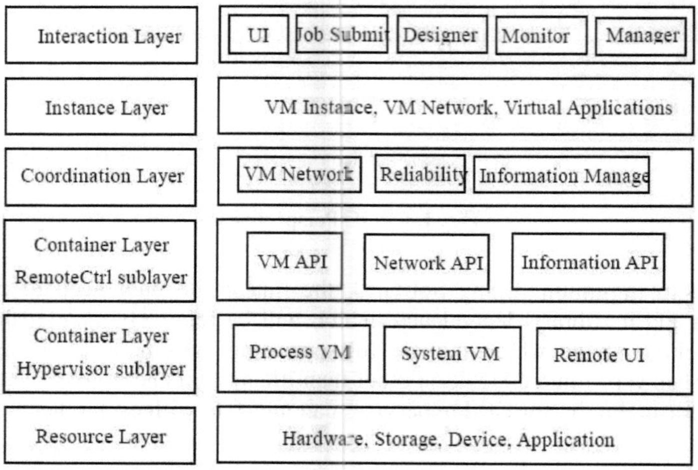

Fig. 2. CIVIC Layers

4.1 Resource Layer

Resource layer is formed by multiple resource nodes. Each resource node is a physical machine. Generally, software is installed directly into these physical machines. However, resource nodes are distributed over the Internet, which may be incompatible with each other; both hardware and software may encounter unexpected failure. All these circumstances make software unreliable and vulnerable to environment changes.

In CIVIC, software will be installed into virtual machines. With the middleware infrastructure established on top of raw hardware resources, CIVIC can provide isolated and reliable hosting environment for virtual machines over dynamic and heterogeneous resource nodes on the Internet.

4.2 Container Layer

The layer on top of the resource layer is the container layer, which is composed of container nodes. Each container node is a physical machine with CIVIC Container component installed. Each container node has hypervisor software deployed that can host multiple virtual machine instances, and these instances can be controlled remotely through container interface.

Container layer can be further divided into two sub-layers:

(1) Hypervisor Sub-layer. CIVIC supports three type of hypervisor. The first one is Xen hypervisor, which can host full functioned virtual machine with operating system and application software installed in the virtual machine instance. The second one is jail virtual machine, which can provide isolated file system hierarchy for software. The third one is Java virtual machine which can host software written in Java.

(2) Remote Control Sub-layer. This sub-layer provides remote management interface, including virtual machine management interface, network configuration interface, and information query interface, etc. Each virtual machine instance can be controlled by other nodes through virtual machine management interface, including operations like booting, halting and hibernating. A tunnel connection can be established between two container nodes using network configuration interface, which can be used to connect multiple virtual machine instances hosted by different container nodes. Administrator can monitor CPU, memory usage in virtual machine instances hosted by container node through information query interface.

4.3 Coordination Layer

Container node has the ability to host virtual machine instance. However it is not efficient to deploy an instance of virtual network containing several virtual machines into a single container node. To keep reliability of the virtual machine instance, it is also important to deploy multiple replicated instances to several container nodes. Therefore, in these cases, it is essential to coordinate several related container nodes to serve one purpose.

CIVIC builds coordination layer over container layer. Coordination layer consists of multiple coordination nodes, which are special container nodes with coordinating function configured. Coordination nodes are responsible for the management of other container nodes.

As mentioned above, there are different kinds of coordination functions in CIVIC, for example, resource management for the resource registration and query, virtual network management for maintaining virtual network over a number of container nodes, etc.

4.4 Instance Layer

There are multiple nodes in instance layer, which are isolated computing environment provided for users. CIVIC support three different kinds of instance nodes: virtual machine instance, virtual machine network instance, and virtual application instance. Virtual machine instance contains a complete operating system, which is isolated by hypervisor like Xen. Virtual machine network consists of multiple virtual machines with a specific network topology. Virtual application instance contains only a single application.

4.5 Interaction Layer

This layer contains two types of interaction modules. First type of interactive modules is CIVIC administrative tools, including CIVIC Monitor, CIVIC Manager and CIVIC Designer. Monitor module can collect and present overall runtime status to administrator. Manager module can provide administrator centralized management for resources registered in CIVIC environment. Designer module provides a visual editor which supports creation of multiple virtual machines with a specific network topology, as shown in Figure 3.

Second type of interaction module provides access interface to CIVIC instance. Users can interact with CIVIC instance nodes through a variety of interactive method, such as command line interface, graphical user interface, and Web Service interface to submit jobs to execute in virtual machine.

5 Implementation and Performance Evaluation

CIVIC is still in its active development. Under the design considerations mentioned above, we have finished part of our systems including CIVIC Designer, which is a GUI application based on Eclipse RCP framework, and CIVIC Job Engine, which implements front-end interface for virtual machine. Performance of our systems is evaluated through carefully designed experiments.

5.1 CIVIC Designer

CIVIC Designer can provide easy to use user interface for creating, modifying, deploying virtual machines and virtual networks. Figure 3 is shows the interface of CIVIC Designer.

There are two ways to create a new virtual machine instance in CIVIC Designer. The first one is to create from scratch, which is to install a new operating system into a virtual machine. Users can customize the software installation and configuration of virtual machine instance via dialogs and wizards in CIVIC Designer. The second one is to create from existing physical machine, also known as Physical-to-Virtual (P2V). CIVIC Designer supports a remote P2V feature, which can turn a physical machine into virtual machine instance remotely with only few requirements from the physical machine. This function is very useful for maintaining runtime environment for legacy software.

CIVIC Designer also provides a visual network topology editor which can be used to create, configure and deploy virtual network with multiple virtual machines. Each instance of virtual machine network created by CIVIC Designer will be treated as an atom element, which can be encapsulated as a single image file, and be deployed and managed as a single instance.

An instance of virtual machine network can be hosted on a single physical machine, and can also be hosted on several different machines, even on two machines that are not attached in the same network link. In case two machines from different links host an instance of virtual network a tunnel will be established between these machines connecting virtual machines among them. In current implementation, OpenVPN is used to establish such network tunnel.

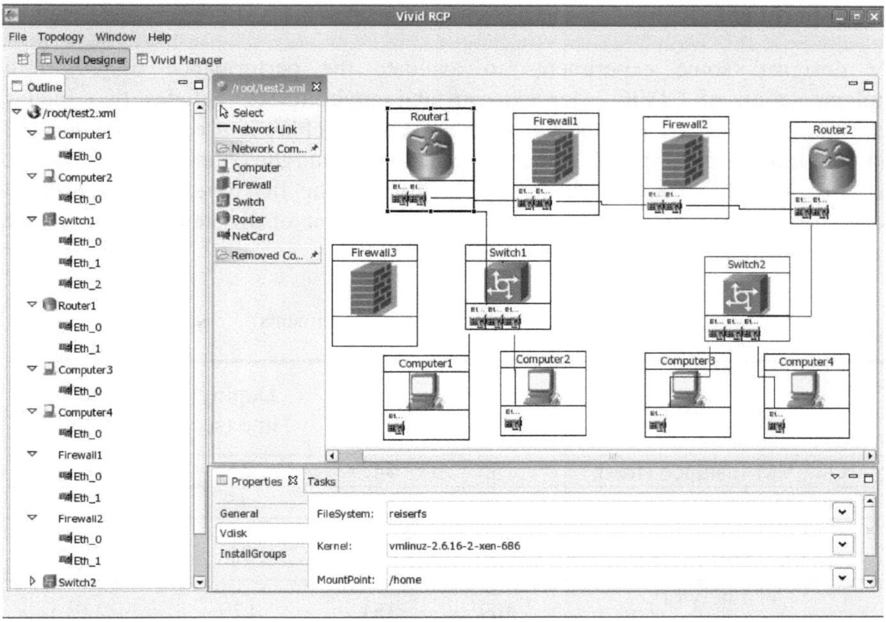

Fig. 3. CIVIC Designer

Performance regarding to creating, deploying virtual machine or virtual network using CIVIC Designer is studied and discussed in section 5.4.

5.2 CIVIC Job Engine

As mentioned, there are several ways for users to use virtual machine instance in CIVIC. Users can log on to the command line interface or graphic user interface to control the virtual machine directly, or use job submission interface to execute a job on virtual machine.

CIVIC Job Engine provides a Web Services interface which support Job Submission Description Language (JSDL) standard[20]. After Job Engine received submitted the job, it will execute the job in the virtual machine instance specified by the JSDL document. Since the runtime environment of job engine and jobs are isolated and protected by hypervisor, and each job is guaranteed by hypervisor with specific allocated computing resources, such as number of processors, and size of memory and disk. Therefore hypervisor can ensure QoS and reliability of job scheduling.

However, since the job engine and job is totally isolated by hypervisor, the communication between engine and job can be inconvenient and inefficient. In current implementation of CIVIC Job Engine, we used SSH as the communication channel between engine and job. In order to improve efficiency, previous SSH session will be cached for future communication.

5.3 Experiment

We designed some experiments to evaluate the performance of our initial implementation of CIVIC. Our experimental environment includes two Intel P4 machine, with 3GHz CPU and 1G memory. One machine is used for the designing and creating of virtual machine and virtual network, and the other one is used to host the virtual machine or virtual network created by the first one. The experimental metric includes install time, deploy time, and boot time of different type of instance. The experimental results are shown in Table 1.

Table 1. CIVIC Designer experiments

Type of instance	Size (MB)	Create Time (sec)	Deploy Time (sec)	Boot Time (sec)
One VM instance (fresh)	136	441	45	79
One VM instance (template)	136	143	45	79
VM network with 2 VM instances	273	246	88	94
VM network with 3 VM instances	409	351	129	113

The virtual machine created in the experiment is a minimized Linux system, including the basic network tools like ftp, ssh, etc. This virtual machine occupied 136 megabytes of disk space. From the experimental results shown in Table 2, Using CIVIC to create a fresh Linux virtual machine only takes about 7 minutes, and if we use previously created virtual machine template to accelerate the installation procedure, we can reduce the install time to 2 minutes. Taking into account that the installation of a physical machine usually takes between 30 to 60 minutes, using CIVIC can effectively speed up the system installation work and minimize manual participation in installation. Furthermore, It only takes about 2 minutes to deploy a virtual network consists of three virtual machines. In practice, it usually takes far more than 2 minutes to install and configure a physical network including several physical machines. This shows that CIVIC can facilitate installation, configuration and deployment of network-oriented applications.

6 Conclusion and Future Work

This paper introduces the emergence and development of virtual machine technology with the conclusion of four characteristics of virtual machine: transparency, isolation, encapsulation and manageability. We also analyzed several related implementations. Finally, we presented the system function design and experimental results of initial implementation of CIVIC, a virtual computing infrastructure based on CROWN.

Our future work includes investigating security access control mechanisms in CIVIC which enforcing resource provision and reservation, as well as improving software availability and resource utilization by supporting virtual machine migration over wide-area network.

Acknowledgments. This work is partially supported by grants from the China National Science Foundation (No.90412011), China 863 High-tech Program (Project No. 2005AA119010), China 973 Fundamental R&D Program (No. 2005CB321803) and National Natural Science Funds for Distinguished Young Scholar (Project No. 60525209). We would also like to thank members in CROWN team in Institute of Advanced Computing Technology, Beihang University for their hard work on CROWN grid middleware.

References

1. Lu, X., Wang, H., Wang, J.: Internet-based virtual computing environment (iVCE): Concepts and architecture. In: Science in China Series F-Information Sciences, vol. 49, pp. 681–701 (2006)
2. Foster, I.: The Physiology of the Grid - An Open Grid Service Architecture for Distributed Systems Integration. In: Open Grid Service Infrastructure WG, Global Grid Forum (2002)
3. Qiu, D., Srikant, R.: Modeling and Performance Analysis of BitTorrent-Like Peer-to-Peer Networks. In: Proceedings of ACM SIGCOMM, ACM Press, New York (2004)
4. Liu, Y., Liu, X., Xiao, L.: Location-Aware Topology Matching in P2P Systems. In: Proceedings of IEEE INFOCOM, IEEE Computer Society Press, Los Alamitos (2004)
5. Goldberg, R.P.: A survey of virtual machine research. IEEE Computer. 7, 34–45 (1974)
6. Keahey, K., Foster, I., Freeman, T., Zhang, X., Galron, D.: Virtual Workspaces in the Grid. In: 11th International Euro-Par Conference Lisbon, Protugal (2005)
7. Ruth, P., Jiang, X., Xu, D., Goasguen, S.: Virtual Distributed Environments in a Shared Infrastructure. IEEE Computer 38, 39–47 (2005)
8. Virtuoso: Resource Management and Prediction for Distributed Computing Using Virtual Machines, http://virtuoso.cs.northwestern.edu/
9. Adair, R.J., Bayles, R.U., Comeau, L.W., Creasey, R.J.: A Virtual Machine System for the 360/40. IBM Corp. Cambridge Scientific Center 320-2007 (1966)
10. Microsoft Virtual PC (2004), http://www.microsoft.com/windows/virtualpc/default.mspx
11. VMware - Virtualization Software, http://www.vmware.com/
12. Nelson, M., Lim, B.H., Hutchins, G.: Fast Transparent Migration for Virtual Machines. In: USENIX Annual Technical Conference, pp. 391–404 (2005)
13. Clark, C., Fraser, K., Hand, S., Hansen, J.G., Jul, E., Limpach, C., Pratt, I., Warfield, A.: Live Migration of Virtual Machines. In: NSDI. 2nd ACM/USENIX Symposium on Networked Systems Design and Implementation, Boston, MA, pp. 273–286. ACM Press, New York (2005)
14. Barham, P., Dragovic, B., Fraser, K., Hand, S., Harris, T., Ho, A., Pratt, I., Warfield, A., Barham, P., Neugebauer, R.: Xen and the Art of Virtualization. In: ACM Symposium on Operating Systems Principles, pp. 164–177. ACM Press, New York (2003)
15. Uhlig, R., Neiger, G., Rodgers, D., Santoni, A.L., Martins, F.C.M., Anderson, A.V., Bennett, M.S., Kägi, A., Leung, F.H., Smith, L.: Intel Virtualization Technology. IEEE Computer 38, 48–56 (2005)
16. Adaballa, S.: From Virtualized Resources to Virtualized Computing Grid: The In-VIGO System. Future-Generation Computing System 21, 896–909 (2005)

17. Kozuch, M., Satyanarayanan, M.: Internet Suspend / Resume. In: Workshop on Mobile Computing Systems and Applications (2002)
18. Greschler, D., Mangan, T.: Networking lessons in delivering 'Software as a Service' Part1. International Journal of Network Management 12, 317–321 (2002)
19. Huai, J., Hu, C., Sun, H., Li, J.: CROWN: A service grid middleware with trust management mechanism. Science in China Series F-Information Sciences 49, 731–758 (2006)
20. Anjomshoaa, A., Brisard, F., Drescher, M., Fellows, D.: Job Submission Description Language (JSDL) Specification Version 1.0 (2005), http://www.ogf.org/documents/GFD.56.pdf

Author Index

Lecture Notes in Computer Science

Sublibrary 1: Theoretical Computer Science and General Issues

For information about Vols. 1–4459
please contact your bookseller or Springer

Vol. 4782: R. Perrott, B.M. Chapman, J. Subhlok, R.F. de Mello, L.T. Yang (Eds.), High Performance Computing and Communications. XIX, 823 pages. 2007.

Vol. 4770: V.G. Ganzha, E.W. Mayr, E.V. Vorozhtsov (Eds.), Computer Algebra in Scientific Computing. XIII, 460 pages. 2007.

Vol. 4763: J.-F. Raskin, P.S. Thiagarajan (Eds.), Formal Modeling and Analysis of Timed Systems. X, 369 pages. 2007.

Vol. 4746: A. Bondavalli, F.V. Brasileiro, S. Rajsbaum (Eds.), Dependable Computing. XV, 239 pages. 2007.

Vol. 4743: P. Thulasiraman, X. He, T.L. Xu, M.K. Denko, R.K. Thulasiram, L.T. Yang (Eds.), Frontiers of High Performance Computing and Networking ISPA 2007 Workshops. XXIX, 536 pages. 2007.

Vol. 4742: I. Stojmenovic, R.K. Thulasiram, L.T. Yang, W. Jia, M. Guo, R.F. de Mello (Eds.), Parallel and Distributed Processing and Applications. XX, 995 pages. 2007.

Vol. 4736: S. Winter, M. Duckham, L. Kulik, B. Kuipers (Eds.), Spatial Information Theory. XV, 455 pages. 2007.

Vol. 4732: K. Schneider, J. Brandt (Eds.), Theorem Proving in Higher Order Logics. IX, 401 pages. 2007.

Vol. 4731: A. Pelc (Ed.), Distributed Computing. XVI, 510 pages. 2007.

Vol. 4711: C.B. Jones, Z. Liu, J. Woodcock (Eds.), Theoretical Aspects of Computing – ICTAC 2007. XI, 483 pages. 2007.

Vol. 4710: C.W. George, Z. Liu, J. Woodcock (Eds.), Domain Modeling and the Duration Calculus. XI, 237 pages. 2007.

Vol. 4708: L. Kučera, A. Kučera (Eds.), Mathematical Foundations of Computer Science 2007. XVIII, 764 pages. 2007.

Vol. 4707: O. Gervasi, M.L. Gavrilova (Eds.), Computational Science and Its Applications – ICCSA 2007, Part III. XXIV, 1205 pages. 2007.

Vol. 4706: O. Gervasi, M.L. Gavrilova (Eds.), Computational Science and Its Applications – ICCSA 2007, Part II. XXIII, 1129 pages. 2007.

Vol. 4705: O. Gervasi, M.L. Gavrilova (Eds.), Computational Science and Its Applications – ICCSA 2007, Part I. XLIV, 1169 pages. 2007.

Vol. 4703: L. Caires, V.T. Vasconcelos (Eds.), CONCUR 2007 – Concurrency Theory. XIII, 507 pages. 2007.

Vol. 4700: C.B. Jones, Z. Liu, J. Woodcock (Eds.), Formal Methods and Hybrid Real-Time Systems. XVI, 539 pages. 2007.

Vol. 4697: L. Choi, Y. Paek, S. Cho (Eds.), Advances in Computer Systems Architecture. XIII, 400 pages. 2007.

Vol. 4688: K. Li, M. Fei, G.W. Irwin, S. Ma (Eds.), Bio-Inspired Computational Intelligence and Applications. XIX, 805 pages. 2007.

Vol. 4684: L. Kang, Y. Liu, S. Zeng (Eds.), Evolvable Systems: From Biology to Hardware. XIV, 446 pages. 2007.

Vol. 4683: L. Kang, Y. Liu, S. Zeng (Eds.), Intelligence Computation and Applications. XVII, 663 pages. 2007.

Vol. 4681: D.-S. Huang, L. Heutte, M. Loog (Eds.), Advanced Intelligent Computing Theories and Applications. XXVI, 1379 pages. 2007.

Vol. 4672: K. Li, C. Jesshope, H. Jin, J.-L. Gaudiot (Eds.), Network and Parallel Computing. XVIII, 558 pages. 2007.

Vol. 4671: V. Malyshkin (Ed.), Parallel Computing Technologies. XIV, 635 pages. 2007.

Vol. 4669: J.M. de Sá, L.A. Alexandre, W. Duch, D. Mandic (Eds.), Artificial Neural Networks – ICANN 2007, Part II. XXXI, 990 pages. 2007.

Vol. 4668: J.M. de Sá, L.A. Alexandre, W. Duch, D. Mandic (Eds.), Artificial Neural Networks – ICANN 2007, Part I. XXXI, 978 pages. 2007.

Vol. 4666: M.E. Davies, C.J. James, S.A. Abdallah, M.D Plumbley (Eds.), Independent Component Analysis and Blind Signal Separation. XIX, 847 pages. 2007.

Vol. 4665: J. Hromkovič, R. Královič, M. Nunkesser, P. Widmayer (Eds.), Stochastic Algorithms: Foundations and Applications. X, 167 pages. 2007.

Vol. 4664: J. Durand-Lose, M. Margenstern (Eds.), Machines, Computations, and Universality. X, 325 pages. 2007.

Vol. 4649: V. Diekert, M.V. Volkov, A. Voronkov (Eds.), Computer Science – Theory and Applications. XIII, 420 pages. 2007.

Vol. 4647: R. Martin, M. Sabin, J. Winkler (Eds.), Mathematics of Surfaces XII. IX, 509 pages. 2007.

Vol. 4646: J. Duparc, T.A. Henzinger (Eds.), Computer Science Logic. XIV, 600 pages. 2007.

Vol. 4644: N. Azémard, L. Svensson (Eds.), Integrated Circuit and System Design. XIV, 583 pages. 2007.

Vol. 4641: A.-M. Kermarrec, L. Bougé, T. Priol (Eds.), Euro-Par 2007 Parallel Processing. XXVII, 974 pages. 2007.

Vol. 4639: E. Csuhaj-Varjú, Z. Ésik (Eds.), Fundamentals of Computation Theory. XIV, 508 pages. 2007.

Vol. 4638: T. Stützle, M. Birattari, H. H. Hoos (Eds.), Engineering Stochastic Local Search Algorithms. X, 223 pages. 2007.

Vol. 4628: L.N. de Castro, F.J. Von Zuben, H. Knidel (Eds.), Artificial Immune Systems. XII, 438 pages. 2007.

Vol. 4627: M. Charikar, K. Jansen, O. Reingold, J.D.P. Rolim (Eds.), Approximation, Randomization, and Combinatorial Optimization. XII, 626 pages. 2007.

Vol. 4624: T. Mossakowski, U. Montanari, M. Haveraaen (Eds.), Algebra and Coalgebra in Computer Science. XI, 463 pages. 2007.

Vol. 4621: D. Wagner, R. Wattenhofer (Eds.), Algorithms for Sensor and Ad Hoc Networks. XIII, 415 pages. 2007.

Vol. 4619: F. Dehne, J.-R. Sack, N. Zeh (Eds.), Algorithms and Data Structures. XVI, 662 pages. 2007.

Vol. 4618: S.G. Akl, C.S. Calude, M.J. Dinneen, G. Rozenberg, H.T. Wareham (Eds.), Unconventional Computation. X, 243 pages. 2007.

Vol. 4616: A. Dress, Y. Xu, B. Zhu (Eds.), Combinatorial Optimization and Applications. XI, 390 pages. 2007.

Vol. 4613: F.P. Preparata, Q. Fang (Eds.), Frontiers in Algorithmics. XI, 348 pages. 2007.

Vol. 4600: H. Comon-Lundh, C. Kirchner, H. Kirchner (Eds.), Rewriting, Computation and Proof. XVI, 273 pages. 2007.

Vol. 4599: S. Vassiliadis, M. Berekovic, T.D. Hämäläinen (Eds.), Embedded Computer Systems: Architectures, Modeling, and Simulation. XVIII, 466 pages. 2007.

Vol. 4598: G. Lin (Ed.), Computing and Combinatorics. XII, 570 pages. 2007.

Vol. 4596: L. Arge, C. Cachin, T. Jurdziński, A. Tarlecki (Eds.), Automata, Languages and Programming. XVII, 953 pages. 2007.

Vol. 4595: D. Bošnački, S. Edelkamp (Eds.), Model Checking Software. X, 285 pages. 2007.

Vol. 4590: W. Damm, H. Hermanns (Eds.), Computer Aided Verification. XV, 562 pages. 2007.

Vol. 4588: T. Harju, J. Karhumäki, A. Lepistö (Eds.), Developments in Language Theory. XI, 423 pages. 2007.

Vol. 4583: S.R. Della Rocca (Ed.), Typed Lambda Calculi and Applications. X, 397 pages. 2007.

Vol. 4580: B. Ma, K. Zhang (Eds.), Combinatorial Pattern Matching. XII, 366 pages. 2007.

Vol. 4576: D. Leivant, R. de Queiroz (Eds.), Logic, Language, Information and Computation. X, 363 pages. 2007.

Vol. 4547: C. Carlet, B. Sunar (Eds.), Arithmetic of Finite Fields. XI, 355 pages. 2007.

Vol. 4546: J. Kleijn, A. Yakovlev (Eds.), Petri Nets and Other Models of Concurrency – ICATPN 2007. XI, 515 pages. 2007.

Vol. 4545: H. Anai, K. Horimoto, T. Kutsia (Eds.), Algebraic Biology. XIII, 379 pages. 2007.

Vol. 4533: F. Baader (Ed.), Term Rewriting and Applications. XII, 419 pages. 2007.

Vol. 4528: J. Mira, J.R. Álvarez (Eds.), Nature Inspired Problem-Solving Methods in Knowledge Engineering, Part II. XXII, 650 pages. 2007.

Vol. 4527: J. Mira, J.R. Álvarez (Eds.), Bio-inspired Modeling of Cognitive Tasks, Part I. XXII, 630 pages. 2007.

Vol. 4525: C. Demetrescu (Ed.), Experimental Algorithms. XIII, 448 pages. 2007.

Vol. 4514: S.N. Artemov, A. Nerode (Eds.), Logical Foundations of Computer Science. XI, 513 pages. 2007.

Vol. 4513: M. Fischetti, D.P. Williamson (Eds.), Integer Programming and Combinatorial Optimization. IX, 500 pages. 2007.

Vol. 4510: P. Van Hentenryck, L.A. Wolsey (Eds.), Integration of AI and OR Techniques in Constraint Programming for Combinatorial Optimization Problems. X, 391 pages. 2007.

Vol. 4507: F. Sandoval, A.G. Prieto, J. Cabestany, M. Graña (Eds.), Computational and Ambient Intelligence. XXVI, 1167 pages. 2007.

Vol. 4501: J. Marques-Silva, K.A. Sakallah (Eds.), Theory and Applications of Satisfiability Testing – SAT 2007. XI, 384 pages. 2007.

Vol. 4497: S.B. Cooper, B. Löwe, A. Sorbi (Eds.), Computation and Logic in the Real World. XVIII, 826 pages. 2007.

Vol. 4494: H. Jin, O.F. Rana, Y. Pan, V.K. Prasanna (Eds.), Algorithms and Architectures for Parallel Processing. XIV, 508 pages. 2007.

Vol. 4493: D. Liu, S. Fei, Z. Hou, H. Zhang, C. Sun (Eds.), Advances in Neural Networks – ISNN 2007, Part III. XXVI, 1215 pages. 2007.

Vol. 4492: D. Liu, S. Fei, Z. Hou, H. Zhang, C. Sun (Eds.), Advances in Neural Networks – ISNN 2007, Part II. XXVII, 1321 pages. 2007.

Vol. 4491: D. Liu, S. Fei, Z.-G. Hou, H. Zhang, C. Sun (Eds.), Advances in Neural Networks – ISNN 2007, Part I. LIV, 1365 pages. 2007.

Vol. 4490: Y. Shi, G.D. van Albada, J. Dongarra, P.M.A. Sloot (Eds.), Computational Science – ICCS 2007, Part IV. XXXVII, 1211 pages. 2007.

Vol. 4489: Y. Shi, G.D. van Albada, J. Dongarra, P.M.A. Sloot (Eds.), Computational Science – ICCS 2007, Part III. XXXVII, 1257 pages. 2007.

Vol. 4488: Y. Shi, G.D. van Albada, J. Dongarra, P.M.A. Sloot (Eds.), Computational Science – ICCS 2007, Part II. XXXV, 1251 pages. 2007.

Vol. 4487: Y. Shi, G.D. van Albada, J. Dongarra, P.M.A. Sloot (Eds.), Computational Science – ICCS 2007, Part I. LXXXI, 1275 pages. 2007.

Vol. 4484: J.-Y. Cai, S.B. Cooper, H. Zhu (Eds.), Theory and Applications of Models of Computation. XIII, 772 pages. 2007.

Vol. 4475: P. Crescenzi, G. Prencipe, G. Pucci (Eds.), Fun with Algorithms. X, 273 pages. 2007.

Vol. 4474: G. Prencipe, S. Zaks (Eds.), Structural Information and Communication Complexity. XI, 342 pages. 2007.

MIX
Papier aus verantwortungsvollen Quellen
Paper from responsible sources
FSC® C105338

If you have any concerns about our products,
you can contact us on
ProductSafety@springernature.com

In case Publisher is established outside the EU,
the EU authorized representative is:
Springer Nature Customer Service Center GmbH
Europaplatz 3, 69115 Heidelberg, Germany

Printed by Libri Plureos GmbH
in Hamburg, Germany